INTERNET OF THINGS AND DATA ANALYTICS HANDBOOK

INTERNET OF THINGS AND DATA ANALYTICS HANDBOOK

Edited by

HWAIYU GENG

Amica Research
Palo Alto, CA, USA

Published by John Wiley & Sons, Inc., Hoboken, New Jersey
Published simultaneously in Canada

***Library of Congress Cataloging-in-Publication data has been applied for*:**

Hardback: 9781119173649

Cover image: Pitju/gettyimages; Oleksiy Mark/gettyimages; Maxiphoto/gettyimage

Set in 10/12pt Times by SPi Global, Pondicherry, India

Printed in the United States of America

10 9 8 7 6 5 4 3 2 1

CONTENTS

LIST OF CONTRIBUTORS

Scott Amyx, Amyx McKinsey, San Francisco, CA, USA

Arun Aryasomajula, Division of Analytics Research and Clinical Informatics, Department of Data Science, Geisinger Health System, Danville, PA, USA

Brenda Bannan, Ph.D., George Mason University, Fairfax, VA, USA

David Bartlett, General Electric, San Ramon, CA, USA

Peter Burnham, FJORD, San Francisco, CA, USA

Alan Carlton, InterDigital Europe Ltd, London, UK

J. Cecil, Ph.D., Co-Director, Computer Science Department, Center for Cyber Physical Systems, Oklahoma State University, Stillwater, OK, USA

Rafael Cepeda, Ph.D., InterDigital Europe Ltd, London, UK

Hubert C.Y. Chan, DBA, The Hong Kong Polytechnic University, Hong Kong, China

Yifan Chen, Ph.D., Ford Research and Advanced Engineering, Dearborn, MI, USA

Dominique Davison, AIA, DRAW Architecture + Urban Design, Kansas City, MO, USA

Abhishek Dubey, Ph.D., Institute for Software Integrated Systems, Vanderbilt University, Nashville, TN, USA

Mumin Tolga Emirler, Ph.D., Ohio State University, Columbus, OH, USA

David Y. Fong, Ph.D., CITS Group, San Jose, CA, USA

Joshua Frank, Ph.D., Intuit Inc., Woodland Hills, CA, USA

Neil Fraser, Ph.D., Macquarie University, Sydney, New South Wales, Australia

Dan Freudberg, Nashville Metropolitan Transport Authority, Nashville, TN, USA

Shane Gallagher, Ph.D., Advanced Distributed Learning, Alexandria, VA, USA

Tim Gammons, ARUP, London, UK

Hwaiyu Geng, P.E., Amica Research, Palo Alto, CA, USA

Girish Ghatikar, Greenlots, San Francisco, CA, USA; Lawrence Berkeley National Laboratory, Berkeley, CA, USA

Francoise Gilbert, J.D., Greenberg Traurig LLP, Silicon Valley, East Palo Alto, CA, USA

Aniruddha Gokhale, Institute for Software Integrated Systems, Vanderbilt University, Nashville, TN, USA

Rikke Gram-Hansen, Copenhagen Solutions Lab, City of Copenhagen, Copenhagen, Denmark

Bilin Aksun Guvenc, Ph.D., Ohio State University, Columbus, OH, USA

Levent Guvenc, Ph.D., Ohio State University, Columbus, OH, USA

Ashley Z. Hand, AIA, CityFi, Los Angeles, CA, USA

David R. Hardoon, Ph.D., Azendian, Singapore, Singapore

David Hindman, FJORD, San Francisco, CA, USA

Michael A. Huff, Ph.D., MEMS and Nanotechnology Exchange (MNX), Corporation for National Research Initiatives, Reston, VA, USA

Rich Hunzinger, B-Scada, Inc., Crystal River, FL, USA

Raj Jain, Ph.D., Department of Computer Science Engineering, Washington University, St. Louis, MO, USA

Karthi Jeyabalan, University of Utah, Salt Lake City, UT, USA

William Kao, Ph.D., Department of Engineering and Technology, University of California, Santa Cruz, CA, USA

Joseph Kimchi, Teledyne Judson Technologies, Montgomeryville, PA, USA

Rafael Laskier, Vale, Rio de Janeiro, Brazil

Martin Lehofer, Siemens Corporate Technology, Princeton, NJ, USA

Jih-Fen Lei, Ph.D., Teledyne Judson Technologies, Montgomeryville, PA, USA

Bridget Lewis, George Mason University, Fairfax, VA, USA

Jennifer Liggett, CH2M, Englewood, CO, USA

George Lu, Ph.D., goodXense, Inc., Edison, NJ, USA

Nicholas Marko, Ph.D., Division of Analytics Research and Clinical Informatics, Department of Data Science, Geisinger Health System, Danville, PA, USA

John Mattison, M.D., Singularity University, Moffett Field, CA, USA; Kaiser Permanente, Pasadena, CA, USA

Jim McKeeth, Embarcadero Technologies, Austin, TX, USA

Pramita Mitra, Ph.D., Ford Research and Advanced Engineering, Dearborn, MI, USA

Gawain Morrison, Sensum, Belfast, UK

Geoff Mulligan, IPSO Alliance, Colorado Springs, CO, USA

Shyam Varan Nath, M.B.A., M.S., Director, IoT at GE Digital, San Ramon, CA, USA

Himanshu Neema, Institute for Software Integrated Systems, Vanderbilt University, Nashville, TN, USA

James Osborne, Microsoft, Redmond, WA, USA

Adrian Pearmine, DKS Associates, Portland, OR, USA

Satyam Priyadarshy, Ph.D., HALLIBURTON Landmark, Houston, TX, USA

Jonathan Reichental, Ph.D., Palo Alto, CA, USA

Oleg Roderick, Ph.D., Division of Analytics Research and Clinical Informatics, Department of Data Science, Geisinger Health System, Danville, PA, USA

Tara Salman, Department of Computer Science Engineering, Washington University, St. Louis, MO, USA

David Sanchez, Division of Analytics Research and Clinical Informatics, Department of Data Science, Geisinger Health System, Danville, PA, USA

Stan Schneider, Ph.D., Real-Time Innovations, Inc., Sunnyvale, CA, USA

Scott Shaw, Hortonworks, Inc., Santa Clara, CA, USA

Shashank Shekhar, Institute for Software Integrated Systems, Vanderbilt University, Nashville, TN, USA

Amy Shi-Nash, Ph.D., Singtel, DataSpark, Singapore, Singapore

Craig Simonds, Ph.D., Ford Research and Advanced Engineering, Dearborn, MI, USA

Brian Skeens, CH2M, Englewood, CO, USA

Danielle Song, University of California, Berkeley, CA, USA

Kai Song, Ph.D., Teledyne Judson Technologies, Montgomeryville, PA, USA

Nick Stein, Indoo.rs GmbH, Brunn am Gebirge, Austria

Gary Strumolo, Ph.D., Ford Research and Advanced Engineering, Dearborn, MI, USA

Fangzhou Sun, Institute for Software Integrated Systems, Vanderbilt University, Nashville, TN, USA

Venkataraman Sundareswaran, Ph.D., Teledyne Judson Technologies, Montgomeryville, PA, USA

Kenneth Thompson, CH2M, Englewood, CO, USA

Stephanie Urbanski, Indoo.rs Inc., Palo Alto, CA, USA

Francesco Valdevies, Selex ES Company, Genova, Italy

Pratik Verma, Ph.D., DB Research Inc., Hopkins, MN, USA

Gene Wang, People Power Company, Redwood City, CA, USA

Y.J. Yang, goodXense, Inc., Edison, NJ, USA

Henry Yuan, Ph.D., Teledyne Judson Technologies, Montgomeryville, PA, USA

FOREWORD

It has been almost 2 years since I met Mr. Hwaiyu Geng for the first time at the SmartAmerica Expo in Washington D.C., a program I established with Geoff Mulligan when we were serving as White House Presidential Innovation Fellows. I still have a vivid memory of Mr. Geng then, as he was one of the few in the audience who sat through more than 8 h of presentations, from 24 teams, without pausing for lunch. At the event, I could see his deep passion for the new technologies—Internet of things (IoT) and cyber–physical systems (CPS)—and his desire to understand how they can help improve everyone's quality of life. Now, I am glad to see his passion for IoT and CPS bear fruit through this book.

IoT is an emerging concept and enabler that has the potential to completely reshape the future of industry. To be exact, IoT is not a completely new concept. It has been around for decades, as can be found in many traditional centralized building-control systems dating back to the 1980s. However, its significance was rediscovered with the emergence of big data analytics, low-cost sensors, and ubiquitous connectivity powered by many modern-age communication technologies. Most importantly, businesses started to realize that new revenue models can be created by adding the IoT concept to their existing product lines, an approach that has fueled the adoption of IoT technologies.

Many people think IoT means "connecting devices." Connectivity is just one piece of the puzzle that defines IoT, which has four layers:

1. At the bottom is the "Hardware" layer, which contains sensors, actuators, chips, and radios—the physical objects that we can touch and feel. Some of the objects are physically small, but others are large, such as cars and airplanes.

2. On top of the Hardware layer is the "Communications" layer, which enables the hardware objects to be connected via wireless or wired communication technologies. It is sometimes misunderstood that IoT is just about these two layers. This is not true.
3. On top of the Communications layer is the "Data Analytics" layer, where the data collected from the bottom two layers are put together and analyzed to extract actionable and useful information. It should be noted that the Data Analytics layer does not necessarily mean big data analytics. For example, the Data Analytics can be a simple sensor data feed into the PID control loop implemented on an 8-bit microcontroller.
4. Finally, there is the "Service" layer on top of the Data Analytics layer, which makes decisions based on the information provided by the Data Analytics layer and takes appropriate actions. The Service layer may include humans as part of the decision process, creating a "human-in-the-loop" system.

It is important to note that the most significant business value of an IoT system is produced at the Service layer where the action is taken. It is quite obvious from the customer's perspective, but it is not widely understood by most of the companies trying to jump into the new wave of IoT phenomena. As more hardware devices become available and connected, the value created by the hardware devices at the bottom layer will continuously decrease as they become gradually commoditized. This is especially true when the cost to manufacture such devices keeps dropping with the growth of the volume. Therefore, the businesses that rely on manufacturing and selling the hardware devices that do not carry a lot of intelligence will likely suffer more. On the other hand, the concentration of the value at the upper layers, such as Data Analytics and Service, will create new lucrative opportunities for the companies that work on extracting useful information from available data sets and monetizing actions based on it.

In this new era of IoT, every company is challenged to come up with new business models while still not only relying on their legacy product lines but also adding new IoT concepts. This is a painful process that requires numerous instances of trial and error, probably including some failures. Moreover, the business models created and validated by a company may not be readily transferred or duplicated by other companies. For example, a new business model created by a jet engine manufacturer using IoT may not be easily adopted by a consumer electronics company. This is a real challenge for many fast followers in the industry, but it is a tremendous opportunity for market leaders who are willing to embrace the new reality and are capable of making investments to create new business cases.

For IoT to be broadly spread, it is important to apply the concept to many applications at scale in our everyday life. Using these advanced technologies, our communities and cities can be made more intelligent, secure, and resilient. The Hardware and Communications layers can serve as part of the city infrastructure, and the Data Analytics and Service layers can provide optimal and synergistic services to the residents. IoT can create tangible benefits to the cities and communities, leading to sustainable smart cities.

The "smart city" concept, by definition, involves many different sectors, including water management, emergency response, public safety, healthcare, energy, transportation, smart home, and even smart manufacturing. Cities strive to coordinate many independent divisions to offer the maximum efficiency and highest quality of service to the residents. However, many smart city solutions are still isolated, fragmented, and built to be a one-off implementation, lacking interoperability, scalability, and replicability. Due to this issue, many communities and cities do not enjoy the level of affordability and sustainability they deserve.

To address this issue, it is important to catalyze the development of new kinds of standards-based, replicable, and interoperable smart city models based on multi-stakeholder involvement and collaboration, so that the cities can leverage each other's investments and the technology providers can create economies of scale. The Global City Teams Challenge (GCTC), a program I lead at the National Institute of Standards and Technology (NIST), is an attempt to encourage just such a transformation of the smart city landscape.

One of the essential elements in the success of IoT and smart city deployment is collaboration and integration among diverse sectors. The value of IoT can be maximized when seemingly unrelated sectors (e.g., healthcare and transportation) get connected and new services are invented using the unique combination of different sectors and businesses. In that sense, successful next-generation IoT and smart city solutions will likely stem from a broad understanding of diverse vertical applications, as well as a fundamental understanding of the cross-sector technical issues.

With over 40 participating authors covering various sectors and applications of IoT, this handbook can provide an overview of many issues and solutions in the complicated IoT playing field. I believe such an interdisciplinary approach is critical in helping readers and the developer community to understand numerous practical issues in IoT and smart cities, and as you examine the contributions of the various authors, I hope you will come to agree with me.

Sokwoo Rhee, Ph.D.
Associate Director of Cyber-Physical Systems Program
National Institute of Standards and Technology
Gaithersburg, MD, USA
February 2016

TECHNICAL ADVISORY BOARD MEMBERS

PREFACE

Designing and implementing a sustainable Internet of Things and data analytics (IoT/DA) project requires core knowledge on a myriad of topics, including invention and innovation, strategic planning, state-of-the-art technologies, security and privacy, business plan, and more. For any successful project, we must consider the following:

- What are the goals?
- What are the givens?
- What are the constraints?
- What are the unknowns?
- Which are the feasible solutions?
- How is the solution validated?

How does one apply technical and business knowledge to optimize a business plan that considers emerging technologies, availability, scalability, sustainability, agility, resilience, best practices, and rapid time to value? Our challenges might include:

- To invent something beneficial
- To design and build using green infrastructure
- To apply best practices to reduce power consumption
- To apply IT technologies, wireless, networks, and cloud
- To prepare a strategic business plan

And this list of challenges is not comprehensive. A good understanding of IoT/DA technologies and their anatomy, taxonomy, ecosystem, and business model will enable one to plan, design, and implement IoT/DA projects successfully.

The goal of this handbook is to provide readers with essential knowledge needed to implement an IoT/DA project. This handbook embraces both conventional and emerging technologies, as well as best practices that are being evolved in the IoT/DA industry. By applying the information encompassed in the handbook, we can accelerate the pace of invention and innovation.

This handbook covers the following IoT/DA topics:

- Business model and strategic planning
- IoT and Industrial IoT
- Data analytics, machine learning, and risk modeling
- Architecture, open source system, security, and privacy
- Microelectromechanical systems and sensor technologies
- Wireless networks and networking protocol
- Wearable designs
- Beacon technology
- Hadoop technology

IoT/DA Handbook is specifically designed to provide technical knowledge for those who are IoT makers and those who are responsible for the design and implementation of IoT/DA projects. It is also useful for IoT/DA decision makers who are responsible for strategic planning. The following professionals and managers will find this handbook to be a useful and enlightening resource:

- C-level executives
- IoT makers and entrepreneurs
- IoT/DA managers and directors
- IoT/DA project managers
- IoT/DA consultants
- Information technology and infrastructure managers
- Network communication engineers and managers

IoT/DA Handbook is prepared by more than 80 world-class professionals from nine countries around the world. It covers the breadth and depth of IoT/DA planning, designing, and implementation and is certain to be the most comprehensive single-source guide ever published in its field.

Hwaiyu Geng, P.E.
Palo Alto, California, USA

ACKNOWLEDGMENTS

The *Internet of Things and Data Analytics Handbook* is a collective effort of an international community of scientists and professionals, with over 70 experts from nine countries around the world.

I am very grateful to the members of the Technical Advisory Board for their diligent reviews for this handbook, confirming technical accuracy while contributing their unique perspectives. Their guidance has been invaluable in ensuring that the handbook can meet the needs of a broad audience.

My sincere thanks to the contributors who took many hours from their professional schedules and personal lives in order to share their wisdom and deep experiences.

Without the Technical Advisory Board members and contributing authors, this handbook could not have been completed. This collective effort has resulted in a work that adds value to the IoT and data analytics communities.

Special thanks go to Dr. Sokwoo Rhee at NIST for the SmartAmerica conference that inspired the creation of this handbook and to Dr. Amy Geng, my daughter, and Todd Park, my son-in-law, for encouraging me to attend the conference.

Many thanks to the following individuals, companies, and organizations for their contributions:

Deborah Geiger, SEMI; Francoise Gilbert, Greenberg Traurig LLP; Dr. Michael A. Huff, MEMS and Nanotechnology Exchange; Dr. Oleg Roderick, Geisinger Health System; Dr. Stan Schneider, Real-Time Innovations, Inc.; Dr. Vanja Subotic, InterDigital; and Dr. Venkataraman Sundareswaran, Teledyne Judson Technologies.

Deloitte University Press, Gartner, Global City Teams Challenge, Harvard Business Review, IBM, IEEE, Industrial Internet Consortium, Intel, International Data Corporation, International Organization for Standardization, McKinsey Global

Institute, MIT Review, NASA, NIST, NOAA, Open Connectivity Foundation, Open Interconnect Consortium, Open Source Hardware Association, Open Source Robotics Foundation, OpenFog Consortium, US Ignite, etc. Thanks are due to many more organizations that are listed in this handbook.

I wish to express my appreciation to the organizers of following events that provided me with comprehensive insights into the IoT and data analytics technologies:

- Big Data Innovation Summit
- Big Data/IoT Summit
- City Innovate Summit
- Global Big Data Conference
- Global IoT Conference
- Hadoop Summit
- Internet of Things Summit
- Internet of Things World
- SEMI
- SmartAmerica Challenge
- SmartGlobal City
- Wearable World Congress

Thanks are due to Brett Kurzman and Alex Castro, Global Research at Wiley, for their support from the beginning of this book project. Thanks also go to Vijayakumar Kothandaraman, K&L Content Management, and his team for the production of this handbook.

Finally, I wish to give my special thanks to my wife, Limei; my daughters, Amy and Julie; and my grandchildren, Abby, Katy, Alex, Diana, and David, for their support and encouragement while I was preparing this book.

Hwaiyu Geng, P.E.
www.AmicaResearch.org
Palo Alto, CA, USA

ABOUT THE COMPANION WEBSITE

The Internet of Things & Data Analytics Handbook is accompanied by a companion website:

www.wiley.com/go/Geng/iot_data_analytics_handbook/

The website includes:

- Figure PPTs
- Case studies
- Web links
- Recommended hardware and/or software for IoT, Data Analytics applications

PART I

INTERNET OF THINGS

1

INTERNET OF THINGS AND DATA ANALYTICS IN THE CLOUD WITH INNOVATION AND SUSTAINABILITY

HWAIYU GENG
Amica Research, Palo Alto, CA, USA

1.1 INTRODUCTION

In January 2016, the US NASA's Global Climate Change reported [1]: "Earth's 2015 surface temperatures were the warmest since modern record keeping began in 1880, according to independent analyses by the National Aeronautics and Space Administration (NASA) and the National Oceanic and Atmospheric Administration (NOAA)."

> The planet's average surface temperature has risen about 1.8°F (1.0°C) since the late 19th century, a change largely driven by increased carbon dioxide and other human-made emissions into the atmosphere (Figure 1.1).

The World Bank issued a report in 2012 [2] describing what the world would be like if it warmed by 4°C (7.2°F): "The 4°C world scenarios are devastation: the inundation of coastal cities, increasing risks of food production potentially leading to higher malnutrition rates; many dry regions becoming dryer, wet regions wetter, unprecedented heat waves in many regions, especially in the tropical substantially exacerbated water scarcity in many region, increase frequency of high-intensity tropical cyclones and irreversible loss of biodiversity, including coral reef system."

Internet of Things and Data Analytics Handbook, First Edition. Edited by Hwaiyu Geng.

Companion website: www.wiley.com/go/Geng/iot_data_analytics_handbook/

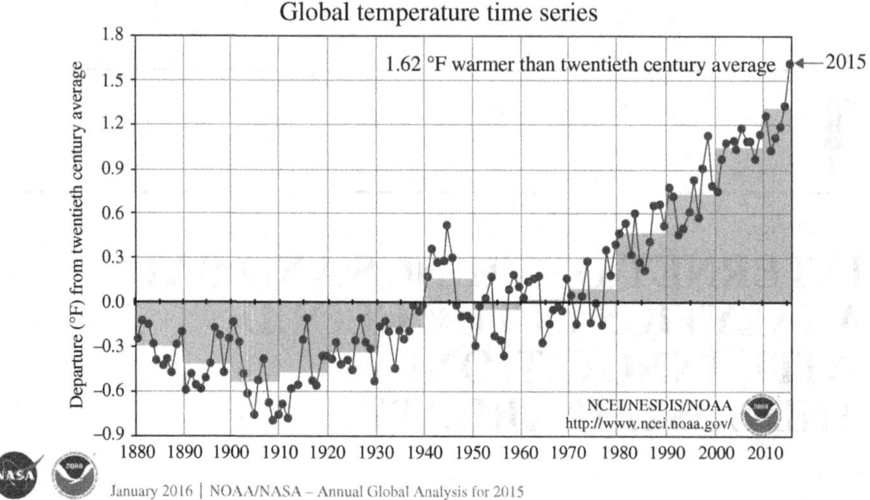

FIGURE 1.1 2015 was planet Earth's warmest year since modern record keeping began in 1880, according to a new analysis by NASA's Goddard Institute for Space Studies. Source: NASA and NOAA.
Note: NASA's data showed that each month in first half of 2016 was the warmest respective month globally in the modern temperature record.

Figure 1.2 shows significant climate anomalies and events in 2015 by major regions or countries. Significant temperature increases and global warming are caused by CO_2 greenhouse effect [3].

Researchers have proven that increasing greenhouse gases (GHGs) and global warming are due to human activities. Human beings generate all kinds of heat from homes, transportations, manufacturing, and data centers that host daunting Internet-related activities [4].

1.2 THE IoT AND THE FOURTH INDUSTRIAL REVOLUTION

The first Industrial Revolution (IR) started in Britain around late eighteenth century when steam power and water power transferred manual labor from homes to powered textile machines in factories. The second IR came in the early twentieth century when internal combustion engines and electrical power generation were invented that led to mass production of T-Model cars by Ford Motor Company. The first two IRs contributed to the higher living standards for mankind. The third IR took place in 1990s when electronics, personal computers, and information technology were used in the automated production systems. Along with technological advancements, the IRs introduced air and water pollution as well as global warming.

The Internet of Things (IoTs) and data analytics are the most significant emerging technologies in recent years that have a disruptive and transformational effect to every industry around the world. The IoT is a technology digitizing the physical world [5].

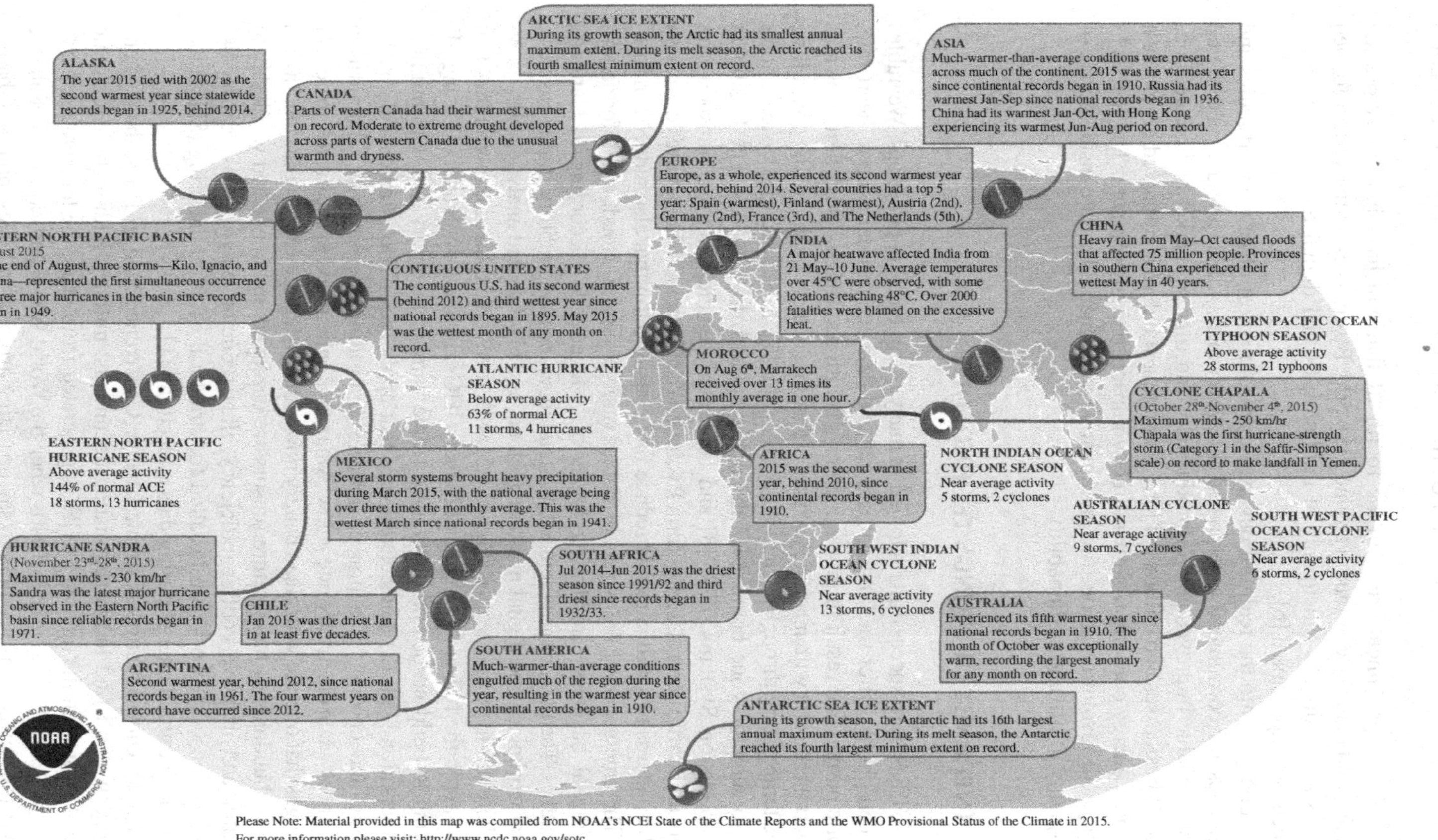
ALASKA
The year 2015 tied with 2002 as the second warmest year since statewide records began in 1925, behind 2014.
ARCTIC SEA ICE EXTENT
During its growth season, the Arctic had its smallest annual maximum extent. During its melt season, the Arctic reached its fourth smallest minimum extent on record.
CANADA
Parts of western Canada had their warmest summer on record. Moderate to extreme drought developed across parts of western Canada due to the unusual warmth and dryness.
ASIA
Much-warmer-than-average conditions were present across much of the continent. 2015 was the warmest year since continental records began in 1910. Russia had its warmest Jan-Sep since national records began in 1936. China had its warmest Jan-Oct, with Hong Kong experiencing its warmest Jun-Aug period on record.
EUROPE
Europe, as a whole, experienced its second warmest year on record, behind 2014. Several countries had a top 5 year: Spain (warmest), Finland (warmest), Austria (2nd), Germany (2nd), France (3rd), and The Netherlands (5th).
EASTERN NORTH PACIFIC BASIN
August 2015
At the end of August, three storms—Kilo, Ignacio, and Jimena—represented the first simultaneous occurrence of three major hurricanes in the basin since records began in 1949.
CONTIGUOUS UNITED STATES
The contiguous U.S. had its second warmest (behind 2012) and third wettest year since national records began in 1895. May 2015 was the wettest month of any month on record.
INDIA
A major heatwave affected India from 21 May–10 June. Average temperatures over 45°C were observed, with some locations reaching 48°C. Over 2000 fatalities were blamed on the excessive heat.
CHINA
Heavy rain from May–Oct caused floods that affected 75 million people. Provinces in southern China experienced their wettest May in 40 years.
WESTERN PACIFIC OCEAN TYPHOON SEASON
Above average activity
28 storms, 21 typhoons
ATLANTIC HURRICANE SEASON
Below average activity
63% of normal ACE
11 storms, 4 hurricanes
MOROCCO
On Aug 6th, Marrakech received over 13 times its monthly average in one hour.
CYCLONE CHAPALA
(October 28th-November 4th, 2015)
Maximum winds - 250 km/hr
Chapala was the first hurricane-strength storm (Category 1 in the Saffir-Simpson scale) on record to make landfall in Yemen.
EASTERN NORTH PACIFIC HURRICANE SEASON
Above average activity
144% of normal ACE
18 storms, 13 hurricanes
MEXICO
Several storm systems brought heavy precipitation during March 2015, with the national average being over three times the monthly average. This was the wettest March since national records began in 1941.
AFRICA
2015 was the second warmest year, behind 2010, since continental records began in 1910.
NORTH INDIAN OCEAN CYCLONE SEASON
Near average activity
5 storms, 2 cyclones
AUSTRALIAN CYCLONE SEASON
Near average activity
9 storms, 7 cyclones
SOUTH WEST PACIFIC OCEAN CYCLONE SEASON
Near average activity
6 storms, 2 cyclones
HURRICANE SANDRA
(November 23rd-28th, 2015)
Maximum winds - 230 km/hr
Sandra was the latest major hurricane observed in the Eastern North Pacific basin since reliable records began in 1971.
SOUTH AFRICA
Jul 2014–Jun 2015 was the driest season since 1991/92 and third driest since records began in 1932/33.
SOUTH WEST INDIAN OCEAN CYCLONE SEASON
Near average activity
13 storms, 6 cyclones
CHILE
Jan 2015 was the driest Jan in at least five decades.
AUSTRALIA
Experienced its fifth warmest year since national records began in 1910. The month of October was exceptionally warm, recording the largest anomaly for any month on record.
SOUTH AMERICA
Much-warmer-than-average conditions engulfed much of the region during the year, resulting in the warmest year since continental records began in 1910.
ARGENTINA
Second warmest year, behind 2012, since national records began in 1961. The four warmest years on record have occurred since 2012.
ANTARCTIC SEA ICE EXTENT
During its growth season, the Antarctic had its 16th largest annual maximum extent. During its melt season, the Antarctic reached its fourth largest minimum extent on record.
NOAA
NATIONAL OCEANIC AND ATMOSPHERIC ADMINISTRATION
U.S. DEPARTMENT OF COMMERCE
Please Note: Material provided in this map was compiled from NOAA's NCEI State of the Climate Reports and the WMO Provisional Status of the Climate in 2015.
For more information please visit: http://www.ncdc.noaa.gov/sotc

FIGURE 1.2 Selected significant climate anomalies and events in 2015. Source: NOAA.

The IoT is a prominent driver to the fourth IR that will have impacts across the business and industry continuum around the world. Business executives and informed citizens are positively anticipating of the fourth IR and digital revolution with low impacts on employment [6].

Applying it into the realm of our lives opens up a host of new opportunities and challenges for consumers, enterprises, and governments. IoT products and digital services enable improvements in productivity and time to market and create thousands of businesses and millions of jobs. Our living standard is improved but at a cost of higher energy consumption that directly impacts our environment. The IoT technology will revolutionize our life and must be implemented with considerations of our quality of life and sustainability.

1.3 INTERNET OF THINGS TECHNOLOGY

The IoT has been successfully adopted in many commercial applications. Wearables and cell phones offer tracking on personalized data such as daily steps, heart rate, calorie burned that result in improving one's health and fitness. Nest, best known as a smart thermostat with machine learning algorithm, centers on the IoT by controlling home temperature through a smartphone from afar [7]. Nest's security system allows you to monitor your home 24/7 through handheld devices. The US Intelligent Transportation Systems (ITS) [8] and connected vehicles improve transportation safety and traffic control. Long-Term Evolution (LTE), or 4G LTE, advises drivers on traffic patterns to avoid jams and reduce fuel consumption that also reduces GHGs.

"The emergence of the Nexus of Forces (mobile, social, cloud, and information) and digital business" are driving forces of "Megatrends" according to Gartner's researches. The megatrends include IoT, smart machines and mobility, digital business, digital workplace and digital marketing, cloud, and big data and analytics [9].

There are hype and reality (Figure 1.3) with many unexplored opportunities that could apply the IoT technology and reduce GHGs. Examples include: (1) using 3D printing to build needed products just-in-time without inventory and with minimal transportation [10]; (2) connected homes that apply the IoT to monitor and control heating and cooling of a house, lighting, entertainment, security, and turn-on appliances when electricity rate is low. Amazon is anticipated to use drones to deliver packages that operate with clean battery power that will reduce CO_2 emission. In data centers, hundreds of temperature sensors are deployed and connected wirelessly to monitor and improve cooling efficiency, thus reducing energy consumption [4]. Wireless sensor networks are installed for structural health monitoring (SHM) that consumes minimal energy during data collection [11].

In 2015 United Nations Climate Change Conference held in Paris, there is one word you won't find in the negotiating documents: military. Although there are no official figures on the amount of GHGs generated from wars, the temperature curve in Figure 1.1 reflects a spike in 1940s when the World War II and explosion of atomic bombs took place. It is needless to say operating armored vehicles, bombers, battleships, and bombings in military actions generate enormous heat and emission that are

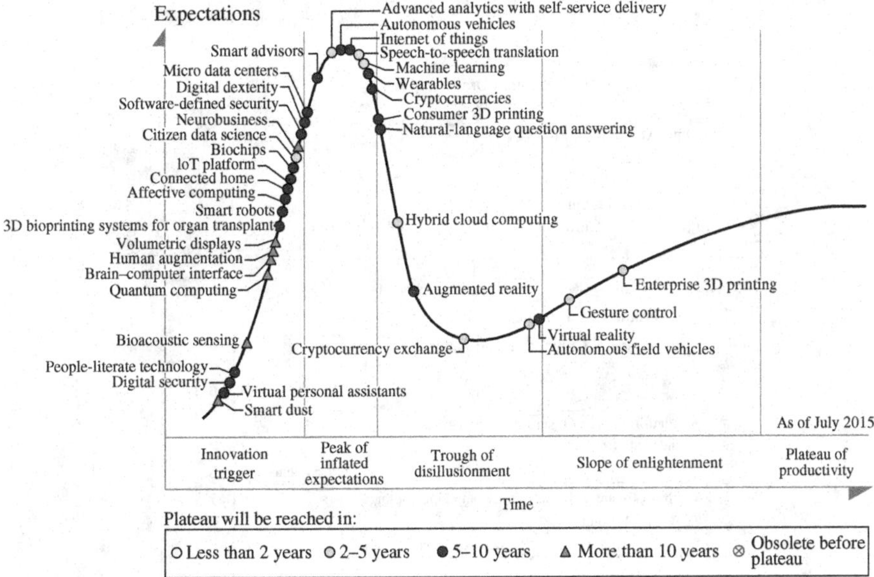

FIGURE 1.3 Hype cycle for emerging technologies, 2015. Source: Gartner (August 2015).

unwarrantedly creating heat to our environment. We have long been taught to help each other, share, keep our hands to ourselves, and work things out. War is an unacceptable behavior that is contradictory to what we have been teaching to our future generations. It creates GHGs that are harmful to our environment.

1.3.1 Definition of IoT/CPS

The term "Internet of Things," also known as "cyber–physical systems (CPS)," at macro level, was first coined by Kevin Ashton in his presentation made to Procter & Gamble (P&G) in 1999. Kevin was working at MIT's AutoID Lab to improved P&G business by linking its supply chain with Radio-Frequency Identification (RFID) information to the Internet.

IEEE has defined it as follows [12]: "Broadly speaking, the Internet of Things is a system consisting of networks of sensors, actuators, and smart objects whose purpose is to interconnect 'all' things, including everyday and industrial objects, in such a way as to make them intelligent, programmable, and more capable of interacting with humans and each other."

1.3.2 Internet of Things Process and Value Chain

A generic way of describing the IoT process was implied in Mark Weiser's ubiquitous computing (ubicomp), and it is well illustrated in the National Institute of Standards and Technology's (NIST) big data reference architecture (Figure 1.4) [13]. The architecture is organized around two axes representing the two Big Data value chains:

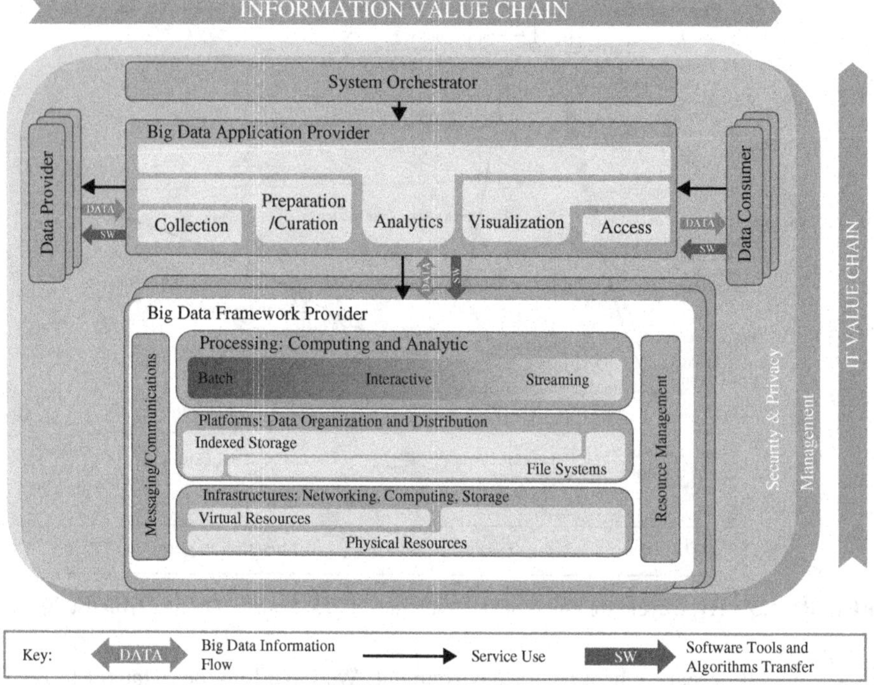

FIGURE 1.4 Big data reference architecture. Source: NIST.

(1) the information (horizontal axis) and (2) the Information Technology (IT) (vertical axis). Along the information axis, the value is created by data collection, preparation/curation, analytics, visualization, and access. Along the IT axis, the value is created by providing infrastructures, platforms, computing, and analytic processing. At the intersection of two axes is the Big Data Application Provider component, indicating that data analytics and its implementation provide the value to Big Data stakeholders in both value chains within "Security and Privacy" and "Management." The names of the Big Data Application Provider and Big Data Framework Provider contain "providers" to indicate that these components provide or implement a specific technical function within the system.

1.3.3 IoT, Pervasive Computing, and Ubiquitous Computing

The IoT, pervasive computing, and ubicomp all share an important trait: computing any time and at any place.

Mark Weiser defined ubicomp as: "enhances computer use by making many computers available throughout the physical environment, but making them effectively invisible to the user."

Both the IoT and pervasive computing use smartphones or handheld devices in the information value loop. "Things" become smarter after being tagged, but ubicomp has horsepower to conduct advanced computing [14] in an environment that is invisible to human beings. Technology advancements over time will narrow the boundaries among IoT, pervasive computing, and ubicomp.

1.3.4 IoT by the Numbers

Gartner research (Table 1.1) shows 4.9 billion connected things worldwide in 2015. A total of 6.4 billion things connected in 2016 (0.9 things per person in the earth), and it will reach to 20.8 billion things (2.7 things per person) in 2020.

In terms of IoT spending, it was $1.2 billion in 2015 and will reach $3.0 billion in 2020 (Table 1.2).

GE estimated that Industrial Internet has the potential to add $15 trillion to global GDP over the next 20 years. IDC's 2015 IoT Global Survey reflects that companies are shifting from planning to execution on IoT plans. The IoT future is here, and we will continue to witness the IoT market in transformation.

1.3.5 Anatomy of the IoT Technology

The concept of IoT is not new. Programmable Logic Controller (PLC) in 1970s is a micro-model of the IoT system that was widely used to control machines and processes within a factory. A PLC consists of inputs (sensors, actuator, and on/off switches), output (digital or analog data), CPU, and communication between components. PLC systems were within a factory, not connected in the Internet and cloud.

Many technologies enable the IoT in connecting products, or things, and services (Table 1.3). The new version of the Internet Protocol (IPv6), supporting 128-bit or 3.4×10^{38} addresses that can connect most atoms in the world (1.33×10^{50} atoms) [15], enables almost unlimited number of devices connected to networks. Sensor prices have declined over the past decades [16]. The size and price of integrated circuit

TABLE 1.1 Internet of Things Units Installed Base by Category (Millions of Units)

Category	2014	2015	2016	2020
Consumer	2,277	3,023	4,024	13,509
Business: cross-industry	632	815	1,092	4,408
Business: vertical specific	898	1,065	1,276	2,880
Grand total	**3,807**	**4,902**	**6,392**	**20,797**

Source: Gartner (November 2015).

TABLE 1.2 Internet of Things End Point Spending by Category (Billions of Dollars)

Category	2014	2015	2016	2020
Consumer	257	416	546	1,534
Business: cross-industry	115	155	201	566
Business: vertical specific	567	612	667	911
Grand total	**939**	**1,183**	**1,414**	**3,010**

Source: Gartner (November 2015).

TABLE 1.3 The Technologies Enabling the Internet of Things

Technology	Definition	Examples
Sensors	A device that generates an electronic signal from a physical condition or event	The cost of an accelerometer has fallen to 40 cents from $2 in 2006. Similar trends have made other types of sensors small, inexpensive, and robust enough to create information from everything from fetal heartbeats via conductive fabric in the mother's clothing to jet engines roaring at 35,000 ft.
Networks	A mechanism for communicating an electronic signal	Wireless networking technologies can deliver bandwidths of 300 megabits per second (Mbps) to 1 gigabit per second (Gbps) with near-ubiquitous coverage.
Standards	Commonly accepted prohibitions or prescriptions for action	Technical standards enable processing of data and allow for interoperability of aggregated data sets. In the near future, we could see mandates from industry consortia and/or standards bodies related to technical and regulatory IoT standards.
Augmented intelligence	Analytical tools that improve the ability to describe, predict, and exploit relationships among phenomena	Petabyte-sized (1015 bytes, or 1,000 terabytes) databases can now be searched and analyzed, even when populated with unstructured (e.g., text or video) data sets. Software that learns might substitute for human analysis and judgment in a few situations.
Augmented behavior	Technologies and techniques that improve compliance with prescribed action	Machine-to-machine interfaces are removing reliably fallible human intervention into otherwise optimized processes. Insights into human cognitive biases are making prescriptions for action based on augmented intelligence more effective and reliable.

processors have dropped with increasing capabilities thanks to Moore's Law. Internet IC chip prices have been declining exponentially. High-bandwidth network technologies such as LTE and LTE-A have arrived. Smartphones become a standard consumer's device that serves as a hub or remote control to IoT. Examples include a personal connected fitness center, connected home, connected car, or connected workplace. Advancements in wireless networking technology and the greater standardization of communications protocols make it possible to collect data ubiquitously from these sensors [17] at very low cost. The IoT technology, in conjunction with big data, has fundamentally transformed how organizations create value to make our lives better.

1.4 STANDARDS AND PROTOCOLS

The Open Systems Interconnection (OSI) model, developed by ISO, is a framework for network communication. The OSI contains seven layers: application, presentation, session, transport, network, data link, and physical. Each layer uses services provided from the layer below it and offers services to the layer above it. The IoT technology concentrates on two broad types of standards, namely, (1) technology standards (network protocols, communication protocols, and data aggregation standards) and (2) regulatory standards related to security and privacy of data [18].

The information collected by sensors must be communicated or transferred over a network, wired or wireless-connected, to other locations for storage and analysis. The process of transferring data from one machine to another needs a unique address for each machine. Internet Protocol (IP) is an open protocol that provides unique addresses to an Internet-connected device such as mobile phones or laptops. For IPv6, there are 128-bit or 3.4×10^{38} (340 undecillion or 340 trillion trillion trillion) Internet addresses.

The Data link layer handles error-free transfer of data frame from one node to another. The data link layer contains two layers: Logical Link Control (LLC) and Media Access Control (MAC). The LLC upper layer controls the multiplexes protocols that provide flow control, acknowledgment, and error notification. The MAC sublayer determines who and when to access media. Some of MAC include IEEE 802.15.4, Wi-Fi, Bluetooth, LTE, ZigBee, NFC, Dash7.

The session layer manages connections, message passing, and termination of a connection between different operating systems. There are many standards and protocols in session layer that include MQTT, XMPP, DDS, SMQTT, OPC UA, CoRE, AMQP, CoAP.

1.5 IoT ECOSYSTEM

We live in a world of cyber and physical things that are fast connecting to each other. They are also ubiquitously connected to our ecological environment that has profound impacts on global warming.

In the IoT ecosystem, all physical things are digitized with digital services and cyber connected, interacted, and functioned to one another and to their physical surroundings. Applying IoT technology will make tremendous economic and environmental impacts that affect global citizens. The following subsections list what potential IoT applications in different sectors that including consumer, government, and enterprise (Figure 1.5) fit together in the IoT ecosystem. Each subsection provide some examples that could be used to facilitate your thoughts to inspire and accelerate the pace of creativity, invention, and innovation.

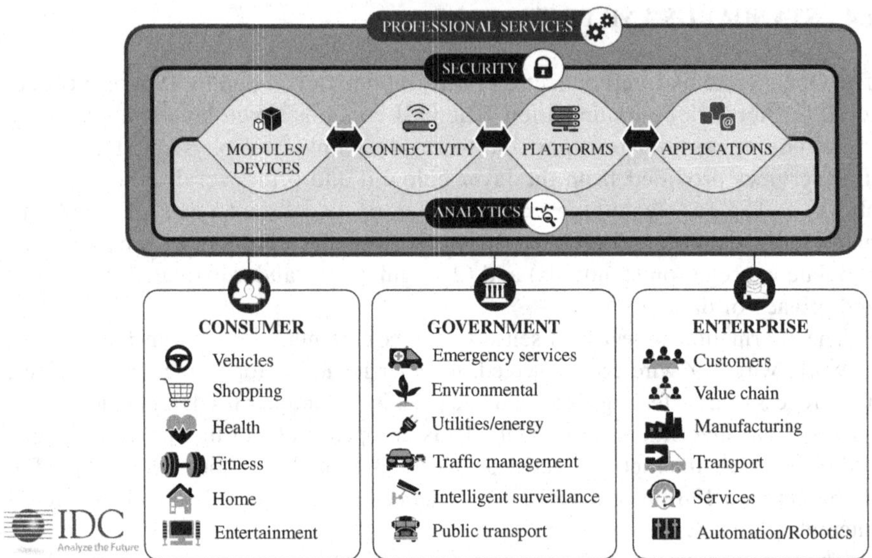

FIGURE 1.5 Internet of Things ecosystem. © International Data Corporation. Use with permission.

1.5.1 Consumer-Facing IoT Applications

Smart home: energy management, water management, home and chore automation, home robots, safety and security, air quality

Connected vehicles: autonomous vehicle, navigation, logistics routing, operations management, condition-based maintenance

Healthcare: illness monitoring and management, personal fitness and wellness

Life and entertainment: hobby, gardening and water, music, smart pet

1.5.2 Government-Facing IoT Applications

Smart city: power and lighting, adaptive traffic management, parking meter, surveillance, events control, natural or human-made disaster management, emergency response system, resource management

Smart transportation: fleet management, connected car, roadway, rail, aviation, port

Smart grid: demand response, power line efficiency

Smart water: domestic waterworks and waste water management

Smart infrastructure: SHM

Environment: environmental monitoring, air quality, landfill and waste management

1.5.3 Enterprise-Facing IoT and Industrial IoT Applications

Energy (Oil/gas, solar, wind, etc.): rigs and wells predictive maintenance, operating management, spill accident management

Smart healthcare: hospital, emergency ambulance service, emergency room, clinic, lab diagnosis, surgery, research, home care, elder care, billing, industrial IoT (IIoT) equipment efficiency, asset management

Smart retails: digital signage, self-checkout, in-store offers, loss prevention, layout optimization, beacon routing, inventory control, customer relationship management

Smart agriculture: wireless sensor on water, tracking cattle, organic food certification

Smart banking: ATM machine, e-statement, online car, or home mortgage

Smart building (office, hotel, airport, education campus, stadium, amusement park, fab and cleanroom, industrial building, data center): energy and water conservation, environment health and safety, security, operating efficiency, equipment maintenance

Smart construction: health, safety, security, inventory control

Smart education: distributed online learning, deep learning

Smart insurance: accident claims, natural disaster claims

Smart logistics: real-time routing, connected navigation, shipment tracking, flight navigation

Smart manufacturing: IIoTs, smart factory, robotics, industrial automation, asset management, energy management, operations management, predictive maintenance, and equipment optimization

1.6 DEFINITION OF BIG DATA

The term "Big data" was first dubbed by Michael Cox and David Ellsworth in 1997 [19]. Big data collects from social networking, video sharing, Internet communication, mobile devices, healthcare (medical records, MRI, CT scan, etc.), science research (astronomy, aerospace, environmental, weather), transportation (land, air, marine traffics), finance and stock market transactions, and sensors and smart devices from IIoTs.

NIST has defined big data as follows: "Big Data consists of extensive data sets—primarily in the characteristics of volume, variety, velocity, and/or variability—that require a scalable architecture for efficient storage, manipulation, and analysis" [20].

1.6.1 Characteristics of Four Vs and the Numbers

Douglas Laney, an analyst at Gartner Research in 2001, introduced the 3Vs data management concept. Volume, velocity, and variety are known colloquially as 3Vs. IBM added one more V, veracity, to the 3Vs. Other Vs that business decision makers, data scientists, and computer scientists have concerns with include value, visualization, validity, volatility. Essential characteristics of 4Vs (volume, velocity, variety, and veracity) are described in the following sections.

1.6.1.1 ***Volume*** Big data implies enormous amount of data at different process that encompass to collect, store, retrieve, process, or update the data. Large data set generated by social media or collected by sensors are large and analyses are massive.

1.6.1.2 ***Velocity*** Big data is often collected in real time at high speed from sources such as businesses, machines, social media, or human interactions via things such as mobile devices. Once received, how fast data could be stored, accessed, processed, analyzed, visualized, and acted becomes crucial.

1.6.1.3 ***Variety*** Big data comes from many sources and can be in homogeneous (structured) or heterogeneous (unstructured) forms. Big data may not be big in number, but may be big in dissimilarity and complexity. For structured data, they could be temperature information from machine tools, mileage information for car maintenance, or turbine blades running hours. For unstructured data, they could be car crash and airbag deployment information, social media with text and video streams, emails with PowerPoint file that contains images, audios, videos, and so on, that are difficult to be sorted. Presenting and rectifying variety of data elements accurately will result in machine learning, quality analytics, precise assessments, and adding value in making informative and accurate business decision.

1.6.1.4 ***Veracity*** The quality of collected data may vary that affects accuracy of analysis. Veracity denotes the completeness and accuracy of the data. There are many uncertainties in the data. The data collected may have "irregularity," "noise," or "dirty" data. It is often said that "garbage in, garbage out."

Big data and the numbers are best illustrated by Figure 1.6 that is presented in the "IBM Big Data and Analytics Hub" [21].

1.6.2 Data Analytics Value Chain

Dan Wagner was the "targeting director" in 2012 President Barack Obama's campaign. He was responsible for collecting voter information, feeding them into his statistical models, and analyzing it. This helped the Democratic National Committee (DNC) to rally individual voters by direct mail and phone. DNC used data analytics technology and successfully aimed and rallied voters, resulting in the return of Obama to office for a second term [22].

NIST describes data life cycle in the Big Data Interoperability Framework [23] having the following components:

- Collection: This stage collects raw data, gathers, and stores data.
- Preparation: This stage screens raw data and cleanses data into cleaned and organized information.
- Analysis: This stage produces knowledge based on cleaned and organized information.
- Action: This stage produces knowledge to generate value for the enterprise.

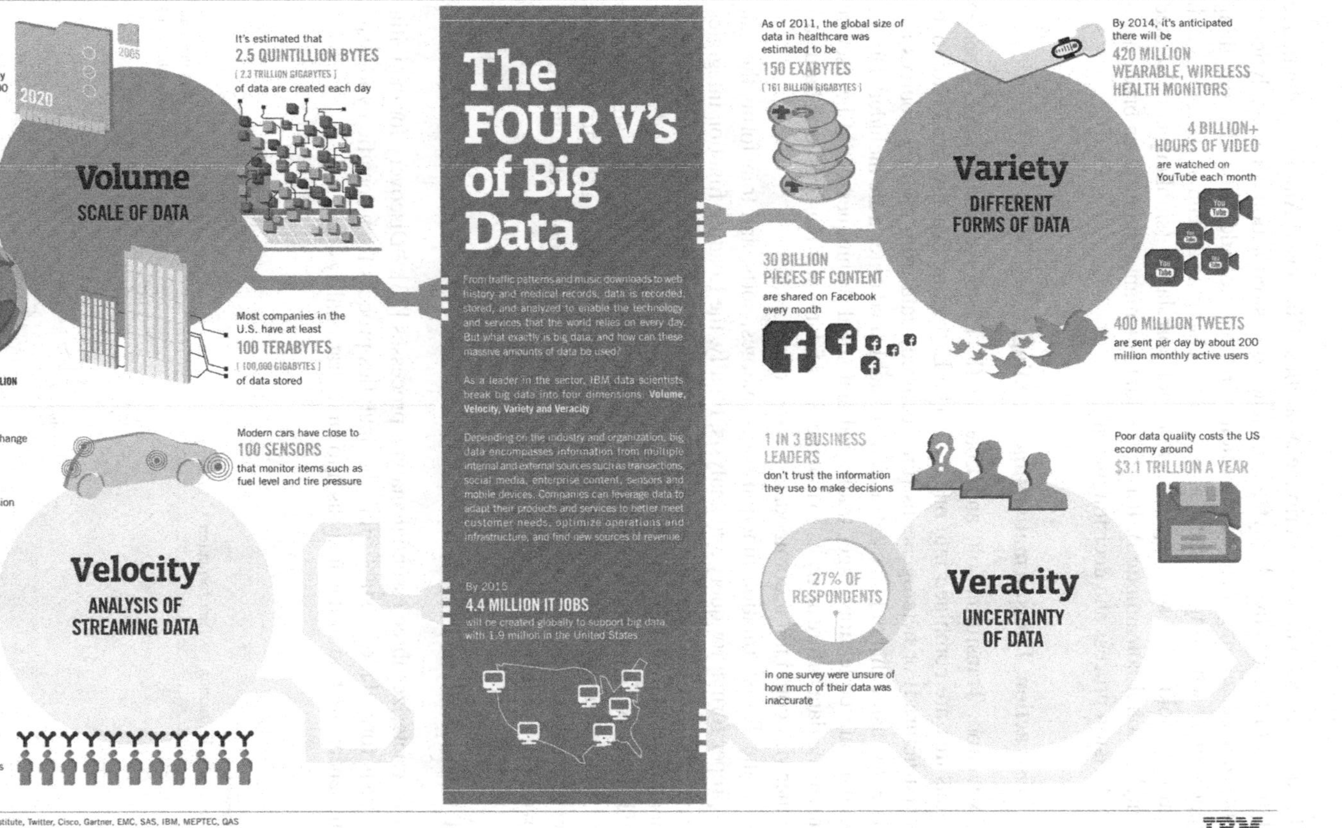

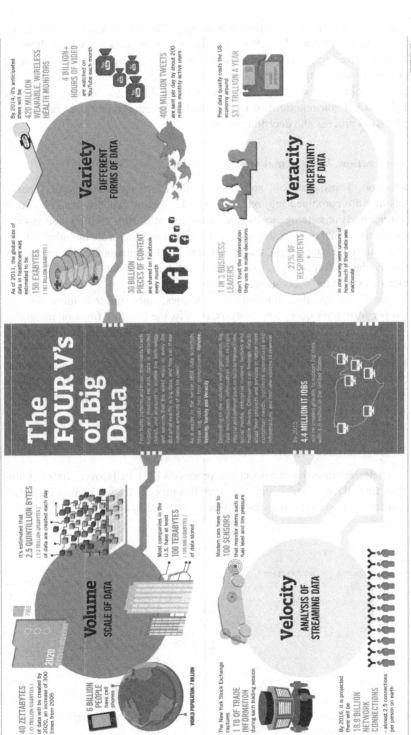

FIGURE 1.6 Big data by the numbers from IBM big data and analytics hub (http://ibmbigdatahub.com). © IBM Analytics. Use with permission.

The analyses encompass Descriptive Analytics, Diagnostic Analytics, Predictive Analytics, and Prescriptive Analytics that lead to desired actions. The analytics in statistics and data mining focus on causation—"being able to describe why something is happening. Discovering the cause aids actors in changing a trend or outcome" [23]. Causation (smoking causes lung cancer) is not correlation (smoking is correlated with high alcohol consumption). It is important for management to embrace data-driven decision process and decision making.

1.6.3 Extraction, Transformation, and Loading

An Extraction, Transformation, and Loading (ETL) tool contains three separate functions that are combined into one tool. The Extraction function reads homogeneous (structured) or heterogeneous (unstructured) data from multiple sources and validates to ensure that only the data meeting established criteria are included. During the initial data collection stage, traditional statistical techniques could be used to downsize the data before analysis so that the size of data set is reasonable on hardware that otherwise could not accommodate the size of data set.

Next, the Transformation function splits, merges, sorts, and transforms the data into a proper format for query and analysis. Finally, the Loading function integrates, arranges, and consolidates data and stores it in a data warehouse ready for analytics applications (Figure 1.7).

ETL tools organize and store the data in relational databases that are easier to query using Structured Query Language (SQL). SQL provides users to modify or retrieve data in centralized or distributed databases. ETL tools can handle data in terabytes (10^{12} bytes) to petabytes (10^{15} bytes).

1.6.4 Data Analytic Process

NIST characterizes three essential analytic processes [23]: "Discovery for the initial hypothesis formulation, establishing the analytics process for a specific hypothesis, and the encapsulation of the analysis into an operational system."

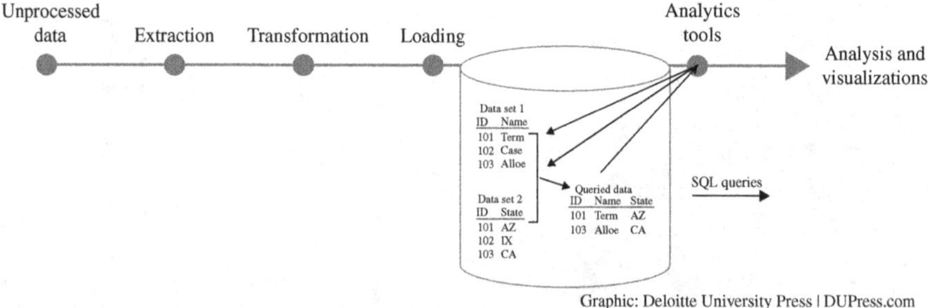

FIGURE 1.7 Typical data aggregation process. © Deloitte. Use with permission. Source: Deloitte analysis. http://dupress.com/collection/internet-of-things/.

Data analytics process can be further detailed in the following value chains:

- Identify problems and determine your business goals.
- Collect, screen, condition, and secure data.
- Explore data and plot the data to see anomalies or patterns.
- Hypothesize and build models.
- Test, validate, and, refine models.
- Deploy model.
- Monitor results and refine model.
- Present for business decision.

"Today, most analytics in statistics and data mining focus on causation" [23]. It is important to emphasize that correlation doesn't imply causation [24].

1.6.5 Data Analytics Tool

There are many data analytics tools available for businesses. The tools cover data cleaning, library, machine learning, statistical analysis, visualization, and Geographic Information System (GIS) mapping. Those data analytics tools enable users to analyze a wide variety of structured and nonstructured information.

Some tools offer free of charge for personal use, such as IBM's Watson Analytics, that get access to cognitive, predictive, and visual analytics. They are "ease of use (no coding and intuitively designed), powerful capabilities (beyond basic excel), and well-documented resources" [25].

A tool interface that supports handling of generic command-line options. Apache Hadoop®, a tool interface that supports handling of generic command-line options, is an open-source framework developed by Apache Software Foundation using Java-based programming framework. The following are some of the tools used in data analytics: Hadoop Distributed File System (HDFS), YARN, MapReduce, Spark.

There is another critical feature of analysis tool: "communicates insight through data visualization." Data visualization and presentation are useful in capturing huge inflow of data and explaining complex results to nontechnical audiences.

1.6.6 Security and Privacy

With the arrival of big data with volume, velocity, variety, and veracity, data security and privacy must be addressed. Security must be given to the retention and use of the data and its metadata (the data describing other data) [18] beyond accessibility. To support big data, the IoT platform must be equipped with protection that has understanding and enforcement of security and privacy requirements (Figure 1.5). The security of the platform must consider distributed computing systems and non-relational data storage as well as various data sets that are increasingly containing with personal identifiable information. Preventing privacy risks on unauthorized access must be considered and designed into IoT applications.

1.7 IoT, DATA ANALYTICS, AND CLOUD COMPUTING

The IoT and big data store and access data and programs over the Internet, or cloud computing, instead of your computer's hard drive or a dedicated Network-Attached Storage (NAS). NIST defines Cloud computing as "A model for enabling ubiquitous, convenient, on-demand network access to a shared pool of configurable computing resources (e.g., networks, servers, storage, applications, and services) that can be rapidly provisioned and released with minimal management effort or service provider interaction."

A cloud model consists of five essential characteristics: on-demand self-service, broad network access, resource pooling, rapid elasticity, and measured service. Cloud computing applies similar business model as utility companies (electricity, natural gas, and water) that is charged as you use.

Mega data centers must have scales in provisioning for peak usage. There is risk of over provisioning due to underutilization.

Cloud computing provides service-oriented architecture, utility computing, virtualization, infrastructure-as-a-service, platform-as-a-service, software-as-a-service [4], and other Web services in cloud. Cloud computing services need no up-front capital investments, pay-as-you-go, no overhead to manage data center, and has speed to deployment. Cloud computing is reliable, scalable, and sustainable.

Cloud computing services are often offered by high-tech companies such as Amazon, Google, Microsoft, and Oracle in the United States. IDC estimates 90% of data for the IoT and Data Analytics will be hosted in cloud in not far future.

Most cloud computing providers experienced crashes and service outages with various recover time. One of the concerns is who own the data that is created in the cloud.

1.8 CREATIVITY, INVENTION, INNOVATION, AND DISRUPTIVE INNOVATION

The IoT is no longer a buzzword; it is a fact of life. The cycle starting from creativity to invention to innovation is a compulsory backbone to flourish products applying IoT technology. A clear understanding will facilitate the discussion of creativity, invention, and innovation.

1.8.1 Creativity, Invention, and Innovation

Creativity is an ability to perceive something unusual and novel as a result of curiosity, inspiration, and imagination (Figure 1.8). Great examples include Sir Isaac Newton and his theory of gravity inspired by a falling apple and Archimedes' Eureka on the volume of water displaced as relating to the volume of his body submerged in a bathtub. Creativity is the capability of conceiving something original. Albert Einstein once said: "Creativity is seeing what everyone else has seen and thinking what no one else has thought."

FIGURE 1.8 Foster creativity. © Amica Research. Use with permission.

Invention is developing something new, satisfying a specific need, and having potential utility value. Bell's telephone and Edison's phonograph and telegraph are well-known examples of invention. Augusta Ada Byron (1815–1852) conceptualized how to instruct a computing machine in binary notation to perform an operation before the first computer was introduced in the 1940. In this way, she was the world's first computer programmer.

"Innovation is making changes to something that already exists by introducing new ideas, new processes, new products, or new business models" [4]. Some good examples include:

1. Flying objects evolved from gliders, propeller airplanes to jet engine airplanes.
2. Light bulbs evolved from incandescent, fluorescent lamps, CFL energy-efficient bulbs to LED lights.

In the article titled "Innovation is not creativity" published in the *Harvard Business Review*, Professor Vijay Govindarajan stated: "Creativity is about coming up with the big idea. Innovation is about executing the idea—converting the idea into a successful business."

1.8.2 Disruptive Innovation

The end of creative, inventive, and innovative cycle is where disruptive (DI) innovation starts.

Professor Clayton Christensen at Harvard Business School coined the term disruptive innovation. "It is a product or service takes root initially in simple applications at the bottom of a market and then relentlessly moves up market, eventually displacing established competitors" [26]. DI is "often centered on customer problems. It is simple yet convincing, accessible, and often cheaper than its competitor."

Classic examples of DI are many: Steve Job's iPod replaced Sony's Walkman; Digital cameras replaced film cameras; LCD terminals replaced CRT terminals; hospitals evolved to clinics in office settings.

There are vast opportunities to applying DI with novel IoT technology in conjunction with drone technology, Uber business model, and many other technologies and business models. "Disruptive" does not have to be novel.

1.9 POLYA'S "HOW TO SOLVE IT"

To brainstorm IoT projects and prepare data analytics model, or business model, the processes explained in the "How to Solve It" by Professor George Polya (1887–1985)—one of the most influential mathematicians of the twentieth century—are extremely helpful. The "How to Solve It" described a heuristic technique in problem solving. The process and steps are excerpted as follows:

- Understanding the problem
- Devising a plan
- Carrying out the plan
- Looking back on your work

1.10 BUSINESS PLAN AND BUSINESS MODEL

What business models are available for your IoT project to be successful? A business model is a simple version and the core concept of a detailed business plan. Peter F. Drucker's theory of business poses a series of assumptions: "assumptions about what a company gets paid for…these assumptions are about markets. They are about identifying customers and competitors, their values and behavior. They are about technology and its dynamics, about a company's strengths and weaknesses…What is the purpose of your business? Who is the customer? What does the customer value? How do you deliver value at an appropriate cost?"

In Professor Michael Porter's "Competitive Strategy," he describes three generic strategies: differentiation, cost leadership, and focus. Those basic strategies still hold true with adding connectivity from the IoT technology.

Clay Christensen suggests a business model should consist of the following elements: a customer value proposition, a profit formula, key resources, and key processes.

Oliver Gassmann's "archetypal business model" illustrates the relationship among key components of a business model (Figure 1.9).

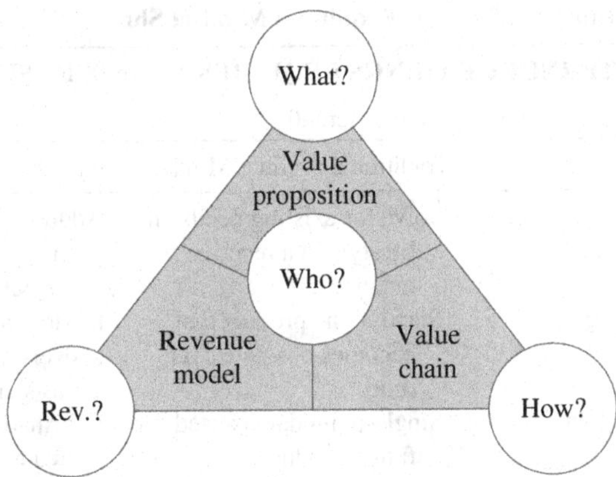

FIGURE 1.9 The archetypal business model. Gassmann O, Frankenberger K, & Csik M, 2014 © Gassmann, Frankenberger, Csik. Use with permission.

1.10.1 Business Model

A business model typically describes a plan on how you transform your creativity, invention, or innovation into a practical product, how you produce, how you market your products, and how you sell them to customers at a profit. The concept of a business model has existed for thousands of years and continues to evolve today. Business models have now propagated with connectivity and sustainability. In Internet age, a product or service can be in the cyber or physical world. By applying the IoT and data analytics technologies along with a "mindset shift" (Table 1.4), many new products with new business models could be created.

1.10.2 Business Model Examples

Exhaustive lists of potential business models with regard to IoT may be found on various websites. One such comprehensive list, "Business Models and the Internet of Things," was prepared by Fleisch, Weinberger, and Wortmann at Bosch IoT Lab, University of St. Gallen, Zurich (www.iot-lab.ch). The following list gives some examples of successful business models for IoT products:

- Add-on model: core is competitive with numerous extras for additional charge such as SAP.
- Affiliation model: someone who helps sell a product in return for commission such as Amazon.
- Digitization: "digitizing physical products" such as Facebook, Dropbox, Netflix.
- Leverage customer data: "making use of what you know" by evaluating customer behavior such as by Amazon, Facebook, Google, Twitter.
- Open source: setting standards by working together and share resources to a community of users, such as the Open Compute Project by Facebook.

TABLE 1.4 The Internet of Things Requires a Mindset Shift

THE INTERNET OF THINGS REQUIRES A MINDSET SHIFT

Because you'll create and capture value differently.

		Traditional Product Mindset	Internet of Things Mindset
Value creation	Customer needs	Solve for existing needs and lifestyle in a reactive manner.	Address real-time and emergent needs in a predictive manner.
	Offering	Stand-alone product that becomes obsolete over time.	Product refreshes through over-the-air updates and has synergy value.
	Role of data	Single-point data is used for future product requirements.	Information convergence creates the experience for current products and enables services.
Value capture	Path to profit	Sell the next product or device.	Enable recurring revenue.
	Control points	Potentially include commodity advantages, Internet Protocol Ownership, and brand.	Add personalization and context; network effects between products.
	Capability development	Leverage core competencies and existing resources and processes.	Understand how other ecosystem partners make money.
SOURCE SMART DESIGN			HBR.org

Source: Hui [27]. © Gordon Hui. Use with permission.

- Pay per use: applying IoT and cloud and pay as you go such as Software as a Service.
- Performance-based contracting: monetizing specialty services based on fee and results by HP, IBM, Oracle.
- Razor and blade (Bait and Hook model): selling razor at low cost with high price on blades for a lifetime of patronage by Gillette razor, Hewlett-Packard's printer and cartridge.
- Servitization: making a product part of a larger service and complete solutions by GE aircraft engine, MRI scanner, Whirlpool washer.

1.10.3 Servitization and Sustainability

Manufacturing companies, for example GE and Whirlpool offer a business model called "servitization." This business model extends product warranty and aftersales service all the way through the end of the product life cycle. By applying cloud computing, this business model can collect big data through embedded sensors, communicate to cloud, analyze the data with machine learning to predict and optimize performance, and even provide prompts for when a component needs maintenance or replacement. Applying IoT with servitization eliminates downtime, extends product life, minimizes overall cost of ownership, saves energy, and maximizes the product performance. The data collected may also be used to improve product design, thus ultimately improving the performance of future products [10]. This business model warrants cradle-to-cradle sustainability.

1.11 CONCLUSION AND FUTURE PERSPECTIVES

IoT technology is here to stay. Big data analytics and machine learning provide intelligence and play a pivotal role in driving IoT devices. These technologies have the potential to improve our economy, productivity, quality of life, and environment if the IoT devices are designed and implemented correctly. But there are many challenges in implementing these technologies as they are new and evolving. Users are inexperienced in selecting IoT platforms, Data Base Management System (DBMS) for multi-Vs data, and IoT services and using complex data analytics tools in modeling with unclear goals. Building a consumer's IoT infrastructure is a complex challenge, and understanding IoT is even a more daunting task. We as a society have an unprecedented opportunity to harness IoT for public good.

To be better prepared and to set the course, it is essential to understand what IoT is, what machine learning and data analytics is, and how to design devices with low power consumption, power harvesting, and sustainability. This chapter introduced the fundamentals and anatomy of IoT and data analytics technologies. The IoT and data analytics technologies interconnect with breadth, depth, and ubiquity. The IoT ecosystem is organized in consumer-facing, government-facing, and enterprise-facing applications. Within each group, different sectors and potential applications are suggested. To fuel new products and new processes in the IoT ecosystem, this chapter also discusses creativity, invention, and innovation as well as DI and "How to Solve It!"

Setting IoT's vision, goals, strategy, and implementation roadmap is challenging due to the nascent nature of technologies, business models, standards, security, and privacy [28]. Huge amounts of heterogeneous data require practitioners from science, mathematics, and statistics to work together with data scientists. This need for cooperation should start with creating interdisciplinary teams from the bottom to the top. Combining big data with machine learning and artificial intelligence for prediction and optimization will provide the greatest value. Pareto's principle, or the 80/20 rule, can be applied as a strategy in selecting IoT projects and prescreening data used in data analytics. Advanced manufacturing with cradle-to-cradle sustainability must be incorporated into IoT products.

To embrace some current conditions and future predictions, the NIST has been actively promoting Global city projects. Many smart city projects have been done in Europe, China, Japan, Korea, Singapore, and other countries [29].

IDC's 2015 IoT Global Survey indicates that "home automation and control represents 35% of consumer IoT installed base by 2020." Connected cars can affect megasized impacts to consumers and society in terms of pollution reduction, improving living standards, increasing social life, and ensuring safety, among many potential benefits. Retail section has the lowest IoT opportunity but is the fastest-growing sector among other sectors that include manufacturing, consumer, and transportation. Standardization will occur in the IoT technology. Security attacks will be handled by automation.

Many topics discussed in this chapter are addressed more fully in other chapters by experts from nine countries around the world. With collective effort, we can apply best practices to accelerate the pace of invention and innovation in IoT products that will improve our life and sustainability.

REFERENCES

[1] Analyses Reveal Record-Shattering Global Warm Temperatures in 2015, NASA's News, January 20, 2016. http://climate.nasa.gov/news/2391/ (accessed January 21, 2016).

[2] A Report for the World Bank by the Potsdam Institute for Climate Impact Research and Climate Analytics. Turn Down the Heat: Why a 4°C Warmer World Must Be Avoid, The World Bank, Washington DC, 2012.

[3] Weart, S., The Discovery of Global Warming, Harvard University Press, second edition, 2008. Timeline (Milestones), https://www.aip.org/history/climate/timeline.htm, American Institute of Physics, February 2015 (accessed August 13, 2016).

[4] Geng, H., *Data Center Handbook*, John Wiley & Sons, Inc., Hoboken, 2014.

[5] Manyika, J., Chui, M., Bisson, P., Woetzel, J., Dobbs, R., Bughin, J., Aharon, D., *The Internet of Things: Mapping the Value Beyond the Hype*, McKinsey Global Institute, McKinsey & Company, New York, 2015.

[6] "2016 GE Global Innovation Barometer," GE Reports, General Electric, January 2016. http://www.gereports.com/innovation-barometer-2016/ (accessed August 13, 2016).

[7] Poindexter, O., The Internet of Things Will Thrive on Energy Efficiency, Future Structure, Government Technology, Washington DC, July 2014. http://www.govtech.com/ (accessed August 13, 2016).

[8] Intelligent Transportation Systems. http://www.its.dot.gov/landing/cv.htm (accessed February 10, 2016).

[9] Burton, B., Willis, D., Gartner's Hype Cycles for 2015: Five Megatrends Shift the Computing Landscape, Gartner, August 2015. https://www.gartner.com/doc/3111522/gartners-hype-cycles-megatrends-shift (accessed August 13, 2016).

[10] Geng, H., *Manufacturing Engineering Handbook*, McGraw-Hill Education, New York, second edition, 2016.

[11] Fu, T., Ghosh, A., Johnson, E., Krishnamachari, B., Energy-efficient deployment strategies in structural health monitoring using wireless sensor networks, *Structural Control Health Monitoring Journal*, Vol. **20**, 971–986, 2012.

[12] IEEE Standards Association (IEEE-SA). *Internet of Things (IoT) Ecosystem Study*, IEEE Standards Association, The Institute of Electrical and Electronics Engineers, Inc., 2015. http://www.sensei-iot.org/PDF/IoT_Ecosystem_Study_2015.pdf (accessed August 13, 2016).

[13] NIST Big Data Public Working Group. NIST Big Data Interoperability Framework: Volume 7, Standards Roadmap, NIST Big Data Public Working Group, NIST Special Publication 1500-7, September 2015.

[14] Fleisch, E., What Is Internet of Thing, Auto-ID-Labs White Paper WP-BIZAPP-053, Zurich, January 2010.

[15] Weisenberger, D., How Many Atoms Are There in the World? https://en.wikipedia.org/wiki/Atom#Earth (accessed February 16, 2016).

[16] Simpson, L., Lamb, R., *IoT: Looking at Sensors*, Equity Research & Strategy, Jefferies, New York, 2014.

[17] Chui, M., Löffler, M., Roberts, R., *The Internet of Things*, McKinsey Quarterly, McKinsey & Company, New York, 2010.

[18] Holdowsky, J., Mahto, M., Raynor, M., Cotteleer, M., *Inside the Internet of Things (IoT)*, Deloitte University Press, New York, 2015.

[19] Cox, M., Ellsworth, D., Application-Controlled Demand Paging for Out-of-Core Visualization, Report NAS-97-010, NASA Ames Research Center, Moffett Field, July 1997.

[20] NIST Big Data Public Working Group. NIST Big Data Interoperability Framework: Volume 1, Definitions, NIST Big Data Public Working Group, NIST Special Publication 1500-1, September 2015.

[21] IBM Big Data and Analytics Hub. http://www.ibmbigdatahub.com/ (accessed February 2, 2016).

[22] Issenberg, S., How Obama's Team Used Big Data to Rally Voters, MIT Technology Review, Cambridge, December 2012. https://www.technologyreview.com/s/509026/how-obamas-team-used-big-data-to-rally-voters/ (accessed February 2, 2016).

[23] NIST Big Data Public Working Group. NIST Big Data Interoperability Framework: Volume 2, Big Data Taxonomies, NIST Big Data Public Working Group, NIST Special Publication 1500-2, September 2015.

[24] Australian Bureau of Statistics. http://www.abs.gov.au/websitedbs/a3121120.nsf/home/statistical+language+-+correlation+and+causation (accessed August 13, 2016).

[25] Jones, A., *Top 10 Data Analysis Tools for Business*, June 2014. http://www.kdnuggets.com/2014/06/top-10-data-analysis-tools-business.html (accessed February 2, 2016).

[26] Clayton Christensen Disruptive Innovation. http://www.claytonchristensen.com/key-concepts/#sthash.UEi5YWYE.dpuf (accessed August 13, 2016).

[27] Hui, G., How the Internet of Things Changes Business Models, Harvard Business Review, July 29, 2014. https://hbr.org/2014/07/how-the-internet-of-things-changes-business-models (accessed August 13, 2016).

[28] Wallin, L., Jones, N., Kleynhans, S., *How to Put an Implementable IoT Strategy in Plan*, Gartner, New York, 2015.

[29] NIST Global City Teams Challenge. http://www.nist.gov/cps/sagc.cfm (accessed August 13, 2016).

FURTHER READING

Altshuller, G., *40 Principles TRIZ Keys to Technical Innovation*, Technical Innovation Center, Inc., Worcester, 2002.

Altshuller, G., *TRIZ—The Theory of Inventive Problem Solving*, Technical Innovation Center, Inc., Worcester, 2004.

Bassi, A., Horn, G., Internet of Things in 2020, European Commission/EPoSS Workshop Report, Brussels, September 2008.

Fleisch, E., What Is the Internet of Things? Auto-ID Labs White Paper WP-BIZAPP-053, University of St. Gallen, Zurich, January 2010.

Fry, A., *Creativity, Innovation and Invention: A Corporate Inventor's Perspective*, Vol. **13**, pp. 1–5, The Creative Problem Solving Group, Inc, CPSB's Communique', Orchard Park, 2002.

Gleeson, A., Have You Optimize Your Business Model? Palo Alto Software, Bplans.co.uk, http://articles.bplans.co.uk/starting-a-business/have-you-optimised-your-business-model/1037 (accessed February 19, 2016).

Govindarajan, V., Innovation Is Not Creativity, Harvard Business Review, Cambridge, August 3, 2010. http://hbr.org/2010/08/innovation-is-not-creativity (accessed August 13, 2016).

Griffith, E., What Is Cloud Computing? PC Magazine, April 17, 2015. http://www.pcmag.com/article2/0,2817,2372163,00.asp (accessed February 22, 2016).

Gupta, P., Trusko, B., *Global Innovation Science Handbook*, McGraw-Hill Education, New York, 2014.

Isaacs, C., 3 Ways the Internet of Things Is Revolutionizing Health Care, Forbes, September 3, 2014.

Kingdon, M., *The Science of Serendipity*, John Wiley & Sons, Ltd, Chichester, 2012.

Kuczmarski, T.D., Innovation Always Trumps Invention, January 19, 2011. http://www.bloomberg.com/news/articles/2011-01-19/innovation-always-trumps-invention (accessed August 23, 2015).

Mack, C., Fifty Years of Moore's Law, *IEEE Transaction on Semiconductor Manufacturing*, Vol. **24**, No. 2, 202–207, 2011.

Manyika, J., Chui, M., Bisson, P., Woetzel, J., Dobbs, R., Bughin, J., Aharon, D., *The Internet of Things: Mapping the Value Beyond the Hype*, McKinsey Global Institute, McKinsey & Company, New York, 2015.

Presser, M., *Internet of Things*, Alexandra Institute, Copenhagen, 2013.

Rifkin, J., *The Zero Marginal Cost Society*, Palgrave MacMillan, New York, 2014.

Schwartz, M., *Internet of Things with the Arduino Yun*, Packt Publishing, Birmingham, 2014.

Spencer, J., *Connected Vehicles Internet of Things World*, U.S. Department of Transportation, Federal Transit Administration, Washington, DC, 2016.

Tidd, J., Bessant, J., *Managing Innovation*, John Wiley & Sons, Ltd, Chichester, fifth edition, 2013.

Vermesan, O., Friess, P., *Internet of Things: Converging Technologies for Smart Environments and Integrated Ecosystems*, River Publishers, Aalborg, 2013.

World Meteorological Organization, Responding to the Challenges of Climate Change, *WMO Bullet*, Vol. **64**, No. 2, 2015, The Journal of the World Meteorological Organization, Geneva, Switzerland, 2015.

Worstall, T., Using Apple's iPhone to Explain the Difference between Invention and Innovation, Forbes, April 20, 2014. http://www.forbes.com/sites/timworstall/2014/04/20/using-apples-iphone-to-explain-the-difference-between-invention-and-innovation/ (accessed August 20, 2015).

USEFUL WEBSITES

3rd Generation Partnership Project (3GPP). http://www.3gpp.org/ (accessed February 28, 2016).

Alliance for Telecommunications Industry Solutions (ATIS). http://atis.org/ (accessed February 28, 2016).

Allseen Alliance. https://allseenalliance.org/ (accessed February 28, 2016).

Beecham Research World of IoT and Wearable. http://www.beechamresearch.com/article.aspx?id=20 (accessed February 28, 2016).

Bluetooth SIG. https://www.bluetooth.com/ (accessed February 28, 2016).

Broadband Forum (BBF). http://www.broadband-forum.org/ (accessed February 28, 2016).

Broadband Forum–TR-069. www.broadband-forum.org/technical/download/TR-069.pdf (accessed February 28, 2016).

Cisco Blog. http://blogs.cisco.com/diversity/the-internet-of-things-infographic (accessed February 28, 2016).

Consumer Electronics Association (CEA). https://www.standardsportal.org/usa_en/sdo/cea.aspx (accessed February 28, 2016).

Deloitte University Press' Internet of Things Collection. http://dupress.com/collection/internet-of-things (accessed February 28, 2016).

Digital Living Network Alliance (DLNA). http://www.dlna.org/ (accessed February 28, 2016).

Eclipse M2M Industry Working Group. http://eclipse.org/org/workinggroups/m2miwg_charter.php (accessed February 28, 2016).

European Telecommunications Standards Institute (ETSI). http://www.etsi.org/ (accessed February 28, 2016).

Gartner IT Glossary. http://www.gartner.com/it-glossary/ (accessed February 28, 2016).

Government Technology. http://www.govtech.com/ (accessed January 21, 2016).

GSM Association (GSMA). http://www.gsma.com/ (accessed February 28, 2016).

Harvard Business Review. https://hbr.org/ (accessed February 28, 2016).

Health Level Seven International (HL7). www.hl7.org/ (accessed February 28, 2016).

Home Gateway Initiative (HGI). http://www.homegatewayinitiative.org/ (accessed August 13, 2016).

IEEE 802.15.4. http://ieee802.org/15/pub/TG4.html (accessed February 28, 2016).

IEEE P2413. http://standards.ieee.org/develop/project/2413.html (accessed February 28, 2016).

IEEE-SA IoT Ecosystem Study. http://standards.ieee.org/innovate/iot/study.html (accessed February 28, 2016).

Industrial Internet Consortium (IIC). www.iiconsortium.org (accessed February 28, 2016).

Intel Big Data Analytics. https://www-ssl.intel.com/content/www/us/en/big-data/big-data-analytics-turning-big-data-into-intelligence.html (accessed February 28, 2016).

International Data Corporation. http://www.idc.com/ (accessed February 28, 2016).

International Electrotechnical Commission (IEC). www.iec.ch (accessed February 28, 2016).

International Organization of Standardization (ISO). www.iso.org (accessed February 28, 2016).

International Society of Automation (ISA). https://www.isa.org/ (accessed February 28, 2016).

Internet Engineering Task Force (IETF). www.ietf.org (accessed February 28, 2016).

Internet Protocol Smart Objects (IPSO) Alliance. www.ipso-alliance.org (accessed February 28, 2016).

IoT European Research Cluster (IERC). http://www.internet-of-things-research.eu/ (accessed February 28, 2016).

IPSO Alliance. http://www.ipso-alliance.org/ (accessed February 28, 2016).

ISO/IEC 7498 Open Systems Interconnection. http://www.iso.org/iso/home/store/catalogue_tc/catalogue_detail.htm?csnumber=25022 (accessed February 28, 2016).

ITU-T Focus Group M2M. http://www.itu.int/en/ITU-T/focusgroups/m2m/Pages/default.aspx (accessed February 28, 2016).

McKinsey Global Institute. http://www.mckinsey.com/mgi/overview (accessed February 28, 2016).

MIT Slogan Management Review. http://sloanreview.mit.edu/ (accessed February 28, 2016).

NIST Big Data. http://www.nist.gov/itl/bigdata/20150406_big_data_framework.cfm (accessed February 2, 2016).

NIST Cyber Physical Systems. http://www.nist.gov/cps/sagc.cfm (accessed February 28, 2016).

OASIS Message Queuing Telemetry Transport (MQTT). http://mqtt.org/ (accessed February 28, 2016).

oneM2M. www.oneM2M.org (accessed February 28, 2016).

Open Connectivity Foundation (OCF). http://openconnectivity.org/ (accessed February 28, 2016).

Open Mobile Alliance (OMA). http://openmobilealliance.org/ (accessed February 28, 2016).

OpenIoT. http://www.openiot.eu/ (accessed February 28, 2016).

Organization for the Advancement of Structured Information Standards (OASIS). https://www.oasis-open.org/ (accessed February 28, 2016).

PC Encyclopedia. http://www.pcmag.com/encyclopedia/index/a (accessed February 28, 2016).

Personal Connected Health Alliance (PCHA). http://www.continuaalliance.org/pchalliance (accessed February 28, 2016).

SAE International (SAE). http://www.sae.org/ (accessed February 28, 2016).

Semiconductor Manufacturing, IEEE Transactions. http://ieeexplore.ieee.org/xpl/RecentIssue.jsp?punumber=66 (accessed April 19, 2015).

Smart Grid Interoperability Panel (SGIP). http://www.sgip.org/ (accessed February 28, 2016).

Smart Manufacturing Leadership Coalition (SMLC). https://www.smartmanufacturingcoalition.org/ (accessed February 28, 2016).

Thread Group. http://www.threadgroup.org/ (accessed February 28, 2016).

Weightless SIG. http://www.weightless.org/ (accessed February 28, 2016).

World Wide Web Consortium (W3C). http://www.w3.org/ (accessed February 28, 2016).

ZigBee Alliance. http://zigbee.org/ (accessed February 28, 2016).

2

DIGITAL SERVICES AND SUSTAINABLE SOLUTIONS

RIKKE GRAM-HANSEN

Copenhagen Solutions Lab, City of Copenhagen, Copenhagen, Denmark

2.1 INTRODUCTION

The majority of the world's population is connected to the Internet via a variety of devices, but the *Internet of Things* (IoT) is not yet a term that most people are familiar with. The IoT does not only connect billions of things but also connect people, services, infrastructure, manufacturing, and many other areas of our physical world and society. To most people, technology and data are not interesting in themselves, but what does interest people is how data and technology can be used. When technology is the answer, what is then the question? Digital services are central in this respect, as they are essentially an answer to what people want: better services that will help them achieve a higher quality of life.

The focus of this chapter is on exactly that: the services that the technological development in IoT and cyber–physical systems makes possible. In the last decades, we have witnessed a fundamental transformation of services to consumers, businesses, and citizens in general. Services from both public sector and commercial players are to a greater extent delivered digitally via the Internet; hence the term "digital services" is now widely used. Digital services are automated, which makes the replication and customization fast and easy, and in a connected society the distribution is nearly free. This goes from banking services, which are available online, to tax filing, to streaming television, to taking an education via massive open online courses. The services offered by IoT go even further than the conventional interfaces

Internet of Things and Data Analytics Handbook, First Edition. Edited by Hwaiyu Geng.

Companion website: www.wiley.com/go/Geng/iot_data_analytics_handbook/

of digital services, namely, computer screens and mobile phones, as data, processing, and connectivity become embedded in everyday objects and everyday situations. IoT takes computing away from the screen and into the everyday life. We interact with IoT without even knowing it, because it is hidden to the eye. The questions this chapter will answer are the following: What are services in a hyperconnected world? What does the IoT impose of opportunities and challenges in regard to delivering digital services to people? What role can the public sector play in ensuring a high quality of service for citizens?

2.2 WHY IoT IS NOT JUST "NICE TO HAVE"

Governments all over the world look to the development in information and communication technology and IoT because of its potential in bringing about new solutions to some of the great challenges of our time, such as challenges to the environment, the quality of life, and the economy. We live in an increasingly urbanized world, where cities house the greatest part of the world's population and economic activities, but cities also account for the greatest pollution and CO_2 emissions. Cities are therefore the central loci which frame the discussions in this chapter, although we should keep in mind that connectivity pervades far beyond city limits, to homes, offices, suburbs, and countryside.

The connected society brings about data that holds the potential to offer citizens with better services while reducing emissions and optimizing city resources, as is illustrated in the concept of Copenhagen Connecting (Figure 2.1). Technologies are not solutions in themselves however. For technologies to actually solve real-world challenges, they need to be contextualized. We need to understand the complexity of cities and urban challenges, and these specific problems should be the driver for innovation in technologies. Technological and social aspects are interlinked, as emphasized by Hajer [1]. This is well illustrated in Hajer's example of the car. The car is not just a technology but part of a broader sociotechnical system. The development of the car has been determining for the structure of Western cities and societies during the twentieth century. Suburbs sprawled, shopping centers thrived, central business districts rose, and express highways cut their ways through cities and landscape. The car means a great deal of freedom as well as physical and social mobility to many people, and the car industry is an important driver in the economy, but it also causes a great deal of problems, from CO_2 emissions, pollution, and congestion to the massive amount of space taken up by roads and parking lots and the numerous people who die in traffic accidents every day.

The exponential development in technology gives us new possibilities on how to look at mobility. Cars have sensors installed that give exact location data, as well as a variety of other information about the car and its environment, and with the access to mobile platforms for sharing this information in real time, we are no longer bound to owning our own private car. We can comfortably share a car with others or buy the mobility we need as a service. Research from MIT shows that by combining ride sharing with car sharing, in a city like New York, would mean 80% fewer cars to satisfy for the needs of all passengers [2].

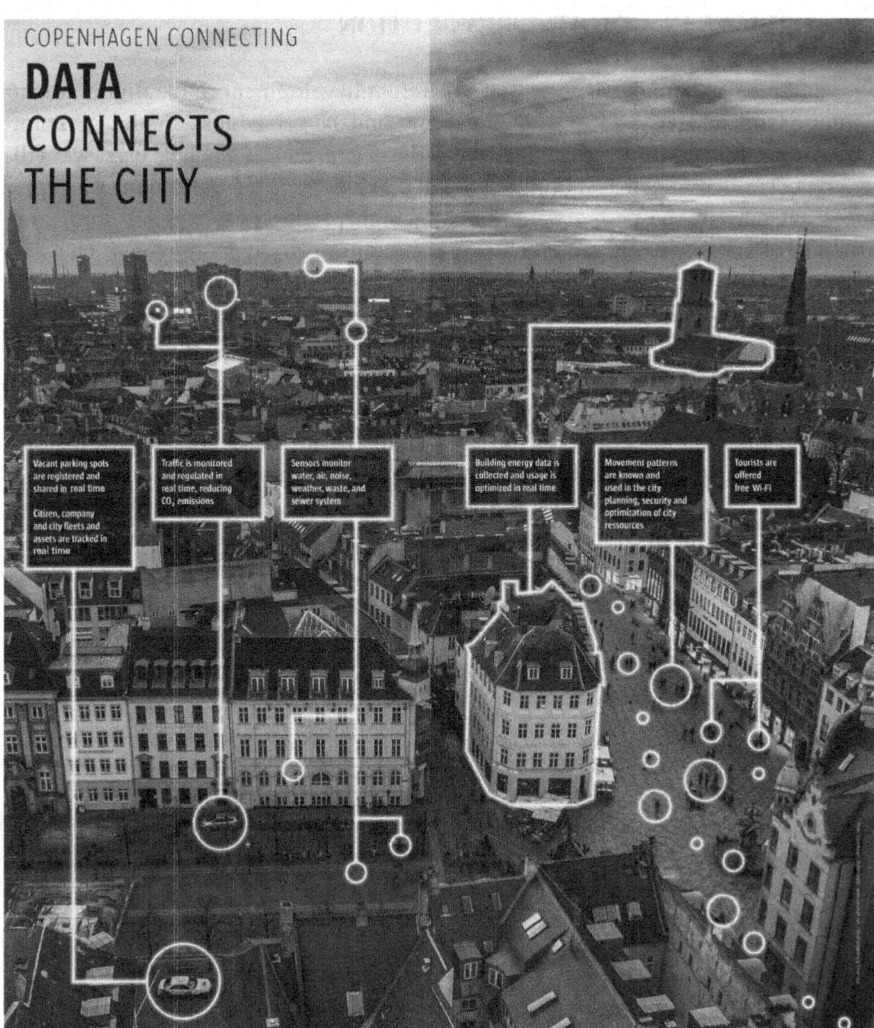

FIGURE 2.1 Data connects the city in ways that offer citizens with better services while reducing emissions and optimizing city resources. Copenhagen Connecting, www.cphsolutionslab.dk, photo: Copenhagen Media Center/www.caecaccph.com/Jacob Schjørring & Simon Lau.

What can we learn from this example of the car? Within the discourse of IoT, there is a lot of "nice to have," but in a broader societal perspective, there is also very much a "need to have" related to IoT. The connection of everything allows for new structuring principles for our society that breaks with the detrimental conduct of the past. It provides data and knowledge for decision makers, as well as ordinary people, to act upon, and it creates a feedback loop of information about the environment, which can perhaps help bring about a more efficient and responsible use of the scarce resources of our planet.

2.3 SERVICES IN A DIGITAL REVOLUTION

The example of the car also shows us a different development in the digital transformation of services, as digital platforms and new business models disrupt traditional markets; for instance, instead of buying products, such as cars, we will be buying services, such as mobility. Digital services have become an increasingly important part of the economy, promising high productivity, transforming the structure of employment, and increasing competition among companies. The service component is a way to distinguish one product from another similar offering to avoid the commoditization of products, and digital technologies are central to this transformation [3]. The transformation of services means a breaking down of barriers between products and services, manufacturing and services, and traditional industrial sectors. The digitization of services is a profound change to society, as the tasks and processes underlying services that were previously performed by people can be expressed as digital information. When services become formalized and codified into computable algorithms, they can be replicated, analyzed, reconfigured, and customized indefinitely, creating new services and new forms of value at a tremendous pace [3].

The walled gardens of early technology, where large capital investments in independent data centers was required to even begin developing digital services, have been broken down. The move toward a connected, open access via cloud computing to storage, computational power, connectivity, distribution, marketing, etc. has made the barriers to entry very low, and this has opened up for new, smaller, and more agile players on the market. The openness and platform-based systems also allow for mash-upping data and existing services into new ones. A prime example of this is Google Maps, which has become the dominant mapping platform, because of its openness and available APIs that allow for developers to reuse the mapping service in ways that were not conceived by Google [4].

2.4 MOBILE DIGITAL SERVICES AND THE HUMAN SENSOR

In 2007, the iPhone was introduced by Apple. This was not only the launch of an innovative technology but also the commencement of a new era of mobile services and mobile connectivity. By now, the Smartphone has become a common property connecting billions of people and giving users instant availability of video, audio, location, motion, voice, (social) media, data, and so much more, proliferating the connectivity of people and things. The Smartphone is an important terminal for most cyber–physical interaction. A Smartphone is, essentially, a bundle of sensors and data communication that is well maintained by the users themselves, who ensure recharging, updating software, etc. and who bring them everywhere.

Mobile digital services are ever present to their users, who have all applications and information at their hand at all times. But the pervasiveness of digital services is not only a downstream phenomenon, as it gives a constant flow of data upstream from all the devices in real time. People become human sensors of their environment, and in many cases this makes additional IoT devices superfluous. This poses some obvious problems in terms of privacy and consent—problems which are treated elsewhere in this volume—but the potential in this information is not just the commercial value of it, but the actual potential in the fact that data can be available when needed for users to act upon; whether it is finding the way, the weather report, or whatever question the user might have, the answer is always there.

2.5 NOT JUST ANOTHER APP

Innovation in digital services does not end at the Smartphone, however, as information in a connected world is ambient and situational, beyond the specific interface of the Smartphone. Regardless of how smart the Smartphone is, IoT opens up for a whole range of different opportunities of human–machine interaction beyond that of the Smartphone screen, which holds a promise for a much more fulfilling provision of services to people. One of the problems with the Smartphone, or any other personal computing device, is that it only provides the user with value as long as the user interacts with it. This makes us spend a lot of time staring into the bright screens of our Smartphones—as nightmarishly depicted by Rose [5]—and it takes up a lot of our cognitive resources. The cognitive overload of conventional cyber–physical interaction has been one of the early motivations for research into and development of IoT and ubiquitous computing. Both information and processing can be embedded in a quiet, invisible, and unobtrusive way in everyday objects, which takes computing from the center stage of our attention to the periphery [6].

MIT Media Lab researcher David Rose and his MIT colleagues have worked with cyber–physical interaction, ubiquitous computing, and tangible media for years. What Rose sees as the true potential in IoT is that it opens up for a more humane and satisfying interaction with technology, because digital services can be integrated in products, which gives everyday things a touch of magic [5]. These "enchanted objects," as Rose argues, give us a much more emotional response than traditional interfaces. His example is Sting, the sword of Bilbo Baggins in *The Hobbit* that lights up blue when there are orcs in the vicinity. This is very smart for Bilbo, who is only a small hobbit; however one of Rose's more mundane examples is also very smart for us: an umbrella with a handle that glows when rain is predicted. That way we don't need to check the weather report on our Smartphone before leaving our homes, as the umbrella will silently communicate whether we will need it during the day. The enchanted objects, or calm technologies as it was discussed already in the mid-1990s, cope with the information overload we are prone to with ICT, by taking tech to the periphery. This motivation for embedding or hiding digital services away from the direct attention of the user is very important because the development is thus an answer to the challenge of coping with information overload. The development has a "why" and not just a "why not": it promises a more human-centered design, and thereby perhaps even more social interaction and presence [7], beyond the interaction with conventional digital interfaces that are in principal asocial.

2.6 THE HIDDEN LIFE OF THINGS

As humans we have an urge to look at things from above. When we arrive in a new city, we seek out the highest spot to enjoy a panorama view over the unknown territory—hence the attraction of ascending the Eifel Tower in Paris or the Chrysler building in New York. The panorama makes it possible for us to see the complex cityscape as a whole, or as the French philosopher Michel de Certeau puts it, it creates a text that is readable to us [8]. Connecting everything gives us a similar pleasure. As ubiquitous computing and data make the invisible visible, all the information that is hidden to the eye becomes comprehensible when automatically collected and visualized. Yet just as de Certeau urges the architect or planner to move down from the ivory tower of the panorama, so must the programmer move away from the screen and look at the *everyday context* of IoT and services. The way we can distance ourselves from

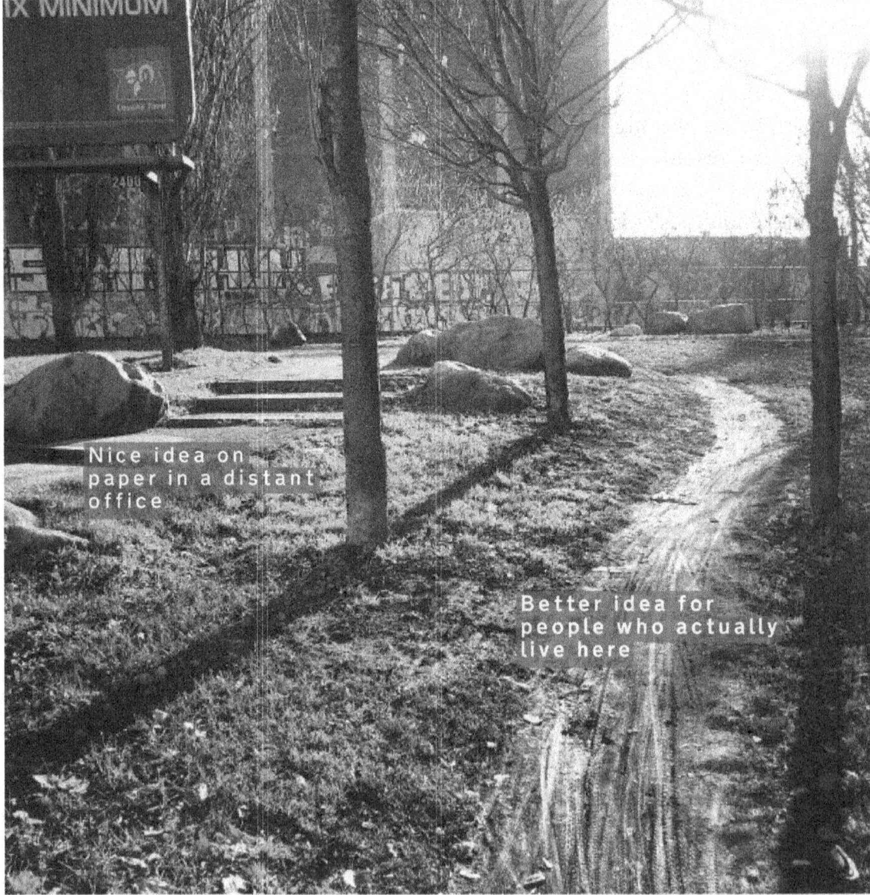

FIGURE 2.2 Desire lines show how people use their own tactics and logic of everyday life, which is impossible to specify by the planner of both physical and virtual spaces. Copenhagenize.eu.

our daily behavior is a "theoretical simulacrum, in short a picture, whose condition of possibility is an oblivion and a misunderstanding of practices" [9]. Both users of physical spaces and users of virtual spaces have their own logic or tactics they employ that are different from the logic of the people planning or programming these spaces. Especially programming proves challenging, as it is necessary to specify and define through the code how a system can be used in advance.

"Desire lines," as shown in Figure 2.2, are not only seen in urban spaces but also in virtual ones. An example could be Twitter, where people are trying to overcome the 140-character limit of tweets by tweeting pictures of text, tweet storming, or number tweets chronologically. The importance of everyday life practices is not only a question of designing for the optimal user experience but actually being able to develop services on top of the massive amounts of data collected and achieve the desired behavioral change. Data in itself is not really that interesting: what is interesting is what data can be used for. Data is a tool that we can use in various situations to make decisions, recognize patterns, and forecast trends. In order to do so, data must be presented to a user. When computing goes beyond traditional interfaces such as PCs and Smartphones, it poses several design challenges both in terms of understanding who is actually the user of a given product or service and how to interact with this user. The interaction with IoT is much more complicated, since there is not only one user who is actively and intentionally interacting with a computer, via a mouse, a keyboard, or a screen, and there is not a one-to-one interaction, but a multiplicity of users and devices all communicating and interacting at the same time [6].

2.7 THE UMBRELLAS ARE NOT WHAT THEY SEEM

IoT is directed toward the everyday and the optimization of our daily behavior. It can even be said to have *colonized* the everyday life [6]. Rose's enchanted umbrella shows how IoT makes data and computing not only silent but also invisible to the user. The umbrella is an example that is easy to comprehend, but as IoT is more profoundly embedded in our environment, it dematerializes and becomes even more hidden in plain sight. From smart, connected homes to offices, buildings, and cities, the physical spaces we inhabit are enhanced by a digital layer, collecting and processing information, and distributing and reacting on this information in ways that are obscure to ordinary people.

The umbrellas are not what they seem, one could say, as their capabilities go beyond their mere appearance (they have been touched by a bit of magic!). IoT is literally hard to see, which makes it difficult for the user to grasp how it functions (how do I know that the umbrella can tell me the weather forecast, when it looks like a normal umbrella? How do I make it glow? Do I speak to it? Touch it? And when it lights up blue, does it mean rain or clear skies?). IoT is also figuratively hard to see, because it is not obvious which sensors are embedded, what data they transmit, and to what purpose [10] (what does my umbrella tell about me? To whom? Why?). As stated previously, technology is an answer to a question, but who actually formulates that question? If it is no longer the users who formulate the question that to what the technologies they employ in their everyday life are answers, a problem with human agency arises. Technology will always reflect the assumptions of the designers who made it, and in conventional interfaces, the user is presented with a choice of how and when to use it, which is not present in IoT and ubiquitous computing [11]. On the other hand, as IoT is not always directly managed by its users, its success depends on being able to function in a chaotic world of volatile everyday tactics.

2.8 INTERACTING WITH THE INVISIBLE

Inherent in IoT is a dialectical relationship between visibility and obscurity at play. For technologies to truly be interactive, they need to be visible—we cannot interact with hidden things [12]. On the other hand, we must consider when the visibility is desired and when it is not, when do we need to direct our attention to the smart objects around us, and when do they serve us better at the periphery, so we can be attentive to other things in life. Do we need to talk to our umbrellas today? The dialectical relation is evident in the development of services, since designers not only have a moral responsibility to question the underlying assumptions and purposes of IoT solutions, but they are also forced to in order to create solutions that anticipate the unpredictability of human behavior. Users of both virtual and public spaces have their own logic, different from the designers'. If we accept that we cannot design or program cities or software to function the way we want, then how should we plan for the unplannable?

2.9 SOCIETY AS OPEN SOURCE

Technologies do not produce their own use or generate their own value; the drivers of the innovation in technology are users and organizations that are open for experimentation [13]. The desire lines in the previous example show us how the designer can look at the user experience to better plan the paths in the park by following the users, and the same could be said of the digital planner. But beyond following users, we need to look at the system itself—physical or virtual. IoT holds a promise for a better understanding of the world, of practices, but in the endeavor of creating a smart connected society, we must not underestimate people, their tactics, as well as the value they add. Connecting things do not make them smart. When it comes to developing services that we want people to use, they must be able to adapt to the actual uses of the people. We need to accommodate for users to be able to hack data, services, and devices to make them meet their needs and desires.

As pointed out by the Dutch-American sociologist Saskia Sassen, the city talks back! This indicates how cities have a resilience in terms of the citizens' tactics, which has made great cities survive great crisis and outlive great nations, because cities are able to change all the time [12]. Cities are essentially open and incomplete. In tech terms it means that resilient cities can be understood as open-source systems, where citizens are able to contribute to, develop, and interact with the code that makes up the city, and the code is transparent to all. The same is valid to resilient digital solutions. For people to actively engage, to interact with the digital, there is a need for us to think it as open.

2.10 LEARN FROM YOUR HACKERS

In his renowned work on Government 2.0, Tim O'Reilly describes how the most creative ideas of using new technologies often do not spring from the minds of the inventors but from the users [4]. O'Reilly gives the example of Google Maps which is, as already mentioned, the leading mapping platform of today. The history behind this leadership lies in Google's ability to learn from its "hackers." When Google Maps was first introduced, an independent developer discovered a hidden feature in the platform that allowed him to use the map coordinate data and mash it up with open data from another source: available apartments from Craigslist.org, creating a whole new service, HousingMaps.com. Instead of dissociating the "hack," Google embraced it by employing the independent developer and created an interface, an API, to Google Maps that easily allowed everyone else to do mash-ups as well.

Opening up technologies for people to collaborate is essential to reach the goals of developing solutions to the major collective problems locally and globally, and open APIs, open data, and open standards are key to achieve collaboration whether it is private sector, public sector, or in civil society. Working closer with people in transparent ways becomes even more urgent as concerns for privacy and digital rights arise, since it is a means to democratize technologies, and make the invisible IoT accessible and comprehensible.

2.11 ENSURING HIGH-QUALITY SERVICES TO CITIZENS

One of the primary objectives for governments is to improve services for citizens. IoT and the massive amounts of data do not only challenge traditional businesses but also greatly affect the role of government. How do we ensure the delivery of high-quality digital services to citizens? And how do we open up government for the advantages within the rapid development and innovation in ICT/IoT? First thing is to learn from those who are the best in doing this, as already illustrated by the examples put forward in this chapter: researchers and businesses. Disruptive innovation seldom occurs within large organizations as they are not able to adapt fast enough to the exponential growth in technologies and they are often characterized by linear thinking and a linear organization of work [14]. Unlike large corporations who are under a constant threat of disruption to their business, governments aren't put out of business so easily. Yet advocating openness in platforms and systems is highly relevant for governments, as well as businesses, to advance the use of the limited public resources. Instead of governments focusing on developing and delivering services to citizens, governments have a great potential in focusing on ensuring that these services are developed and delivered by the best players in the market.

However, open government should not entail a closed vendor shop for a few large vendors of solutions, but an open platform, which with the words of Tim O'Reilly means "government stripped down to the essentials. A platform provider builds essential infrastructure, creates core applications that demonstrate the power of the platform and inspire outside developers to push the platform even further, and enforces 'rules of the road' that ensure that applications work well together" [4].

Government as a platform is a powerful way of interacting with and encouraging the market to build the digital services that enhances the quality of life for the citizens. The key enablers for this interaction are open data, open APIs, and open standards, and it is necessary for governments to continuously collaborate on this across cities and borders to create a large enough market for digital services, to make it relevant for companies to create the solutions we need, that are interoperable between systems, situations, and places [15].

2.12 GOVERNMENT AS A PLATFORM

Opening up government means a democratization of information and technology, and to keep it democratic, it is important for governments to consider which services are core to the public sector. Government as a platform should not entail that core welfare activities are externalized to people themselves or to the private sector. Open data and government as a platform should not become an excuse for not investing in collective goods, and if citizens are to "spark innovation" as O'Reilly writes, they need the proper tools and knowledge to work with data and digital services.

Investment in collective goods is central to the role of government, especially in building the right infrastructure. We have already visited the example of the car: government investment in infrastructure was foundational for the massive sprawl of car-dominated transportation. For society to take full advantage of the promise of IoT, there is also a great need for public infrastructure, not as in highways, but in networks and open platforms. Government as a platform means to get the right actors to meet and collaborate across sectors and develop closer partnerships with businesses and universities in a triple-helix fashion, because there is a shared interest in investing in the infrastructure necessary for the digital transformation of society [16].

Collaboration is also a question of cross-fertilization from different sectors to the mutual benefit for the collaborators. Working across verticals of traditional business domains and learning how IoT is employed in other industries, and which data are collected, have a transformative potential for all sectors. Taking the car again as an example, the amount of data collected by the embedded sensors in cars does not only provide analytics on the car, its functioning, maintenance, environment, etc. that are important for the car driver or the manufacturer, but it can also be of great value in urban planning—from providing data on traffic flows and potholes to lacking road marking and free parking spots.

2.13 CONCLUSION

Mid-twentieth-century science fiction literature and films already imagined several of the great services that could help us live a better life, which are now made possible by IoT. But many of the challenges we face today are consequences of the technologies developed in the twentieth century. This chapter began with the example of the car, which served to show how the development of a specific technology gave way to a range of new possibilities but entailed grave problems too. To bring a happy ending to the development in IoT, we need to reflect critically on the technologies and services that are made possible thereby and to consider carefully which possibilities we want to make use of and why.

The present challenges of our planet are to a great extent the effects of human activities up until today, in particular on top of the industrial revolution and the great exploitation of resources, which led way to the high degree of life quality that we benefit from in the developed world. The possibilities presented to us now with the exponential development of technology need to not only make a flawed existing system more efficient but also to look at completely new ways of organizing society that keeps up with a high degree of life quality without the dire consequences for the planet.

We need to envision a new future of the kind of city and community we want to live in that doesn't replicate the bad structures of the past and that takes the everyday life to the center stage, with technologies as enablers for a good life of citizens and not an end in its own.

REFERENCES

[1] Hajer, M. & Dassen, T. (2014) *Smart about Cities: Visualising the Challenge for 21st Century Urbanism*. nai010 publishers/PBL Publishers, Rotterdam.

[2] Claudel, M. & Ratti, C. (2015) Full speed ahead: How the driverless car could transform cities. Available at: http://www.mckinsey.com/insights/sustainability/full_speed_ahead_how_the_driverless_car_could_transform_cities (accessed August 9, 2016).

[3] Zysman, J., Feldman, S., Kushida, K., Murray, J., & Nielsen, N.C. (2013) Services with Everything. In: Breznitz, D. & Zysman, J. *The Third Globalization: Can Wealthy Nations Stay Rich in the Twenty-First Century?* Oxford University Press, New York.

[4] O'Reilly, T. (2010) Government as Platform. In: Lathrop, T. & Ruma, L. *Open Government*. O'Reilly Media, Inc., Sebastopol. Available at: http://chimera.labs.oreilly.com/books/1234000000774/ch02.html (accessed August 9, 2016).

[5] Rose, D. (2014) *Enchanted Objects: Innovation, Design and the Future of Technology*. Scribner, New York.

[6] Greenfield, A. (2006) *Everyware: The Dawning Age of Ubiquitous Computing*. New Riders, Berkeley.

[7] Weiser, M. & Brown, J.S. (1995) Designing Calm Technologies, Xerox PARC. Available at: http://www.ubiq.com/hypertext/weiser/calmtech/calmtech.htm (accessed August 9, 2016).

[8] De Certeau, M. (1984) *The Practice of Everyday Life*. University of California Press, Berkeley.

[9] De Certeau, M. (1984) *The Practice of Everyday Life*. University of California Press, Berkeley, p. 93.

[10] Greenfield, A. (2006) *Everyware: The Dawning Age of Ubiquitous Computing*. New Riders, Berkeley, p. 93.

[11] Greenfield, A. (2006) *Everyware: The Dawning Age of Ubiquitous Computing*. New Riders, Berkeley, p. 61.

[12] Sassen, S. (2012) The Future of Smart Cities. Available from Nancy Rubin at: http://nancy-rubin.com/2012/11/10/saskia-sassen-the-future-of-smart-cities/ (accessed August 9, 2016).

[13] Zysman, J., Feldman, S., Kushida, K., Murray, J., & Nielsen, N.C. (2013) Services with Everything. In: Breznitz, D. & Zysman, J. *The Third Globalization: Can Wealthy Nations Stay Rich in the Twenty-First Century?* Oxford University Press, New York, p. 114.

[14] Ismail, S., Malone, M.S., & van Geest, Y. (2014) *Exponential Organizations: Why New Organizations Are Ten Times Better, Faster and Cheaper Than Yours (and What to Do about It).* Diversion Books, New York.

[15] Goldstein, B. & Dyson, L. (2013) *Beyond Transparency: Open Data and the Future of Civic Innovation.* Code for America Press, San Francisco.

[16] Lathrop, T. & Ruma, L. (2010) *Open Government.* O'Reilly Media, Inc., Sebastopol. Available at: http://chimera.labs.oreilly.com/books/1234000000774 (accessed August 9, 2016).

3

THE INDUSTRIAL INTERNET OF THINGS (IIoT): APPLICATIONS AND TAXONOMY

STAN SCHNEIDER
Real-Time Innovations, Inc., Sunnyvale, CA, USA

3.1 INTRODUCTION TO THE IIoT

The Internet of Things (IoT) is the name given to the future of connected devices. There are two clear subsets. The "Consumer IoT" includes wearable computers, smart household devices, and networked appliances. The "Industrial IoT" includes networked smart power, manufacturing, medical, and transportation. Technologically, the Consumer IoT and the Industrial IoT are more different than they are similar.

The Consumer IoT attracts more attention, because it is more understandable to most people. Consumer systems typically connect only a few points, for instance, a watch or thermostat to the cloud. Reliability is not usually critical. Most systems are "greenfield," meaning there is no existing infrastructure or distributed design that must be considered. There are many exciting new applications that will change daily life. However, the Consumer IoT is mostly a natural evolution of connectivity from human-operated computers to automated things that surround humans.

While it will grow slower than the Consumer IoT, the Industrial Internet of Things (IIoT) will eventually have much larger economic impact. The IIoT will bring entirely new infrastructures to our most critical and impactful societal systems. The opportunity to build truly intelligent distributed machines that can greatly improve function and efficiency across virtually all industries is indisputable. The IIoT is the strategic future of most large companies, even traditional industrial manufacturers and infrastructure providers. The dawn of a new age is clear.

Internet of Things and Data Analytics Handbook, First Edition. Edited by Hwaiyu Geng.

Companion website: www.wiley.com/go/Geng/iot_data_analytics_handbook/

Unlike connecting consumer devices, the IIoT will control expensive mission-critical systems. Thus, the requirements are very different. Reliability is often a huge challenge. The consequences of a security breach are vastly more profound for the power grid than for a home thermostat. Existing industrial systems are already networked in some fashion, and interfacing with these legacy "brownfield" designs is a key blocking factor. Plus, unlike consumer devices that are mostly connected on small networks, industrial plants, electrical systems, or transportation grids will encompass many thousands or millions of interconnected points.

Building a technology stack for any one of these applications is a challenge. However, the real power is a single architecture that can span sensor-to-cloud, interoperate between vendors, and span industries. The challenge is to evolve from today's mash-up of special-purpose standards and technologies to a fast, secure, interoperable future.

In the long term, there is an even larger opportunity. The future of the IIoT will include enterprise-class platforms that guarantee real-time delivery across enterprises and metro or continental areas. This will become a *new utility* that enables reliable distributed systems. This utility will support twenty-first-century infrastructure like intelligent transportation with autonomous vehicles and traffic control, smart grids that integrate distributed energy resources, smart healthcare systems that assist care teams, and safe flying robot air traffic control systems. This utility will be as profound as the cell network, GPS, or the Internet itself.

There are many consortia of companies targeting the IIoT. The largest and fastest growing is the Industrial Internet Consortium (IIC) [1]. The IIC was founded in 2014 by global industrial leaders: GE, Intel, Cisco, AT&T, and IBM. As of this writing in 2016, it includes over 250 members. The German government, along with several large German manufacturers, has an active effort called Industrie 4.0 [2]. There is also a smaller start-up consortium called the OpenFog Consortium [3]. Of these, the IIC is by far the broadest. It addresses end-to-end designs in all industries. Industrie 4.0 is focused only on manufacturing. And OpenFog targets "intelligence at the edge," meaning the movement of powerful elastic computing out of data centers into neighborhoods and customer premises. However, all share many common members and goals. They are working together in many ways.

Because of its size and growth, the IIC gets by far the most attention. The goal of the IIC is to develop and test an architecture that will span all industries. Just as Ethernet, Linux and the Internet itself grew as general-purpose technologies that pushed out their special-purpose predecessors, the IIC will build a general-purpose Industrial Internet architecture that can build and connect systems such as transportation, medical, power, factory, industrial controls, and others. The IIC's unique and powerful combination of leaders from both government and industry gives it the necessary platform to make this huge impact.

The IIC was the first to create a venue for users across industries with similar challenges. This actually created the IIoT as a true market category and changed the landscape dramatically. Suddenly, hundreds of companies are deciding their strategy for this new direction. Gartner, the large analyst firm, predicts that the Smart Machine era will be the most disruptive in the history of IT. That disruption will be led by smart distributed infrastructure called the IIoT.

3.2 SOME EXAMPLES OF IIoT APPLICATIONS

The author is the CEO of Real-Time Innovations, Inc. (RTI) [4]. RTI is the largest embedded middleware company and the leading vendor of middleware compliant with the Data Distribution Service (DDS) standard [5]. All applications in this section are operational RTI Connext DDS systems. The DDS standard is detailed in later sections; the applications are presented first to provide background and highlight the breadth of the IIoT challenge. These are only a few of nearly 1,000 applications that form RTI's experience base. Further examples can be found at www.rti.com.

3.2.1 Connected Medical Devices for Patient Safety

Thirty years ago, healthcare technologists realized a simple truth: monitoring patients improves outcomes. That epiphany spawned the dozens of devices that populate today's hospital rooms: pulse oximeters, multiparameter monitors, ECG monitors, Holter monitors, and more. Over the ensuing years, technology and intelligent algorithms improved many other medical devices, from infusion pumps (IV drug delivery) to ventilators. Healthcare is much better today because of these advances (Figure 3.1).

However, hospital error is still a leading cause of death; in fact, the Institute of Medicine named it the third leading cause of death after heart disease and cancer. Thousands and thousands of errors occur in hospitals every day. Many of these errors are caused by false alarms, slow responses, and inaccurate treatment delivery.

Today, a new technology disruption is spreading through patient care: intelligent, distributed medical systems. By networking devices, alarms can become smart, only

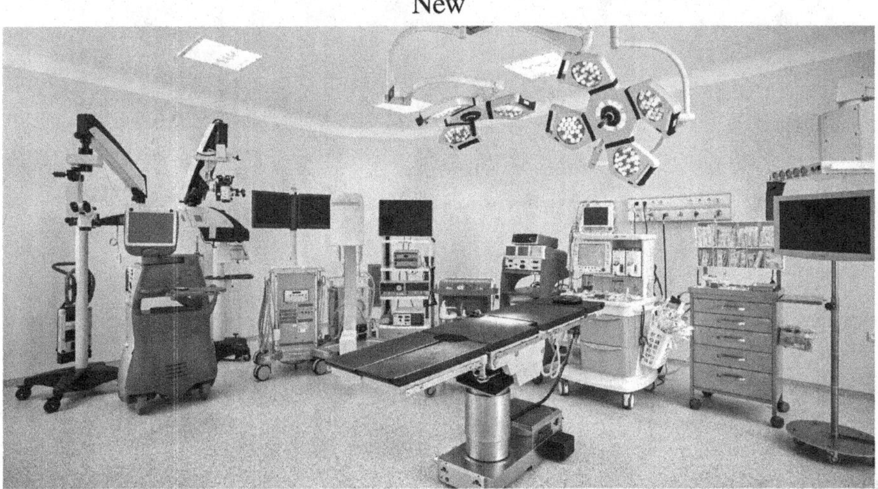

FIGURE 3.1 Connected medical devices will intelligently analyze patient status, create "smart alarms" by combing instrument readings, and ensure proper patient care. An intelligent, distributed IIoT system will help care teams prevent hundreds of thousands of deaths.

sounding when multiple devices indicate errant physiological parameters. By connecting measurements to treatment, smart drug delivery systems can react to patient conditions much faster and more reliably than busy hospital staff. By tracking patients around the hospital and connecting them to cloud resources, efficiency of care can be dramatically improved. The advent of true Internet of Things networking in healthcare will save costs and lives (Figure 3.2).

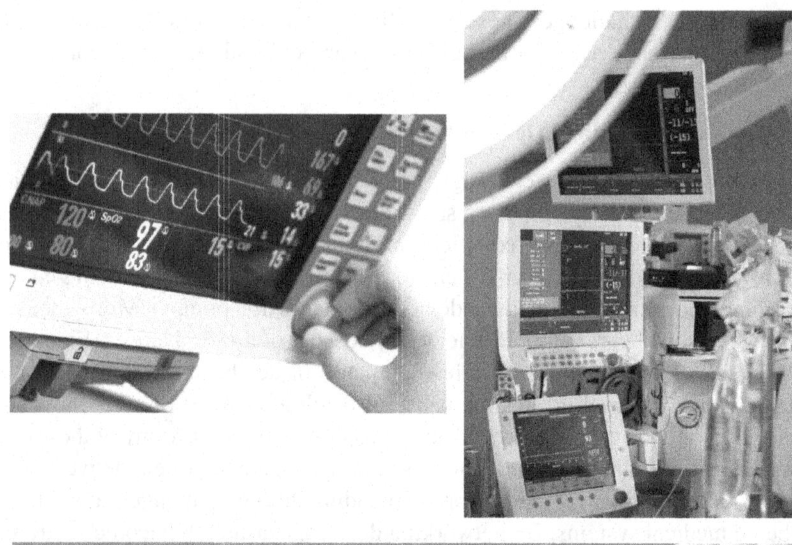

FIGURE 3.2 A modern hospital needs hundreds of types of devices. These must communicate to improve patient safety and outcome, to aid resource deployment and maintenance, and to optimize business processes. RTI Connext DDS adapts to handle many different types of data flows, different computing platforms, and transports.

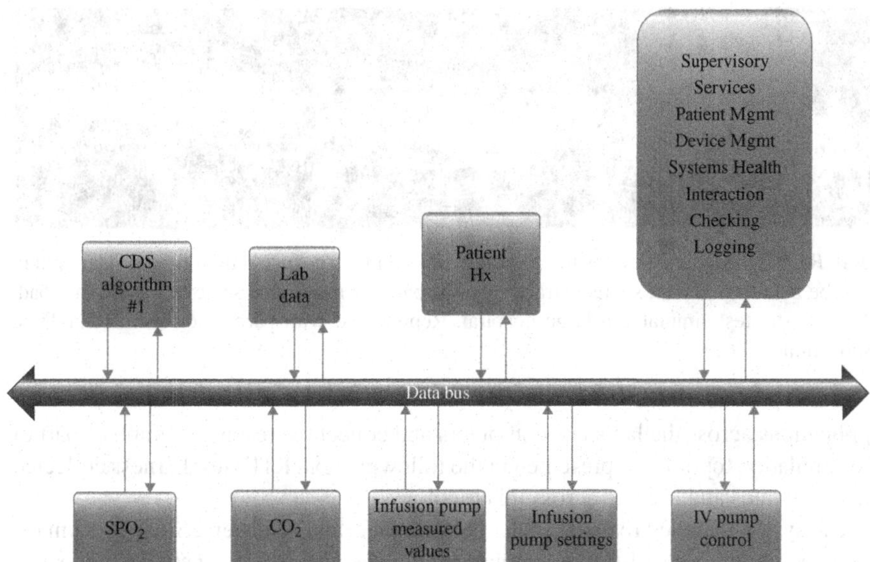

FIGURE 3.3 An intelligent Patient Controlled Analgesia system. The supervisor combines oximeter and respirator readings to reduce false alarms and stop drug infusion to prevent overdose. The RTI DDS databus connects all the components with appropriate real-time reliable delivery. Reproduced with permission from Real-Time Innovations.

3.2.1.1 The Integrated Clinical Environment (ICE) Researchers and device developers are making quick progress on medical device connectivity. The Integrated Clinical Environment (ASTM F2761) standard [6] is one key effort to build such a connected system. ICE combines standards. It takes data definitions and nomenclature from the IEEE 11073 (×73) standard for health informatics. It specifies communication via the DDS standard. ICE then defines control, data logging, and supervisory functionality to create a connected, intelligent substrate for smart clinical connected systems (Figure 3.3).

Like most standards, large organizations may take time to adapt to a standards-driven environment. ICE nonetheless represents an excellent example of how to build smarter systems.

3.2.1.2 Patient Monitoring Modern hospitals use hundreds of types of devices for patient care and monitoring. These systems must work in a large hospital environment. Integrating whole hospitals with thousands of devices presents challenges for scalability, performance, and data discovery.

To prove the design viable for GE Healthcare, RTI built a simulation to prove that DDS could handle a thousand-bed hospital with over 100,000 devices. The simulation ran in RTI's networking lab. It sent realistic data flows between hundreds of applications, instances of RTI's Connector product. RTI services developed a matrix (Excel spreadsheet) to configure Connector to send the mix of data types and rates expected from real devices. RTI developed an automated test harness to deploy these

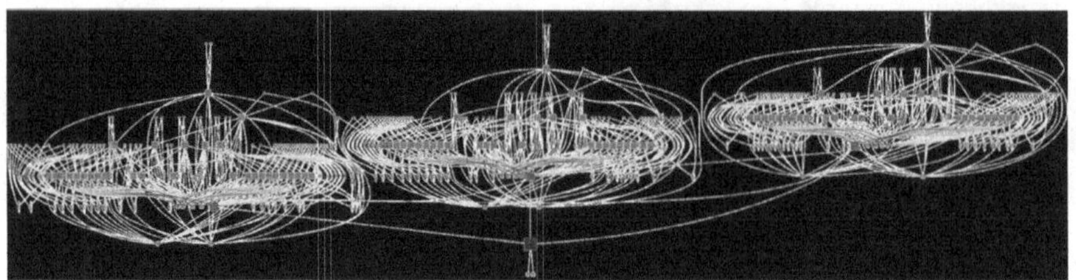

FIGURE 3.4 Medical devices must operate in a complex hospital environment. The system must be able to find data sources, track them as patients move, and scale to handle the load. This realistic test simulated a large hospital. Reproduced with permission from Real-Time Innovations.

applications across the lab's test computers and collect the results. A graph of part of the simulation topology is presented in the following text. RTI's test harness collected data flow rates and loading across this topology.

The system handled realistic scale, performance, and discovery. Since it is important to communicate real-time waveforms and video, the potential network-wide data flow is large. However, the need is "sparse"; most data is only needed at relatively few points. As explained in the succeeding text, DDS can propagate specifications to indicate exactly what each receiver needs from the senders. The senders then filter the information to send only what's needed, thereby eliminating wasted bandwidth. Discovering data sources is also critical, since 62% of hospital patients move every day. So, the system also tested transitions between network locations (Figure 3.4).

When deployed, the new system will ease patient tracking. It will coordinate devices in each room and connect rooms into an integrated whole hospital. Information will flow easily and securely to cloud-based Electronic Health Records (EHR) databases. The hospital of the future will become an intelligent distributed machine in the IIoT.

3.2.2 Microgrid Power Systems

The North American electric power grid has been described as the biggest machine in the world. It was designed and incrementally deployed based on centralized power generation, transmission, and distribution concepts.

Times have changed. Instead of large centralized power plants burning fossil fuels that drive spinning masses, Distributed Energy Resources (DERs) have emerged as decentralized local alternatives to bulk power. DERs are typically clean energy solutions (solar, wind, thermal) that take advantage of local environmental and market conditions to manage the local generation, storage, or consumption of electricity.

3.2.2.1 The DER Time Challenge Most renewable energy sources are not reliable producers. Solar and wind can change their power output very quickly.

Unfortunately, that dynamic behavior is not compatible with today's grid. Today's grid uses local power substations to convert high-voltage power to neighborhood distribution voltage levels. Those stations estimate power needs and report back to the utility. The utility then needs up to 15 minutes to spin up (or down) a centralized generation plant to match the estimate.

So, since a solar array can lose power in a matter of seconds with a fast-moving cloud, the grid cannot react. An alternate source has to be available and ready to pick up the load immediately. If there isn't sufficient backup, the voltage on the grid can drop and the grid can fail. The only way to provide that backup today is to provide "spinning reserve" capacity, meaning the generators use more energy than the grid needs.

As solar energy resources grow in a utility's service area, the utility has to have more excess spinning reserve ready as backup. While the sun is shining, power may be flowing from these distributed solar arrays back to the grid. However, the fossil fuel generators need to be running and spun up sufficiently to quickly take over if the solar arrays stop producing. So, with every solar array pushing power on to the grid, there is an equivalent fossil fuel generator spinning in the background to take over. Thus, little fossil fuel is saved. Even worse, driving the generators without load makes them overheat and even prematurely wears out bearings.

To fix this, the utility needs 15–30 minutes of extra time to ramp up the generators. Then, they would not need to have the spinning reserve. The only way to provide the time needed is to implement energy storage or load reduction.

3.2.2.2 Microgrid Architecture Microgrids are the leading way to provide that time. Microgrids combine intermittent energy sources, energy storage systems like batteries, and some local control capability. This allows the microgrid to smooth out the changes in DER power. A microgrid can even "island" itself from the main power grid and run autonomously.

Microgrids usually encompass a well-defined, relatively small geographic region. College campuses have been proving grounds for this technology, as have military bases. A microgrid can respond rapidly and locally to a loss of power from solar arrays or local wind turbines using backup energy sources like batteries. Many proof-of-concept microgrid projects are active; they range from small demos to utility-class pilot test beds. All seek to incorporate energy storage and load reduction techniques into the grid.

Two key capabilities for microgrids are intelligent control at the edge of the grid and peer-to-peer, high-performance communications for local autonomy (see Figure 3.5). With these, a local battery energy storage system can receive a message in milliseconds from the solar arrays when backup energy is needed. The local controller on the battery can then quickly switch the battery from charge to source mode. This keeps the local energy consumers powered and gives the utility time to spin up central power resources as needed.

The OpenFMB™ Framework [7] is the first field system addressing the need for reliable, safe, upgradeable distributed intelligence on the grid. OpenFMB directly addresses the decentralization issue facing utilities and regulators by leveraging existing electricity information models (e.g., IEC 61968/61970, IEC 61850, MultiSpeak, and SEP 2) and creating a data-centric "bus" on the grid to allow devices to talk directly to one another. The initial use cases targeted by

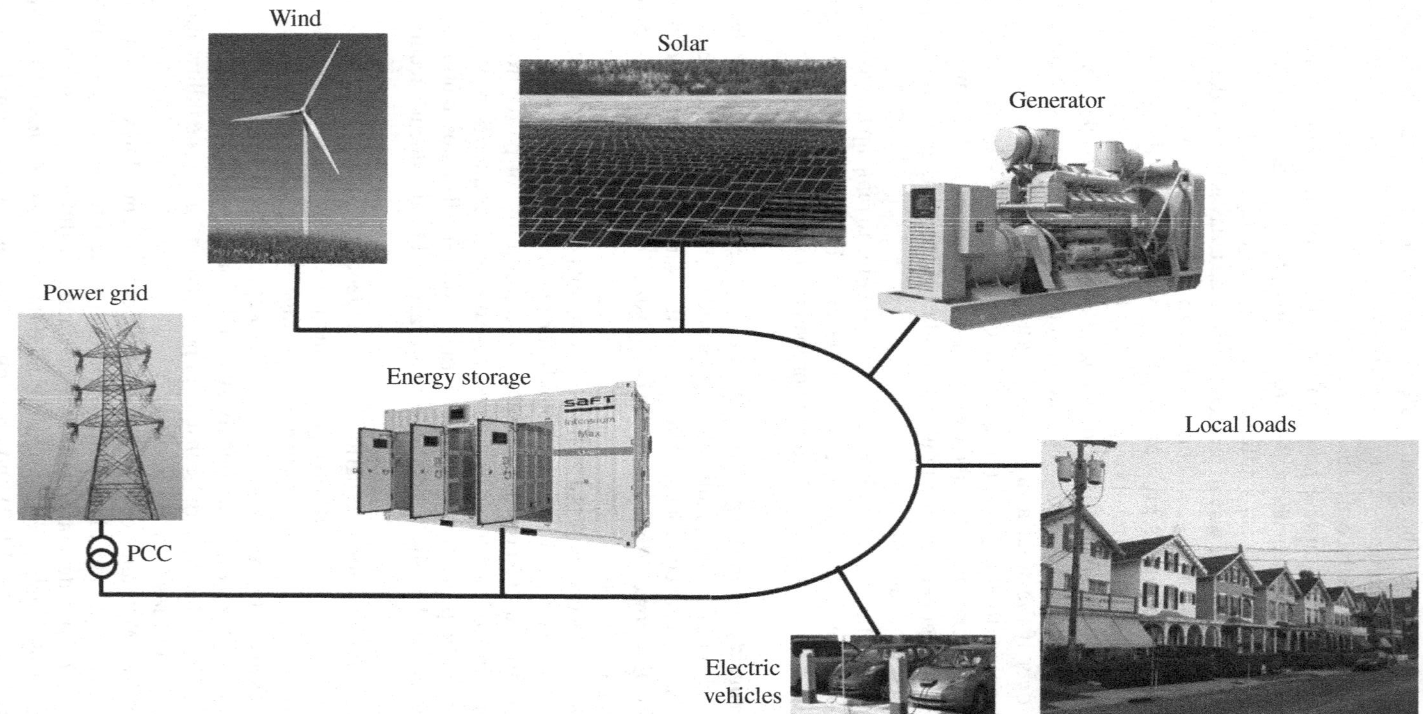

FIGURE 3.5 A Microgrid uses peer-to-peer data communication and edge intelligence to automate local power generation and balance against the power load. Microgrids help integrate intermittent energy sources like solar and wind. Reproduced with permission from Real-Time Innovations.

OpenFMB are microgrid focused, and the OpenFMB framework is closely adhering to the IIC's Industrial Internet Reference Architecture (IIRA).

The OpenFMB team held a major demonstration in February 2016 with 25 different companies. Many parts of the implementation use RTI's Connext DDS platform. DDS interfaces were developed for the Optimization Engine, Load Simulators, the Point of Common Coupling (PCC) transition logic, and other required simulators to drive the demonstration. To test nonproprietary interoperability, the system built a cross-platform solution with multiple operating system targets and CPU architectures. The demonstration proves that IIoT interoperability is a practical, achievable path for fielded utility devices and systems.

3.2.3 Large-Scale SCADA Control

NASA Kennedy Space Center's launch control system is the largest SCADA (Supervisory Control And Data Acquisition) system in the world. With over 400,000 control points, it connects together all the equipment needed to monitor and prep the rocket systems. Before launch, it pumps rocket fuels and gasses, charges all electrical systems, and runs extensive tests. During launch, a very tightly controlled sequence enables the main rocket engines, charges and arms all the attitude thrusters, and monitors thousands of different values that make up a modern space system. It must also adapt to the various mission payloads, some of which need special preparation and monitoring for launch (Figures 3.6 and 3.7).

The launch control system has very tight and unique communication requirements. The system is distributed over a large area and the control room. It must be secure. Data flow is "tidal": activity cycles through the surge of preparation, spikes during the actual launch, and then ebbs afterward. During the most critical few seconds, it sends hundreds of thousands of messages per second. Connext DDS intelligently batches updates from thousands of sensors, reducing traffic dramatically. Everything must be stored for later analysis. All information is viewable (after downsampling) on HMI stations in the control room. After launch, all the data must be available for replay, both to analyze the launch and to debug future modifications in simulation.

3.2.4 Autonomy

RTI was founded by researchers at the Stanford Aerospace Robotics Laboratory (ARL) [8]. The ARL studies complex electromechanical systems, especially those with increasing levels of autonomy.

Unmanned air and defense vehicles have long relied on DDS for deployments on land, in the air, and under water. DDS is a key technology in many open architecture initiatives, including the Future Airborne Capability Environment (avionics), Unmanned Air Systems (UAS) Control Segment Architecture (UAS ground stations), and the Generic Vehicle Architecture (military ground vehicles).

UAS have complex communication requirements, with flight-critical components distributed across the air and ground segments. Further, to operate in the U.S. National Airspace System (NAS), the system must be certified to the same safety standards as civil aircraft. RTI recently announced a version of DDS with full DO-178C level A safety certification evidence. It was developed to meet the needs of the Ground Based Sense and Avoid (GBSAA) system pictured in Figure 3.8.

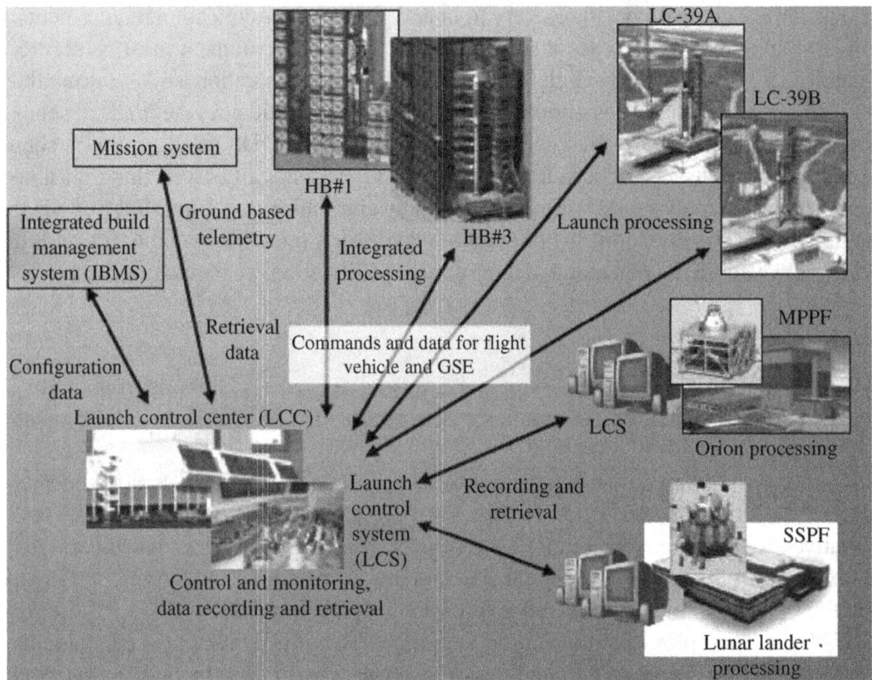

FIGURE 3.6 NASA KSC's launch control is a massive, reliable SCADA system. It comprises over 400,000 points, spread across the launch platform and the control room. The launch control system integrates many thousands of devices, from tiny sensors to large enterprise storage systems. It spreads over many miles. Photo: NASA/Bill Ingalls.

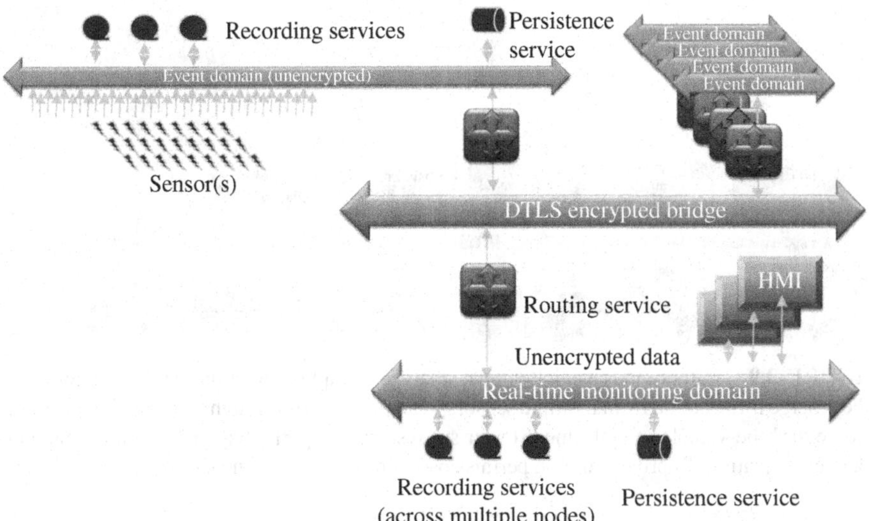

FIGURE 3.7 NASA's launch control SCADA system is massive. It captures sensor data to both recording services (for forensic use) and persistence service (for durability). The Routing Service-to-Routing Service bridge encrypts data between the event platform and control room. Reproduced with permission from Real-Time Innovations.

FIGURE 3.8 The Ground Based Sense and Avoid (GBSAA) system includes many distributed radars. It will soon allow unmanned vehicles to fly in the U.S. National Airspace System (NAS). Applications of unmanned vehicles will include operator training, repositioning, search and rescue, and disaster relief. US Army photo by Sofia Bledsoe.

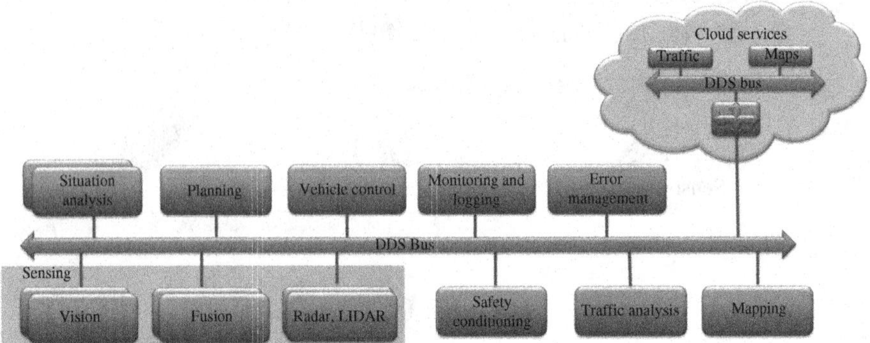

FIGURE 3.9 Autonomous vehicles must analyze complex situations and react quickly. They merge information from multiple sensors, plan trajectories through traffic and road lanes, and control the vehicle in real time. Slower subsystems support navigation, monitoring, and route optimization. Reproduced with permission from Real-Time Innovations.

This technology is now also being applied aggressively to autonomous cars for consumer use. This market is perhaps the most disruptive of the IIoT applications. Because the ground is a much more complex environment, autonomous cars face even greater challenges than air systems. They must coordinate navigation, traffic analysis, collision detection and avoidance, high-definition mapping, lane departure tracking, image and sensor processing, and more (Figure 3.9).

Safety certification for the entire system as whole is prohibitively expensive. Dividing the system into modules and certifying them independently reduces the cost dramatically. "Separation kernels" are operating systems that provide guaranteed separation of tasks running on one processor. "Separation middleware" provides a similar function to applications that must communicate, whether they are running on one processor or in a distributed system. A clean, well-controlled interface eases certification by enabling modules to work together.

3.3 TOWARD A TAXONOMY OF THE IIoT

There is today no organized system science for the IIoT. We have no clear way to classify systems, evaluate architectural alternatives, or select core technologies. To address the space more systematically, we need to develop a "taxonomy" of IIoT applications based on their system requirements.

This taxonomy will reduce the space of requirements to a manageable set by focusing only on those that drive significant architectural decisions. Based on extensive experience with real applications, we suggest a few divisions and explain why they impact the architecture. Each of these divisions defines an important dimension of the IIoT taxonomic model. We thus envision the IIoT space as a multidimensional requirement space. This space provides a framework for analyzing the fit of architectures and technologies to IIoT applications.

FIGURE 3.10 Environment does not indicate architecture. Dividing animals by "land, sea, and air" environment is scientifically meaningless. The biological taxonomy instead divides by fundamental characteristics.

FIGURE 3.11 Industry does not indicate architecture. Dividing IIoT applications by "medical, power, or transportation" environment is as scientifically meaningless as dividing animals by their environments. To make progress, we need an IIoT taxonomy that instead divides by fundamental characteristics.

A taxonomy logically divides types of systems by their characteristics. The first problem is to choose top-level divisions. In the animal kingdom, you could label most animals "land, sea, or air" animals. However, those environmental descriptions don't help much in understanding the animal. For instance, the "architecture" of a whale is not much like an octopus, but it is very like a bear. To be understood, animals must be divided by their characteristics and architecture, such as heart type, reproductive strategies, and skeletal structure (Figure 3.10).

It is similarly not useful to divide IIoT applications by their industries like "medical, transportation, and power." While these environments are important, the requirements simply do not split along industry lines. For instance, each of these industries has some applications that must process huge data sets: some that require real-time response and others that need life-critical reliability. Conversely, systems with vastly different requirements exist in each industry. The bottom line is that fundamental system requirements vary by application and not by industry and these different types of systems need very different approaches (Figure 3.11).

Thus, as in biology, the IIoT needs an *environment-independent* system science. This science starts by understanding the key system challenges and resulting requirements. If we can identify common cross-industry requirements, we can then logically specify common cross-industry architectures that meet those requirements. That architecture will lead to technologies and standards that can span industries.

There is both immense power and challenge in this statement. Technologies that span industries face many challenges, both political and practical. Nonetheless, a clear fact of systems in the field is the similarity of requirements and architecture across industries. Leveraging this fact promises a much better understood, better connected future. It also has immense economic benefit: over time, generic technologies offer huge advantage over special-purpose approaches. Thus, to grow our understanding and realize the promise of the IIoT, we must abandon our old industry-specific thinking.

3.3.1 Proposed Taxonomic Criteria

So, what can we use for divisions? What defining characteristics can we use to separate the mammals from the reptiles from the insects of the IIoT?

There are far too many requirements, both functional and nonfunctional, to consider in developing a "comprehensive" set to use as criteria. As with animals, we need to find those few requirements that divide the space into useful major categories.

The task is simplified by the realization that *the goal is to divide the space so we can determine system architecture*. Thus, good division criteria are (1) unambiguous and (2) impactful on the architecture. That makes the task easier, but still nontrivial. The only way to do it is through experience. We are early on our quest. However, significant progress is within our collective grasp.

This work draws on extensive experience with nearly 1,000 real-world IIoT applications. Our conclusion is that an IIoT taxonomy is not only possible but also critical to both the individual system building and the inception of a true cross-industry IIoT.

While the classification of IIoT systems is very early, we do suggest a few divisions. To be as crisp as possible, we also chose numeric "metrics" for each division. The lines, of course, are not that stark. And those lines evolve with technology over time at a much faster pace than biological evolution. Nonetheless, the numbers are critical to force clarity; without numerical metrics, meaning is often too fuzzy.

3.3.1.1 Reliability

Metric: Continuous availability must exceed "99.999%" to avoid severe consequences.

Architectural Impact: Redundancy

Many systems describe their requirements as "highly reliable," "mission critical," or "minimal downtime." However, those labels are more often platitudes than actionable system requirements. For these requirements to be meaningful, we must be more specific about the reasons we must achieve that reliability. That requires understanding how quickly a failure causes problems and how bad those problems are.

Thus, we define "continuous availability" as the probability of a temporary interruption in service over a defined system-relevant time period. The "five 9s" golden specification for enterprise-class servers translates to about 5 minutes of downtime per year. Of course, many industrial systems cannot tolerate even a few milliseconds of unexpected downtime. For a power system, the relevant time period could span years. For a medical imaging machine, it could be only a few seconds.

The consequences of violating the requirement are also meaningful. A traffic control system that goes down for a few seconds could result in fatalities. A web site that goes down for those same few seconds would only frustrate users. These are fundamentally different requirements.

Reliability thus defined is an important characteristic because it greatly impacts the system architecture. A system that cannot fail, even for a short time, must support redundant computing, sensors, networking, storage, software, and more. Servers become troublesome single-point-of-failure weak points. When reliability is truly critical, redundancy quickly becomes a—or perhaps *the*—key architectural driver (Figure 3.12).

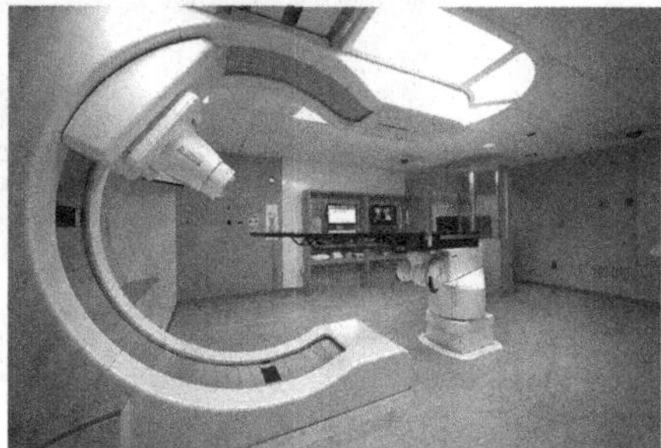

FIGURE 3.12 IIoT reliability-critical applications. Hydropower dams can quickly modulate their significant power output by changing water flow rates and thus help balance the grid: even a few milliseconds of unplanned downtime can threaten stability. Air traffic control faces a similar need for continuous operation: a short failure in the system endangers hundreds of flights. A proton-beam radiation therapy system must guarantee precise operation during treatment: operational dropouts threaten patient outcomes. Applications with severe consequences of short interruptions in service require a fully redundant architecture, including computing, sensors, networking, storage, and software.

3.3.1.2 Real Time

Metric: Response < 100 ms

Architectural Impact: Peer-to-peer data path

There are many ways to characterize "real time." All systems should of course be "fast." However, for these requirements to be useful we must specifically understand which timing requirements drive success.

Thus, "real time" is much more about guaranteed response than it is about fast. Many systems require low average latency (delivery delay). However, true real-time systems succeed only if they always respond "on time." This is the maximum latency, often expressed as the average delay plus the variation or "jitter." Even a fast server with low average latency can experience large jitter under load (Figure 3.13).

In a distributed system, the most important architectural impact is the potential jitter imposed by a server or broker in the data path. An architecture that can satisfy a human user annoyed by a wait longer than 8 seconds for a web site will never satisfy an industrial control that must respond in 2 milliseconds. We find that the "knee in the curve" that greatly impacts design occurs when the speed of response is measured in a few tens of milliseconds or even microseconds. We choose 100 ms, simply because that is about the unpredictable delay of today's servers. Systems that must respond faster than this usually must be peer to peer, and that is a huge architectural impact (Figure 3.14).

3.3.1.3 Data Item Scale

Metric: More than 10,000 addressable data items

Architectural Impact: Selective delivery filtering

Scale is a fundamental challenge for the IIoT. It is also complex; there are many dimensions of scale, including number of "nodes," number of applications, number

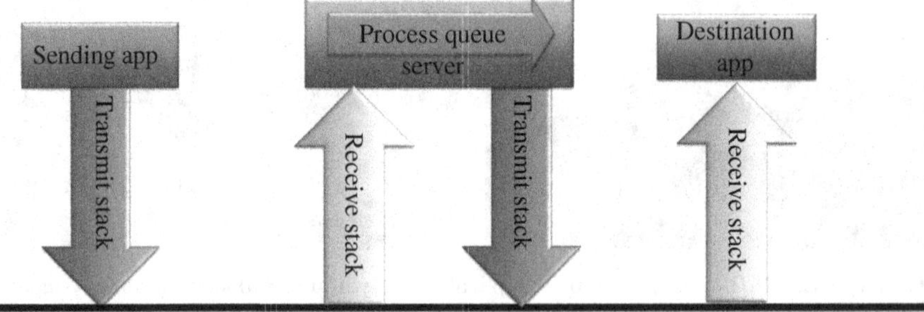

FIGURE 3.13 Added server latency. Although the hardware transmit time is often negligible, sending data through a server "hop" requires traversing the sending machine's transmit stack, the server's receive stack, the server's processing queue, the server's transmit stack, and finally the destination's receive stack. Each of these has threads, queues, and buffers that add uncontrolled latency. Worse, the server cannot easily prioritize traffic as easily as the end points. Thus, systems that are sensitive to maximum latency often cannot use data servers. Reproduced with permission from Real-Time Innovations.

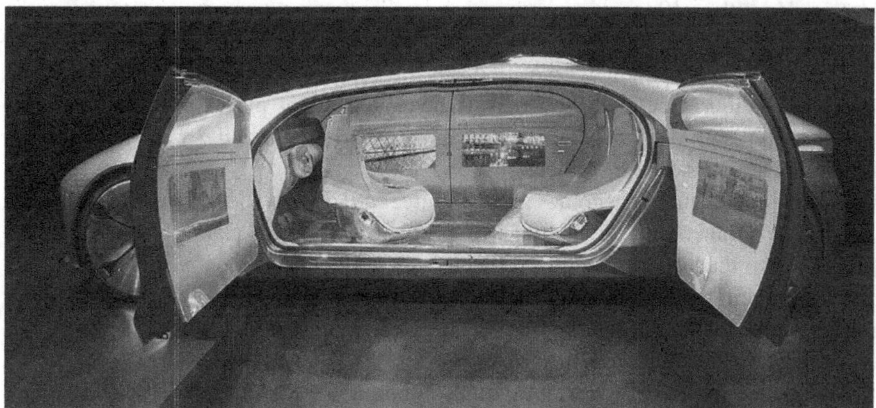

FIGURE 3.14 IIoT real-time applications. To provide quality feel to surgeons, distributed control loops for medical robotics must run at rates up to 3 kHz and control the "jitter" to only tens of microseconds. Similarly, autonomous cars must react fast enough to safely control the vehicle and prevent collisions. These fundamental performance needs imply a system architecture that does not send data through intermediaries. Robotics photo: DLR CC-BY 3.0.

of developers on the project, number of data items, data item size, total data volume, and more. We cannot divide the space by all these parameters.

In practice, however, they are related. For instance, a system with many data items probably has many nodes.

Despite the broad space, we have found that two simple metrics correlate well with architectural requirements.

The first scale metric is addressable "data item scale," defined as the number of different data instances that could be of interest to different parts of the system. Note that this is not the same as the size of a single large data set, such as a stream of data from a single fast sensor. The key scale parameter is the existence of many different data items that could potentially be of interest to different consumers. So, a few fast sensors create only a few addressable data items. Many sensors or sources create many data items. A large number of addressable data items implies difficulty in sending the right data to the right place (Figure 3.15).

When systems get "big" in this way, it is no longer practical to send every data update to every possible receiver. We find that the challenge is significant for as few as 100 data items. It is extreme for systems with more than 10,000 addressable data items. Above this limit, managing the data itself becomes a key architectural need. These systems need an architectural design that explicitly understands the data, thereby allowing selective filtering and delivery. There are two approaches in common use: runtime introspection that allows consumers to choose data items themselves and "data-centric" designs that empower the infrastructure itself to understand and actively filter the data system wide.

3.3.1.4 Module Scale

Metric: More than 10 teams or interacting applications
Architectural Impact: Interface control and evolution

The second scale parameter we choose is the number of "modules" in the system, where a module is defined as a reasonably independent piece of software. Each module is typically an independently developed application built by an independent team of developers on the "project."

Module scale quickly becomes a key architectural driver. The reason is that system integration is inherently an "*n*-squared" problem. Each new team presents another interface into the system. Smaller projects built by a cohesive team can easily share interface specifications without formality. Larger projects built by many independent groups of developers face a daunting challenge. System integration can occupy half of the delivery schedule and most of its risk.

In these large systems, interface control dominates the interoperability challenge. It is not practical to expect interfaces to be static. Modules, or groups of modules, that depend on an evolving interface schema must somehow continue to interoperate with older versions of that schema. Communicating all the interfaces becomes hard. Forcing all modules to "update" on a coordinated time frame to a new schema becomes impossible. Thus, interacting teams quickly find that they need tool, process, and eventually architectural support to solve the system integration problem.

Of course, this is a well-studied problem in enterprise software systems. In the storage world, databases ease system integration by explicitly modeling and controlling "data tables," thus allowing multiple applications to access information in a controlled manner. Communication technologies like enterprise service buses (ESBs), Web services, enterprise "queuing" middleware, and textual schema like XML and JSON all provide evolvable interface flexibility. However, these are often not appropriate for industrial systems, usually for performance or resource reasons.

FIGURE 3.15 IIoT applications with many data items. IIoT systems often produce far too much data to send everything to every possible consumer. "Gust control" in a wind turbine farm, for instance, needs weather updates from the turbines immediately "upwind," a specification that changes with time. Traffic control systems are very interested only in vehicles approaching an intersection. These applications require the architecture to provide selective data availability, so only the right information loads the network and the participants.

Data-centric systems expose and control interfaces directly, thus easing system integration. Databases, for instance, provide data-centric storage and are thus important in systems with many modules. However, databases provide storage for data at rest. Most IIoT systems require data in motion, not (or in addition to) data at rest.

Data-centric middleware is a relatively new concept for distributed systems. Similar to a database data table, data-centric middleware allows applications to interact through explicit data models. Advanced technologies can even detect and manage differences in interfaces between modules and then adapt to deliver to each end point in the schema what the end point expects [9]. These systems thus decouple application interface dependencies, allowing large projects to evolve interfaces and make parallel progress on multiple fronts (Figure 3.16).

3.3.1.5 Runtime Integration

Metric: More than 20 "devices," each with many parameters and data sources or sinks that cannot be configured at development time

Architectural Impact: Must provide a discoverable integration model

Some IIoT systems can (or even must) be configured and understood before runtime. This does not mean that every data source and sink is known, but rather that this configuration is relatively static. Others, despite a potentially large size, have applications that implement specific functions that depend on knowing what data will be available. These systems can or must implement an "end point" discovery model that finds all the data in the system directly.

However, other systems cannot easily know what devices or data will be available until runtime. For instance, when IIoT systems integrate racks of field-replaceable machines or devices, they must often be configured and understood during operation. For instance, a plant controller HMI may need to discover the device characteristics of an installed device or rack so a user can choose data to monitor.

The key factor here is not addition or changes in which device is used. It is more a function of not knowing which types of devices may be involved.

These systems must implement a different way to discover information. Instead of searching for data, it is more efficient to automate the process by building runtime maps of devices and their data relationships. The choice of "20" different devices is arbitrary. The point is that when there are many different configurations for many devices, mapping them at runtime becomes an important architectural need. Each device requires some sort of server or manager that locally configures attached subdevices and then presents that catalog to the rest of the system. This avoids manual gymnastics (Figure 3.17).

3.3.1.6 Distribution Focus

Metric: Fan-out > 10

Architectural Impact: Must use one-to-many connection technology

We define "fan-out" as the number of data recipients that must be informed upon change of a single data item. Thus, a data item that must go to 10 different destinations each time it changes has a fan-out of "10."

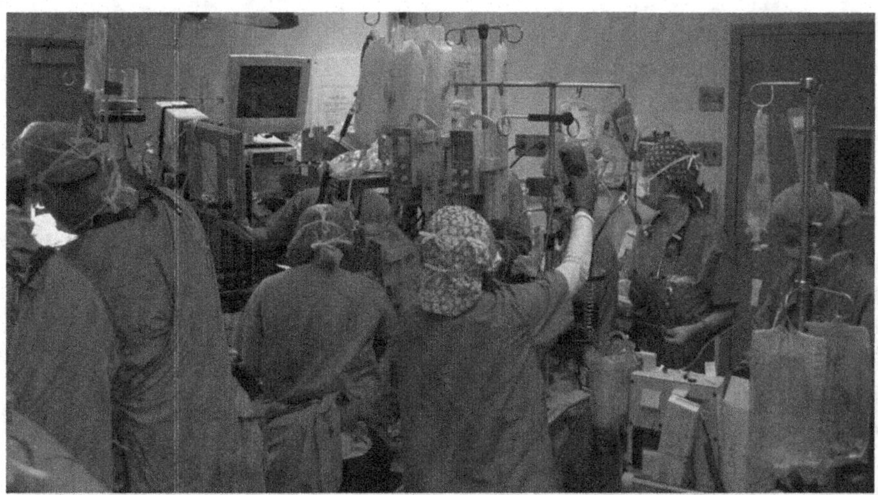

FIGURE 3.16 IIoT applications built by large teams. Hundreds of different types of hospital medical devices, from heart monitors to ventilators, must combine to better monitor and care for patients. Similarly, ship systems integrate dozens of complex functions like navigation, power control, and communications. When a complex "system of systems" integrates many complex interfaces, the system architecture itself must help to manage system integration and evolution.

Fan-out impacts architecture because many protocols work through single 1 : 1 connections. Most of the enterprise world works this way, often with TCP, a 1 : 1 session protocol. Examples include connecting a browser to a Web server, a phone app to a backend, or a bank to a credit card company. While these systems can achieve significant scale, they must manage a separate connection to each end point.

FIGURE 3.17 IIoT device integration challenge. Large systems assembled in the field from a large variety of "devices" face a challenge in understanding and discovering interacting devices and their relationships. The most common example applications are in manufacturing. These applications benefit from a design that offers the ability for remote applications and human interfaces to "browse" the system, thus discovering data sources and relationships.

FIGURE 3.18 IIoT applications needing data distribution. Many applications must deliver the same data to many potential end points. Coordinated vehicle fleets may update a cloud server, but then that information must be delivered to many distributed vehicles. An emergency service communication system must allow many remote users access to high-bandwidth distributed voice and video streams. Many industries use "hardware in the loop" simulation to test and verify modules during development. Across all these industries, an efficient architecture must deliver data to multiple points easily.

When many data updates must go to many end points, the system is not only managing many connections, but it is also sending the same data over and over through each of those connections.

IIoT systems often need to distribute information to many more destinations than enterprise systems. They also often need higher performance on slower machines. Complex systems even face a "fan-out mesh" problem, where many producers of information must send it to many recipients. When fan-out exceeds 10 or so, it becomes impractical to do this branching by managing a set of 1 : 1 connections. An architecture that supports efficient multiple updates greatly simplifies these systems (Figure 3.18).

3.3.1.7 Collection Focus

Metric: One-way data flow from more than 100 sources

Architectural Impact: Local concentrator or gateway design

Data collection from field systems is a key driver of the IIoT. Many systems transmit copious information to be stored or analyzed in higher-level servers or the cloud. Systems that are essentially restricted to the collection problem do not share significant data between devices. These systems must efficiently move information to a common destination, but not between devices in the field.

This has huge architectural impact. Collection systems can often benefit from a hub-and-spoke "concentrator" or gateway. Widely distributed systems can use a cloud-based server design, thus moving the concentrator to the cloud (Figure 3.19).

FIGURE 3.19 IIoT collection and monitoring applications. Collecting and analyzing field-produced data is perhaps the first "killer app" of the IIoT. The IIC's "track and trace" test bed, for instance, tracks tools on a factory floor so the system can automatically log use. Other applications include monitoring gas turbines for efficient operation, testing aircraft landing gear for potentially risky situations, and optimizing gas pipeline flow control. Since there is little interdevice flow, "hub and spoke" system architectures that ease collection work well for these systems.

3.3.2 Dimensional Decomposition and Map to Implementation

The analogy with a biological taxonomy only goes so far. Industrial systems do not stem from common ancestors and thus do not fall into crisply defined categories. As implied previously, most systems exhibit some degree of each of the characteristics. This is actually a source of much of the confusion and the reason for our attempt to choose hard metrics at the risk of declaring arbitrary boundaries. In the end, however, the goal is to use the characteristics to help select a single system architecture. Designs and technologies satisfy the previously mentioned goals to various degrees. With no system science to frame the search, the selection of a single architecture based on any one requirement becomes confusing.

Perhaps a better analysis is to consider each of the key characteristics as an axis in an n-dimensional space. The taxonomical classification process then places each application on a point in this n-dimensional space.

This is not a precise map. Applications may be complex and thus placement is not exact. The metrics mentioned before delineate architecturally significant boundaries that are not in reality crisp. So, the lines that we have named are somewhat fuzzy. However, an exact position is often not important. Our classification challenge is really only to decide on which side of each boundary our application falls.

In this framework, architectural approaches and the technologies that implement them can be considered to "occupy" some region in this n-dimensional space. For instance, a data-centric technology like the Object Management Group (OMG) DDS provides peer-to-peer, fully redundant connectivity with content filtering. Thus, it would occupy a space that satisfies many *reliable*, *real-time* applications with significant numbers of *data items*, the first three challenge dimensions previously mentioned. The Message Queuing Telemetry Transport (MQTT) protocol, on the other hand, is more suited to the data collection focus challenge. Thus, these technologies occupy different regions of the solution space. Figure 3.20 represents this concept in three dimensions.

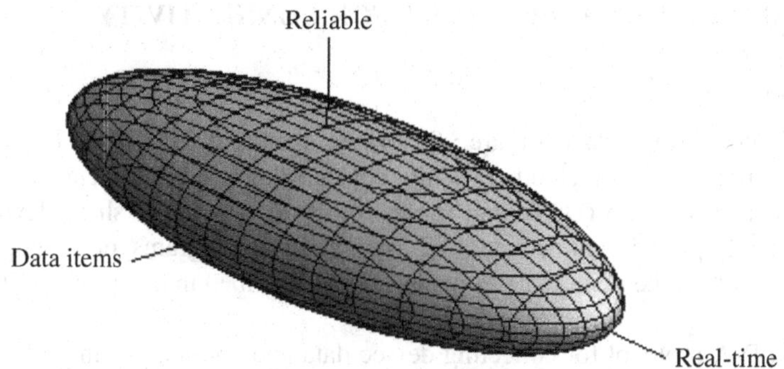

FIGURE 3.20 *n*-Dimensional requirement space. Architectural approaches and their implementing technologies satisfy some range of each of the dimensions above and thus occupy a region in an *n*-dimensional "requirement space." The value of a taxonomy is to help designers decompose their problem into relevant dimensions so they can then select an appropriate approach.

Thus, the application can be placed in the space, and the architectural approaches represented as regions. This reduces the problem of selecting an architecture to one of mapping the application point to appropriate architectural regions.

Of course, this may not be a unique map; the regions overlap. In this case, the process indicates options. The trade-off is then to find something that fits the key requirements while not imposing too much cost in some other dimension. Thinking of the system as an *n*-dimensional mapping of requirements to architecture offers important clarity and process. It greatly simplifies the search.

3.3.3 Taxonomy Benefits

Defining an IIoT taxonomy will not be trivial. The IIoT encompasses many industries and use cases. It encompasses much more diversity than applications for specialized industry requirements, enterprise IT, or even Consumer IoT. Technologies also evolve quickly, so the scene is constantly shifting. This present state just scratches the surface.

However, the benefit of developing a taxonomical understanding of the IIoT is enormous. Resolving these issues will help system architects choose protocols, network topologies, and compute capabilities. Today, we see designers struggling with issues like server location or configuration, when the right design may not even require servers. Overloaded terms like "real time" and "thing" cause massive confusion between technologies despite the fact that they have no practical use-case overlap. The industry needs a better framework to discuss architectural fit.

Collectively, organizations like the IIC enjoy extensive experience across the breadth of the IIoT. Mapping those experiences to a framework is the first step in the development of a *system science* of the IIoT. Accepting this challenge promises to help form the basis of a better understanding and logical approach to designing tomorrow's industrial systems.

3.4 STANDARDS AND PROTOCOLS FOR CONNECTIVITY

3.4.1 IoT Protocols

The IoT Protocol Roadmap in Figure 3.21 outlines the basic needs of the IoT. Devices must communicate with each other (D2D), device data must be collected and sent to the server infrastructure (D2S), and server infrastructure has to share device data (S2S), possibly providing it back to devices, to analysis programs, or to people.

From 30,000 ft, the main IoT protocols can be described in this framework as:

- MQTT: A protocol for collecting device data and communicating it to servers (D2S) [10]
- XMPP: A protocol best for connecting devices to people, a special case of the D2S pattern, since people are connected to the servers [11]
- DDS: A fast bus for integrating intelligent machines (D2D) [12]
- Advanced Message Queuing Protocol (AMQP): A queuing system designed to connect servers to each other (S2S) [13]
- Open Platform Communications (OPC) Unified Architecture (OPC UA): A control plane technology that enables interoperability between devices [14]

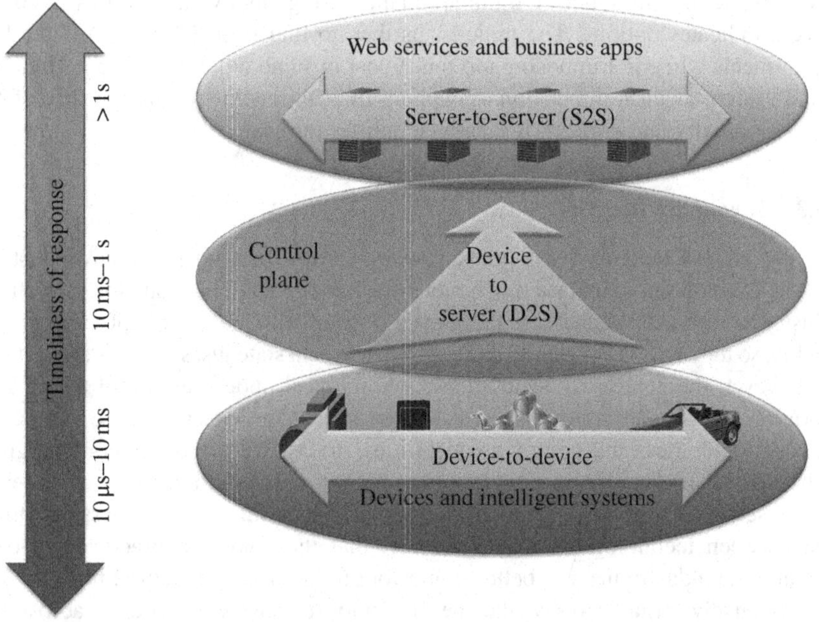

FIGURE 3.21 IoT protocol road map. Devices communicate with each other (D2D) and send data to the IT infrastructure (D2S). The IT infrastructure servers use the data (S2S), communicating back to devices or to people. Reproduced with permission from Real-Time Innovations.

Each of these protocols is widely adopted. There are at least 10 implementations of each. Confusion is understandable, because the high-level positioning is similar. In fact, the first four all claim to be real-time publish–subscribe IoT protocols that can connect thousands of things. Worse, those claims are true, depending on how you define "real time," "publish–subscribe," and "thing."

Nonetheless, all five are very different indeed! They do not in fact overlap much at all. Moreover, they don't even fill strictly comparable roles. For instance, OPC UA and DDS are best described as information systems that *have* a protocol; they are much more than just a way to send bits.

The previously mentioned simple taxonomy frames the basic protocol use cases (Figure 3.21). Of course, it's not really that simple. For instance, the "control plane" represents some of the complexity in controlling and managing all these connections. Many protocols cooperate in this region.

Today's enterprise Internet supports hundreds of protocols; the IoT will support hundreds more. It's important to understand the class of use that each of these important protocols addresses.

3.4.1.1 MQTT MQTT targets device data collection (Figure 3.22). As the name states, the main purpose is telemetry, or remote monitoring. Its goal is to collect data

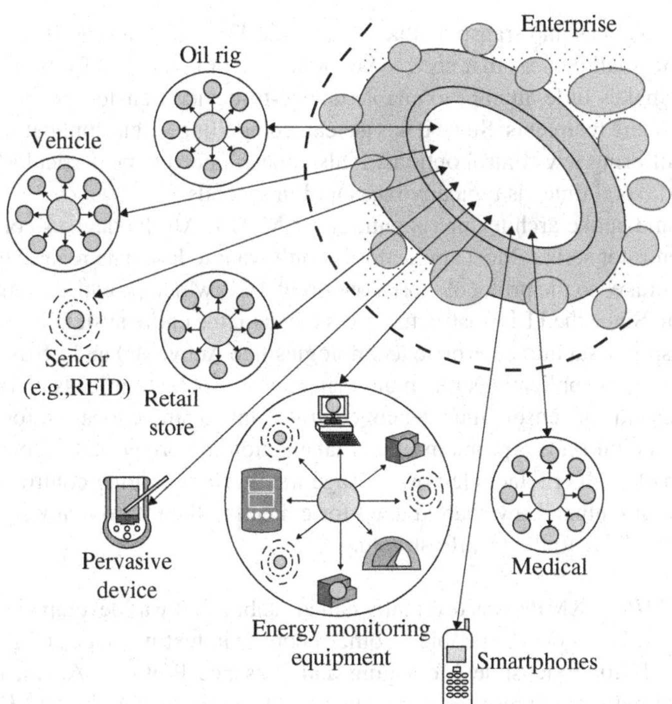

FIGURE 3.22 Message Queuing Telemetry Transport (MQTT) implements a hub-and-spoke data collection system.

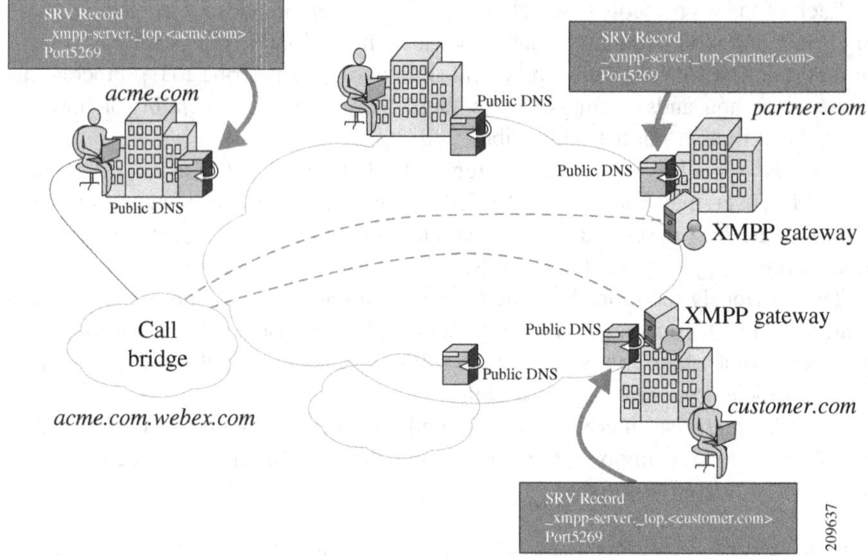

FIGURE 3.23 Extensible Messaging and Presence Protocol (XMPP) provides text communications between diverse points.

from many devices and transport that data to the IT infrastructure. It targets large networks of small devices that need to be monitored or controlled from the cloud.

MQTT makes little attempt to enable device-to-device transfer, nor to "fan out" the data to many recipients. Since it has a clear, compelling single application, MQTT is simple, offering few control options. It also doesn't need to be particularly fast. In this context, "real time" is typically measured in seconds.

A hub-and-spoke architecture is natural for MQTT. All the devices connect to a data concentrator server. Most applications don't want to lose data regardless of how long retries take, so the protocol works on top of TCP, which provides a simple, reliable stream. Since the IT infrastructure uses the data, the entire system is designed to easily transport data into enterprise technologies like ActiveMQ and ESBs.

MQTT targets applications like monitoring an oil pipeline for leaks or vandalism. Those thousands of sensors must be concentrated into a single location for analysis. When the system finds a problem, it can take action to correct that problem. Other applications for MQTT include power usage monitoring, lighting control, and even intelligent gardening. They share a need for collecting data from many sources and making it available to the IT infrastructure.

3.4.1.2 XMPP XMPP was originally called "Jabber." It was developed for instant messaging (IM) to connect people to other people via text messages (Figure 3.23). XMPP stands for Extensible Messaging and Presence Protocol. Again, the name belies the targeted use: presence—meaning people are intimately involved.

XMPP uses the XML text format as its native type, making person-to-person communications natural. Like MQTT, it runs over TCP, or perhaps over HTTP on top

FIGURE 3.24 Data Distribution Service (DDS) connects devices at physics speeds into a single distributed application.

of TCP. Its key strength is a name@domain.com addressing scheme that helps connect the needles in the huge Internet haystack.

In the IoT context, XMPP offers an easy way to address a device. This is especially handy if that data is going between distant, mostly unrelated points, just like the person-to-person case. It's not designed to be fast. In fact, most implementations use polling or checking for updates only on demand. A protocol called Bidirectional-streams over Synchronous HTTP (BOSH) lets servers push messages. But "real time" to XMPP is on a human scale, measured in seconds.

XMPP provides a great way, for instance, to connect your home thermostat to a Web server so you can access it from your phone. Its strengths in addressing, security, and scalability make it appropriate for Consumer IoT applications.

3.4.1.3 DDS In contrast to MQTT and XMPP, DDS targets devices that directly use device data. It distributes data to other devices (Figure 3.24). DDS's main purpose is to connect devices to other devices. It is a data-centric middleware standard with roots in high-performance defense, industrial, and embedded applications. DDS can efficiently deliver millions of messages per second to many simultaneous receivers.

Devices demand data very differently than the IT infrastructure demands data. First, devices are fast. "Real time" may be measured in milliseconds or microseconds. Devices need to communicate with many other devices in complex ways, so TCP's simple and reliable point-to-point streams are far too restrictive. Instead, DDS offers detailed quality-of-service (QoS) control, multicast, configurable reliability, and pervasive redundancy. In addition, fan out is a key strength. DDS offers powerful ways to filter and select exactly which data goes where, and "where" can be thousands of simultaneous destinations. Some devices are small, so there are lightweight versions of DDS that run in constrained environments.

Hub-and-spoke is completely inappropriate for device data use. Rather, DDS implements direct device-to-device "bus" communication with a relational data model. This is often termed a "databus" because it is the networking analog to a database. Similar to the way a database controls access to stored data, a databus controls data access and updates by many simultaneous users. This is exactly what many high-performance devices need to work together as a single system.

High-performance integrated device systems use DDS. It is the only technology that delivers the flexibility, reliability, and speed necessary to build complex, real-time applications. DDS is very broadly used. Applications include wind farms, hospital integration, medical imaging, autonomous planes and cars, rail, asset tracking, automotive test, smart cities, communications, data center switches, video sharing, consumer electronics, oil and gas drilling, ships, avionics, broadcast television, air traffic control, SCADA, robotics, and defense.

RTI has experience with nearly 1,000 applications. DDS connects devices together into working distributed applications at physics speeds.

3.4.1.4 AMQP AMQP is sometimes considered an IoT protocol. AMQP is all about queues (Figure 3.25). It sends transactional messages between servers. As a message-centric middleware that arose from the banking industry, it can process thousands of reliable queued transactions.

AMQP is focused on not losing messages. Communications from the publishers to exchanges and from queues to subscribers use TCP, which provides strictly reliable point-to-point connection. Further, end points must acknowledge acceptance of each message. The standard also describes an optional transaction mode with a formal multiphase commit sequence. True to its origins in the banking industry, AMQP middleware focuses on tracking all messages and ensuring each is delivered as intended, regardless of failures or reboots.

AMQP is mostly used in business messaging. It usually defines "devices" as mobile handsets communicating with back-office data centers. In the IoT context, AMQP is most appropriate for the control plane or server-based analysis functions.

3.4.1.5 OPC UA OPC UA is an upgrade of the venerable OPC (OLE for Process Control) protocol. OPC is operational in thousands of factories all over the world. Traditionally, OPC was used to configure and query plant-floor servers (usually Programmable Logic Controllers (PLCs)). Actual device-to-device communications were then effected via a hardware-based "fieldbus" such as ModBus or PROFINET.

OPC UA retains some of that flavor; it connects and configures plant-floor servers. The UA version adds better modeling capabilities. Thus, a remote client (e.g., a graphical interface) can "browse" the device data controlled by a server on the floor. By allowing this introspection across many servers, clients can build a model of the "address space" of all the devices on the floor.

OPC UA specifically targets manufacturing. It connects applications at the shop-floor level as well as between the shop floor and the enterprise IT cloud. In the taxonomy previously mentioned, it targets systems that require runtime integration (Figure 3.26).

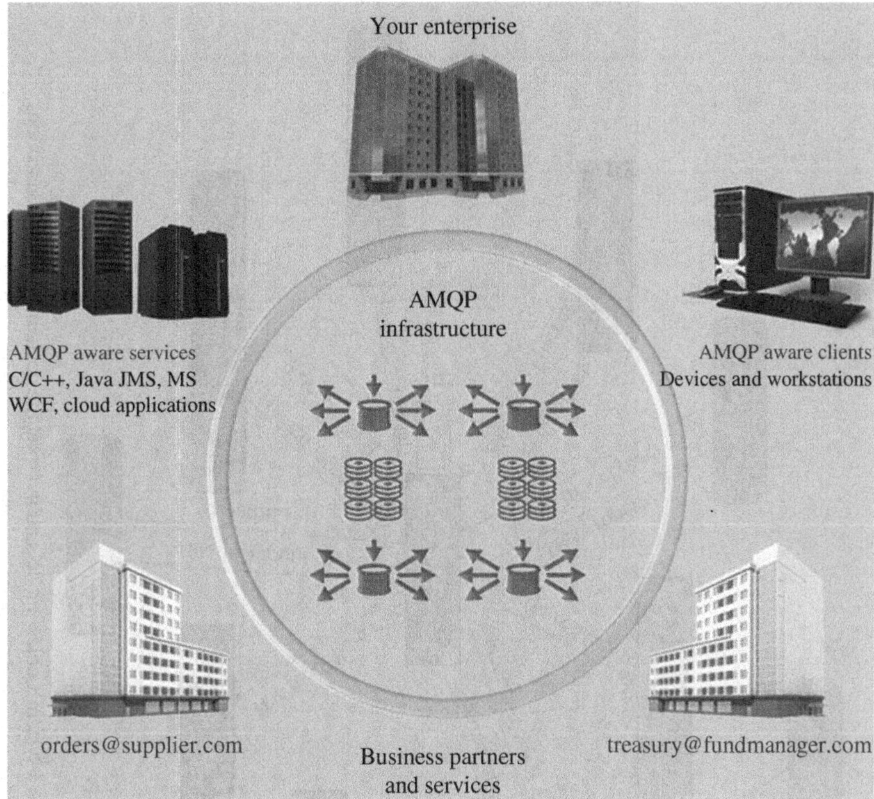

FIGURE 3.25 Advanced Message Queuing Protocol (AMQP) shares data reliably between servers.

3.4.2 The Bottom Line: How to Choose?

The IoT needs many protocols. Those outlined here differ markedly. Perhaps it's easiest to categorize them along a few key dimensions: QoS, addressing, and application.

QoS control is a much better metric than the overloaded "real-time" term. QoS control refers to the flexibility of data delivery. A system with complex QoS control may be harder to understand and program, but it can build much more demanding applications.

For example, consider the reliability QoS. Most protocols run on top of TCP, which delivers strict, simple reliability. Every byte put into the pipe must be delivered to the other end, even if it takes many retries. This is simple and handles many common cases, but it doesn't allow timing control. TCP's single-lane traffic backs up if there's a slow consumer.

Because it targets device-to-device communications, DDS differs markedly from the other protocols in QoS control. In addition to reliability, DDS offers QoS control of "liveliness" (when you discover problems), resource usage, discovery, and even timing.

Next, "discovery"—finding the data needle in the huge IoT haystack—is a fundamental challenge. XMPP shines for "single item" discovery. Its "user@domain"

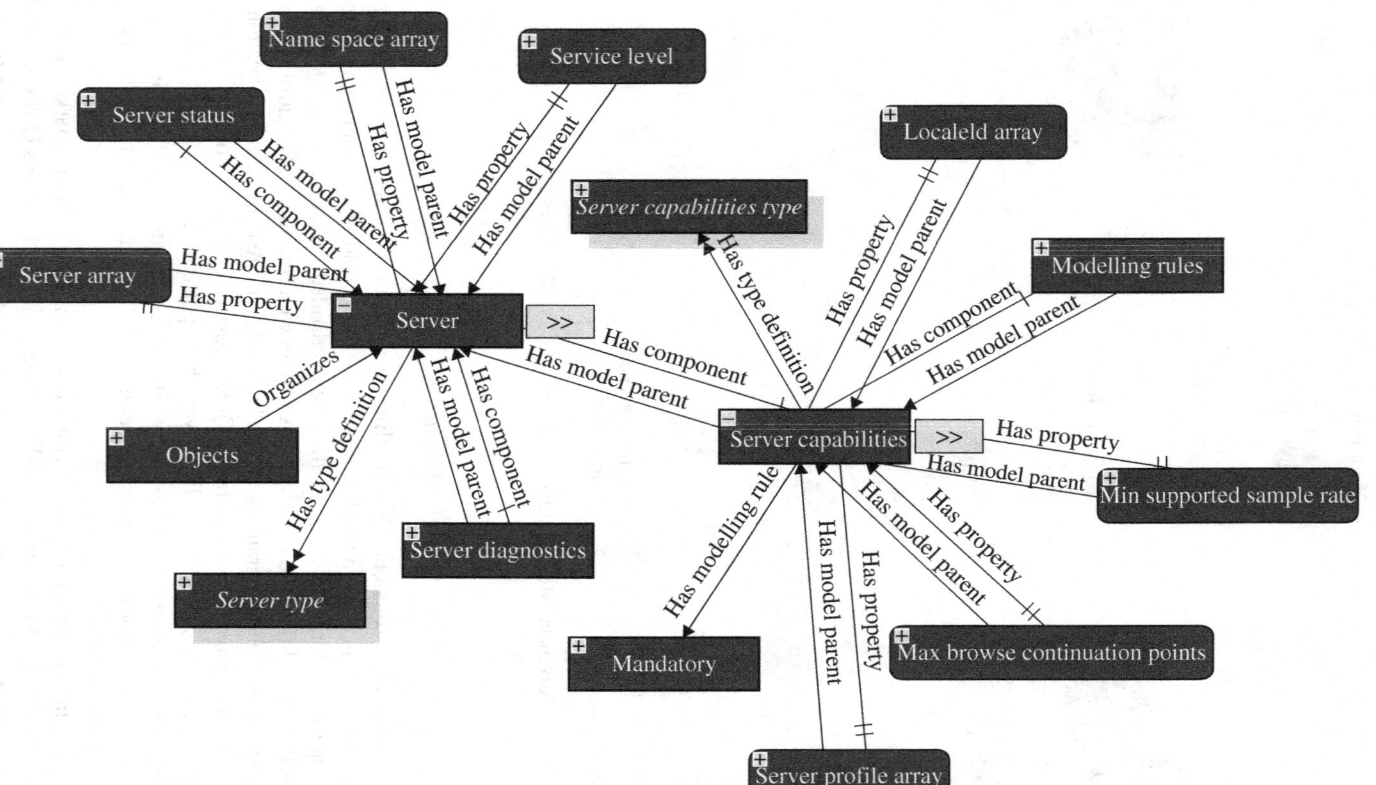

FIGURE 3.26 OPC UA allows applications to browse a system's object model and interpret the devices and their connections.

addressing leverages the Internet's well-established conventions. However, XMPP doesn't easily handle large data sets connected to one server. With its collection-to-a-server design, MQTT handles that case well. If you can connect to the server, you're on the network. AMQP queues act similarly to servers, but for S2S systems. Again, DDS is an outlier. Instead of a server, it uses a background "discovery" protocol that automatically finds data. DDS systems are typically more contained; discovery across the wide-area network (WAN) or huge device sets requires special consideration.

OPC UA specializes in communicating the information about the system, its configuration, topology, and data context (the "metadata"). These are exposed in the collective address space of the individual OPC UA servers. This data can be accessed by OPC UA clients; they can see what is available and choose what to access. OPC UA is not designed for flexible device-to-device interaction.

Perhaps the most critical distinction comes down to the intended applications. Interdevice data use is a fundamentally different use case from device data collection. For example, turning on your light switch remotely (best for XMPP) is worlds apart from generating that power (DDS), monitoring the transmission lines (MQTT), or analyzing the power usage back at the data center (AMQP).

Of course, there is still confusion. For instance, DDS can serve and receive data from the cloud, and MQTT can send information back out to devices. Nonetheless, the fundamental goals of all five protocols differ, the architectures differ, and the capabilities differ. All of these protocols are critical to the (rapid) evolution of the IoT. And all have a place; the IoT is a big place with room for many protocols. To make confusion even worse, many applications integrate many subsystems, each with different characteristics.

With this variety, what's the best path? Experience suggests that designers first identify the application's toughest challenge and then choose the technology that best meets that single challenge. This is a critical decision; choose carefully and without prejudice of what you know. The aforementioned requirements-based dimensional decomposition can help. After this step, the choice is usually fairly obvious; most applications contain a key challenge that clearly fits better with one or the other.

Once the hardest challenge is met, the best way to cover the rest of the application is usually to push your initial choice as far as it will go. There are many bridging technologies, so it is possible to mix technologies. However, it's usually easier to avoid multiple protocol-integration steps when possible.

In the long term, the technologies will offer better interoperability. The IIC's "connectivity core standard" design described later is a key approach. So, regardless of your initial choice, the vendor communities are working to provide a nonproprietary path to interoperability.

3.5 CONNECTIVITY ARCHITECTURE FOR THE IIoT

There is no way to build large distributed systems without connectivity. Enterprise and human-centric communications are too slow or too sparse to put together large networks of fast devices. These new types of intelligent machines need a new technology.

That technology has to find the right data and then get that data where it needs to go on time. It has to be reliable, flexible, fast, and secure. Perhaps not as obviously, it also must work across many types of industries. Only then can it enable the efficiencies of common machine-based and cloud-based infrastructure for the IIoT.

Connectivity faces two key challenges in the IIoT: interoperability and security. Interoperability is a challenge because the IIoT must integrate many subsystems with different designs, vendor equipment, or legacy infrastructures. Security is a challenge because most enterprise security approaches target hub-and-spoke designs with a natural center of trust. Those designs cannot handle vast networks of devices that must somehow trust each other.

The IIC's IIRA [15] addresses both. Ultimately, the IIoT is about building distributed systems. Connecting all the parts intelligently so the system can perform, scale, evolve, and function optimally is the crux of the IIRA. The IIoT must integrate many standards and connectivity technologies. The IIC architecture explicitly blends the various connectivity technologies into an interconnected future that can enable the sweeping vision of a hugely connected new world.

3.5.1 The *n*-Squared Challenge

When you connect many different systems, the fundamental problem is the "*n*-squared" interconnect issue. Connecting two systems requires matching many aspects, including protocol, data model, communication pattern, and QoS parameters like reliability, data rate, or timing deadlines. While connecting two systems is a challenge, it is solvable with a special-purpose "bridge." But that approach doesn't scale; connecting *n* systems together requires *n*-squared bridges. As *n* gets large, this becomes daunting.

One way to ease this problem is to keep *n* small. You can do that by dictating all standards and technologies across all systems that interoperate. Many industry-specific standards bodies successfully take this path. For instance, the European Generic Vehicle Architecture (GVA) specifies every aspect of how to build military ground vehicles, from low-level connectors to top-level data models. The German Industrie 4.0 effort takes a similar pass at the manufacturing industry, making choices for ordering and delivery, factory design, technology, and product planning. Only one standard per task is allowed.

This approach eases interoperation. Unfortunately, the result is limited in scope because the rigidly chosen standards cannot provide all functions and features. There are simply too many special requirements to effectively cross industries this way. Dictating standards also doesn't address the legacy integration problem. These two restrictions (scope and legacy limits) make this approach unsuited to building a wide-ranging, cross-industry Industrial Internet.

On the other end of the spectrum, you can build a very general bridge point. Enterprise Web services work this way, using an "ESB" or a mediation bus like Apache Camel. However, despite the "bus" in its name, an ESB is not a distributed concept. All systems must connect to a single point, where each incoming standard is mapped to a common object format. Because everything maps to one format, the

ESB requires only one-way translation, avoiding the n-squared problem. Camel, for instance, supports hundreds of adapters that each convert one protocol or data source to and from Camel's internal object format. Thus, any protocol can in principal connect to any other.

Unfortunately, this doesn't work well for demanding industrial systems. The single ESB service is an obvious choke and failure point. ESBs are large, slow programs. In the enterprise, ESBs connect large-grained systems executing only a few transactions per second. Industrial applications need much faster, reliable, smaller-grained service. So, ESBs are not viable for most IIoT uses.

3.5.2 The IIRA Connectivity Core Standard

The IIRA takes an intermediate approach. The design introduces the concept of a "Connectivity Core Standard." Unlike an ESB, the core standard is very much a distributed concept. Some end points can connect directly to the core standard. Other end points and subsystems connect through "gateways." The core standard then connects them all together. This allows multiple protocols without having to bridge between all possible pairs. Each needs only one bridge to the core.

Like an ESB, this solves the n-squared problem. But, unlike an ESB, it provides a fast, distributed core, replacing the centralized service model. Legacy and less capable connectivity technologies transform through a gateway to the core standard. There are only n transformations, where n is the number of connectivity standards (Figure 3.27).

Obviously, this design requires a very functional connectivity core standard. Some systems may get by with slow or simple cores. But most industrial systems need to

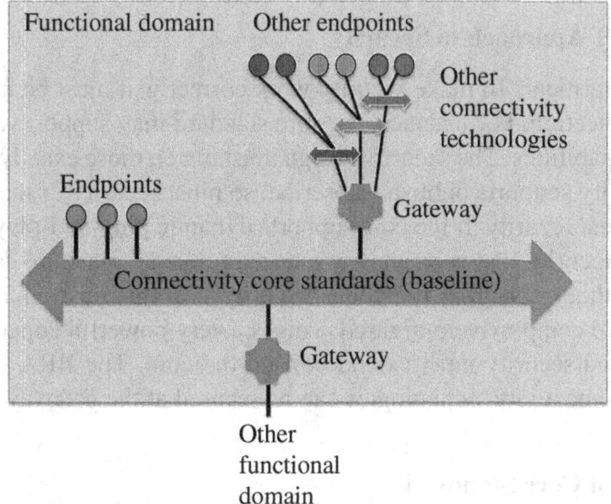

FIGURE 3.27 The IIRA connectivity architecture specifies a Quality-of-Service controlled, secure connectivity core standard. All other connectivity standards must only bridge to this one core standard.

identify, describe, find, and communicate a lot of data with demands unseen in other contexts. Many applications need delivery in microseconds or the ability to scale to thousands or even millions of data values and nodes. The consequences of a reliability failure can be severe. Since the core standard really is the core of the system, it has to perform.

The IIRA specifies the key functions that connectivity framework and its core standard should provide data discovery, exchange patterns, and "QoS." QoS parameters include delivery reliability, ordering, durability, lifespan, and fault tolerance functions. With these capabilities, the core connectivity can implement the reliable, high-speed, secure transport required by demanding applications across industries.

DATA QOSS

1. *Delivery*: Provide reliability and redelivery
2. *Timeliness*: Prioritize and inform when information is "late"
3. *Ordering*: Deliver in the order produced or received
4. *Durability*: Support late joiners and survive failures
5. *Lifespan*: Expire stale information
6. *Fault Tolerance*: Enable redundancy and failover
7. *Security*: Ensure confidentiality, integrity, authenticity, and nonrepudiation

The IIRA outlines several data QoS capabilities for the connectivity core standard. These ensure efficient, reliable, secure operation for critical infrastructure.

3.5.3 The IIoT Approach to Security

Security is also critical. To make security work correctly, it must be intimately married to the architecture. For instance, the core standard may support various patterns and delivery capabilities. The security design must match those exactly. For example, if the connectivity supports publish–subscribe, so must security. If the core supports multicast, so must security. If the core supports dynamic plug-and-play discovery, so must security. Security that is intimately married to the architecture can be imposed at any time without changing the code. Security becomes just another controlled QoS, albeit more complexly configured. This is a very powerful concept.

The integrated security must extend beyond the core. The IIRA allows for that too; all other connectivity technologies can be secured at the gateways.

3.5.4 DDS as a Core Standard

The IIRA does not currently specify standards; the IIC will take that step in the next release. However, it's clear that the DDS standard is a great fit to the IIRA for many applications. DDS provides automated discovery, each of the patterns specified in the IIRA, all the QoS settings, and intimately integrated security.

This is no accident. The IIRA connectivity design draws heavily on industry experience with DDS. DDS has thousands of successful applications in power systems (huge hydropower dams, wind farms, microgrids), medicine (imaging, patient monitoring, emergency medical systems), transportation (air traffic control, vehicle control, automotive testing), industrial control (SCADA, mining systems, PLC communications), and defense (ships, avionics, autonomous vehicles). The lessons learned in these applications were instrumental in the design of the IIRA.

3.5.5 DDS Unique Features

DDS is not like other middleware. It directly addresses real-time systems. It features extensive fine control of real-time QoS parameters, including reliability, bandwidth control, delivery deadlines, liveliness status, resource limits, and (new) security. It explicitly manages the communication "data model," or types used to communicate between end points. It is thus a "data-centric" technology. Like a database, which provides data-centric storage, DDS understands the contents of the information it manages. DDS is all about the data. This data-centric nature, analogous to a database, justifies the term "databus."

At its core, DDS implements a connectionless data model with the ability to communicate data with the desired QoS. Originally, DDS focused on publish–subscribe communications; participants were either publishers of data or subscribers to data. Later versions of the specification added request–reply as a standard pattern. Currently, RTI also offers a full queuing service that can implement "one of *n*" patterns for applications like load balancing. The key difference between DDS and other approaches is not the publish–subscribe pattern. The key difference is data-centricity.

A DDS-based system has no hard-coded interactions between applications. The databus automatically discovers and connects publishing and subscribing applications. No configuration changes are required to add a new smart machine to the network. The databus matches and enforces QoS.

DDS overcomes problems associated with point-to-point system integration, such as lack of scalability, interoperability, and the ability to evolve the architecture. It enables plug-and-play simplicity, scalability, and exceptionally high performance.

3.5.6 The DDS Databus

The core architecture is a "databus" that ties all the components together with strictly controlled data sharing. The infrastructure implements full QoS control over reliability, multicast, security, and timing. It supports fully redundant sources, sinks, networks, and services to ensure highly reliable operation. It needs no communication servers for discovery or configuration; instead, it connects data sources and sinks through a background "meta traffic" system that supports massive scale with no servers.

The databus technology scales across millions of data paths, ensures ultrareliable operation, and simplifies application code. It does not require servers, greatly easing configuration and operations while eliminating failure and choke points. DDS is by far the most proven technology for reliable, high-performance,

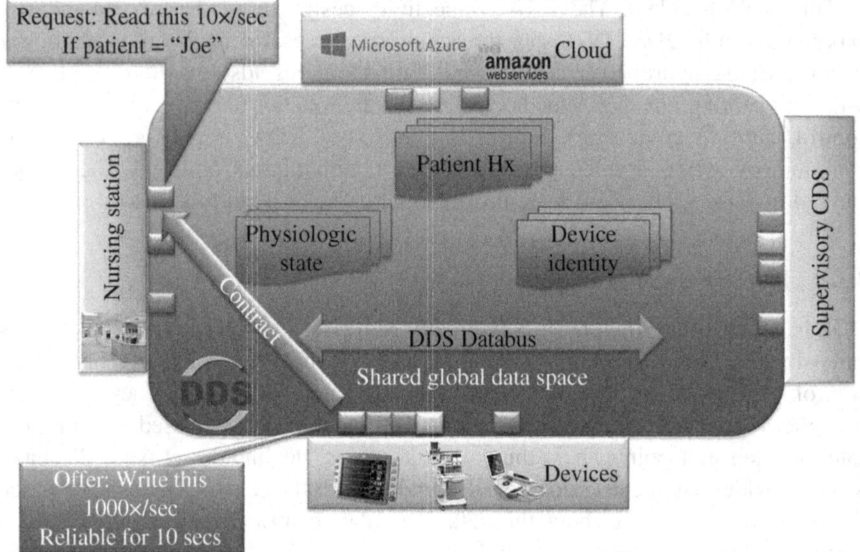

FIGURE 3.28 Data-centric communications. The databus links any language, device, or transport. It automatically discovers information sources, understands data types, and communicates them to interested participants. It scales across millions of data paths, enforces submillisecond timing, ensures reliability, supports redundancy, and selectively filters information. Each path can be unicast, multicast, open data, or fully secure. In the figure, a medical device that produces heart waveforms will send only one patient's information to the nursing station, at a rate it can handle, if it has permission to receive the information. Reproduced with permission from Real-Time Innovations.

large-scale IIoT systems. As of this writing, there are at least 12 implementations. Several of these are backed by commercial vendors. RTI provides the leading implementation.

Conceptually, DDS is simple. Distributed systems must share information. Most middleware works by simply sending that information to others as messages.

DDS sends information too, of course. However, DDS conceptually has a local store of data, which looks to the application like a simple database table. When you write, it goes into your local store, and then the messages update the appropriate stores on remote nodes. When you read, you just read locally. The local stores together give the applications the illusion of a global data store. Importantly, this is only an illusion. There is no global place where all the data lives; that would be a database. In a *databus*, each application stores locally only what it needs and only for as long as it needs it. Thus, DDS deals with data in motion; the global data store is a virtual concept that in reality is only a collection of transient local stores (Figure 3.28).

DDS also matches producers and consumers to ensure proper operation. The matching includes data flow rates, remote "liveliness," filtering, security, and performance. Once matched, the middleware enforces the "contract." For instance, if a subscriber needs updates 1,000 times per second, the middleware will ensure that a fast producer is available. The standard defines more than 20 of these QoS policies.

3.6 DATA-CENTRICITY MAKES DDS DIFFERENT

Systems are *all about the data*. Distributed systems must also share and manage that data across many processors and applications. The strategy to understand and manage this state is a fundamental design decision.

Data-centricity can be defined by these properties:

- The interface *is* the data. There are no artificial wrappers or blockers to that interface like messages, objects, files, or access patterns.
- The infrastructure understands that data. This enables filtering/searching, tools, and selectivity. It decouples applications from the data and thereby removes much of the complexity from the applications.
- The system manages the data and imposes rules on how applications exchange data. This provides a notion of "truth." It enables data lifetimes, data model matching, CRUD interfaces, etc.

An analogy with the database, a data-centric *storage* technology, is instructive. Before databases, storage systems were files with application-defined (ad hoc) structure. A database is also a file, but it's a very *special* file. A database has known structure and access control. A database defines "truth" for the system; data in the database can't be corrupted or lost.

By enforcing structure and simple rules that control the data model, databases ensure consistency. By exposing the structure to all users, databases greatly ease system integration. By allowing discovery of data and schema, databases also enable generic tools for monitoring, measuring, and mining information.

Like a database, data-centric middleware imposes known structure on the transmitted data. The databus also sends messages, but it sends very *special* messages. It sends only messages specifically needed to maintain state. Clear rules govern access to the data, how data in the system changes, and when participants get updates. Importantly, the infrastructure sends messages. To the applications, the system looks like a controlled global data space. Applications interact directly with data and data properties like age and rate. There is no application-level awareness or concept of "message" (Figure 3.29).

With knowledge of the structure and demands on data, the infrastructure can do things like filter information, selecting when or whether to do updates. The infrastructure itself can control QoS like update rate, reliability, and guaranteed notification of peer liveliness. The infrastructure can discover data flows and offer those to applications and generic tools alike. This knowledge of data status, in a distributed system, is a crisp definition of "truth." As in databases, this accessible source of truth greatly eases system integration. The structure also enables tools and services that monitor and view information flow, route messages, and manage caching.

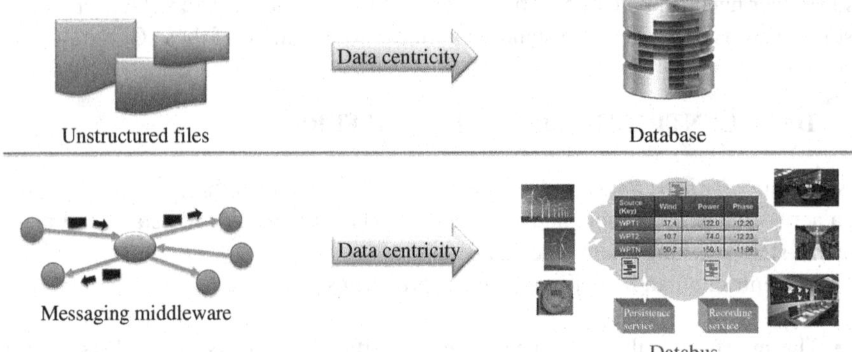

FIGURE 3.29 Data-centric middleware does for data in motion what a database does for data at rest. The database's data-centric storage fundamentally enables the simplified development of very complex information systems. Analogously, the databus offers data-centric networking that fundamentally enables the simplified development of very complex distributed systems. Both move much of the complexity from the application (user code) to the infrastructure. Reproduced with permission from Real-Time Innovations.

3.7 THE FUTURE OF THE IIoT

The IIoT is clearly in its infancy. Like the early days of the Internet, the most important IIoT applications are not yet envisioned. The "killer application" that drove the first machine-to-machine connections for the Internet was email. However, once connected, the real power of distributed systems created an entirely new ecosystem of value. This included web pages, search, social media, online retail and banking, and so much more. The real power of the Internet was barely hinted in its early days.

The IIoT will likely follow a similar pattern. Today, many companies are most focused on collecting data from industrial systems and delivering it to the cloud for analysis. This is important for predictive maintenance, system optimization, and business intelligence. This "killer app" is driving the initial efforts to build connected systems.

However, the future holds much more promise than optimizing current systems. By combining high-quality connectivity with smart learning and machine intelligence, the IIoT future holds many new systems that will revolutionize our world. It will, for instance, save hundred of thousands of lives a year in hospitals, make renewable energy sources truly practical, and completely transform daily transportation. Projections of the economic and social benefits range greatly, but all agree that the impact is measured in the multiple trillions of dollars in a short 10 years. That is a daunting but inspiring number. The IIoT will be a daunting but inspiring transformation across the face of industry.

REFERENCES

[1] http://iiconsortium.org (accessed August 24, 2016).
[2] https://en.wikipedia.org/wiki/Industry_4.0 (accessed August 24, 2016).
[3] http://www.openfogconsortium.org (accessed August 24, 2016).
[4] http://www.rti.com (accessed August 24, 2016).
[5] http://www.omg.org/dds (accessed August 24, 2016).
[6] http://www.icealliance.org/ (accessed August 24, 2016).
[7] http://sgip.org/Open-Field-Message-Bus-OpenFMB-Project (accessed August 24, 2016).
[8] http://arl.stanford.edu (accessed August 24, 2016).
[9] http://www.omg.org/spec/DDS-XTypes/ (accessed August 24, 2016).
[10] http://mqtt.org/ (accessed August 24, 2016).
[11] http://xmpp.org/ (accessed August 24, 2016).
[12] http://portals.omg.org/dds/ (accessed August 24, 2016).
[13] https://www.amqp.org/ (accessed August 24, 2016).
[14] http://opcfoundation.org (accessed August 24, 2016).
[15] http://www.iiconsortium.org/IIRA.htm (accessed August 24, 2016).

4

STRATEGIC PLANNING FOR SMARTER CITIES

Jonathan Reichental
Palo Alto, CA, USA

4.1 INTRODUCTION

It's 2016 and our planet is undergoing rapid urbanization. Today there are over 3.5 billion people living in cities, and by 2050 the United Nations says three billion more will join them [1]. Remarkably, this means that by then, urban infrastructure for 1.4 million people will need to be built each week [2]. The city challenges of population growth, inadequate infrastructure, generating new economic opportunities, climate change, and more are daunting and require the creation of bold and innovative urban development strategies. Without immediate and major shifts in our current trajectory, a good quality of life for city dwellers and the preservation of the environment are not sustainable.

Fortunately, a growing global consensus is beginning to emerge to tackle a seemingly intractable list of urban issues. To build and grow our cities in a sustainable manner, we're going to have to think and act differently. In particular, we're going to need to use innovative technology and a greater degree of civic engagement to move forward [3].

The architecture for a new generation of sustainable and high-performing cities is starting to take shape. These *smart cities* are using mobile apps and connectivity, data and cloud, an Internet of Things (IoT), people engagement and process redesign, and much more to help redefine possibilities.

Urban design and development must be a deliberate process and by extension so must be the creation of smarter cities. While a small set of cities are already moving

Internet of Things and Data Analytics Handbook, First Edition. Edited by Hwaiyu Geng.

Companion website: www.wiley.com/go/Geng/iot_data_analytics_handbook/

forward with projects that deploy smart capabilities, most cities have not yet begun to think deeply about what a plan might look like. Soon these cities will start the important work of creating a smart city strategy. This strategic planning—a systematic process of envisioning a desired future—is essential and high-value work that will guide the long and complex journey ahead. The purpose of this chapter is to introduce this planning process.

4.2 WHAT IS A SMART CITY?

At time of writing, there exists no universally agreed definition for a smart city. As with any emerging domain, agreement will take time as experts, practitioners, and other stakeholders slowly coalesce around an acceptable definition. There are some consistent high-level themes which are emerging that reflect the motivation for developing smart cities. These include creating more efficient and sustainable city services and infrastructure, enabling new economic opportunities for residents, making government more open and accessible, and increasing civic engagement. I've reviewed many sources that range from academia to industry, and in addition to my own experiences, I've developed the following definition for the purposes of this chapter:

What is a Smart City? Urbanization that uses innovative technology to enhance community services and economic opportunities, improves city infrastructure, reduces costs and resource consumption, and increases civic engagement.

While it could be argued that most cities aspire to these goals, it's the specific focus on the role of innovative technology that differentiates smart cities strategies from other approaches. Rather than being relegated to a supporting role, technology is central to designing and building a smart city. That said, it remains important to stress that the outcomes of successful initiatives within a smart city strategy are not the attendant technologies, but rather improved human experiences. Despite an emphasis on technology, the software and systems should largely dissipate into the fabric of everyday urban life, remaining largely invisible relative to our day-to-day human needs [4].

An increasing number of cities across the world are pursuing smart city initiatives. These can range from completely new cities such as Masdar [5] in the United Arab Emirates (UAE) to existing metropolitan areas like Barcelona [6] and Singapore [7]. In each instance, the emphasis differs from such diverse focus areas as transportation to sustainability to a combination of areas. The Smart Cities Council [8] provides a good list of global case studies [9]. That said, it is still rare to find a city that has a cohesive, integrated strategy that traverses the complexity of the modern city environment. A smart city is a system of systems.

4.2.1 Common Smart City Focus Areas

While we recognize that every city has its own specific needs, it's possible to begin to categorize common areas of interest. To gain a better understanding of the scope and complexity of smart cities, I've identified nine core areas of focus in the smart city

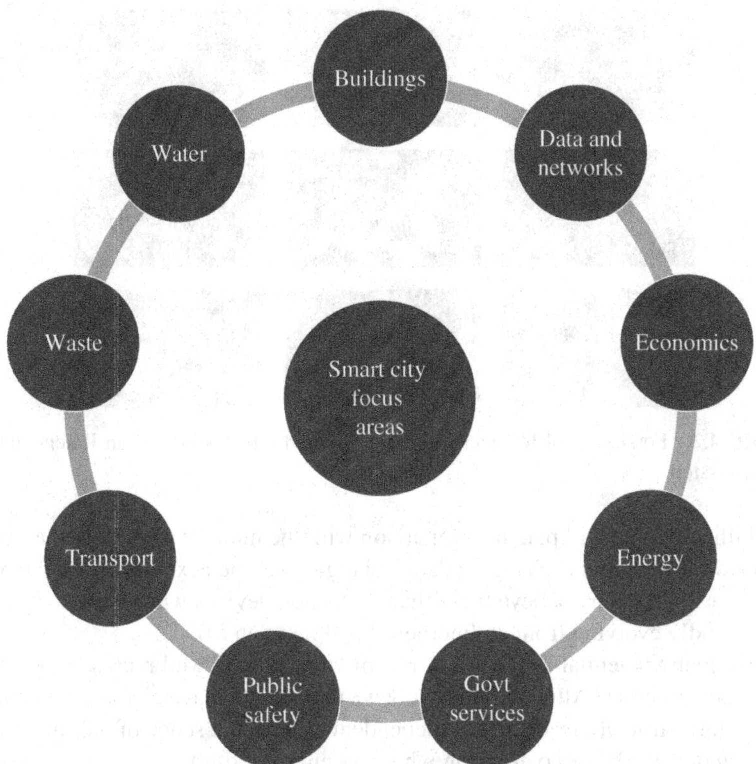

FIGURE 4.1 Smart city focus areas. © Jonathan Reichental. Use with permission.

domain that effectively represent much of the current literature and global work being done (Figure 4.1). With these core areas, a number of quick observations can be made. These categories alone are complex, specialized areas that require unique skills and experience. The diversity of domains illustrate both the enormous challenges and the depth and breadth of opportunity that lies ahead. Finally, it will become clear over time that there are benefits to integration across these diverse mostly siloed areas and that will require the invention of completely new technologies and the emergence of new skills.

4.3 SMART CITIES AND THE INTERNET OF THINGS

If the opening act of the Internet revolution was defined as connecting people to information and then people to people, this next emergent phase involves connecting devices to devices. For the past 20 years, the focus has been on finding ways to get as many people as possible online. Surprisingly, even today, only around 40% of the global population has access to the Internet [10].

Fortunately, there is a concerted effort by governments and several private enterprises to bring the Internet to the majority of the remaining 60%. That said, connecting

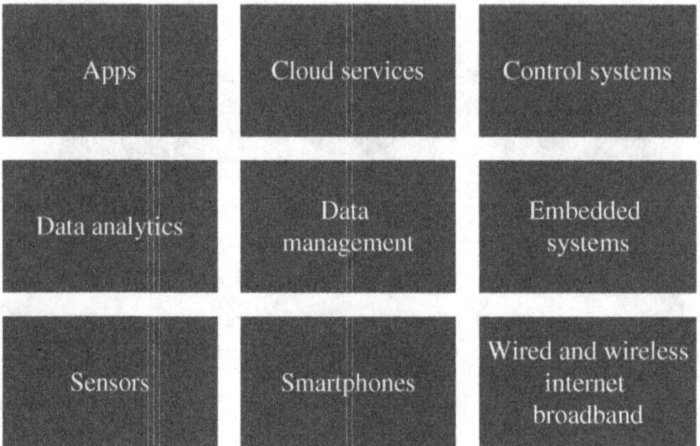

FIGURE 4.2 Foundational IoT technologies for smart cities. © Jonathan Reichental. Use with permission.

seven billion people will pale in comparison with the number of devices that will be connected together and exchanging data and logic over the next 10 years and beyond. As soon as 2020, many believe more than 25 billion devices will be connected [11]. We are rapidly evolving from an Internet of people to an *IoT*.

Of the many potential uses for an IoT, smart cities may well represent one of the greatest beneficiaries. After all, intuitively we understand that a city is a collection of complex infrastructure made up of independent things. The scope of our imagination will expand as we begin to envision what a connected, digitally enabled city infrastructure might make possible.

Today, we already see advantages at just the hint of what is to come. For example, our smartphones know the exact location and arrival of buses and trains, government open data initiatives power apps that provide insights on crime and power usage, trash cans can alert when they are full in order to optimize collection, sensors in parking spaces help drivers find available spots, and small sensors detect and report water leaks.

Without a doubt, the IoT will represent a core foundation and enabler of smart cities. As city leaders and others begin planning their smart city strategy, they'll need to understand, at least conceptually at first, the emerging technical characteristics that IoT brings to their efforts. To provide some further context as we begin to work on the strategic planning processes, I've included some of the current foundational technology here (Figure 4.2).

4.4 WHY STRATEGIC PLANNING MATTERS

Benjamin Franklin said, "By failing to prepare, you are preparing to fail." It's a quote that I refer to often as a reminder that enthusiasm to make progress should not trump the important work of planning. It's particularly relevant when the work ahead is complex, extensive, high risk, and interdependent. By its nature, this largely defines the work that's required to achieve success in smart city initiatives.

While planning and executing each specific smart city initiative is the approach taken by most cities today, our emphasis here is on planning at a higher level. We're concerned with *strategic planning*. This work elevates planning efforts to focus on the big picture and the set of major steps that must be taken to realize a vision. It begins with the end in mind and then works backward to today, identifying what needs to happen along the way:

> *What is strategic planning? A systematic process of envisioning a desired future, and translating this vision into broadly defined goals or objectives and a sequence of steps to achieve them* [12].

As more leaders begin to articulate a vision for the future of their cities and commit to the smart city journey, there will be a greater need to develop a strategic plan. In addition, in many places where smart city initiatives are already underway, leaders are realizing that they have to retrofit a strategic plan that unifies and optimizes efforts. In these cases it's not necessarily because city leaders were negligent of the effort, but it's more likely because we've now begun to label existing initiatives as smart city efforts and recognized the amazing integration opportunity.

4.5 BEGINNING THE JOURNEY: FIRST THINGS FIRST

The logical milestones in the pursuit of most goals follow a similar pattern: define, design, develop, and deploy. Two additional components should almost always be considered: a form of ongoing and final measurement and a capacity for iteration along the way, the latter being essential to contemporary strategy execution as needs and capabilities increasingly change over time.

I've used these concepts to create the high-level steps to becoming a smart city—a somewhat misleading notion given that the work will never really be done (cities are continually evolving)—which is summarized as follows (Figure 4.3).

Since we're focused on strategic planning, this chapter only covers steps 1 and 2 from Figure 4.3.

As Figure 4.3 illustrates, a strategic plan begins with a vision of the future. It's an essential first step because it will guide every decision that comes next. It is also an essential vehicle to create alignment with a large group of stakeholders and to attain agreement on purpose and desired outcomes. It will help to clarify at a high level what must remain the same and what must change.

FIGURE 4.3 Steps to create and deliver a smart city. © Jonathan Reichental. Use with permission.

A vision of the future is not the same as a vision statement. While they share core characteristics, the envisioning exercise is broader and more enumerated than the traditional 1–2 sentences of a vision statement. That said, you'll likely include a vision statement in the introduction of the vision document. Finally, a vision of the future will need to be encapsulated into an artifact that can be used as a communication tool throughout the smart city strategy.

4.5.1 Steps to Creating a Vision

The literature is replete with methods to create a vision. All of them have value. I've created the following approach that distills the core ideas into a simple set of steps (Table 4.1):

TABLE 4.1 Steps in an Envisioning Process

Envisioning Steps	Description
1. Define what a smart city means in the context of your city	This may be the most difficult step but also the most fun. It's likely you and your team already have some ideas on this. You'll want to elicit input and use different tools like brainstorming or design thinking to help In the envisioning phase you're not necessarily identifying challenges (although you may choose this route). I recommend that you focus on what you want to become
2. Create a short list of goals	In step 1, in addition to an overarching vision for your Smart City, you've coalesced around many important ideas that support the vision. Group those ideas into meaningful, categorized goals. Here's an example, "Goal 1: Create a transportation environment that supports climate change management, reduces traffic, and improved parking capabilities" The right amount of goals is in the range of 6–10
3. Determine a time frame	By definition executing on a vision takes a long time. You're certainly looking at several years, but not too long such that it becomes impractical. Agreeing on a time frame around the defined goals creates an important boundary and helps to sharpen everyone's focus. While recognizing that a smart city strategy is never really finished, you must articulate a time frame for this round of visionary goals
4. Identify strengths	This step requires some careful and honest introspection. Articulate your city qualities that lend themselves to the work ahead. Knowing these strengths will help focus efforts, understand potential risks, optimize for those strengths, and assist in prioritizing objectives. Some organizations go further in this step and complete a SWOT (strengths, weaknesses, opportunities, and threats) analysis [13]

TABLE 4.1 (Continued)

Envisioning Steps	Description
5. Create a first draft	Combine steps 1–4 into a cohesive narrative. This is not an essay. It should take the form of a high-level vision statement, followed by a short paragraph that supports the statement in terms of some background, motivation, and strengths. Next, each goal should be listed with descriptions, outcomes, approximate timelines, and high-level metrics If the document is more than 20 pages, it's probably already too long. The vision exercise should result in a relatively simple elevator pitch that stakeholders can summarize when requested
6. Circulate	The next few steps are what I like to call *rinse and repeat*. The draft vision of the future for your smart city must be circulated with a broad and diverse community. Create a mechanism to make it easy to elicit feedback and track changes
7. Reviews and redraft	The first round of feedback will likely elicit a high volume of comments. In subsequent circulations you should expect reduced volume
8. Circulate	You'll need to decide how many rounds of rinse and repeat make sense. Do it too little and people will not feel like their voices are being heard. Do it too much and people will tune out
9. Finalize	With several iterations completed, it's time to lock down the vision. It will be clear at this point what has resonated with the stakeholders. I suggest you get the right talent to create the final vision document. You'll want this to be easy to consume and something that everyone will be proud of referencing and sharing for many years to come
10. Socialize	You've reached the end of a major milestone in the strategic planning process. You have a completed vision document. Now share it widely and often. With so many channels available for both analog and digital, use them all. For the core online presence—possibly a microsite—consider a way for people to provide comments

4.6 FROM VISION TO OBJECTIVES TO EXECUTION

Well done. In many ways, the envisioning exercise is one of the hardest phases of creating a smart city strategy. At this point your team have articulated a bold future state and you've gained agreement from a large set of stakeholders. The vision identifies the strengths of the city, the areas that should continue as is, and the key areas that need to change to accommodate the needs of the future.

It's time to move from the vision to beginning the process of articulating the work ahead. The complex work of creating a smart city from the vision will involve a master plan with a large number of objectives. Figure 4.4 illustrates the relationship between our vision, the goals, and now the objectives.

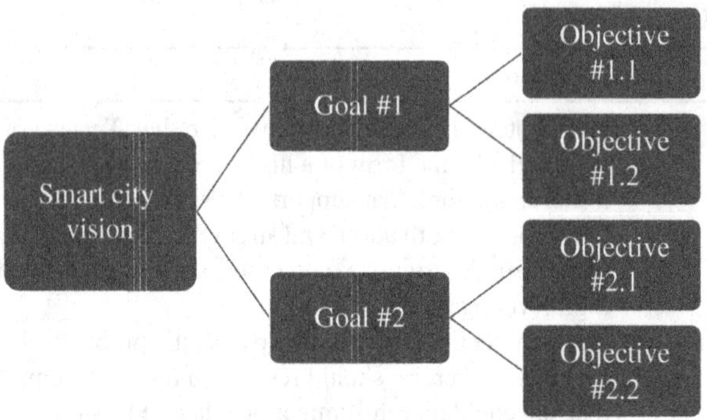

FIGURE 4.4 Relationship between vision, goals, and objectives.

TABLE 4.2 Extracting Objectives from Goals

Extracted Smart City Vision Goal	Supporting Objectives
1. A transportation environment that supports climate change management, reduces traffic, and improved parking capabilities by 2020	1.1 Climate change management 1.1.1 Provide electric car chargers at 60% of city-provided parking spaces by 2019 1.2 Reduce traffic 1.2.1 Create restricted driving zones that dynamically change dependent on conditions by 2020 1.3 Improved parking capabilities 1.3.1 Enable drivers to easily find an available parking space by 2018

Objectives will lead to projects that will ultimately result in reaching your vision goals:

What is an objective? The act of achieving a specific result in a defined time frame. They are the building blocks of a strategic plan. Unlike a goal which is long term and aspirational, an objective is short-term and clearly defined.

How might we begin to convert our vision to objectives? Let's use transportation—a core smart city theme (Figure 4.1)—as an example. Keep in mind that I've summarized both the vision goal and objectives information in the table for the purposes of succinctness.

We arrive at Table 4.2 through one or more ideation sessions [14]. I recommend assigning teams to each goal. Team members should be diverse, but certainly include subject matter experts. Teams should report up to some form of decision authority such as a steering committee [15]. Many objectives can be determined, but they will need to be reduced to a manageable number. Objectives will become projects, so dependent on capacity and funding; effort must be made to ensure that the strategy is achievable. This may be a good opportunity to determine a prioritization process [16, 17].

TABLE 4.3 Mapping Projects to Common IoT Technology

	App	Cloud Services	Control Systems	Data Analytics	Data Management	Embedded Systems	Sensors	Smartphones	Wired Internet	Wireless Internet
(a) Install parking space detection technology		x		x	x		x			x
(b) Deploy parking space finder app	x	x		x	x			x		x
(c) Install digital signage that displays available parking spaces in multi-story parking lots			x	x	x	x	x		x	x

Let's take a look at how one of these objectives may manifest in one or more projects. Again, I've simplified the project list for this chapter. Keep in mind that your projects require significant definition for them to succeed.

Objective	Potential Projects
1.3.1 Transportation: improved parking capabilities—enable drivers to easily find an available parking space by 2018	1.3.1.1 Install parking space detection technology 1.3.1.2 Deploy parking space finder app 1.3.1.3 Install digital signage that displays available parking spaces in multistory parking lots

Finally, let's take a look at how common IoT technology might underlie the projects in this single objective. While technical architecture is outside of the scope of this chapter, the following is an example used to illustrate how technical architects may look to the IoT to build a common scalable architecture for many of your smart city objectives. In this example, data analytics, data management, and wireless Internet span all three projects. We've now identified common components that when viewed against a larger set of categorized projects may form the foundation of some core IoT infrastructure. We've successfully taken the vision right down to the underlying technology to enable it. (Note: IoT categories are taken from Figure 4.1 and do not represent the full degree of technology possible (Table 4.3).)

4.7 PULLING IT ALL TOGETHER

Strategic planning is a stepwise process. Each step builds on the previous. Importantly later steps can be traced backward to source. For example, done right, any stakeholder should be able to understand the genesis of the parking space finder app by tracing it back through the strategy. This is also a validation mechanism. If a project cannot be easily traced back to the strategy—this can happen as the strategic plan becomes large and complex—then it may call for an analysis of validity. A project without a clear strategic alignment and a measureable contribution to goals may likely be unnecessary. I think we'd agree it's best to know that prior to starting the project.

The strategic plan for your smart city should not be a 300-page document. In fact, great strategic plans are best when they are just a few pages. It should be just enough to communicate the vision, the objectives, timelines, and desired outcomes with high-level metrics.

I suggest that projects—which are typically defined after the strategic plan is completed and ratified—are contained in a different document. Of course I'm using the terminology of documents, but you're likely to use some form of online presence. That said, even in 2016, many of your stakeholders will want to view and use the strategic plan in physical form.

If you've generally followed my guidance to create a strategic plan, you and the larger stakeholder group will now have a clear vision of what a smart city means for your city. This will provide both the elevator pitch and the deeper articulation of value that the upcoming efforts will contribute to. This vision will have enabled you to identify objectives within the different areas of the vision. Having high-level objectives will help with prioritization, assessing potential costs, timelines, and detailed outcomes. There is significant substance in this exercise alone to generate detailed discussion, create answers for many people, and make the intention real.

Finally, with the vision, objectives, and potential projects identified, technologists can begin to determine what common technologies—in particular the IoT—may be needed to provide a scalable foundation. These common technologies such as a sensor network and city Wi-Fi will be a justified investment, help to integrate many different objectives, and form an integrative architecture for the future of your smart city.

REFERENCES

[1] https://www.un.org/development/desa/en/news/population/world-urbanization-prospects.html (accessed November 29, 2015).

[2] http://2014.newclimateeconomy.report/cities/ (accessed November 29, 2015).

[3] Townsend, A. M. (2014). *Smart Cities*. New York: W. W. Norton & Company.

[4] Newsom, G. & Dickey, L. (2014). *Citizenville*. London: Penguin Books.

[5] http://www.masdar.ae/en/masdar-city/live-work-play (accessed November 29, 2015).

[6] http://smartcity.bcn.cat/en (accessed November 29, 2015).

[7] https://www.ida.gov.sg/Tech-Scene-News/Smart-Nation-Vision (accessed November 29, 2015).

[8] http://smartcitiescouncil.com/ (accessed December 4, 2015).

[9] http://smartcitiescouncil.com/smart-cities-information-center/examples-and-case-studies (accessed December 4, 2015).

[10] http://www.internetlivestats.com/internet-users/ (accessed November 29, 2015).

[11] http://www.gartner.com/newsroom/id/2905717 (accessed November 29, 2015).

[12] http://www.businessdictionary.com/definition/strategic-planning.html (accessed November 29, 2015).

[13] http://pestleanalysis.com/how-to-do-a-swot-analysis/ (accessed December 4, 2015).

[14] Gray, D., Brown, S., & Macanufo, J. (2010). *Gamestorming: A Playbook for Innovators, Rulebreakers, and Changemakers*. Sebastopol: O'Reilly Media.

[15] http://www.businessdictionary.com/definition/steering-committee.html (accessed November 29, 2015).

[16] Kodukula, D. P. (2014). *Organizational Project Portfolio Management: A Practitioner's Guide*. Plantation, FL: J. Ross Publishing.

[17] http://www.cio.com/article/3007575/project-management/6-proven-strategies-for-evaluating-and-prioritizing-it-projects.html (accessed December 4, 2015).

5

NEXT-GENERATION LEARNING: SMART MEDICAL TEAM TRAINING

BRENDA BANNAN[1], SHANE GALLAGHER[2] AND BRIDGET LEWIS[1]

[1] *George Mason University, Fairfax, VA, USA*

[2] *Advanced Distributed Learning, Alexandria, VA, USA*

5.1 INTRODUCTION

Big data analytics, next-generation networking, and the Internet of Things (IoT) technologies have been shown to have positive impact on many areas related to transportation and smart home applications as well as other environments; however, there are currently few examples or case studies available in formal education, informal learning, and structured training, particularly related to complex situations involving teams and teamwork. The demands of new skill sets and tools to deal with the variety and velocity of data multiplying at an exponential rate may be part of the challenge. The integration of physical networked objects into learning design and research and meaningful use of learning analytics (LA) that can result may require new mindset and processes for those involved in education and training. For example, a learning action or path can be represented as a "smart" service and can be encapsulated by a software object that can be discovered and integrated into high-level analysis and potentially leveraged for reflection and learning. IoT technologies that incorporate this capability are rapidly becoming a mainstream data source to improve our services and infrastructure; however, these technologies also have the capacity to potentially improve teaching, learning, and training by providing a new window on a learner's own behavior or learning path in real-world contexts. In the space of healthcare or emergency training simulations, which have historically been limited either to

Internet of Things and Data Analytics Handbook, First Edition. Edited by Hwaiyu Geng.

Companion website: www.wiley.com/go/Geng/iot_data_analytics_handbook/

fully computer-based simulations or live-action human-observable field-based simulations, IoT technologies can open up innovative, hybrid digital–physical opportunities both for delivering and for understanding the outcomes of education and training efforts in a much more comprehensive way. Team-based training efforts have traditionally been difficult to employ, study, and build on individual and team-based experience across many types of learning contexts, and IoT technologies may offer new insights and methods into these challenges.

For example, one context where it is imperative to learn from human action and experience is medical emergencies. Learning from applied experiences and working as a team are critical components of a safe healthcare system. Simulation-based training holds significant promise to reduce errors and promote performance improvement in high-stakes, high-criticality medical emergency situations. Simulation is a highly complex intervention [1]. However, simulation team training is only as effective as the learning that results, and research studies indicate that most of the learning occurs at the end of the simulation in the debriefing process [2]. Facilitating reflective observation on individual and team performance factors concluding a complex medical simulation is key for the experiential learning cycle that promotes active reflection and revision of mental models for future beneficial actions and experiences [3].

This prototype of learning design and development effort described in this chapter leverages a smart medical team training IoT system employed across multiple technologies that dynamically collects and standardizes experiential learning data (e.g., proximity, time, and location) for the purposes of increasing our understanding of how enhanced information in the simulation debriefing process may impact emergency medical team training (and potentially team training in other contexts). By providing dynamic, detailed information on human activity with devices and objects through the employment of a wireless software protocol, during and after a complex, fast-moving high-fidelity medical simulation, this system may provide a new window of information into education and training outcomes. The importance of learning and analytics for IoT and the learning design, prototyping, and initial deployment of the system is described in the succeeding text (see Sections 5.3.1, 5.3.2, 5.3.3, and 5.3.4).

5.2 LEARNING, ANALYTICS, AND INTERNET OF THINGS

5.2.1 Learning Design

Mor et al. [4] speak to the potential and synergy of Learning Design, Design Inquiry, and Design Science in relation to LA in stating:

> Learning analytics provide the instrument for making this design inquiry of learning truly powerful [4].

Learning design, design inquiry, and analytics are inherently part of the design research process in attempting to enact and investigate "…the creative and deliberate act of devising new practices, plans of activity, resources and tools aimed at achieving particular educational aims in a given context" [4]. LA are often connected to the emerging movement or term the Internet of Things (IoT) in potentially representing the pervasive connected devices that collect fine-grain behavioral data through sensors and other technologies that can be immediately analyzed and visualized. There are very few examples related to improving learning and performance in real-world settings through the use of emergent IoT technologies such as sensors, networked devices, and LA. This project describes an initial complex exploratory design research effort that integrated characteristics of learning design in attempting to devise a new way of improving team-based behavior in a high-fidelity medical simulation context leveraging IoT technologies and LA.

Learning design characteristics in this effort may be described as design principles that attempt to promote or enculturate learning processes. Generating, identifying, and modeling technology-mediated social learning and behaviors or learning characteristics in order to design tools that support, promote, and study the phenomena of interest are core goals of education design research and this work [5]. These learning characteristics or design principles are generated through integrating information and applied research results to postulate what characteristics, principles, or requirements are linked to learning. In the case of medical team training, we uncovered and generated learning design characteristics linked to the analytics we collected such as (1) visualizing the timing of response, location of the patient at all times, presence of professionals in particular environments, and the exchange of the patient between teams; (2) detecting the patient and several key medical events over time and location feeding that information forward from en route in the ambulance to the medical professionals; and (3) automatically collecting real-time behavioral data about proximity, location and time of the patient, and professional interaction for immediate display back to the simulation participants in the debriefing session. These characteristics or principles are linked to learning theory and relevant to this particular learning design context. The educational design research process that generated and modeled them through a prototyping process is described in the succeeding text.

5.2.2 Education Design Research

Education design research provides a systematic and thoughtful approach to improve teaching, learning, or training interventions while also generating new knowledge in iterative cycles of applied research. Bannan et al. [5] describe the education design research process as "a form of inquiry into how individuals and groups use digital resources to support educational and other forms of cultural/social processes incorporating best practices" [5]. These digital resources may involve IoT technologies and LA to investigate and improve complex teaching, learning, and training situations. Education design research is somewhat distinct from other applied design professional's (e.g., engineering) conceptualization of design research in the following ways:

> Design research is distinct from these approaches by integrated rigorous, long-term cycles of applied and empirical research as part of a complex, evolving design process

attempting to positively influence and effect change in a learning context through the building of a design intervention through which we uncover pedagogical principles that may be applicable and researchable in similar situations. This is often conducted through identifying and investigating a learning problem, the design and development of an educational innovation and its trial, and iteration in multiple contexts over time. Determining a particular phenomenon to focus on in design research is often based on an identified need and the selective perception of the researcher who becomes intimately familiar with the learning context and activity as he or she closely examines and analyzes real-world practices and settings for the purposes of design (or re-design) of an educational innovation [5].

These following sections provide more detail on the initial cycle and process of design research to provide a case study for a type of next-generation learning involving smart medical team training.

5.3 IoT LEARNING DESIGN PROCESS

5.3.1 Problem Identification and Contextual Analysis

To understand the existing needs and context related to this work, we conducted a contextual analysis that included observational research as well as interviews and focus groups and literature review. The problem context included a multiteam simulation involving emergency response and medical teams engaged together in a live-action, high-stakes simulated medical emergency in a staged car accident scene and transport of the patient via ambulance to the hospital emergency room setting.

According to the literature, simulation-based training holds significant promise to reduce errors and promote learning and performance improvement in high-stakes, high-criticality medical emergency situations. Simulation is also a highly complex intervention employed in healthcare and emergency response situations [1]. However, simulation team training is only as effective as the learning that results and research studies indicate that most of the learning occurs concluding the simulation in the debriefing process [2]. Facilitating reflective observation on individual and team performance factors concluding a complex medical simulation is key for the experiential learning cycle. The learning design proposition that evolved from this initial design inquiry incorporated ideas for engineering networked tracking devices for smart medical and emergency response training to potentially promote active reflection and revision of mental models for future beneficial actions and experiences in real-world simulation contexts [2]. Attempting to engender effective team-based reflection within and across teams at the moment of need through innovative data collection, dynamic analysis, and immediate display of meaningful LA became the core focus of this research effort.

In this effort, in a review of best practices and performance measures related to simulation-based team training theory, the researchers identified factors such as adaptability, situation awareness, leadership, and team orientation and coordination as important components of teamwork [6]. Stress in demanding situations such as medical emergency simulations can also impact team performance by narrowing attention to focus more prominently on individual tasks rather than team-oriented information [7]. These factors encompass relevant research constructs that can be examined in new ways through the deliberate collection and analysis of explicit individual and team activity stream and other data automatically generated across multiple individuals, devices, and objects in real-time within a complex, team-based simulation scenario.

However, before intervening in the context, we attempted to increase our understanding of each context through observing emergency response simulations in the field and medical simulations in the hospital. Interviewing the emergency medical technicians (EMTs), Fire and Rescue department personnel, and hospital medical staff was crucial to understanding how these teams accomplished simulations and their perspectives and behaviors to establish what they perceived as important contextual information that are aligned with their practices, culture, and needs in improving their learning and performance from simulations. This information directly informed our learning design and exploratory design research cycle.

Medical and emergency response teams are required to quickly comprehend a complex array of factors including time, situational awareness, coordination of team/individual actions, as well as manage physiological stress, any of which can impair performance in high-stakes situations [8]. Often, basic needs are identified to address complex situations and establish important criteria including time, location, and interaction of professionals. In the emergency response and medical simulation carried out in this exploratory research and written by an emergency room trauma surgeon, the core elements/criteria for learning and performance of the multiteam simulation involved timing of response, location of the patient at all times, presence of professionals in particular environments, and the exchange of the patient between teams. As was stated by the trauma surgeon during the contextual analysis, "If you can just tell me who was in the room [trauma bay], of who was supposed to be in the room and the location of the patient at all times, that is good enough for me."

Learning design involves employing empathy and observation to "...understand where learners are, and creating the things that will help them get to where you want them to be, be those tasks, resources, social configurations or tools" [4]. The previously articulated statement by the trauma surgeon became our learning design goal to enable enhanced reflection by the teams in the debriefing by attempting to engineer an IoT sensor-based system to immediately visualize the timing and presence of the team members in their natural setting. The design phase is described in the next section.

5.3.2 Prototyping

To provide conditions for the multiteam simulation participants to improve their reflection and potentially their learning and performance in the debriefing session, we conceptualized, designed, and built an initial prototype system to strive toward that goal. The simulation participants and medical practitioners informed our work in a continuous interactive dialog throughout the conceptualization, design, and enactment of the system.

5.3.3 Technology System Design

A technology system was derived that would automatically track the proximity of each of the emergency response and medical professional to the simulated (mannequin) patient across the contexts of the accident scene and transport in the ambulance to the hospital emergency room. The system would also detect the patient and several key medical events over time and location feeding that information forward from en route in the ambulance to the medical professionals at the hospital prior to the patient's arrival. Display of this information was provided in real time as well as visual display of the proximity sensor data of each team member to the simulated patient during the debriefing session.

The learning design prototype specifically involved leveraging an API software specification that allowed Bluetooth proximity beacons/sensors, cell phones, and minicomputing devices to communicate with each other, dynamically collecting and standardizing experiential learning data (e.g., time, location, and proximity) for immediate display and visualization to increase our understanding of how this enhanced information for the simulation debriefing process might impact emergency medical team training (see Figure 5.1). We hypothesized that by providing dynamic, detailed LA on human activity with IoT sensor-based devices through this specification, during and concluding a complex, fast-moving, high-fidelity medical simulation, this type of system might provide a new window into participant understanding of and reflection on their own behavior in the simulation.

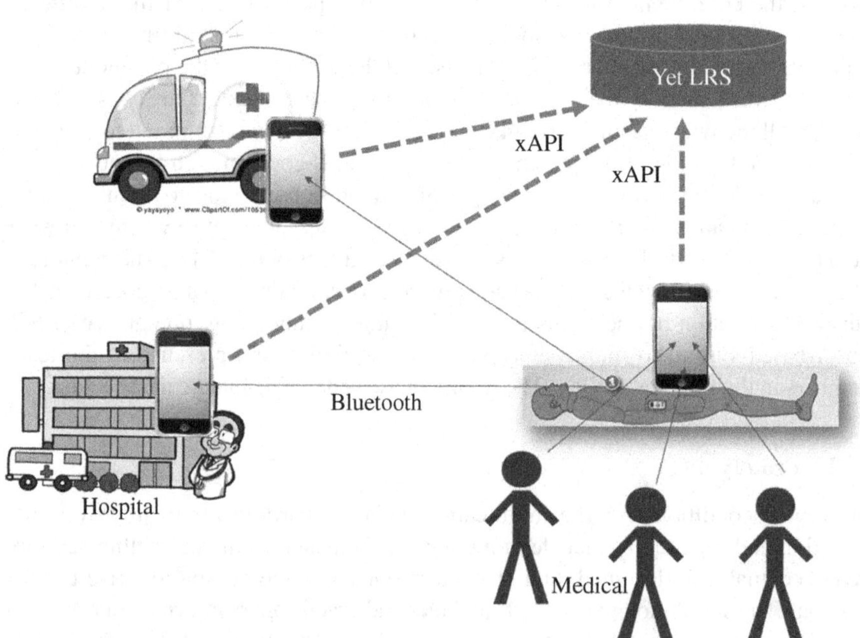

FIGURE 5.1 Initial proximity learning system design.

Designed by the Advanced Distributed Learning Initiative, the Experience API (xAPI) is a learning data specification created to allow interoperability and scale in next-generation training environments. Rather than focus on merely the technical practice of linking devices, xAPI provides a way to standardize the formative data of human experience and activity captured through digital wireless connectivity of devices. The xAPI specification provides a way to capture information and formalize human experience from multiple and varied networked devices in standardized, English-like statements (e.g., JSON) that can inform both human and machine learning through leveraging big data analysis and interoperability of the IoT technologies.

Long and Siemens [9] define LA as the measurement, collection, analysis, and reporting of data about learners and their contexts, for purposes of understanding and optimizing learning and the environments in which it occurs. By automatically collecting real-time behavioral data about proximity, location and time of the patient, and professional interaction for immediate display back to the simulation participants in the debriefing session, we also wanted to understand their impressions of how this technology might optimize their learning. Minicomputers and cell phones acted as "listening" devices were integrated into the patient mannequin and strategic locations such as in the ambulance and in the emergency room trauma bay, while proximity beacons/sensors provided by Radius Networks, Inc. were attached to each individual team member in the emergency response and medical teams. These proximity beacons provided a continuous stream of data related to the proximity of each team member to the patient across time and setting.

Detailed data regarding proximity of professionals to the simulation mannequin was generated and presented back to the participants through a specialized database and visualization system designed by Yet Analytics, Inc. called the YetCore Learning Record Store (LRS). This system provided dynamic and automatic data collection, processing, and immediate visualization of more than 12,000 data points from the proximity sensors during the hour and a half simulation.

5.3.4 Simulation Results

Concluding the debriefing session the participants confirmed that the data demonstrated who was present in what location and proximity to the patient at what point in time (see Figure 5.2).

The 12,000 points of data generated from continuous streams of information from each proximity beacon/sensor attached to each individual team member across the emergency response and medical teams generated potential information for learning and reflection on individual and team-based behavior. The system generated useful data for later analysis also related to proximity by team member role, event, time, and space (see Figure 5.3).

The proximity of the team members to the "patient" and other listening devices was also analyzed to provide insights into cross-team behavior and efficiency such as who was near the patient or in which location at what point in the trauma bay context (see Figure 5.4).

FIGURE 5.2 Data collection in emergency response and trauma bay and visualization in debrief.

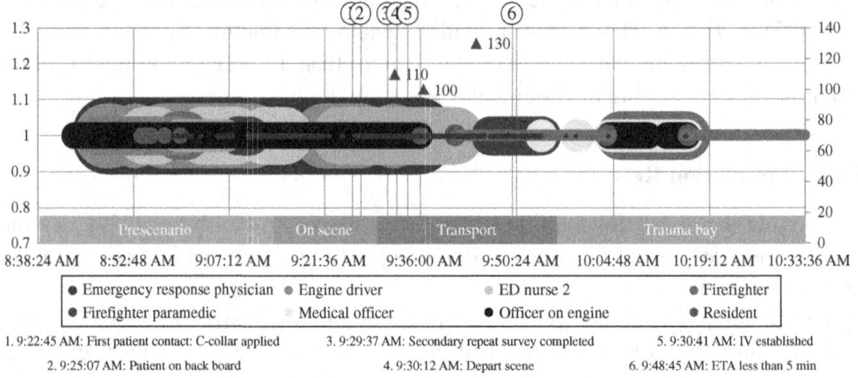

FIGURE 5.3 Detection of proximity of sensors by team member role, event (number), mannequin heart rate (triangle), and time (horizontal axis).

The data collection system demonstrated the stated needs of the practitioners including (1) the location of patient at all times, (2) the presence or absence of emergency response and medical professionals in each context including the accident scene, ambulance, and trauma bay and, (3) the proximity of the patient to individual team members across time and setting.

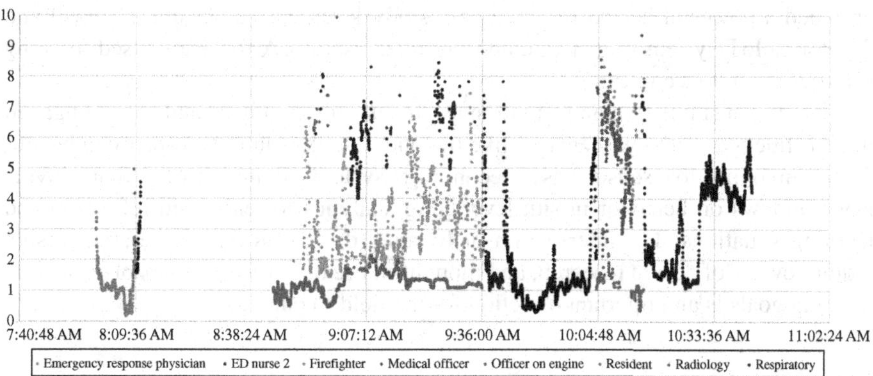

FIGURE 5.4 Trauma bay listener distances as measured through Radius Network Bluetooth beacons (experiment performed in May, 2015). Time of day is indicated in *x*, and distance in meters is the *y* axis.

The participants also informed our assumptions about the potential for the system including statements such as how this system may (1) enhance learning about team dynamics, (2) provoke awareness of individual and team behavior in situ including important information about transferring the patient from the emergency response to the medical teams in a patient "handoff," (3) track fine-grain behavior of individuals, objects, and teams in real time. Successfully achieving our goals of providing enhanced information in real time for reflection during the simulation debrief and informed feedback on the system design, the research provided information for the next cycle of design inquiry.

Through this exploratory cycle of design research, we were able to demonstrate what Mor et al. [4] refer to "the option of designing learning experiences which assume the availability of LA and thus incorporate checkpoints for self-reflection using the analytics output." The emergency response and medical teams participating in this high-fidelity complex simulation were able to view their behavior (e.g., proximity to the patient, time, and location) to provide enhanced information for self-reflection and team-based reflection in the debriefing session. The teams were also able to project forward and discuss how the system could potentially improve their situation awareness, reflection, and learning in future cycles of research.

5.4 CONCLUSION

5.4.1 The Intersection of Learning Design, IoT, and Analytics

The multiteam medical simulation exercise described in this chapter provided important insights into the learning design of the IoT, sensors, and analytics. Engaged in a design inquiry process, we were able to engineer a system that met the stated need of a trauma surgeon to provide real-time analytics on team-based behavior during a complex live-action medical simulation that provided new information for enhanced reflection as perceived by the teams. This demonstration and initial prototype then provided a powerful design research case study leveraging a custom-designed and developed IoT system to generate and visualize useful LA for team-based learning and training in this context.

Live-action simulation and situation awareness of individuals and teams engaged in high-fidelity, complex scenarios like the one described in this chapter can benefit from employing IoT sensor-based technology systems along with LA that provide new windows on behavior in situ leveraging analytics for enhanced reflection and learning situations. The intersection of learning design processes, iterative design inquiry cycles of design research, and thoughtful use of LA for participant-generated learning goals is an important direction for the field to pursue.

REFERENCES

[1] Haji, F. A., Da Silva, C., Daigle, D.T., & Dubrowski, A.: From bricks to buildings: adapting the medical research council framework to develop programs of research in simulation education and training for the health professions. *Society for Simulation in Healthcare*, **9**(4), 249–275 (2014).

[2] Sawyer, T. L. & Deering, S.: Adaptation of the US Army's after-action review for simulation debriefing in healthcare. *Simulation in Healthcare: Journal of the Society for Simulation in Healthcare*, **8**(6), 388–393 (2013).

[3] Sawyer, T. L. & Deering, S.: Adaptation of the US Army's after-action review for simulation debriefing in healthcare. *Simulation in Healthcare: Journal of the Society for Simulation in Healthcare*, **8**(6), 388–397 (2013).

[4] Mor, Y., Craft, B., & Mania, M. (2015). Introduction: learning design: definitions, current issues and grand challenges. In: M. Maina, B. Craft, & Y. Mor (Eds.), *The Art & Science of Learning Design*. Rotterdam, The Netherlands: Sense Publishers, pp. 9–26.

[5] Bannan, B., Cook, J., & Pachler, N.: Reconceptualizing design research in the age of mobile learning. *Interactive Learning Environments*, **24**(5), 938–953 (2016).

[6] Salas, E., Rosen, M. A., Held, J.D., & Weissmuller, J.J.: Performance measurement in simulation-based training. *Simulation & Gaming*, **40**(3), 328–376 (2009).

[7] Driskell, J. E., Salas, E., & Jonston, J.: Does stress lead to a loss of team perspective? *Group Dynamics: Theory, Research and Practice*, **3**(4), 291–302 (1999).

[8] Arora, S., Sevdalis, N., Nestel, D., Woloshynowych, M., Darzi, A., & Kneebone, R.: The impact of stress on surgical performance: a systematic review of the literature. *Surgery*, **147**(3), 318–330 (2009).

[9] Long, P. D. & Siemens, G.: Penetrating the fog: analytics in learning and education. *EDUCAUSE Review*, **46**(5), 31–40 (2011).

FURTHER READING

Bannan, B., Gallagher, P. S., & Blake-Plock, S. (2015). Ecosystem for smart medical team training. Presentation at the US Ignite Global City Teams Challenge, National Institutes for Standards and Technology, Gaithersburg, MD.

Bannan-Ritland, B.: The role of design in research: the integrative learning design framework. *Educational Researcher*, **32**(1), 21–24 (2003).

Bannan-Ritland, B. & Baek, J. (2008). Investigating the act of design in design research: the road taken. In: A. E. Kelly, R. A. Lesh, & J. Baek (Eds.), *Handbook of Design Research Methods in Education: Innovations in Science, Technology, Mathematics and Engineering*. Mahway, NJ: Taylor & Francis.

GovLoop Guide. (2015). The internet of things—what the IoT means for the public sector. Accessed at http://direct.govloop.com/IoT-Public-Sector (accessed August 17, 2016).

Middleton, J., Gorard, S., Taylor, C., & Bannan-Ritland, B. (2008). The "compleat" design experiment: from soup to nuts. In: A. E. Kelly, R. A. Lesh, & J. Baek (Eds.), *Handbook of Design Research Methods in Education: Innovations in Science, Technology, Mathematics and Engineering*. Mahway, NJ: Taylor & Francis.

Steffens, K., Bartolome, A., Estev, V., Bannan, B., & Cela-Ranilla, J.: Recent developments in technology-enhanced learning: a critical assessment. *RUSC Universities and Knowledge Society Journal*, **12**(2), 73–86 (2015).

6

THE BRAIN–COMPUTER INTERFACE IN THE INTERNET OF THINGS

JIM MCKEETH

Embarcadero Technologies, Austin, TX, USA

6.1 INTRODUCTION

We all have a brain, and we keep it with us at all times, but how connected is our brain? We typically connect our brain through our conscious perceptions, senses, and physical interactions: we see, hear, touch, talk, gesture, and so on, but a new generation of technologies is opening the path to connect our brains directly, bypassing our usual input and output mechanics. What is this technology, how does it work, what does it do, and where is it headed?

Neuroscience is a complex field, and this chapter only offers to serve as an overview and an introduction to a wide variety of topics within neuroscience and how they pertain to the Internet of Things (IoT), specifically connecting our brains. The hope is your eyes will be open to technology and achievements you were not aware of, and you will be inspired to research these topics more on your own.

6.1.1 What Is BCI

The Brain–Computer Interface (BCI) is a technology that allows a direct interface between the brain and the computer. This interface bypasses our usual input and output mechanics of senses, voice, gestures, and so on. This allows for greater understanding of how our brain works and has the possibility to unlock greater efficiency in communication and cognition.

Internet of Things and Data Analytics Handbook, First Edition. Edited by Hwaiyu Geng.

Companion website: www.wiley.com/go/Geng/iot_data_analytics_handbook/

BCI is advanced to the point where it is practical, and there are consumer devices available. Let's take a look at the science of BCI, some of the significant breakthroughs, and some of the consumer devices available to make BCI available to the masses today. Finally, we will look at some of the practical aspects of working with a BCI device today.

6.1.2 BCI in Science Fiction

The BCI is a popular element of science fiction where it is shown as the best way to communicate and interface with a computer or other people. It bypasses the limitations of our physical existence and elevates our consciousness to new levels of existence. While the science fact is not to that level yet today, some of science fiction is becoming a reality, and it is easy to see how this science has the potential to change our lives as we know it in the very near future.

Here is a look at examples of some significant different themes from science fiction and comments about how they may show up as reality. Later we will see how some steps are already taking place to do exactly that.

***6.1.2.1 1982: Thought-Controlled Weapons in* Firefox** Hollywood loves the idea of the government using BCI to fire weapons. In the movie *Firefox*, Clint Eastwood steals a Russian airplane by the same name. One of its many advanced features is the weapons are controlled by thought.

While BCI can serve as another input channel for pilots, it is unlikely that it will replace traditional controls completely for an able-bodied pilot. Using BCI to augment these traditional controls is however more likely.

***6.1.2.2 1983: Brain-to-Brain Communication via the "Hat" in* Brainstorm** Sometimes words get in the way and we just can't communicate effectively. Wouldn't it be great if we could share our thoughts and feelings without the need to reduce them to words which could be misunderstood? All of that and more was accomplished in the Christopher Walken movie *Brainstorm*, thanks to their technology called the "hat." Naturally the government wanted to take it over and weaponize it.

6.1.2.3 1991—Star Trek: The Next Generation, Episode 93 (4×19): The Nth Degree In Episode 93 of *Star Trek: The Next Generation*, the character Reginald Barclay finally became interesting when he was possessed by a super intelligent alien. In an effort to save the ship, he built a BCI on the holodeck because the traditional computer interface was just too inefficient to make the changes necessary in the time available.

Today, just as in Star Trek, the primary input mechanism for the computer is a combination of voice, pointing devices or gestures, and keyboards (including specialized ones like those that you would find on a grocery checkout kiosk). While arbitrary data input via a BCI is certainly possible, its potential to be more efficient and effective than these other mechanisms is very far off.

6.1.2.4 1999: Accelerated Learning with the "Headjack" in The Matrix Learning can be a tedious process, but in *The Matrix* Neo learns kung fu in a matter of seconds when the data is uploaded directly to his brain. While that granular of a skill upload is still a way off, experiments today are improving training, learning, and studying by 50–100%.

6.1.2.5 2013: Seeing and Hearing through the "Monitor" in Ender's Game Forget surveillance cameras; in the movie *Ender's Game*, the students become the cameras via an implanted "monitor." It allows the staff to see what the student sees and hear what the student hears. This is a huge invasion of privacy but an effective way to monitor students remotely.

While current technology doesn't allow for BCI to replace the resolution of a digital camera or microphone, it is showing that it is possible to intercept these standard sensory input channels for viewing by a third party. I don't think the movie and music industry needs to start worrying about this being used to pirate movies and live concerts, yet.

***6.1.2.6 2014: Uploading Our Consciousness in* Transcendence** While the movie *Transcendence* was plagued with technical issues, the idea of uploading consciousness isn't as crazy as it may seem. Research is taking place today to find a way to map the connectome that represents all of the neural connections within the brain. Once that is completed then creating a computerized model of an individual's memories and personality isn't too far off.

6.2 THE SCIENCE BEHIND READING THE BRAIN

We subconsciously transmit our thoughts through our body language, buying habits, and any number of other physical indications. Science today allows more direct and immediate access to reading out thoughts directly from the brain. The most common, controversial, and perhaps most coveted technique is the polygraph or lie detector. Other techniques rely on blood flow and oxygen levels within the brain as well as the electrical impulses that flow through our neurons and synapses within our brains.

6.2.1 Polygraph and Stress

The polygraph is commonly referred to as the lie detector, since that is its primary use. If we break down the word, "poly" means many, while "graph" means message, and that is exactly how the polygraph works. It collects some of the many messages or physical manifestations of stress that our body produces. The idea is that when someone lies, they feel stress about the lie [1].

Some of the typical methods of measuring stress include blood pressure, pulse, respiration, and skin conductivity [2]. A baseline for these readings is determined by asking simple, not contentious questions. Then those readings are compared against questions that may result in greater stress.

One note on the effectiveness of polygraphs is that since they only measure changes in the level of stress, they are easy to defeat and just about as likely to give a false positive [3]. So the reality is all they report is stress, which could come from any number of sources. For our purposes we are interested in the technology as an indication of stress and not truthfulness. And most all of these physical indicators of stress are easily measured through connected devices.

Take, for example, the Bluetooth Special Interest Group which publishes a Generic Attribute Profile (GATT) for the Heart Rate service [4] that all compliant heart rate monitors implement (there are many). This makes it easy to develop an app that can communicate with a typical Bluetooth heart rate monitor. Such an app could be used to help determine the subjects' stress levels.

Possible uses include monitoring your performance during the day and knowing when it would be a good time to take a break. Or maybe an employer could gather aggregate data of employees to know the overall stress levels and office moral. Perhaps a movie studio could use it to determine stress and excitement during a test screening.

This is the simplest way to tap into the brain from a computer to understand stress levels in our brain. It can easily be combined with other techniques to create a better picture of our brain, our thoughts, and our feelings.

6.2.2 Functional Magnetic Resonance Imaging (fMRI)

The use of large electromagnets combined with radio waves to form images of internal body systems is Magnetic Resonance Imaging (MRI). When a computer is added to analyze changes in blood flow and oxygen levels to different areas of the brain, then it is functional MRI (fMRI).

Different regions of the brain represent different functions and specialties. Changes in the level of utilization in these regions result in changes in the blood flow to that area of the brain. This is referred to as hemodynamic response (*hemo* meaning blood and *dynamic* meaning motion, so the "motion of blood").

fMRI is a very effective way of studying the function of the brain. It is mostly for research purposes but also has clinical and diagnostic uses. It requires the use of large powerful electromagnets and permanent magnets. Because of the size and magnetic fields, it is not portable and cannot be used around ferromagnetic materials.

6.2.2.1 Reconstructing Visuals In 2010, a team led by Shinji Nishimoto partially reconstructed YouTube videos based on brain recordings. The reconstruction wasn't a perfect representation but was typically enough that you could make out the basic form or composition of what was viewed.

This was accomplished with fMRI analysis of the visual cortex within the brain while a video was viewed. That was used to build a dictionary or regression model of the brain activity for the video. That model was then used to rebuild the images from a different library of videos.

The paper is available from *Current Biology* [5], and you can also view YouTube videos showing the clips presented to the subject along with the video reconstructed from the computer models: http://youtu.be/nsjDnYxJ0bo [6].

6.2.2.2 Identifying Thoughts and Intentions Marcel Just and Tom Mitchell of Carnegie Mellon University successfully used fMRI to show that the similar thoughts show up similarly in different subject's brains. This allowed them to identify which object a subject was thinking about [7].

In 2007, John-Dylan Haynes of the Bernstein Center in Berlin, Germany, used fMRI to read intentions from the subject's mind. The subject was asked to decide if they wanted to add or subject a series of numbers. They successfully detected the intention up to 4 seconds before the subject was conscious of deciding [8].

6.2.3 Brain Waves

Brain waves, or neural oscillations, are the rhythmic patterns of neural activity found within our brains. These brain waves reflect different mental states (Table 6.1). When you look at the location of the waves, they unlock information from within the depth of the brain.

Unlike hemodynamics, brain waves can be read with much simpler technology.

6.2.3.1 Electroencephalography (EEG) Electroencephalography (EEG) (*electro* = electrical, *encephalo* = brain, *graphy* = writing) is a method of using very sensitive electrodes placed along the scalp to read signals from the brain. The neurons of the brain communicate via very small electrical impulses—ionic current. Electrical impulses are found throughout the body. When they are detected in the heart, it is an Electrocardiogram (ECG or EKG), in other muscles it is an Electromyogram (EMG), and in the skin it is Galvanic Skin Response (GSR).

You can detect these electrical signals yourself with a simple voltage meter. Simply set it to its most sensitive direct current (DC) setting, and place the electrodes on your skin. The display should give you a value that changes over time. This value is from a combination of sources. The electrodes in an EEG are much more sensitive and use specific configurations to look into specific areas of the brain.

Whenever there is a moving electrical charge, there is also a magnetic field. Magnetoencephalography detects these magnetic fields as a way of imaging the functionality of the brain and is able to provide a different mechanism for looking into the brain.

While EEG has been used for years for research, clinical, and diagnostic purposes, the recent advent of consumer grade EEG devices opens the door for a new generation of BCI applications. Many of these devices support Bluetooth LE connectivity. We will explore the different EEG devices and their uses in detail later in this chapter.

TABLE 6.1 Brain Wave Frequencies and Associated Mental States

Name	Frequency (Hz)	Associated With
Alpha	8–13	Relaxed wakefulness
Delta	1–4	Deep stage 3 of NREM sleep
Theta	4–8	Alertness and motor activity
Beta	12.5–30	Normal waking consciousness
Gamma	25–100	Conscious perception and meditation

6.3 THE SCIENCE OF WRITING TO THE BRAIN

Reading directly from the brain is pretty useful, but writing to the brain unlocks accelerated learning, direct brain input, and rewriting memories.

6.3.1 Transcranial Direct Current Stimulation (tDCS)

EEG picks up the slight electrical currents that propagate to the scalp. Transcranial direct current stimulation (tDCS) runs that process in reverse. It applies slight electrical current to the scalp, in specific locations, so that the current can cross the cranium and enter the brain and nervous system noninvasively.

Unlike electroshock therapy, tDCS uses very low current. Current safety protocol restricts the current to 2 mA. Subjects occasionally report a tingling or slight skin irritation as a result. Other side effects include nausea and headaches. It is also common to see a phosphene, which is the perception of light where no external source is present.

US Defense Advanced Research Projects Agency (DARPA) did some work with tDCS in military training. The brain locations for the tDCS placement were based on fMRI analysis of the parts of the brain active during training. The results were that those with tDCS treatment learned twice as quickly [9].

Beyond accelerated learning, tDCS is also used in studies to reduce depression, anxiety, risk taking, cravings, migraine pain, and addictive behaviors and to increase attention, memory, problem solving, socialization, and mathematical abilities. The effects are typically most intense during the treatment but often times continue after the treatment is complete as well.

There is still a lot of controversy and research going into the use of tDCS to enhance healthy brains. While tDCS has existed for over 100 years, it is just recently with the advent of fMRI that tDCS interest has peaked. Other similar neurostimulation techniques include Transcranial Alternating Current Stimulation (tACS) and Transcranial Random Noise Stimulation (tRNS). All are considered safe, painless, and noninvasive.

The similar neurostimulation technique of Transcranial Magnetic Stimulation (TMS) places an electromagnet near the target area of the brain. It applies either an alternating or a constant magnetic field to induce an electrical current in the brain. It has the advantage of not requiring contact with the scalp but also produces different results.

Within the past few years, a few companies and individuals have started producing low-cost tDCS and related systems for individuals who wish to take neurostimulation into their own hands as a form of DIY brain hacking. One such company, Foc.us [10], has a Bluetooth LE-enabled tDCS stimulator, allowing wireless control of the neurostimulation process from a connected device.

Both EEG and tDCS use the same international 10–20 system for the placement of electrodes. It is conceivable that a system could be built that contains both EEG and tDCS functionality. The EEG system could monitor brain performance during a task or throughout the day, and then tDCS could be applied as necessary to make adjustments for optimal performance.

6.3.2 Transcranial Pulsed Ultrasound

A limitation with TMS, tDCS, and similar is their low spatial resolution—it is hard to accurately target the specific areas of the brain necessary to be most effective. Ultrasound on the other hand is much more accurate while still remaining noninvasive.

DARPA started work with Dr. William Tyler of Arizona State University to develop a helmet with ultrasonic neurostimulation built-in [11]. The system would help soldiers deal with stress, boost alertness, and relieve pain [12]. In 2014 Dr. Tyler went on to show that targeted ultrasound could be used to boost sensory performance and accurately target an area about 1 cm in diameter [13].

The system could be controlled by the soldier either directly, remotely, or automatically. Easily combined with other biometric signals like heart rate monitoring and EEG to collect stress and fatigue information, the soldier's performance could be tweaked as necessary to maintain optimal performance in whatever situation arrives.

6.4 THE HUMAN CONNECTOME PROJECT

The ultimate in BCI would be uploading of the brain into a computer. The Human Connectome Project [14] represents the potential to do that. Similar to how the Human Genome Project sequenced the human genome—all the genes that make us who we are as an individual—the Human Connectome Project seeks to build a comprehensive map of the neural connections in the brain.

As we learn, gain memories, and experiences, we create new neural connections within our brain. This wiring diagram would represent those connections at a fixed point in time. Once the diagram is created, it could be simulated on computer hardware.

Before mapping the human connectome, the nematode worm, with its 302 neurons, was an easier target. Timothy Busbice of the OpenWorm [15] project built a LEGO Mindstorms representation of a worm and connected it to a software model of the worm connectome [16]. This resulted in a robot that behaved like a worm, without any specific programming to do so. The behavior was apparent in the connectome as a result of the sensory input the robot collected.

A human connectome is certainly much more complicated than a worm, but it is an interesting prospect to consider.

6.5 CONSUMER ELECTROENCEPHALOGRAPHY DEVICES

The advance of consumer EEG devices opens the technology up to a whole new world of BCI options. This isn't an exhaustive list but represents a few devices to provide an overview.

Generally, the consumer devices have four target usages: Wellness, Games, Alternative Input, and Research.

Wellness uses the EEG devices as biofeedback, allowing the wearer to monitor their stress, concentration, and other mental factors. They frequently come with exercises helping the wearer to reduce stress, improve concentration, and so on while all receive input via the EEG sensors about the effectiveness of the process.

When used for Games the EEG headsets are mostly capitalizing on the novelty of a BCI. This could be either an alternative input for a computer or mobile game or sometimes even with dedicated hardware allowing the wearer to control some physical object. Some are branded as Star Wars Force Trainers or similar. These are keyed off of the wearers' ability to concentrate.

The use as alternative input is especially applicable for individuals who are unable to use traditional input for computers. This could include people without use of their hands or people whose hands are otherwise occupied. It can also be useful to augment traditional input methods.

Research is the catch-all bucket for other uses. EEG's been used in research for years, and with the development of new EEG and BCI uses, this might be the most interesting use case for most readers. It typically includes reading raw EEG data.

6.5.1 NeuroSky's MindWave

The MindWave from NeuroSky is widely available in a number of different configurations, everything from a bundle for EEG research to Star Wars-related toys for kids.

The MindWave has one dry electrode (without the need for conductive gel) placed on the forehead, and the reference electrode attaches with a clip to the earlobe. This EEG electrode measures attention and meditation. It also provides EMG signals from the muscles in the forehead to detect eyeblinks.

The MindWave connects via Bluetooth wireless connection to PC, Mac, Android, or iOS devices. There are SDKs available for Java, Objective-C, .NET, and Python [17]. It provides the following outputs: attention, meditation, eyeblinks, brain wave bands, and raw output.

Because of its low price, it makes a great entry-level device, is popular in educational and maker communities, and is also commonly found in toys and games.

6.5.2 Interaxon's Muse

The Muse by Interaxon is specifically focused on wellness and providing biofeedback to improve the effectiveness of meditation. It is used in combination with a mobile app that it communicates with via Bluetooth. It makes use of seven dry sensors: two forehead sensors, two sensors behind the ears, and three reference sensors on the forehead.

For research and development purposes, the Muse also provides research tools to view and work with EEG data, as well as a mobile app development SDK for iOS and Android. Desktop platform SDKs are coming soon [18].

6.5.3 OpenBCI

With hardware based on the open Arduino platform, OpenBCI is an open-source approach to the BCI. Each 32 bit board supports eight wet (using electroconductive gel) electrodes, and you can daisy chain two boards together for a total of 16 electrodes [19].

OpenBCI is targeted toward high-end research use. It most closely replicates clinical EEG hardware of all the other consumer EEG devices. In addition to brain activity, it also samples muscle activity (EMG) and heart activity (EKG). It supports Bluetooth LE connectivity to mobile devices and desktops.

As an open-source project, OpenBCI provides downloads of all their software and hardware firmware with source code to make it easy to modify and extend. They also provide SDKs for Java (Processing), Python, and Node.js.

6.5.4 Emotiv EPOC and Insight

There are two different headsets available from Emotiv. Their original headset is the EPOC and it comes with 14 sensors with two reference electrodes. Instead of electroconductive gel, it uses saline-soaked felt pads. Their next-generation headset is the Insight, with five "semidry polymer" sensors and two reference electrodes. The new semidry polymer sensors don't require rewetting, which simplifies the application process. The Insight makes up for less sensors with better processing.

Both the EPOC and Insight provide both EEG information from the brain and EMG for facial expressions, with the EPOC detecting a couple more expressions. They divide the EEG signals into two classifications. One is mental commands, and the other is mental states. The mental states include excitement, frustration/stress, engagement, meditation/relaxation with the Insight adding interest, and focus [20].

The mental commands allow the user to train the software to recognize specific thought patterns as a command. The commands it provides are direction oriented, but the Insight also adds user definable commands. Once it is trained, then it will recognize this command when repeated. In the provided software the user can use these commands to move 3D objects on the screen, but with the SDK it can be expanded to control real-world objects, including drones.

The Emotiv SDK support Windows, Linux, OS X, iOS, and Android with samples in Java, C#, C++, Objective-C, and Python. Then there are third party samples that also work with Embarcadero Delphi [21].

6.6 SUMMARY

The BCI is an evolving technology, just like the IoT. The ability to interface directly with the brain is here today and will continue to improve in accuracy and function. Combining EEG technologies with other IoT technologies like heart rate monitoring, facial emotion recognition, and data mining will further enhance the picture describing the brain and its functions.

REFERENCES

[1] Bonsor, K. "How Lie Detectors Work." *HowStuffWorks*. InfoSpace LLC, July 16, 2001. http://people.howstuffworks.com/lie-detector.htm (accessed January 6, 2016).

[2] Rosenfeld, J. P. (1995). "Alternative Views of Bashore and Rapp's (1993) Alternatives to Traditional Polygraphy: A Critique." *Psychological Bulletin*, 117: 159–166.

[3] Stromberg, J. "Lie Detectors: Why They Don't Work, and Why Police Use Them Anyway." *Vox*. Vox Media, Inc., December 15, 2014. http://www.vox.com/2014/8/14/5999119/polygraphs-lie-detectors-do-they-work (accessed January 6, 2016).

[4] Bluetooth Special Interest Group. "Heart Rate GATT Specification." https://developer.bluetooth.org/gatt/services/Pages/ServiceViewer.aspx?u=org.bluetooth.service.heart_rate.xml (accessed August 20, 2015).

[5] Nishimoto, S., Vu, A. T., Naselaris, T., Benjamini, Y., Yu, B., and Gallant, J. L. (2011). "Reconstructing Visual Experiences from Brain Activity Evoked by Natural Movies." *Current Biology*, 21(19): 1641–1646.

[6] Gallant, J. "Movie Reconstruction from Human Brain Activity." *YouTube*, September 21, 2011. http://youtu.be/nsjDnYxJ0bo (accessed February 13, 2016).

[7] Shinkareva, S. V., Mason, R. A., Malave, V. L., Wang, W., Mitchell, T. M., and Just, M. A. (2008). "Using fMRI Brain Activation to Identify Cognitive States Associated with Perception of Tools and Dwellings." *PLoS ONE*, 3(1): e1394.

[8] Smith, K. (2011). "Neuroscience vs Philosophy: Taking Aim at Free Will." *Nature*, 477(7362): 23–25.

[9] Batuman, E. "Electrified." *The New Yorker*. Condé Nast, April 16, 2015. http://www.newyorker.com/magazine/2015/04/06/electrified (accessed February 20, 2016).

[10] Oxley, M. "The Foc.us V2 Stimulator." *Foc.us*. Transcranial Ltd, February 20, 2016. http://www.foc.us/v2/ (accessed August 19, 2016).

[11] Tyler, W. J. "Remote Control of Brain Activity Using Ultrasound." *Armed with Science*. U.S. Department of Department, September 1, 2010. http://science.dodlive.mil/2010/09/01/remote-control-of-brain-activity-using-ultrasound/ (accessed February 20, 2016).

[12] Dillow, C. "DARPA Wants to Install Transcranial Ultrasonic Mind Control Devices in Soldiers' Helmets." *Popular Science*. Bonnier Corporation, September 9, 2010. http://web.archive.org/web/20160614220851/http://www.popsci.com/technology/article/2010-09/darpa-wants-mind-control-keep-soldiers-sharp-smart-and-safe (accessed February 21, 2016).

[13] Virginia Tech. "Ultrasound Directed to the Human Brain Can Boost Sensory Performance." *EurekAlert!* American Association for the Advancement of Science (AAAS), January 12, 2014. http://www.eurekalert.org/pub_releases/2014-01/vt-udt011014.php (accessed February 22, 2016).

[14] The Human Connectome Project. *NIH Blueprint for Neuroscience Research*. National Institutes of Health, January 14, 2016. http://www.humanconnectome.org/ (accessed February 22, 2016).

[15] OpenWorm. *OpenWorm*, November 28, 2015. http://www.openworm.org/ (accessed February 22, 2016).

[16] Black, L. "A Worm's Mind in a Lego Body." Web log post. *I Programmer*, November 16, 2014. http://www.i-programmer.info/news/105-artificial-intelligence/7985-a-worms-mind-in-a-lego-body.html (accessed February 22, 2016).

[17] "NeuroSky Developer Program." *MindWave*. NeuroSky, September 30, 2014. http://developer.neurosky.com/ (accessed February 28, 2016).

[18] "Muse." *Muse Developers*. InteraXon, 2015. http://developer.choosemuse.com/ (accessed February 28, 2016).

[19] "Open-Source Human-Computer Interface Technologies." *OpenBCI*, 2015. http://openbci.com/ (accessed February 29, 2016).

[20] "Wearables for Your Brain." *Emotiv Bioinformatics*. Emotiv, Inc., 2015. https://emotiv.com/ (accessed February 29, 2016).

[21] McKeeth, J. "Delphi Emotiv EPOC." *GitHub*, February 29, 2016. https://github.com/jimmckeeth/Delphi-Emotiv-EPOC (accessed February 29, 2016).

7

IoT INNOVATION PULSE

JOHN MATTISON
Singularity University, Moffett Field, CA, USA
Kaiser Permanente, Pasadena, CA, USA

7.1 THE CONVERGENCE OF EXPONENTIAL TECHNOLOGIES AS A DRIVER OF INNOVATION

The playing field for IoT technical and business innovation is expanding rapidly in every dimension. We are witnessing many phenomena unprecedented in all of human history. Various technologies, each evolving at exponential rates, drive the expansion of innovation opportunities. One way of organizing these opportunities is to consider each of the six genres of platforms in this multidimensional set of Platform Ecosystems as a *PLECOSYSTEM:* The Plecosystem consists of six dimensions of platforms and a set of corresponding principles.

7.2 SIX DIMENSIONS OF THE PLECOSYSTEM

1. *Service platforms*
   Internet, Cloud, Smartphone, BioCurious (for bio hackers), qPCR (DNA sequencer for $200), TensorFlow (for machine learning hackers), BlockChain, Deep Learning Analytics, IoT, personal real-time IOT (aka PRIOT), Robotics, Smartwatch, Avatars, Social Avatar networks, Augmented Reality, Virtual Reality (>35,000 certified developers in healthcare VR), CRISPR/

Internet of Things and Data Analytics Handbook, First Edition. Edited by Hwaiyu Geng.

Companion website: www.wiley.com/go/Geng/iot_data_analytics_handbook/

cas9-GeneDrive, Cpf1-…, Stem Cell therapy, 3D Printers, Drones, self-driving cars, and so on.

2. *Data platforms*

   Panarome (gen-, transcript-, proteo-, lipid-ome, etc.), Digital Phenome, Microbiome, Immunome, Exposome, Socialome, Neurobiome, Quantified Self, Wearable Sensors, Pervasive fixed and mobile sensors, and so on.

3. *API Platforms*

   SMART on FHIR, Open MHealth, Apple, Samsung, Google, EHR vendors, and a wide variety of vertical-specific and cross-vertical PaaS vendors, and so on.

4. *Experience platforms*

   UX, empathic design, VR Physical Transfer with Empathic Training, Social Connected, Avatar social network (intraindividual and interindividual, locally autonomous to globally networked), for example, FaceSpace [1].

5. *Financial platforms*

   From Angels to institutional investors to IPOs, fit for purpose and stages of life cycle, startup incubators and accelerators.

6. *Sociocultural/geopolitical/value/ethical platforms*

   Failure Fault Tolerance? Data sharing/Access? Balancing individual versus collective benefits. Many Spinozan trade-offs. Opportunity for global treaties to establish and monitor compliance, reciprocity, and net producers versus consumers of shared commons through blockchain technologies.

7.3 FIVE PRINCIPLES OF THE PLECOSYSTEM

1. *Exponential growth*: 10 years from first smartphone to first billion phones, similar patterns for each of the components within the Plecosystem.
2. *Synergy*: Value rises exponentially with both the number of nodes accessible and the number of platforms within each of the six platform dimensions of the Plecosystem linked across the Plecosystem, aka Matticalfe's law for both big data and small data.
3. *Data liquidity* is critical, hence federation of data, respecting local privacy values/policy. Blockchain and smart contracts, e.g. Cancer Genome Trust of GA4GH [2].
4. *Person-centricity and X-centricity (Matticalfe's law)*: The data liquidity is important in its own right, but the value of confluent data is dramatically enhanced with reliable "indexing" to a single entity, for example, a single individual person, a single 777 airliner, a single set of GPS coordinates, a single storm system, a single genetic lineage, and so on.
5. *Accordion model for learning within the plecosystem*: Ideally, we can continue to refine how we isolate the signal from the noise across the entire sensing network. The final section of this chapter outlines how we can use the accordion model for continuous learning across sensor networks, data stores, and collaborations.

The unprecedented metaexponential opportunities for innovation in this plecosystem are still subject to several tried and true principles of design and managing product life cycles.

Several quotes inform that approach:

1. Begin with the end in mind

   Steven Covey Habit #2 in *Seven Habits*

2. Vision without action is a daydream. Action without vision is a nightmare.

   Japanese Proverb (permutations attributed to Edison)

3. Start with Minimum Viable Product (MVP) then learn and innovate quickly through iteration (paraphrased)

   Eric Ries, *The Lean Startup*

4. Future success of the IoT depends upon three things:
   a. Empathic Design to understand and address needs of individuals and the larger community.
   b. Accordion model of learning through analytics and transparency.
   c. An urgent focus on international policy and diplomacy that recognizes the ability of many IoT networks to cross all geographic and political boundaries. This is perhaps our biggest opportunity to innovate in the IoT space.

   John Mattison, Singularity University, 2013

Successful innovation in the emerging universe of IoT will follow some key principles and strategies. This chapter begins with the end in mind using empathic design and evolves toward a road map. There is no crystal ball for the road map, but guiding principles and values can be articulated. The largest "butterflies" impacting the road map and the destination revolve around value adjudication and governance. Governance over the IoT will evolve through market forces, international consortiums, open-source collaborations, and to a lesser extent digital and cyber treaties between nation states.

7.4 THE BIOLOGIC ORGANISM ANALOGY FOR THE IoT

The IoT has a digital analog in the living organism and can be reduced to three key components in a closed loop:

1. Sensing (afferent perception)
2. Analytics (cognition)
3. Action (efferent neuromuscular motion)

The context for all three of these closed loop functions is social relationships as communities evolve, interact, and transform. While these social aspects of the evolution of IoT become increasingly important, they warrant an entirely separate and expansive treatment beyond the scope of this chapter. Suffice it to say that the extension of teleconferencing capabilities into the realm of photorealistic avatar interactions with haptics during real-time conversations and activities will utterly transform some of our concepts of human relationships, create entirely new concepts, and reinforce many of our preexisting social constructs.

The analogy to the human organism reveals that IoT sensing and action are virtually identical to their analogs in nature and will likely converge on a path of commoditization. In contrast the fundamental differences between machine analytics and human cognition represent more significant areas for innovation and market differentiation as AI and deep learning continue to surpass the capacity of the human brain in challenge after challenge.

7.5 COMPONENTS FOR INNOVATION WITH THE ORGANISMAL ANALOG

1. *Sensing*

   The physicochemical–social sensors that acquire new information are under rapid innovation cycles involving each of the following:
   - The modality of physical, chemical, or social probe used
   - The progressive noninvasiveness of those probes (especially for the human sensor network)
   - Cross modality sensing to enhance precision, reproducibility, reliability
   - Miniaturization
   - Reducing power demand
   - Raising power capacity over space and time
   - Embedding filtering, rectification, anomaly detection
   - Extending the range of transmission of output at low energy cost
   - Embedding capacity for early fault detection of any aspect of the sensing cycle and self-healing with escalation to the "network authority" on early failure mode and response
2. *Output from sensor networks*

   The array of outputs includes all actions perceptible by humans and many well beyond that perceptual capacity. General Classes of output will be optimized through innovation across each of the following parameters:
   - Specific targeting of action (physical, chemical, informational, emotional sentiment, social interaction, and responsiveness) within space and time
   - Autovalidation of guaranteed delivery/guaranteed receipt
   - Fidelity of intended effect with actual effect observed
   - Transparency and broadcasting of the effect (from deliberately obscure as in Stuxnet to deliberately conspicuous as providing public aid during a disaster)
   - Coordination of outputs to achieve desired set points for different objectives
   - AI and machine learning to tune the coordination real time and as a cumulative science

3. *Analytics*

   Since analytics fundamentally encroaches on the most human of functions, cognition, and the application of values and objectives, it is useful to first frame the human factors and heuristics guiding the innovation in this final and most significant engine of innovation.

7.6 SPINOZAN VALUE TRADE-OFFS

The context of human values and objectives will profoundly shape how a given input results in a given action. Behavioral science, behavioral economics, and social networks drive this transformation of information into behavioral action. The philosopher, Baruch Spinoza, elaborated how human values and virtues often conflict with each other, and frequently there is not a "correct set point" for how we adjudicate values and virtues in conflict. The emergence of context-specific human adjudication requires efficient and transparent visualization tools that help individuals, communities, and societies "tune" the IoT ecosystem according to extent views of balancing conflicting values and virtues.

A classic example is the FBI versus Apple dispute over Apple's creation of an uncrackable operating system, and the FBI demands that Apple creates software to crack that code and extract data from a deceased terrorist's cell phone to better protect the public. It is easy to imagine a myriad of similar value/context-specific adjudications to be managed through human machine interfaces. A recurring challenge/opportunity will be in managing when public transparency of IoT generated information outweighs the virtue of individual privacy rights.

A similar Spinozan dilemma is illustrated in autonomous vehicles. These vehicles use algorithms to optimize net flux of vehicles over time against acceptable levels of safety. Where in the curve of flux versus safety is the optimal set point? Should an ambulance take more risk of collision to optimize speed? Will it divert competing traffic? Who decides what clinical situations warrant additional risk? The ability to embed value-sensitive empathic design and controls into complex, otherwise autonomous systems represents a design challenge and opportunity for innovative differentiation. While this issue is fodder for many sci-fi novels or movies, it's no longer sci-fi, it's here, and it's now. Today's generation is the "Sci-Fi Generation" (copyright John Mattison, TM pending) because "if we can imagine it, we will build it in our lifetime."

7.7 HUMAN IoT SENSOR NETWORKS

The ultimate application of the analogy to a living organism is the human individual itself. Massive sensing of individuals (significantly beyond current quantified selfing) will allow for more informed decisions on the part of each individual and, by extension, social networks.

The ability to represent how different value choices, for example, variable risk tolerance of an individual (near term vs. long term) will be critical to informing those decisions. The role of the plecosystem in shaping innovation is the interaction of the IoT with millions of increasingly quantified lives, each of whose genome is known, and their transcriptome, microbiome, proteome, and so on are monitored for diurnal patterns. Understanding patterns and archetypes that generate different algorithmic rule sets that span microcohorts of individuals based upon analytics of large populations that cross all of these massive data sets. Then coming full circle will be the deviations detected in individuals that fall outside not just their personal diurnal norms but also outside of the norms of their microcohort.

7.8 ROLE OF THE IoT IN SOCIAL NETWORKS

As the volume of digital social interaction grows, personal digital avatars will become key brokers to manage these interactions. The algorithms that drive how personal avatars interact with the avatars of friends, family, and colleagues will grow in complexity. That complexity will require a simplified human machine interface that allows for efficient tuning and training of those algorithms. FaceSpace and others are bringing these services to market now, and the potential commoditization of such advanced services combined with advanced haptics will finally deliver on the promise of providing an "in person" experience for conversations around the globe that today are held hostage to physical travel. The implications are indeed profound ranging from the preservation of closer family ties and friendships to the better opportunity to establish and maintain more trusting relationships between individuals and nation states in disputes.

7.9 SECURITY AND CYBERTHREAT RESILIENCE

Current sensors and transmissions are infamously insecure. As sensors become more critical to various utilities, including power generation and transmission, transportation, supply chain management, the social impact of cyberthreats grows. While it is difficult to predict which threats are most likely, an atmospheric electromagnetic pulse (EMP) could permanently disable entire sensor networks and cripple an entire economy and the health and safety of large populations. Network topologies will require innovation with hot failover devices, and Faraday shields as critical design features for large IoT networks in public and private infrastructures.

7.10 IoT OPTIMIZATION FOR SUSTAINABILITY OF OUR PLANET

There will be many opportunities for innovation in how we design and deploy sensor networks to maximize values to communities while minimizing impact on the planet. This innovation opportunity includes every aspect of sensing, analytics, and action and extends into a collective consciousness around both physical and aesthetic impacts. Smart Communities, Smart Cities, Smart Nations will all drive design toward sustainability and a balance of social values.

In agriculture, water management, and natural ecosystem sustainability, there is a growing debate about the sustainability implications of large monocultural farms. Could IoT innovation support high levels of agricultural productivity while managing more biological diversity?

John Muir's advocacy for the preservation of wilderness will resurrect political debate as to whether to embed sensor networks in our wilderness to protect against degradation. That policy debate will manifest as innovation in a "light footprint" for sensor networks and call for judgments about what is "natural evolution" of an ecosystem, what is human-inflicted change, and how to decide when human-inflicted change is net negative, neutral, or positive and who gets to decide. The ultimate example of this is that we know multiple sources of greenhouse gases and the impact they are having on climate change. We will likely soon have multiple mechanisms (physical catalytic, biologic, etc.) to sequester greenhouse gases and begin to offset human and natural causes, perhaps even reverse them. When we have the ability to regulate the net production or sequestration of greenhouse gases, who gets to decide what the ideal set point is for climate at a global level? There is little prospect anytime soon that we can isolate the impact of free-ranging gases on climate to specific geographies, so we may well emerge with one regional IoT network autonomously sequestering greenhouse gases in accordance with homeostatic algorithms, while a neighboring regional IoT network will directly compete by autonomously producing more greenhouse gases. These opportunities will clearly force our planet to become much more global in its thinking, governance, action, and, in the context of this chapter, its treaties are numerous issues that regional IoT networks might regulate with futile cycles of conflict. Cyberthreats and conflicts between such conflicting regional IoT networks would be an undesirable outcome. We desperately need imagination and inspiration at the global diplomatic and multinational scale to take on these issues while they are still future and theoretical before we find ourselves in conflict over established implementations and "assertions of heritable rights."

A major collateral benefit of having both capacity and governance over the "greenhouse gas set point knob" will be when the inevitable supervolcano eruption occurs under Yellowstone Park and threatens a global mass extinction similar to what happened with the last such event in Indonesia some 80,000 years ago. That collective, well-governed opportunity should enable the global community to rapidly and efficiently govern the restoration and modulation of a more inhabitable global climate. IoT networks will play a key role but only if our innovation in the policy and diplomatic spheres enjoys more imagination and inspiration.

7.11 MAINTENANCE OF COMPLEX IoT NETWORKS

Billions of sensors will need to be continually identified, managed, upgraded, replaced, and repaired before failure events through early detection of prefailure patterns of output. Technologies such as block chain and implementations like blockstack may generate entire ancillary industries.

7.12 THE ACCORDION MODEL OF LEARNING AS A SOURCE OF INNOVATION

As the exponential proliferation of IoT sensors continues, we will have the ability to know all measurable things about anything, anywhere, anytime. We will also begin to impact our environment and the people and cultures that inhabit our planet and beyond in increasingly profound and deliberate ways. That array includes environmental conditions, social conditions, and personal activities and biologics. Extending this analogy of a human organism, many IoT-based systems already reliably maintain homeostasis, comparable to many complex physiologic processes. Water treatment plants automate conversion of sewage inputs into clean water outputs, analogous to kidney function. Full autonomy of these homeostatic closed loop systems will likely become commodity services. The role of innovation in these systems emerges from knowledge on process optimization based upon new understanding of the processes involved (aka the science) and new sensing capabilities (aka the technology). One of the key principles on how this innovation will occur is what can be described as "The Accordion Model of Learning."

Autonomously functioning systems are built according to the prevailing science and technology. Any period of stability in the science and technology permits a localization of the analytics closer to the data so that the "meaningful signal" is escalated to control centers and the noise is left or discarded closer to the local sensors. However, whenever new science or technology emerges, the data needs to be less filtered and more centrally analyzed with deep learning tools to retune the entire autonomous functioning based on more advanced understanding. Every autonomous service will have different components with varying periodicities that determine reconsideration of what is signal and what is noise. This creates multiple accordions opening to new signals (previously considered noise or silence) then closing once a new set of signals is established.

A third source for opening the accordion will be anomaly detection of outlying events. This is an ongoing learning and design process that proceeds independently of new science and new technology.

7.13 SUMMARY

The growth of the IoT is truly exponential, and innovation will be catalyzed by the convergence with many other technologies, each of which is subject to similar exponential growth. While innovation on both the sensing and the output side will continue, it is likely that the analytic aspects of the IoT will generate more sustained opportunities for innovation that will create new markets and differentiation within those markets. Most importantly, we need to have a much more deliberate and transparent discussion of the need for more effective international dialog to begin to lay a foundational approach for policy affecting large IoT networks with homeostatic or other objectives. When regional or conflicting multinational IoT networks operate toward conflicting goals or set points in what is increasingly a global planetary impact, we need to urgently draw upon the best imagination, inspiration, and diplomacy to lay the foundation for global treaties on how to adjudicate these imminently approaching conflicts through peaceful and diplomatic means. Without those innovations in policy, the promise of the IoT will be tightly constrained by the conflicts of the past bleeding into the opportunities of the future.

REFERENCES

[1] https://ww2.kqed.org/futureofyou/2016/04/22/stanfords-virtual-reality-lab-turned-me-into-a-cow-then-sent-me-to-the-slaughterhouse/ (accessed 10/1/16).

[2] https://github.com/ga4gh/cgtd (accessed 10/1/16).

FURTHER READING

Blockstack. http://continuations.com/post/139970467265/announcing-blockstack-decentralized-dns-and; https://blockstack.org/blockstack.pdf (accessed August 19, 2016).

Sudha Jamthe. The Internet of Things Business Primer, 2015, self-published. https://www.amazon.com/IoT-Disruptions-Internet-Things-Innovations-ebook/dp/B00Y0CPV5I (accessed September 13, 2016).

Joseph Kvedar, et al. The Internet of Healthy Things, 2015, Health Care Partners, El Segundo, CA.

PART II

INTERNET OF THINGS TECHNOLOGIES

8

INTERNET OF THINGS OPEN-SOURCE SYSTEMS

SCOTT AMYX
Amyx McKinsey, San Francisco, CA, USA

8.1 INTRODUCTION

Open source will accelerate the innovation and adoption of the Internet of Things (IoT or Internet of Everything), fundamentally altering the way businesses, organizations, communities, and individuals interact and operate [1]. Since open source will have such a broad impact, it is important for innovators to have an understanding of the current open-source projects available for use. This work briefly examines the background of open source, the motivation behind it, and the benefits of it. Using the results of an industry study, the importance of open source to the IoT is explored, and lists of consortiums and useful open-source projects are provided.

8.2 BACKGROUND OF OPEN SOURCE

Sharing technological information is vital to advancement. The current state of sharing information is not new: Henry Ford successfully challenged the patent for the gasoline engine, which led to the creation of a partnership that would help distribute information throughout the burgeoning automotive industry. Open source is following in the grand tradition of free information sharing: it is free code (or economical) that performs a certain function. Learner and Tirole noted that the threat of lawsuits over the use of UNIX may have been a motivator in developing open-source code [2].

Internet of Things and Data Analytics Handbook, First Edition. Edited by Hwaiyu Geng.

Companion website: www.wiley.com/go/Geng/iot_data_analytics_handbook/

The Open Source Initiative (OSI, founded in 1998) notes that code sharing has been an on-going event, with its status raised in the public square by the growing popularity of Linux and the release of Netscape's source code [3]. The OSI defined open source through a list of ten items, most notably that the code must be free, must not be technology specific (it cannot depend on the purchase of some other product), and has to be technology neutral (it cannot depend on any certain product or interface) [4]. Deshpande and Riehle noted that there has been an explosive growth in the number of open-source projects available, with more being added [5].

8.3 DRIVERS FOR OPEN SOURCE

Fundamentally, the motivation behind open-source projects is utility: people have a need for the types of deliverables open source can provide. Consumers want to use any device in the marketplace and want no restrictions on access; for example, a consumer may like one company's smartphone and another company's wearable. The issue arises from the incompatibility of the devices: proprietary devices and software lock buyers into only one choice, which frustrates consumers. The same holds true for enterprises; companies need to be able to incorporate a range of technologies into their operations but can run into difficulties by being forced to go with a single vendor for all of their technology requirements. Application developers struggle to support a wide range of devices and may be forced to only work with vendor-specific code. If a company changes its software provider, application development needs to start fresh.

8.4 BENEFITS OF USING OPEN SOURCE

According to Bailetti, businesses can benefit from open source if it can "solve a problem or fill a need," "increase access or control over the key resources, processes, and norms required to deliver value to customers, channel partners, and complementors," and "increase the company's ability to protect its intellectual property" [6]. Open source allows for scalability, increased velocity of innovation, and flexibility. While there are some issues with open-source technologies (e.g., the documentation and support options vary wildly from one project to another), open source provides a strong foundation for the expansion of the IoT.

8.4.1 Scalability

CISCO estimates that there will be 50 billion connected devices within the next 5 years, potentially creating a massive network of online objects [7]. Supporting all of these devices will require a significant investment in hardware and software—as well as other resources, such as office space and tech employees. Millions of routers, gateways, and data servers will be needed. Open-source frameworks and platforms are the best solution to this problem, as they are already available to scale to any infrastructure.

8.4.2 Velocity of Innovation

The IoT will require rapid innovation, so vendors will find it difficult to create and deploy quality products using only proprietary solutions.

8.4.3 Cost

Open-source code's natural advantage over its counterparts is its cost. Since it is (usually) free, small start-ups, mid-sized companies, and large enterprises all have access. Open source is typically easy to download and begin working with, saving on implementation costs. Companies can use open source to significantly enhance their technology resources without significant monetary investment.

8.4.4 Royalty-Free

Litigation, or the threat of litigation, damages innovation and costs companies billions in court costs or insurance fees. Open source is by definition permissionless, saving individuals and companies from the stress related to using proprietary software. Use of open source encourages innovation, since developers can integrate a variety of components into a solution that targets a company's specific needs.

8.4.5 Vibrant Developer Community

Markus noted that open-source software tends to innovate faster than proprietary solutions; this is largely because of its volunteer army of developers that provide tweaks, insights, and troubleshooting [8]. Since developers are typically using the open-source code for enterprise projects, the development of targeted, company-centered programs is enhanced. In this way, open source beats proprietary software with its limited list of out-of-box solutions. Weber highlighted the fact that open source is empowering to coders and aids in technology development [9].

8.4.6 Interoperability

Open source offers a new level of interoperability across devices and networks. Interoperability speeds innovation and reduces costs further. Open standards can provide the framework for open-source implementation.

8.4.7 No Lock-In

A key advantage of open source is that it does not lock-in an organization to a proprietary solution provider. Almeida et al. noted that vendor lock-in is particularly irritating to consumers [10]. The cost of switching to another provider keeps enterprises and start-ups hostage to proprietary service providers. This freedom also extends to support, mitigating the risk of proprietary solutions being discontinued or no longer being supported. Numerous firms have purchased proprietary software and invested in costly customizations only to discover the system is no longer supported or now requires a costly upgrade.

8.5 IoT OPEN-SOURCE CONSORTIUMS AND PROJECTS

Amyx McKinsey examined a broad range of IoT open-source projects in order to garner insight into the open source available to enterprises. It is impossible to provide complete coverage of all available open-source projects, as new ones are being added at a rapid pace. However, the survey did capture important information concerning the types of open source currently being used. The research methodology employed a survey questionnaire, video interviews, and secondary sources from websites, industry reports, and other sources.

One finding of particular note was that some of the open-source projects defied the OSI's definition, namely, they were not entirely free. Some projects offer a freemium model or free open-source code with the option of professional services for support and advanced tools.

8.5.1 Open-Source Projects in Scope

Some projects may fall under more than one category.

8.5.1.1 Industry Consortiums Industry consortiums provide a voice to enterprises, groups, and individuals within their respective communities. They facilitate interactions with the public and provide platforms for information transfer; becoming involved with a consortium offers a number of benefits, including up-to-date information in an industry, on-going education opportunities, and ability to network with others. Table 8.1 shows the top 5 consortiums reported by respondents.

8.5.1.2 Protocols and Operating Systems Protocols such as TCP, HTTP, and SSL provide a familiar foundation for managing and sending data; there are a number of open-source protocols that can provide value. The operating system (OS), formerly an area of little competition, is now ripe with growth in the open-source arena (Table 8.2).

8.5.1.3 APIs, Horizontal Platforms, and Middleware While Google and Facebook provide some of the most well-known APIs, open source has a variety of offerings. Horizontal platforms target users with a variety of backgrounds and skill sets, and middleware is become increasingly important in effecting reliable communication (Table 8.3).

TABLE 8.1 Industry Consortiums

Consortium	URL
AllSeen Alliance	allseenalliance.org
Open Interconnect Consortium (OIC)	iotivity.org
COMPOSE	compose-project.eu
Eclipse	eclipse.org
Open Source Hardware Association (OSHA)	oshwa.org

Courtesy of Amyx McKinsey.

TABLE 8.2 Protocols and Operating Systems

	URL
Protocols	
Advanced Message Queuing Protocol (AMQP)	amqp.org
Constrained Application Protocol (CoAP)	coap.technology
Extensible Messaging and Presence Protocol (XMPP)	en.wikipedia.org/wiki/XMPP
OASIS Message Queuing Telemetry Transport (MQTT)	oasis-open.org
Very Simple Control Protocol (VSCP)	vscp.org
Operating system (OS)	
ARM mbed	mbed.org
Canonical Ubuntu and Snappy Ubuntu core	developer.ubuntu.com/en/snappy
Contiki	contiki-os.org
Raspbian	raspbian.org
RIOT	riot-os.org
Spark	spark.github.io
webinos	webinos.org

Courtesy of Amyx McKinsey.

TABLE 8.3 APIs, Horizontal Platforms, and Middleware

	URL
API	
BipIO	bip.io
Qeo Tinq	github.com/brunodebus/tinq-core
Zetta	zettajs.org
1248.io	wiki.1248.io/doku.php
Horizontal platforms	
Canopy	canopy.link
Chimera IoT	https://chimeraiot.com/
DeviceHive	github.com/devicehive
Distributed Services Architecture (DSA)	iot-dsa.org
Pico labs (Kynetx open source assigned to Pico labs)	github.com/Picolab
M2MLabs mainspring	m2mlabs.com
Nimbits	nimbits.com/index.jsp
Open Source Internet of Things (OSIOT)	osiot.org
prpl Foundation	prplfoundation.org
RabbitMQ	rabbitmq.com
SiteWhere	sitewhere.org
webinos	webinos.org
Yaler	yaler.net/download
Middleware	
IoTSyS	code.google.com/p/iotsys
OpenIoT	openiot.eu
OpenRemote	openremote.org/display/HOME/ OpenRemote
Kaa	kaaproject.org

Courtesy of Amyx McKinsey.

8.5.1.4 Node Flow Editors, Toolkits, Data Visualizations, and Search As can be seen from the previous tables, popular open-source projects tend to focus on areas of high interest with little market saturation. Table 8.4 is a kind of "miscellaneous" category for open-source projects.

8.5.1.5 Hardware and In-Memory Data Grids The IoT relies on a firmware network that is flexible and relatively cost-effective (there are no free pieces of hardware, unlike the open-source coding projects). In-memory data grids provide for quick, efficient communication that skirts the typical network relays (Table 8.5).

8.5.1.6 Home Automation, Robotics, Mesh Network, Health, Air Pollution, and Water The IoT will have a revolutionary impact on systems not already digitized. The dearth of data in Table 8.6 indicates significant room for growth in these areas.

TABLE 8.4 Node Flow Editors, Toolkits, Data Visualizations, and Search

	URL
Node flow editor	
Node-RED	nodered.org
ThingBox	thethingbox.io
Toolkit	
KinomaJS	github.com/kinoma
IoT toolkit	iot-toolkit.com
Data visualization	
Freeboard	github.com/Freeboard/freeboard
ThingSpeak	thingspeak.com
Search	
Thingful	thingful.net

Courtesy of Amyx McKinsey.

TABLE 8.5 Hardware and In-Memory Data Grids

	URL
Hardware	
Arduino ethernet shield	arduino.cc/en/Main/ArduinoEthernetShield
BeagleBone	beagleboard.org/getting-started
Intel Galileo	arduino.cc/en/ArduinoCertified/IntelGalileo
openPicus FlyportPRO	http://openpicus.com/
Pinoccio	pinocc.io
WeIO	we-io.net/hardware
In-memory data grid	
Ehcache	ehcache.org
Hazelcast	hazelcast.org

Courtesy of Amyx McKinsey.

TABLE 8.6 Home Automation, Robotics, Mesh Network, Health, Air Pollution, and Water

	URL
Home automation	
Home gateway initiative (HGI)	http://www.homegatewayinitiative.org/
Ninja blocks	http://shop.ninjablocks.com/
openHAB	openhab.org
Eclipse SmartHome	eclipse.org/smarthome
PrivateEyePi	projects.privateeyepi.com
RaZberry	razberry.z-wave.me
Robotics	
Open-source robotics foundation	osrfoundation.org
Mesh network	
Open garden	opengarden.com
OpenWSN	openwsn.org
Health	
e-Health sensor platform	cooking-hacks.com/documentation/tutorials/ ehealth-v1-biometric-sensor-platform-arduino-raspberry-pi-medical
Air pollution	
HabitatMap Airbeam	takingspace.org
Water	
Oxford flood network	http://flood.network/

Courtesy of Amyx McKinsey.

8.6 FINDING THE RIGHT OPEN-SOURCE PROJECT FOR THE JOB

With so many open-source possibilities, the only issue that remains is how to find the right open-source project for your needs. Key questions to ask involve the interoperability and scalability required for a given enterprise.

8.6.1 Interoperability

Enterprise-IoT involvement requires standards and protocols that can operate with the broad range of technologies already available. Interoperability is foundational to transitioning to what Korzun, Balandin, and Gurtov referred to as the IoT's "ubiquitous interconnections of highly heterogeneous networked entities" [11] environment. Standards and protocols provide an infrastructure that allows for reliable communication for connected devices, including mobile devices, wearables, automotive devices, TVs, video and digital cameras, printers and scanners, and household appliances.

8.6.2 Standards

Standards groups provide a necessary architecture for navigating the open-source ecosystem. These organizations provide a measure of standardization as well as educational opportunities like reference materials and training. Standards bodies like the AllSeen Alliance (AllJoyn framework), the Open Interconnect Consortium (OIC) (IoTivity open-source software framework), Industrial Internet Consortium, OASIS IoT/M2M, and Eclipse work to ensure the current and future interoperability of networked items.

8.6.2.1 AllSeen Alliance: AllJoyn The AllSeen Alliance consortium has more than 120 members, including Qualcomm, Microsoft, Panasonic, Sharp, Sony, LG, Cisco, Honeywell, HTC, Lenovo, Haier, TP-Link, D-Link, ADT, Netgear, Symantec, and Verisign. AllJoyn is its open-source development framework, and over 10 million devices currently use it. Since AllSeen Alliance enjoys such broad enterprise support, it hosts a robust technical community and targets devices, platforms, and networks across all verticals. As a motivator for the advancement of the IoT, AllSeen Alliance has noted the growing rise in the IoT applications in healthcare, education, automotive, and enterprise [12].

8.6.2.2 OIC: IoTivity The OIC, which was founded in July 2014, sponsors the IoTivity framework. The group has a somewhat smaller membership, with more than 50 members, such as Intel, Samsung, Cisco, Acer, Dell, GE, Honeywell, HP, Siemens, Lenovo, and McAfee. IoTivity allows for customization across all verticals and is especially of interest to those in the areas of Smart Homes, Industrial Automation, Healthcare, Automotive, Smart Cities, and Smart Grids [13].

8.6.2.3 Eclipse Foundation Founded in 2004 as a natural outgrowth of IBM's Eclipse Project that began in 2001, Eclipse has 228 members, including Google, Oracle, SAP, Siemens, Texas Instruments, Research in Motion, BMW, Cisco, Dell, Ericsson, HP, Intel, Nokia, and Bosch. It has a number of working groups that target areas such as Automotive, LocationTech, OpenMDM, and the IoT [14].

8.6.3 Scalability

Scalability is of concern for any business or organization capitalizing on the spread of the IoT. Current and projected business plans need to incorporate projections for the IoT. Here, we first examine some of the more popular protocols and then take note of platforms that can support large-scale deployments.

8.6.3.1 Protocols and Platforms Protocols are an alphabet of tools that allow connected devices to communicate. Message Queue Telemetry Transport (MQTT), Constrained Application Protocol (CoAP), Extensible Messaging and Presence Protocol (XMPP), and Advanced Message Queuing Protocol (AMQP) are the more popular protocols. Platforms allow for creative development and smoother integrations.

8.6.3.2 MQTT MQTT is a "lightweight" messaging protocol that works with TCP/IP. It is particularly geared toward resource-constrained IoT devices, as well as low-bandwidth, high-latency, or unreliable networks. MQTT works well for devices that operate in remote locations where network bandwidth is restricted.

8.6.3.3 CoAP CoAP is an application layer protocol that was defined by the Internet Engineering Task Force (IETF) to target the small electronics, particularly the technology of the IoT and its associated networks [15]. CoAP can work with a variety of internet devices that only consume limited resources, like wireless sensor network nodes, low power sensors, and other types of components [16, 17].

8.6.3.4 XMPP XMPP began its life as Jabber, an open instant messaging technology [18, 19]. XMPP is based on XML, and it has been extended for use in publish-subscribe systems, video, and IoT applications like smart grids.

8.6.3.5 AMQP AMQP is a binary wire protocol designed for message-oriented middleware known for its interoperability and reliability. Its key features are message orientation, queuing, routing (including point to point and publish and subscribe), reliability, and security. It provides for a great deal of control. (This protocol is used by a number of large organizations, such as Bank of America and Barclays.) [20]

8.6.3.6 Platform: SiteWhere SiteWhere is a platform that provides for a number of functionalities, including a system framework, data management (MongoDB and Apache HBase servers), and asset management. SiteWhere supports MQTT, JSON, AMQP, XMPP, Stomp, JMS, and WebSockets and provides published APIs [21]. It allows for either local or cloud data storage. Since it allows for bidirectional and asynchronous communication, it can handle enterprise operations that include a large volume of devices.

8.6.3.7 Platform: Xively The Xively platform (acquired by LogMeIn) offers a range of features to allow for the management of data and communication. The platform also uses MQTT, WebSockets, and HTTP [22]. Xively is built on the cloud platform Gravity, which can handle a tremendous number of devices.

8.6.3.8 webinos The webinos platform (Secure WebOS Application Environment) promotes the idea of a "single service for every device" and utilizes a number of standards such as those from the W3C and IETF [23]. webinos has largely focused on the areas of TV, Automotive, Health, and Home Automation Gateways. It provides for secure data management and easy application development. Supported standards and protocols include HTTPS and JSON-RPC. It has the support of 30 project partners, including W3C, University of Oxford, Samsung, and BMW.

8.6.4 Smart Cities Technologies

Perera et al. noted the increasing movement of governments toward the implementation of the IoT into populated areas [24]. Smart cities are not a vision of the future—they are already in production because governments want to be able to manage populations more effectively. The ASC defines a "smart city" as an urban environment where "social and technological infrastructures and solutions facilitate and accelerate sustainable economic growth" [25]. Clearly, large communal living areas with either complete or partial IoT implementations are fragmented environments that require a variety of technological approaches to successfully integrate networks and devices into everyday life.

8.6.4.1 RIOT RIOT is "a microkernel-based OS matching the various software requirements for IoT devices" [26].

The open-source RIOT OS can be used with sensors with minimal processing and memory, is "developer-friendly" (programming can be done in C or C++), and supports multiple chip architectures, such as MSP430, ARM7, Cortex-M0, Cortex-M3, and Cortex-M4. RIOT utilizes RPL mesh network protocols [27, 28].

8.6.4.2 Contiki Veteran OS Contiki boasts a robust community (it has been around for more than a decade) and targets very small devices [29]. It supports 6lowpan, Ripple, CoAP, TCP, HTTP, MQTT, DNS, and JSON; it also utilizes RPL mesh network protocols. Contiki offers full IP networking, low-power consumption, and dynamic module loading.

8.6.4.3 OpenRemote: A Case Study OpenRemote is middleware that targets the IoT in home automation, commercial buildings, and healthcare [30]. OpenRemote's architecture enables autonomous intelligent buildings.

OpenRemote offers end-user control interfaces for iOS, Android, and web browser-based devices. It also provides insight into how an open-source solution can be implemented into an urban environment. Eindhoven, a city in the Netherlands, used OpenRemote to create a crowd management system that implemented workflow and messaging capabilities. Data visualization software was integrated with devices, sensors, and subsystems through a local controller.

The OpenRemote system helped with crowd management metrics by tallying the number of pedestrians and gauging the sound level of the area. A simple UI dashboard for mobile devices allowed for managers and team members to assess the crowd metrics.

8.6.5 Smart Homes

Davidoff et al. noted that smart homes, or homes that implement the IoT, are designed to allow families to "have more control of their lives" [31]. This means that the technology needs to work around the actual users, so companies investing in the IoT in this area need to have a broad understanding of family needs and interactions. Competitors like NEST, Icontrol Networks, and People Power are competing in a market that is experiencing rapid growth. Taking advantage of open-source projects will enhance the competitiveness of smart home enhancements and improve the users' quality of life.

8.6.5.1 Smart Home Platforms: openHAB and Eclipse SmartHome The vendor-neutral openHAB platform can incorporate a number of home automation systems and technologies and provides for "over-arching automation rules" [32]. The platform's framework rests on Open Services Gateway Initiative (OSGi)'s architecture. Notably, openHAB can be used within the Eclipse SmartHome project, an abstraction and translation architecture designed for heterogeneous smart home automation systems.

8.6.5.2 Home Gateway Initiative (HGI) The Home Gateway Initiative (HGI), a consortium of broadband service providers (Deutsche Telekom, Telecom Italia, and NTT) and home consumer electronics manufacturers, works to create specifications for the features, diagnostic solutions, and performance of home networking systems. HGI also provides test programs for home gateways, as well as publications targeted at developing the home automation ecosystem [33].

8.6.6 Rapid Time to Benefit

Businesses need to realize rapid time to benefit. While open-source projects offer numerous benefits, the amount of time needed to successfully implement myriad solutions to complex business problems is a drawback. There are, however, a growing number of open-source solutions that address the issue of fast integration.

8.6.6.1 Zetta Zetta, a platform built on Node.js, lays claim to the ability to "run everywhere" since its servers can operate in the cloud or on devices. Zetta allows for used APIs and supports most device protocols, enabling a broad range of impact for the IoT. Apigee, a company that develops and maintains APIs, uses Zetta because of its extensive toolkit [34].

8.6.6.2 OpenIoT OpenIoT is a middleware architecture for IoT sensors and applications [35]. The framework allows for the integration of virtual and physical sensors, and it allows for the retrieval of data from sensor clouds. OpenIoT is relevant for managing cloud environments for IoT entities and resources such as sensors, actuators, and smart devices and offers a cloud-based and utility-based sensing-as-a-service model. OpenIoT has use cases for smart agriculture, intelligent manufacturing, urban crowdsensing, smart living, and smart campuses [36].

8.6.7 Limited Network Coverage

In the near future, IoT devices and networks will be ubiquitous but will still face the same structural issue confronting communications today: certain areas will suffer from either a lack of cellular coverage or limited access to a Wi-Fi network.

8.6.7.1 Mesh Network Mesh networks are wireless and arranged so that every node in the network can distribute data. This type of topology is obviously the best network for the IoT, since the network's size and speed will increase as devices are added. Mesh networks use dynamic routing and will not experience major disruptions if one of the nodes fails or has its signal blocked. Rural areas are one possible area of use, as are concert venues, amusement parks, or even very remote locations.

8.6.7.2 Open Garden Open Garden uses peer-to-peer mesh networking connectivity to quickly transfer data. It offers its Open Garden SDK IoT to developers to create apps that make use of its mesh networking technology [37]. Open Garden, known for its popular FireChat app that allows users to go "off the grid," the appeal of creating a network from whatever devices are around cannot be overlooked—especially in an age when eavesdropping—has become an issue. Open Garden's networking application is currently supported on Mac, Android, and Windows.

A number of companies and organizations are already taking advantage of Open Garden and an LPWA in order to create networks that can operate within a 20 mile radius.

8.6.7.3 OpenWSN Open Wireless Sensor Network (OpenWSN) is a network protocol stack for low-power devices (like IoT sensors) [38]. It uses C and supports RPL, 6LoWPAN, and CoAP. OpenWSN has been acknowledged as the "the first open-source implementation of the IEEE802.15.4e standard," which refers to time-synchronized channel hopping (TSCH) [39].

8.6.8 Utilization-Based Model

Monetization, or the ability to generate revenue from a product or service, is fundamental to the successful implementation of IoT devices, networks, and strategies. For the IoT, this monetization will most likely take the form of utilization; users will pay a fee based on usage of a product or service [40].

8.6.8.1 Chimera IoT Chimera IoT's platform is at the forefront of monetizing the utility of enterprise applications [41]. The platform supports the development of mobile and web applications, as well as the reliable storage and transfer of data. Its messaging service is based on RabbitMQ (an implementation of AMQP), which runs on the top OS [42].

8.6.9 High-Volume, Real-Time Analytics: In-memory Data Grids

The increased volumes of data produced by apps and the necessity of real-time analytics require significant processing power and memory, areas that are currently suffering from the ubiquity of legacy systems.

One possible solution to these issues lies in in-memory data grids, structures that exist only in RAM that are distributed among a number of servers. This allows for the rapid accessibility of data and for the scalability of the system, making in-memory data grids vital for industrial and other uses [43].

Ehcache and Hazelcast both support enterprise-grade, in-memory computing solutions for high-volume transactions, real-time analytics, and hybrid data processing.

8.6.9.1 Ehcache Ehcache is a robust, performance-enhancing cache [44]. Touted as the most popular Java-based cache, it boasts a simple API and is scalable. Ehcache's newest version allows for the persistent storage of data after an app has been shut down, allowing for access to the cached data once the app is restarted. It can be used with ColdFusion, Google App Engine, and Spring.

8.6.9.2 Hazelcast The Hazelcast platform is used by companies like CISCO and AT&T and has been implemented by over 8,000 other entities. Hazelcast is based on Java and provides in-memory NoSQL, Java caching, data grid, messaging, application scaling, and clustering [45].

8.7 CONCLUSION

Open source will enjoy growing influence as the IoT expands. Its benefits, including interoperability, scalability, and cost, make open source a useful component of the IoT as well as a driver for it. Organizations interested in taking advantage of the available open-source technologies have numerous options. Which open-source solutions are best for any particular enterprise or group depend upon the technical needs and business model. Since open-source projects are able to more easily interact with each other than technologies offered through proprietary vendors, utilizing a mix to target each individual need makes sense. The benefits of open source far outweigh its potential limitations. Companies can utilize these open-source projects to accelerate product development and enhance their technology resources.

GLOSSARY[1]

API: Application program interface, an abbreviation application program interface, is a set of routines, protocols, and tools for building software applications. The API specifies how software components should interact, and APIs are used when programming graphical user interface (GUI) components.

Asynchronous communication: The term asynchronous is usually used to describe communications services, software programs, or mobile apps that are offered to users free of charge but typically with limited functionality, advertiser support, or additional features that are only available for a premium charge.

Binary wire protocol: A protocol that carries operation-invocation requests from the client to the server and operation-result replies from the server to the client across a stateful connection. The stateful connection provides extra efficiency [46].

Freemium: An amalgamation of the words *free* and *premium* that refers to services, software programs, or mobile apps that are offered to users free of charge but typically with limited functionality, advertiser support, or additional features that are only available for a premium charge.

Middleware: Software that connects two otherwise separate applications.

[1] All definitions, unless otherwise stated, are direct quotes from http://www.webopedia.com.

Operating system (OS): The most important program that runs on a computer. Computers and mobile devices must have an OS to run programs.

Protocol: an agreed-upon format for transmitting data between two devices. It determines type of error checking and data compression used.

REFERENCES

[1] Worldwide Internet of Things (IoT) 2013–2020 Forecast: Billions of Things, Trillions of Dollars. Market Analysis 243661, IDC; 2013.

[2] Lerner J., and Triole J. The simple economics of open source (No. w7600). National Bureau of Economic Research; 2000. Available at http://www.nber.org/papers/w7600.pdf (accessed August 24, 2016).

[3] http://opensource.org (accessed August 24, 2016).

[4] http://opensource.org/osd (accessed August 24, 2016).

[5] Deshpande A., and Riehle D. The total growth of open source. Proceedings of the Fourth Conference on Open Source Systems (OSS 2008). Springer Verlag; 2008. Available at: http://dirkriehle.com/wp-content/uploads/2008/03/oss-2008-total-growth-final-web.pdf (accessed August 24, 2016).

[6] Bailetti T. How Open Source Strengthens Business Models. *Technology Innovation Management Review* February: 4–10; 2009.

[7] http://www.cisco.com/web/solutions/trends/iot/portfolio.html (accessed August 24, 2016).

[8] Markus M., Manville B., and Agres C. What Makes a Virtual Organization Work: Lessons from the Open-Source World. MIT Sloan Management Review; October 15, 2000. Available at: http://sloanreview.mit.edu/article/what-makes-a-virtual-organization-work-lessons-from-the-opensource-world/ (accessed August 24, 2016).

[9] Weber S. *The Success of Open Source*. Harvard University Press, Cambridge, MA; 2004.

[10] Almeida F., Oliveira J., and Cruz J. Open Standards and Open Source: Enabling Interoperability. *International Journal of Software Engineering & Applications* **2** (1); 2011. Available at http://airccse.org/journal/ijsea/papers/0111ijsea01.pdf (accessed August 24, 2016).

[11] Korzunn D., Balandin S. I., and Gurtov, A. V. Deployment of Smart Spaces in Internet of Things: Overview of the Design Challenges. In *Internet of Things, Smart Spaces, and Next Generation Networking* (vol. **8121**, pp. 48–59) Springer, Berlin Heidelberg; 2013.

[12] https://allseenalliance.org/ (accessed August 24, 2016).

[13] https://www.iotivity.org/ (accessed August 24, 2016).

[14] https://eclipse.org/org/foundation/ (accessed August 24, 2016).

[15] https://tools.ietf.org/html/draft-ietf-core-coap-18 (accessed August 24, 2016).

[16] Kovatsch M., Lanter M., Shelby Z. Californium: scalable cloud services for the Internet of Things with CoAP. 2014 International Conference on the Internet of Things (IOT). Cambridge, MA: IEEE; October 6–8; 2014.

[17] Cirani S., Picone M., and Veltri L. mjCoAP: An Open-Source Lightweight Java CoAP Library for Internet of Things Applications. *Interoperability and Open-Source Solutions for the Internet of Things Lecture Notes in Computer Science: Volume* **9001**: 118–133; March 2015.

[18] http://www.jabber.org/faq.html (accessed August 24, 2016).

[19] http://xmpp.org/ (accessed August 24, 2016).

[20] https://www.amqp.org/ (accessed August 24, 2016).

[21] http://www.sitewhere.org/ (accessed August 24, 2016).

[22] https://xively.com/whats_xively/ (accessed August 24, 2016).

[23] http://webinos.org/ (accessed August 24, 2016).

[24] Perera C., Zaslavsky A., Christen P., and Georgakopoulos D. Sensing as a Service Model for Smart Cities Supported by Internet of Things. *Emerging Telecommunications Technologies* **25** (1): 81–93; 2014.

[25] http://amsterdamsmartcity.com/about-asc (accessed August 24, 2016).

[26] http://www.riot-os.org/#about (accessed August 24, 2016).

[27] Hahm O., Baccelli E., Petersen H., Wählisch M., and Schmidt T. Demonstration abstract: simply RIOT: teaching and experimental research in the Internet of Things. IPSN'14 Proceedings of the 13th International Symposium on Information Processing in Sensor Networks. IEEE Press: 329–330; 2014.

[28] Baccelli E., Hahm O., Wählisch M., Günes M., and Schmidt T.. RIOT: One OS to Rule Them All in the IoT. [Research Report] RR-8176, 2012.

[29] http://www.contiki-os.org/ (accessed August 24, 2016).

[30] http://www.openremote.org/display/HOME/OpenRemote (accessed August 24, 2016).

[31] Davidoff S., Lee M., Yiu C., Zimmerman J., and Dey A. *Principles of Smart Home Control in UbiComp 2006: Ubiquitous Computing* (vol **4206**, pp. 19–34) Springer, Berlin Heidelberg; 2006. Available at: 10.1007/11853565_2 (accessed August 24, 2016).

[32] http://www.eclipse.org/smarthome/ (accessed August 24, 2016).

[33] http://www.homegatewayinitiative.org/ (accessed August 24, 2016).

[34] http://www.zettajs.org/ (accessed August 24, 2016).

[35] http://open-platforms.eu/library/openiot-the-open-source-internet-of-things/ (accessed August 24, 2016).

[36] Kim J., and Lee J. OpenIoT: An open service framework for the Internet of Things. In Internet of Things (WF-IoT), 2014 IEEE World Forum (pp. 89–93). IEEE; 2014.

[37] http://www.opengarden.com/ (accessed August 24, 2016).

[38] https://openwsn.atlassian.net (accessed August 24, 2016).

[39] Watteyne T., Vilajosana X., Kerkez B., Chraim F., Weekly K., Wang Q., Glaser S., and Pister, K. OpenWSN: A Standards-Based Low-Power Wireless Development Environment. *European Transactions on Telecommunications* **00**: 1–13; 2013. Available at: http://glaser.berkeley.edu/glaserdrupal/pdf/openWSN%20Euro%20Trans%20Telecom.pdf (accessed August 24, 2016).

[40] Ferreira E., and Tanev S. How Companies Make Money Through Involvement in Open Source Hardware Projects. Open Source Business Resource; February 2009.

[41] https://www.chimeraiot.com/ (accessed August 24, 2016).

[42] https://www.rabbitmq.com/ (accessed August 24, 2016).

[43] Williams J.W., Aggour K.S., Interrante J., McHugh J., and Pool E., Bridging high velocity and high volume industrial big data through distributed in-memory storage & analytics in *Big Data (Big Data)*, 2014 IEEE International Conference (pp. 932–941, 27–30); October 2014. Available at: http://ieeexplore.ieee.org/document/7004325/ (accessed August 24, 2016).

[44] http://ehcache.org/ (accessed August 24, 2016).

[45] https://hazelcast.com/ (accessed August 24, 2016).

[46] https://www.safaribooksonline.com/library/view/http-the-definitive/1565925092/ch10s09.html (accessed August 24, 2016).

FURTHER READING

Bell J. Government Transparency via Open Data and Open Source. Open Source Business Resource; February 2009.

Ding A., Korhonen J., Savolainen T., Kojo M., Ott J., Tarkoma S., and Crowcoft J. Bridging the Gap between Internet Standardization and Networking Research. *ACM SIGCOMM Computer Communication Review* **44** (1): 56–62; 2014.

Dunkels A., and Vasseur J. IP for smart objects, Internet Protocol for Smart Objects (IPSO) Alliance. White Paper, N. 1, IPSO Alliance; 2008.

Fambon O., Fleury E., Harter G., Pissard-Gibollet R., and Saint-Marcel F. FIT IoT-LAB tutorial: hands-on practice with a very large scale testbed tool for the Internet of Things. INRIA Grenoble Rhone-Alpes, Montbonnot, France; 2014.

Gershenfeld N., and Vasseur J. As Objects Go Online; The Promise (and Pitfalls) of the Internet of Things. Foreign Aff. 93; 2014.

Jara A., Lopez P., Fernandez D., Castillo J., Zamora M., and Skarmeta A. Mobile Digcovery: Discovering and Interacting with the World Through the Internet of Things. *Journal Personal and Ubiquitous Computing* **18** (2): 323–338; 2014.

Jara A., Zamora M., and Skarmeta A. GLoWBAL IP: An Adaptive and Transparent IPv6 Integration in the Internet of Things, Mobile Information Systems. *Mobile Information Systems* **8** (3): 177–197; 2012.

Shelby Z., and Chauvenet C. The IPSO application framework, draft-ipso-app-framework-04. IPSO Alliance: Interop Committee; 2012.

Smith M., Reilly K., and Benkler Y. *Open Development: Networked Innovations in International Development.* MIT Press, Cambridge, MA; January 10, 2014.

Soldatos J., Kefalakis N., Serrano M., and Hauswirth M. Design Principles for Utility-Driven Services and Cloud-Based Computing Modelling for the Internet of Things. *International Journal of Web and Grid Services* **10** (2–3):139–167; 2014.

Tarkoma S., and Ailisto H. The Internet of Things Program: The Finnish Perspective. *IEEE Communications Magazine*, **51** (3):10–11; 2013.

Tian L. Lightweight M2M (OMA LWM2M), OMA device management working group (OMA DM WG). Open Mobile Alliance (OMA); 2012.

9

MEMS: AN ENABLING TECHNOLOGY FOR THE INTERNET OF THINGS (IoT)

MICHAEL A. HUFF

MEMS and Nanotechnology Exchange (MNX), Corporation for National Research Initiatives, Reston, VA, USA

The *Internet of Things* (*IoT*) is defined as a collection of physical objects (i.e., "things") and their interconnected communication networks that allow the physical objects to gather, store, process, and exchange information. Importantly, the physical objects can be almost anything, from the smallest devices or products to the largest systems. In the most general form, the IoT is a world where everything and everyone is connected together (Figure 9.1). Additionally, the physical objects may also make decisions about the amassed, processed, and exchanged information, as well as take actions to control the physical objects and the environment in which they are embedded. The capabilities that enable the physical objects to participate in the IoT are usually composed of an assemblage of different types of advanced technologies including electronics, sensors, actuators, and software. These capabilities are either connected to or integrated into conventional products and systems, such as vehicles; appliances; heating, ventilation, and air conditioning systems (HVAC); consumer electronics; entertainment systems; security systems; power generators; medical devices; sports and recreation equipment; commercial building controls; tools; industrial manufacturing equipment; health monitoring devices; etc.

There are several important elements that are needed in the realization of the IoT. The first are the communication networks that enable devices, perhaps using different operating systems, to communicate with one another. This element leverages the existing Internet standards of the TCP/IP protocol suite. The second is the enormous and inexpensive information storage and processing power available in

Internet of Things and Data Analytics Handbook, First Edition. Edited by Hwaiyu Geng.

Companion website: www.wiley.com/go/Geng/iot_data_analytics_handbook/

FIGURE 9.1 An illustration of the *Internet of Things* (IoT) showing a myriad of physical objects represented as devices, products, and systems (including common household and business appliances) that are all connected together and where information can be gathered by the devices and then stored, processed, and accessed from the cloud. Reproduced with permission from MEMS and Nanotechnology Exchange.

modern integrated circuits. The third are inexpensive and unobtrusive devices that can sense and actuate to control the things and/or environment in which the things are embedded. Microminiature devices that can sense and actuate are commonly called "MEMS[1]" in the semiconductor industry, which is an acronym for "Micro-ElectroMechanical Systems." MEMS is an emerging technology that is an important enabler for the IoT and is the focus of this chapter.

9.1 THE ABILITY TO SENSE, ACTUATE, AND CONTROL

As noted earlier, a key element of the realization of the IoT is the ability to *sense, actuate, and control.* The ability to sense is enabled through the use of sensors, and the ability to control is enabled through the use of both sensors and actuators, combined with decision-making capabilities enabled by integrated circuit (IC) devices such as microcontrollers and microprocessors.

Sense and Actuate

Sensors are devices whose purpose is to monitor some physical parameter of interest (e.g., temperature, pressure, force, etc.) and provide a suitable output signal that

[1] Sometimes MEMS are also referred to as "microsystems."

is in the form of information that is an accurate representation of that parameter. In general, sensors are a type of transducer, which is a device that converts energy from one form to another; often the sensor measures a mechanical, electrical, magnetic, chemical, biological, optical, or other parameter (i.e., in one form of energy) and converts it into an electrical signal (a different form of energy). Having the output of the transducer in the form of an electrical signal is convenient since it can be directly inputted into other electronic devices that act as information storage or communication devices or decision-making devices or systems, such as microcontrollers or microprocessors.

Actuators, like sensors, are another form of transducer device. Actuators take an energy input, usually in the form of an electrical signal, and usually convert this energy into a mechanical physical motion. This physical motion can be used to undertake actions to control physical objects and potentially modify the environments in which they are located.

Control

The concept of control is slightly more complicated. Control involves several elements including (Figure 9.2) the knowledge of a desired state of a system (also sometimes called the "set point" or "target state"), the ability to actively determine the current state of the system, and the ability to direct and cause the system to move toward the desired state. The control system as described previously is often called a *control loop*. Automatic control is when this process is performed without human intervention. Control loops are used in vast numbers of products and systems. Control systems usually involve one or more sensors to measure the state of a system as well as one or more actuators to direct the system to the desired state. Both sensors and actuators have been around for a very long time. However, traditionally they have been only available as discrete components that are relatively large, consume significant levels of power, and are relatively expensive. However, the recent advent of MEMS technology has resulted in a revolution in the implementation of sensors and actuators as explained in the following text.

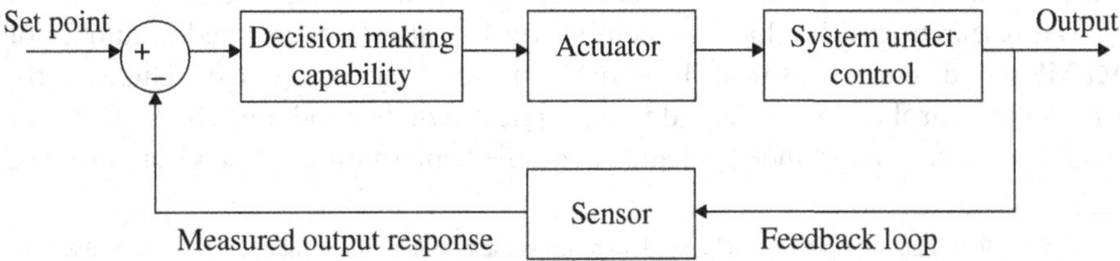

FIGURE 9.2 An illustration of a control loop that employs a decision-making capability that based on the difference between the set point and the measured output directs an actuator to make a physical motion; an actuator that makes a physical motion; a system under control that based on the motion of the actuator changes its state; and a sensor that measures the output parameter of the system and converts it into a form that can be compared to the set point. Reproduced with permission from MEMS and Nanotechnology Exchange.

9.2 WHAT ARE MEMS?

MEMS is a technology that in its most general form can be defined as microminiaturized electromechanical devices that are made using the techniques of micro- and nanofabrication.[2] That is, MEMS are made using many of the same manufacturing technologies as used in the production of ICs. However, instead of being purely electronic in function as are ICs, MEMS typically have some type of electromechanical functionality. It is this electromechanical functionality that enables MEMS technology to be used for the implementation of the most common device types of MEMS, which are microminiaturized sensors and actuators.

Over the past two decades, researchers and developers have demonstrated an extremely large diversity and number of MEMS-based sensors for almost every possible sensing application. Examples of the types of MEMS sensors reported to date have included temperature, pressure, inertial forces, tactile forces, chemical and biologic species, sound, magnetic fields, radiation in several bands of the electromagnetic spectrum, proximity, fluid level, flow rate, seismic, and many more [1–4]. Importantly, many of these MEMS sensors have demonstrated performances far exceeding those of their discrete large-dimensional-scale device counterparts.

Not only is the performance of MEMS sensors exceptional, but also their method of manufacturing allows them to be produced in high volume with exceptionally low cost levels. This is because MEMS manufacturing leverages the same *batch fabrication* techniques used in the IC industry whereby hundreds to thousands of individual devices are fabricated simultaneously on each substrate and each manufacturing lot contains a number of substrates that are processed together (i.e., as a batch). This method of manufacturing translates into low per-device production costs, similar to what is seen in the IC industry where the number of transistors per microprocessor approximately doubles every 2 years (this is known as "Moore's Law") and the quality-adjusted price improvement of microprocessors halves every 2 years[3] [5, 6].

Consequently, it is possible to not only obtain exceptional device performance, but this high level of performance can be obtained at relatively low cost levels. Not surprisingly, the pace of commercially exploiting MEMS sensors has been accelerating, and the markets for these sensors is growing at a very rapid rate.

The research and development community has also demonstrated a number of MEMS-based actuators as well. The MEMS-based actuators reported include microvalves for control of gas and liquid flows; optical switches and mirrors to redirect or modulate light beams; independently controlled micromirror arrays for displays;

---

[2] As noted before, nanofabrication can be used in the implementation of microminiaturized electromechanical devices. Often electromechanical devices having nanodimensional critical features are called nanoelectromechanical systems (NEMS). For the purposes of our discussion, we will only use the term MEMS to refer to both since MEMS is the term most commonly used in the industry.

[3] The rate of quality-adjusted microprocessor price improvement likewise varies and is not linear on a log scale. Microprocessor price improvement accelerated during the late 1990s, reaching 60% per year (halving every 9 months) versus the typical 30% improvement rate (halving every 2 years) during the years earlier and later.

microresonators for a number of different applications such as communication filters and chemical sensors; micropumps to develop positive fluid pressures for drug delivery and other fluidic applications; microflaps to modulate airstreams on the airfoils of airplane wings; radio-frequency (RF) switches; relays; as well as many others [3]. While these MEMS actuators are extremely small, remarkably they often have been able to produce consequences on much larger dimensional scales, even at our size scales. That is to say, these microscopic-sized actuators can perform mechanical feats far larger than their microminiature sizes would suggest [7].

The true potential of MEMS technology really only begins to become fully realized when MEMS sensors and actuators of any conceivable type are merged with ICs onto the same microchip substrate. This is a compelling paradigm shift in technology systems. The sensors are able to monitor parameters of interest in the environment and provide this sensory information to the electronics; the electronics communicate with other devices and the cloud, process this sensory information, and make decisions about how to influence the environment; and the actuators act to influence the environment. This represents the major components of the control loop concept we described in the preceding text, but enabled by tiny microscopic devices potentially all on a single substrate that can be attached or integrated into anything, and having remarkable levels of functionality combined with exceptional performance and low cost.

By bringing together the computational capability of microelectronics with the perception and control capabilities of microminiaturized sensors and actuators, MEMS technologies are enabling *smart systems on a chip* to be mass-produced. The use of smart systems that can actively and autonomously *sense and control* their environments has far-reaching implications for a tremendous number of applications.

For example, MEMS technologies have become some of the most important advanced technologies in the automotive safety markets. Initially, the first MEMS sensors for safety on automobiles were the crash airbag sensors. These devices were smaller, were lower in cost, and provided much higher performance levels compared to the macroscale technologies they replaced (the older technology was a metal ball held with a tube attached to a magnet, and the impact of the car would cause the ball to overcome the magnetic attractive forces allowing the metal ball to travel down the tube to close a switch to initiate the charge to allow the airbag to deploy).

The MEMS crash airbag sensing devices were lower in cost, smaller, and higher in performance than the technology they replaced. Importantly, the previous crash airbag technology was so expensive that it was only deployed on very expensive vehicles (it was not found on moderately priced cars and trucks) and typically only used to provide protection to the driver and maybe the front seat occupant.

As a result of the lower-cost MEMS airbag sensors, crash airbag technology quickly expanded to be placed on nearly every vehicle. Additionally, the use of these devices has expanded from protection of only the driver in frontal crashes on very expensive vehicles to the protection of all of the passengers in front, side, rear, and rollover accidents on every vehicle sold.

Subsequent advancements in MEMS technology enabled vehicles to progress from merely protecting the occupants in a crash to being able to avoid crashes altogether. These MEMS devices have included steering stabilization sensors; tire air pressure sensors; brake pressure sensors; rollover protection sensors; sensors to detect other vehicles, pedestrians, and potential obstacles; detection sensors and warning systems for nonalert drivers; and many more.

Now the automakers are moving toward driverless vehicles wherein a number of MEMS are being deployed on vehicles to provide sufficient information to the control system of the vehicle to correctly and safely navigate as well as avoid potential problems and obstacles. Undoubtedly, the automobile industry will be more and more dependent on MEMS technologies in the future. Figure 9.3 shows some of the sensors commonly used on vehicles at the present time.

Despite all these revolutionary changes in vehicle technologies enabled by MEMS devices, the auto industry has yet to fully introduce vehicles that encompass the IoT and embrace the importance of MEMS to the implementation of IoT. Therefore, it is expected that MEMS devices will have an even larger role to play in vehicle technologies as the concept of IoT rolls out into the wider markets in the future.

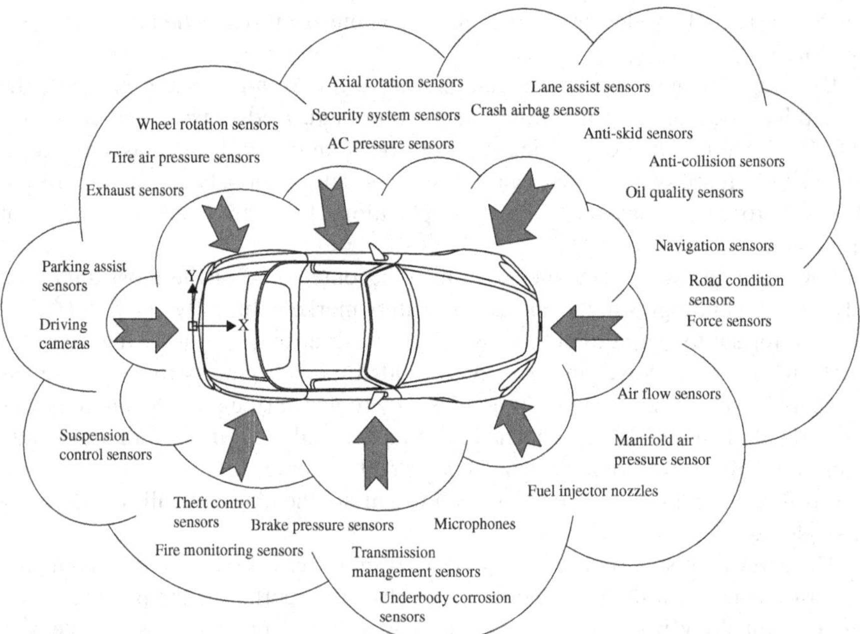

FIGURE 9.3 Illustration of some of the variety of sensors being deployed on vehicles. Automakers originally only employed MEMS-based manifold air pressure sensors and Hall effect sensors early on, but the number and diversity of MEMS sensors have radically increased over the past decade and will continue to increase as consumers demand increased safety, drivability, and reliability. The dramatic growth of MEMS sensors on vehicles is expected to significantly increase with the future introduction of autonomous vehicle technology. Reproduced with permission from MEMS and Nanotechnology Exchange.

Products of every type are increasingly dependent on a large diversity of different MEMS sensor and actuator technologies to successfully control and operate these products as well as perceive the user and their surrounding environments. Increasingly, MEMS technologies are being exploited in more and more products due to the advantages of their extremely small size and weight, ability to have extremely high levels of functionality, high levels of reliability, low cost, and reduced power consumption.

With MEMS technology every conceivable product and system can be outfitted with any type of MEMS device, thereby enabling every product and system to become smart wherein it constantly is monitoring any and all parameters of interest to the user, processes this information, and, based on desired or optimal conditions, can perform functions to alter or modify these parameters to the desired or optimal state. Therefore, MEMS devices are a key element to enable the *Internet of Everything.*

9.3 MEMS AS AN ENABLING TECHNOLOGY FOR THE IoT

As discussed in the preceding text, MEMS is an enabling technology for the IoT because MEMS manufacturing makes possible small, low-cost, high-performance sensors and actuators. Nevertheless, there are other unique and important benefits that can be derived from MEMS that have not been discussed that are very beneficial and impactful for the deployment and advancement of the IoT. This section discussed some of these less obvious but immensely useful and valuable aspects of MEMS technology for the IoT.

Humans sense and interact with their environments and the world through their major senses. The most commonly cited human senses include vision, hearing, touch, smell, and taste. These sensing capabilities are the result of millions of years of evolution and have allowed humans to build relatively technically advanced, safe, healthy, comfortable, and wealthy societies.

While human senses have evolved to be quite good at specific functional capabilities, they are very limited compared to MEMS sensor technologies. For example, we are limited in the number of things going on simultaneously that we can pay attention to at any moment of time. That is, when confronted with monitoring multiple stimuli going on at the same time, humans can easily suffer sensory overload. The result is that some important sensory information may be overlooked completely or detected too late for an adequate response.

MEMS sensors on the other hand, particularly when coupled with enormous information processing power, do not get overwhelmed and can continuously and simultaneously monitor a very large number of important parameters of interest in the environment (and on the user) without suffering sensory overload situations. This makes for a safer, more productive, and rewarding environment.

Second, while humans can detect changes in some parameters in the environment, we tend to have difficulty with detecting modest rates of change in parameters over longer periods of time.

In contrast, MEMS sensors can continuously monitor parameters of interest to us over both short and long periods of time and analyze this collected data to give the user information about trends, warn the user about anomalies or unexpected events occurring, prevent failures and accidents, prevent disruptions and failures of activities and services, as well as maintain higher-quality and safer services.

Third, human senses can be limited by the amount of additional information (i.e., above and beyond the information provided by the senses) that is available and can be received and processed along with the sensory information for decision-making purposes. This means that we may miss important clues from the environment that would otherwise assist us.

In comparison, MEMS sensors and their associated vast processing power can continuously exchange information with the environment and thereby give us increased contextual awareness and perceptive capabilities about our environments. For example, when we walk into a local environment, the MEMS sensors can notify or alert us to opportunities and dangers in that environment that we might be oblivious to otherwise.

Fourth, most of the human senses have limited levels of sensitivities. Humans have a limited sense of smell compared to some animals such as dogs. It is well known that scent hounds have a sense of smell sensitivity orders of magnitude higher than that of humans. For example, the bloodhound dog has a sense of smell that is reportedly 10–100 million times more sensitive than a person. And bears, such as the silvertip grizzly bear, have a sense of smell seven times more sensitive than even the bloodhound.

MEMS sensors, on the other hand, can have sensitivities far higher than any human, or even some animals, thereby enabling the ability to detect far smaller levels of an olfactory parameter, or any other parameter, than otherwise would be possible.

Fifth, most of the human senses also have limited dynamic ranges. For example, the human ear (before age-related degradation or noise-related hearing loss) can detect sound in the frequency band from about 20 and 20,000 Hz, but we mostly cannot hear anything outside this limited spectrum. However, some animals have evolved the capabilities to hear sound waves far outside the human bandwidth range. For instance, dogs can detect sound from about 60 to over 60,000 Hz. Some other animals have even larger ranges; the porpoise marine animal has a range from about 75 to over 150,000 Hz, and bats have the ability to determine their relative location by detection and processing of reflected sound signals within a dynamic range from about 10,000 to over 200,000 Hz.

Similarly, while the human eye can detect reflected light from objects within the optical wavelengths from about 400 to 700 nm, we cannot see anything outside this range, specifically in the infrared or ultraviolet parts of the spectrum as well as other electromagnetic bands. In fact, the part of the spectrum visible to humans is quite a tiny portion of the electromagnetic spectrum. Some animals have evolved the ability to see in the visible as well as the infrared and ultraviolet spectrums. For example, snakes can detect heat (i.e., infrared signals) from their prey. Arctic caribou, bees, and other animals have the ability to see in the ultraviolet. The caribou using its ultraviolet vision has the ability to detect wolves that are virtually invisible above 400 nm in the arctic winter landscape. Bees use their ultraviolet vision to find nectar in flowers.

MEMS sensors can be engineered to have dynamic ranges far exceeding that of humans and even exceeding that of animals as well. For example, MEMS acoustical sensors can be engineered and deployed to measure virtually any part of the acoustical spectrum. It is well known that rotating machinery will emit a very-high-frequency acoustical signal if the bearings are just beginning to wear. While a human cannot hear this phenomenon, a MEMS sensor can detect sound waves at these frequencies and provide an alert to check the system before a catastrophic event were to occur. Additionally, MEMS sensors have been reported that can "see" in many of the portions of the spectrum nonvisible to humans, including far outside the known detectable spectrum of any animal.

Sixth, humans do not have sensing capabilities for many parameters of interest. For example, some animals, such as sharks and dolphins, can detect changes in nearby electrical fields, while no humans have any known ability to sense this parameter.

In contrast, MEMS sensors that can sense almost any known parameter have been developed and reported in the literature.

Seventh, humans and even the most capably trained animals fatigue in sensing duties after some relatively short period of time. However, MEMS sensors as inanimate devices can operate almost indefinitely without tiring.

Obviously, these are extremely powerful benefits that can provide considerably more capability for the IoT than is currently possible using human senses alone, or large-scale discrete sensor devices.

9.4 MEMS MANUFACTURING TECHNIQUES

MEMS manufacturing uses many of the same fabrication processes that are used in the IC industry (e.g., photolithography, oxidation, diffusion, ion implantation, LPCVD, sputtering, etc.) and combines these fabrication methods with specialized fabrication processes that are often collectively called "micromachining" processes. In this section, we briefly review some of the more widely known and commonly used MEMS fabrication processes. We will highlight a small number of the most popular methods of micromachining in this chapter. Readers interested in a more comprehensive discussion of MEMS fabrication techniques and manufacturing methods, including the challenges of custom manufacturing process development, are referred to [2], and readers interested in material covering conventional microelectronics manufacturing technologies are referred to [8].

9.4.1 Wet Chemical Bulk Silicon Micromachining

One of the first technologies specifically developed for MEMS manufacturing is known as *wet chemical bulk silicon micromachining* and involves the selective removal of the silicon substrate material to implement MEMS devices. The ability to

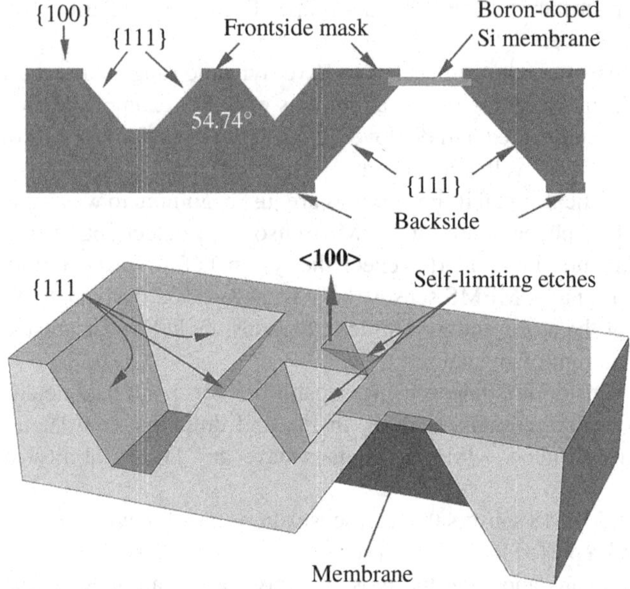

FIGURE 9.4 Illustration of shape of the etch profiles of a (100) oriented silicon substrate after immersion in an anisotropic wet etchant solution. Copyright CNRI/MNX, used with permission.

delineate the different crystal planes of the silicon lattice in wet chemical etching provides the ability to form three-dimensional features in silicon substrates with a reasonable level of dimensional control. Figure 9.4 shows an illustration of some of the shapes that are possible using anisotropic wet etching of a <100> oriented silicon substrate.

9.4.2 Deep Reactive Ion Etching Bulk Micromachining

Deep reactive ion etching (DRIE) is a highly anisotropic plasma etch process that can be used to etch very deep features into silicon with high aspect ratios. It was first introduced in the mid-1990s and has been widely adopted by the MEMS community [9]. The sidewalls of the etched features are nearly vertical, and the depth of the etch can be tens of microns, hundreds of microns, or even completely through the entire silicon substrate. The etch is a dry plasma etch and uses a high-density plasma to repeatedly alternate between an etch cycle of the silicon and a deposit cycle wherein an etch resistant polymer layer is deposited on the sidewalls. The protective polymer layer is deposited on the sidewalls as well as the bottom of the etch pit, but the anisotropy of the etch removes the polymer at the bottom of the etch pit faster than the polymer being removed from the sidewalls. Figure 9.5 shows a cross-sectional SEM of a silicon microstructure fabricated using DRIE

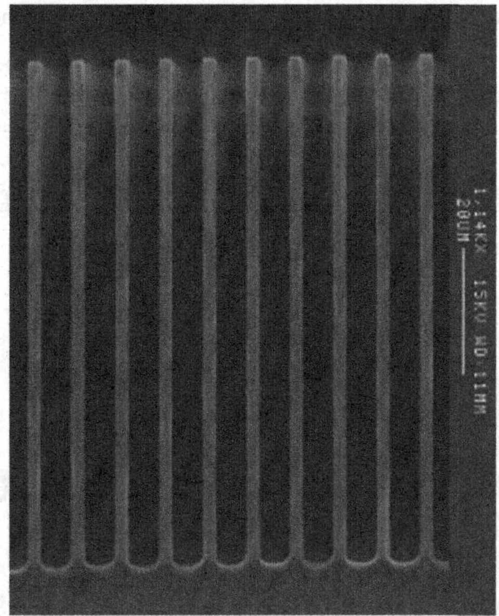

FIGURE 9.5 SEM of the cross section of a silicon wafer demonstrating high aspect ratio and deep trenches that can be fabricated using DRIE technology. Copyright CNRI/MNX, used with permission.

technology. As can be seen, the etch is very deep into the silicon substrate and the sidewalls are nearly vertical.

9.4.3 Surface Micromachining

Surface micromachining is another very popular technology for the fabrication of MEMS devices. Surface micromachining involves a sequence of steps starting with the deposition of some thin-film material to act as a temporary sacrificial layer onto which the actual device layers are built, followed by the deposition and patterning of the thin-film device layer of material which is referred to as the structural layer and followed by the removal of the temporary sacrificial layer to release the mechanical structural layer from the constraint of the underlying sacrificial layer, thereby allowing the structural layer to move [10]. An illustration of a surface micromachining process is given in Figure 9.6, wherein an oxide layer is deposited and patterned. This oxide layer is temporary and is commonly referred to as the sacrificial layer. Subsequently, a thin-film layer of polysilicon is deposited and patterned and this layer is the structural mechanical layer. Lastly, the temporary sacrificial layer is removed and the polysilicon layer is now free to move as a cantilever. Figure 9.7 shows an SEM of a polysilicon microresonator structure made using surface micromachining.

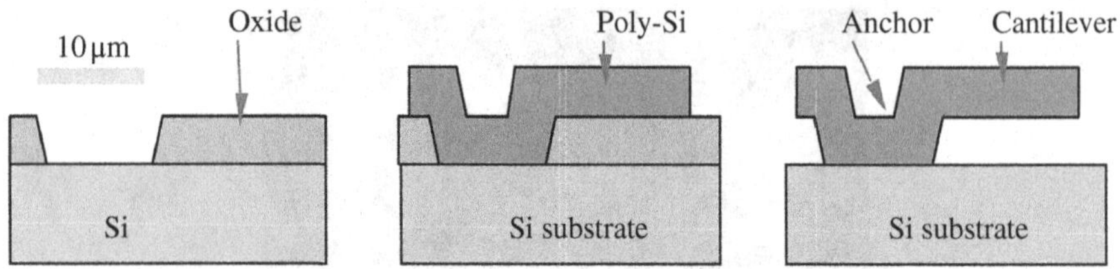

FIGURE 9.6 Illustration of a surface micromachining process. Copyright CNRI/MNX, used with permission.

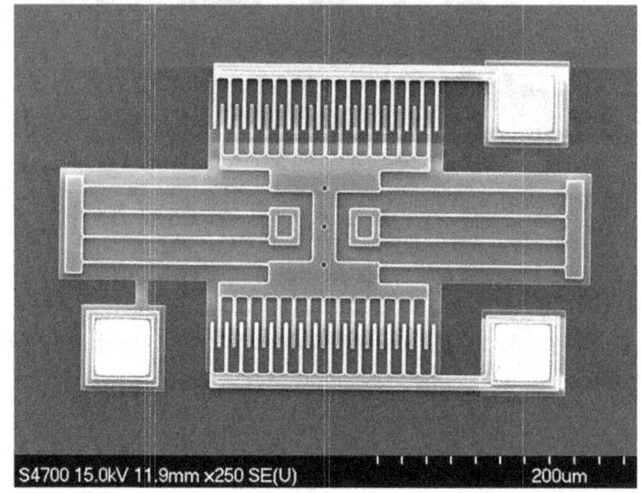

FIGURE 9.7 Polysilicon resonator structure fabricated using a surface micromachining process. Copyright CNRI/MNX, used with permission.

9.4.4 Other Micromachining Technologies

In addition to wet chemical bulk micromachining, DRIE bulk micromachining, and surface micromachining, there are a number of other techniques used to fabricate MEMS devices, including wafer bonding, XeF2 dry-phase isotropic silicon etching, LIGA, electrodischarge micromachining, laser micromachining, focused ion beam micromachining, and others. The reader is referred to [1–4] for an exhaustive catalog of MEMS fabrication methods.

9.5 EXAMPLES OF MEMS SENSORS

As noted earlier, sensors are a type of transducer that converts a form of energy (that represents a parameter of interest) into another form of energy. Over the recent past almost every imaginable type of MEMS sensor has been demonstrated including pressure, acoustic, temperature (including infrared focal plane arrays), inertia (including acceleration and rate rotation sensors), magnetic field (Hall, magnetoresistive, and magnetotransistors), force (including tactile), strain, optical, radiation, and chemical and biological sensors [1]. In this section we review a few selected MEMS sensor devices that have been developed successfully for the commercial market. There are far too many types of MEMS devices to provide a comprehensive review of all the MEMS sensors developed, and therefore we will only review a few selected examples. The reader is referred to [1–4] for more information.

9.5.1 MEMS Integrated Piezoresistive Pressure Sensor by Freescale

Piezoresistivity is one of the oldest and most common material properties used for the implementation of transduction in MEMS sensors. A piezoresistive material is a material wherein an applied mechanical strain to the material results in a change in the resistance across the material. This material property has been most widely used in MEMS pressure sensors.

A notable example of a MEMS pressure sensor that employs the piezoresistive effect is the *Integrated Pressure Sensor* (IPS) process technology that was originally developed and put into production by Motorola (now Freescale Semiconductor) and represents one of the most successful high-volume MEMS products (Figure 9.8). The manufacturing of this device employs bulk micromachining to make a thin pressure-sensitive diaphragm onto which the piezoresistive strain sensors are positioned (Figure 9.9). The sensor employs the piezoresistive effect to measure the mechanical deflection of a thin silicon membrane and combines bipolar microelectronics for signal conditioning and calibration on the same silicon substrate as the sensor device. Freescale employs an electrochemical etch stop to precisely control the pressure-sensing membrane thickness [11].

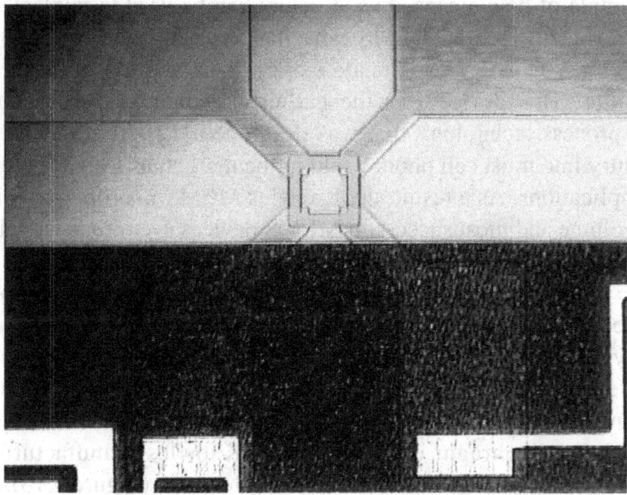

FIGURE 9.8 An overhead optical photograph of the MEMS integrated pressure sensor device that employs a piezoresistor configuration. Reprinted with permission, Copyright Freescale Semiconductor Inc.

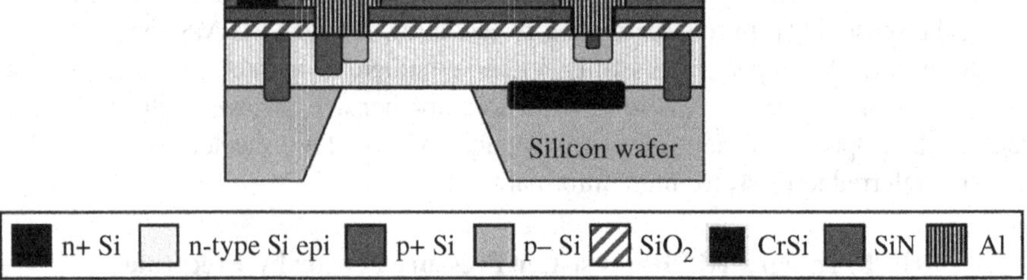

FIGURE 9.9 Cross-sectional illustration of the Freescale Pressure Sensor. The materials used in the fabrication of this device are given in the legend shown earlier. Reproduced with permission from MEMS and Nanotechnology Exchange.

9.5.2 MEMS Capacitance-Based Microphone Sensor by Knowles

Capacitive sensing is very commonly utilized in MEMS sensors due to its inherent simplicity and high sensitivity. In general, the capacitance, *C*, of a two terminal device is given by

$$C = (\varepsilon_0 \varepsilon_r A) / d(\text{Farads}),$$

where ε_0 is the dielectric constant of free space, ε_r is the relative dielectric constant of any material between the electrodes, *A* is the area of the capacitor, and *d* is the separation of the electrodes. Capacitors can be used as sensors in several different ways with the most common method varying the distance between electrodes.

As an example of how a capacitive structure can be used to implement a microphone device, we consider the MEMS capacitance-based microphone sensor developed by Knowles. This device is made using a combination of surface and bulk micromachining. This device was the earliest commercially successful MEMS microphone process technology that was developed [12]. This device has gained successful entry into most cell phones sold in the market, as well as other consumer electronic applications. As a result, the Knowles MEMS microphone is now one of the highest-volume and most successful MEMS devices ever produced with volumes in excess of five billion components.

The basic design requirement for a microphone is the construction of a low mass and mechanically compliant diaphragm offsetting a short distance from a mechanically rigid "backplate." The diaphragm and backplate form the electrodes of a variable capacitance-type microphone. The combination of the flexible diaphragm and the stiff backplate makes a variable capacitor whose capacitance is a function of the diaphragm deflection. The Knowles manufacturing process fabricates the diaphragm and backplate on a single wafer (Figure 9.10). The sensor is assembled, along with an associated readout ASIC, in a package. Figure 9.11 shows an optical photograph of the Knowles microphone looking downward on the device.

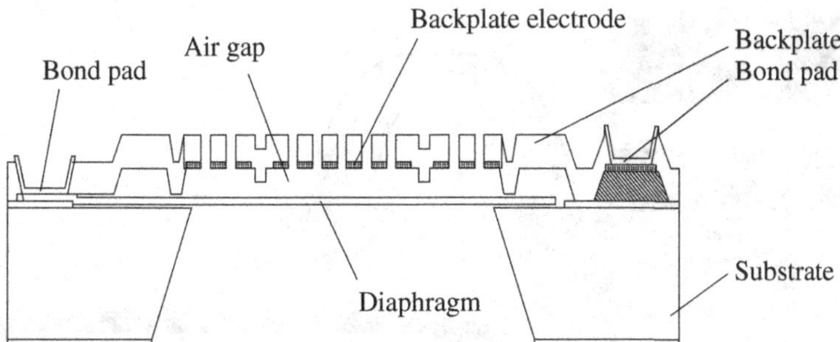

FIGURE 9.10 Cross-sectional diagram of the Knowles, Inc. microphone sensor structure. Reprinted with permission. Copyright Knowles Inc.

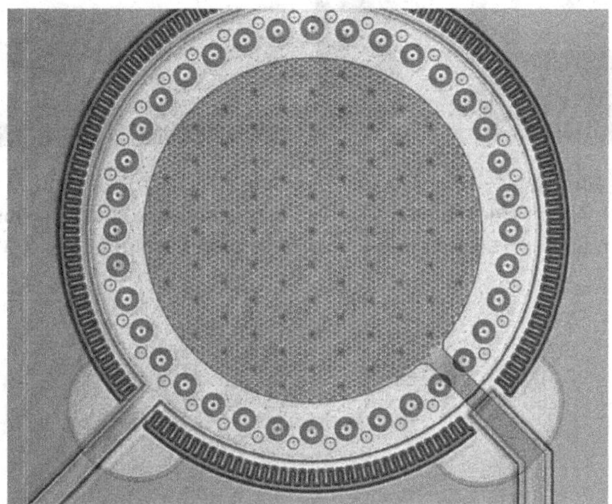

FIGURE 9.11 Top-down optical micrograph of the Knowles, Inc. microphone sensor. Reprinted with permission. Copyright Knowles Inc.

9.5.3 MEMS Capacitance-Based Accelerometer by STMicroelectronics

Another example of capacitive-based transduction used for implementation of a MEMS sensor is the STMicroelectronics accelerometer product series. STMicroelectronics, Inc. has very quickly captured a significant portion of the consumer electronics MEMS-based inertial sensor market by introducing a line of accelerometer devices that have excellent performance levels and are sold at an extremely attractive price: this is an important requirement for most consumer applications [13].

This strategy has been very successful and has allowed STMicroelectronics to rapidly become one of the largest producers of MEMS devices. STMicroelectronics' MEMS devices are currently used in such notable and highly recognizable consumer products such as Nintendo's Wii, Apple's iPhones (Figure 9.12) and iPod Touch, and others [14].

The STMicroelectronics MEMS inertial products are based on their Thick Epitaxial Layer for Microactuators and Accelerometers (THELMA) process technology [15].

FIGURE 9.12 Photographs of products employing STMicroelectronics' inertial MEMS sensor technology that provide the customer with increasing capabilities in consumer products, in this case for properly orientating the screen to the user. Copyright CNRI/MNX, used with permission.

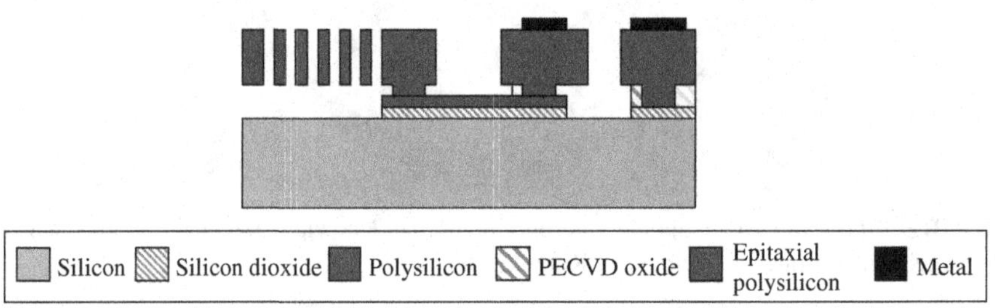

FIGURE 9.13 Cross-sectional illustration of MEMS device structure on substrate made using the STMicroelectronics' THELMA process technology that is used to manufacture their line of inertial sensors. Reproduced with permission from MEMS and Nanotechnology Exchange.

The THELMA process is a nonintegrated MEMS process technology that has the distinct advantage of allowing thicker structures to be implemented, which is extremely useful for capacitive-based inertial sensors. The THELMA process technology is used to implement capacitive-based inertial devices but is sufficiently flexible to be used for the production of other MEMS device types.

The THELMA process uses a thick layer of polysilicon deposited using epitaxial deposition as the structural material for the accelerometer proof mass. The thick polysilicon layer is patterned and etched using DRIE. The DRIE allows very high aspect ratio structures to be made in the thick polysilicon layer (Figure 9.13).

9.6 EXAMPLE OF MEMS ACTUATOR

9.6.1 MEMS Electrostatically Actuated Digital Light Processor (DLP) by Texas Instruments

A number of different basic physical principals are used to implement MEMS actuators including electrostatic, piezoelectric, magnetic, magnetostrictive, bimetallic, shape memory alloy, and others. Each of these has its respective advantages and disadvantages, and therefore careful consideration of the specific application requirements must be part of the selection process. We shall review one of the most popular methods for MEMS actuators called electrostatic actuation in this section. The reader is referred to [1–4] for a more comprehensive review of MEMS actuators.

Digital Light Processor (DLP™) technology developed by Texas Instruments Corporation is used in various large-volume commercial markets such as projection televisions and projection display systems. DLP technology is a disruptive technology for movie theaters since it replaces the 100-year-old celluloid film with a completely digital format. Already, DLP is currently used in more than 100,000 theaters. The key component of DLP® technology is the micromirror array chip (called the Digital Micromirror Device (DMD) [16]). Essentially the DMD chip is a large array of individually and digitally controlled MEMS micromirrors—there are up to millions of mirrors in each chip array. The mechanism for actuation of each mirror is electrostatic actuation.

Each mirror in the array measures just over 10 μm by 10 μm and is electrostatically actuated by the microelectronics physically located underneath the mirror array [17]. This enables the fill factor of the DMD to reach levels of over 90%, thereby allowing very high optical efficiency and contrast. Integrated microelectronics is a necessity in DLP technology since the number, density, and size of the mirrors combined with individual addressability would preclude having all the electronics off-chip. The DMD fabrication is made using a surface micromachining process on top of a microelectronics substrate [18].

Figure 9.14 shows an SEM of a portion of the DMD device showing the center pixel in the actuated state and the surrounding pixels in the unactuated state. For a description of how DMDs are used in projection optical systems, see [18]. The DMD pixel consists of an SRAM cell fabricated in the silicon substrate with a MEMS superstructure implemented on top (Figure 9.15). The DMD is made using a low-temperature, surface micromachining, MEMS-last, monolithic, integrated MEMS process technology.

9.7 THE FUTURE OF MEMS FOR THE IoT

As discussed in the preceding text, MEMS devices as they are currently embodied are being used to implement the IoT. However, it is important to point out that the MEMS devices that are currently available on the market are mostly either discrete devices, either an individual sensor or an actuator, or an array of the same device type

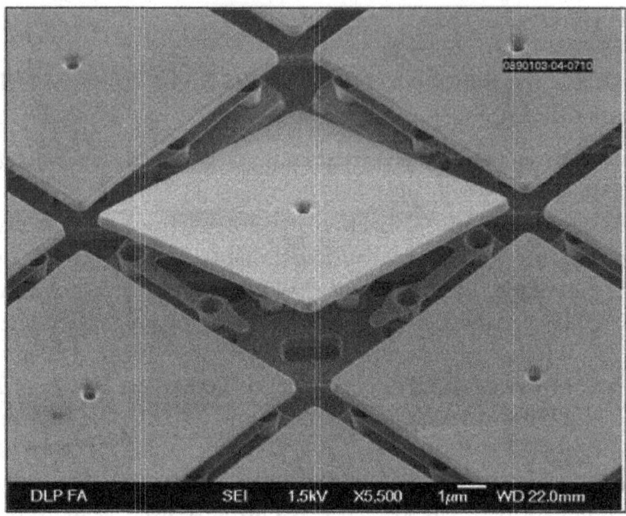

FIGURE 9.14 Magnified SEM image of Texas Instrument's DMD device with center pixel in actuated (i.e., rotated) state. Reprinted with permission, Copyright Texas Instruments, Inc.

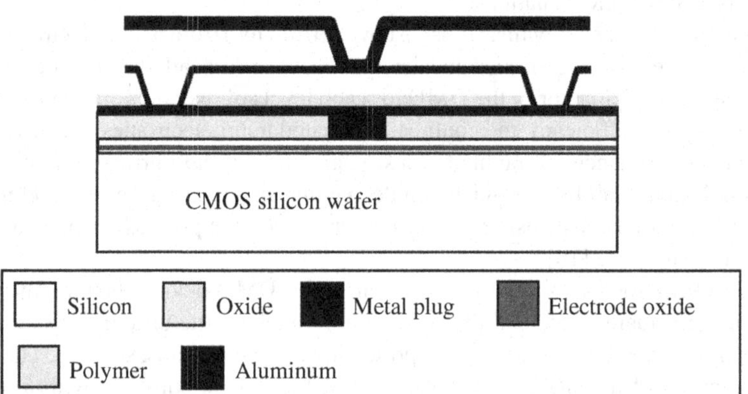

FIGURE 9.15 Cross section of one pixel of the Texas Instruments DMD. Reproduced with permission from MEMS and Nanotechnology Exchange.

replicated many times over the surface of a substrate. That is, the MEMS devices currently available do not combine any type of sensor, actuator, and IC all on the same substrate. Nevertheless, as MEMS technology matures, specifically as the manufacturing methods continue to develop, it is expected that combining different MEMS devices onto a single die will become not only feasible but also desirable and cost effective. Moreover, this ability will provide an enormous catalyst to the further development and proliferation of the IoT.

9.8 CONCLUSION

MEMS are revolutionizing the implementation of sensors, actuators, and control systems through miniaturization, batch fabrication, and integration with electronics. MEMS technology is enabling smart systems on a chip with high levels of functionality, performance, and reliability to be available in a small microsized chip and at very low cost levels. Presently, the largest market drivers for MEMS industry include silicon-based pressure sensors for automotive, medical, and industrial control applications; crash airbag inertial sensors for automotive applications; inertial sensors for consumer electronics; microphones for cell phones and computers; DLP for displays and projectors; and ink-jet cartridges for printers. In the industrial, commercial, medical, and defense sectors, MEMS devices are already emerging as product performance differentiators in numerous applications and markets. Because of these success stories and the applicability of this technology in so many other products, the market potential for MEMS is very promising. Nevertheless, since MEMS is a nascent and highly synergistic technology, it is expected that many new applications will emerge, thereby expanding the markets beyond that which are currently known or identified, particularly as the IoT develops and matures. In short, MEMS is a new and extremely important technology that has a very promising future. The diversity, economic importance, and extent of potential applications of MEMS for the implementation of the IoT make it the hallmark technology of the future.

REFERENCES

[1] Madou, M., *Fundamentals of Microfabrication*, CRC Press, Boca Raton, FL, 1997.

[2] Huff, M.A., Bart, S.F., and Lin, P., MEMS Process Integration, Chapter 14 of the *MEMS Materials and Processing Handbook*, editors R. Ghodssi and P. Lin, Springer Press, New York, 2012.

[3] Huff, M.A., Fundamentals of Microelectromechanical Systems, Chapter 23 of the *Semiconductor Manufacturing Handbook*, editor H. Geng, McGraw-Hill, New York, 2005.

[4] Kovacs, G.T.A., *Micromachined Transducers Sourcebook*, McGraw-Hill, New York, 1998.

[5] Aizcorbe, A., "Why are semiconductor prices falling so fast?," U.S. Department of Commerce Bureau of Economic Analysis. Retrieved 2005.

[6] Liyang, S., "What are we paying for: A quality adjusted price index for laptop microprocessors," Wellesley College. Accessed July 11, 2014.

[7] Huff, M.A., Mettner, M.S., Lober, T.A., and Schmidt, M.A., "A Wafer-Bonded Electrostatically-Actuation Microvalve," Solid-State Sensor and Actuator Workshop, 4th Technical Digest IEEE, 1990.

[8] Jaeger, R.C., *Introduction to Microelectronic Fabrication: Volume 5 of Modular Series on Solid-State Devices*, 2nd Edition, Prentice Hall, Upper Saddle River, 2001.

[9] Larmar, F., and Schilp, P., "Method of Anisotropically Etching of Silicon," German Patent DE 4,241,045, 1994.

[10] Howe, R.T., and Muller, R.S., "Polycrystalline and amorphous silicon micromechanical beams: Annealing and mechanical properties," *Sensors and Actuators*, vol. **4**, p. 447, 1983.

[11] G. Bitko, A. McNeil, and R. Frank, "Improving the MEMS pressure sensor," *Sensors Magazine*, vol. **17**, no. 7, July 2000.

[12] Loeppert, P.V., and Sung, B.L., "SiSonic—The first commercialized MEMS microphone," Solid-State Sensors, Actuators, and Microsystems Workshop, Hilton Head Island, SC, June 4–8, 2006, pp. 27–30.

[13] Vigna, B., "MEMS Epiphany," MEMS 2009 Conference, Sorrento Italy, January 26, 2009.

[14] Source, iSuppli Corporation, See: http://www.isuppli.com. Accessed August 13, 2016.

[15] De Masi, B., and Zerbini, S., "Process builds more sensitive structures," *EE Times*, November 22, 2004.

[16] Hornbeck, L.J., "From cathode rays to digital micromirrors: A history of electronic projection display technology," *Texas Instruments Technical Journal*, vol. **15**, no. 3, 1998, pp. 7–46.

[17] Grimmett, J., and Huffman, J., "Advancements in DLP® Technology—The New 10.8 μm Pixel and Beyond," IDW/AD'05, Proceedings of the 12th International Display Workshops, (in conjunction with Asia Display 2005) Vol. **2**, pp. 1879–1882 (2005).

[18] Hornbeck, L.J., "Combining Digital Optical MEMS, CMOS, and Algorithms for Unique Display Solutions," IEEE International Electron Devices Meeting Technical Digest, Plenary Session, pp. 17–24 (2007).

OTHER INFORMATION

The reader is referred to three very popular additional sources of information concerning MEMS technology. The first source is a website called the MEMS and Nanotechnology Clearinghouse which is located at http://www.memsnet.org and is a general informational portal about MEMS technology and includes events, news announcements, directories of MEMS organizations, and a MEMS material database. The second source is the MEMS and Nanotechnology Exchange (MNX) located at http://www.mems-exchange.org. This website represents a large MEMS foundry network and offers MEMS design, fabrication, packaging, product development, and related services as well as considerable information about MEMS and nanotechnologies. Lastly, the reader is referred to several electronic discussion groups concerning MEMS technology that have very active participation from several thousand MEMS developers and researchers from around the world. These groups can be accessed through the following URL: http://www.memsnet.org/memstalk/archive.

10

ELECTRO-OPTICAL INFRARED SENSOR TECHNOLOGIES FOR THE INTERNET OF THINGS

Venkataraman Sundareswaran, Henry Yuan, Kai Song, Joseph Kimchi and Jih-Fen Lei
Teledyne Judson Technologies, Montgomeryville, PA, USA

10.1 INTRODUCTION

Sensors at the network edge are essential to most Internet of Things (IoT) devices. While there are a number of ways of sensing the environment using electromechanical, electrochemical, and chemical sensors, optical sensing plays a dominant role due to its versatility and accuracy. A large variety of interesting phenomena can be sensed with optical sensors. Some examples are temperature, gas, food quality, chemical species, humidity, moisture, liquid level, position, distance, displacement, velocity, acceleration, laser power, pressure, strain, and flow. Electro-Optical (EO) sensing denotes the process by which optical indicators are converted to electrical signals.

Simple EO devices are found everywhere in the form of photodiodes that sense a break in a beam of infrared (IR) light—as used in elevator and garage doors to prevent closing on a person or in a water faucet to sense the presence of a hand. Noncontact temperature monitoring, or Remote Thermal Sensing, is dominated by EO devices in industrial process control applications such as in steel, aluminum, glass, cement, plastic, and semiconductor processing. EO sensors provide the temperature feedback to control these processes. The EO sensors operate in the IR wavelength applicable to the process temperature being sensed: Near Infrared (NIR) for steel manufacturing processes, Short Wave Infrared (SWIR) for semiconductor

Internet of Things and Data Analytics Handbook, First Edition. Edited by Hwaiyu Geng.

Companion website: www.wiley.com/go/Geng/iot_data_analytics_handbook/

growth and processing, and Mid Wave Infrared (MWIR) for cement temperature monitoring are examples. Gas sensing using optical spectroscopy is a widely used approach, though electrochemical methods offer a competitive alternative. Many interesting gases are absorbed, and therefore optically detected, in the MWIR. Food quality inspection, which is a relatively new application, exploits numerous optical sensing phenomena such as transmission and reflection in SWIR. We will see later in more detail what the different IR wavebands are and how sensors in these bands are useful in a large number of sensing applications. While sensors in the visible range (400–700 nm) have some applicability in IoT devices, the real impact of EO sensors for IoT is in the IR range (>750 nm), and this chapter is focused on IR devices.

There are several key benefits to optical sensing:

- *Remote sensing*: When noncontact sensing, or remote sensing, is desired due to moving parts, very high process temperature, or harsh environment, optical sensing approaches offer the most flexibility. A simple example of remote sensing is the handheld laser IR thermometers (see Ref. [1]) that use built-in laser beam and a photodiode to measure temperature remotely. Commercial aircraft, helicopters, and satellites use remote sensing to observe ocean and land temperature, monitor crop moisture and health, survey ground mineral composition, monitor pipeline gas leaks, and determine oil seepage on the ocean surface.
- *Ease of digitization*: A key benefit of electro-optical sensing is the conversion of optical energy (photons) to electrical charge (electrons and holes). For example, materials used in typical optical IR sensor have specific structures that result in electron–hole pairs being generated when photons impinge upon them. The resulting electrical energy (called minority carriers) can be collected, amplified, digitized, and processed, before being transmitted.
- *No moving parts*: A typical optical sensor uses either passive lighting (object temperature sensing) or active lighting (an LED, laser, or other light source used along with the sensor), which illuminates the area/object of interest, with the resulting optical signal impinging upon the sensor generating an electrical signal. No moving parts are involved. Contrast this with electromechanical or chemical sensors, which involve motion of sensor components or chemical reagents. The absence of moving parts translates to lower power requirements and higher structural robustness and reliability.
- *Miniature size*: Optical sensing uses semiconductor photon detectors in sensing the incoming optical signal. Photodetector chips can be made in miniature size and can be potentially integrated with other integrated circuits (ICs) to form complex function analog or digital photodetector ICs.
- *High sensitivity*: IR optical sensing using photon detectors usually achieves the highest sensitivity of all sensors (typically 100–1,000 times more sensitive than IR thermal sensors). This natural characteristic makes optical sensing very useful in applications such as sensing low-concentration poisonous gases below the short-term exposure level or explosive gas below the low explosive level.

These benefits make optical sensing an ideal candidate for IoT applications, which often require digital miniature remote sensors with no moving parts.

However, there are some limitations to optical sensing:

- *Line of sight*: Optical IR sensing requires light to travel from the sensed area/object to the sensor. Since light travels in a straight line, optical sensing requires line of sight to the sensed area/object. In some applications, it may be possible to route light through fiber optics or via an optical bench, but in general, line of sight is necessary.
- *Loss of energy*: If polarizing or other special absorptive/reflective optics is used, some part of the optical energy may be lost in the optics before reaching the sensor. Usually, sensors can be designed to account for these losses.
- *Optical interference*: Related to line of sight are challenges posed by obscurants such as dust, smoke, rain, and clutter. These are of particular importance when the sensor operates in outdoor or uncontrolled environments. By proper placement of the sensors and proper selection of the sensor wavelength, most of the obscurant-related problems can be circumvented.
- *Accuracy*: IR sensors are sensitive to the ambient temperature, and ambient temperature stabilization may be needed to achieve stability and accuracy, especially when operating in the longer wavelengths. This is usually achieved using a thermoelectric cooler.
- *Cost*: IR sensors are relatively more expensive due to the lower-yield, higher-complexity manufacturing processes used.

10.2 SENSOR ANATOMY AND TECHNOLOGIES

The core of an optical sensor is a detector material element, such as a photodiode, that converts incoming photon energy into electrical energy or the modulation of electrical energy. Electrical energy can be then amplified, digitized, and transmitted (see Figure 10.1). There are a wide variety of materials that can be employed to perform this energy conversion.

Photodetectors cover a broad spectral wavelength range, from X-ray (0.01–10 nm), ultraviolet (UV) (10–400 nm), visible (400–750 nm), IR (750 nm–100's μm) to millimeter

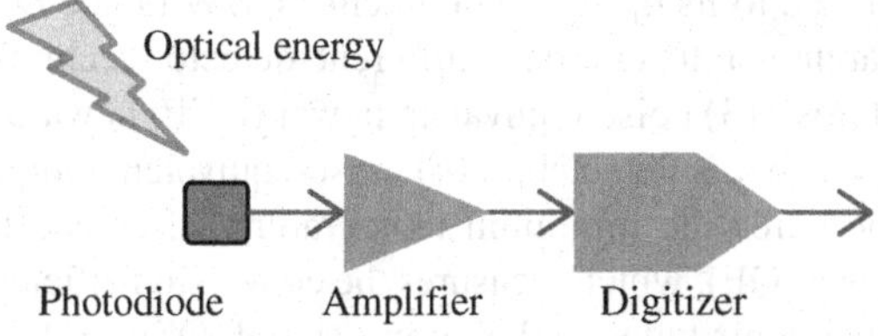

FIGURE 10.1 Digital detector schematic. Courtesy of Teledyne Judson Technologies. Use with Permission.

waves. For IoT, IR detectors play a critical role due to the relevant wavelength range of interest. IR photodetectors can be divided into several groups based on their spectral response wavelength range, such as NIR (0.75–1 μm), SWIR (1–3 μm), MWIR (3–8 μm), long wavelength IR (LWIR) (8–14 μm), very long wavelength IR (VLWIR) (14–24 μm), and far IR (FIR) (24–100's μm). In general, there are two main types of IR detectors, thermal detectors and photon detectors.

Thermal IR detectors are based on temperature increase of IR materials from absorption of IR illumination, which can cause change of certain material characteristics, such as resistance change effect (thermistor and bolometer), pyroelectric effect, or thermoelectric effect. Thermal detectors have the following advantages: (1) flat spectral response, where the output signal does not change with spectral wavelength; (2) operation (usually) at room temperature; (3) lower cost, compared to most of photon detectors. Disadvantages include (1) lower sensitivity or detectivity and (2) slower response speed (normally in milliseconds). Thermal detectors can be made of semiconductors, metals, organic materials, etc.

Photon detectors are based on absorption of photons by special semiconductor materials due to IR illumination, which generates electron–hole pairs (called photocarriers) with the output signal being either photocurrent or photovoltage. The materials used are narrow band-gap semiconductors, where the band gap, E_g, is a critical parameter of the semiconductor energy band structure and defines the energy threshold of photons that can be absorbed. The spectral response cutoff wavelength (50%) of a photon detector is defined as λ_c (μm) = $1.24/E_g$ (eV). Therefore, IR semiconductor materials have band gaps smaller than that of silicon (1.11 eV at 300K), which is the most important semiconductor material used in visible and UV detectors. Unlike thermal detectors with flat spectral response, photon detectors show linear spectral response under ideal conditions. That is, the detector *Responsivity*, defined as the ratio of output current or voltage to the input light power (A/W or V/W), increases with spectral wavelength linearly up to the cutoff wavelength. In reality, the spectral response of a photon detector is not linear around the cutoff wavelength. See Figure 10.2 for a typical spectral response curve, where peak wavelength λ_p at peak responsivity R_p and a 50% cutoff wavelength are shown.

In addition to Responsivity, there are a number of other important detector parameters. They include (1) detector noise, usually measured in frequency bandwidth (BW) units (A/√Hz for current noise, generally used for photovoltaic (PV) detectors, or V/√Hz for voltage noise, generally used for photoconductive (PC) detectors); (2) detectivity, or D^* (pronounced D star), which is the signal-to-noise ratio (SNR) of the detector normalized to its optical area, in cm-√Hz/W (since D^* is normalized, it is a convenient parameter to compare different detector technologies, sizes, and operating temperatures); (3) noise-equivalent power (NEP) in watts, a measure of the minimum light power that is detectable; (4) noise-equivalent temperature difference (NETD), which measures the minimum temperature difference that is detectable; (5) quantum efficiency (QE), which measures the conversion efficiency from a photon to a photocarrier that contributes to the output signal, $\text{QE}(\lambda) = 1.24 * R_\lambda / \lambda$. Note that for an ideal detector with a linear spectral response, QE would be constant (flat) over the entire wavelength range. In reality, QE is wavelength dependent, resulting in

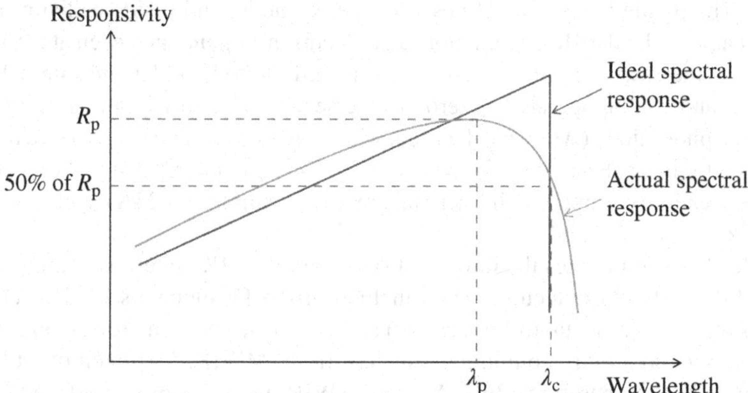

FIGURE 10.2 Relative spectral response curve of a photon detector. Courtesy of Teledyne Judson Technologies. Use with Permission.

nonlinear spectral response curve, and determined by IR material properties, detector structure design, and detector fabrication processes. QE can be enhanced by using a properly designed antireflection (AR) coating, which can be either a single-layer AR coating, optimized to a certain narrow wavelength band, or a broadband (multilayer) AR coating, optimized to a certain broad wavelength band or even to the entire wavelength range of the detector.

There are several sources of noise, including intrinsic thermal noise and background photon noise. Note that the spectral D^* curve of a photon detector has the same spectral shape as a spectral responsivity curve, because the detector noise is usually independent of wavelength. Other detector performance parameters include (1) linearity, which measures how well the output signal increases linearly with the input light power over a wide input power range, and (2) response speed, an indicator of how fast the output signal can respond to the change in input power, measured in either frequency BW or response time constant. Note that photon detectors have much faster response speed than thermal detectors, the former usually in microseconds to nanoseconds. For high-speed photodiodes, which require special structure designs, it can reach picoseconds or less.

There are two types of photon detectors, PC detectors (or photoconductors) and PV detectors (or photodiodes). A photoconductor is an electric resistor where photocarriers change the resistance due to IR illumination and generate an output voltage under an externally applied bias current. Its responsivity is in V/W. Photoconductors have an internal gain, which increases with the bias across the detector until it becomes saturated at large biases. As the detector noise also increases with the bias, there is an optimal bias for a given detector, producing an optimal SNR. The output voltage signal of a photoconductor is generally amplified by a voltage preamplifier with a certain gain.

PV detectors are based on a barrier-like structure, such as a p–n junction, with a built-in electric field. A p–n junction is formed at an interface between p-type semiconductor doped with acceptor dopant and n-type semiconductor doped with donor dopant. The p-side is also called positive side or anode, and n-side called negative side or cathode. Under IR illumination, a photocurrent is generated even at zero bias. Therefore, the responsivity is in A/W. A standard photodiode has no internal gain (gain = 1) and can be operated at zero or reverse bias. If internal gain is desired, an avalanche photodiode (APD) can be used, which operates at a relatively large reverse bias close to the avalanche breakdown voltage. The output current signal of a photodiode is generally amplified by a transimpedance amplifier (TIA) with a certain feedback resistance (gain).

Table 10.1 summarizes the important semiconductor IR materials commercially available for photon detectors. They include group IV elements of the Periodic Table, such as Si (NIR, up to 1 μm cutoff) and Ge (NIR/SWIR, up to 1.7 μm cutoff); III–V binary and ternary compounds, such as InSb (MWIR, 5.4 μm cutoff at liquid nitrogen (LN_2) temperature), InGaAs (NIR/SWIR, up to 2.6 μm cutoff), and InAs (SWIR/MWIR, up to 3.5 μm cutoff); II–VI ternary compounds, such as HgCdTe (SWIR/MWIR/LWIR/VLWIR, up to 26 μm cutoff); and IV–VI binary compounds, such as PbS (SWIR, up to 3 μm cutoff) and PbSe (MWIR up to 5 μm cutoff)—which are often referred to as Lead Salt (PbX) detectors. Most of these detectors are photodiodes, except for PbS and PbSe (which are mainly PC) and HgCdTe, which can be made as both photodiodes and photoconductors. Note that the single-element and binary compound semiconductor IR materials have a fixed cutoff wavelength (or energy band gap) for a given temperature. In contrast, ternary compound semiconductors, which are combination of two binary compound semiconductors, have a tunable cutoff wavelength, which is a function of the mixing composition, denoted by the parameter x. For $In_xGa_{1-x}As$, the cutoff wavelength

TABLE 10.1 Summary of Important Commercially Available Photon IR Materials

Material	Detector Type	Wavelength Range	Operating Temperature (Typical)	Temperature Coefficient of λ_c
Si	PV	NIR, up to ~1 μm	Room temperature (RT)	Positive
Ge	PV	NIR/SWIR, up to 1.7 μm	RT	Positive
$In_xGa_{1-x}As$	PV	NIR/SWIR, λ_c tunable, up to 2.6 μm	RT and thermoelectric cooler (TEC)	Positive
InAs	PV	SWIR/MWIR, up to 3.5 μm	RT and TEC	Positive
PbS	PC	SWIR, up to 3 μm	RT and TEC	Negative
PbSe	PC	MWIR, up to 5 μm	RT and TEC	Negative
InSb	PV	MWIR, up to 5.4 μm	LN_2	Positive
$Hg_{1-x}Cd_xTe$	PV & PC	SWIR/MWIR/LWIR/VLWIR, λ_c tunable, up to 24 μm	RT, TEC, and LN_2	Negative

Reproduced with permission from Teledyne Judson Technologies.

can be tuned from 1.7 μm (x=0.53) to 2.6 μm (x=0.82). For $Hg_{1-x}Cd_xTe$, its cutoff wavelength can be tuned from SWIR (1–3 μm) to VLWIR (up to 24 μm) with x adjustable in a wide range, from >0.7 to <0.18. In addition, cutoff wavelength is a function of temperature. For group IV elements and III–V compounds, λ_c decreases with decreasing temperature. In contrast, for II–VI and IV–VI compounds, λ_c increases with decreasing temperature.

IR semiconductor materials can be grown using a variety of techniques, including bulk crystal pulling, liquid-phase epitaxy (LPE), metalorganic chemical vapor deposition (MOCVD), molecular beam epitaxy (MBE), chemical deposition, etc. A substrate material is used for epitaxy growth or deposition. For example, InP substrate is used for InGaAs epi-growth, CdZnTe for HgCdTe epi-growth, and quartz plate for PbX deposition. These bulk or epi-grown semiconductor wafers are then processed into PC or PV detectors using a variety of semiconductor processing techniques.

In general, NIR/SWIR detectors operate at or near room temperature, SWIR/MWIR detectors operate at thermoelectrically cooled (TEC) temperatures in the range of −20 to −80°C, and LWIR detectors operate at cryogenic cooled temperatures such as LN_2 or cooled with closed cycle coolers (CCCs). LWIR detectors operating at TEC temperatures (−85°C) are also becoming available. They are typically designed with an integrated lens to increase signal collection area in order to improve the SNR. For IoT, detectors operating at or near room temperature are desired in order to reduce size, weight, power and cost (SWaPC).

Teledyne Judson Technologies (TJT) is one of very few IR EO sensor suppliers in the world that offer a broad range of IR materials and photon detectors covering a broad IR wavelength range from NIR to FIR (see Figure 10.3), in which the spectral D^* plots are shown for various detector materials, including Ge (J16), 1.7 μm InGaAs (J22), extended-wavelength (EW) InGaAs (J23), InAs (J12), PbS (J13), PbSe (J14), InSb (J10), PV HgCdTe (J19), and PC HgCdTe (J15).

Photodetectors are manufactured in various formats, including discrete (single-element) detectors; multielement, such as quadrant detectors (Quad); position sensors (PS); linear (1D) arrays; and two-dimensional (2D) arrays (see Figure 10.4). Discrete detectors have the lowest cost and are widely used. The detector active area size can be as small as <50 μm diameter and as large as >10 mm diameter (see Figure 10.4a). Quad detectors and position sensors are used for position alignment (see Figure 10.4 b and c), where Quad has four equally divided elements with small gaps between them, while PS has four outputs from either the four sides or four corners of the same detector. 1D array expands object detection length and greatly improves resolution. This is a format suitable for a variety of applications, from sorting objects on a moving conveyer belt to polar-orbiting Earth observation satellites. It can either have a parallel output, such as for 32- and 64-element arrays, where each element has its own pre-amp, or have a multiplexer (mux) output, such as for 128-, 256-, 512-element arrays and up to 1K/2K arrays and more. The pixel shape can be rectangular or square, with highly customized [3] pitch (center to center) size from >1 mm for small format down to 10 μm for large format arrays [4]. A mux is a 1D silicon readout integrated circuit (ROIC) and can be placed next to the detector array die to reduce SWaP significantly (see Figure 10.4d). 2D arrays are used in large-area imaging

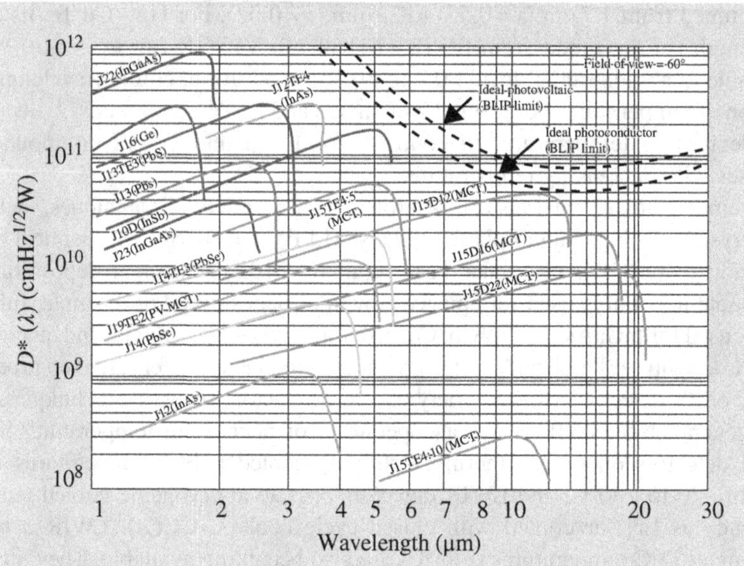

FIGURE 10.3 Graph showing a typical performance parameter (D^*) for a number of detector classes offered by Teledyne Judson Technologies (TJT), operating in the 1–20 μm spectral range [2]. See Teledyne Judson website for more information and detailed specification sheets. Courtesy of Teledyne Judson Technologies. Use with Permission.

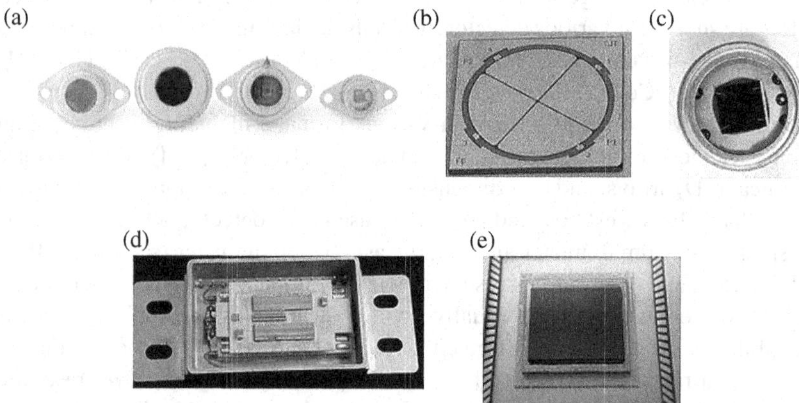

FIGURE 10.4 Various detector formats: (a) discrete detectors of various sizes, (b) Quad detector, (c) position sensor, (d) linear array, and (e) two-dimensional focal plane array (FPA). Courtesy of Teledyne Judson Technologies. Use with Permission.

systems, such as IR cameras, and are integrated to a 2D Si ROIC to form a focal plane array (FPA) (see Figure 10.4e). 2D FPAs have formats from 256×256 to 4K×4K, with (usually square) pixel size from >50 μm down to <10 μm.

Discrete detectors and 1D arrays can be either PV or PC type and are usually front-side illuminated (FSI), where the detectors are connected to pre-amp or ROIC through wire bonding (see Figure 10.4d). 2D arrays/FPAs are usually PV type and backside illuminated (BSI), where the detector array is flip-chip-bonded on an ROIC die through metal bumps [5] (see Figure 10.4e). The corresponding pixels, on the detector and ROIC, are connected through metal bumps for each pixel. Backside illumination is also used for high-speed detectors, lensed detectors (detector with an immersion lens), integrated detector/electronics assembly, and digital detectors.

Most PV detectors are p-on-n structure, that is, the active area is anode and the substrate is cathode. For PV arrays, all elements or pixels share a common contact, such as a common cathode. Both anode and cathode can be made on the front side of the detector, as done in BSI detectors. For discrete detectors and 1D arrays of FSI type, the common contact is usually made on the backside of the detector. Table 10.2 summarizes the general selection criteria for IR photon detectors.

TABLE 10.2 General Considerations in the Selection of an Infrared Photon Detector

	Parameter	Selection Criteria
1	Cutoff wavelength, λ_c	λ_c is determined by the upper end of the IR wavelength band of interest for a given application (IR source or target). It is also often driven by the IR system design requirements. Since detector noise increases with longer λ_c, λ_c should be kept as short as possible in order to achieve the lowest noise and hence maximum detectivity or SNR. For temperature sensing application, the Wien displacement law [6] provides some guidance for detector peak wavelength selection as $\lambda_p(\mu m) = 2{,}898/T(K)$
2	Detector size, A_d	For discrete detectors, A_d is determined by the illumination light spot size. Since detector noise is proportional to the square root of A_d, A_d should be kept as small as possible while still be able to capture all or most of illumination light, in order to achieve the lowest noise and hence maximum detectivity or SNR. For multielement detectors and 1D and 2D arrays, A_d is in general driven by IR system design requirements, such as resolution, optics, cost, and so on
3	Detector format	Depending on application and IR system design requirements, as well as cost budget, one may select either discrete detectors (single element or multielement) or 1D or 2D arrays. Cost increases with number of elements and detector size
4	Detector performance: detectivity, responsivity, noise, etc.	Detector performance is driven by application and IR system design requirements. Two primary parameters that one must know by design and modeling are (1) how much IR energy (light intensity) that the detector can capture and (2) what is the detector noise that is acceptable. A system designer or an application engineer should clearly know the SNR requirement for a given application. There will always be trade-offs between performance and cost, which will determine the selection of detector material, format, and operating temperature

(Continued)

TABLE 10.2 (Continued)

	Parameter	Selection Criteria
5	Detector type, PV versus PC	PC detectors usually have lower cost due to simpler detector fabrication processes, readily available for certain materials such as PbX, and simpler electronics. PV detectors usually have higher response speed, wider linearity range, and lower power consumption due to low or zero detector bias needed, compared to high bias required for PC. Selection of one over the other is in general driven by application, IR system design requirements, and cost
6	IR material	For the same wavelength band, there are usually two or more IR materials available. The selection of one material over other materials is driven by detector performance, cost, detector type and format, operating temperature and package, etc. Examples include Ge versus standard InGaAs for NIR/SWIR, EW InGaAs versus PV HgCdTe for SWIR, PbX versus InAs versus HgCdTe for MWIR, etc.
7	Operating temperature	Although room temperature operation is always desired, especially for IoT, some applications may require higher performance and longer wavelength. As a result, some kind of cooling of the detector may be still needed, which drives up cost, size, and power consumption, and also requires package design to accommodate the cooling temperature
8	Package	Detector package is driven by detector size and format, as well as detector operating temperature. It can be surface mounts, room temperature TO packages, TE-cooled TO packages, cryogenic cooled dewar packages, etc.
9	Cost	Cost is always driven by detector performance requirements and often plays a big role in the selection of IR material, detector type, and format
10	Field of view (FOV)/cold shield	For background limited performance (BLIP) detectors such as InSb and HgCdTe, proper cold shield design can enhance detector performance by limiting background noise. For BLIP detectors, the improvement in D^* is related to 1/sin(cold shield half angle) [7]; thus for 60° FOV, the D^* improvement can be up to a factor of 2 compared to 180° FOV

Reproduced with permission from Teledyne Judson Technologies.

10.3 DESIGN CONSIDERATIONS

When utilizing a photodetector in an EO system for sensing, one needs to consider all key aspects at the complete system level. The key aspects are shown in Figure 10.5:

- Optical sensing mechanism
- Object characteristics
- Photodetector

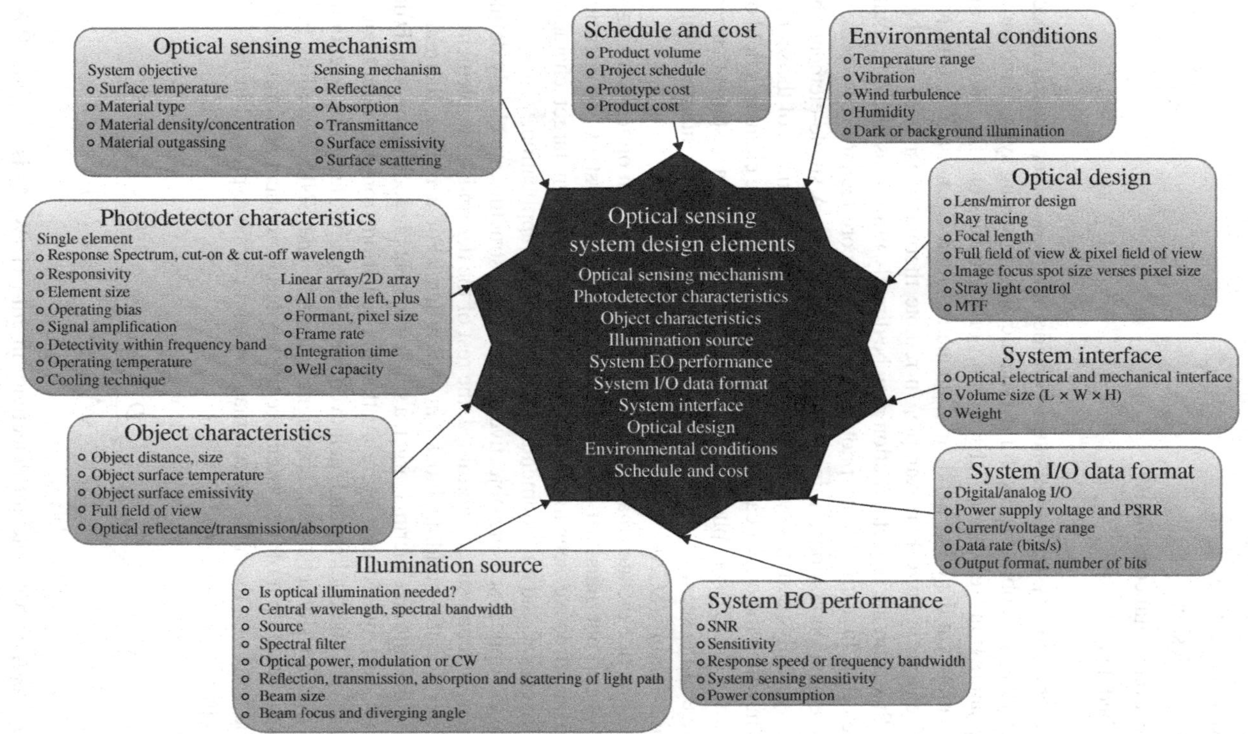

FIGURE 10.5 Key system aspects to be considered when designing an optical sensing system. Courtesy of Teledyne Judson Technologies. Use with Permission.

- Illumination source
- System EO performance
- System I/O data format
- System interface
- Optical design
- Environmental conditions
- Schedule and cost

More complex systems will need considerations of other special aspects and parameters such as object speed for a dynamic object sensing system and multispectral signatures in reflection, absorption, or transmission for remote material identification. On the other hand, an illumination source is not a necessity for every optical sensing system. For example, in remote thermal sensing of an object, the photon flux from the object radiation may be directly utilized as photon signal to the photodetector. In each system, photodetector is the core technology whether it is a single-element photodiode, a linear array, or a 2D array. A single-element photodetector is typically used in a system that does not require an image of the scene or in a system in which the object can be fully covered by the single element's field of view. Linear array or 2D array photodetectors are typically used in systems that require a full image of the scene. A photodetector and a Si ROIC can be integrated into a sensor chip assembly (SCA) using flip chip bonding techniques.

Good practice dictates that at the start of the design of an optical sensing system or subsystem, one must set up a clear objective of the system and then identify/develop an effective sensing mechanism based on known target characteristics, system interface, and environmental conditions in the application.

Once the optical sensing objective and interface structure are clearly defined and the sensing mechanism is determined, the rest of the design is to implement the right technology for the building blocks in the system. For a remote thermal sensing system with finite size thermal target, a single-element photodetector or a 2D detector array may be required to capture all thermal radiation from the target but not the radiation from the background, achieved through careful design or selection of the optical transmission channel. The signal intensity and SNR as well as frequency response at the output of the system are the key electro-optical performance parameters. The SNR of a thermal optical image system can be expressed in the following formula [6]:

$$\mathrm{SNR} = \Delta T \times \frac{D^*}{\sqrt{\Delta f}} \times \frac{\partial L}{\partial T} \times \frac{\pi}{4} \times \frac{\sqrt{A_\mathrm{d}}}{(F/\#)^2} \tag{10.1}$$

where D^* is the photodetector detectivity in cm-rtHz/W, Δf is the 3 dB BW of the signal in Hz, L is the radiance of the target in W/cm^2·sr, T is the temperature in Kelvin, ΔT is the resolvable temperate difference between the target and the background for a certain SNR, $F/\#$ is the optical system f-number which is defined as the ratio of the focal length to the clear aperture diameter of the system optics, A_d is the detector photosensitive area in cm^2.

Noise-equivalent temperature difference (NETD) is a widely used terminology that defines the minimum resolvable temperature difference that produces unity SNR by setting SNR = 1 in Equation (10.1) and solving the equation for ΔT [6]:

$$\text{NETD} = \frac{4}{\pi}\left[\frac{(F/\#)^2\sqrt{\Delta f}}{D^*(\partial L/\partial T)\sqrt{A_d}}\right]$$

For a planar Lambertian source-type target, the source target radiance gradient is given by the following expression:

$$\frac{\partial L}{\partial T} = \frac{1}{\pi T} M_e(\lambda, T)\frac{x}{1-e^{-x}}$$

where $M_e(\lambda, T)$ is spectral radiant exitance of the target in W/cm^2·μm, x is a parameter [7] defined as $x = c_2/\lambda T$, $c_2 = 14{,}388$ μm·K is a radiation constant.

For a given NETD, we can easily calculate the required D^* by the system

$$D^* = \left[4 * (F/\#)^2\sqrt{\Delta f}\right] / \left\{\sqrt{A_d} \times \text{NETD} \times \left[\int_{\lambda_1}^{\lambda_2} R(\lambda) * T_s(\lambda) * \left(dM_e(\lambda, T)/dT\right)d\lambda\right]\right\} \tag{10.2}$$

Now, one can see that the system design reduces to the physical layer which specifically calls out the photodetector detectivity and spectral responsivity $R(\lambda, T)$ as well as the optical channel transmission $T(\lambda, T)$. Equation (10.2) provides useful guideline at the system level in choosing a photodetector out of many different choices. If a photodetector has to be cooled down to a certain operating temperature to achieve the required detectivity and responsivity, then an appropriate cooling technique must be considered in photodetector packaging design. Typically, a single-stage or a multistage thermoelectric cooler (TEC) is used in cooling a single-element photodetector down to temperatures around −40 to −85°C from room temperature in a hermetically sealed package.

System-level design includes consideration and trade-offs among various performance parameters while satisfying the constraints posed by the end application. Almost in every single circumstance, a system will often be required to fit in a volume size defined in length, width, and height and not to exceed certain weight and power consumption. Additional system-level considerations such as thermal management and mechanical interface may need to be addressed through design/selection.

10.4 APPLICATIONS

EO–IR sensors can be used in numerous applications. Figure 10.6 summarizes the choice of sensors for different typical applications. In the next section, we look at three IoT applications: thermal sensing, gas sensing, and food inspection.

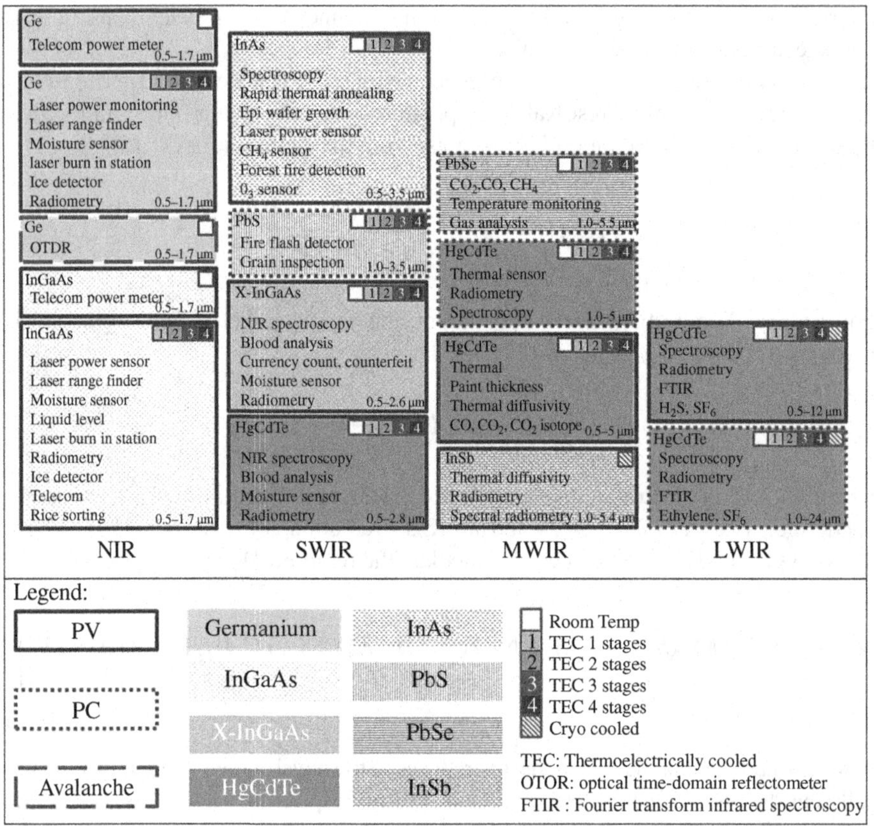

FIGURE 10.6 Choice of sensors in the four different bands—near, short wave, mid, and long wave infrared—for typical infrared sensor applications. Courtesy of Teledyne Judson Technologies. Use with Permission.

10.4.1 Thermal Sensing

Thermal IR energy is emitted from all objects that have temperature greater than absolute zero (−273°C). Human eyes cannot detect thermal IR energy because our eyes are sensitive only to visible light (400–700 nm). An object's internal kinetic heat is converted into radiant energy in the form of thermal IR emission. Some part of it may appear as visible radiation (especially in very-high-temperature situations, such as molten steel), but the invisible part of the spectrum, however, contains up to 100,000 times more energy.

Due to the high correlation between the emitted energy and the internal heat, by measuring the emitted energy using an IR detector, the internal heat can be estimated fairly accurately. This is a remote sensing approach, in contrast to other types of sensors that directly measure the internal heat through contact methods. As we have seen before, there are several materials that are sensitive to thermal radiation in the 3–10 µm range, where thermal radiation dominates over reflected IR energy.

TABLE 10.3 Key Criteria for Selecting a Thermal Sensor for an Application

Criterion	Thermocouple	Thermistor	Resistance Temp Detectors (RTD)	IR Photodiode
Temp range	–260–2,320°C	–100–500°C	–240C–650°C	Up to 3,000°C
Accuracy	Good	Good	Better	Good
Linearity	Better	Good	Best	Best
Sensitivity	Good	Best	Better	Best
Cost	Low	Low medium	Medium	High
Noncontact	No	No	No	Yes
Field of view control	None	None	None	Yes
Response Speed	Slow	Slow	Slow	Fast

Reproduced with permission from Teledyne Judson Technologies.

Table 10.3 summarizes criteria that may be used to select the appropriate thermal sensor for an application.

Note that IR photodiodes are best suited for noncontact, high-speed, high-sensitivity, and high-linearity applications. There are several advantages due to these features:

- High-speed measurements allow for more measurements and accumulation of data.
- Can be used to measure temperature of moving objects (conveyer belts).
- Can be used to measure temperature of hazardous and physically inaccessible processes.
- For higher temperatures, noncontact devices have longer life.
- Better than alternatives for measuring temperatures of poor heat conducting surfaces (plastic, wood).
- No risk of contamination or physical damage to the object surface.

However, it should also be noted that in general, IR measurements require line of sight, with no obscurants (smoke, dust, etc.); only surface temperature can be measured; optics in the IR devices need to be kept clean.

The following design parameters impact the selection of IR temp sensors to measure the temperature of an object:

- Temperature range to be measured
- The size of the object
- How close to the object can the detector be installed
- Does the detector fill the field of view?
- Object material
- Speed of movement of the object

- Discrete object or continuous process
- Ambient temperature
- Ambient conditions (dust, smoke, steam)
- Compatibility with existing control system

10.4.2 Gas Sensing

Gas sensing is receiving increasing attention in many applications: industrial production, automotive industry, medical applications, indoor air quality supervision, environmental studies, etc. A number of sensing technologies are available for gas sensing [8]. Electromagnetic energy, including IR, is absorbed by gas as the energy passes through gas. The complexity of the gas molecule determines the number of wavelengths in which energy is absorbed. The more atoms that form a gas molecule, the more are the absorption "bands" or sets of wavelengths. The band in which the absorption occurs, the amount of absorption, and specific character of the absorption curve are unique to each gas. These characteristics form a "fingerprint" of the gas. In typical gas sensing applications, only one specific absorption region is used to quantitatively determine the gas concentration (even if the gas has multiple absorption regions or bands). IR gas sensors are widely used for methane, carbon monoxide, carbon dioxide, etc.

A typical configuration has a "Gas Cell," also referred to as "Light Path," designed in such a way as to allow the light path to interact with the sample gas (Figure 10.7). This is constructed with a tube that allows light to enter from one end and exit to the other, where the detector is positioned. Inlet and outlet ports allow the sample gas to circulate through the tube. A band-pass filter is used to select the band being characterized. The filter may be at the source end or the detector end. Often a reference gas cell or a reference detector may be used to minimize drift due to contamination or other factors.

This type of detector is used to detect saturated hydrocarbons (methane, ethane, propane, etc.), cycloalkanes (cyclopropane, cyclohexane, etc.), unsaturated hydrocarbons

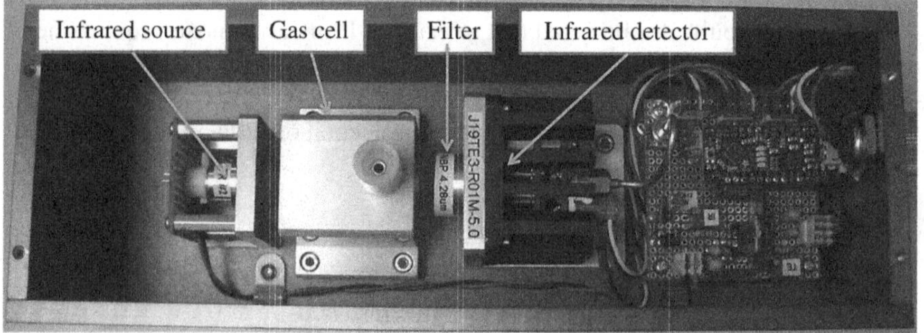

FIGURE 10.7 Demonstration model of a typical gas sensing device using an infrared source and detector to measure absorption in specific bands determined by a filter. Courtesy of Teledyne Judson Technologies. Use with Permission.

TABLE 10.4 Comparison of Various Sensing Methods for Gas Detection

Criterion	Metal Oxide Semi-conductors	Electro-chemical	Thermal Conductive	IR Sensors	Catalytic Sensor	Photo-ionization
Sensitivity	Low	High	Low	High	High	High
Selectivity	Poor	High	High	High	Poor	High
Response time	Short	Short	High	Short	High	High
Energy consumption	High	Low	Low	Mid	High	High
Lifetime	Long	Long		Long	Short	Short
Variety of gas detection	Broad	Good	Limited	Limited	Limited	Good
Noncontact	No	No	No	Yes	No	Yes
Fabrication cost	Low	High	Mid	High	Low	High

Reproduced with permission from Teledyne Judson Technologies.

(ethylene, propylene, butane, etc.), aromatics (benzene, toluene, xylene), alcohols (methanol, ethanol, propanol, etc.), amines, ethers, ketones, and aldehydes. Absorption of carbon dioxide at 4.3 μm and carbon monoxide at 4.6 μm has very little interference by other gases.

Remote gas sensing techniques have been used in sensing gas concentration in free air. In this case, the gas sensing mechanism is based on reflectance and scattering of the "light path" from sensor to the gas to be sensed and the target gas absorption or reflectance of the IR light signal. Depending on application, the light path could range up to a kilometer. A reference light emitted from the sensing system is often used to detect the reflectance and scattering of the light path.

There are a variety of gas detection technologies in use today. Among the most commonly employed are IR Sensors, Electrochemical Sensors, Metal Oxide Semiconductor (also known as "solid state"), Thermal Conductivity Detectors, Catalytic Gas Sensor, and Photoionization Sensors. Each technology has certain advantages and disadvantages as shown in Table 10.4.

10.4.3 Food Inspection

Food quality has gained wide interest since it can affect the health and safety of vast populations. In recent years, due to globalization of food products and food ingredients, people everywhere are more exposed to potential cases of contaminated food products and cases of intentional substitute of ingredients for financial profit, but with devastating consequences to the consumer. Traditionally, food quality inspections have been done by certified laboratories offline, and the wet chemical methods involved are expensive and require additional time for sample removal; further, there is concern with hygiene. In recent years, the need to inspect food ingredients at the source and through the supply chain demands alternate inspection methods that are fast, on-site (online and in-line) noncontact, deterministic, and noncontaminating. Electro optic IR sensors offer a promising solution.

IR photon detectors are being used in food inspection by taking advantage of optical phenomenon such as reflection and transmission. Reflection methodology is prevalent when solids are inspected and transmission methodology is used when liquid is inspected. Both reflection and transmission methodologies are used for paste inspection. Solids such as grains and seeds are inspected by reflection, and the absorbance level is an indication of the constituent of interest, such as grain oil, moisture, and protein content. The common configuration is a dish with grains illuminated by a light source in the wavelength of interest, and the reflected light from the grain is detected by a SWIR photon detector such as extended InGaAs or PbS detector. In this methodology, the grain's absorbance is an indication of the moisture and protein content. A similar approach is used in grain (e.g., rice) sorting. In this configuration, a light source illuminates grains as they move through a vertical chute and an array of detectors measure the spectral signature of the grain. When a difference in signature is detected, a nozzle blows a puff of air to remove the nonrice material. This methodology is also used for sorting nuts such as pecan sorting and separation from the shell. A similar method is used when sorting peaches and other fruits for bruising. This approach is also effective to screen livestock feed and pet food for quality and adulteration.

Liquid food items are mostly inspected by the transmission methodology. Light in the wavelength of interest passes through the liquid of interest, and the resultant spectral signature is detected by a SWIR detector. The location of the absorption peak and difference in the absorption amplitude are indicators of the chemical constituents and their relative concentration. For example, the inspection of virgin olive oil for additives and substitutions with less expensive oils can reveal the purity of the oil. Milk and milk products are inspected for fat and protein content through the manufacturing process that requires separation and labeling of the fat content. SWIR detectors are used for in-line measurement of fat and protein. A combination of reflection and transmission techniques is used for food products in semisolid state, such as butter, yogurt, ice cream. In the gas phase, it is possible to examine food freshness and ripening based on photon absorption of outgassing. It is well known that the presence of ethylene (C_2H_4) triggers fruit ripening [9]. Using an LWIR gas sensor that detects ethylene concentration (near 10.6 μm) in a food container, it is possible to prevent maturing and spoilage of stored fruits like apples.

We have summarized some of the benefits of utilizing IR sensor technology for food inspection. The trend is toward increasing access to food safety and quality inspection instrumentation through the entire supply chain. We anticipate the proliferation of mobile accessories for the consumer to select and inspect food items prior to purchase.

10.5 CONCLUSION

In this chapter, we have presented an overview of IR electro-optical sensing technologies. Due to multiple beneficial properties of electro-optical sensors, they are currently used widely in numerous industrial, military, and space applications.

We anticipate widespread incorporation of the sensors in IoT devices for the consumer, industry, and smart cities. IR sensors are already available in sizes and packages ready to be incorporated in IoT applications, and as the sensor usage expands, we expect further miniaturization and reduction of cost to follow.

REFERENCES

[1] Infrared Thermometers, Visual IR Thermometers, and Non-Contact Infrared Thermometers. http://www.fluke.com/fluke/inen/products/thermometers. Accessed October 9, 2016.

[2] Teledyne Judson Technologies. http://www.teledynejudson.com. Accessed October 9, 2016.

[3] T.A. Ellis, J. Myers, P. Grant, S. Platnick, D.C. Guerin, J. Fisher, K. Song, J. Kimchi, L. Kilmer, D.D. LaPorte, and C.C. Moeller, "The NASA enhanced MODIS airborne simulator," SPIE Earth Observing Systems XVI, 8153ON (August 23–25, 2011).

[4] H. Yuan, M. Meixell, J. Zhang, P. Bey, J. Kimchi, L. Kilmer, "Low Dark Current Small Pixel Large Format InGaAs 2D Photodetector Array Development at Teledyne Judson Technologies," Proceedings of SPIE 8353, Infrared Technology and Applications XXXVIII, 835309 (May 1, 2012).

[5] H. Yuan, G. Apgar, J. Kim, J. Laquindanum, V. Nalavade, P. Beer, J. Kimchi, T. Wong, "FPA Development: From InGaAs, InSb to HcCdTe," Proceedings of SPIE 8353, Infrared Technology and Applications XXXIV, 69403C (2008).

[6] E.L. Dereniak, G.D. Boreman, *Infrared Detector and Systems*, Wiley-Interscience Publication, 1996. Hoboken, NJ.

[7] J. Vincent, *Fundamentals of Infrared Detector Operation and Testing*, Wiley-Interscience Publication, 1990. Hoboken, NJ.

[8] X. Liu, S. Cheng, H. Liu, S. Hu, D. Zhang, and H. Ning, "A survey on gas sensing technology," *Sensors* 2012, 12, 9635–9665. doi:10.3390/s120709635.

[9] The Origin of Fruit Ripening. https://www.scientificamerican.com/article/origin-of-fruit-ripening/. Accessed October 9, 2016.

FURTHER READING

J. Fraden, *Handbook of Modern Sensors: Physics, Design, and Applications*, fourth edition, Springer, 2010. Berlin.

Y. Ozaki, F. McClure, and A. Christy, Editors, *Near-Infrared Spectroscopy in Food Science and Technology*, Wiley Publication, 2006. Hoboken, NJ.

G. Rakovic, "Overview of sensors for wireless sensor networks," *Transactions on Internet Research*, 2009, 5, 13–18.

G. Zissis, Editor, *The Infrared & Electro-Optical System Handbook*, Volume 1, ERIM and SPIE Copublication, 1993. Ann Arbor, MI and Bellingham, WA.

11

IPv6 FOR IoT AND GATEWAY

GEOFF MULLIGAN

IPSO Alliance, Colorado Springs, CO, USA

11.1 INTRODUCTION

While many have declared the Internet of Things (IoT) revolutionary, it should instead be considered in the context of *evolution* of the current Internet. The Internet of today provides connectivity between people—sending email, sharing social information, exchanging pictures—whereas the main purpose of the IoT is and will be to interconnect devices and things, just as the name would suggest. Whether called the Internet of Things, the Internet of Everything, Industrial Internet, Cyber–Physical Systems, or Machine to Machine, utilizing open and interoperable standard protocols is critically important for the success of this next generation of the Internet. The reason for the success of today's Internet was Interoperability. Prior to this there were many incompatible proprietary protocols that made interconnecting devices from different vendors a real nightmare and had stalled the expansion of local area and long-distance networking. The advent and adoption of the Internet Protocol (IP) changed all of this, and such is necessary for the realization of the potential for the IoT.

11.2 IP: THE INTERNET PROTOCOL

On May 5, 1974, Vinton G. Cerf (known as the "Father of the Internet") and Robert E. Kahn published the paper "A Protocol for Packet Network Interconnection" which described a new protocol that provided the capability to connect disparate network to

Internet of Things and Data Analytics Handbook, First Edition. Edited by Hwaiyu Geng.

Companion website: www.wiley.com/go/Geng/iot_data_analytics_handbook/

TABLE 11.1 The ISO Seven-Layer Model

Application Layer	Message Format and Interfaces
Presentation Layer	Data Translation, Encryption, Compression
Session Layer	Authentication, Permission
Transport Layer	End to End control, retransmission
Network Layer	Network addressing, routing
Data Link Layer	Error detection, physical link
Physical Layer	Physical medium

create a "network of networks" which was the basis for the Transmission Control Protocol (TCP/IP) deployed across the ARPANET in 1983. One of the fundamental new concepts that came from work on the ARPANET was the idea of protocol layering. Rather than having just one single monolithic protocol to do everything, the idea was that there should be separate layers for different functions. This eventually became the ISO Seven-Layer Model (see Table 11.1).

TCP and IP are used to break apart and transport larger data items, like email, pictures, documents, into and using smaller packets. In the ISO model shown previously, TCP and IP are examples of the Transport and Network Layers. [The actual mapping of the protocols used on the Internet is not as clean as this (see Table 11.1) but is close enough for this description.] IP assigns addresses to computers and provides a mechanism to "route" the packets from the source node to the destination node through a number of intermediary nodes. This design also allows for the packets to take different paths and therefore route around outages in order to make a more robust system—if there is congestion in one part of the network, the packets can be routed through a less busy path automatically. IP version 4 (IPv4) was the first version of the IP and supported up to 2^{32} addresses (about four billion addresses). The TCP makes it possible for the packets being moved by IP to take different paths, to arrive at the destination out of order, or even to be lost. The TCP ensures that at the destination node the packets are delivered to the application in order and complete. Since TCP/IP was agnostic to the type of connectivity, this meant that devices on different types of networks (phone line, Ethernet, token ring, etc.) could all cooperate and could share data. The IP makes very few assumptions about the underlying network and allows it to work across nearly any type of network. These "lower layer" networks fit into the ISO model in the Data Link and Physical Layers with TCP/IP "over" them. With a foundation of TCP/IP, many new application protocols such as Simple Mail Transfer Protocol (SMTP) for email, File Transfer Protocol (FTP) for files, Simple Network Management Protocol (SNMP) for managing devices connected to the network, Session Initiation Protocol (SIP) for Voice, and a number of others have been developed and standardized over the intervening 40 years. To complete the description of the ISO model, these protocols occupy the Session, Presentation, and Application layers and sit "atop" TCP/IP. Quite often the complete set of protocols used on the Internet are called the "Internet Protocol Stack" or "Internet Protocol Suite." Table 11.2 shows the approximate mapping of the IP Model (sometime called the ARPANET model) and the ISO Model.

TABLE 11.2 ISO Model Versus Internet Model

ISO Model	Internet Model
Application Layer	Application Layer
Presentation Layer	
Session Layer	
Transport Layer	Transport Layer
Network Layer	Internetwork Layer
	Network Layer
Data Link Layer	
Physical Layer	

This layered network design and the definition of the protocol interfaces make it possible to use different protocols at different layers without having to change all of the rest of the protocols. Besides TCP for the transport layer, there is also the User Datagram Protocol (UDP). UDP differs from TCP in that it does not try to maintain the order of delivery nor does it ensure that lost packets are retransmitted. As a result the protocol is very simple and very small, and for applications that need to send the same information to many many different destinations, UDP is the right choice. The design of the IP stack then seems to be complete, but there was one major flaw that was exposed because of the success of the Internet itself. The initial seemingly gigantic number of address support by IPv4 (four billion addresses) would in fact be severely limiting as the growth and demand for connecting to the Internet and the wealth of services grew—network engineers started worrying about running out of addresses.

11.3 IPv6: THE NEXT INTERNET PROTOCOL

By the early 1990s it became obvious that some of the basic design of IPv4 was causing problems with expansion of the Internet and some action needed to be taken. The Internet Engineering Task Force—a group of engineers from companies, universities, and governments from around the world who are charged with standardizing protocols for use on the Internet—started work on the design of the next version of the IP, IP version 6 (IPv6).

What happened to IP version 5

Somewhat in parallel to the deployment of IPv4 for data sharing and computer interconnection, engineers started working on a protocol that could be used to stream voice, video and simulations. This was called the Internet Stream Protocol and was IPv5 and was designed to provide a "guaranteed quality of service (qos)". It was implemented by a number of companies, but never widely deployed.

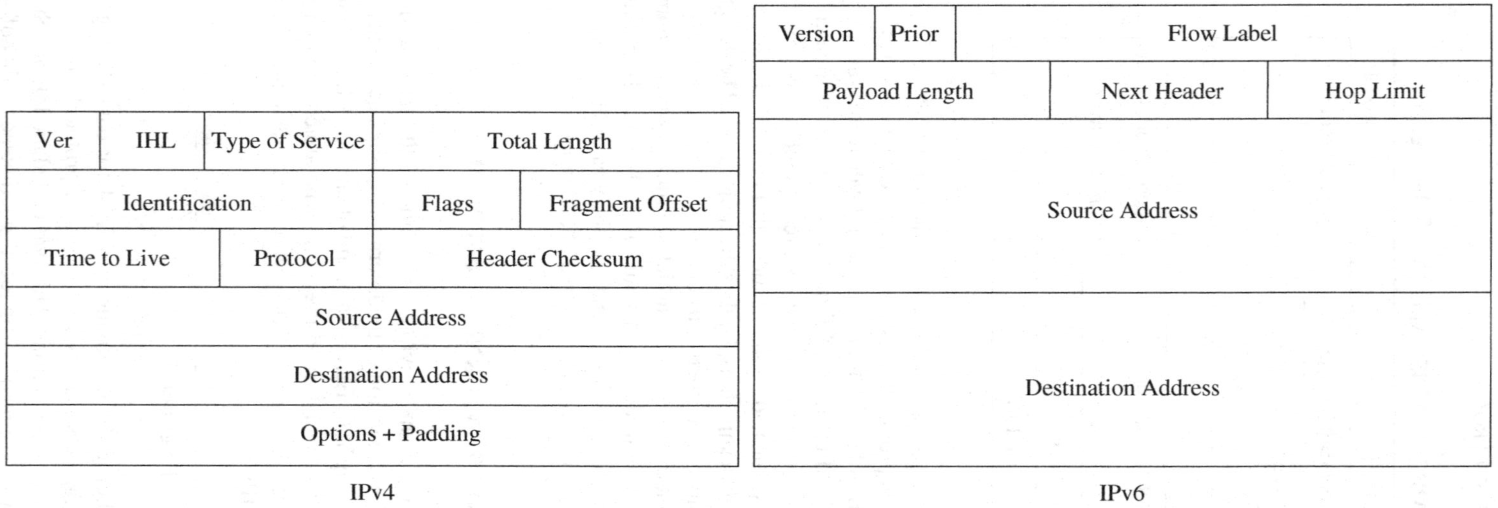

Ver	IHL	Type of Service	Total Length	
Identification			Flags	Fragment Offset
Time to Live		Protocol	Header Checksum	
Source Address				
Destination Address				
Options + Padding				

IPv4

Version	Prior	Flow Label	
Payload Length		Next Header	Hop Limit
Source Address			
Destination Address			

IPv6

FIGURE 11.1 IP version 4 and IP version 6.

There were two main goals for the design of IPv6: (1) increase the number of available addresses and (2) simplify the protocol to make it faster for newer networking technologies. In Figure 11.1, you can see the differences between IPv4 on the left and IPv6 on the right.

With IPv6 the available address space is increased from 2^{32} to 2^{128}. This is not 4 times larger (32 vs. 128) but instead 2^{96} (or 79 billion billion billion) times larger. To try to put this in some perspective, since this number is so so large, if a person could count all of the IPv4 addresses in 1 second, it would take them 25 million trillion years to count all of the IPv6 addresses. Another way to visualize just how large this number is that if all of the IPv4 addresses were represented as a stick 1 m long, that same stick would be 92 times the width of the Universe to accommodate all of the IPv6 address. But this expansion of the address space comes at a cost. The IPv6 header, shown in Figure 11.1, is twice as large as the original IPv4 address (40 bytes vs. 20 bytes). While this might not seem like much, the extra memory necessary to store the addresses and the required time to transmit the extra bytes have a huge impact on small embedded devices to be used in the IoT.

11.4 6LoWPAN: IP FOR IoT

When working with small sensors or simple appliances—devices like light switches, door locks, thermostats, garage door openers, blood pressure monitors, motion sensors, and the like, adding extra memory adds cost and having to transmit extra bytes reduces the battery life. To deal with these problems, the author Geoff Mulligan who Vint Cerf referred to as the "Father of the *Embedded* Internet" created the 6LoWPAN protocol. It "compresses" the IPv6 header from 40 bytes down to as little as 1 byte. This increase in efficiency can extend the battery life of one of these small devices from what would have been a few months to over 5 years, or more importantly it makes it possible for these tiny embedded devices to make use of energy-harvesting technology and eliminate batteries all together. 6LoWPAN accomplishes this by eliminating some of the fields in the IPv6 header that for sensor and embedded networks are not necessary and by deleting redundant fields. This compression is not possible with IPv4 because of the more complex header (again see Figure 11.1). The use of IPv6 provides additional benefits including the elimination of extra devices and complex setup, such as Dynamic Host Configuration Protocol (DHCP) servers, and the extra address space means there is no need for Network Address Translators (NAT). For these reasons the success and future of the IoT rest squarely on the shoulders of IPv6.

Just as with IPv4 for the Person-to-Person Internet today, the use of IPv6 for the Device-to-Device (D2D) IoT provides the necessary interoperability. Without the IPv4 interoperability today, we might need completely different applications to talk to different websites or worse completely different computers. For example, imagine if AT&T cell phones could only call or text other AT&T phones or Verizon could only call or text Verizon. An alternative for dealing with the incompatibility is to insert other devices—Gateways—to translate between the incompatible protocols.

11.5 GATEWAYS: A BAD CHOICE

While it might appear simple to put in a device that can talk to two different networks using two different protocols, this translation is fraught with problems. Things like maintaining the proper versions, imprecise translation, management, and updates can cause huge issues not only for the vendor and OEM but also for the end customer. Additionally these devices introduce a single point of failure and a single point of attack. In order to translate protocols, a gateway must, in general, decrypt the message or data coming from one network and then reencrypt the data or message to send to the other network. This means that there is a point in the process where the data is vulnerable to be read or changed or new data inserted unbeknown to either the sender or receiver, and therefore this is little or no security or privacy.

Privacy vs. Security

Often today these terms are used interchangeably or conflated, but they are very different. Privacy is a policy and Security is a technology. A privacy policy defines who should be able to have access to the data and how it should be protected and the security technology is used to enforce that policy via techniques like encryption.

By using open standards the sender can protect the data from transmission to reception such that only the receiver can read it, and both can be assured that the data was not changed—known as end-to-end integrity. Another benefit of using open standards and avoiding gateways and proprietary protocols is the ability to leverage existing and future developments—no need to reinvent the wheel over and over again. For example, if the device needs a software upgrade mechanism, it is likely that this can be accomplished securely with protocols already available. While open standards, such as the IP Suite, are not a panacea for problems, they do provide common "plumbing" to interconnect these devices and applications in a tested and interoperable way.

11.6 EXAMPLE IoT SYSTEMS

An example of a use case for Smart Cities combined with intelligent transportation systems for the IoT is automobiles "talking" to traffic lights. In this scenario vehicles would communicate with the traffic light system so that the lights could be controlled to more effectively control the flow of traffic. This sort of design would be nearly impossible if ever different car manufacturer had to know about the inner workings and protocols of the numerous different traffic light systems in different cities or if every traffic light system needs to understand different protocols in each different auto manufacturer. But with a common set of protocols and interoperability, it is possible to eliminate waiting at a light when there are no other cars on the road.

Another example closer to home would be the interaction of home smoke detectors and home appliances. By using IP standards a smoke detector would be able to broadcast a message across the home network when it detects smoke. Locally the appliances would "hear" this message and make decisions about how to react. For example, all of the gas appliances (stove top, oven, water heater) could turn off—no point in adding fuel to the fire. The house fan depending on design might either turn off or turn on. If the house fan vents to the outside, it would know to turn on and exhaust any smoke out of the house. If instead the fan vents inside the house, it would know to turn off and not move the smoke throughout the house. Additionally the message from the smoke detector could be routed to services and devices on the Internet to provide warnings to first responders and the homeowner. All of this is more easily accomplished using common open protocols rather than inserting unnecessary translating gateways.

In a final scenario, "Crash to Care," at the scene of an accident, the cars automatically send a message to first responders and the traffic light system. The traffic light system reacts by working to redirect traffic around or away from the accident while at the same time managing to bring the first responders efficiently to the accident. Once on scene the medical information about the injured would be transmitted to nearby emergency rooms, and those rooms with the necessary equipment and space would respond (within the Emergency Room of the near future, the equipment and room exchange information to ensure the right devices are in the right place at the right time). The traffic system would now reverse to ensure the first responders can move efficiently to the not just the closest hospital but the best hospital with the staff and equipment required to deal with the specific trauma. During the transport from the accident to the emergency room, the patients' vital health information is constantly relayed to the hospital.

For all of these views of the future, the technology already exists. The protocols and devices are already available, but at this moment in time, the problem is interoperability. For the few years companies have been attempting to build system to translate and interconnect this various systems in one-off and unique ways. It has turned out to be much more complicated and generally not cost effective. During the demonstrations from the White House Smart America Challenge in 2014, these concepts were put into practice. With the impetus of the White House, companies came together to do work with open standards and build systems to deliver these ideas.

Not all of the issues have been solved. IPv6 is not yet fully deployed and available. As of September 2014, the United States "ran out" of IPv4 address. To be clear, this does not mean that the Internet stops working, but it does mean that adding new systems to the Internet is a bit more difficult. Addresses are still available but much more expensive—like real estate in New York City or any large downtown. But, as of October 2014, only about 20% of the world's top networks are running IPv6, and only about 6% of the top one million websites are using IPv6, so clearly much work needs to be done to expand the deployment of this next generation of IP. As the deployment of the IoT and small embedded devices expands, the demand for addresses and therefore networks supporting IPv6 will increase. The IoT will likely be the tipping point for IPv6 use. But it is not just about connectivity at the network layer. As mentioned, the useful connection of devices requires an interoperable stack, the protocols at all of the various layers. Alliances such as the Internet Protocol for Smart Objects (IPSO) have been working to educate not only about the benefits of the adoption of open standards but also the best practices for building embedded systems and embedded networking for the IoT.

11.7 AN IoT DATA MODEL

In 2013, IPSO published the IPSO Application Framework, which defined a data model for the IoT. IPSO has been clear to point out that this is just one path for interoperability, but this data model is protocol agnostic, meaning that if an application requires specific services (real time, quality of service, acknowledgements, multicasting), it is free to utilize the most appropriate—the data objects are not tied to any specific protocol. In 2014 IPSO published the Smart Object Starter Pack to provide examples of how to create interoperable data objects. These examples included objects for controlling lights or controlling home temperature or measuring energy usage or temperature. With these and based on common open standard protocol, the product engineer can build truly interoperable devices and applications and can easily connect them to other devices or the cloud.

And IPSO is not the only Alliance on the IoT landscape. As of Q4 2015 there are at least 10 groups vying for thought leadership of the IoT—IPSO, Thread Group, AllSeen, Industrial Internet Consortium, Open Connectivity Forum formerly known as the Open Interconnect Consortium, oneM2M, ZigBee, Z-Wave, UPnP, and LoRa. Each group is working on some piece of the IoT puzzle with some focused on specific layers or on trying to define a complete solution and some utilizing open standards and others not. It is important, though, to see that as time continues many are moving from a closed proprietary design to an open system embracing IP.

In order to achieve ubiquity for the IoT, any data model needs to deal with Identity and Life Cycle Management and the previously mentioned data privacy. These are critical pieces of any complete IoT system. As of mid-2015 work has been started on Identity Management to understand concepts like immutable identity and to hammer out concepts like data ownership. The latter must be understood.

11.8 THE PROBLEM OF DATA OWNERSHIP

If a utility company places an electric meter on the side of house and this "smart meter" collects data about energy usage in the home, who owns that data? Is the information about energy consumption the property of the homeowner? After all it is personal information about how much air conditioning is used or how long lights are left on. Is the information owned by the utility? They actually own the smart meter and manage the energy provided to the home and generate the energy being used. Or do they? Maybe the data is owned by the smart meter manufacturer because they are leasing the meters and provide the actual energy management system. And how does this all change when the home owner generates their own energy via renewables such as solar or wind? Now consider more personal information, such as health information or daily habits, that might be collected by IoT devices and the privacy and data ownership rights. These are still open questions. As we move from IoT devices on our walls to IoT devices on our wrist (wearables) to IoT in our bodies (ingestables), these issues of privacy rights and data ownership multiply.

11.9 MANAGING THE LIFE OF AN IoT DEVICE

No one thinks twice about throwing out a burned-out light bulb—what else would you do with it? But what if it is a "smart bulb" that is connected to a home network? It therefore contains security information, perhaps encryption keys, that allows that bulb to talk to other devices on the network. If that bulb were to fall into the wrong hands and the security information was extracted, then the attacker could communicate with devices in the home including door locks. Putting a bulb in the trash might be the same as putting the keys to the front door in the trash. As with disposing of a computer, everyone is now advised to "wipe" the disk; it will be necessary to "wipe" these devices to protect the networks they were connected to. Additionally, adding 10 or 100 or 1000 devices to a network is pretty well understood, but what about 1 or 10 or 100 million. Today it takes some time and effort to activate a new phone or set up a new computer, and each person generally has only one or a few of these, but a home could have 100.

11.10 CONCLUSION: LOOKING FORWARD

Mark Weiser, widely considered to be the father of ubiquitous computing, said, "the most profound technologies are those that disappear … they weave themselves into the fabric of everyday life until they are indistinguishable from it." Even with issues and problems left to be solved, the onward march of technology and innovation continues. It will not wait for all questions to be answered. Henry Ford didn't stop building cars because there were no roads everywhere or gas stations or auto insurance and Vint Cerf and the early Internet pioneers didn't wait to build out the original ARPANET, just as the IoT can't and won't wait for full deployment of IPv6 and solutions to privacy, but will instead push these to happen. The benefits of these systems will be profound in increasing the efficiency of our transportation and healthcare systems, making our homes safer and making our cities more livable and sustainable, and steps can be taken today without all of the answer, but it is important not to ignore these questions and to keep an eye on the design and consideration of solutions.

FURTHER READING

Cicileo, G., Gagliano, R., O'Flaherty, C., et al. (October 2009). *IPv6 for All: A Guide for IPv6 Usage and Application in Different Environments*, ISOC Argentina chapter. Accessed March 6, 2016.

Google IPv6 Statistics. http://www.google.com/intl/en/ipv6/statistics.html. Accessed March 6, 2016.

IPv6 Address Allocation and Assignment Policy, RIPE Network Coordination Centre. https://www.ripe.net/publications/docs/ripe-512. Accessed March 6, 2016.

12

WIRELESS SENSOR NETWORKS

DAVID Y. FONG
CITS Group, San Jose, CA, USA

12.1 INTRODUCTION

Wireless Sensor Networks (WSN) play a critical role in Internet of Things (IoT). They are the interface of the IoT system to the physical world, collecting information from the terminals and sending it back to the system. Each node of the network is responsible for sensing and collecting certain physical properties, sending and receiving information to and from other nodes in the network, and aggregating the information. The information collected is sent to a gateway and fed into the Internet.

In the past decade, the cost of sensors, actuators, microprocessors, and network infrastructures has come down rapidly, making the WSN a practical reality. Many applications that were too expensive to operate are now within reach with reasonable cost.

In many cases, the sensor nodes are deployed in remote areas or difficult-to-access locations, drawing operating power from a battery, and communicate wirelessly. Due to this nature, the sensor nodes need to consume very little power and be able to self-organize in the context of network configuration in case of individual node failure.

However, even in cases where sensors are sufficiently small, smart, and inexpensive, challenges remain. In addition to power consumption, they include data security and interoperability.

Internet of Things and Data Analytics Handbook, First Edition. Edited by Hwaiyu Geng.

Companion website: www.wiley.com/go/Geng/iot_data_analytics_handbook/

In this chapter, we describe the characteristics of the WSN and analyze the power constraints, the network operation, the operating system, and the computing paradigm resulted from these characteristics.

12.2 CHARACTERISTICS OF WIRELESS SENSOR NETWORKS

12.2.1 Power Consumption Constraints

Sensors can be powered through either connected power or batteries [1, 2]. Connected power sources are constant but may be impractical or expensive in remote applications or hard-to-reach locations. Batteries represent a viable alternative, but battery life, charging, and replacement represent significant issues that cannot be ignored [3].

There are two dimensions to the issue of power:

1. Efficiency: Modern sensors are very efficient in terms of power consumption due to the advancement of silicon technologies. In some cases, sensors can stay operational for over 10 years on batteries [3]. Yet due to the lower cost of deploying sensors, the number of sensors used in a network increases, thus eliminating the low power advantage and even increasing power consumption.
2. Source: Most of the time, the sensor nodes are deployed at a remote place or in hard-to-reach locations. The availability of power is limited and replacing the battery is difficult. Under this condition, the power consumption of the sensor nodes must be restricted. While sensors often depend on batteries, energy harvesting of various energy sources may provide some alternatives [4]. However, energy-harvesting technologies that are currently available are expensive, and companies are hesitant to make the investment given the unreliability associated with the supply of alternative power [4].

Because of this constraint, the operating system of the sensor nodes, the organization of the network, the protocol used in network communication, and the computing paradigm all have to be designed or constructed to conserve power.

As alluded, an option in replenishing the battery is harvesting the energy from ambient environment. Various power-harvesting mechanisms have been proposed [5]. These include photoelectric conversion, piezoelectric, magnetic induction, and thermoelectric. The harvested power can be used directly, and excess power can be used to recharge the battery.

12.2.2 Energy-Harvesting Sensor Nodes

Due to the nature of deploying IoT sensor nodes, many sensor nodes are placed at remote areas or difficult-to-reach places. Also due to the untethered nature of the sensor nodes, most of the power source of the sensor nodes has to be the battery. Yet the use of battery alone at times does not fulfill the design goals of lifetime, cost, sensing reliability, and sensing and transmission coverage [5]. Thus energy-harvesting mechanisms that convert ambient energy into electricity were proposed as an augmented source of energy. If the harvested energy source is sufficient and stable, available either continuously or with predictable period, the sensor node can be powered for an extremely lengthy time [6].

In addition, the sensor node can optimize its energy usage during the period when the energy source is not available, making the operation viable for periodically available energy source. The challenge lies in estimating the periodicity and magnitude of the harvestable source and avoiding premature energy depletion before the next harvesting cycle [5].

A typical energy-harvesting system has three components: the energy source, the harvesting architecture, and the load. The energy source can be classified into two categories: ambient energy sources, such as solar, wind, and RF energy; and human power, energy harvested from body movements of human [7–9].

The energy sources are usually intermittent. The amount of power harvested is usually measured in the microwatt range. In the following table, the typical amount of energy harvested is shown [5, 10, 11].

Energy Sources and the Energy They Can Generate

Energy source	Typical energy level generated	Typical application
Small solar panels	100s of mW/cm^2 (direct or indirect sunlight)	Handheld electronic devices
Wind	1,200 mWh/day	Handheld electronic devices or remote wireless actuators
Seebeck devices (which convert heat energy into electrical energy)	10s of μW/cm^2 (from body heat)	Remote wireless sensors
	10s of mW/cm^2 (from furnace exhaust stack)	Remote wireless actuators
Piezoelectric devices (which produce energy by either compression or deflection of the device)	100s of mW/cm^2	Handheld electronic devices or remote wireless actuators
RF energy from an antenna	100s of pW/cm^2	Remote wireless sensors
Finger motion	2 mW	Small electronic devices
Footfalls	5 W	Small electronic devices
Breathing	100s of μW	Remote wireless sensors
Blood pressure	100s of μW	Remote wireless sensors

Energy-harvesting architecture can be broadly divided into two categories [5]: (1) *harvest-use*, where energy is harvested and used immediately; and (2) *harvest-store-use*, where energy is harvested whenever possible and stored for future use. The energy storage devices can be capacitors or thin-film batteries.

Two storage technologies, nickel–metal hydride (NiMH) and lithium based, emerge as good choices for energy-harvesting nodes [5]. Lithium batteries are characterized by high output voltage, energy density, efficiency, and moderately low self-discharge rate. They do not suffer from *memory effect*—loss of energy capacity due to repeated shallow recharge. However, lithium batteries require *pulse charging* for recharge—a high pulsating charging current. Usually an auxiliary battery or a charging circuit is required for this purpose. On the other hand, NiMH batteries can be *trickle* charged, that is, directly connected to an energy source for charging, and do not need complex pulse charging circuits. They have reasonably high energy, power density, and number of recharge cycles though NiMH batteries do suffer from memory effect. Additionally, the charge–discharge efficiency of NiMH batteries is lower than Lithium-based batteries. Both storage technologies have pros and cons, and the choice depends on the trade-off dictated by the application requirements and constraints.

Alternatively, supercapacitors can be used as storage components or along with rechargeable batteries. Like batteries, they self-discharge but at a much higher rate than batteries, as much as 5.9% per day [12]. Additionally, a supercapacitor's weight-to-energy density is very low, only 5 Wh/kg as compared to 100 Wh/kg of NiMH batteries. However, supercapacitors have high charge–discharge efficiency (97–98%) and also do not suffer from memory effect. Supercapacitors can also be trickle charged like NiMH batteries and hence do not need complex charging circuitry. Theoretically, supercapacitors have infinite recharge cycles [13]. Thus, supercapacitors are useful storage elements in locations where ample energy is available at regular intervals. They can also be used to buffer the available energy if the energy source is jittery, that is, the supercapacitor is trickle charged and a stable discharge from the capacitor charges the battery.

Energy-harvesting WSN is not suitable for every situation. The applications where microenergy-harvesting system makes sense are aircraft corrosion sensors, autodimming windows, bridge monitors, building automation, electricity usage meters, gas sensors, health monitors, HVAC controls, light switches, remote pipeline monitors, and water meters [10].

12.2.3 Ability to Cope with Node Failure

It is important to recognize that the nodes in a wireless sensor network will fail. When failure of a node occurs, the connections around the node in the network are broken, and services the node was providing cease to exist. Any self-organization algorithm has to be cognizant of device failures and capable of managing service recovery after the failures occur. The effectiveness of service failure recovery depends on structure of the network and the positions of devices in the network [14].

12.2.4 Security of Sensor Nodes

Security is definitely a key feature when considering IoT deployment [1, 15]. Within the IoT system, data integrity and authenticity are major security concerns at the sensor node level. Yet due to limitations such as low processing power, memory capacity, and power availability at the sensor level, light-weight communication protocols are preferable [16]. As a result, this might preclude the use of complex cryptographic algorithms for data integrity.

One such light-weight protocol, Constrained Application Protocol (CoAP) [17], seeks to apply the same application transfer paradigm and basic features of HTTP to constrained networks, while maintaining a simple design and low overhead. While CoAP is well suited for energy-constrained sensor systems, it does not come with built-in security features. CoAP is vulnerable to the usual Internet attacks, including attacks from the external networks such as distributed deny of service (DDoS). These attacks not only bring new security problems but also could drain the battery of constrained nodes. The problem is not the lack of available security solutions but that the implementations often do not support those since the lightness is preferred over security [1, 18, 19].

More research is needed in the preventive measures and the disaster recovery mechanisms [15].

12.2.5 Interoperability

Many sensor systems are designed for specific application and, when interconnected, lead to interoperability issues in areas of communication, exchange, storage and security of data, and scalability [1]. To resolve this issue, communication protocols are needed to enable communications between heterogeneous sensor systems.

Yet the conventional communication protocols such as HTTP are not suitable for sensor networks due to the constraints of sensor systems such as limited power availability, small amount of available memory, and limited processing power. As such, light-weight communication protocols such as CoAP are preferable. It uses the same REST model [1] as HTTP. From a developer's point of view, CoAP is much like HTTP, and that makes the integration process easier. Obtaining a value from a sensor is just like obtaining a value from a Web API. Since both HTTP and CoAP share the same REST model, they can be connected using application-agnostic cross-protocol proxies.

12.3 DISTRIBUTED COMPUTING

We can identify four major components in IoT applications: sensors, communications, computation, and service. Large amount of data are collected and analyzed to reveal information and create knowledge. To support such large-scale data size and computation tasks, it is not feasible to send every bit of data to centralized cloud servers. A feasible solution is to distribute computation to every node in IoT [20].

One solution is what HP termed Distributed Mesh Computing [21] and what Cisco termed Fog Computing [22]. Distributed mesh computing consists of small servers "out on the edge," medium-sized servers and storage in the middle that store data, and big servers in a central location where the heavy analytics and long-term storage take place.

The local computing may determine that some data should be transmitted to the cloud for more analysis, but in this scenario not all data will need to go back and forth to the cloud for processing.

All of these servers and nodes will work together in a distributed mesh computing system to serve the massive growth in data and connected devices expected in an IoT system.

Key principles of distributed mesh computing include the following [21]:

1. Distribute, compute, and storage everywhere.

   Each node has the computing power and local storage for distributed computing. The data and processing is close to where the event is, thus promoting efficiency.
2. Move data and/or analysis as needed.

   The locally collected data can be aggregated to different extent and send to other nodes as needed. Each device is low cost, with some computing power. Further analysis can be done in “upper” level.
3. Assume intermittent network connectivity.

   Network connectivity will fail. However, with distributed mesh computing, the network can be reorganized and data can be routed differently up the level.
4. Ultralow power requirements.

   Since each sensor node requires low energy and less space, the cost of the system is lowered.

12.4 PARALLEL COMPUTING

In developing IoT and the concept of smart world, we are essentially solving many complex optimization problems. How to efficiently use the embedded, distributed, or hosted computing power in IoT is the fundamental principle of addressing the challenges in constructing a smart world [23]. In other words, IoT needs High-Performance Computing (HPC) to resolve its optimization problems.

New approaches to HPC are made possible by advances in microprocessor architectures which include proliferation of parallelism and high-bandwidth networks. Several solutions exist, such as Graphics Processing Unit (GPU) computing, Many Integrated Core (MIC) computing, heterogeneous computing, Peer-to-Peer (P2P) computing, Cloud computing, and Grid computing [23].

12.4.1 GPU Computing

GPUs [24, 25] are highly parallel, multithreaded many-core architectures. They are better known for image processing. Nevertheless, NVIDIA introduced in 2006 Compute Unified Device Architecture (CUDA), a parallel programming platform and technology that enables users to use GPU accelerators in order to address general-purpose parallel applications.

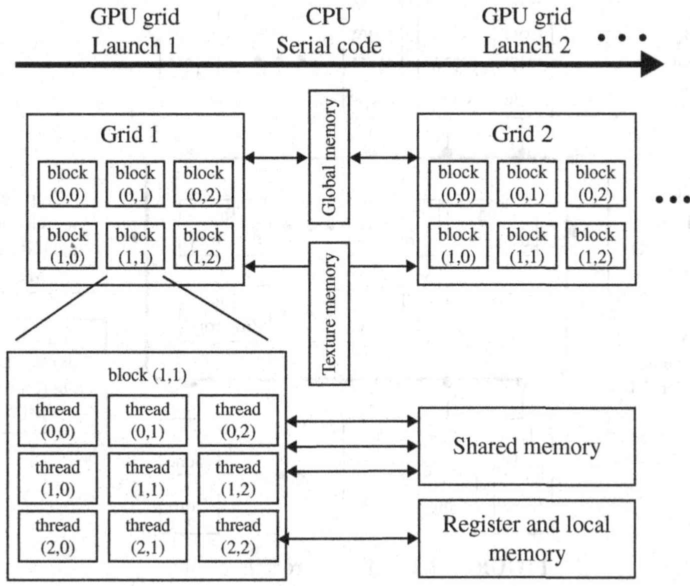

FIGURE 12.1 Thread and memory hierarchy in a GPU.

As shown in Figure 12.1 [23], a parallel code on GPU situated in the device is interleaved with a serial code executed on the CPU at the host. The parallel threads are grouped into blocks which are in turn organized in a grid. The grid is launched via a single CUDA program.

GPUs function as accelerators with many cores. Overall, they are low cost and run on less energy than other computing devices. As we deploy IoT applications, the low cost and low energy usage feature is an enabling factor. GPUs can combine with CPUs to build efficient heterogeneous computing platforms.

12.4.2 MIC

In 2013, Intel released the Xeon Phi, a coprocessor for HPC. Different configurations of the coprocessor contain 57, 60 or 61, ×86 processor cores, interconnected by a high-speed bidirectional ring (Figure 12.2). The architecture of a core is based on the Pentium architecture for ease of programming. Each core can hold four hardware threads (two per clock cycle and per ring's direction).

The Xeon Phi is connected to the CPU via the PCIe connector. The memory controllers and the PCIe client logic provide a direct interface to the GDDR5 memory on the coprocessor and the PCIe bus, respectively. The design of the coprocessor permits one to run existing applications parallelized via OpenMP or MPI; the MIC can be used either in offload mode or native mode [26, 27]. The peak double-precision floating-point [24] performance of the MIC is 1.2 TFLOPS which makes it a powerful computing accelerator [28].

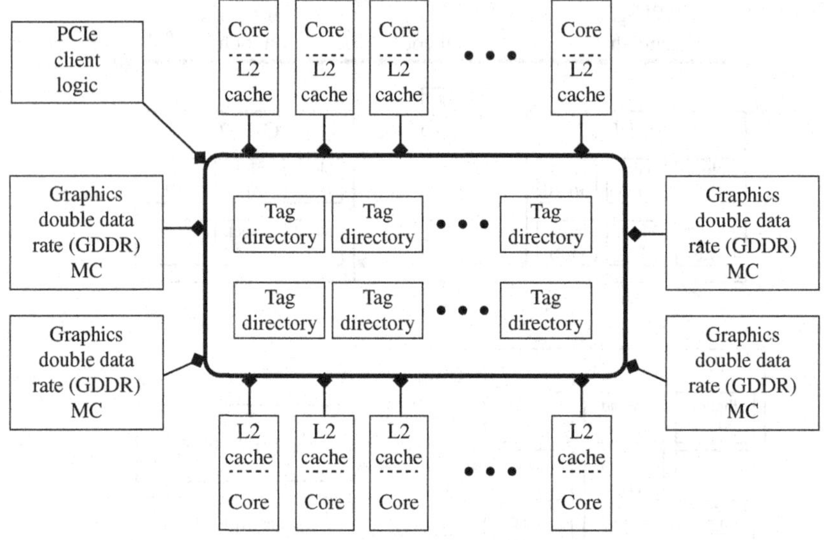

FIGURE 12.2 MIC microarchitecture.

12.4.3 Peer-to-Peer Computing

The P2P concept has been considered recently for parallel and distributed iterative methods with application to optimization and numerical simulation (see Refs. [29–31]). The Peer-to-Peer Distributed Computing (P2PDC) environment is a decentralized and robust environment for the implementation of HPC applications on peer-to-peer networks [23, 31].

P2PDC is intended for solving optimization problems in parallel via distributed iterative methods that lead to frequent direct data exchanges between peers [23]. P2PDC relies on the use of the P2P self-adaptive communication protocol P2PSAP [29] and a reduced set of communication operations (P2Psend, P2Preceive, and P2Pwait) in order to facilitate programming. From the programmer's point of view, he needs to concern only about the choice of distributed iterative scheme of computation (synchronous or asynchronous) that he wants to be implemented and does not need to be aware of the communication mode between any two machines. The programmer also has the option of selecting a hybrid iterative scheme of computation, whereby computations are locally synchronous and globally asynchronous.

P2PSAP chooses dynamically the most appropriate communication mode between any two peers according to decision taken at application level like scheme of computation and elements of context like network topology at transport level. In the hybrid case, the communication mode between peers in a group of machines that are close and that present the same characteristics is synchronous, and the communication mode between peers in different groups is asynchronous. The decentralized environment P2PDC is based on a hybrid topology manager and a hierarchical task allocation mechanism which make P2PDC scalable [23].

The reader is referred to Ref. [32] for a previous study on P2P computing.

12.4.4 Cloud Computing

Cloud computing [24] occupies an important position in IoT, as the main high-speed computing resources for processing data collected from the WSN. Two HPC concepts exist, and it is not clear which solution will prevail in the future [23]: Cloud HPC or HPC in the cloud. Nevertheless, one can receive HPC as a service like others in the cloud.

The cloud has started to be used for HPC applications. There exist now solutions like Amazon Web Services (AWS) cloud including EC2 for HPC, SGI Cyclone cloud, and IBM RC2. Hybrid solutions have also been proposed whereby cloud computing is supported by volunteer computing.

12.4.5 Summary

The fact that each sensor node has a processor unit embedded in it calls for parallel computing in order to exploit the available computing resources. All these small processors embedded in the sensor nodes do not locate within a box, but rather scattered within some area. They do not share the same I/O but need to communicate through single memory and disk. For applications of sensors and actuators, longer wait in communications might be fine.

The programming should use “declarative programming” and not imperative, threaded programming.

Due to the more prevalent availability of GPU, parallel computing using GPUs is a viable option for IoT programming. However, programming infrastructure like CUDA is needed to make the tasks easier.

12.5 SELF-ORGANIZING NETWORKS

In IoT, the sensor nodes and devices rely on networks to interact in order to achieve common goals. However, this infrastructure is subject to deterioration, disaster, and other adverse conditions. In order to mitigate against the situation that renders the original network infrastructure less effective or useless, the ability to self-organize among these devices is needed to maintain communication resilience [14].

When we look at the IoT from a system perspective, all the devices envisioned in the IoT paradigm are part of a heterogeneous network. This heterogeneity includes device’s computation capability, network and communication technologies, and the services being offered by various devices in the IoT. Thus we treat the computation and the communication operations of these devices toward the common goal in IoT as one system.

The key properties that are critical for the efficient self-organization in the IoT include cooperative communication model, situational awareness, and automated load balancing. With the perspective of these properties, five key components of self-organization in the IoT can be identified:

1. Neighbor discovery
2. Medium access control
3. Local connectivity and path establishment
4. Service recovery management
5. Energy management

Challenges exist. They are the following:

- Cross-layer design for self-organization
- Heterogeneity for self-organization
- Multiradio multichannel communications
- Low power computing and load balancing
- Delay-tolerant networking over self-organized networks

12.6 OPERATING SYSTEMS FOR SENSOR NETWORKS

Due to the nature of sensor networks, which potentially consists of thousands of small sensor nodes operating in unfriendly environments, the operating system has to be able to handle event-centered concurrency without consuming substantial power and occupy a large footprint in storage. Several operating systems emerged in the research of sensor networks are described in Sections 12.6.1 and 12.6.2.

12.6.1 TinyOS

TinyOS fulfills the requirements for the operating systems for sensor networks in the following way [33]:

1. Limited resources: The application program image occupies small size with resolution at compile time: footprint optimization with whole-program compilation and cross-component optimization to reduce overhead.
2. Reactive concurrency: TinyOS can handle large numbers of potential concurrency by its component model; the compiler detects potential racing condition and modify the code by inlining.
3. Flexibility: The fine-grained components allow applications to be constructed from a large number of very fine-grained components; Bidirectional interfaces

and explicit support for events enable any component to generate events; hardware/software transparency makes the shifting of hardware–software boundary easy.

4. Low power: The application-specific nature of TinyOS ensures that no unnecessary functions consume energy.

12.6.2 LiteOS

LiteOS is an open-source, interactive, UNIX-like operating system designed for WSN [34]. With the tools that come with LiteOS, you can operate one or more WSN in a UNIX-like manner, transferring data, installing programs, retrieving results, or configuring sensors. You can also develop programs for nodes and wirelessly distribute such programs to sensor nodes.

The LiteOS 2.0 is the latest version of LiteOS. It runs on the following platforms: Windows XP/Vista/7, MICAz as target board, and MIB510/MIB520 as programming boards. Unlike 1.0, LiteOS 2.0 is closely integrated with AVR Studio 5.0. This brings multiple advantages, such as IDE editing, debugging, and built-in JTAG support.

12.7 WEB OF THINGS (WoT)

The Web of Things (WoT) is a software and programming architecture that allows real objects to be part of the World Wide Web.

The architecture has four layers [35]:

Accessibility layer

This layer makes accessing of the things via the Internet possible. It exposes the services of things via Web APIs. This is critical as it ensures things have a Web-accessible API, thus transforming things into programmable things.

Two main mechanisms exist in the access layer of WoT: the first one exposes their services through a RESTful API either directly or through a gateway [36]. Since REST is implemented in HPPT 1.1, for things offering RESTful APIs over HTTP, they get a URL and become integrated to the World Wide Web and browsers, hyperlinked HTML pages, and JavaScript applications [37].

On the other aspect, the request–response nature of HTTP is seen as one of the limitations for IoT use cases as it does not match the event-driven nature of applications that are common in the WSN. To overcome this shortcoming while keeping a focus on fostering integration with the Web, several authors have suggested the use of HTML5 WebSockets either natively or through the use of translation brokers (e.g., translating from MQTT or CoAP to WebSockets). This complements the REST API of things with a publish–subscribe mechanism that is largely integrated with the Web ecosystem [38–40].

Due to the fact that battery-powered devices such as wireless sensor nodes have to conserve energy and thus would prefer not to connect directly to the Internet, devices can access the Internet through Smart Gateways. Smart Gateways are protocol translation gateways at the edge of the network [41, 42].

Findability layer

The function of this layer is to provide a way to find and locate things on the Web. It is influenced by the semantic Web [38, 40].

One reasonable thought is to reuse Web semantic standards to describe things and their services, in an approach that reuses the standards in a different context. Specifically, this includes HTML5 Microdata Integration, RDF/RDFa, JSON-LE and EXI [38, 40]. This enables searching for things through search engines and other Web indexes as well as enabling machine-to-machine interaction based on a small set of well-defined formats and standards.

Sharing layer

The WoT is largely based on the idea of things pushing data to the Web where more intelligence and big data patterns can be applied as an example to help us manage our health (Wearables), optimize our energy consumption (Smart Grid), etc.

The motivation for establishing the WoT is to enable things to push data to the Web and use the richer resources on the Web for data aggregation and big data analysis. This can only happen in a large scale if some of the data can be efficiently shared across services. The sharing layer ensures that data generated by things can be shared in an efficient and secure manner.

Several approaches toward a granular and social context-based sharing have been proposed such as the use of social network to build a Social Web of Things [43, 44].

Composition layer

Through the composition layer, the services and data of things can be integrated into higher-level Web tools, for example, analytics software, and mash-up applications such as IFTTT [45], making it simple to create applications involving objects, things, and virtual Web services.

Tools in the composition layer include Web toolkits (e.g., JavaScript SDKs offering higher-level abstractions) [39], dashboards with programmable widgets [46], and Physical Mash-up tools [38]. Inspired by Web 2.0 participatory services and in particular Web mash-ups, the Physical Mash-ups offer a unified view of the classical Web and WoT and enable users to build applications using the WoT services without programming skills [38].

12.8 WIRELESS SENSOR NETWORK ARCHITECTURE

Most common architecture for WSN follows the OSI Model. Basically in sensor network we need five layers: application layer, transport layer, network layer, data link layer, and physical layer. Added to the five layers are the three cross-layers/planes as shown in Figure 12.3 [47].

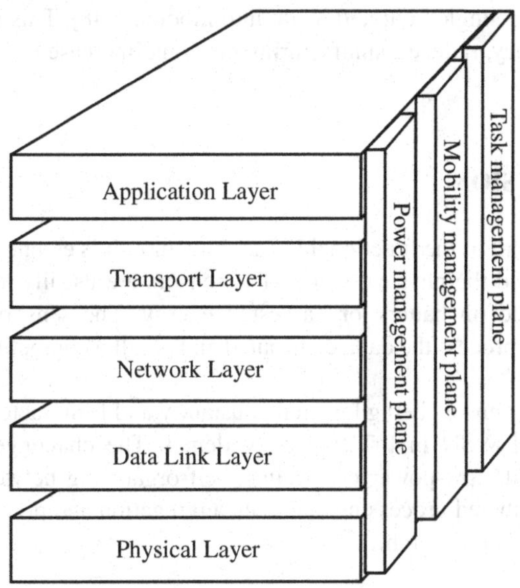

FIGURE 12.3 WSN architecture.

TABLE 12.1 Difference of Architecture between WSN, WLAN, and OSI

Wireless Sensor Network	WLAN	OSI Model
WSN application	Application program	Application layer
WSN middleware	Middleware	Presentation layer
	Socket API	Session layer
WSN transport protocols	TCP/UDP	Transport layer
WSN routing protocols	IP	Network layer
Error control	WLAN adaptor and device driver	Data link layer
WSN MAC protocols	WLAN MAC protocols	
Transceiver	Transceiver	Physical layer

The three cross-planes or layers are power management plane, mobility management plane, and task management plane. These layers are used to manage the network and make the sensors work together in order to increase the overall efficiency of the network [47].

The difference of architectures between OSI, WLAN, and WSN are shown in Table 12.1 [47].

12.9 MODULARIZING THE WIRELESS SENSOR NODES

The wireless sensor nodes currently are manufactured with specific sensors integrated into the sensor node. With the expansion of WSN, the large-scale manufacturing of sensor nodes is necessary. When we examine the sensor nodes, we will see that portions of the sensor node design are identical in various sensor nodes. These identical designs can then be singled out and made into modules [48]. This will improve the production efficiency, reduce manufacturing cost, and increase the flexibility of the sensor node usage.

12.10 CONCLUSION

WSN is a critical part of the functional IoT architecture. It is essentially the interface between the IoT and the physical world. The sensors are usually located in remote areas and depended on battery or harvested energy. The sensors generate large amount of data. To process the data, distributed and parallel computing methods have to be employed.

With the cost of sensors being lowered constantly and form factor being reduced, the applications of WSN in IoT will be widened. The challenges ahead in IoT development include low-power constraints, self-organizing networks, security of sensor nodes, distributed processing, and data aggregation and processing. All these areas will need further research.

REFERENCES

[1] J. Holdowsky, M. Mahto, M. Raynor, and M. Cotteleer, *Inside the Internet of Things*, Deloitte University Press, Westlake, August 21, 2015, accessed February 15, 2016.

[2] S.K. Singh, M.P. Singh, and D.K. Singh, Routing protocols in wireless sensor networks: A survey, *International Journal of Computer Science and Engineering Survey (IJCSES)* **1**, 2, 2010, http://www.airccse.org/journal/ijcses/papers/1110ijcses06.pdf accessed February 23, 2015.

[3] Powercast Corporation, Lifetime Power Wireless Sensor System, http://www.powercastco.com/PDF/wireless-sensor-system.pdf, 2012, accessed January 29, 2015.

[4] G. Zhou, L. Huang, W. Li, and Z. Zhu, Harvesting ambient environmental energy for wireless sensor networks: A survey, *Journal of Sensors* 2014, article ID 815467, 2014. 10.1155/2014/815467 accessed February 23, 2015.

[5] S. Sudevalayam and P. Kulkarni, Harvesting sensor nodes: Survey and implications, *IEEE Communications Surveys and Tutorials* **13**, 3, pp. 443–461, THIRD QUARTER 2011, Hoboken, New Jersey.

[6] A. Kansai, J. Hsu, S. Zahedi, and M.B. Srivastava, Power management in energy harvesting sensor networks, *Transactions on Embedded Computing Systems* **6**, 4, p. 32, 2007.

[7] J.A. Paradiso and M. Feldmeier, A Compact, Wireless, Self-Powered Pushbutton Controller, Proceedings *of the 3rd International Conference on Ubiquitous Computing*, Atlanta, GA, September 2001, pp. 299–304, Springer-Verlag, Berlin/Heidelberg.

[8] N. Shenck and J. Paradiso, Energy scavenging with shoe-mounted piezoelectrics, *IEEE Micro* **21**, 3, pp. 30–42, May/Jun 2001.

[9] J. Kymissis, C. Kendall, J. Paradiso, and N. Gershenfeld, Parasitic Power Harvesting in Shoes, *Second International Symposium on Wearable Computers*, Pittsburgh, PA, October 1998, pp. 132–139.

[10] T. Armstrong, Harvesting Energy for Wireless Sensors, *Machine Design*, March 24, 2014 http://machinedesign.com/batteriespower-supplies/harvesting-energy-wireless-sensors accessed August 23, 2016.

[11] T. Starner, Human-powered wearable computing, *IBM Systems Journal* **35**, 3–4, pp. 618–629, 1996.

[12] J. Taneja, J. Jeong, and D. Culler, Design, Modeling, and Capacity Planning for Micro-Solar Power Sensor Networks, *Proceedings of the 7th International Conference on Information Processing in Sensor Networks*, St. Louis, MO, April 22–24, 2008, pp. 407–418, IEEE, Washington, DC.

[13] X. Jiang, J. Polastre, and D. Culler, Perpetual Environmentally Powered Sensor Networks, *Fourth International Symposium on Information Processing in Sensor Networks*, Los Angeles, CA, April 15, 2005, pp. 463–468.

[14] A.P. Athreya and P. Tague, Network Self-Organization in the Internet of Things, *IEEE International Workshop on Internet-of-Things Networking and Control* (IoT-NC 2013), New Orleans, LA, June 24, 2013.

[15] H. Suo and C. Zou, Security in the Internet of Things: A Review, *Proceedings of the 2012 International Conference on Computer Science and Electronic Engineering (ICCSEE 2012)*, Hangzhou, Zhejiang China. March 23, 2012, pp. 648–651, doi: 10.1109/ICCSEE.2012.373.

[16] M.O. Farooq and T. Kunz, Operating Systems for Wireless Sensor Networks: A Survey, Department of Systems and Computer Engineering, Carleton University Ottawa, Canada, May 31, 2011, http://www.mdpi.com/1424-8220/11/6/5900/pdf, accessed February 2, 2015.

[17] LiteOS User's Guide, Version 2.1, last updated: October 2, 2011, http://lanterns.eecs.utk.edu/software/liteos/LiteOS_User_Guide.pdf, accessed September 8, 2016.

[18] A. Ludovici, P. Moreno, and A. Calveras, TinyCoAP: A novel Constrained Application Protocol (CoAP) implementation for embedding RESTful Web services in wireless sensor networks based on TinyOS, *Journal of Sensor and Actuator Networks* **2**, 2, pp. 288–315, 2013.

[19] M. Ilaghi, T. Leva, and M. Komu, Techno-Economic Feasibility: Analysis of Constrained Application Protocol, *Proceedings of the IEEE World Forum on Internet of Things* (WF-IoT), Seoul, Korea, March 6–8, 2014, http://www.leva.fi/wp-content/uploads/CoAP-feasibility_wf-iot_accepted_manuscript.pdf accessed April 2, 2015.

[20] S.-Y. Chien, W.-K. Chan, Y.-H. Tseng, C.-H. Lee, V.S. Somayazulu, and Y.-K. Chen, Distributed Computing in IoT: System-on-a-Chip for Smart Cameras as an Example, *Proceedings of 20th Asia and South Pacific Design Automation Conference* (ASP-DAC), Chiba, Japan, January 19–22, 2015, pp. 130–135.

[21] B. Patrick, IoT and Distributed Mesh Computing, *Connected Cloud Summit*, Boston, Massachusetts, September 18, 2014.

[22] K. Lee, Fog Computing, Paradigm Shift or Same Old Same Old? IT Knowledge Exchange, TotalCIO, http://itknowledgeexchange.techtarget.com/total-cio/fog-computing-paradigm-shift-or-same-old-same-old/ accessed February 10, 2016.

[23] D. El Baz, IoT and the Need for High Performance Computing, *Proceedings of 2014 International Conference on Identification, Information and Knowledge in the Internet of Things,* Beijing, China, October 17–18, 2014.

[24] K. Hwang, G. Fox, and J. Dongarra, *Distributed and Cloud Computing, from Parallel Processing to the Internet of Things*, Morgan Kaufmann, Boston, MA, 2012.

[25] R. Fernando, *GPU Gems: Programming Techniques, Tips and Tricks for Real-Time Graphics*, Addison Wesley Professional, Boston, MA, 2004.

[26] J. Jeffers and J. Reinders, *Intel Xeon Phi Coprocessor High-Performance, Programming*, Morgan Kaufmann/Elsevier Science & Technology Books, Boston, MA, 2013.

[27] R. Rahman, *Intel Xeon Phi Coprocessor Architecture and Tools: The Guide for Application Developers*, 1st ed., Apress, Berkeley, CA, 2013.

[28] B. Plazolles, D. El Baz, M. Spel, and V. Rivola, Comparison between GPU and MIC on Balloon Envelope Drift Descent Analysis, LAAS Report No. 14549, Toulouse, France, November 2014.

[29] D. El Baz and T.T. Nguyen, A Self-Adaptive Communication Protocol with Application to High Performance Peer to Peer Distributed Computing, *Proceedings of 18th International Conference on Parallel, Distributed and network based Processing*, Pisa, Italy, February 17–19, 2010, pp. 323–333.

[30] T.T. Nguyen, D. El Baz, P. Spiteri, G. Jourjon, and M. Chau, High Performance Peer-to-Peer Distributed Computing with Application to Obstacle Problem, *Proceedings of HOTP2P in conjunction with the Symposium IEEE IPDPS 2010*, Atlanta, USA, April 19–23, 2010.

[31] T. Garcia, M. Chau, T.T. Nguyen, D. El Baz, and P. Spiteri, Asynchronous Peer-to-Peer Distributed Computing for Financial Applications, *Proceedings of the* 25th *IEEE Symposium IPDPSW* 2011*/PDSEC 2011*, Anchorage, USA, May 19–23, 2011.

[32] G. Jourjon and D. El Baz, Some Solutions for Peer to Peer Global Computing, *Proceedings of the 13th Conference on Parallel, Distributed and Network-based Processing, PDP* 2005, Lugano, Suisse, February 9–11, 2005, IEEE CPS, pp. 49–58.

[33] P. Levis, S. Madden, J. Polastre, R. Szewczyk, K. Whitehouse, A. Woo, D. Gay, J. Hill, M. Welsh, E. Brewer, and D. Culler, TinyOS: An Operating System for Sensor Networks, Computer Science Division, Electrical Engineering and Computer Sciences Department, University of California Berkeley, https://people.eecs.berkeley.edu/~culler/papers/ai-tinyos.pdf accessed September 8, 2016.

[34] Q. Cao, T. Abdelzaher, J. Stankovic, and T. He, The LiteOS Operating System: Towards Unix-Like Abstractions for Wireless Sensor Networks, *IEEE International Conference on Information Processing in Sensor Networks 2008*, St. Louis, Missouri, April 22–24, 2008, pp. 233–244.

[35] Wikipedia. https://en.wikipedia.org/wiki/Web_of_Things accessed August 23, 2016.

[36] D. Guinard, V. Trifa, F. Mattern, and E. Wilde, *From the Internet of Things to the Web of Things: Resource Oriented Architecture and Best Practices*. Springer, Berlin/Heidelberg/New York, 2011, pp. 97–129.

[37] D. Guinard and V. Trifa, Towards the Web of Things: Web Mashups for Embedded Devices, Workshop on Mashups, Enterprise Mashups and Lightweight Composition on the Web (MEM 2009), *Proceedings of WWW (International World Wide Web Conferences)*, Madrid, Spain, April 2009.

[38] D. Guinard, *A Web of Things Application Architecture—Integrating the Real-World into the Web*, ETH, Zurich, 2011.

[39] V. Trifa, *Building Blocks for a Participatory Web of Things: Devices, Infrastructures, and Programming Frameworks*, ETH, Zurich, 2011.

[40] Web of Things Interest Group Charter. https://www.w3.org/2014/12/wot-ig-charter.html accessed August 23, 2016.

[41] D. Guinard, V. Trifa, and E. Wilde, A Resource Oriented Architecture for the Web of Things. *Internet of Things 2010 International Conference (IoT 2010)*, Tokyo, Japan, November 29–December 1, 2010.

[42] V. Trifa, S. Wieland, D. Guinard, and B. Thomas, Design and Implementation of a Gateway for Web-Based Interaction and Management of Embedded Devices. *International Workshop on Sensor Network Engineering (IWSNE 09)*, Marina del Rey, CA, June 2009.

[43] D. Guinard, V. Trifa, and M. Fischer, Sharing Using Social Networks in a Composable Web of Things. *First IEEE International Workshop on the Web of Things (WOT2010)*, IEEE PerCom 2010, Mannheim, Germany, March 2010.

[44] T.-Y. Chung, I. Mashal, O. Alsaryrah, V. Huy, W.-H. Kuo, and D.P. Agrawal, Social Web of Things: A Survey. *International Conference on Parallel and Distributed Systems (ICPADS)*, Seoul, December 15–18, 2013, pp. 570–575. doi:10.1109/ICPADS.2013.102.

[45] About IFTTT. https://ifttt.com/wtf accessed August 23, 2016.

[46] D. Guinard, M. Mathias, and P. Jacques, Giving RFID a REST: Building a Web-Enabled EPCIS. *Internet of Things 2010 International Conference (IoT 2010)*, Tokyo, November 29 to December 1, 2010.

[47] A.A.A. Alkhatib and G.S. Baicher, Wireless Sensor Network Architecture, *2012 International Conference on Computer Networks and Communication Systems (CNCS 2012) IPCSIT* vol. 35, 2012, IACSIT Press, Singapore.

[48] M. Grisostomi, L. Ciabattoni, M. Prist, L. Romeo, Modular Design of a Novel Wireless Sensor Node for Smart Environment, *Proceedings of 2014 IEEE/ASME 10th International Conference on Mechatronic and Embedded Systems and Applications* (MESA), Senigallia, Italy, September 10–12, 2014.

FURTHER READING

A. Kansal, J. Hsu, S. Zahedi, and M.B. Srivastava, Power management in energy harvesting sensor networks, *Transactions on Embedded Computing Systems* **6**, 4, p. 32, 2007.

Y. Shu et al., Internet of Things: Wireless Sensor Networks, IEC White Paper, http://www.iec.ch/whitepaper/pdf/iecWP-internetofthings-LR-en.pdf, accessed February 15, 2016.

13

NETWORKING PROTOCOLS AND STANDARDS FOR INTERNET OF THINGS

Tara Salman and Raj Jain

Department of Computer Science Engineering, Washington University, St. Louis, MO, USA

13.1 INTRODUCTION

Internet of Things (IoT) and its protocols are among the most highly funded topics in both industry and academia. The rapid evolution of the mobile Internet, mini-hardware manufacturing, microcomputing, and machine-to-machine (M2M) communication has enabled the IoT technologies. According to Gartner, IoT is currently on the top of their hype cycle, which implies that a large amount of money is being invested on it by the industry. Billions of dollars are being spent on IoT enabling technologies and research, while much more is expected to come in the upcoming years [1].

IoT technologies allow things, or devices that are not computers, to act smartly and make collaborative decisions that are beneficial to certain applications. They allow things to hear, see, think, or act by allowing them to communicate and coordinate with others in order to make decisions that can be as critical as saving lives or buildings. They transform "things" from being passively computing and making individual decisions to actively and ubiquitously communicating and collaborating to make a single critical decision. The underlying technologies of ubiquitous computing, embedded sensors, light communication, and Internet protocols allow IoT to provide its significance; however, they impose lots of challenges and introduce the need for specialized standards and communication protocols.

Internet of Things and Data Analytics Handbook, First Edition. Edited by Hwaiyu Geng.

Companion website: www.wiley.com/go/Geng/iot_data_analytics_handbook/

In this chapter, we highlight IoT protocols that are operating at different layers of the networking stack, including Medium Access Control (MAC) layer, network layer, and session layer. We present standards protocols offered by the Internet Engineering Task Force (IETF), Institute of Electrical and Electronics Engineers (IEEE), International Telecommunication Union (ITU), and other standard organizations. These standards were proposed over the past half decade to meet IoT current and future needs.

The rest of the chapter is organized as follows: Section 13.2 describes the first layer of networking protocols, which is the data link layer and MAC protocols. Following that, Section 13.3 handles the network layer routing protocols, while Section 13.4 presents network layer encapsulation protocols and Section 13.5 handles the session layer protocols. Section 13.6 briefly summarizes the management, and Section 13.7 describes security mechanisms in key protocols. Section 13.8 gives some discussion points about IoT challenges. Finally, Section 13.9 summarizes our discussion and highlights the main points presented.

13.1.1 Related Works

There are several survey papers that handle different aspects of standardization in IoT. Examples of such surveys include a survey of IETF standards in Ref. [2], security protocols in Ref. [3], and application, or transport, layer standards in Ref. [4]. Other papers discuss a specific layer of standardizations such as communication protocols or routing. Most importantly, we recommend Ref. [5] to readers interested in more details. That paper summarizes some of the most important standards that are offered by different standards organizations. It also provides a discussion of different IoT challenges including mobility and scalability. In this chapter, we aim to provide a comprehensive survey of newly arising standards including some other drafts and protocols that were not discussed in Ref. [5]. This allows us to discuss more standards, add some of the recent standard drafts offered in IETF, and discuss state-of-the-art protocol that are expected to go for standardization in the future.

13.1.2 IoT Ecosystem

Figure 13.1 shows a seven-layer model of IoT ecosystem. At the bottom layer is the market or application domain, which may be smart grid, connected home, or smart health. The second layer consists of sensors that enable the application. Examples of such sensors are temperature sensors, humidity sensors, electric utility meters, or cameras. The third layer consists of interconnection layer that allows the data generated by sensors to be communicated, usually to a computing facility, data center, or a cloud. There the data is aggregated with other known data sets such as geographical data, population data, or economic data. The combined data is then analyzed using machine learning and data mining techniques. To enable such large distributed applications, we also need the latest application-level collaboration and communication software, such as software-defined networking (SDN), service-oriented architecture (SOA), etc. Finally, the top layer consists of services that enable the market and may include energy management, health management, education, transportation, etc. In addition to these seven layers that are built on the top of each other, there are security and management applications that are required for each of the layers and are, therefore, shown on the side.

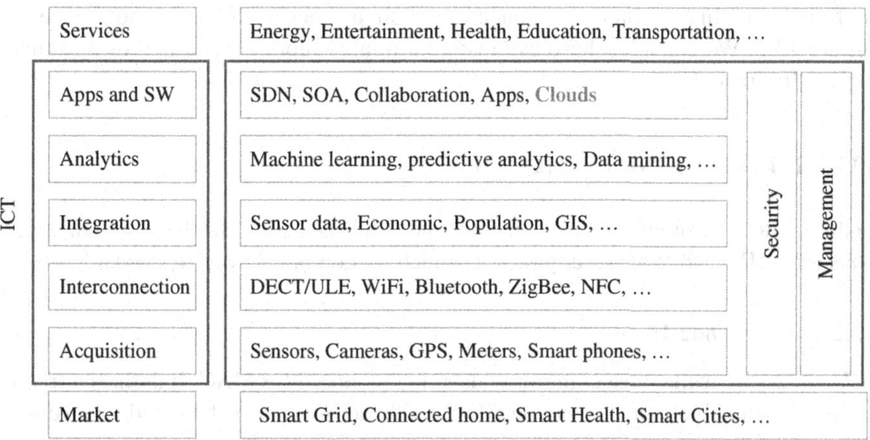

FIGURE 13.1 IoT ecosystem. © Tara Salman and Raj Jain. Use with permission.

Session		**MQTT, SMQTT, CoRE, DDS, AMQP, XMPP, CoAP, ...**
Network	**Encapsulation**	**6LowPAN, 6TiSCH, 6Lo,** Therad, ...
	Routing	**RPL, CORPL, CARP, ...**
Datalink		WiFi, **Bluetooth Low Energy, Z-Wave, ZigBee Smart, DECT/ULE,** 3G LTE, NFC, **Weighless, HomePlug GP, 802.11ah, 802.15.4e, G.9959, WirelessHART, DASH7,** ANT+, **LTE-A, LoRaWAN, ...**

Security	**Management**
TCG, Oath 2.0, SMACK, SASL, ISASecure, ace, DTLS, Dice, ...	**IEEE 1905, IEEE 1451, ...**

FIGURE 13.2 Protocols for IoT. © Tara Salman and Raj Jain. Use with permission.

In this chapter, we concentrate on the interconnection layer. This layer itself can be shown in a multilayer stack as shown in Figure 13.2. We have shown only the data link, network, and transport/session layers. The data link layer connects two IoT elements which generally could be two sensors or the sensor and the gateway device that connects a set of sensors to the Internet. Often there is a need for multiple sensors to communicate and aggregate information before getting to the Internet. Specialized protocols have been designed for routing among sensors and are part of the routing layer. The session layer protocols enable messaging among various elements of the IoT communication subsystem. A number of security and management protocols have also been developed for IoT as shown in the figure.

Many different standards and protocols for IoT have been proposed by various standards organizations. Prominent among them are IEEE, IETF, and ITU. These protocols are listed in Figure 13.2. Although we have tried to make the list as current as possible, new protocols are continuously being proposed and may appear in future. In this chapter, we concentrate on protocols shown in bold face in Figure 13.2. We consider these as most commonly recommended and/or designed especially for IoT.

13.2 IoT DATA LINK PROTOCOLS

In this section, we discuss the data link layer protocol standards. The discussion includes physical (PHY) and MAC layer protocols which are combined by most standards.

13.2.1 IEEE 802.15.4e

IEEE 802.15.4 is the most commonly used IoT standard for MAC. It defines a frame format, headers including source and destination addresses, and how nodes can communicate with each other. The frame formats used in traditional networks are not suitable for low-power multihop networking in IoT due to their overhead. In 2008, IEEE802.15.4e was created to extend IEEE802.15.4 and support low-power communication. It uses time synchronization and channel hopping to enable high reliability and low cost and meet IoT communications requirements. Its specific MAC features can be summarized as follows [6]:

- *Slotframe structure*: IEEE 802.15.4e frame structure is designed for scheduling and telling each node what to do. A node can sleep, send, or receive information. In the sleep mode, the node turns off its radio to save power and stores all messages that it needs to send at the next transmission opportunity. When transmitting, it sends its data and waits for an acknowledgment (ACK). When receiving, the node turns on its radio before the scheduled receiving time, receives the data, sends an acknowledgement, turn off its radio, delivers the data to the upper layers, and goes back to sleep.
- *Scheduling*: The standard does not define how the scheduling is done, but it needs to be built carefully such that it handles mobility scenarios. It can be centralized by a manager node which is responsible for building the schedule and informing others about the schedule, and other nodes will just follow the schedule.
- *Synchronization*: Synchronization is necessary to maintain nodes' connectivity to their neighbors and to the gateways. Two approaches can be used: acknowledgment-based or frame-based synchronization. In acknowledgement-based mode, the nodes are already in communication and they send ACK for reliability guarantees; thus they can be used to maintain connectivity as well. In frame-based mode, nodes are not communicating, and hence, they send an empty frame at prespecified intervals (about 30 second typically).
- *Channel hopping*: IEEE802.15.4e introduces channel hopping for time-slotted access to the wireless medium. Channel hopping requires changing the frequency channel using a predetermined random sequence. This introduces frequency diversity and reduces the effect of interference and multipath fading.

Sixteen channels are available which adds to network capacity as two frames over the same link can be transmitted on different frequency channels at the same time.

- *Network formation*: Network formation includes advertisement and joining components. A new device should listen for advertisement commands, and upon receiving at least one such command, it can send a join request to the advertising device. In a centralized system, the join request is routed to the manger node and processed there, while in distributed systems, they are processed locally. Once a device joins the network and it is fully functional, the formation is disabled and will be activated again if it receives another join request.

13.2.2 IEEE 802.11ah

IEEE 802.11ah is a light (low-energy) version of the original IEEE 802.11 wireless medium access standard. It has been designed with less overhead to meet IoT requirements. IEEE 802.11 standards (also known as wireless fidelity (Wi-Fi)) are the most commonly used wireless standards. They have been widely used and adopted for all digital devices including laptops, mobiles, tablets, and digital televisions (TVs). However, the original Wi-Fi standards are not suitable for IoT applications due to their frame overhead and power consumption. Hence, IEEE 802.11 working group initiated 802.11ah task group to develop a standard that supports low-overhead, power-friendly communication suitable for sensors and motes [7]. The basic 802.11ah MAC layer features include:

- *Synchronization frame*: A station is not allowed to transmit unless it has valid medium information that allows it to capture the medium and stop packet exchange by others. It can know such information if it receives the duration field packet correctly. If it does not receive it correctly, then it should wait for a duration called *Probe Delay*. Probe Delay can be configured by the access points in 802.11ah and announced by transmitting a synchronization frame at the beginning of the time slot.
- *Efficient bidirectional packet exchange*: This feature allows the sensor device to save more power by allowing both uplink and downlink communication between the access point and the sensor and allowing it to go to sleep as soon as it finishes the communication.
- *Short MAC frame*: The normal IEEE 802.11 frame is about 30 bytes, which is too large for IoT applications. IEEE 802.11ah mitigates this problem by defining a short MAC frame with about 12 bytes.
- *Null data packet*: In IEEE 802.11 the control frames, such as ACK frames, are about 14 bytes and have no data, which adds a lot of overhead. IEEE 802.11ah mitigates this problem by replacing the ACK frame with a preamble, a tiny signal.
- *Increased sleep time*: 802.11ah is designed for low-power sensors and, hence, it allows a long sleep period of time and waking up infrequently to exchange data only.

13.2.3 WirelessHART

Wireless Highway Addressable Remote Transducer (WirelessHART) protocol is a data link protocol that operates on the top of IEEE 802.15.4 PHY and adopts time division multiple access (TDMA) in its MAC. It is a secure and reliable MAC protocol that uses advanced encryption to encrypt the messages and calculate the integrity in order to offer reliability. The architecture, as shown in Figure 13.3, consists of a network manager, a security manager, a gateway to connect the wireless network to the wired networks, wireless devices as field devices, access points, routers, and adapters. The standard offers end-to-end, per-hop, or peer-to-peer security mechanisms. End-to-end security mechanisms enforce security from sources to destinations, while per-hop mechanisms secure it to next hop only [8, 9].

13.2.4 Z-Wave

Z-Wave is a low-power MAC protocol designed for home automation and has been used for IoT communication, especially for smart home and small commercial domains. It covers about 30 m point-to-point communication and is suitable for small messages in IoT applications, like light control, energy control, wearable healthcare control, and others. It uses carrier sense multiple access with collision avoidance (CSMA/CA) for collision detection and ACK messages for reliable transmission. It follows a master/slave architecture in which the master controls the slaves, sends them commands, and handles scheduling of the whole network [10].

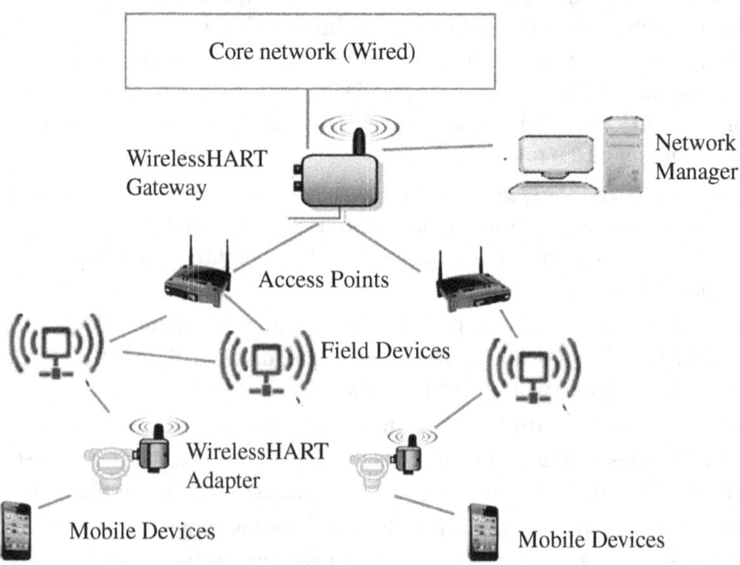

FIGURE 13.3 WirelessHART architecture. © Tara Salman and Raj Jain. Use with permission.

13.2.5 Bluetooth Low Energy

Bluetooth low energy or Bluetooth smart is a short-range communication protocol with PHY and MAC layer widely used for in-vehicle networking. Its low energy can reach ten times less than the classic Bluetooth, while its latency can reach 15 times. Its access control uses a contention-less MAC with low latency and fast transmission. It follows master/slave architecture and offers two types of frames: advertising and data frames. The Advertising frame is used for discovery and is sent by slaves on one or more of dedicated advertisement channels. Master nodes sense advertisement channels to find slaves and connect them. After connection, the master tells the slave its waking cycle and scheduling sequence. Nodes are usually awake only when they are communicating and they go to sleep otherwise to save their power [11, 12].

13.2.6 ZigBee Smart Energy

ZigBee smart energy is designed for a large range of IoT applications including smart homes, remote controls and healthcare systems. It supports a wide range of network topologies including star, peer to peer, or cluster tree. A coordinator controls the network and is the central node in a star topology or the root in a tree or cluster topology and may be located anywhere in peer to peer. ZigBee standard defines two stack profiles: ZigBee and ZigBee Pro. These stack profiles support full mesh networking and work with different applications allowing implementations with low memory and processing power. ZigBee Pro offers more features including security using symmetric key exchange, scalability using stochastic address assignment, and better performance using efficient many-to-one routing mechanisms [13].

13.2.7 DASH7

Named after last two characters in ISO 18000-7 (DASH7) is a wireless communication protocol for active radio-frequency identification (RFID) that operates in globally available Industrial Scientific Medical (ISM) band and is suitable for IoT requirements. It is mainly designed for scalable, long-range outdoor coverage with higher data rate compared to traditional ZigBee. It is a low-cost solution that supports encryption and Internet Protocol version 6 (IPv6) addressing. It supports a master/slave architecture and is designed for burst, lightweight, asynchronous, and transitive traffic. Its MAC layer features can be summarized as follows [14]:

- *Filtering*: Incoming frames are filtered using three processes: cyclic redundancy check (CRC) validation, a 4-bit subnet mask, and link quality assessment. Only the frames that pass all three checks are processed further.
- *Addressing*: DASH7 uses two types of addresses: the unique identifier (ID) which is the extended unique identifier 64-bit (EUI-64) ID and dynamic network ID which is 16-bit address specified by the network administrator.
- *Frame format*: The MAC frame has a variable length of maximum 255 bytes including addressing, subnets, estimated power of the transmission, and some other optional fields.

13.2.8 HomePlug

HomePlug Green PHY (HomePlug GP) is another MAC protocol developed by HomePlug Powerline Alliance that is used in home automation applications. HomePlug suite covers both PHY and MAC layers and has three versions: HomePlug audio visual (HomePlug AV), HomePlug AV2, and (HomePlug GP). HomePlug AV is the basic power line communication protocol which uses TDMA and CSMA/CA as MAC layer protocol, supports adaptive bit loading which allows it to change its rate depending on the noise level, and uses Orthogonal Frequency Division Multiplexing (OFDM) and four modulation techniques.

HomePlug GP is designed for IoT generally and specifically for home automation and smart grid applications. It is basically designed to reduce the cost and power consumption of HomePlug AV while keeping its interoperability, reliability, and coverage. Hence, it uses OFDM, as in HomePlug, but with one modulation only. In addition, HomePlug GP uses Robust OFDM coding to support low rate and high-reliability transmission. HomePlug AV uses only carrier sense multiple access (CSMA) as a MAC layer technique, while HomePlug GP uses both CSMA and TDMA. Moreover, HomePlug GP has a power-save mode that allows nodes to sleep much more than HomePlug by synchronizing their sleep time and waking up only when necessary [15].

13.2.9 G.9959

G.9959 is a MAC layer protocol from ITU, designed for low-bandwidth and low-cost, half-duplex reliable wireless communication. It is designed for real-time applications where time is really critical, reliability is important, and low power consumption is required. The MAC layer characteristics include unique network IDs that allow 232 nodes to join one network, collision avoidance (CA) mechanisms, backoff time in case of collision, automatic retransmission to guarantee reliability, and dedicated wake-up pattern that allows nodes to sleep when they are out of communication and hence saves their power. G.9959 MAC layer features include unique channel access, frame validation, acknowledgments, and retransmission [16, 17].

13.2.10 LTE-A

Long-Term Evolution Advanced (LTE-A) is a set of standards designed to fit M2M communication and IoT applications in cellular networks. LTE-A is a scalable, lower-cost protocol compared with other cellular protocols. LTE-A uses Orthogonal Frequency Division Multiple Access (OFDMA) as a MAC layer access technology, which divides the frequency into multiple bands, and each one can be used separately. The architecture of LTE-A consists of a core network (CN), a radio access network (RAN), and the mobile nodes. The CN is responsible for controlling mobile devices and to keep track of their Internet protocols. RAN is responsible for establishing the control and data planes and handling the wireless connectivity and radio access control. RAN and CN communicate using S1 link, as shown in Figure 13.4 where RAN consists of the E-UTRAN Node B (4G base station) (eNBs) to which other mobile nodes are connected wirelessly [18].

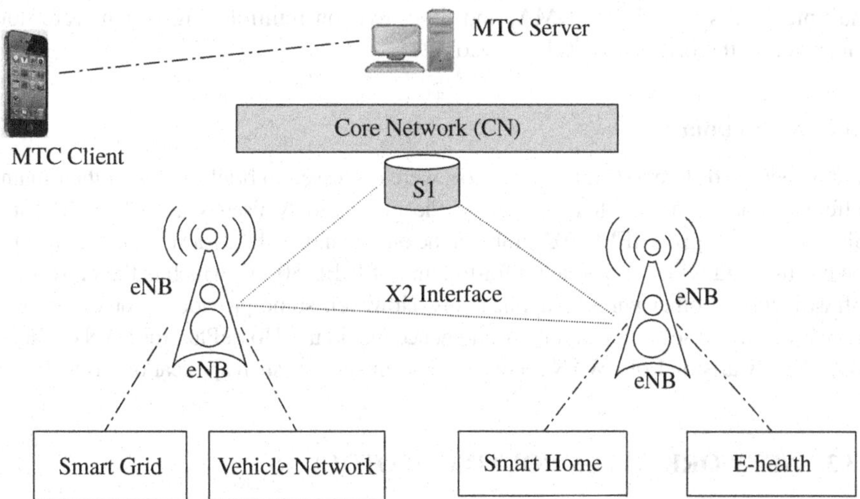

FIGURE 13.4 LTE-A architecture. © Tara Salman and Raj Jain. Use with permission.

13.2.11 LoRaWAN

Long-Range Wide Area Network (LoRaWAN) is a newly arising wireless technology designed for low-power WAN networks with low cost, mobility, security, and bidirectional communication for IoT applications. It is a low-power-consumption optimized protocol designed for scalable wireless networks with millions of devices. It supports redundant operation, location-free, low-cost, low-power, and energy-harvesting technologies to support the future needs of IoT while enabling mobility and ease-of-use features [19].

13.2.12 Weightless

Weightless is another wireless WAN technology for IoT applications designed by the Weightless Special Interest Group (SIG)—a nonprofit global organization. It has two sets of standards: Weightless-N and Weightless-W. Weightless-N was first developed to support low-cost, low-power M2M communication using TDMA with frequency hopping to minimize the interference. It uses ultranarrow bands in the sub-1 GHz ISM frequency band. On the other hand, Weightless-W provides the same features but uses TV band frequencies [20].

13.2.13 DECT/ULE

Digital enhanced cordless telecommunications (DECT) is a universal European standard for cordless phones. In their latest extension DECT/Ultra Low Energy (ULE), they have specified a low-power and low-cost air interface technology that can be used for IoT applications. Due to its dedicated channel assignment, DECT does not suffer from congestion and interference. DECT/ULE supports frequency division multiple access (FDMA), TDMA, and time division multiplexing which were not supported in the original DECT protocol [21].

13.2.14 Summary

In this section, different data link protocols were discussed in brief to present their main differences and usage in IoT. Generally, the most widely used standards in IoT are Bluetooth and ZigBee. IEEE 802.11ah, on the other hand, is the easiest to be used due to the existing and widely separated infrastructure of IEEE 802.11 which is the most used infrastructure in other wireless applications. However, some providers would seek for more reliable and secured technology and hence would use HomePlug for LAN connectivity. Newly arising LoRaWAN seems to be promising for such applications as well.

13.3 NETWORK LAYER ROUTING PROTOCOLS

In this section, we discuss some standard and nonstandard protocols that are used for routing in IoT applications. It should be noted that we have partitioned the network layer in two sublayers: routing layer that handles the transfer the packets from source to destination and an encapsulation layer that forms the packets. Encapsulation mechanisms will be discussed in the next section.

13.3.1 RPL

Routing Protocol for Low-Power and Lossy Networks (RPL) is distance vector protocol that can support a variety of data link protocols, including the ones discussed in the previous section. It builds a Destination-Oriented Directed Acyclic Graph (DODAG) that has only one route from each leaf node to the root in which all the traffic from the node will be routed to. At first, each node sends a DODAG Information Object (DIO) advertising itself as the root. This message is propagated in the network and the whole DODAG is gradually built. When communicating, the node sends a Destination Advertisement Object (DAO) to its parents, the DAO is propagated to the root, and the root decides where to send it depending on the destination. When a new node wants to join the network, it sends a DODAG Information Solicitation (DIS) request to join the network and the root will reply back with a DAO Acknowledgment (DAO-ACK) confirming the join. RPL nodes can be stateless, which is most common, or stateful. A stateless node keeps tracks of its parents only. Only root has the complete knowledge of the entire DODAG. Hence, all communications go through the root in every case. A stateful node keeps track of its children and parents, and hence when communicating inside a subtree of the DODAG, it does not have to go through the root [22].

13.3.2 CORPL

An extension of RPL is CORPL, or (cognitive RPL), which is designed for cognitive networks and uses DODAG topology generation but with two new modifications to RPL. CORPL utilizes opportunistic forwarding to forward the packet by choosing multiple forwarders (forwarder set) and coordinates between the nodes to choose the best next hop to forward the packet to. DODAG is built in the same way as RPL. Each node maintains a forwarding set instead of its parent only and updates its neighbor with its changes using DIO messages. Based on the updated information, each node dynamically updates its neighbor priorities in order to construct the forwarder set [23].

13.3.3 CARP

Channel-Aware Routing Protocol (CARP) is a distributed routing protocol designed for underwater communication. It can be used for IoT due to its lightweight packets. It considers link quality, which is computed based on historical successful data transmission gathered from neighboring sensors, to select the forwarding nodes. There are two scenarios: network initialization and data forwarding. In network initialization, a HELLO packet is broadcasted from the sink to all other nodes in the networks. In data forwarding, the packet is routed from sensor to sink in a hop-by-hop fashion. Each next hop is determined independently. The main problem with CARP is that it does not support reusability of previously collected data. In other words, if the application requires sensor data only when it changes significantly, then CARP data forwarding is not beneficial to that specific application. An enhancement of CARP was done in E-CARP by allowing the sink node to save previously received sensory data. When new data is needed, E-CARP sends a *Ping* packet which is replied with the data from the sensors nodes. Thus, E-CARP reduces the communication overhead drastically [24].

13.3.4 Summary

Three routing protocols in IoT were discussed in this section. RPL is the most commonly used one. It is a distance vector protocol designed by IETF in 2012. CORPL is a nonstandard extension of RPL that is designed for cognitive networks and utilizes the opportunistic forwarding to forward packets at each hop. On the other hand, CARP is the only distributed hop-based routing protocol that is designed for IoT sensor network applications. CARP is used for underwater communication mostly. Since it is not standardized and just proposed in literature, it is not yet used in other IoT applications.

13.4 NETWORK LAYER ENCAPSULATION PROTOCOLS

One problem in IoT applications is that IPv6 addresses are too long and cannot fit in most IoT data link frames which are relatively much smaller. Hence, IETF is developing a set of standards to encapsulate IPv6 datagrams in different data link layer frames for use in IoT applications. In this section, we review these mechanisms briefly.

13.4.1 6LoWPAN

IPv6 over Low-power Wireless Personal Area Network (6LoWPAN) is the first and most commonly used standard in this category. It efficiently encapsulates IPv6 long headers in IEEE802.15.4 small packets, which cannot exceed 128 bytes.

The specification supports different length addresses, low bandwidth, different topologies including star or mesh, power consumption, low cost, scalable networks, mobility, unreliability, and long sleep time. The standard provides header compression to reduce transmission overhead, fragmentation to meet the 128-byte maximum frame length in IEEE802.15.4, and support of multihop delivery. Frames in 6LoWPAN use four types of headers: No 6LoWPAN header (00), Dispatch header (01), Mesh header (10), and Fragmentation header (11). In No 6LoWPAN header case, any frame that does not follow 6LoWPAN specifications is discarded. Dispatch header is used for multicasting and IPv6 header compressions. Mesh headers are used for broadcasting, while Fragmentation headers are used to break long IPv6 header to fit into fragments of maximum 128-byte length.

13.4.2 6TiSCH

IPv6 over Time Slotted Channel Hopping Mode of IEEE 802.15.4e (6TiSCH) working group in IETF is developing standards to allow IPv6 to pass through Time-Slotted Channel Hopping (TSCH) mode of IEEE 802.15.4e data links. It defines a channel distribution usage matrix consisting of available frequencies in columns and time slots available for network scheduling operations in rows. This matrix is portioned into chucks where each chunk contains time and frequencies and is globally known to all nodes in the network. The nodes within the same interference domain negotiate their scheduling so that each node gets to transmit in a chunk within its interference domain. Scheduling becomes an optimization problem where time slots are assigned to a group of neighboring nodes sharing the same application. The standard does not specify how the scheduling can be done and leaves that to be an application-specific problem in order to allow for maximum flexibility for different IoT applications. The scheduling can be centralized or distributed depending on application or the topology used in the MAC layer [25].

13.4.3 6Lo

IPv6 over Networks of Resource-constrained Nodes (6Lo) working group in IETF is developing a set of standards on transmission of IPv6 frames on various data links. Although 6LowPAN and 6TiSCH, which cover IEEE 802.15.4 and IEEE 802.15.4e, were developed by different working groups, it became clear that there are many more data links to be covered and so 6Lo working group was formed. At the time of this writing most of the 6Lo specifications have not been finalized and are in various stages of drafts. For example, IPV6 over Bluetooth Low Energy Mesh Networks, IPv6 over IEEE 485 Master–Slave/Token Passing (MS/TP) networks, IPV6 over DECT/ULE, IPV6 over near-field communication (NFC), IPv6 over IEEE 802.11ah, and IPv6 over Wireless Networks for Industrial Automation–Process Automation (WIA-PA) drafts are being developed to specify how to transmit IPv6 datagrams over their respective data links [26]. Two of these 6Lo specifications—“IPv6 over G.9959” and “IPv6 over Bluetooth Low Energy”—have been approved as request for comments (RFC) and are described next.

13.4.4 IPv6 over G.9959

RFC 7428 defines the frame format for transmitting IPv6 packet on International Telecommunications Union—Telecommunications (ITU-T) G.9959 networks. G.9959 defines a unique 32-bit home network ID that is assigned by the controller and 8-bit host ID that is allocated for each node. An IPv6 link local address must be constructed by the link layer derived 8-bit host ID so that it can be compressed in G.9959 frame. Furthermore, the same header compression as in 6LowPAN is used here to fit an IPv6 packet into G.9959 frames. RFC 7428 also provides a level of security by a shared network key that is used for encryption. However, applications with a higher level of security requirements need to handle their end-to-end encryption and authentication using their own higher-layer security mechanisms [26].

13.4.5 IPv6 over Bluetooth Low Energy

Bluetooth Low Energy is also known as Bluetooth Smart and was introduced in Bluetooth V4.0 and enhanced in V4.1. RFC 7668 [27], which specifies IPv6 over Bluetooth LE and reuses most of the 6LowPAN compression techniques. However, since the Logical Link Control and Adaptation Protocol (L2CAP) sublayer in Bluetooth already provides segmentation and reassembly of larger payloads in to 27-byte L2CAP packets, fragmentation features from 6LowPAN standards are not used. Another significant difference is that Bluetooth Low Energy does not currently support formation of multihop networks at the link layer. Instead, a central node acts as a router between lower-powered peripheral nodes.

13.4.6 Summary

In this section, encapsulation protocols for IPv6 in IoT MAC frame were discussed. First, two standards for IPv6 over 802.15.4 and 802.15.4e were discussed. Such protocols are important as 802.15.4e is the most widely used encapsulation framework designed for IoT. Following that, 6Lo specifications are briefly and broadly discussed just to present their existence in IETF standards. These drafts handle passing IPv6 over different channel access mechanisms using 6LoWPAN standards. Then, two of 6Lo Specifications which became IETF RFCs are discussed in more details. The importance of presenting these standards is to highlight the challenge of interoperability between different MAC standards which is still challenging due to the diversity of protocols.

13.5 SESSION LAYER PROTOCOLS

This section reviews standards and protocols for message passing in IoT session layer proposed by different standardization organizations. Most of the IP applications including IoT applications use Transmission Control Protocol (TCP) or User Datagram Protocol (UDP) for transport. However, there are several message distribution functions that are common among many IoT applications; it is desirable that these functions be implemented in an interoperable standard way by different applications. These are the so-called "Session Layer" protocols described in this section.

13.5.1 MQTT

Message Queuing Telemetry Transport (MQTT) was introduced by International Business Machines Corporation (IBM) in 1999 and standardized by Advancing Open Standards for the Information Society (OASIS) in 2013 [28]. It is designed to provide embedded connectivity between applications and middlewares on one side and networks and communications on the other side. It follows a publish–subscribe architecture, as shown in Figure 13.5, where the system consists of three main components: publishers, subscribers, and a broker. From IoT point of view, publishers are basically the lightweight sensors that connect to the broker to send their data and go back to sleep whenever possible. Subscribers are applications that are interested in a certain topic, or sensory data, so they connect to brokers to be informed whenever new data are received. The brokers classify sensory data in topics and send them to subscribers interested in the topics.

13.5.2 SMQTT

An extension of MQTT is Secure MQTT (SMQTT) which uses encryption based on lightweight attribute-based encryption. The main advantage of using such encryption is the broadcast encryption feature, in which one message is encrypted and delivered to multiple other nodes, which is quite common in IoT applications. In general, the algorithm consists of four main stages: setup, encryption, publish, and decryption. In the setup phase, the subscribers and publishers register themselves to the broker and get a master secret key according to their developer's choice of key generation algorithm. Then, when the data is published, it is encrypted, published by the broker

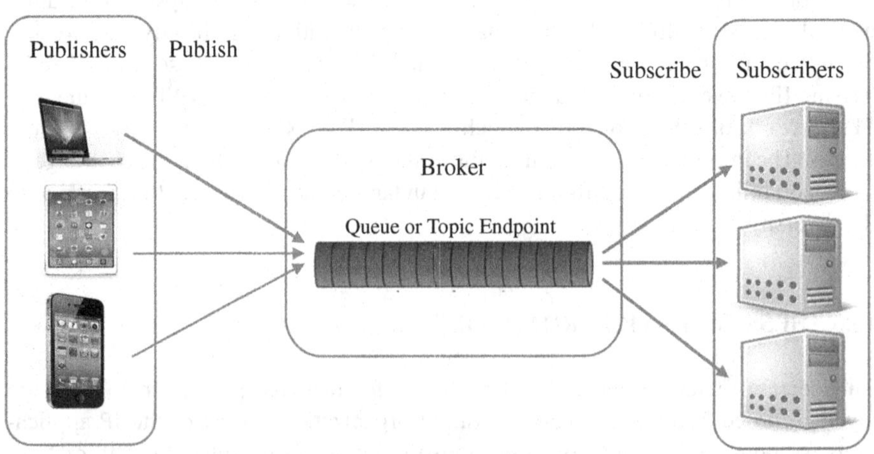

FIGURE 13.5 MQTT architecture. © Tara Salman and Raj Jain. Use with permission.

which sends it to the subscribers, and finally decrypted at the subscribers which have the same master secret key. The key generation and encryption algorithms are not standardized. SMQTT is proposed only to enhance MQTT security feature [29].

13.5.3 AMQP

The Advanced Message Queuing Protocol (AMQP) is another session layer protocol that was designed for financial industry. It runs over TCP and provides a publish–subscribe architecture which is similar to that of MQTT. The difference is that the broker is divided into two main components, exchange and queues, as shown in Figure 13.6. The exchange is responsible for receiving publisher messages and distributing them to queues based on predefined roles and conditions. Queues basically represent the topics and subscribed by subscribers which will get the sensory data whenever they are available in the queue [30].

13.5.4 CoAP

The Constrained Application Protocol (CoAP) is another session layer protocol designed by IETF Constrained RESTful Environment (CoRE) working group to provide lightweight representational state transfer based (RESTful) (HTTP) interface. Representational State Transfer (REST) is the standard interface between HTTP client and servers. However, for lightweight applications such as IoT, REST could result in significant overhead and power consumption. CoAP is designed to enable low-power sensors to use RESTful services while meeting their power constrains. It is built over UDP, instead of TCP commonly used in HTTP, and has a light mechanism to provide reliability. CoAP architecture is divided into two main sublayers: messaging and

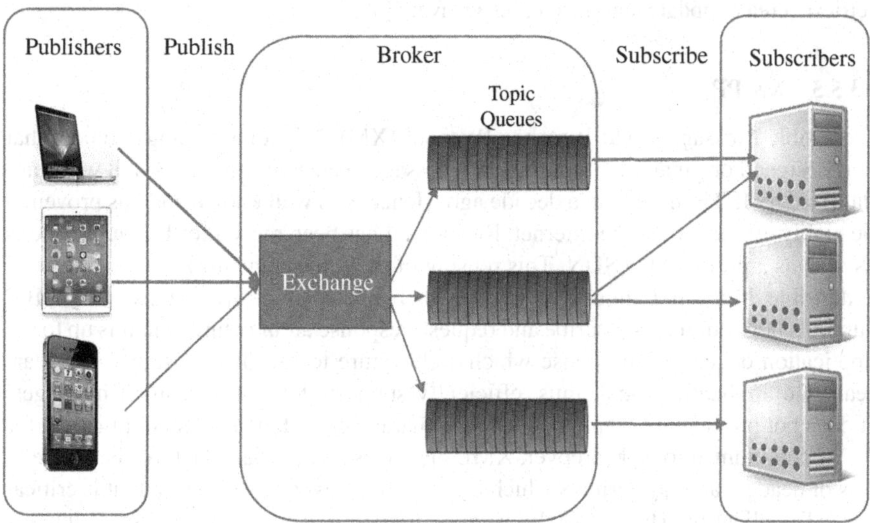

FIGURE 13.6 AMQP architecture. © Tara Salman and Raj Jain. Use with permission.

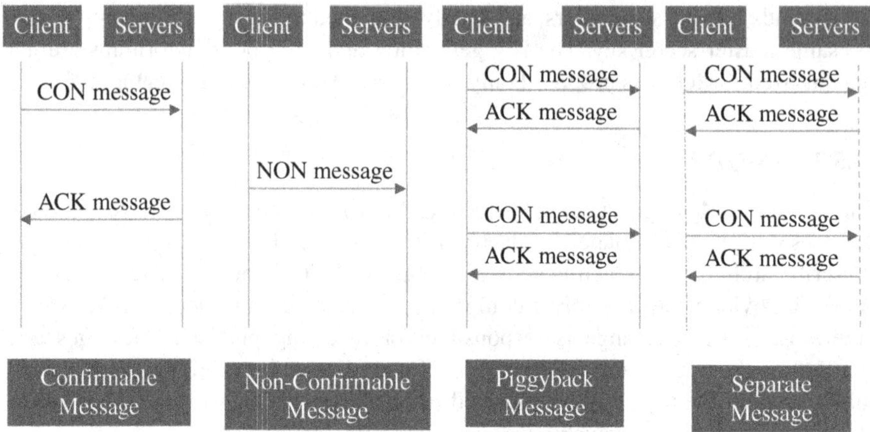

FIGURE 13.7 CoAP messages. © Tara Salman and Raj Jain. Use with permission.

request–response. The messaging sublayer is responsible for reliability and duplication of messages, while the request–response sublayer is responsible for communication. As shown in Figure 13.7, CoAP has four messaging modes: confirmable, nonconfirmable, piggyback, and separate. Confirmable and nonconfirmable modes represent the reliable and unreliable transmissions, respectively, while the other modes are used for request/response. Piggyback is used for client/server direct communication where the server sends its response directly after receiving the message, that is, within the ACK message. On the other hand, the separate mode is used when the server response comes in a message separate from the ACK and may take some time to be sent by the server. As in HTTP, CoAP utilizes GET, PUT, PUSH, and DELETE message requests to retrieve, create, update, and delete, respectively [4, 31].

13.5.5 XMPP

Extensible Messaging and Presence Protocol (XMPP) is a messaging protocol that was designed originally for chatting and message exchange applications. It was standardized by IETF more than a decade ago. Hence, it is well known and has proven to be highly efficient over the Internet. Recently, it has been reused for IoT applications as well as a protocol for SDN. This reusing of the same standard is due to its use of Extensible Markup Language (XML) which makes it easily extensible. XMPP supports both publish–subscribe and request–response architecture, and it is up to the application developer to choose which architecture to use. It is designed for near-real-time applications and, thus, efficiently supports low-latency small messages. It does not provide any quality-of-service guarantees and, hence, is not practical for M2M communications. Moreover, XML messages create additional overhead due to lots of headers and tag formats which increase the power consumption that is critical for IoT application. Hence, XMPP is rarely used in IoT but has gained some interest for enhancing its architecture in order to support IoT applications [4, 32].

13.5.6 DDS

Data Distribution Service (DDS) is another publish–subscribe protocol that is designed by the Object Management Group (OMG) for M2M communications [33]. The basic benefit of this protocol is the excellent quality-of-service levels and reliability guarantees as it relies on a brokerless architecture, which suits IoT and M2M communication. It offers 23 quality-of-service levels which allow it to offer a variety of quality criteria including security, urgency, priority, durability, reliability, etc. It defines two sublayers: data-centric publish–subscribe and data-local reconstruction sublayers. The first takes the responsibility of message delivery to the subscribers, while the second is optional and allows a simple integration of DDS in the application layer. Publisher layer is responsible for sensory data distribution. Data writer interacts with the publishers to agree about the data and changes to be sent to the subscribers. Subscribers are the receivers of sensory data to be delivered to the IoT application. Data readers basically read the published data and deliver it to the subscribers and the topics are basically the data that are being published. In others words, data writers and data reader take the responsibilities of the broker in the broker-based architectures.

13.5.7 Summary

IoT has many standardized session layer protocols which were briefly highlighted in this section. These session layer protocols are application dependent, and the choice between them is very application specific. It should be noted that MQTT is the most widely used in IoT due to its low overhead and power consumption. It's an organizational application specific to choose between these standards. For example, if an application has already been built with XML and can, therefore, accept a bit of overhead in its headers, XMPP might be the best option to choose among session layer protocols. On the other hand, if the application is really overhead and power sensitive, then choosing MQTT would be the best option; however, that comes with the additional broker implementation. If the application requires REST functionality as it will be HTTP based, then CoAP would be the best option if not the only one. Table 13.1 summarizes comparison points between these different session layer protocols.

TABLE 13.1 A Comparison of IoT Session Layer Standards

Protocols	UDP/TCP	Architecture	Security and Quality of Service (QoS)	Header Size (bytes)	Max Length(bytes)
MQTT	TCP	Pub/Sub	Both	2	5
AMQP	TCP	Pub/Sub	Both	8	–
CoAP	UDP	Req/Res	Both	4	20 (typical)
XMPP	TCP	Both	Security	–	–
DDS	TCP/UDP	Pub/Sub	QoS	–	–

Reproduced with permission of Tara Salman and Raj Jain, Washington University.

13.6 IoT MANAGEMENT PROTOCOLS

This section discusses two main management standards for IoT that provide heterogeneous—communication between different data links. Management protocols play an important role in IoT due to the diversity of protocols and standards at different layers of networking. The need for heterogeneous and easy communication between different protocols at the same or different layers is critical for IoT applications. Existing standards mainly facilitate communication between protocols at the same layer; however, it is still a challenge to facilitate communication at different layers in IoT.

13.6.1 Interconnection of Heterogeneous Data Links

As IoT environments rely on many different MAC protocols, interoperability among all these technologies is a challenge that needs to be handled. IEEE 1905.1 standards offer such interoperability by providing an abstraction layer that is built in top on all these heterogeneous MAC protocols [34]. This abstraction hides the diversity of the different protocols without requiring any change to the design of each MAC, as illustrated in Section 13.2. The basic idea behind this protocol is the abstraction layer which is used to exchange messages, called Control Message Data Units (CMDUs), among all standards compatible devices. As shown in Figure 13.8, all IEEE 1905.1 compliant devices understand a common "Abstraction Layer Management Entity (ALME)" protocol which offers different services including neighbor discovery, topology exchange, topology change notification, measured traffic statistics exchange, flow forwarding rules, and security associations.

13.6.2 Smart Transducer Interface

IEEE 1451 is a set of standards developed to allow management of different analog transducers and sensors. The basic idea of this standard is the use of plug-and-play identification using standardized transducer electronic data sheets (TEDSs). Each

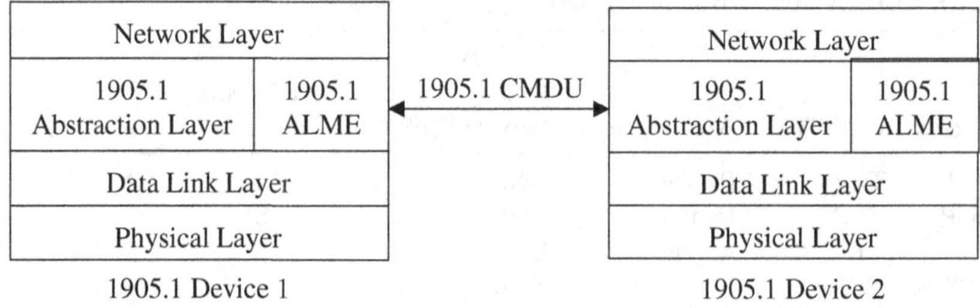

FIGURE 13.8 IEEE 1905.1 protocol structure. © Tara Salman and Raj Jain. Use with permission.

transducer contains a TEDS which includes all the information needed by the measurement system including device ID, characteristics, and interface besides the data coming from the sensors. Data sheets are stored embedded memory within the transducer or the sensor and have a defined encoding mechanism to understand a broad number of sensor types and applications. The memory usage is minimized by utilizing the small XML-based messages which are understood by different manufactures and different applications [35].

13.7 SECURITY IN IoT PROTOCOLS

Security is another aspect of IoT applications which is critical and can be found in all almost all layers of the IoT protocols. Threats exist at all layers including data link, network, session, and application layers. In this section, we briefly discuss the security mechanisms built in the IoT protocols that we have discussed in this survey.

13.7.1 MAC 802.15.4

MAC 802.15.4 offers different security modes by utilizing the “Security Enabled Bit” in the Frame Control field in the header. Security requirements include confidentiality, authentication, integrity, access control mechanisms, and secured Time-Synchronized Communications.

13.7.2 6LoWPAN

6LoWPAN by itself does not offer any mechanisms for security. However, relevant documents include discussion of security threats, requirement, and approach to consider in IoT network layer. For example, RFC 4944 discusses the possibility of duplicate EUI-64 interface addresses which are supposed to be unique [36]. RFC 6282 discusses the security issues that are raised due to the problems introduced in RFC 4944 [37]. RFC 6568 addresses possible mechanisms to adopt security within constrained wireless sensor devices [38]. In addition, a few recent drafts in Ref. [26] discuss mechanisms to achieve security in 6LoWPAN. See also Refs. [39, 40].

13.7.3 RPL

RPL offers different level of security by utilizing a “Security” field after the 4-byte Internet Control Message Protocol version 6 (ICMPv6) message header. Information in this field indicates the level of security and the cryptography algorithm used to encrypt the message. RPL offers support for data authenticity, semantic security, protection against replay attacks, confidentiality, and key management. Levels of security in RPL include Unsecured, Preinstalled, and Authenticated. RPL attacks include Selective Forwarding, Sinkhole, Sybil, Hello Flooding, Wormhole, Black hole, and Denial-of-Service attacks.

13.7.4 Application Layer

Applications can provide additional level of security using transport layer security (TLS) or secure sockets layer (SSL) as a transport layer protocol. In addition, end-to-end authentication and encryption algorithms can be used to handle different levels of security as required. For further discussion on security, see Ref. [3].

It should be noted that a number of new security approaches are also being developed that are suitable for resource-constrained IoT devices. Some of these protocols are listed in Figure 13.2.

13.8 IoT CHALLENGES

Developing a successful IoT application is still not an easy task due to multiple challenges. These challenges include mobility, reliability, scalability, management, availability, interoperability, and security and privacy. In the following, we briefly describe each of these challenges.

13.8.1 Mobility

IoT devices need to move freely and change their IP address and networks based on their location. Thus, the routing protocol, such as RPL, has to reconstruct the DODAG each time a node goes off the network or joins the network which adds a lot of overhead. In addition, mobility might result in a change of service provider which can add another layer of complexity due to service interruption and changing gateway.

13.8.2 Reliability

System should be perfectly working and delivering all of its specifications correctly. It is a very critical requirement in applications that requires emergency responses. In IoT applications, the system should be highly reliable and fast in collecting data, communicating them, and making decisions, and eventually wrong decisions can lead to disastrous scenarios.

13.8.3 Scalability

Scalability is another challenge of IoT applications where millions and trillions of devices could be connected on the same network. Managing their distribution is not an easy task. In addition, IoT applications should be tolerant of new services and devices constantly joining the network and, therefore, must be designed to enable extensible services and operations.

13.8.4 Management

Managing all these devices and keeping track of the failures, configurations, and performance of such large number of devices is definitely a challenge in IoT. Providers should manage Fault, Configuration, Accounting, Performance and Security (FCAPS) of their interconnected devices and account for each aspect.

13.8.5 Availability

Availability of IoT includes software and hardware levels being provided at anytime and anywhere for service subscribers. Software availability means that the service is provided to anyone who is authorized to have it. Hardware availability means that the existing devices are easy to access and are compatible with IoT functionality and protocols. In addition, these protocols should be compact to be able to be embedded within the IoT constrained devices.

13.8.6 Interoperability

Interoperability means that heterogeneous devices and protocols need to be able to interwork with each other. This is challenging due to the large number of different platforms used in IoT systems. Interoperability should be handled by both the application developers and the device manufacturers in order to deliver the services regardless of the platform or hardware specification used by the customer.

13.9 SUMMARY

In this chapter, we have provided a comprehensive survey of protocols for IoT. Many such protocols have been developed by IETF, IEEE, ITU, and other organizations and many more in development. Due to their large number, the discussion of each protocol is brief and references for further information have been provided. The aim of this chapter is to give an insight to developers and service providers of different layers of protocols in IoT and how to choose between them.

We categorized the standards based on their layer of operation to: data link layer, network routing standards, network encapsulation layer, session layer, and management standards. At each layer, we presented most of the finalized standards and some drafts. In addition, we briefly reviewed IoT management protocols and current state of security issues related to these protocols. We also provided a brief comparison between different IoT protocols and how to choose between them. Finally, we discussed some challenges that still exist in IoT systems and researchers are trying to solve them.

REFERENCES

[1] Gartner, Gartner's 2014 Hype Cycle for Emerging Technologies Maps the Journey to Digital Business, August 2014, http://www.gartner.com/newsroom/id/2819918 (accessed August 24, 2016).

[2] Z. Sheng, S. Yang, Y. Yu, A. Vasilakos, J. McCann, and K. Leung, A survey on the IETF protocol suite for the internet of things: Standards, challenges, and opportunities, *IEEE Wireless Communications*, **20**, 6, 91–98, December 2013, http://ieeexplore.ieee.org/xpl/articleDetails.jsp?arnumber=6704479 (accessed August 24, 2016).

[3] J. Granjal, E. Monteiro, and J. Sa Silva, Security for the internet of things: A survey of existing protocols and open research issues, *IEEE Communications Surveys Tutorials*, **17**, 3, 1294–1312, 2015, http://ieeexplore.ieee.org/xpl/articleDetails.jsp?arnumber=7005393 (accessed August 24, 2016).

[4] V. Karagiannis, P. Chatzimisios, F. Vazquez-Gallego, and J. Alonso-Zarate, A survey on application layer protocols for the internet of things, *Transaction on IoT and Cloud Computing*, **3**, 1, 11–17, 2015, https://jesusalonsozarate.files.wordpress.com/2015/01/2015-transaction-on-iot-and-cloud-computing.pdf (accessed August 24, 2016).

[5] A. Al-Fuqaha, M. Guizani, M. Mohammadi, M. Aledhari, and M. Ayyash, Internet of things: A survey on enabling technologies, protocols and applications, *IEEE Communications Surveys Tutorials*, **17**, 4, 2347–2376, 2015, http://ieeexplore.ieee.org/xpl/articleDetails.jsp?tp=&arnumber=7123563 (accessed August 24, 2016).

[6] IEEE 802.15.4-2011, IEEE Standard for Local and Metropolitan Area Networks—Part 15.4: Low-Rate Wireless Personal Area Networks (LR-WPANs), 314 pp., September 5, 2011, http://standards.ieee.org/getieee802/download/802.15.4-2011.pdf (accessed August 24, 2016).

[7] M. Park, IEEE 802.11ah: Sub-1-GHz license-exempt operation for the internet of things, *IEEE Communications Magazine*, **53**, 9, 145–151, September 2015, http://ieeexplore.ieee.org/xpl/articleDetails.jsp?arnumber=7263359 (accessed August 24, 2016).

[8] A. Kim, F. Hekland, S. Petersen, and P. Doyle, When HART Goes Wireless: Understanding and Implementing the WirelessHART Standard, in Proceedings of 13th IEEE International Conference on Emerging Technologies and Factory Automation (ETFA 2008), Hamburg, Germany, September 15–18, 2008, pp. 899–907, https://library.e.abb.com/public/eb20fe80a391ca8485257bc600667573/When%20HART%20Goes%20Wireless%20Understanding%20and%20Implementing%20the%20WirelessHART%20Standard.pdf (accessed August 24, 2016).

[9] S. Raza and T. Voigt, Interconnecting WirelessHART and Legacy HART Networks, in 6th IEEE International Conference on Distributed Computing in Sensor Systems Workshops (DCOSSW), Santa Barbara, June 21–23, 2010, pp. 1–8, http://ieeexplore.ieee.org/xpl/articleDetails.jsp?arnumber=5593285 (accessed August 24, 2016).

[10] Z-Wave, Z-Wave Protocol Overview, v. **4**, May 2007, https://wiki.ase.tut.fi/courseWiki/images/9/94/SDS10243_2_Z_Wave_Protocol_Overview.pdf (accessed August 24, 2016).

[11] J. Decuir, Bluetooth 4.0: Low Energy, Presentation slides, 2010, http://chapters.comsoc.org/vancouver/BTLER3.pdf (accessed August 24, 2016).

[12] C. Gomez, J. Oller, and J. Paradells, Overview and evaluation of Bluetooth low energy: An emerging low-power wireless technology, *Sensors*, **12**, 9, 11734–11753, 2012, http://www.mdpi.com/1424-8220/12/9/11734 (accessed August 24, 2016).

[13] ZigBee Standards Organization, ZigBee Specification, Document 053474r17, January 2008, 604 pp., http://home.deib.polimi.it/cesana/teaching/IoT/papers/ZigBee/ZigBeeSpec.pdf (accessed August 24, 2016).

[14] O. Cetinkaya and O. Akan, A DASH7-Based Power Metering System, in 12th Annual IEEE Consumer Communications and Networking Conference (CCNC), Las Vegas, NV, January 9–12, 2015, pp. 406–411, http://ieeexplore.ieee.org/xpl/articleDetails.jsp?reload=true&arnumber=7158010 (accessed August 24, 2016).

[15] HomePlug Alliance, Homeplug GreenPHY v1.1, 2012, http://www.homeplug.org/tech-resources/resources/ (accessed August 24, 2016).

[16] A. Brandt and J. Buron, Transmission of IPv6 Packets over ITU-T G.9959 Networks, IETF RFC 7428, February 2015, http://www.ietf.org/rfc/rfc7428.txt (accessed August 24, 2016).

[17] ITU-T, Short Range Narrow-Band Digital Radio Communication Transceivers—PHY, MAC, SAR and LLC Layer Specifications, 2015, https://www.itu.int/rec/T-REC-G.9959-201501-I/en (accessed August 31, 2016).

[18] M. Hasan, E. Hossain, and D. Niyato, Random access for machine-to-machine communication in LTE-advanced networks: Issues and approaches, *IEEE Communications Magazine*, **51**, 6, 86–93, June 2013, http://ieeexplore.ieee.org/xpl/articleDetails.jsp?reload=true&arnumber=6525600 (accessed August 24, 2016).

[19] LoRa Alliance, LoRaWAN Specification, 2015, https://www.lora-alliance.org/portals/0/specs/LoRaWAN%20Specification%201R0.pdf (accessed August 24, 2016).

[20] I. Poole, Weightless Wireless—M2M White Space Communications—Tutorial, 2014, http://www.radio-electronics.com/info/wireless/weightless-m2m-white-space-wireless-communications/basics-overview.php (accessed August 24, 2016).

[21] S. Bush, DECT/ULE Connects Homes for IoT, *Electronics weekly*, September 8, 2015, http://www.electronicsweekly.com/news/design/communications/dect-ule-connects-homes-iot-2015-09/ (accessed August 24, 2016).

[22] T. Winter, P. Thubert, A. Brandt, J. Hui, R. Kelsey, P. Levis, K. Pister, R. Struik, J.P. Vasseur, and R. Alexander, RPL: IPv6 Routing Protocol for Low-Power and Lossy Networks, IETF RFC 6550, March 2012, http://www.ietf.org/rfc/rfc6550.txt (accessed August 24, 2016).

[23] A. Aijaz and A. Aghvami, Cognitive machine-to-machine communications for internet-of-things: A protocol stack perspective, *IEEE Internet of Things Journal*, **2**, 2, 103–112, April 2015, http://ieeexplore.ieee.org/xpl/articleDetails.jsp?tp=&arnumber=7006643 (accessed August 24, 2016).

[24] Z. Zhou, B. Yao, R. Xing, L. Shu, and S. Bu, E-CARP: An energy efficient routing protocol for UWSNs in the internet of underwater things, *IEEE Sensors Journal*, **16**, 11, 4072–4082, 2015, http://ieeexplore.ieee.org/xpl/articleDetails.jsp?arnumber=7113774 (accessed August 24, 2016).

[25] D. Dujovne, T. Watteyne, X. Vilajosana, and P. Thubert, 6TiSCH: Deterministic IP-enabled industrial internet (of things), *IEEE Communications Magazine*, **52**, 12, 36–41, December 2014, http://ieeexplore.ieee.org/xpl/articleDetails.jsp?arnumber=6979984 (accessed August 24, 2016).

[26] IETF, IPv6 over Networks of Resource-Constrained Nodes (6lo), https://datatracker.ietf.org/wg/6lo/documents/ (accessed August 24, 2016).

[27] J. Nieminen, T. Savolainen, M. Isomaki, B. Patil, Z. Shelby, and C. Gomez, IPv6 over Bluetooth Low Energy, IETF RFC 7668, October 2015, http://www.ietf.org/rfc/rfc7668.txt (accessed August 24, 2016).

[28] D. Locke, MQ Telemetry Transport (MQTT) v3. 1 Protocol Specification, IBM Developer Works Technical Library, 2010, http://www.ibm.com/developerworks/webservices/library/ws-mqtt/index.html (accessed August 24, 2016).

[29] M. Singh, M. Rajan, V. Shivraj, and P. Balamuralidhar, Secure MQTT for Internet of Things (IoT), in 5th International Conference on Communication Systems and Network Technologies (CSNT 2015), Gwalior, India, April 4–6, 2015, pp. 746–751, http://ieeexplore.ieee.org/xpl/articleDetails.jsp?arnumber=7280018 (accessed August 24, 2016).

[30] OASIS, OASIS Advanced Message Queuing Protocol (AMQP) Version 1.0, 2012, http://docs.oasis-open.org/amqp/core/v1.0/os/amqp-core-complete-v1.0-os.pdf (accessed August 24, 2016).

[31] Z. Shelby, K. Hartke, and C. Bormann, The Constrained Application Protocol (CoAP), IETF RFC 7252, June 2014, http://www.ietf.org/rfc/rfc7252.txt (accessed August 24, 2016).

[32] P. Saint-Andre, Extensible Messaging and Presence Protocol (XMPP): Core, IETF RFC 6120, 2011, https://tools.ietf.org/html/rfc6120 (accessed August 24, 2016).

[33] Object Management Group, Data Distribution Service V1.4, April 2015, http://www.omg.org/spec/DDS/1.4 (accessed August 24, 2016).

[34] IEEE 1905.1-2013, IEEE Standard for a Convergent Digital Home Network for Heterogeneous Technologies, 93 pp., April 12, 2013, http://ieeexplore.ieee.org/xpl/articleDetails.jsp?arnumber=6502164 (accessed August 24, 2016).

[35] K. Malar and N. Kamaraj, Development of Smart Transducers with IEEE 1451.4 Standard for Industrial Automation, in 2014 International Conference on Advanced Communication Control and Computing Technologies (ICACCCT), Ramanathapuram, India, May 8–10, 2014, pp. 111–114, http://ieeexplore.ieee.org/xpl/articleDetails.jsp?&arnumber=7019280 (accessed August 24, 2016).

[36] G. Montenegro, N. Kushalnagar, J. Hui, and D. Culler, Transmission of IPv6 Packets over IEEE 802.15.4 Networks, IETF RFC 4944, September 2007, https://tools.ietf.org/html/rfc4944 (accessed August 24, 2016).

[37] J. Hui and P. Thubert, Compression Format for IPv6 Datagrams over IEEE 802.15.4-Based Networks, IETF RFC 6262, September 2011, https://tools.ietf.org/html/rfc6282 (accessed August 24, 2016).

[38] E. Kim, D. Kaspar, and J. Vasseur, Design and Application Spaces for IPv6 over Low-Power Wireless Personal Area Networks (6LoWPANs), IETF RFC **6568**, April 2012, http://www.ietf.org/rfc/rfc6568.txt (accessed August 24, 2016).

[39] P. Pongle and G. Chavan, A Survey: Attacks RPL and 6LowPAN in IoT, in International Conference on Pervasive Computing (ICPC 2015), Pune, India, January 8–10, 2015, pp. 1–6, http://ieeexplore.ieee.org/xpl/articleDetails.jsp?arnumber=7087034 (accessed August 24, 2016).

[40] L. Wallgren, S. Raza, and T. Voigt, Routing attacks and countermeasures in the RPL-based internet of things, *International Journal of Distributed Sensor Networks*, **9**, 8, 794326, 2013, http://www.hindawi.com/journals/ijdsn/2013/794326/ (accessed August 24, 2016).

14

IoT ARCHITECTURE

Shyam Varan Nath

IoT at GE Digital, San Ramon, CA, USA

14.1 INTRODUCTION

Internet of Things (IoT) is an emerging area where multiple vendors are developing their offerings, either as IoT Platform or specific part of the stack. According to the Gartner's Hype Cycle of Emerging Technologies, while IoT remains at the peak of the hype curve, IoT Platform is on the emerging side or as Gartner calls it "Innovation Trigger" [1]. Since IoT brings together several different technologies, such as sensor technologies, data collection, connectivity, data store, analytics, visualization, business logic, and mobile, IoT Architecture provides a "glue" to align the different pieces within an IoT Platform. At a high level, the physical world converges with the computing hardware, software, and the interconnecting systems. To derive the business value from the IoT systems, people have to be connected to consume the information and monetize it. Hence, any IoT application cannot ignore the User Experience (UX). IoT Platforms should account for the UX aspects, as well.

14.2 ARCHITECTURAL APPROACHES

Software design patterns have been used to provide reusable solutions for generic problems, to guide the software engineers. An example of such a software pattern can be a sorting algorithm. As a resource, it allows the software engineer to get jump-started with the coding in a specific programming language. Likewise, an architecture pattern helps to jump-start the software architecture of the proposed solution, without

Internet of Things and Data Analytics Handbook, First Edition. Edited by Hwaiyu Geng.

Companion website: www.wiley.com/go/Geng/iot_data_analytics_handbook/

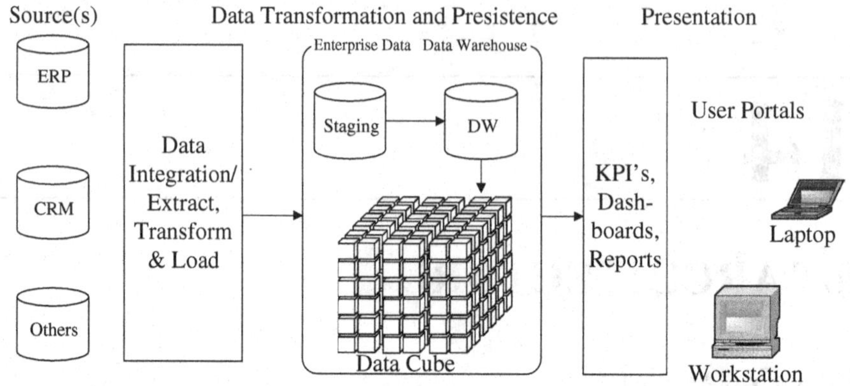

FIGURE 14.1 Reference architecture of data warehouse. Reproduced with permission from Object Mgt Group/llC.

the need to reinvent the wheel. While the architecture pattern is a concept, the IoT Architecture can be used as a reference for the application architecture. Such reference architecture provides a common terminology and concepts, to ensure that the different stakeholders are communicating at the same level.

Let us start with simple reference architecture from a well-understood area, such as data warehousing. At a glance, we can see in Figure 14.1 that a data warehousing and Business Intelligence solution would generally have three main parts. The starting point is the source system(s). This stage would be similar to the devices or the "Things" in the IoT space. The next phase is the data processing or the transformation phase which determines how the data needs to be stored. The final stage is where the human users consume the information.

Most IoT systems have a similar purpose, namely:

- Connect to the device(s) and ambient sources, to gather the data.
- Store/transform the data in a central system in which data can be analyzed.
- Consume the information through the application that helps to solve a specific business problem, for example, condition-based maintenance of the machine.

We will briefly look at the area of Enterprise Architecture (EA) here which has been around for 25–30 years now. The four well-respected EA Methodologies are as follows:

1. The Zachman Framework for EA
2. The Open Group Architecture Framework (TOGAF)
3. Federal Enterprise Architecture (FEA)
4. Gartner

A detailed comparison of these EA approaches has been provided at this site [2]. The reason we bring up the existence of the different EA approaches for last two to three decades is that same problems can be dealt in many ways. Likewise, there will be a flurry of IoT Reference Architectures as it is an emerging field. Each of these

Consortium	Members	Prominent companies
AllSeen alliance	200+	Microsoft, Cisco, Qualcomm, IBM
IEEE P2413	25+	GE, Qualcomm, Cisco, Intel
Industrial internet consortium	235+	GE, IBM, Intel, AT&T, Cisco
OCF (include AllSeen alliance*)	170+	GE, Intel, HP, Dell, Samsung, IBM
Thread Group	160+	Samsung, Google (Nest), ARM,

FIGURE 14.2 The state of IoT-related consortiums for standardization. Reproduced with permission from Object Mgt Group/llC.
Note: *AllSeen alliance and OCF merged as of 10/10/16.

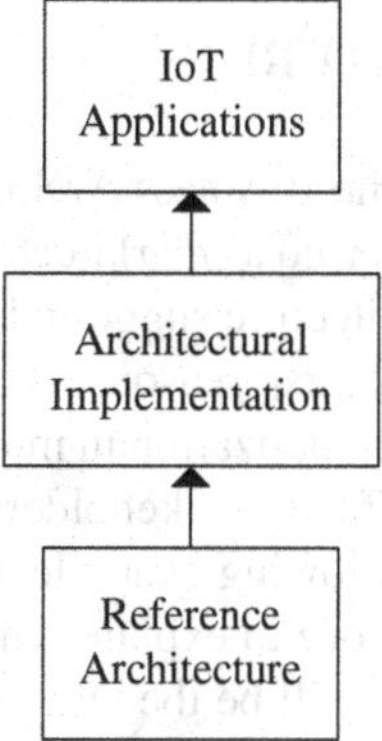

FIGURE 14.3 Hierarchy of IoT architecture. Reproduced with permission from Object Mgt Group/llC.

IoT Architectures will try to address the same class of problem with a slightly different perspective. This does not undermine one approach versus the other (Figure 14.2).

The previous table clearly shows that there are several efforts to self-organize into different groups and consortiums to bring some degree of "regulation" in the field of IoT and Industrial Internet. It is interesting to note that some of the prominent vendors are part of multiple such groups. As a result of these movements, we will see several IoT Reference Architectures emerge representing different perspective. For instance, the Industrial Internet Consortium (IIC), where AT&T, Cisco, GE, IBM, and Intel are at the driving seat, will primarily focus on the use IoT for industrial use cases. The consumer use cases such as wearables to monitor a human body or commercial use cases such as connected cars will not be the front and center of IIC's reference architecture. However, on a positive note, several such efforts to bring standards and regulation to the broader IoT space will invite a lot of investments and attentions and provide guidelines to emerging vendors and potential consumers of IoT.

Figure 14.3 shows the hierarchic view of the IoT Architectures. The reference architecture is a reusable concept that is used to create the architecture for a specific company or line of IoT solutions. While the reference architecture is independent of the IoT Platform, the final IoT solution may be based on use of a specific IoT Platform for deployment. Using the architecture implementation, the specific project of specific application may create a solution. It is important for the IoT reference architectures and the IoT Platform providers to ensure that there is sufficient degree of interoperability between the IoT Platforms, to allow hybrid solutions. The end customers may choose the best-of-breed components for their solution. In some cases, the expertise may be spread among different IoT Platform providers, and the fastest way to implement a specific solution may be by interoperating two or more such IoT Platforms. The IIC recognizes the need for such cooperation among the different stakeholders and encourages creation of test beds with multiple contributors. Most details of the test beds are provided here [3]. Cutting edge and emerging concepts like blockchain [4] will impact the IoT architectures at the device level, very soon. However, for the current book chapter, it is out of scope.

14.3 BUSINESS MARKITECTURE

The following is an example of the Business Architecture or the Marketing view of the Architecture or commonly referred as markitecture [5]. Markitecture is defined as "a computing or networking architecture suggested by the marketing department for sales purposes rather than for technical reasons." It is important to plan a real-world project with a business goal and monetization in mind. A markitecture is a quick way to initiate the dialog with the different stakeholders; some of them may not have a technology background. In the following example, the concept of Industrial Internet can be easily explained using a story to explain what kind of industrial devices will be connected and eventually who will be the target set of users who will monetarily benefit from it (Figure 14.4).

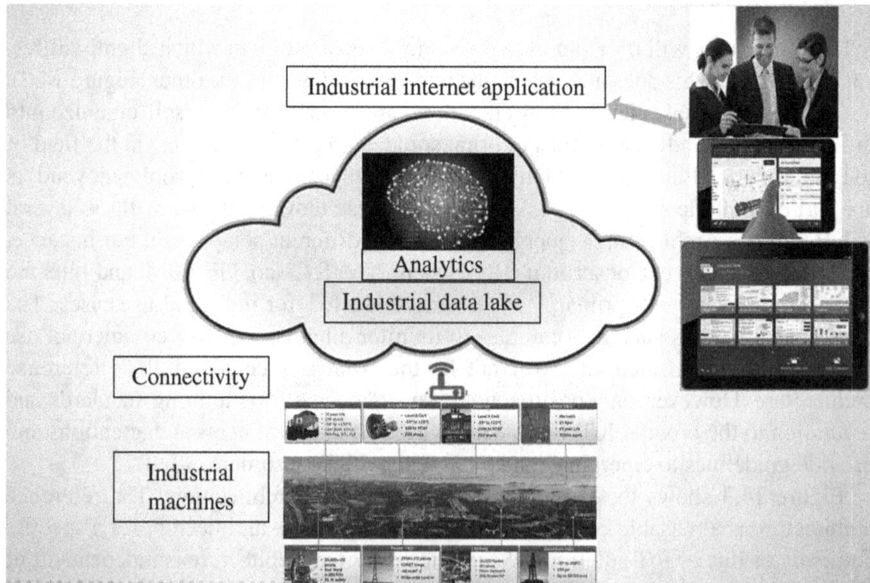

FIGURE 14.4 Conceptual view of IoT Cloud. Reproduced with permission from Object Mgt Group/IIC.

14.4 FUNCTIONAL ARCHITECTURE

The functional IoT architecture is a model that helps to map the subsystem functions, stakeholders, interactions between them, and corresponding technology needs. Such a functional architecture can be used by the different experts such as the business subject matter experts (SMEs); software engineers and IoT architects are able to create the blueprint of the IoT architecture solution. This is the big picture stage that helps to model the functional landscape (Figure 14.5).

The previous visual shows a functional view from the health IT Federal domain. Such functional view of business could be a starting point for the development of the IoT functional architecture for a health IT-related IoT solution such as Connected Care. The next stage would be the application architecture.

14.5 APPLICATION ARCHITECTURE

There are three different commonly used patterns for IoT (Figures 14.6 and 14.7):

1. Cloud/Data Center centric
2. Edge + Gateway with Cloud
3. Edge intelligence + Cloud (Cloud vs. fog computing)

The Cloud or the Data-Centric pattern relies heavily on the use of Cloud environment or private Data Center for majority of the IoT data and analytics processing. The device data and any other external data sources are persisted in the Cloud tier.

FIGURE 14.5 Functional view for healthcare. https://etrij.etri.re.kr/etrij/images/2014/v36n5/ETRI_J001_2014_v36n5_730_f002.jpg. Reproduced with permission from Object Mgt Group/llC.

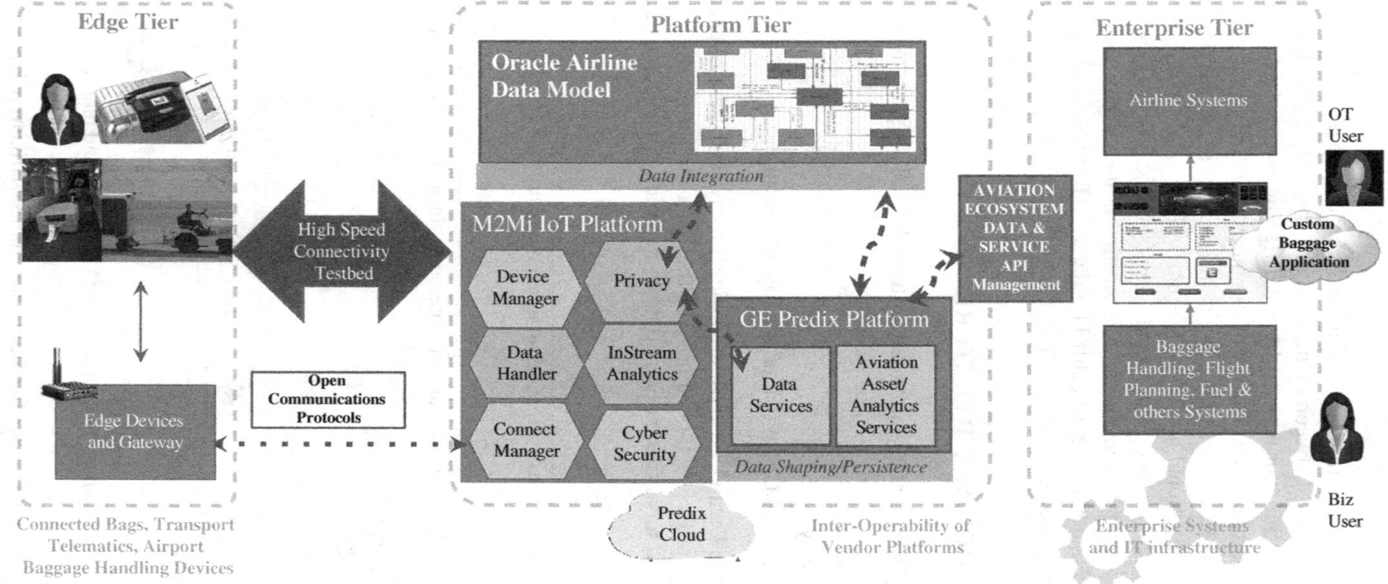

FIGURE 14.6 IoT application architecture. Reproduced with permission from Object Mgt Group/IIC.

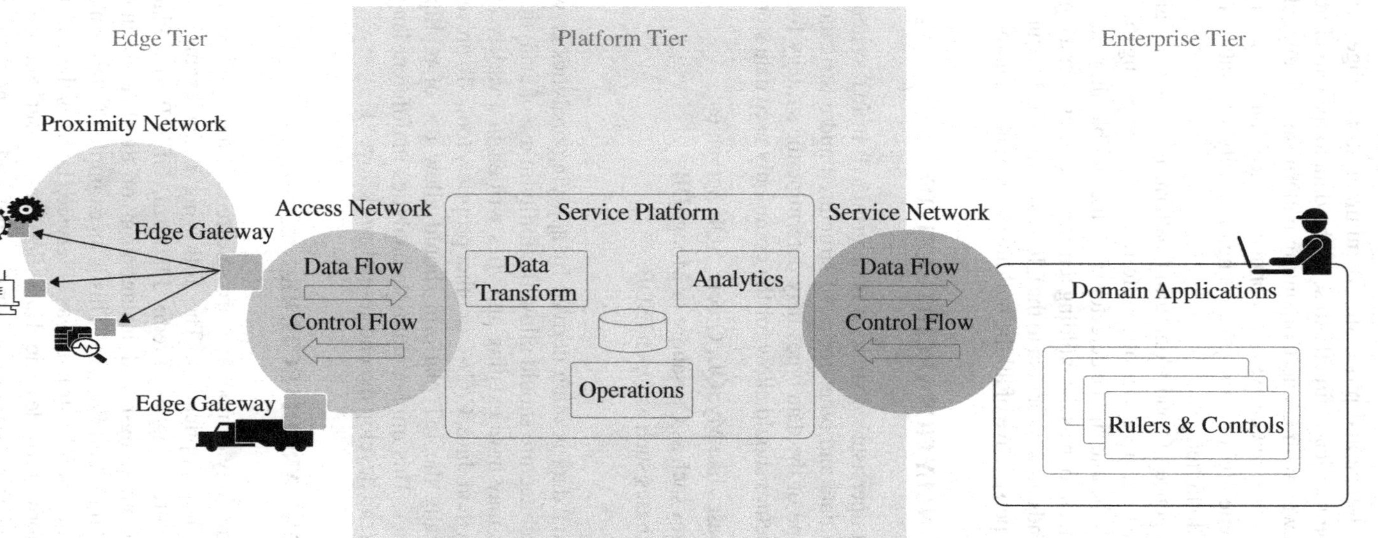

FIGURE 14.7 Industrial Internet Consortium's Industrial Internet Reference Architecture (IIRA). http://www.iiconsortium.org/IIRA-1-7-ajs.pdf. Reproduced with permission from Object Mgt Group/llC.

The amount of data storage and processing is minimal at the Edge tier. The Edge devices may often communicate directly to send the data to the Cloud tier.

In the Edge + Gateway with Cloud Computing pattern, several edge devices may be connected via one or more Gateways to connect to the Cloud. While the Gateway devices could do some lightweight processing, majority of Data and Analytics processing is in the Cloud tier.

In the Edge intelligence + Cloud Computing pattern, there is a certain degree of processing logic in the Edge tier and is often referred as fog computing [6]. An example of such a pattern would be that smoke detector starts the audible alarm and turns on the water sprinkler system as soon as the building temperature turns over a high threshold. At the same time it sends the sensor data to the Cloud system which in turn can analyze the data with a larger perspective of all the buildings in certain geography.

14.6 DATA AND ANALYTICS ARCHITECTURE

At this stage, the data persistence strategy is determined. In IoT systems, we often deal with time series data and other forms of structured and unstructured data. The variety and the volume of the data types may determine the strategy for persistence data. The IoT architecture should allow for the commonly used data stores such as:

- Relational databases like MySQL, Oracle, SQL Server, etc.
- NoSQL Databases such as Cassandra or MongoDB
- Time series databases such as OpenTSDB
- Hadoop framework

The value from IoT data is often derived by applying analytics or workflow of analytics. The IoT architecture should allow flexibility in use of analytics and workflows. Organizations may have existing analytics written in a variety of languages such as MATLAB, Python, Java, C/C++, or other propriety tools. To prevent rewriting of the analytics, ideally the IoT platform should allow use of multiple of these languages for analytics. This implies multiple analytics in different languages have to be chained together to develop the end-to-end workflow.

14.7 TECHNOLOGY ARCHITECTURE

At this stage the technology components are selected for the implementation of the instantiation of the IoT application. The following is an example of the IoT Technology Architecture using the Predix Platform. GE's Predix Platform is an example of Cloud Computing-based instantiation of the IoT Architecture. The Platform tier is based on Cloud Foundry. This is an example of Platform as a Service (PaaS) that in turn requires Infrastructure as a Service (IaaS) as the underlying tier.

The Predix Platform provides the Data Services, Asset Services, Analytics Orchestration, and other supporting constructs like end-to-end security services. Such Platforms can help to build IoT solutions using reference architecture in a rapid and repeatable fashion. Each application does not have to solve the scale and security problem, rather the Platform incorporates the best practices from the IoT Reference Architecture (Figure 14.8).

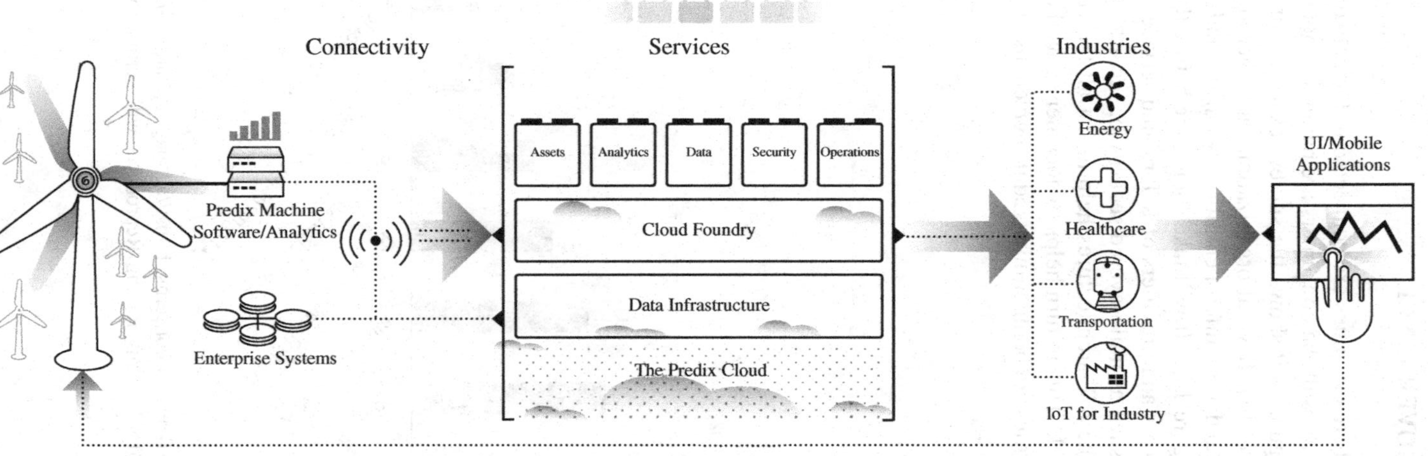

FIGURE 14.8 Predix—technology view of the architecture. Reproduced with permission from Object Mgt Group/IIC.

14.8 SECURITY AND GOVERNANCE

IoT is an emerging area where machines are connected to the network in ways it was never done before. As a result the consumers are paranoid about security till this area fully matures. Figure 14.9 shows the survey results about the top programmer concerns related to IoT. Several organizations are emerging to address the related topics like Standards, Interoperability, Security, and Governance around IoT as shown in Figure 14.2. It will take some time before this landscape stabilizes. Till then consumers and providers of IoT-related services will have to use current best practices such as multilayered security or security at each layer to minimize the threats. The IoT architectures help to list all the attack vectors within the full-solution stack. This in turn helps to document how the risks to the different system vulnerabilities are being minimized in a given instantiation of the IoT system.

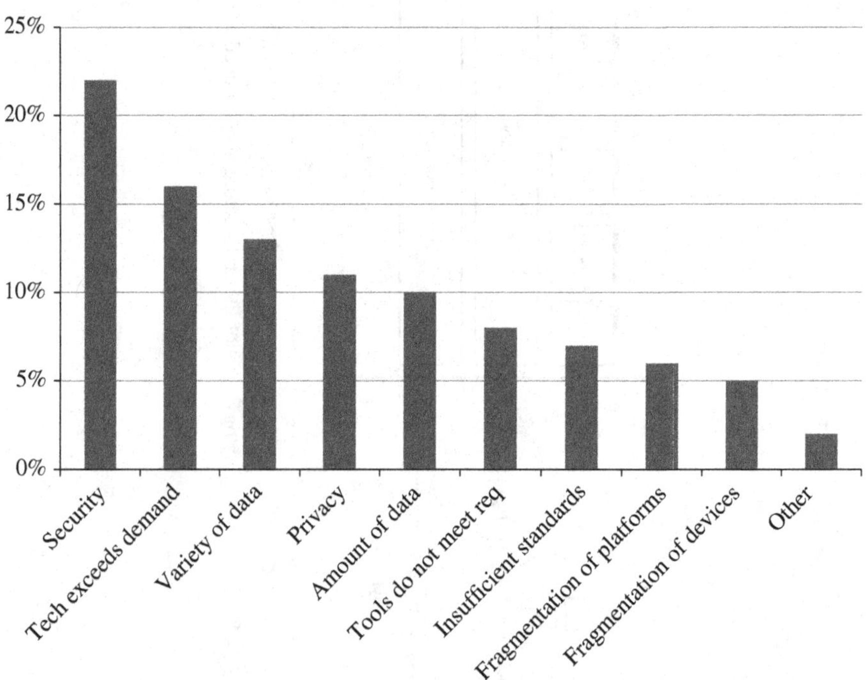

FIGURE 14.9 Survey on IoT-related concerns. https://adtmag.com/articles/2015/07/29/~/media/ECG/adtmag/Images/2015/07/evans_iot.png. Reproduced with permission from Object Mgt Group/llC.

REFERENCES

[1] http://www.gartner.com/technology/research/hype-cycles/ (accessed August 29, 2016).

[2] https://msdn.microsoft.com/en-us/library/bb466232.aspx (accessed August 29, 2016).

[3] http://www.iiconsortium.org/test-beds-faq.htm (accessed August 29, 2016).

[4] http://fortune.com/2015/08/18/filament-blockchain-iot/?xid=soc_socialflow_twitter_FORTUNE (accessed August 29, 2016).

[5] http://www.urbandictionary.com/define.php?term=markitecture (accessed August 29, 2016).

[6] https://www.cisco.com/c/dam/en_us/solutions/trends/iot/docs/computing-overview.pdf (accessed August 29, 2016).

15

A DESIGNER'S GUIDE TO THE INTERNET OF WEARABLE THINGS

DAVID HINDMAN AND PETER BURNHAM
FJORD, San Francisco, CA, USA

15.1 INTRODUCTION

The wearable technology industry is germinating conceptually new formats at a lightning pace and in a million disparate directions at once. These new form factors and approaches to wearable technologies are unique expressions of an industry still finding its footing as it innovates toward a few common wearable formats. Think pre-iPhone and pre-Android, back to 2005 when dozens of different mobile phone form factors existed somewhat independently from each other. There was no agreed upon design framework that united the industry. Everything from Blackberries, to flip phones, to Sidekicks existed independently, and semiunique operating systems were baked into each manufacturer's device lineup. Now, in 2015 we've more or less settled on a single form factor and a few operating systems that the success of the iPhone is responsible for solidifying across the industry.

As the Internet of wearable things matures, we can expect a similar increase in consistency to shape the market into a few known expressions. Currently, all signs point to the watch form factor coming out on top as the industry converges on this trend. Again, though relatively late to the game, the Apple Watch has cemented the watch form factor into the public's perception of wearables as a category and into the technology market's product lineups. However, even though the competition for the wrist is heating up, the wearable technology market is still in the foggy exploration era where shrinking technology, new types of sensors, batteries, and

Internet of Things and Data Analytics Handbook, First Edition. Edited by Hwaiyu Geng.

Companion website: www.wiley.com/go/Geng/iot_data_analytics_handbook/

computing power are forming entirely new concepts of what a wearable technology can do. To make sense of this ever-changing landscape, a design framework is needed that considers the microcontexts that underpin the Internet of wearable things.

The scope of this chapter includes an actionable set of design principles that serve to unify these disparate product ecosystems into a congruent whole. The chapter also provides a checklist for designers, entrepreneurs, product managers, and inventors to make sure that the wearable user interface (UI) of the future is focused around the needs of the user and their behaviors.

15.2 INTERFACE GLANCEABILITY

Traditionally, digital services and products have been designed to encourage users to engage with them as often as possible. Each engagement is a potential conversion or additional data point to sell to marketers and keeps the user fully engaged. However, Gartner explains that the goal of glanceable design is that it "enables users to absorb the information without having their attention distracted from foreground tasks" [1]. When we look at the use cases common in wearable technology, this attention-monopolizing paradigm undermines a service's ability to meet the needs of the user from a core level. Since the majority of consumer wearables are focused around data collection and its periodic review, the service model necessitates that the wearable should recede into the background of its owners' life and be called upon only when needed or for key inputs.

When discussing glanceability it is necessary to evaluate key moments and unpack why they are important to the user's immediate context, what type of information is being delivered, and how it is delivered. Take, for instance, the runner who wants to see their average speed and set an interval timer. This interaction is tricky at best on a smartphone midstride. After unlocking the device and tapping and swiping through applications and menus, the runner's attention is anywhere but on the course ahead of them. Wearable devices, however, are in a unique position to provide custom-tailored interfaces that focus on concise interactions that are both contextually relevant and task focused.

To help the runner out, we first break down the scenario into two separate activities. First, there is the "review" of the data, which in this case is the average speed. Then there is the "action" when the runner sets a timer. Each of these requires a different approach to make them truly glanceable activities that don't distract the user from their goal at hand.

15.2.1 Reviewing Activity

Reviewing information is one of the most common wearable tasks. Most often this information is the current system status of progress toward a goal. To make reviewing as quick and impactful as possible, focus on the most important data and simplify its display as drastically as possible. When reviewing data, the first decision you'll need to make is whether or not the user needs to see multiple data points or whether a single data point will suffice. For the runner in this example, a single data point will suffice to show their average speed. However, if the runner wanted to check multiple data points like distance and time, creating a glanceable interface becomes much more difficult. In this case the best option is often to aggregate multiple data points and award them an abstract score. Once the user understands the mental model of the score, a single data point can be presented that lets them understand the summation of several data points at once. So now we have a simple numerical score, but how can we simplify the information even further to make the glance as quick as possible? Depending on the capability of the device, we can use graphics to show an image representative of the data set. The Misfit Shine device, for instance, shows a set of a dozen simple LEDs that light up to indicate the quantity of physical activity that the user has accomplished during a 24 hour period. Misfit states that "health and body information on a wearable device like the Shine should be able to be conveyed in a second or two" [2]. Nike+ FuelBand similarly uses an array of different colored LEDs to show progress from green (low) to red (high). Other wearables use a range of traditional screen-based graphics from growing vines, to hearts, to Tomodachi-esque character development that ultimately lets the user understand a large data set with the least amount of time spent looking at a screen and the least amount of mental effort needed to decipher a complex set of information.

15.2.2 Glanceable Actions

When we look at glanceable Actions, the goal is similar but slightly more complex because we need to help the user complete a task on the device while not interrupting their flow. To do this we focus on "single-serving activities," which are actions that a user can complete with a single tap, swipe, or shake. This lets the user interact with the service without stopping what they are doing to interact with their device. Framing these interactions into yes or no questions helps keep the interface as simple as possible. "Start new lap?" for the runner lets them tap yes or no and trigger an action within the service with minimal effort. For more complex interactions, let the graphical nature of the UI help guide the task. In the Tesla Apple Watch application, the user can unlock their car's doors by tapping the door itself. They don't need to swipe to a "door-unlocking screen" to take this simple interaction. Instead they are able to interact with the graphic content itself, which is a more immediate mental model of the action in the first place.

In summary, the overarching goal when crafting glanceable interfaces is to simplify information and actions down to their core components ultimately reducing the user's cognitive load.

For reviewing data:

- Aggregate and summarize several data points into a general activity score.
- Create an abstract representation of the activity score.
- Use hardware (LEDs, smart materials, etc.) or graphic interfaces like shape, color, size, and movement to indicate progress.

For actions:

1. Focus on "single-serving actions."
2. Frame interactions into yes or no questions.
3. Let users interact with graphical content instead of interacting with menus, buttons, or forms.

15.3 THE RIGHT DATA AT THE RIGHT TIME

Unwanted personal information on a wearable is the equivalent of a banner ad at best. How can new technologies in wearable devices sense their situational surrounds and adjust content accordingly to provide a better experience? First we need to decipher what is important within the service that the wearable technology is providing. Does the user need to know key information now to make a decision, or can they wait to do this later, perhaps on another device entirely?

15.3.1 Context Is Always King

The wearable ecosystem in its current state is complex and often involves multiple devices to provide data connections to or support management capabilities of wearable devices. This lets us choose the right time and the right place to provide essential information to the user. To do this a thorough examination of the user's context is essential. Using traditional service design techniques, we can understand all of the various inputs that affect context including time, speed, proximity, environmental data like weather, bio data like heart rate, historical data, and more. Each of these inputs can act as a trigger to the system, but a combination of multiple context flags can pinpoint key interactions that tell us when to present information to the user and what form it should take.

15.3.2 Methodology

To serve the right data at the right time, we focus on three service design fundamentals: customer journey mapping, service blueprinting, and interaction prioritization. Customer journey mapping involves substantial research with end users of the wearable service but uncovers each individual interaction that they have with the service. Each of these interactions or "moments" is documented including what they did and why they did it. This lets us understand not only the tactical action but also their underlying motivations. Then the service is blueprinted. Service blueprinting takes a deep look into all the business and technology processes that are needed to support the user's journey. This uncovers the moments that business needs to communicate with the customer or when the customer needs to communicate with the business to keep everything running smoothly. Looking at both the journey and the blueprint gives service designers insight into the crucial communication moments that are needed.

These are then prioritized by value to the user. This point can't be understated because many services overprioritize things that may be important from a business perspective but inconsequential to the end user. To make sure that we are prioritizing the right moments, we ask ourselves:

- Does it support the user's goals?
- Did the user ask to see this information?
- What happens to the service if we don't show the user this information?
- Does the user get this information somewhere else?
- How often do they need this information?
- How can we reduce the amount of information they need to see?

Answering these few questions—and the flowchart (Figure 15.1)—pays dividends to the user who ultimately gets a more tactical and useful service that doesn't feel needy or obtrusive or clumsy.

In summary, to serve only the right information at the right time, always consider the user first and foremost and prioritize the needs of the marketing and advertising as often as possible.

The key tools to find the right information are:

1. Customer journey maps
2. Service blueprints
3. Information/interaction prioritization

15.4 CONSISTENCY ACROSS CHANNELS

Developing seamless transitions that account for changing data connectivity and multiple platforms is table stakes when designing for wearables. Customers expect services to work each and every time in every single situation. "Service drag" happens when a service requires too much effort to use and starts to become literally a drag. The runner who decides to leave their phone at home loses the data connection to their smartwatch, and it fails to track their run. This is an example of a service failure because there was a clear misunderstanding of a key use case that the user expected the device to fulfill. When we design for consistency across channels, what we are aiming for is a service that is flexible and predicts pain points in the service and uses smart defaults and baseline services to continue operating. To design within these constraints, we focus on two key areas, offline mode and transition points.

15.4.1 Offline Mode

Until our wearables have dedicated data connections, which may be fairly soon, we need to design for two states, paired and offline. The paired mode assumes full connectivity between the smartphone or other device and the wearable. In this case we

(a)

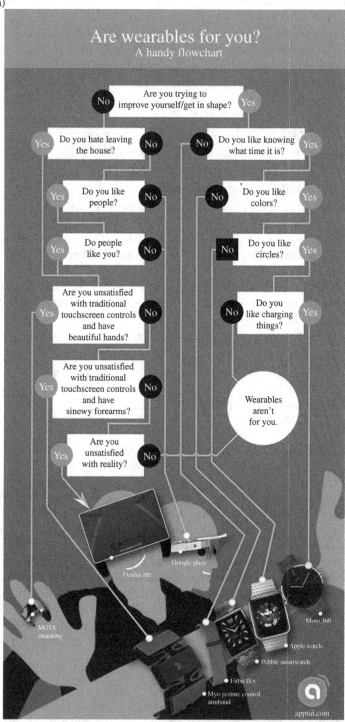

FIGURE 15.1 (a) Are wearables for you? Permission by Apptui.com or (b) choosing a pedometer. Permission by Livescience.com. (c) Method-controlling charge within a portable device. Permission by Apple, Inc.

(b)

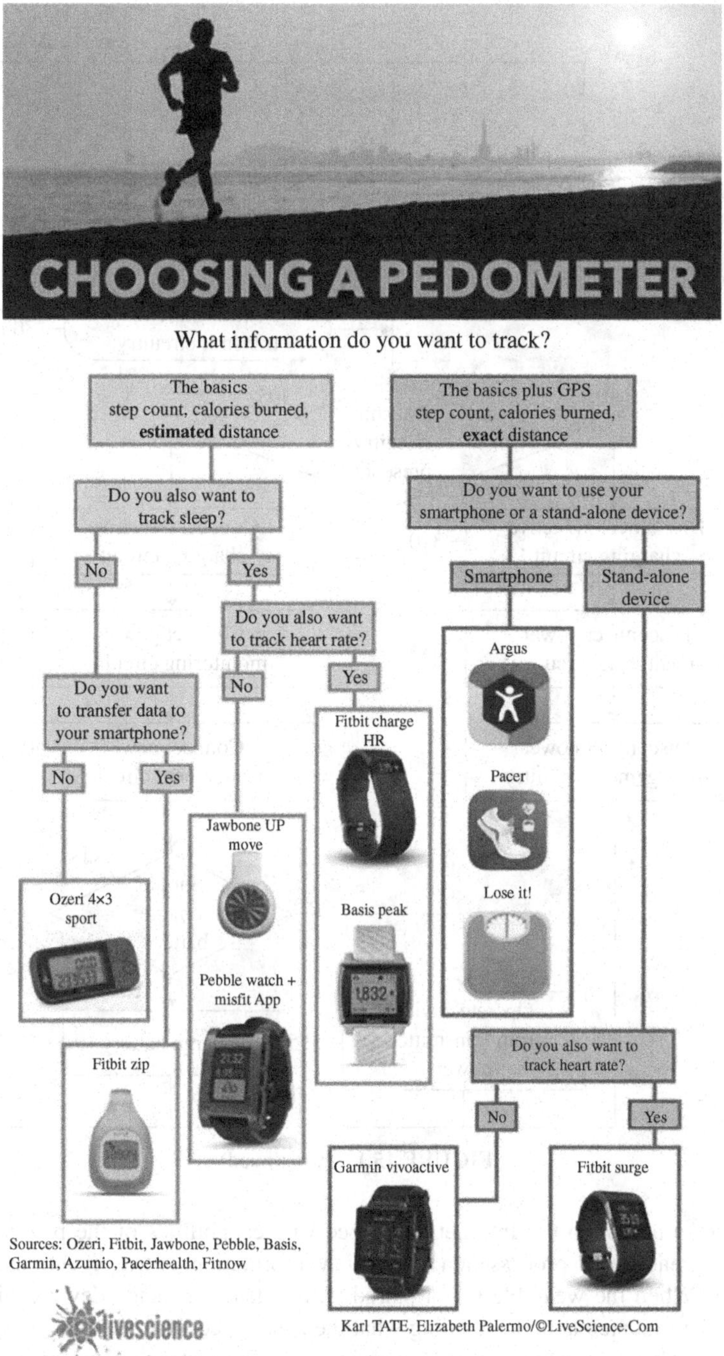

What information do you want to track?

The basics
step count, calories burned,
estimated distance

Do you also want to track sleep?

No

Do you want to transfer data to your smartphone?

No

Ozeri 4×3 sport

Yes

Fitbit zip

Yes

Do you also want to track heart rate?

No

Jawbone UP move

Pebble watch + misfit App

Yes

Fitbit charge HR

Basis peak

The basics plus GPS
step count, calories burned,
exact distance

Do you want to use your smartphone or a stand-alone device?

Smartphone

Argus

Pacer

Lose it!

Stand-alone device

Do you also want to track heart rate?

No

Garmin vivoactive

Yes

Fitbit surge

Sources: Ozeri, Fitbit, Jawbone, Pebble, Basis, Garmin, Azumio, Pacerhealth, Fitnow

livescience

Karl TATE, Elizabeth Palermo/©LiveScience.Com

FIGURE 15.1 (Continued)

(c)

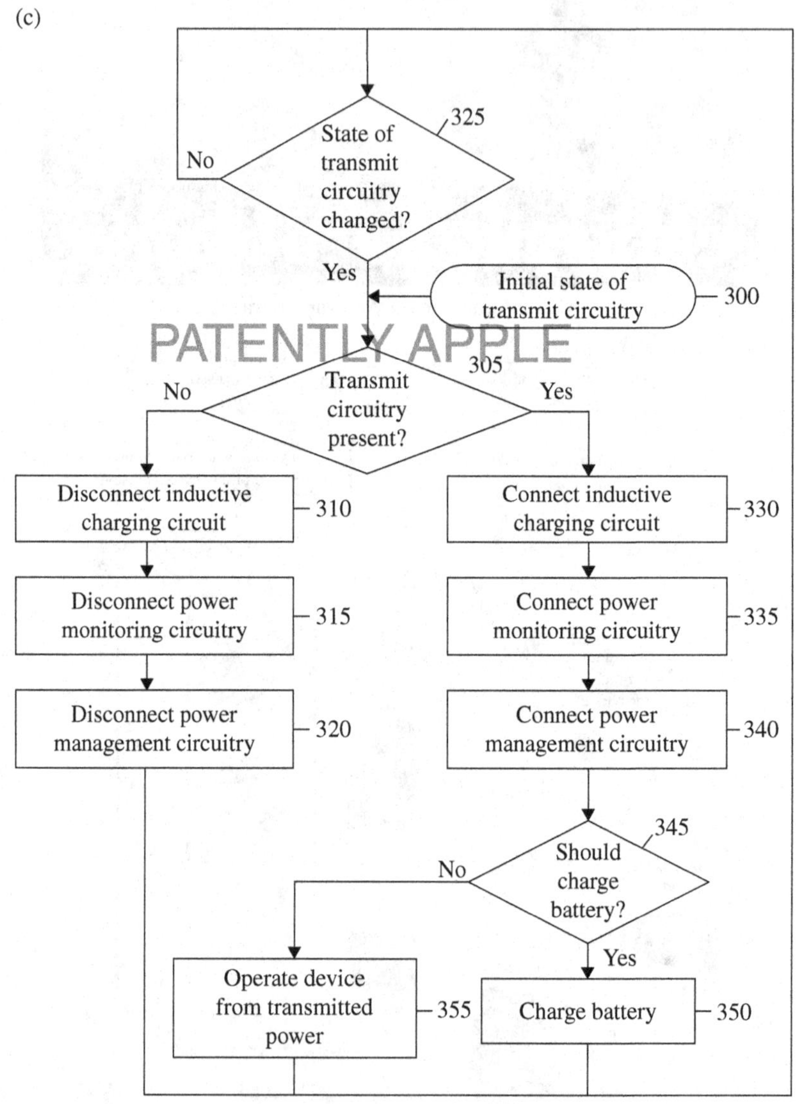

FIGURE 15.1 (Continued)

can assume access to the Internet and processing capabilities of the paired device. This is a great time to process data sets, review information, or trigger actions in other services. When the wearable is not paired with a data-connected device, things get tricky. The first step is understanding what the user expects the wearable to continue doing and what the device's offline capabilities are. For instance, the runner still may expect the device to record distance and heart rate but understand that full GPS tracking is unavailable. This information must be transparently communicated to the user at all times including the exact moment that a connection is lost. The three most important items to maintain in offline mode are continued tracking and reviewing of raw data, goal progress, and data from a previously synced state.

15.4.2 Transition Points

The moment that someone transitions from one application to another, from one service to another, or from one device to another is common in wearable ecosystems but often results in annoying and time-consuming work-arounds to manage the transition. To craft a seamless experience, focus on the user's cognitive resources at the time of the transition point, utilize visual anchors between services, and try to let smart integrations do the heavy lifting:

- **Cognitive resources** change depending on device type. When the runner glances at their smartwatch to see their pace, their cognitive resources are low because they're actively engaged in an activity. Cognitive resource theory, although focusing on a different area, states that "*stress* is the enemy of *rationality*, damaging leaders' ability to think logically and analytically" [3]. Our goal is to design products that let people make better informed and stress-free decisions in an instant. To that end, only show them the critical information and remove all distractions that undermine clarity and immediacy. When they stop their run and review their activity on their smartphone, they have more cognitive resources. Show them more information like historical trends but keep it simple and to the point. The mobile device still limits the user's cognitive ability to take in large amounts of data. For the runner who reviews their run on their computer, this may be a good time to show them everything and give them multiple views of data sets. Match information density to devices and you'll help people get the most from each platform.
- **Visual anchors** are used when transitioning between two services of platforms to make the transition feel congruent. For instance, the runner who first looks at their distance on a wearable like a Fitbit should also be presented with this information as the first piece of data on the mobile or desktop application, so there is data parody anchoring them in the new experience.
- **Smart integrations** between services underpin nearly every successful wearable service. Apple HealthKit may become a successful platform simply because the user is not required to do any work to input their data into the system. Similarly, there are few successful calorie-counting applications because of the overwhelming amount of effort it takes to manually input every single item we eat. Do the heavy lifting for your users and your service can truly become seamless.

For offline mode:

- Continue tracking raw data and let users review it.
- Show the user's progress toward previously set goals.
- Show data from a previously synced state.

For transition points:

- Shift the amount of information displayed on a given device to match the user's cognitive resources.
- Use visual anchors to make data feel congruent between services.
- Integrate backend services to make the experience seamless.

15.5 FROM PUBLIC TO PERSONAL

There are a number of differences between wearables and other mobile devices, but the biggest difference is their persistent attachment to our body. With these new devices, we may find ourselves "wearing" some of the most personal aspects of ourselves: our conversations, relationships, and even health. Wearables are simultaneously the most private and public devices yet. Not only must designers account for a user's context and surroundings, but also the angle and position of their wrist or head may make or break an experience. When designing for this paradox, we should keep in mind this precarious tipping point between Public and Personal.

To approach designing for the shift from public to personal, we can apply the following framework of categorizing shifts as "Planned or Unplanned." Planned shifts include user-initiated transitions: checking the time, sending a message, making a payment, and so on. Unplanned shifts include most incoming communication, notifications, or alerts.

15.5.1 Planned Shifts

In the majority of the time, our smart watches won't in fact be doing anything, so to speak. They will be in ambient mode, similar to a screen saver. When users need to wake up the device and perform a task, this constitutes a planned shift. The goal for us as designers is facilitating this shift for users and making it as easy as possible and ideally one handed or no handed. To this end, we should consider the following:

1. **Recognize Orientation**. A simple rotation of the wrist from outward facing to inward facing has become an industry standard, especially with the recent release of the first Apple Watch. This trigger allows a one-handed mechanism for, at the very least, viewing the time. Moving forward, we would expect and hope to be able to sense device orientation with a greater degree of subtlety, without having to make a dramatic shift or arm movement to wake up the device.
2. **Recognize Gestures**. Similar to recognizing the direction that the watch is facing, we could think about having the device learn custom gestures that are assigned to awake the device from its slumber. With gesture-recognition technology increasing rapidly, we can expect to be able to assign choreography to waking triggers, creating a custom mechanism for users to purposefully engage with their devices.

3. **Respond to voice**. Another way we might want to purposefully shift wearables from "standby" mode is to design voice recognition so that we may keep the transition one handed or, in this case, no handed. Currently there are a number of wearables that support voice recognition, but the key will be to be able to engage with the devices without pressing a key first (which renders the interaction two handed). Looking at examples like "Ok Google" as initial waking triggers can be a model that will enable this transition.

15.5.2 Unplanned Shifts

When considering unplanned shifts, we must take into account user context and be as considerate as possible because these shifts are by definition interruptions. In order to minimize the potentially disruptive nature of these transitions, we can consider the following principles:

1. **Vibrate first, display second**. This is perhaps a no-brainer but as such is something we should not allow to be overlooked. Because wearables rest on our skin, vibration could actually take center stage in being even more noticeable than typical audible alerts. It is also important to vibrate to alert the user of incoming messaging but wait until the device is inward facing to display content. Those riding the bus with you may not want to come face to face with your friend's cat selfie or even something more embarrassing.
2. **Recognize orientation**. This principle holds for both planned and unplanned shifts but in this case is more of a defensive mechanism. For incoming communications, especially if it's visual content, we will want to recognize if the device is facing outward (public) or inward (personal). This can be used in concert with vibration to alert the user without accidentally displaying content that was meant to be private.
3. **Use considerate defaults**. Like we just mentioned, the key when designing new service with wearables will be starting the user off on the right foot. We can do so by always asking ourselves if the default settings for the shifts from ambient to personal are considerate. Just because an option is buried in a settings menu doesn't mean we have solved the problem—we should anticipate as much as possible and bake these into the default configuration of our services.

15.5.3 Summary

When designing for the shifts from public to personal, we should think of the transitions in one of two categories: Planned shifts and Unplanned shifts.

For Planned shifts:

- Consider device orientation, voice, and gestures to enable the transition.

For Unplanned shifts:

- Be aware of user context by launching services with considerate defaults.
- Vibrating first and recognizing device orientation are a few key examples of keeping the defaults empathetic.

15.6 NONVISUAL UI

One of the most exciting developments accompanying many wearables is the numerous possibilities for interaction using more of our five senses. Wearables are constrained by lack of screen real estate which not only limits the amount of display area but also reduces the region for interaction. But with constraint comes opportunity. Limiting visual UI opens up a world of possibilities for interactions using touch, sound, gestures, vibration, health data, and more.

15.6.1 Challenges with Visual UI and Wearables

The biggest challenge when designing for wearables is that if it does have a screen, it's a lot smaller than we're used to. Traditional touch-screen paradigms like pinch to zoom don't translate due to lack of screen "runway." At the same time, screen real estate is drastically limited even compared with the smallest smartphones. It's clear that wearables' time to shine will not be a result of screen size or interactivity but rather the way it incorporates other interaction methods. In this section we will look at some of the opportunities provided by nonvisual UI.

15.6.2 Opportunities

When considering interaction methods beyond the screen, it helps to categorize them as either Inputs or Feedback. Inputs include actions that feed information into the system, such as vocal control, tapping, or scrolling. Feedback includes information the user receives from the device, such as audible alerts and vibrations.

15.6.3 User Inputs

- **Voice input** is probably the most obvious input method that doesn't involve a screen. We already see capabilities such as voice to text, Siri, and Ok Google on our mobile devices. The recently launched Amazon Echo further proves voice as a valid input method due to its total lack of screen. In the near future, designers that are trained in traditional UI tools and methods will soon need to augment their skill set with those possessed by voice designers. We will also need to consider the types of information and tasks that lend themselves better to audible or visual input. As voice recognition technology continually improves, we should count on voice as a crucial, if not primary, input method for wearables.

- **Gestures**: Many of you reading this chapter will remember the "Shake To" capability provided by and touted on many of the early iPhones. Where shaking is a fitting input method for some specific apps, this gesture has largely not been adopted beyond the gimmick stage. The reason is that this gesture is limited and not specific enough. With the launch of the apple watch and its gesture-recognition technology, the promise of being able to assign your own gestures to specific tasks becomes more of a reality. A world in which we can pound our fist to send a tweet or wave to email goodbye is a very real possibility. The most interesting will be the potential to teach wearables. Each of us may have our own personal input gestural language as well as use common gestures to trigger specific events. As designers, we will be challenged with devising a standard way of documenting these types of inputs.
- **Tapping**, or haptic input, is also an exciting frontier with regard to information entry. The apple watch, for example, allows force touch and recognizes taps. As capabilities for haptic input on additional devices become commonplace, we should consider tapping as a new language for outbound communication. While pattern-based communication have existed for a while (Morse Code), we should think of providing the ability for users to assign their own tapping patterns to their communications. Maybe two taps means "I'm on the way" or "Be there soon"; three taps in a row could mean "Running late." The possibilities are endless, and the best part is that none requires taking attention away or looking at a screen.
- **Health data** is powering many wearable experiences these days, and many devices now come equipped with a heart monitor. While this data has specific uses, the input is a totally new paradigm. Might there be other ways that this input mechanism could be used beyond monitoring specific health data?

15.6.4 System Feedback

- **Vibration**: Since wearables often rest against our skin, the opportunity for subtle and complex vibrational patterns becomes a real possibility for translating inbound messages. In the same way that we can expect tapping languages to send outbound messages, we can expect that haptic touch feedback for inbound messages can emerge as well. Vibration can also continue to be used to alert users.
- **Sound** is a valuable feedback mechanism in numerous systems and has been for a long time. It will still be a contender when it comes to nonscreen-based feedback of wearables. The key will be to minimize or find the right moments for this distraction. Often, sound is used in conjunction with a visual input, to give the user the assurance that the commands have been heard. In this case we can expect that sound may continue to provide users with alerts for incoming notifications. With connection to headphones, the distraction factor is eliminated and we can begin to imagine more sonic possibilities for this feedback.

15.6.5 Summary

Screens are small and not very useful on wearables, but what wearables lack in screen-based interaction capability they make up for in unlocking the power of our other senses. As designers, we can make sense of the expanded capabilities by thinking of them as either user inputs or system feedback.

For user input:

- Design using voice, touch, gesture, and even health input.

For system feedback:

- Consider touch (vibrational) feedback as well as sonic feedback.

As we forge a path and create new and enhanced existing services, we should consider all the interactive capabilities available to us. Looking forward, it's fair to think that the future for wearable interactions is largely a screenless one.

15.7 EMERGING PATTERNS

Remember when phones first got smart? There were flip phones, feature phones, phones with keyboards and some without.

Now, interestingly, we more or less agree on what a connected mobile device looks like. It contains an interactive touch screen and a home or back button and uses apps to distribute services.

Right now, the wearable device landscape is at least as, if not more, varied than early smartphone varieties. There are devices for fitness, activity tracking, those with screens, and those without. It's true that there is a lot of experimentation and that we haven't converged on the archetypal "Wearable" as we finally have with the smartphone. But moving forward, there are enough truths that are emerging that we can hang our hat on. As we continue to forge our way through the emerging wearables landscape, we should be able to count on these trends as being constants in a sea of uncertainty.

15.7.1 Wearables = Fashion

While the look and aesthetic appearance of wearables is of course the responsibility of product designers, those of us in charge of designing the screens and interactions must remember that wearables are essentially an interactive item of clothing. Screens we design for these devices must not only communicate the brand of the particular service but also support the user's fashion sense—our UI can't be ugly even if the brand is. As UI designers, we have a new dimension to consider, and we will have the new challenge of designing for brand accuracy but also fashionability.

15.7.2 Get Ready for New Languages

As we discussed in Section 15.4, new interactive capabilities will give rise to new forms of communication. For example, new tapping languages will emerge, either that are specific to the user's personal preferences or that fit into a larger more widely approved standard. Will two taps mean "yes" and three mean "no" universally? Or will we each have our own private haptic languages with those we care about? The answer is not certain, but we can bet on new languages and new forms of communication beyond the written digital or spoken word. We'll be able to communicate through touch in ways never possible before.

15.7.3 Count on Voice

Voice control will become more reliable both on how well it understands you and the degree to which it can plug into other services. With wearables, the ability to say "Order the usual" and have my typical takeout food arrive in 30 minutes without having to open my laptop is one promise of vocal interaction.

15.7.4 Speed Dial 2.0

While wearables aren't great as content creation tools (yet), they are effective at quick simple tasks. It is already common to build a list of common responses to incoming communications. Phrases like "I'm on the way home" or "Can't talk now" can be built-in and provide users with stock responses. We can bet on this Speed Dial paradigm being more prevalent, not only for responding to but also for initiating conversations.

15.7.5 Get Ready for Gestures

In the same way that we can expect new haptic languages, we can prepare for gestures as being a legitimate way for inputting information. Whether it's waving to tweet "hi" or raising your arms to turn the volume up, the gesture-recognition technology will only improve, creating more opportunity for the emergence of our custom gestural languages as well as a foundation for new standardized gesture-triggered messages.

15.8 CONCLUSION

The wearables product landscape is still in its infancy—the current market is replete with devices of all shapes, sizes, capabilities, and interaction paradigms. From simple activity trackers to full visual UI products like the Apple watch and everything in between, we haven't yet decided on what the archetypal "wearable" should be. As such, creating design rules for a vast and varied market is not without its challenges. And we expect that in a few years, wearables will evolve beyond some elements of this publication. With that in mind, we have attempted to present design principles that should be as timeless as possible. Regardless of the technological leaps that wearables make in the next few years, we should always try to design for glanceability, context, channel consistency, considerate defaults, and new interaction methods.

REFERENCES

[1] Gartner, Ambient and Glanceable Displays: http://www.gartner.com/it-glossary/ambient-and-glanceable-displays/ (accessed August 20, 2015).

[2] Misfit Shine: https://gigaom.com/2013/04/08/wearable-design-misfit-and-the-age-of-the-glanceable-ui/ (accessed August 20, 2015).

[3] Cognitive Resource Theory: https://en.wikipedia.org/wiki/Cognitive_resource_theory (accessed August 20, 2015).

FURTHER READING

Amazon Echo: http://www.amazon.com/Amazon-SK705DI-Echo/dp/B00X4WHP5E (accessed August 20, 2015).

Apple Gesture Recognition: https://developer.apple.com/library/ios/documentation/EventHandling/Conceptual/EventHandlingiPhoneOS/GestureRecognizer_basics/GestureRecognizer_basics.html (accessed August 20, 2015).

Apple Watch: http://www.apple.com/watch/?afid=p238%7CsrjqK1qvv-dc_mtid_20925qtb42335_pcrid_79417951813_&cid=wwa-us-kwg-watch-slid- (accessed August 20, 2015).

Google Voice Control: https://support.google.com/websearch/answer/6031948?hl=en (accessed August 20, 2015).

SIRI: http://www.apple.com/ios/siri/?cid=wwa-us-kwg-features (accessed August 20, 2015).

Voice to Text: https://play.google.com/store/apps/details?id=com.impressive4.voicetotext&hl=en (accessed August 25, 2016).

16

BEACON TECHNOLOGY WITH IoT AND BIG DATA

NICK STEIN[1] AND STEPHANIE URBANSKI[2]

[1] *Indoo.rs GmbH, Brunn am Gebirge, Austria*
[2] *Indoo.rs Inc., Palo Alto, CA, USA*

16.1 INTRODUCTION TO BEACONS

The history of beacons starts a long time ago as a way to signal over long distances using fire or later light houses. Beacons, as pertains to the Internet of Things (IoT), started with Bluetooth and accelerated with advancement to Bluetooth Low Energy (BLE), now referred to as Bluetooth Smart.

The Bluetooth protocol (invented by telecom vendor Ericsson in 1994) [1] has been around for many years enabling wireless transfer of data including both audio and video by the creation of a sharable wireless personal network.

In 2010 BLE, as part of Bluetooth 4.0, was released with a focus on the transfer of simple links URLs, and similar small amounts of data, instead of video and audio directly. It is called Low Energy as the protocol was adapted to be less battery intensive, which opened up a lot of possibilities.

Beacons are a piece of hardware that emits BLE signal to a Bluetooth-enabled receiver "which allows another device to determine its proximity to the broadcaster" [2]. This advance in, or simplification of the Bluetooth technology has opened up many possibilities for small amounts of data to be transferred. This allows devices to communicate with one another, either with updates, commands, or other information, which we now refer to as IoT (see Figure 16.1).

Internet of Things and Data Analytics Handbook, First Edition. Edited by Hwaiyu Geng.

Companion website: www.wiley.com/go/Geng/iot_data_analytics_handbook/

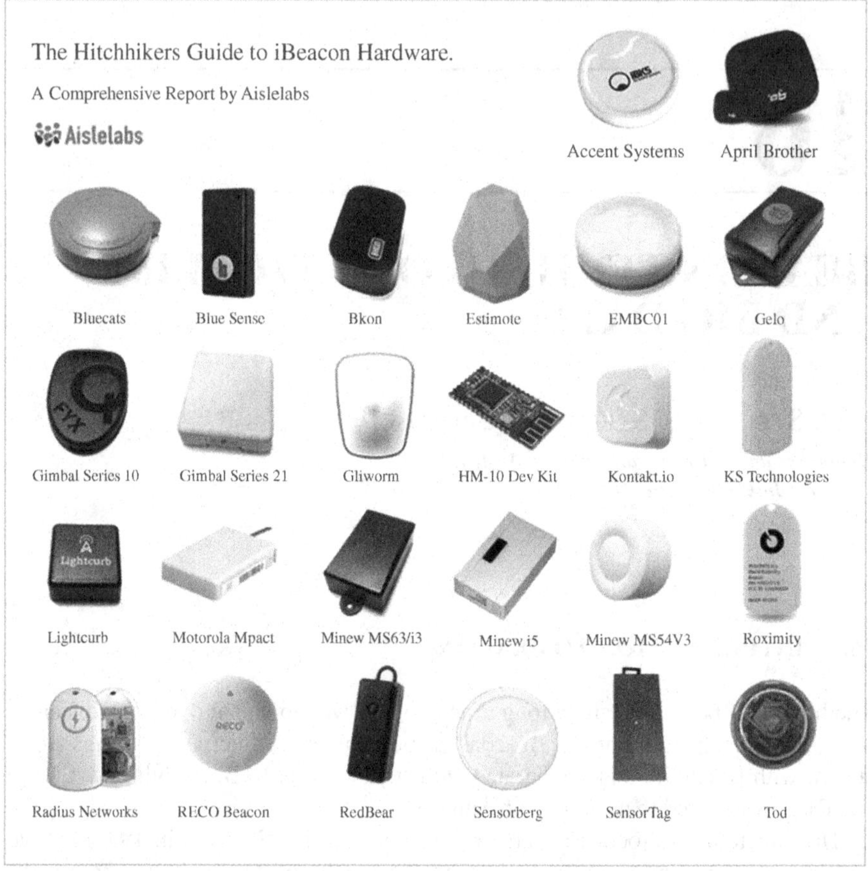

FIGURE 16.1 The Hitchhikers guide to iBeacon hardware. Courtesy of Aislelabs [3].

When we talk about machine-to-machine (M2M) communication, here is what is really meant in regard to BLE beacon communication:

> A beacon doesn't transmit *content*; it simply transmits a signal that lets a user's phone or tablet figure out what its proximity to the beacon is. The content (a coupon, for example) is delivered separately to the user's app [4].

The limitations of GPS in an indoor environment mean that navigation or positioning is impossible with the use of GPS, and therefore beacons and BLE combine to fill that void;

- *We expect beacons to directly influence over $4 billion worth of US retail sales this year at top retailers (0.1% of the total)* [5], and that number will climb 10-fold in 2016.

- *Many early adopters who opt in to receive beacon-triggered messages will likely be coupon clippers.* Half of beacon-triggered messages sent currently are some form of coupon, according to Shopkick. Mobile coupons are a significant part of this market. Mobile couponing app company RetailMeNot claims that its offers alone influenced $3.5 billion in retail sales in 2013.
- *Loyalty programs will also be important drivers.* Beacons and loyalty apps could be used together to reward customers for all sorts of location-based actions, even just walking into a store. Brick-and-mortar stores will use mobile payment apps and in-store technology to establish integrated online and offline loyalty programs among their own customers.
- *Since beacon-powered apps will collect valuable data on consumers' in-store activity*, they could result in highly personalized and targeted offers, which will reinforce the aforementioned programs. Once a consumer opens an app in-store, the data on their clicks and location can help retailers target them with offers customized to their location in a store or based on past in-store shopping behavior.
- *Some retailers are explicitly using in-store signs and promotions to encourage shoppers to download apps and opt in to beacon campaigns.* GameStop is experimenting with prominently displaying beacons in stores so that shoppers can voluntarily wave their smartphone in front of them to receive some type of offer or information [6].

16.2 WHAT IS BEACON TECHNOLOGY

There are currently two main types of software protocols that are used in beacons today, enabling a beacon to communicate with an app or with a browser.

These can be rudimentarily categorized as:

- The iBeacon protocol, which is App based
- The Eddystone protocol which is Browser based

There is an overlap where the Eddystone Universally Unique IDentifier (UUID) actually works with apps as well, but generally each protocol is confined to working with either apps or browsers.

The beacons themselves can be configured to work with either of these protocols and be used for a number of purposes:

- iBeacon protocol allows a notification to be sent in the background or foreground of an app, thus giving the ability to "wake up" an app for the user to receive a message.
- Eddystone protocol talks with a web browser, which needs to be open for the user to receive messages/URL. Background notifications are supported on Android but not on iOS.

Both protocols require the receiver to have Bluetooth switched on, to be able to interact with the beacons.

16.3 BEACON AND BLE INTERACTION

Focusing on the more common iBeacon protocol, a beacon transmits/transfers one packet of data, consisting of four types of information.

UUID, Major, Minor, cal Tx Power—the combination of this information enables position, navigation, and also a way to open dialog with a user.

Here all the information is broken down:

UUID: This is a 16 byte string used to differentiate a large group of related beacons. For example, if Coca-Cola maintained a network of beacons in a chain of grocery stores, all Coca-Cola beacons would share the same UUID. This allows Coca-Cola's dedicated smartphone app to know which beacon advertisements come from Coca-Cola-owned beacons.

Major: This is a 2 byte string used to distinguish a smaller subset of beacons within the larger group. For example, if Coca-Cola had four beacons in a particular grocery store, all four would have the same Major. This allows Coca-Cola to know exactly which store its customer is in.

Minor: This is a 2 byte string meant to identify individual beacons. Keeping with the Coca-Cola example, a beacon at the front of the store would have its own unique Minor. This allows Coca-Cola's dedicated app to know exactly where the customer is in the store.

cal Tx Power: Calibrated Tx power is used to determine proximity (distance) from the beacon. How does this work? cal Tx power is defined as the strength of the signal exactly 1 m from the device. This has to be calibrated and hardcoded in advance. Devices can then use this as a baseline to give a rough distance estimate [7]. cal Tx power and Tx power should not be confused; the former is specifically from Apple, while the latter is the real transmission power value of the beacon, which mostly depends on the settings of the beacon and the physical abilities.

This differs a little with Eddystone, three data packets are sent UID, URL, and TLM:

Eddystone-UID is 16 bytes long and split into two parts:

Namespace (10 bytes), similar in purpose to iBeacon's UUID. In iBeacon, you'd usually assign a unique UUID to all of your beacons to easily filter them out from other people's beacons. In Eddystone-UID, you can do the same with the namespace.

Instance (6 bytes), similar in purpose to iBeacon's major and minor numbers, that is, to differentiate between your individual beacons. With Estimote beacons broadcasting Eddystone-UID, instance is represented as a string up to 12 characters long.

Eddystone-URL

Eddystone-URL packet contains a single field: URL. The size of the field depends on the length of the URL.

Eddystone-TLM

Eddystone-TLM packet is designed to be broadcast by the beacon alongside the "data" packets (i.e., UID and/or URL) for the purposes of fleet management. Nearby Bluetooth-capable devices can read these packets and relay them to a fleet management service. This service can then notify the owner of the beacon that, for example, the battery is running out.

The telemetry packet consists of:

- Battery voltage, which can be used to estimate the battery level of a beacon
- Beacon temperature
- Number of packets sent since the beacon was last powered up or rebooted, beacon uptime, that is, time since last power up or reboot [8]

There are many other different types of beacon protocols available, but these are the two main commercially used protocols that are the focus of this chapter. Besides iBeacon and Eddystone, there is also Altbeacon (evolved from iBeacon by Radius Networks after a patent infringement case, so there is an open standard).

Prior to Eddystone Google worked on URI beacons, that standard later led to the Eddystone-URL packet.

At the moment the Eddystone protocol is still relatively new and therefore is likely to improve rapidly in its applicability and functionality.

16.4 WHERE BEACON TECHNOLOGY CAN BE APPLIED/USED

Static beacons serve as digital landmarks. They are bound to their physical location and serve a purpose of context, as well as bridge the gap between material and digital (this applies for all beacons really). When surrounded by beacons, we can perceive the mobile device like our sixth sense. The beacons act as stimuli for this sense. This allows us to observe our surroundings (and space) in ways that it wasn't possible before.

> *Indoor Location and Positioning*, enabled by Smartphones, will become a part of many existing apps, and enable new services. Think about how web maps started as a stand-alone application and were later embedded into all sorts of web sites. The same will happen for Indoor Location on mobile apps [9].

Beacon technology is very much in its infancy, so uses for it are still being discovered. Currently the main focuses are in the fields of navigation and proximity. This is where you see companies investing heavily at the moment.

For navigation, the verticals that are investing in beacons infrastructure are:

- Retail Centers
- Hospitals
- Transportation Hubs
- Museums/Galleries/Exhibition halls
- Offices
- Others

Onboarding—a good onboarding can determine future performance. Onboarding new employees effectively can increase retention by 25% and improve employee performance by 11%. It can also lead to greater employee engagement [10].

For Proximity, the list is similar but the uses are different.

Though there is some overlap in the way they are used, the information for proximity purposes is much more focused on what is in your surroundings and not just getting you from A to B.

- Shops—location-based special offers
- Managing Agile teams—instant room booking
- Public Safety—overcrowding situations, location of Emergency First Responders
- Office-finding a colleague, finding office hotelling
- Catastrophic event—Emergency evacuation
- Others

Other areas of applications include:

Trigger events

- Turning your oven on when you are unloading your food shopping in the garage.
- Checking into a queue and then enjoying something else while waiting.
- Contactless payments more fluid than NFC.
- Enter a lecture hall and get the notes sent to your phone.

Asset management

- Supply chain management
- Fleet management
- Asset tracking
- Ticketing
- Baggage tracking

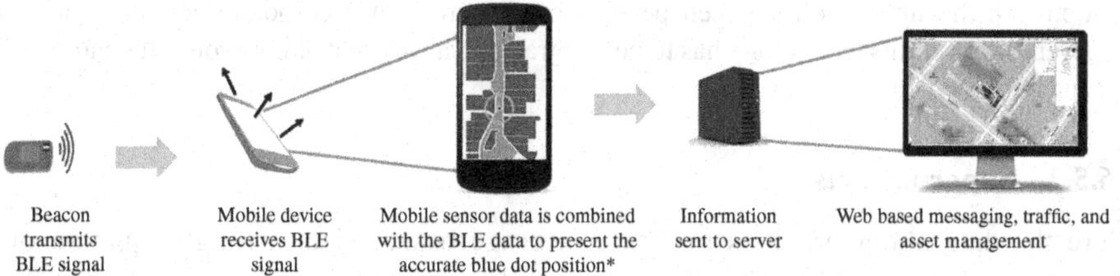
Beacon transmits BLE signal
Mobile device receives BLE signal
Mobile sensor data is combined with the BLE data to present the accurate blue dot position*
Information sent to server
Web based messaging, traffic, and asset management
All buffering and computation done locally on the mobile device, locating the device on the map or delivering message sent by the beacon.
*This step has been simplified for this graphic.
indoors

FIGURE 16.2 Interaction between beacons, mobile devices, and a server.

16.5 BIG DATA AND BEACONS

Big data is about using large data sets to predict and reveal patterns in human behavior. This is where indoor location and positioning can help everything from Marketers to UN peace keepers understand situations better and react accordingly.

Figure 16.2 describes the interaction between Beacons, mobile devices, and a server all happening in real time.

A few examples to show how beacons feed into the Big Data picture are discussed in the following text.

16.5.1 Corporate Campuses

An employee is issued with an ID card, which can be used for a number of different purposes:

- Timecard—checking in and out
- Charging lunch to their account
- Sick leave
- Room bookings

The data attained from a person's corporate ID can be combined with information about their movement, who they are meeting with, and cross referenced against other members of their team.

It would then be possible to find out if this person might require to move desks to be closer to another team member, if their sick leave might be reduced if their diet was changed, if they might need a taxi as they are working late, if they might require a bigger meeting budget, and so on.

> Integrating digital wayfinding applications with Integrated Workplace Management Systems (IWMS), scheduling applications, providing indoor navigation and directory services are becoming commonplace. Over the past decade, advances in location aware technologies, the Internet of Things (IoT) and enterprise wayfinding are helping

optimize the interaction between people and resources in a corporate environment. Every layer of this technology has to be integrated in a way that allows you full control and enhanced data security [11].

16.5.2 Supermarkets

Here the data about previous purchase, average basket size, a rough equation of dietary restrictions could be combined with aisle precision location or quickest route navigation to help more fluid journey through a supermarket.

Two examples:

A person in a household, where one or more people might be lactose intolerant, has stopped in the dairy aisle of a supermarket. That event could trigger a message to be sent with a special offer for a lactose-free item.

A person could enter their entire shopping list before they enter a supermarket and the quickest route, including the best offers, is then calculated out, minimizing the effort and time spent in the supermarket and enhancing the customer's experience.

16.5.3 Public Safety

Here Emergency First responders can be tracked so that the dispatchers as well as the captain know where all of the firefighters are at all times, making it easier to coordinate the safe control of the fire or emergency situation.

This can also be used to evacuate people out of the building safely and quickly too. An alarm can pop up on a phone, and then the person can be shown where the nearest fire exit is, and, if that one is blocked, it can be recalculated as well.

In both of these scenarios, there would be someone outside of the pressure-filled situation to calmly monitor and control the flow of people.

This is not an exhaustive list for how beacon technology can be applied to real-world situations, just an example of a few current use cases.

The pressing needs from the end customers for location-based services are playing the major role in defining the future of the indoor location market. The advent of hybrid positioning systems, location-based analytics and real-time tracking, and navigation solutions are the products of huge demand for indoor location-based products. Companies are putting sincere efforts in building customer-specific products. Major players in this market are Apple, Broadcom, Cisco, Ericsson, Google, Nokia, Microsoft, Motorola, Qualcomm, and STMicroelectronics [12].

16.6 SAN FRANCISCO INTERNATIONAL AIRPORT (SFO)

Navigating foreign and unfamiliar indoor spaces can be challenging for everyone, but it is even more challenging for blind and visually impaired people. That's why the San Francisco International Airport (SFO) paired with indoo.rs to create a prototype real-time indoor navigation solution, specifically tailored for blind and visually impaired travelers.

16.6.1 The Infrastructure

SFO's 640,000 ft^2 Terminal 2 was equipped with around 300 battery-powered iBeacons. The beacons are installed about 2–4 m above ground so that smartphones can establish a line of sight with 3–5 beacons at a time. Depending on the beacons' configuration, they transmit a Bluetooth low-energy signal at a certain strength and interval. The beacon batteries are expected to last 4 years and can then easily be replaced.

After installation of the iBeacons, all radio signals (bluetooth signals, Wi-Fi access points) at the venue were measured in a process called "fingerprinting." With the help of a mobile app, a person walks around the venue and takes a 15 second measurement every few meters within the venue. This map of all the radio signals then serves as a reference for determining a user's position inside the venue.

16.6.2 The App Development Process

The app was developed in very close cooperation with the Lighthouse for the Blind. To kick off the project, indoo.rs met with visually impaired members of the Lighthouse for the Blind to better understand their needs when it comes to exploring and navigating an unknown venue. We learned that turn-by-turn navigation is not necessary since blind people rely on their cane or a guiding dog. An airport further provides challenges for blind people when it comes to luggage or groups of people standing around. Blind people wanted a way of exploring the airport and its facilities as well as reliable guidance in the direction of important points of interest (e.g., the dog relief areas in airports are especially important point of interests for blind people traveling with their guide dogs).

At the heart of the app is the indoo.rs SDK, which allows the localization of a user (or technically speaking a user's smartphone) inside a venue.

16.6.3 The App and How It Works

After several iterations and focus groups with Lighthouse for the Blind, indoo.rs came up with a working prototype of the app after only 6 weeks of development time.

Upon opening the app, the app knows exactly where the user is located and on which floor and which direction he or she is facing. Based on the user's location, the app announces point of interests in the vicinity, their direction, and distance. How much information is provided also depends on the traveler's movement context and change in location. Results are refreshed when the user moves a certain distance or changes direction.

For the first time travelers can now walk through the airport and be informed which shops/restaurants/places are around them. The point of interests will change automatically while the user is walking or changing his/her direction. The traveler is

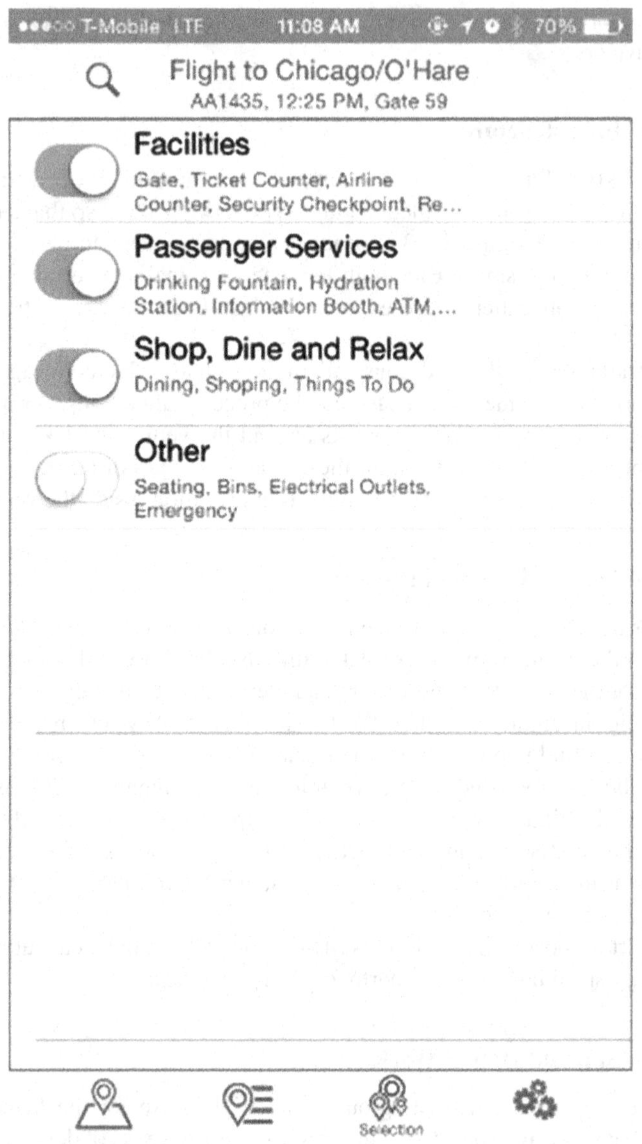

FIGURE 16.3 Screenshot of indoo.rs prototype at SFO—user preferences.

also able to customize the point of interests based on their interests, for example, they can only have airport facilities read out loud to them so that they can easily find the airline counter, security check, or the gate. Or they could just opt for "shop, dine, and relax" facilities while they are waiting for their flight as shown in Figure 16.3. With over 500 points of interest included in the app, a traveler can even find electrical outlets at the airport as seen in Figure 16.4.

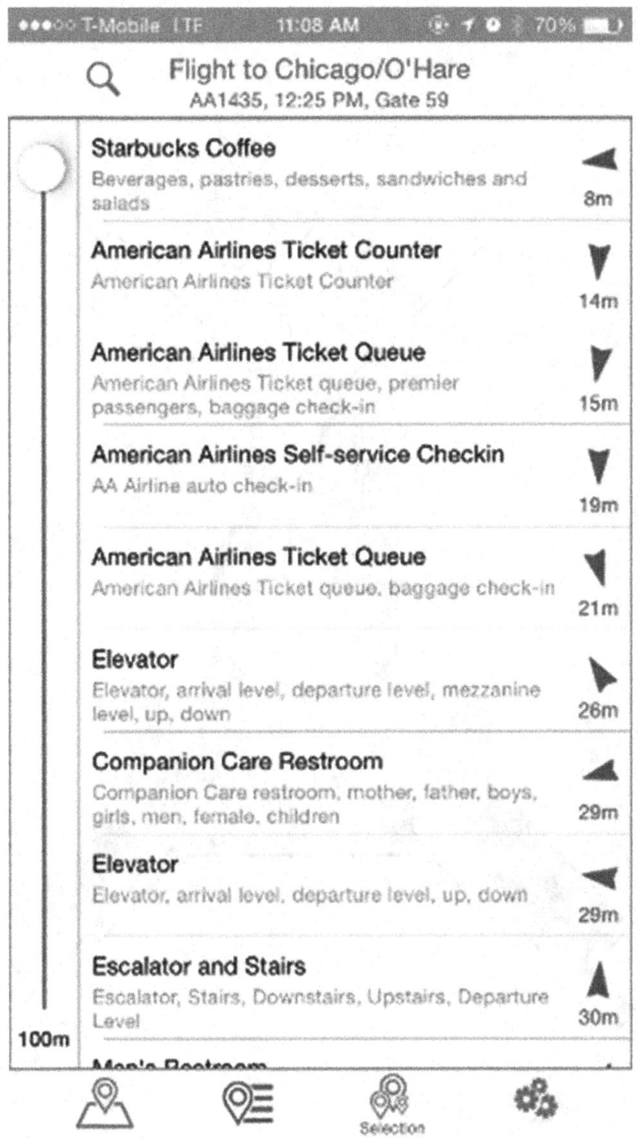

FIGURE 16.4 Screenshot of indoo.rs prototype at SFO—points of interest.

A traveler not only can explore the airport facilities but also can search for specific point of interests. The user can select any point of interest and have the app calculate a route from the current location to the destination on the map (even across floors). Figure 16.5 shows the navigation from one point to another.

For visually impaired users, we built a hybrid solution that combines features of strict directional orientation and turn-by-turn navigation. A routing sound will

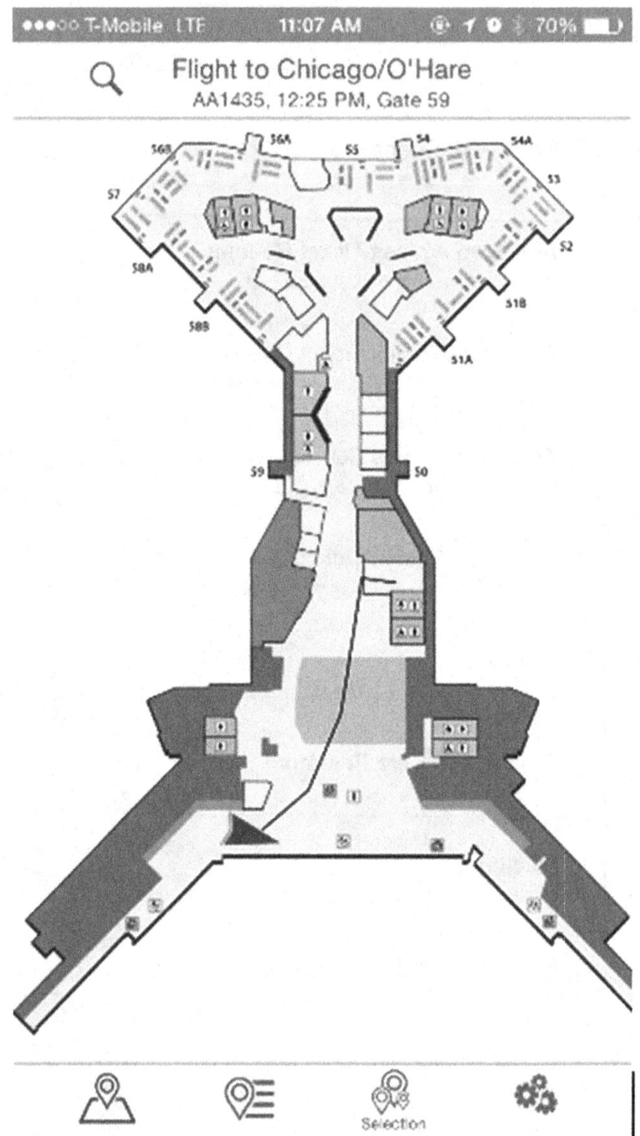

FIGURE 16.5 Screenshot of indoo.rs prototype at SFO—navigation on map.

provide orientation assistance and guide the user to their destination and will keep them on track. The routing sound provides audio cues to guide a traveler in the correct direction and along a given path. When a route is set, the app will play a subtle continuous tone. The pitch of the tone changes if the user is getting off track from the recommended path. The bigger the offset, the higher the pitch of the tone. A very low offset means the traveler is heading in the correct direction, and the continuous tone will be paused.

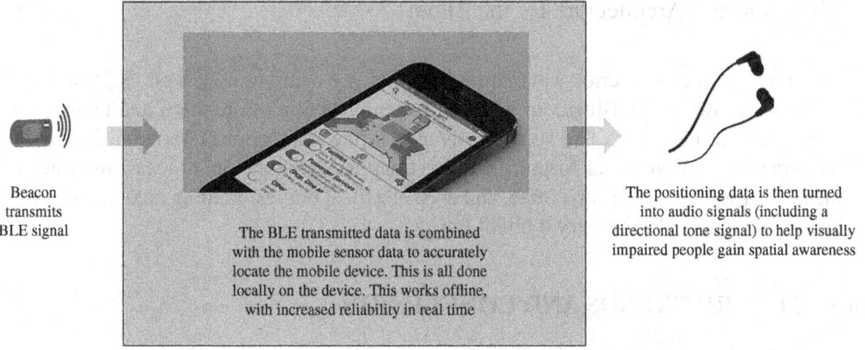

Beacon transmits BLE signal
The BLE transmitted data is combined with the mobile sensor data to accurately locate the mobile device. This is all done locally on the device. This works offline, with increased reliability in real time
The positioning data is then turned into audio signals (including a directional tone signal) to help visually impaired people gain spatial awareness

Indoo.rs unique sensor fusion postprocessing methodology uses the data provided by the sensors of the mobile devices (e.g., compass, gyroscope, accelerometer, barometer) to improve the floor level accuracy and deliver a smooth navigation experience.

indoors

FIGURE 16.6 Showing how the SFO app works for visually impaired passengers. Courtesy of indoo.rs.

Apart from point of interest, travelers can also search for their flight information, pin their flight in the app, and monitor their flight updates right in the app (Figure 16.6).

16.6.4 The Future of Indoor Navigation for Blind and Visually Impaired

The application is still a prototype, and we keep improving its features and functionalities as we receive feedback from the blind and visually impaired community. Another community, MD Support, is now actively using the application coined LowViz Guide at conferences and conventions throughout the United States. We are receiving amazing feedback, and seeing how our technology impacts the life and self-reliability of so many people is truly motivating to keep working on this application of our technology.

16.6.5 Praise/Advocates

Lisamaria Martinez, Director of Community Services, Lighthouse for the Blind:

> Both indoo.rs and SFO care about things that matter, and what matters is the freedom to move about one's environment independently with confidence and grace. With this solution, indoo.rs and SFO will change the way blind and low vision people travel.

Dan Roberts, Director MD Support:

> The LowViz Guide app has not only been proven in the field, but it displays an elegant concept design which works well for both visual and non-visual people.

Chris Downey, Architecture for the Blind:

> The most promising interior wayfinding [tool], in my opinion, is beacon technology, which uses low-energy Bluetooth [signals] to send location information and notifications to a smartphone. .... I've successfully used it [the installation at Terminal 2 of the San Francisco International Airport], with no training or orientation, to locate my gate, the men's room, a bite to eat, coffee, and even electrical outlets. Finding an outlet in an airport is a real challenge—try it blind! [13].

16.7 FUTURE TRENDS AND CONCLUSION

The future is hard to predict; who would have thought that in such a short space of time, a cell phone would have the same computing power as a computer that used to fill a whole room?

In beacon technology that is no difference; with all the big players, such as Apple, Google, Facebook, to name just a few, starting to heavily invest in this area, the rate of advancements and uptake will be exponential.

2015 has been a phenomenal year for beacons as Figure 16.7 of US retailers shows.

The prediction is that in 2016/17 this trend will only continue and in fact gather pace. Beacons will no longer be for the bigger department stores, the prices will continue to drop, the technology to set up a beaconized environment will get quicker, and therefore this will become the everyman's tool to interact better with customers, users, visitor, and so on.

MarketsandMarkets estimates that the global indoor location market is estimated to be $935.05 million in 2014 and is expected to grow to $4,424.1 million by 2019. This represents an estimated Compound Annual Growth Rate (CAGR) of 36.5% from 2014 to 2019 [15].

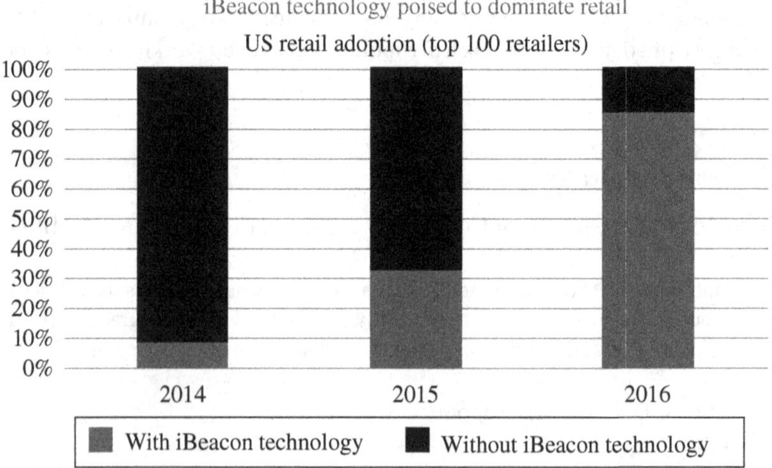

FIGURE 16.7 iBeacon technology poised to dominate retail. Courtesy of echidna Inc. [14].

The recent release of SLAM Engine™ will be quickly followed up by SLAM Crowd Engine™ from indoo.rs and has already sped up mapping of an indoor environment by 10-fold [16] without adversely affecting the accuracy. This will revolutionize the implementation of indoor navigation and proximity especially with the event industry.

As the price and setup barriers are removed, more and more uses will be found, with the M2M communication automating more tasks. The drop in price will lead to more applications to focus on home uses, a dynamic shopping list that pops up every time you open the fridge, a reminder that you have a show ready to watch on Netflix when you are in the living room, and so on.

Triggering events that act as reminders helping you optimize what you do or what your appliances do will be all possible. The IoT is something that will happen; in the vicinity of the Bluetooth-enabled device, lights will turn on, coffee will be made, stolen bikes will be found, special personalized offers will be sent, educational material will be sent, lives will be saved, disabilities will be eased.

This is just the start of automation; the physical world is the new digital channel!

IoT is basically the idea that everything from your car to your refrigerator to your thermostat should be connected to the Internet. With a simple app on your phone, you should be able to preheat your oven, get an alert that you're out of milk, and have the temperature in the house set perfectly before you even get home.

But it's not just consumer benefits that IoT has. With this technology, you'll be able to take your blood pressure, pulse, temperature, and other basic tests for your doctor, so that they can analyze the data at their office without you needing to go in.

All of this data, though, will require a significant amount of security. The last thing you want is your personal information, such as health documents, to be easily taken by hackers. And once something is connected to the Internet, it is available for hack.

REFERENCES

[1] https://en.wikipedia.org/wiki/Bluetooth (accessed December 14, 2015).

[2] http://beekn.net/guide-to-ibeacons/ (accessed December 14, 2015).

[3] http://www.aislelabs.com/reports/beacon-guide/ (accessed December 14, 2015).

[4] https://en.wikipedia.org/wiki/Bluetooth (accessed December 14, 2015).

[5] https://intelligence.businessinsider.com/welcome (This is not the correct link) (accessed December 14, 2015).

[6] http://uk.businessinsider.com/beacons-impact-billions-in-reail-sales-2015-2?r=US&IR=T (accessed December 14, 2015).

[7] http://www.ibeacon.com/what-is-ibeacon-a-guide-to-beacons/ (accessed December 14, 2015).

[8] http://developer.estimote.com/eddystone/ (accessed December 14, 2015).

[9] http://dondodge.typepad.com/the_next_big_thing/2015/01/my-5-predictions-for-2015-and-beyond.html (accessed December 14, 2015).

[10] http://www.jibestream.com/blog/how-wayfinding-helps-with-onboarding-new-employees (accessed December 14, 2015).

[11] http://www.jibestream.com/blog/transitioning-to-the-digital-workplace (accessed December 14, 2015).

[12] http://www.marketsandmarkets.com/PressReleases/indoor-location.asp (accessed December 14, 2015).

[13] http://www.dwell.com/profiles/article/architect-lost-his-sight-and-kept-working-thanks-breakthrough-technologies-blind (accessed December 14, 2015).

[14] http://echidnainc.com/ibeacon-technology-2015-adoption-statistics/ (accessed December 14, 2015).

[15] http://www.marketsandmarkets.com/PressReleases/indoor-location.asp (accessed December 14, 2015).

[16] https://slaminfo.indoo.rs/slam-revolution/ (accessed August 26, 2016).

17

SCADA FUNDAMENTALS AND APPLICATIONS IN THE IoT

RICH HUNZINGER
B-Scada, Inc., Crystal River, FL, USA

17.1 INTRODUCTION

Coming from a background firmly rooted in industrial software technology—in particular Supervisory Control and Data Acquisition (SCADA) software—I may have different ideas about the Internet of Things (IoT) than you or your colleagues. As revolutionary as the end results may be, the truth is that the IoT is just a new name for a bunch of old ideas. In fact, in some ways the IoT is really just a natural extension and evolution of SCADA. It is SCADA that has burst free from its industrial trappings to embrace entire cities, reaching out over our existing Internet infrastructure to spread like a skin over the surface of our planet, bringing people, objects, and systems into an intelligent network of real-time communication and control.

17.1.1 The IoT Is a Constellation of Different Technologies

Not entirely unlike a SCADA system—which can include Programmable Logic Controllers (PLC), Human–Machine Interface (HMI) screens, database servers, large amounts of cables and wires, and some sort of software to bring all of these things together—an IoT system is also composed of several different technologies working together (Figure 17.1). That is to say you can't just walk in to the electronics section of your local department store, locate the box labeled "IoT," and carry it up to the counter to check out.

Internet of Things and Data Analytics Handbook, First Edition. Edited by Hwaiyu Geng.

Companion website: www.wiley.com/go/Geng/iot_data_analytics_handbook/

FIGURE 17.1 The IoT is composed of various contemporary technologies working together. Courtesy of B-Scada, Inc.

It also means that your IoT solution may not resemble your neighbor's IoT solution. It may be composed of different parts performing different tasks. There is no such a thing as a "one-size-fits-all" IoT solution. There are, however, some common characteristics that IoT solutions will share:

- *Data access*

  It's obvious, but there has to be a way to get to the data we want to work with (i.e., sensors).
- *Communication*

  We have to get the data from where it is to where we are using it—preferably along with the data from our other "things."
- *Data manipulation*

  We have to turn that raw data into useful information. Typically, this means it will have to be manipulated in some way. This can be as simple as placing it in the right context or as complex as running it through a sophisticated algorithm.
- *Visualization*

  Once we have accessed, shared, and manipulated our data, we have to make it available to the people who will use it. Even if it's just going from one machine to another to update a status or trigger some activity, we still need some kind of window into the process in order to make corrections or to ensure proper operation.

There could be any number of other elements to your IoT system—alarm notifications, workflow, and so on—but these four components are essential and will be recognized from one IoT system to the next. Coincidentally (or not so coincidentally), these are technologies that all cut their teeth in the world of SCADA [1].

17.2 WHAT EXACTLY IS SCADA?

The term SCADA was originated in the late 1960s. It is the technology used to remotely monitor and control physical devices and processes, typically for heavy industry or utilities. SCADA has improved the efficiency of manufacturing processes, reducing unscheduled downtime and waste. SCADA has allowed water treatment facilities to detect leaks and increase the output of accountable water. SCADA has completely changed the way we operate our factories, refineries, power plants, subway systems, and more. In fact, the advent of SCADA technology is often referred to as the dawn of the third industrial revolution (Figure 17.2). According to this model, the first industrial revolution occurred near the end of the eighteenth century when the steam engine was applied to industrial processes, greatly increasing the scale of production. The second industrial revolution occurred near the beginning of the twentieth century with the introduction of electrically powered production and the moving assembly line. The third revolution began with the invention of the PLC and the beginning of distributed control and automation, or SCADA.

It is noteworthy here to take account of the fact that each one of these revolutionary advances was necessary to the next. For example, electric power and moving assembly lines could not have revolutionized industry if steam power had not already changed the way factories operated and individuals worked. The industrial infrastructure created by steam engines paved the way for successful companies to quickly adopt electricity and moving assembly lines. Likewise, the electrical infrastructure that was in place and the predictability of moving assembly lines made it possible for companies to quickly adopt SCADA technology and implement remote control and automation of production processes.

In this same way, the advances brought about by SCADA have paved the way for the fourth industrial revolution that we are entering right now. There are differing schools of thought on industrial revolutions and the manner in which they transpire. Some leading minds consider the current revolution to be a continuation of the original

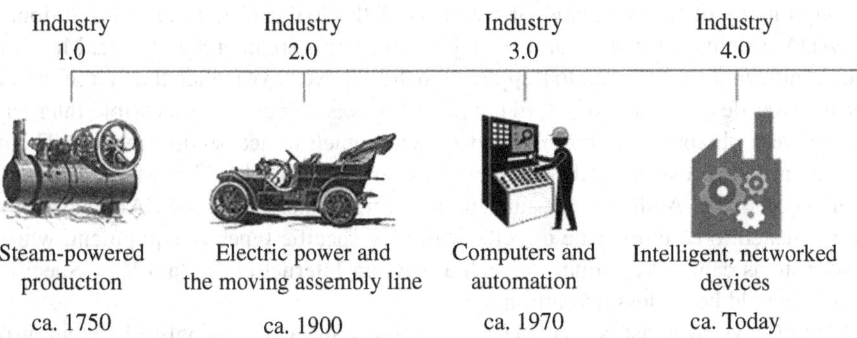

FIGURE 17.2 Industry has evolved through a series of revolutionary leaps. Courtesy of B-Scada, Inc.

digital revolution that began in the 60s and 70s, while others argue that the current wave of technology is yet another revolutionary advance on top of the third. Industry 4.0 (or Industrie 4.0) is a term coined in Germany as a policy directive for shaping the future of their industrial infrastructure [2].

17.2.1 The IoT Is the Beginning of the Fourth Industrial Revolution

The effects of the IoT on industry are already making the same kind of revolutionary waves that earlier technologies did. In the same way that the changes brought about by steam power allowed successful organizations to quickly adopt electricity and moving assembly lines and the changes brought about by electricity and moving assembly lines allowed successful organization to quickly adopt SCADA technology, the changes brought about by SCADA technology are allowing successful organizations to adopt IoT technology.

Of course, the IoT is not restricted to industrial applications, and it promises a bold new world for commercial and residential consumers as well. Fortunately, though, SCADA's time in the heavy industrial world has allowed the technology to address several of the problems posed by the IoT:

- Several decades of tangible results in high-volume, high-velocity industrial applications
- Codified standards for communication and security
- Interoperability and hardware-agnostic integration of multiple data systems

These advances and more have made it possible for industrial enterprises to adopt the IoT with much greater ease than they would have otherwise. In fact, some of the key concepts associated with the IoT—like real-time data visualization and machine-to-machine (M2M) communication—were born in the world of SCADA.

17.2.2 How Does the IoT Differ from SCADA?

Though many of the foundational elements of the IoT can be found in traditional SCADA systems, it is not accurate to say that the terms are interchangeable. Many of the significant features of an IoT application did not even exist when the first SCADA system was devised. First of all, of course, there was no publicly accessible Internet. There were obviously no mobile devices with which to access information. There was no need to cross firewalls, and there was certainly no "cloud" in which solutions could be hosted. Additionally—and perhaps most importantly—SCADA systems were created to communicate directly with very specific types of equipment, while IoT systems cannot be limited in such a way; an Internet of Certain Very Specific Things would be far less revolutionary.

An IoT system must be capable of communicating with not only PLCs but also sensors of different types, databases, web-based resources, and users. An IoT system is also likely to involve analytic tools to transform and manipulate data into actionable information. And since the IoT system is not confined to a factory floor or oil refinery, there are likely to be very different answers for how, where, and why an IoT system is being deployed.

At B-Scada, we were fortunate that we began to develop our latest SCADA software system at a time when most of the new technologies were already available, and our experience with end users of our earlier SCADA product taught us that the future of SCADA was in fact something quite different from SCADA as it was known years earlier. As we continued to research and develop the new system, it was with an understanding that people were expecting certain features, like the ability to consume data from different types of hardware from different vendors, the ability to include data from other enterprise systems (ERP, CMMS, etc.), the ability to visualize data on mobile devices, and the ability to include sensor data as part of a solution. It didn't take long to realize that our latest SCADA application was moving us in a new direction.

This realization was very important. Once we realized we were developing the foundation of an industrial IoT application, our focus changed. We were able to build on that SCADA foundation knowing that we had to reach out to a larger audience with a wider variety of needs. Had we kept ourselves trapped in the SCADA "box," only thinking in terms of industrial process control, we would have missed an opportunity to build a platform that can solve the problems people are truly facing today. So, while our SCADA background provided us with the knowledge and experience needed to develop our IoT platform, it took a change in focus to get from there to here.

17.3 WHY IS SCADA THE RIGHT FOUNDATION FOR AN IoT PLATFORM?

To revisit our introduction, the IoT is a natural extension and evolution of SCADA. It is essentially SCADA plus the new technology that has evolved since SCADA was first devised. Just like the previous revolutionary periods, it was catalyzed by one major transformative technology. Just like steam power put a hook in all other industrial technology and pulled it forward into a new era, electric power did the same thing a century later. Several decades later, with the advent of microchips and computer technology, once again industry was swept forward into a new era by the gravity of a single revolutionary technology. As we sit here today, well aware of the revolutionary power of what we call the "Internet," we are now feeling that gravity once again pulling us toward a new era.

In a drastic oversimplification, we can consider the evolution of industry to have proceeded something like:

Labor + steam power = industry
Industry + electricity = automation
Automation + computers = distributed control and M2M
Distributed control and M2M + Internet = IoT

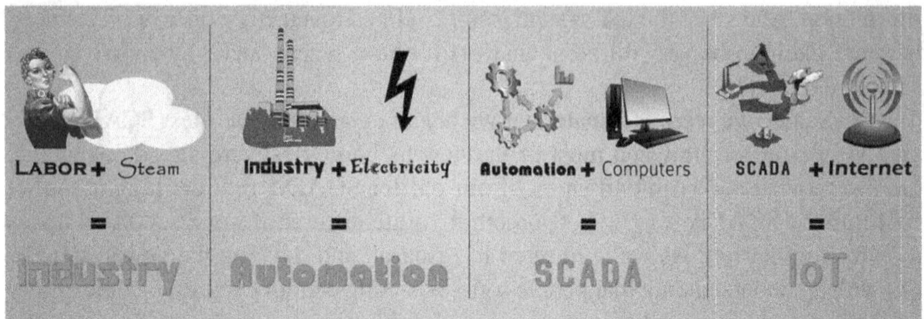

FIGURE 17.3 Major technological advances have triggered revolutionary changes in industry. Courtesy of B-Scada, Inc.

Obviously, adding Internet to a SCADA system doesn't automatically make it an IoT system any more than adding steam to manual labor automatically makes it industrial. The step-up requires a change in focus, a new production model. The technological advance simply prepares the foundation—just like how the communication standards and business models that have been connecting production processes with business processes in industrial enterprises have laid the foundation for our connected world of intelligent "things" (Figure 17.3).

17.3.1 How Does SCADA Work in IoT Applications?

One of the unfortunate things about SCADA technology is that it was historically very expensive and time-consuming to implement. For that reason, many of the SCADA systems in place today were installed 10–20 years ago—if not more. People who use SCADA technology everyday may not even be aware of the many ways that SCADA software technology has evolved over the years. Long before anyone was using the term "IoT," SCADA had been working toward today's inevitable reality.

Early SCADA systems were deployed to monitor specific production processes, and separate systems were employed to manage assets, coordinate maintenance operations, optimize supply chains and other business operations (ERP, CMMS, etc.). Today's enterprises are frequently composed of a patchwork of these legacy information systems [3].

Later generation SCADA systems came to incorporate data from plant floor controllers, databases, and other data systems in a single solution. Some modern SCADA systems employ concepts like information modeling and data virtualization to consolidate and organize data from these disparate devices and systems. Modern SCADA software also leverages the power of the Internet to provide even greater data visibility. If you were to install a modern SCADA system today, taking advantage of all of the many advances SCADA has made over the years, the system you end up with would likely be indistinguishable from an industrial IoT system. The more specifically we try to define IoT, the more difficult it becomes to separate it from SCADA.

17.3.2 Machine to Machine, Machine to System, System to Man

One of the concepts central to the IoT, M2M communication, is not a new idea. It is one of the advances that made the PLC such a revolutionary invention. By giving machines the ability to make decisions and perform actions based on nothing more than information provided to them by other machines, SCADA technology completely redefined the processes involved in industries like manufacturing, water treatment, oil and gas, and much more. Task automation driven by mechanically autonomous devices has improved the speed, efficiency, and quality of industrial processes.

In IoT applications, we begin to encounter not only M2M communication but also Business-to-Machine (B2M) and Business-to-Business (B2B) communications. Now, machines can make decisions not only based on information about themselves or other machines but also based on business conditions. This is the realm of the autonomous factory—a notion not far removed from science fiction, yet surprisingly realistic.

Imagine the following scenario:

At ABC Company's factory, a certain machine's throughput drops below what's expected. At this time, a workflow task is triggered to generate a work order; the work order is automatically assigned to the technician who is closest and best able to perform the work (the technician is notified in real time on his smartphone or tablet). While the maintenance is performed, the lights in that part of the facility are automatically adjusted to help him do his work; thermostats are automatically adjusted to accommodate the reduced ambient temperature caused by the downtime; and all of the day's numbers related to production, profit, warehousing, shipping, and so on are automatically adjusted to accommodate the downtime. Meanwhile, the technician is able to perform the maintenance, test the machine, and update the work order right there on the spot (on his phone, for instance). All other systems are automatically aware of the work order's completion in real time, and everything is adjusted again accordingly. If you imagine that the technician in this scenario is a robot, then we have a completely self-aware, self-managing facility.

Also, the technology deployed to enable this vision would allow a sort of predictive analysis. For instance you could create a real-time report that displays a current "cost per production unit" based on material cost, labor, energy cost (based on real-time kWh rates pulled from the web). You could then test scenarios where a certain piece of equipment is replaced by a more energy-efficient unit and measure the difference in "cost-per-production unit" and discover in seconds the true amount of time it would take to recover the investment. Scenarios of this sort could undergo constant testing to ensure that processes are always optimized to current conditions.

From simple M2M communication, machines began to communicate with entire systems, and systems began to communicate with other systems. When this communication takes place over the Internet and includes a visualization component that brings real-time human input into the fold, we begin to recognize that something revolutionary is happening. This is SCADA that has evolved beyond SCADA and into something else.

Consider the example of Algae Lab Systems (ALS) in Colorado. Their story calls attention to the space where SCADA meets IoT.

17.4 CASE STUDY: ALGAE LAB SYSTEMS

ALS provides Monitoring and Control systems for two industries: the commercial algae growing industry and the aquaculture industry. Their systems enable customers to optimize their productivity and save them time by automating their quality control processes.

They provide a turnkey Photobioreactor Monitoring and Control system designed to help optimize algae yield. Their AlgaeConnect™ platform integrates the data from sensors: pH, temperature, DO, density, flow rate, water level, and many others into a cohesive software/hardware platform. The system monitors, stores, and displays all status parameters of the photobioreactors and also automates their behavior with controlled CO_2 or nutrient dosing, as well as activation of other devices like harvesting pumps and valves.

The AlgaeConnect platform has proven to be a wonderful tool to help small- and medium-sized algae growers improve their processes and increase their yield and profitability. This has been a great success for ALS, but they weren't able to provide the same type of insight and control to larger customers who have dozens—if not hundreds—of tanks farms and photobioreactors spread out over different locations.

ALS contacted B-Scada when one such customer came to them looking for a more far-reaching solution. The customer was looking for a way to aggregate data from several hundred AlgaeConnect systems, each with several photobioreactors/ponds, in a single system (Figure 17.4). The customer wanted a centralized interface through which to access their data, and they wanted it to be composed of graphics they chose.

Custom data access and visualization have long been a hallmark of B-Scada, and ALS had very little trouble creating a solution that allows their customer to remotely monitor and control their algae farms from anywhere at any time.

So, ALS has provided a system that uses the Internet to communicate real-time data about various sensors and devices (things) to relevant parties—wherever they may be. This is one of the many examples of how traditional methods of remote monitoring and control are leveraging modern sensing, communication, and networking technology to create something entirely new to our world. This is the natural, organic evolution of SCADA and the fountainhead of the IoT.

17.5 THE FUTURE OF SCADA AND THE POTENTIAL OF THE IoT

While SCADA and the IoT share many common traits, it is not the case that the terms are becoming interchangeable. In the world of heavy industry, where SCADA technology has been used for decades and the term is both familiar and understood, we are seeing the beginnings of Industrial IoT. In these ecosystems, SCADA will not disappear. It will continue to function as a subset of a larger solution. This is particularly true where sensitive proprietary data is being communicated and must remain behind a firewall.

The industrial space is also where the earliest seeds of the IoT were planted. There is less reluctance to employ IoT technology because industrial enterprises already

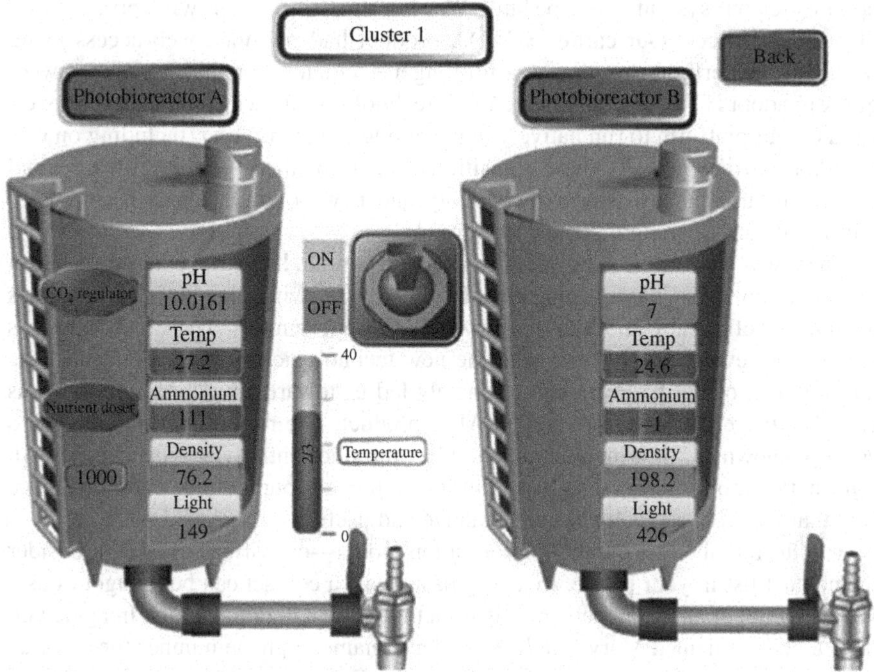

FIGURE 17.4 With the help of B-Scada, Algae lab systems was able to enhance their product portfolio and appeal to a new class of customer. Courtesy of B-Scada, Inc.

have a SCADA foundation in place. While the IoT tries to deal with issues related to standards, security, reliability, and cross-platform compatibility, industrial users are well aware of how these issues have been addressed by SCADA developers, and the natural inclination is to begin building the IoT system on that foundation. This is, of course, the way it has always been. Those who are best positioned to take advantage of revolutionary technology are the ones who have already leveraged existing technology to its full extent.

17.5.1 Beyond SCADA

Maybe the best way to describe the transition from SCADA to IoT is to describe the transition as it occurred in my own place of business. My company wasn't always known as B-Scada. We were founded in 2003 as Mobiform Software, and we developed a successful SCADA application that came to be used all over the world. Portions of that SCADA technology were actually adopted and licensed into the products of some well-known Fortune 500 companies. Meanwhile, technology marched along, and when we set to work developing the system that would ultimately become our Status Device Cloud Platform, some things had already changed.

The OPC Foundation had introduced their Unified Architecture (UA) specification [4], providing an open, cross-platform, service-oriented architecture upon which our system would be built. We also witnessed the growing proliferation of mobile devices. Our earlier SCADA product had provided web access using Microsoft Silverlight, but that was no longer adequate. For that reason we were quick to adopt HTML5 as a preferred web technology, allowing applications developed on our platform to run natively in any modern web browser, including on iOS and Android devices. As these transitions occurred, there was a growing trend toward the use of new sensor technology and new communication protocols to allow software to communicate with sensors.

There was also a shift in business models at this time. Industrial enterprises began to realize new value in integrating plant floor processes with other business processes to improve efficiency and identify new growth opportunities. These new business models that evolved in parallel with the new technological advances informed the development of our platform and ultimately led us toward an architecture that was entirely different from our earlier SCADA product. The new platform employed a concept known as "information modeling" to introduce entirely new capabilities. An information model allows data from multiple unrelated sources to be normalized in a way that it becomes much more informative and useful.

To illustrate the concept of an information model—or "virtual" model—consider the contact list in your phone. In a very basic way, a contact can be thought of as a virtual model of an actual person. It is something like a digital identity. Imagine you have a contact named Mary Smith. Mary has a name, a phone number (or two), an e-mail address, maybe a photo. Mary can have a Facebook profile, a Twitter alias—you can even assign Mary a special ringtone. All of these things combine to create a virtual model of Mary stored in your phone.

Now, to make your model of Mary a bit more intelligent and useful, you could add her date of birth, her hair color, her favorite book, her pet cat's name, or any number of different properties of Mary. If we connected some sensors to Mary, we may ascertain things like her current location, current body temperature, her heart rate, her blood pressure. If this information is communicated to your model in real time, you have an active, living representation of Mary that tells you more about her than she may know herself.

Imagine applying this same process to a particular department of your organization or a specific piece of equipment. Your "virtualized" assets (or "things") are able to share information with one another in real time, and this information can be used to trigger automated tasks or perform analysis. Object virtualization allowed for a new type of automation, allowing seemingly unrelated objects to speak the same language. Now, a value change in a database could trigger a certain machine to power off, or a particular sensor reading could trigger the opening of a valve or switch. Because a virtual model is completely indifferent to the source of a particular piece of data, the potential for advanced analytics and automation is nearly unlimited.

While this new software platform shared many traits with our SCADA software, it was also abundantly clear that this was not SCADA as it had come to be known. It was, in fact, a powerful and sophisticated platform for the IoT. Of course, once that fact was realized, development proceeded in the direction of IoT instead of SCADA, and we ended up with an open software platform that allows users to easily develop custom applications for thousands of potential use cases.

The fact that the platform was built on a standards-based SCADA foundation gives it stability and reliable performance. It also lends confidence to those industrial customers who are looking to leverage new sensor technology and the promise of the IoT.

The story of B-Scada's transition from SCADA to IoT is very similar to that experienced in the rest of the world. It was less a conscious decision to deviate and strike new ground than it was a natural and necessary response to consumer demand and the march of progress.

REFERENCES

[1] J. Holler, V. Tsitsis, C. Mulligan, S. Avesand, S. Karnouskos, D. Boyle, *Machine-to-Machine to the Internet of Things: Introduction to a New Age of Intelligence*, First Edition, 2014, Academic Press, Oxford.

[2] G. Schuh, C. Reuter, A. Hauptvogel, C. Dolle, Hypothesis for a Theory of Production in the Context of Industrie 4.0. *Advances in Production Technology*, C. Brecher, 2015, Springer, Berlin.

[3] J. He, B. Seppelt, Automation Under Service-Oriented Grids, *Springer Handbook of Automation*, S. Y. Nof, 2009, Springer, Heidelberg.

[4] W. Mahnke, S. Leitner, M. Damm, *OPC Unified Architecture*, 2009, Springer, Heidelberg.

FURTHER READING

F. Behmann, K. Wu, *Collaborative Internet of Things (C-IoT)*, First Edition, 2015, John Wiley & Sons, Ltd., Chichester.

S. Umeda, M. Nakano, H. Mizyuma, H. Hibino, D Kiritsis, G. van Cieminski, *Advances in Production Management Systems, Part* 2, First Edition, 2015, Springer International Publishing, Cham.

H. Zhou, *The Internet of Things in the Cloud*, First Edition, 2012, CRC Press, Boca Raton, FL.

PART III

DATA ANALYTICS TECHNOLOGIES

18

DATA ANALYSIS AND MACHINE LEARNING EFFORT IN HEALTHCARE: ORGANIZATION, LIMITATIONS, AND DEVELOPMENT OF AN APPROACH

OLEG RODERICK, NICHOLAS MARKO, DAVID SANCHEZ AND ARUN ARYASOMAJULA

Division of Analytics Research and Clinical Informatics, Department of Data Science, Geisinger Health System, Danville, PA, USA

18.1 INTRODUCTION

In the last decade, we are seeing a steady increase in automated, data-driven analysis applied to the needs of industry [1]. Reviews occasionally use the term "third wave": of data sharing, of scientific approach, of impact on individual lives [2–4]. We may state that we are entering a third wave of interest in data science.

Techniques of data mining were successfully applied in finance, marketing, and social media [5]. In its third wave, data science is spreading to the more traditional, predigital fields: journalism, urban planning, political process, education, sports, art and design, fashion, even culinary arts, and physical fitness. The need to share data for research purposes is slowly transforming the practices of data stockpiling [6, 7]. Ideas and methods that previously belonged in engineering and academic fields are now used to deal with massive data in almost every type of human activity.

Healthcare occupies an interesting position: it is *both* cutting-edge and very traditional. It makes use of the most modern research in development of means of

Internet of Things and Data Analytics Handbook, First Edition. Edited by Hwaiyu Geng.

Companion website: www.wiley.com/go/Geng/iot_data_analytics_handbook/

diagnosis and intervention. At the same time, its goals—increased safety, health, and informed freedom of choice—are old indeed. Medical clinics, research organizations, and medical insurance companies have always had access to specialists in biostatistics, medical actuarial science, and to a collection of generic problem-solving recipes lately known as operation research. Presently, healthcare industry is hiring specialists with degrees in mathematics and computer science. In this material, we provide some insight as to why this is happening, explain how to organize a research and development effort in data science for healthcare, and review a medical case study of intermediate complexity.

The material is intended for medical professionals interested in the predictive analytics, aspects of healthcare, and also for readers with background in statistics or computer science willing to diversity their professional skills.

The material is organized as follows:

In Section 18.2, we overview the ideology of data-driven healthcare. This will help the reader to understand how new projects are defined and what is delivered as a proof of concept (POC). Note that we do not discuss the use of human subjects in research and clinical implementation; such issues lie outside of the scope of a data science group.

In Section 18.3, we discuss prerequisites and personnel qualifications for data science in healthcare; with a contributed subsection on the role of high-performance computing (HPC).

In Section 18.4, we provide a description of medical data and list the issues arising during data acquisition and transformation; a contributed subsection discusses de-identification data sharing.

In Section 18.5, we overview the main themes of machine learning; the intention is to provide the readers with ideas and keywords for future independent exploration.

In Section 18.6, we present a case study: prediction of rare adverse events based on traditional, nonspecific medical data.

Concluding remarks are in Section 18.7. Also, see a list of suggested additional reading.

18.2 DATA SCIENCE PROBLEMS IN HEALTHCARE

Data science is defined as process of extracting knowledge from externally generated data [8, 9]. This implies an ability to make true, verifiable statements about subject matter without complete understanding of first principles and without additional physical experiments and observations.

It is a composite field of study, freely borrowing from mathematics, statistics, computer science, and applied fields of study.

Modern healthcare can be characterized as personalized [10, 11] and evidence-driven [12]. Planners and practitioners are required to base their decisions on the best, most relevant knowledge from the entire body of human medical experience. The enabling technology is unified electronic medical records (EMRs) that allow access, search, and simple sorting or comparison of hundreds of standard clinical features for millions of patients. Specialized data (such as sequenced DNA) and massive data (such as fine-resolution real-time reading of vitals) are also becoming available for an increasingly large part of the population [13].

Database systems alone are not sufficient to address the questions of risk prediction and optimal resource allocation. Researchers in data-driven healthcare have to continuously improve the analytic methods used to predict the future of a patient, or to recover crucial information based on incomplete, sometimes incorrectly measured evidence, and answer many other questions that cannot be resolved by descriptive statistics, or by a medical expert's insight. Modern healthcare industry needs analytic techniques of modern data science.

One notable feature of data science is its subject blindness: for example, very similar inference techniques can be used to predict increase in toxicity of a patient on dialysis and to predict the workload of a busy emergency response department. Very different types of data can be used to achieve the same goal: for example, biometric measurements can be used to diagnose influenza but so can sentiment analysis of patient's posts on social media. The modern paradigm of transfer learning allows prediction using information from disjoint populations.

Thus, it is not informative to classify data science projects in healthcare by the diagnostic group, by the department, or by data contents. We suggest a method-centric view, in which every need of a clinician, patient, or a medical insurance provider is represented as a problem in applied mathematics.

Guidance via mathematical insight cuts across the healthcare organization structure, accepts only limitations of the available development effort and computational power, and produces work that is both personalized (aware that different groups of patients have different rules and needs) and modular (allows replacement or reuse of parts of workflow).

Of course, there is no such thing as a universal approach to data analysis, just as there is no truly universal specialist in data analysis. In each project, it is important to understand the scope and limitations. While a qualified research group, with access to large sets of medical data, may model and analyze virtually anything, it is advisable to conserve and focus the effort via development of a research group identity.

18.2.1 Areas of Interest and Research Group Identity

Data science is driven by empiricism: confidence in results of study comes not from the complete theoretical explanation but from overwhelming evidence observed in data. It is a composite field of study, combining academic rigor of mathematics, power of modern technology, and accumulated narratives of applied knowledge. The relationship between the themes around the use of data in healthcare can be visualized with Venn diagram [14]. Our version (Figure 18.1) has labels chosen to emphasize the role of all modern technology, not just practices of programming; and of all formal academic research, not just mathematics or statistics. The center of the

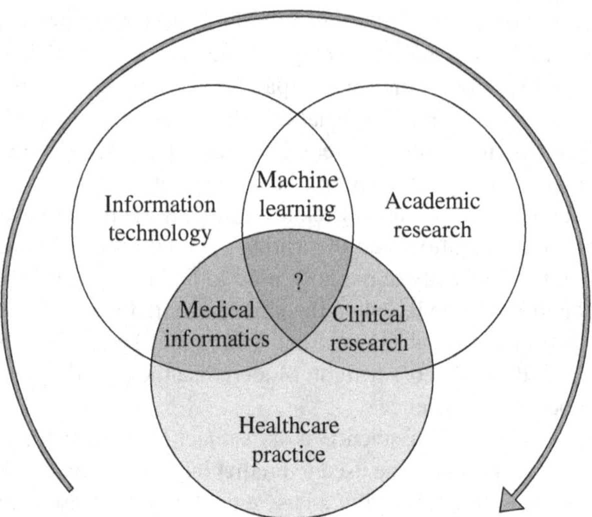

FIGURE 18.1 Venn diagram of data science in healthcare. Courtesy of Geisinger Health System.

diagram is reserved for a hypothetical universal specialist, familiar with all aspects of data storage and delivery, academic research, and medical practice. This is too much to expect from one person or even in a small group of specialists, hence a question mark (or a metaphorical "unicorn").

In healthcare, new questions are motivated by clinical and medical insurance practice. The workflow then follows the diagram clockwise. The role of medical informatics is to extract, interpret, and deliver the relevant records. The methods of machine learning are then used to model the abstract relationship between contents of data and outcomes of interest (this is a focus of present chapter). The results of abstract analysis are then validated as appropriate for clinical research; this may involve study with human subjects, or investment into software development and training.

The same healthcare phenomena can be discussed in multiple ways. For example, suppose that a medical database contains comprehensive records for patients diagnosed with forms of schizophrenia disorder; there is an open-ended request to research solutions that would result in improvement of service to such mental health patients. The condition is still incompletely understood (in fact, even its definition is controversial); almost any new result would be of practical interest. Possible tasks include:

- Classify patients based on their response to treatment (any new patient, even before the treatment starts, would be placed, with maximal possible confidence, into one of the classes).
- Predict time of recidivism to provide timely, preemptive support to medical practitioners and the families of patients.

- Generate a new method to track and document the progression of condition, in an objective manner, with massive observational evidence used in place of preconceived assumptions.

To generalize the topics of interest in predictive analytics research, we have identified four large areas of interest, each with its own take on what "analysis" means. They are:

A. Inductive reasoning, generalizations from experiences of a single provider, familiar with perhaps a hundred of relevant medical cases, to massive data sets describing millions of patients
B. Diagnostic reasoning, discovery of several main factors contributing to a medical condition from a list of potential candidates
C. Causal reasoning, formal description of causes, effects, and rules in healthcare phenomena
D. Prognostic reasoning, extrapolation from existing data to future events, and interactions

The resulting variety of research themes is visualized in Figure 18.2 (we submitted a related version to ODBMS Industry Watch [15]). The intersections are labeled with keywords that often come up when two types of reasoning are used together. This includes statistical inference as taught in college coursework, various types of modeling, and uncertainty analysis, that is, propagation of levels of confidence in the inputs to the level of confidence in the outputs.

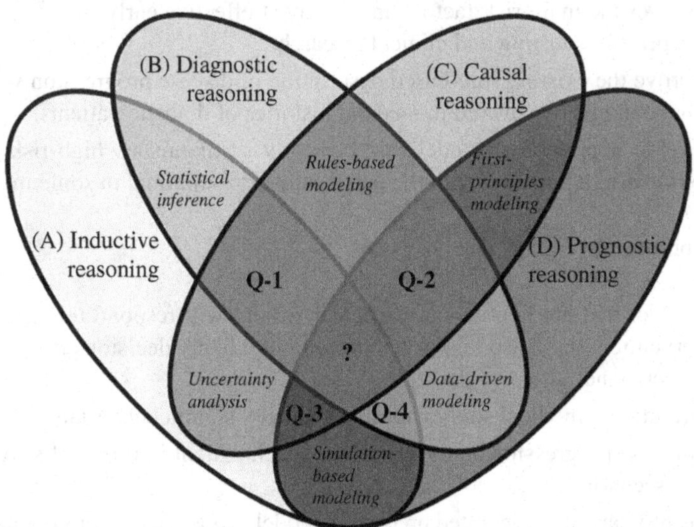

FIGURE 18.2 Venn diagram of analysis interests. Reproduced with permission of Oleg Roderick.

The quadrants (Q-1–Q-4) correspond to significant multidisciplinary effort, with many kinds of knowledge working together. Contradictions and knowledge gaps can become very obvious:

- Q-1 has no objective estimation of quality, so there is no strategy for incremental improvement of results; instead, there are competing narratives.
- Q-2 operates on one scale: either only big data, with the relatively trivial operations that are possible at such magnitude or only local data, with limited ability to generalize.
- Q-3 has no discrimination ability: all data is potentially an input into the model.
- Q-4 has no causal, didactic narrative; a scientific problem was solved for its own sake but may not produce an impact.

Overall, we advise against projects that use too many different conceptual systems.

A new problem formulated on medical data initially falls into one of the four areas. In addition to a mental health example in the preceding text, consider the needs of patients potentially diagnosed with a common, debilitating condition, for example, diabetes. Susceptibility to this condition and responsiveness to treatment are linked to hundreds of medical features; lifestyle choices and socioeconomic conditions also play a role. A data science team may be asked to:

A. Describe effectiveness of a particular prevention and treatment plan when applied to a larger patient population, based on detailed knowledge about a few patients and general knowledge about thousands of patients.
B. Discover the main risk factors and the most effective early warning signs, to help prioritize testing and clinical research.
C. Improve the existing rule-based description of disease progression with additional patterns discovered in medical histories of diabetic patients.
D. Develop a predictive model that correctly distinguishes high-risk patients from low-risk patients, with the prognostic error minimal in some metric.

Appropriate deliverables are:

A. A model that predicts how a particular patient will respond to each stage of intervention plan, also taking into account the likely decisions made by medical personnel.
B. A traditional medical study listing statistically significant variables.
C. A disease progression study that places each patient into one of several possible scenarios.
D. An automatically generated prediction model that achieves optimal match between predicted and final outcomes on a trial set; the intermediate outcomes may be in a tradeoff between quality and level of detail.

In a method-centric view, we treat every project as a combination of a predictive model and an optimization metric. We record this schematically as: $Y = F(X; P), J := J(Y, P): \min_P J$.

Here, Y is an outcome (or outcomes) of interest; X is a set of features characterizing a particular patient or a group of patients; P are independent parameters—of a treatment plan, of lifestyle choices made by the patient, or of externally influencing forces. J is a metric of quality on the possible outcome: For example, the task may be to minimize the cost of treatment, to minimize the difference between expected and observed personnel working hours, to maximize self-reported patient satisfaction. The core predictive model F may not be explicitly reported in the study, as it is often a standard, reuseable tool, and not itself a subject of research.

Schematically, the most common activity within data science research is this sequence: data acquisition and transformation to define X, P, choice of classification or extrapolation methods to define F, choice of optimization methods to characterize best possible J given P.

Data science always seeks to establish a formal, reproducible relationship; schematics can be different depending on the area of interest:

(A) $(X,P) \rightarrow Y$	(B) $P \rightarrow Y,J$	(C) $(X,Y) \rightarrow F$	(D) $(X,P) \rightarrow J$

We see that even the most unambiguous medical questions, for example, "will patient A experience medical condition B?," can be discussed in multiple ways. Attempting to address a variety of forms of analysis, both transformative and incremental improvement, may be detrimental to the quality of scientific work. In practice, this will consume the available development resources and result in a disorganized and unremarkable collection of studies. To prioritize effort, a data science team should develop a strategic vision. We suggest several options for strategic development:

I. An internally focused data science effort, supporting a particular clinical department. In this case, impactful problems may be already defined; additional effort is needed for transfer of knowledge to other applications.

II. An actively networking data science group, sharing ideas and sources of support with other research entities. The effort is defined by several main collaborative relationships, with university or grant coapplicants. This approach emphasizes the academic aspect of data science work. The amount of development resources is increased (to the entire scientific field of study). Many of the projects are now externally defined; it is harder to control their impact on the home organization.

III. A data science group can view itself as a career cornerstone for its members. The projects are matched to the already developed skills and interests of subject matter experts, data scientists, and programmers on the team. The idea is to attract professional talent, and to place researchers in the position of unique access to intellectual resources (very relevant for healthcare industry that cannot freely disclose its data and practices). A disadvantage of this approach is isolation from the values of the larger organization.

We note that even a volatile and relatively disorganized predictive analytics effort makes a positive impact on performance of healthcare organization. In contrast, the absence of such effort leads to ad hoc allocation of resources and extremely conservative estimation of risks.

18.2.2 Precision or Personalized, Data or Big Data?

This subsection is prompted by a question that is answered ambiguously in popular representations of healthcare data science: how massive is the body of medical data used in routine studies? Is it correct to expect terabytes [13, 16] (as in DNA sequence analysis) or merely megabytes [17, 18] (as in review of history of clinic visits and medications)? The ambiguity comes from the fact that both genomic data and traditional data are subjects of active interest in healthcare data science.

In the previous material, we mentioned that modern medicine is *personalized* [19] and that modern medical informatics is geared for *big data* [20]. Controversy exists around definitions of both terms. The twin concepts of personalized and precision healthcare are based on an understanding that different individuals and groups of individuals have different needs and responses to treatment. In general, it is not possible to produce unique medical devices or prescription drugs for a single individual (although it is possible to fine-tune exact dosage or physical dimensions). Instead, medical decisions can be personalized at the level of a small group of patients. Thus, analysis for the needs of personalized medicine can be thought of as series of classification problems, where patients are clustered into groups so that patients in the same group will have different responses to treatment and different risk outcomes. The ambiguity comes from the choice of patient features to be used in personalized decision-making. The term "personalized" is commonly used in connection with traditional clinical data, that is, biometric features and patterns observed in patient-provider-insurer interactions. The term "precision" medicine, sometimes replaced with "molecular" medicine, is used in connection with analysis on modern genetic and fine-resolution biological features. The blanket term "-omics" covering genomics, proteomics, and metabolomics is also used.

-Omics data is almost ideally suited to illustrate the idea of individual precision: it opens up a very detailed biochemical map for every individual [21]. Molecular markers have been successfully used to detect health conditions in very early stages of disease progression. In the recent past, only a limited amount of such data was available for research, with not enough observation points to represent the variety of patterns. However, modern massive sequencing efforts are changing this as we write [13]. Thus, there is merit to opinion that only molecular data can be truly precise.

At the same time, human behavior is not chemically deterministic. There is not clear connection between -omics and personal behavior, choices, and socioeconomic condition. Recent work shows importance of medical and sociological factors at the macrolevel for prediction of needs and outcomes of a patient. Thus, it is possible to treat genetic data as simply another set of variables for modeling purposes.

Precision versus personalized discussion is an example of tradeoff between medical technology (reliance on sequencing) and academic research (reliance on multidisciplinary insight). This example is a good introduction into issues surrounding Big Data as it is applied to healthcare industry.

The scope and scale of typical business challenges have expanded with the increased volume, velocity, and value of enterprise data. This has forced a larger fraction of businesses to adopt "big data" methodologies (we use the phrase synonymously with "distributed computing," although special mention in the following text is made to HPC workflows). Although this practice has yet to find large-scale adoption within the healthcare industry [22, 23, 24], trailblazers in the "biomedical Big Data" space are enjoying increased curation of vital metadata, superior throughput for complex data queries, reduced barrier to the analysis of unstructured data, and a substantially reduced hardware/software financial profile than with conventional architectures. In virtually every field that has deployed big data solutions, data scientists share common complaints of their experience: interfaces which are designed for the conventional business analysts or the software developers do little to support the data scientist. These interfaces are currently excellent resources for users at the two extremes (local data and very massive data),but remain insufficient for routine tasks of data science research.

While these systems can offer extremely powerful opportunities for developers to leverage distributed compute and storage resources in order to tackle larger, more dynamic challenges more efficiently (and with commodity hardware!) than before, many data scientists are not skilled at leveraging MapReduce [24], other distributed computing abstractions, or higher-level Big Data services in ways that improve their own workflows. Moreover, because many enterprises maintain data governance standards which tend to restrict access to scientific computing tools which can properly leverage the distributed nature of Big Data solutions, the data scientist is reduced to treating the Hadoop cluster [25] (or comparable system) as any other SQL database. The synergistic relationship between data scientist and Big Data is denied.

Consider the data landscape of a typical healthcare system. The most operationally useful data is recorded as transactional logs of the various services delivered to the patient over the course of an encounter, often to be flattened and stored in a secondary database for reporting. Much of the supplementary narrative data exists as RTF-encoded strings. Radiological images are captured and traditionally made available through a PACS server [26] (if at all), with the accompanying narrative documents (by either the provider or an imaging technician) stored for archive, not consumption. Laboratories will maintain semistructured records, especially with regard to pathology or sequencing results.

In a Relational DataBase Management System [27] (RDBMS), such as the ones in use for many conventional reporting and analytics purposes, the representation of data on disk is optimized for rapid retrieval and querying. This can make integrating the various data sources behind common query interfaces challenging (even though most RDBMS vendors have well-established methodologies for, say, implementing inverted indices appropriate for document stores), inefficient, and expensive. The data scientist is then responsible for integrating these various source systems themselves, according to the various query or bulk-extract interfaces available across the spectrum of devices. Not only does this put the burden on data scientists to discover how to perform the integration but also, within a healthcare environment where specialized IT groups service, maintain, and provision access to source systems, obtaining the access itself can be challenging and time-consuming.

Accordingly, the first role Big Data systems often perform for data scientists is an aggregator of data sources. In healthcare, this effect is even more pronounced than in other industries, as the act of deploying and integrating a Big Data solution encourages an enterprise to adapt more flexible and comprehensive data access/governance policies, which is a particular weakness of healthcare. Moreover, it also provides extremely convenience fall-back mechanisms for obtaining data when query interfaces are insufficiently performant or expressive for the task at hand. Unlike RDBMS, Big Data systems quite often sacrifice efficiency of local IO for generality of the underlying file format, which invites the opportunity for enterprise IT to engineer source data capture methodologies from the file layer. This ensures that, at the very least, it will almost always be technically possible to work with "raw" data.

18.3 QUALIFICATIONS AND PERSONNEL IN DATA SCIENCE

The degree of specialization needed in a healthcare data science effort depends on the size of the group and ease of access to resources. A one-person team will likely access one main source of data for each project: for example, by analyzing scheduling of surgeries and timestamps of specific stages of each operation, biometric data on the patient is easily available, but career and professional skill information on medical personnel is not. On the other end of the range, a large and well-supported analytics team does not need to include focused specialists inside the group, as missing skills can be outsourced to other departments or external analytics companies.

18.3.1 Organization and Workflow of Predictive Analytics Projects

Projects given to a medium-size team can still be very ambitious and require coordinated, competent effort. Medium-scale (i.e., most common) research in healthcare is characterized by (1) the available data and how it is prepared for analysis, and (2) specialized predictive algorithms. We suggest an idealized distribution of roles in the team capable of doing most of the development internally, and, at the same time, having specialists responsible for interaction with: (1) sources of data; (2) cutting-edge computational research; (3) needs of main customers or home healthcare organization.

We visualize the distribution of roles in a data science team in Figure 18.3. This is not a complete organizational flowchart: the reporting structure can be fairly flat, and we are defining roles in relationship with the development effort, not sorting them by status and rank. Depending on the available personnel and resources, it is entirely possible that the same person will play multiple roles, although we do not recommend allocating less than 0.25 full-time effort (FTE) to any given task.

To define a problem, a senior manager (e.g., chief data officer) who represents the home organization, or the client, works with a senior scientist who would lead

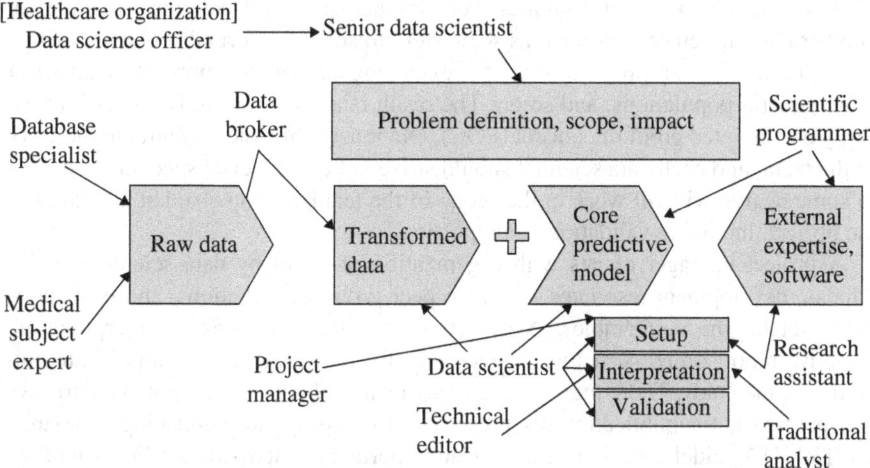

FIGURE 18.3 Medium-scale data science team: sample organizational chart. Reproduced with permission of Oleg Roderick.

the project. They discuss the scope, final deliverables, and expected impact of the problem.

The following is needed for a data pull:

- A database specialist organizes storage and effective search in data. Healthcare organizations are bound by strict confidentiality rules: legal and reputation consequences of security breach are dire. You can expect a multilayer structure of data storage, with research teams only able to access the highest level of abstraction. A person familiar with the entire acquisition, storage, and transformation process is crucial for effective data pulls.
- A data broker is responsible for anonymization of information. For most research purposes, the team must be insulated from patient features that uniquely identify an individual.
- A subject-matter expert, in our case, a practicing doctor, a laboratory technician, or a person responsible for collecting socioeconomic information, prepares and interprets a data dictionary, stays in contact with the team to clarify, and assists with additional data pulls (the situation "this feature does not mean what we thought it means" is very common).

If a team is using external intellectual resources, such as collaboration with an academic research group, it is helpful to include a junior member of that group in the research process; for increased outreach and scientific impact, this can be a coordinated part of the graduate student's education. Thus, we allocate a position for an academic research assistant: perhaps not as independent researcher but familiar with the algorithm or process.

A scientific programmer uses standard libraries and algorithms to produce and maintain code—for internal purposes (testing and validation of the method) and for external use.

A data scientist uses techniques from mathematics and computer science (or another data-driven science such as sociology) to solve the formal problem: writing the best diagnostic or prognostic model, extending an existing approach to larger or more general populations, and so on. The result is a scientific study, but not necessarily a completed proof of concept (POC). Obviously, this role is central to the work of the team, and each data scientist should strive to be a universal specialist, familiar to some degree with all work in the scope of the team, and involved at all stages of the project, including validation and reporting.

A project manager assists with organization of effort by data scientists, helps allocate development resources, organize peer group consultations, and ultimately, helps set up the deliverables. A (traditional) analyst prepares an interpretation: narrative report on the structure of data used for the study and the immediate outcomes of the study. There may be a separate technical editor to review the narrative and prepare it for publication (as prescribed by appropriate publishing standards, see TRIPOD guidelines). Publication is an important counterpart to validation of the model: in both cases, the main concern is reproducibility.

18.3.2 The Role of High-Performance Computing Resources in Data Science for Healthcare

The analytical arm of data science is involved with the exploration and consumption of numerical techniques for diagnosis, transformation, and modeling of data relative to a scientific analytical workflow. However, the product of these activities is typically constrained by the flexibility and suitability of a computing environment which is often beyond the skill or authority of the data scientist to modify in their own favor. In many businesses, this stems from the segregation of enterprise IT from analyst groups, which often roll up under separate business units and are invested with competing priorities (security, reliability, and management of data versus flexible and complete access to resources).

This separation often has consequences to the data scientist. The impediments to the success of a practicing data scientist are concerned with the capture, featurization, and standardization of data, as well as the dissemination of analysis results to a broader community of business stakeholders and leaders. This competes with the conventional workflow of enterprise IT, which is prioritized to deploy conventional Business Intelligence (BI) tools for business analysts, often constraining interactions with the underlying enterprise data warehouse in exchange for providing convenient abstractions for organizing and creating reports and simple dashboards. Accordingly, the design and maintenance of the computing environment in which data science takes place is a driving factor in the value derived from a data science program.

This is especially true in enterprises such as healthcare, in which the flexibility of devices and services on any given business network context is heavily constrained by the particular operating characteristics of that industry. For example, in healthcare enterprise, there is additional complexity introduced by federal oversight organs, a cultural and business imperative to keep the clinical network extremely stable and secure, unnecessarily sophisticated data models which originate from flattening data stored in transactional systems (as are many modern EMR systems), and general resistance to the idea that data science can be a valuable contributor to the patient's experience, quality of care, or cost of care. These forces manifest as extremely aggressive network security policies which completely inhibit the deployment of scientific or statistical computing tools to workstations, extremely restricted access to data (especially data-bearing Protected Health Information—PHI—or similar identifiers), a lack of data lineage and comparable metadata establishing the originating context of data (e.g., record of the kind of device from which a given biometric reading was taken), and a general kind of plodding, methodological culture of "do it once, do it right," which conflicts with the scientific approach of success through rapid, incremental changes.

We propose a DevOps-inspired model for the organization of a data science computing environment. DevOps, a combination of the words for development and operations, is a software development paradigm which has proven successful for technology-focused enterprises whose software development practices are tightly involved with the provisioning and implementation of infrastructure. A DevOps environment stresses collaboration between cross-functional team members, automation of core processes and business policies to reduce human overheads, and integration of commonly reused activities into common pipelines. Ultimately, the goal of such an environment is to facilitate and prioritize the methodologies that improve and hasten the delivery of a final "product" to the client (be it internal or external).

In a "data science-DevOps" environment, analysts and developers are paired with infrastructural and system engineers who can deploy, create, and maintain the tools and systems critical to success. This is an especially valuable design methodology in healthcare, as it allows the entire data science effort to operate on its own subnetwork separate from the rest of the enterprise, and manage its own IT needs. This enables the relaxation of software security policies for that particular subnetwork; it facilitates the deployment of infrastructure (which, especially for testing and prototyping, is more rapidly configured in a relaxed environment than in one for which there is opportunity to interfere with production systems facing the clinical operations). This can also lead to synergistic prioritization of data science infrastructure along with enterprise IT, as the DevOps layer integrates mature systems and workflows into the broader enterprise offering.

Ultimately, a data science program will generate a demand for new kinds of data, novel processing techniques, exotic hardware configurations, and general pressure for any external IT department. If such a program is in any way predictive of the future business or technical needs of an enterprise, then a data science-DevOps environment is a proving ground for new technologies and solutions, without much of the risk of broadly deploying new services to a large user community. In this sense, the perspective given here is not representative merely of a particular pattern for deploying a data science team within a larger organization but is a working model for how to increase the general agility and flexibility of healthcare IT to meet the rapidly changing business landscape in which hospitals exist.

HPC systems are characterized by high-throughput interconnects between nodes and by workloads which tend to be scientific in nature. Although there is no technical definition of either HPC or Big Data which draws a line between the two infrastructures, in practice they tend to suit substantially different purposes, receive attention from different kinds of teams, and even fall under separate business divisions, with HPC typically closer to research and Big Data landing closer to the heart of the enterprise data effort. Because data scientists often leverage techniques and tools originating from academic research—especially from the hard sciences, applied mathematics, and computational statistics—it is common to find overlap between data science and HPC, in design and philosophy if not in practice.

HPC clusters provide a convenient mechanism for IT departments to flatten the computational demand profile of a data science program and offer superior computational resources without investing in high-powered workstations. A cluster which includes heterogeneous hardware partitions—groups of nodes consisting of systems with extreme memory capacity, more modern or parallelized CPUs, or powerful accelerator boards such as Intel Xeon Phi or compute GPUs—can satisfy the most intense needs of data scientists. Depending on the software licensing landscape, it may also be possible to efficiently undersubscribe to licenses, which can often represent enormous financial savings.

However, HPC hardware is not always necessary for these tasks. Consider the example of ensemble learning [28] (or another massively parallelizable task). On a single machine, each model will be trained sequentially (or at least, up to the degree of parallelism which can be recovered from a single node). A cluster can train multiple models simultaneously, but only needs to transfer model coefficients (a few floating-point numbers) between nodes, obviating the need for high-performance interconnects. Large HPC deployments can expect to spend as much on interconnects as on computational hardware, so this is a scenario in which a suitably-configured Big Data system would be a competitive, if not superior, option.

Of course, HPC, just like Big Data, is no panacea. There exist many powerful numerical libraries for distributing sparse or dense linear algebraic payloads across HPC clusters, but the improved distributed processing capacity comes at a cost. Not only are these systems more difficult to utilize effectively (and occasionally require knowledge of advanced semaphore concepts, which can present a steep learning curve to the novice distributed-computing developer), but because communication overhead, even along a high-performance interconnect, can be orders of magnitude more expensive than RAM-to-CPU transactions or even local storage IO, true HPC workflows are generally slower for a given problem size than the corresponding duration on a single-node system. Distributed computing makes possible extraordinary efforts which would be impossible on a single machine, but there are few turnkey solutions for that power.

18.4 DATA ACQUISITION AND TRANSFORMATION

Healthcare is a popular subject of discussion in media, in connection with dominant role of informatics, in modern society. Both traditional clinical data and genomics data are associated with massive acquisition and processing efforts. We know that healthcare organizations are storing and analyzing medical histories for millions of patients, using thousands of variables to better allocate resources, better cure thousands, and improve life of hundreds of thousands of individuals. There are common order-of-magnitude claims, such as "one terabyte of data is collected every day" [29].

This view is incomplete. While it is true that billions of records are available, covering the entire variety of socioeconomic and medical events, medical practice is motivated by violation of the norm. Trauma, chronic disease, and unprevented death are, fortunately, rare events. Thus, for any given subject of study, billions of records also have no direct relevance. Patients (even hospital in-patients) are not observed in real time, the amount of allowable testing is finite and most forms of medical surveillance miss critical events.

Very different levels of motivation and precision in recording data result in incomplete records, random, and systematic errors and misinterpretations in the definition of patient features. Records are not stored in the same format—conversion from a system (or systems!) of billing codes, insurance claims, narratives, and images requires development effort, with proportionally large chance of error. It is more correct to expect that medical data will be massive, but also partially irrelevant, possibly noisy, and incomplete.

The saving trace of healthcare data is, indeed, high volume, and its affinity for transfer learning (building predictive models for rare events of interest by establishing connections with seemingly different, better observed, phenomena). Data science research in healthcare operates under an unspoken agreement that many different sets of variables contain almost the entire information of interest. Clinical interactions, social behavior, financial data, and -omics markers provide thousands of potentially relevant inputs into the model, and we can try to use *almost any* subset of these thousands.

18.4.1 Healthcare Data Types

To describe the foundations of our work in an organized fashion, we sorted types of healthcare data by practical ease of access: from the most commonly available, to the most rare and effort-consuming. They are:

I. Census information (such as age or ethnicity) and simplest biometrics (such as weight).
II. Clinical information: timestamped labels of transactions, assigned according to a medical coding system. These record instances of medical decisions, not detailed outcomes.
III. Financial information: depending on conditions of data sharing with insurance providers, this may include claims data, payments received, and aggregate metrics of subscriber loyalty or risk.
IV. Narratives, including unstructured text documents such as doctors' notes.
V. Multiformatted measurements and test results, including numerical data and images.

VI. Self-reported behavior, such as questionnaires of patient satisfaction.
VII. Inferred and miscellaneous socioeconomic features, such as frequency of use of social media or likelihood to purchase an insurance plan.
VIII. Miscellaneous external data that is not collected by the hospital, such as academic grades, property prices, involvement in criminal activity.
IX. -Omics data. We place this in the last category because of specialized effort required to collect this data for an arbitrary group of patients. We note that with the development of technology and computational power, this type of data is becoming much more available for almost any kind of study.

Some elements of comprehensive patient records allow unambiguous identification of an individual (mostly type I). Such features are, in most cases, obscured or completely removed from study. Some are not directly threatening patient's privacy, but contain business-sensitive information (mostly type III); they will also be removed or available under restrictions in use and publication. In general, a researcher on medical data should be prepared to deal with the situation when a subset of requested data is not available, or is stored in a very inconvenient manner.

We overview the types of data and the expected volume, in Table 18.1. The volume of data in each type is given per 100,000 patients, reported correct to order of magnitude. The estimation is based on our recent work on data collected by Geisinger Health System and Geisinger Health Plan in 2010–2014 [30–32]. These numbers should not be applied to general population, since hospital in-patients with severe debilitating conditions represent a disproportionately large part of medical data.

In Figure 18.4, we (informally) visualize the basic quality of different types of data by the volume available for research and the integrity of data (i.e., the number of missing and unreliable entries).

The rightmost column of Table 18.1 lists standard preprocessing tasks required to transform multiformat data into annotated, reproducible numerical tables that will be used in statistical models. While external software exists for the purpose, large portion of the data transformation work will be performed internally. A few comments on the listed tasks:

I. Recovery of missing entries is actually required for every data type. For local sources of data, "cleanup" may mean detection of obviously erroneous entries and reacquisition. At massive scale, some error detection is possible, but recovery has to be done by statistical imputation.
II. Transactional labels are either binary (a test was either ordered or not) or trinary (ordered, not ordered, or status unknown, event possibly recorded elsewhere). The dimensionality and sparsity are very high; it may be useful to merge codes with similar medical contents into one feature.
IV and VI. Natural language processing is still a relatively rare scientific specialization; without it, only trivial operations are possible, such as counting frequencies and cooccurrences of words.

TABLE 18.1 Sources of General Healthcare Data

Class of Data	Contents	Volume Per 100k Patients, Per Year	Postprocessing
I. Biometric information	Basic census facts: age, sex, ethnicity	~100k individual records ~750k vital measurements taken	Recover missing entries
II. Clinical transactional information	ICD-10 diagnostic codes, outreach codes, CPT laboratory test codes	~2 million physical encounters, each can generate multiple codes ~250k problems diagnosed ~2 million procedures ordered ~750k medications prescribed	Convert categorical to binary, merge ambiguous features according to data dictionary
III. Financial transactions	Claims, payments, insurance records	~10 million records	Scramble/obscure business sensitive information
IV. Narratives	Doctors' notes, misc. reports	~2 million text documents including ~750k progress notes	Term frequency search, sentiment analysis
V. Laboratory results	Quantitative information, images, vitals printouts	~1 million images including ~100k CT scans >5 million records on laboratory test outcomes	Signal processing, image analysis, real-time data compression, representation learning
VI. Self-reported behavior	Questionnaires, surveys, social media posts	<1 million filled out surveys; ~10,000 social media posts	Manual processing, narrative interpretation
VII. Socioeconomic and inferred behavior	Municipal, educational, political data	~1 million relevant records exist, only a small fraction (~10k) is readily available	Attribution of records to patients
VIII. External data	Misc. life events not recorded in clinical data	~10k records	Ad hoc tasks of transfer learning
IX. -Omics data	Genetic, proteomic, metabolomics markers	A goal-oriented public database may have ~1,000 records; this is rapidly changing	-Omics annotation and analysis, discussed in specialized literature

Reproduced with permission of Oleg Roderick.

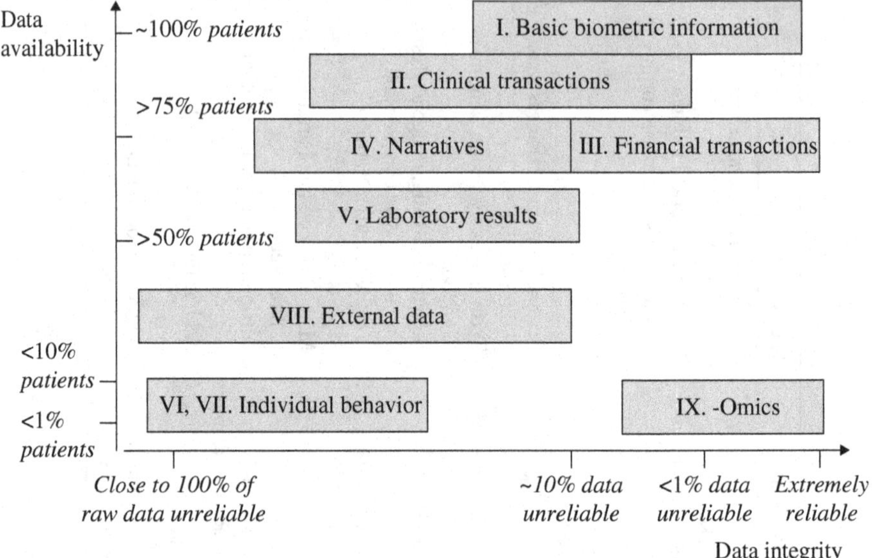

FIGURE 18.4 Types of healthcare data, volume versus reliability per year. Reproduced with permission of Oleg Roderick.

VII and VIII. Using data flows that are initiated and controlled by patients is a new and exciting direction of development, with many recent successes. The main technical task here is data attribution: not every published narrative is easily linked to a particular patient, and not every patient discloses their life choices in the same way.

18.4.2 Protected Health Information Data

Healthcare is like many other industries in that federal standards regulate the fair use of data, even within an organization itself. In particular, HIPAA [33] identifies eighteen different kinds of PHI for which there are significant consequences should they be treated inappropriately. In order to make certain business and research workflows possible, honest broker (elsewhere in this chapter and in popular vernacular called a "data broker," although the phrase could refer to various conflicting concepts) systems have emerged. These programs allow a certified individual or entity can de-identify PHI-laden medical data so that it is not (reasonably) possible for a recipient of that data to identify the individuals about whom the data was collected.

The following identifiers of the individual, relatives, employers, or household members of the individual are considered PHI [34]:

1. Names
2. Geographic subdivisions smaller than a state (minor modifications to this rule exist)

3. All elements of dates, years allowed if not indicative of extremely old age
4. Telephone numbers
5. Fax numbers
6. Electronic mail address
7. Social security numbers
8. Medical record numbers
9. Health plan beneficiary numbers
10. Account numbers
11. Certificate/license numbers
12. Vehicle identifiers
13. Device identifiers
14. Web Universal Resource Locators (URLs)
15. Internet Protocol (IP) address numbers
16. Biometric identifiers, for example, fingerprints
17. Full face photographic images and any comparable images
18. Any other unique identifying number, characteristic, or code

These restrictions are essentially invisible for research on traditional clinical data (Types I, II, III, V) and in precision medicine (Type IX). However, multiformatted socioeconomic records may include identifiable information in ways that become obvious only after manual inspection.

To protect PHI data, a data broker removes identifiable elements, and, where appropriate, replaces them with temporary randomized and/or time shifted records. Examples include adding an undisclosed time shift to dates in the study, replacing medical record numbers with study-specific identification numbers using undisclosed linkage file.

It is not always necessary for the honest broker systems to be utilized. For example, if a healthcare organization is capable of proving that it satisfies certain Meaningful Use criteria (including protection and auditing), business analysts are free to leverage identified patient information for business purposes. However, because IRB must impose higher standards for the consumption of identified patient information, and because the transition of data into lower-security environments might break certain security requirements, it is common for honest broker systems to support research workflows. For data science programs following the DevOps model discussed elsewhere in this chapter, it is also necessary for such teams to abide by strict data de-identification standards for the majority of their work.

Additional stipulations can add complexity to this process, depending on the host site. For example, unstructured documents and images may contain PHI in subtle ways. PHI may hide in the body of a text document, written as text on an image, as components of RTF metadata, within header data, or in other ways. Programmatic de-identification tools need to be quite sophisticated to win approval by a hospital system for these purposes. At many institutions, it results in an expensive manual de-identification process.

18.5 BASIC PRINCIPLES OF MACHINE LEARNING

As healthcare industry is becoming more integrated with data science, we have to continuously address the question of best choice of machine learning methods for medical data that is inherently sparse, noisy, and has imbalanced classes of interest. On this data, standard predictors and classifiers may show only very modest quality and robustness.

When a standard data analysis tool systematically fails, a modern data scientist (as opposed to a more traditional business analyst) has to be able to solve theoretical problems in mathematics and computer science to respond to the discovered knowledge gap. Historically, establishment of global communications raised new questions in network analysis; ubiquitous use of personal computers inspired advances in hardware development and scalable algorithms. Similarly, expensive and complex knowledge gaps now exist in the best use of data for healthcare applications. In response to this challenge, we expect new and exciting research in data science.

This section is written in a manner of academic lecture notes. For readers with experience with statistical inference, this is an attempt to put familiar but disjoint topics into a system that is convenient in context of healthcare analytics. For readers mostly concerned with management and scientific workflow, this material lists keywords for independent exploration.

We return to the representation, $Y = F(X; P)$; $J := J(Y, P) : \min_P J$, used in Section 18.1. The central task of analysis is to relate inputs X to outputs Y, or to metric J. This requires a development effort in two parts:

1. Data acquisition: from distributed, multiformat raw data to formally defined X, Y, P, J
2. Development of a core predictive model, that is, choice of mathematical format for F.

We discussed aspects of (1) in previous material. Here, we would like to reinforce the attitude of representation equivalence: *almost any* robust, repeatable transformation of raw data into numerical tables has predictive potential. And several, sometimes many different representations should be examined during the study. We recommend viewing data as raw material to be transformed, as opposed to as an immutable source of knowledge. In fact, the process of data acquisition itself can be subjected to mathematical analysis. We would schematically represent it as $(X, Y) = Q(R, P)$ where R is a body of raw data and Q is a transformation process, described by parameters P. The issue of the best choice of Q is of particular interest in fields with large cost of data acquisition and cleanup.

In part (2), we use the term "core predictive model" very informally. The outcome of analysis does not have to be a simulation model (as in engineering studies or computer visualizations); it does not have to predict anything when description is sufficient, does not even have to be an explicit model.

Machine learning imitates cognition process in that it deals with patterns [35, 36]. It *decomposes* and *synthesizes* information. Decomposition leads to creation of

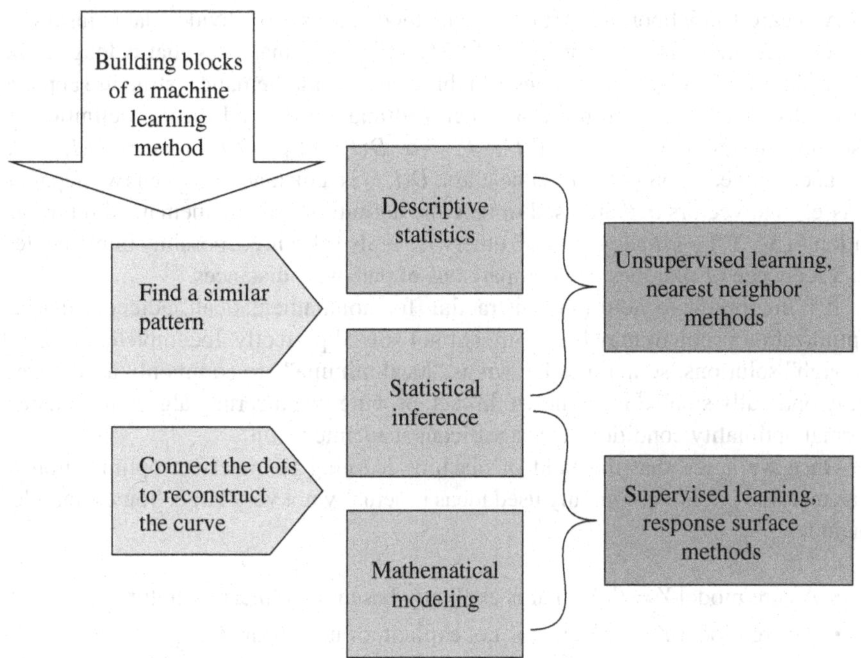

FIGURE 18.5 Building blocks of machine learning. Reproduced with permission of Oleg Roderick.

patterns that will be recognized when observed again. Synthesis allows completion of partially observed data into familiar patterns. In simpler words, an automatically learning computer program either searches for objects, or completes objects, or both. To search in data, we use various aggregating and summarizing methods. To interpolate, we create idealized mathematical representations. To do both, we create idealized representations and adjust them if it produces aggregate quantities that are very different from those observed in data. These building blocks of almost any machine learning method are visualized in Figure 18.5.

The terms supervised and unsupervised learning are sometimes used to distinguish methods based on whether an output Y is explicitly known or even defined. Supervised learning aims at achieving the best representation $Y = F(X)$ and is usually extrapolative. Unsupervised learning attempts to best group subsets of X, according to the proximity metric J, and tends to be represented as a classification problem. The two types of learning may overlap.

We suggest that every machine learning task can be understood as an optimization problem of fitting an algebraic formula to the available data. The differences between models are in the choice of the mathematical representation, and, of course, in the way data was chosen and prepared in (1).

For example, a linear regression method uses a (matrix–vector) representation $F(X) = X \cdot B + E$ with optimization criteria $J = |Y - F(X)|$ in some metric $|..|$, for example, Euclidean distance.

A nearest-neighbor classification method seeks to divide data instances $X_1, X_2, \ldots, X_n$ into discrete classes "1", "2", "3", … so instances that belong to the same class are similar in some sense. In this example, mathematical structure appears to be disguised. The optimization criteria ultimately depend on the definition of "similar." It may have a form $J(\text{"}I\text{"}, \text{"}J\text{"}) = \alpha \cdot D(I,J) + \beta \cdot D(I,I)$ where $D(I,I)$ is a distance between points in the same class, $D(I,J)$ is a distance between two separate classes, and vectors α, β are used to tune the definition. The mathematical representation $F(X_i, X_j) = 1$ if same class, 0 otherwise is also obscured, possibly implemented as a sequence of searches and comparisons of pair-wise distances.

It is important to note (and distracting for nonmathematical audience) that an optimization problem may be set up, but not solved perfectly. Incomplete, or "good enough" solutions, sometimes known as "local minima" are commonly used where truly optimally small J is unknown. In fact, proving that a certain algorithm satisfies formal optimality conditions is a significant academic result.

Once we agree that the field of machine learning is imperfect optimization in disguise, the pool of commonly used ideas is actually not very large. Some examples include:

- A core model $Y = F(X)$ that is explicitly based on a linear operator
- A core model $Y = F(X)$ that is not explicit, or not linear
- A network representation, in which F describes relationships, not outputs
- A stochastic representation, in which F describes statistical distributions, not specific values
- A representation in which F is not explicit, and never tuned, for example, there is no "mathematical formula," only a set of distances
- A hierarchical representation, in which an output of a core model is fed into another model or method

Of course, the ideas may overlap and can be combined to improve multiple steps of (1) and (2). We present some of the resulting variety in Figure 18.6. The keywords we included in the image are very differently distributed in terms of popularity. Some of them are taught in academic courses of statistics for medical professionals, some are more familiar to specialists in computational science. Even more variety may have been included, but, in the scope of this material, we only provide a basic idea of how the field of machine learning is organized.

We now follow with remarks on every method mentioned. An impression that choice of mathematical model is an art rather than a science is at least partly correct: scientific objectivity comes from optimization, not representation aspect of our studies.

Linear regression is the basic technique of statistical inference [37]. While very few relationships even in mechanical systems are strictly linear, a locally linear relationship very often suffices to model as unknown phenomena on variables that are mostly independent and recorded in an error-free manner. Linear relationship

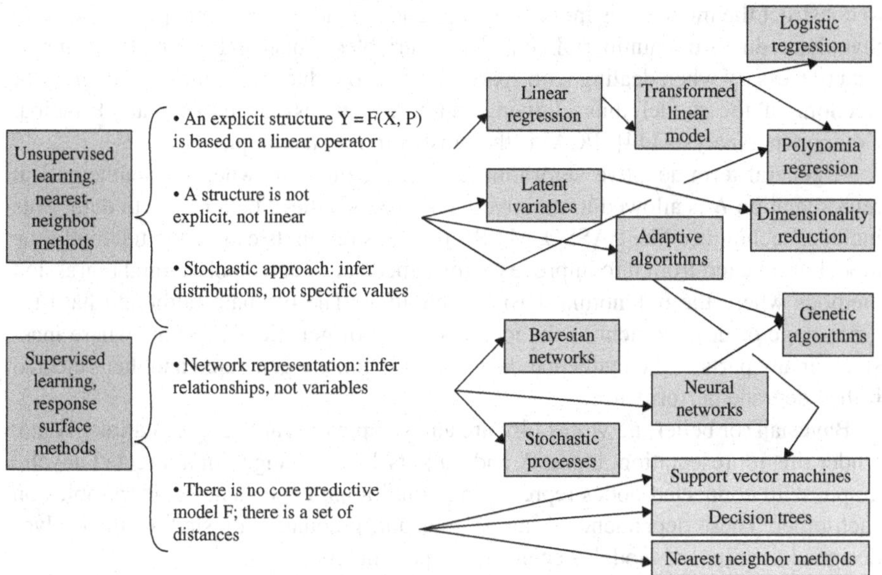

FIGURE 18.6 Building blocks of machine learning: continued. Reproduced with permission of Oleg Roderick.

between variables is also assumed in many experimental design considerations (planning sample size, estimating error, trending, i.e., estimating sensitivity).

A linear model can be passed to an a priori-defined nonlinear operator [38], $F(X) = \Pi(X \cdot B + E)$. An operator Π may be a variant of logistic function, leading to logistic regression [39]; a single most popular classification method of healthcare analytics. Another choice of the transforming operator may create a model that is linear in patient features but nonlinear in time. Expecting some quantity to exponentially increase or to decay leads to a family of survival models (such as Cox proportional hazards model [40]), also very popular in medical literature.

Polynomial regression (also see polynomial chaos [41]) is another modification on linear regression: it uses an expansion in multivariate, nonlinear polynomials $\Psi(X)$: $F(X) = \Psi(\mathrm{X}) \cdot B + E$. Including high-dimensional polynomials allows to capture complex dependencies on variables, but runs the risk of overfitting. Time-dependent variations are possible: $F(X, T) = (\Psi(X) \cdot B + E) \cdot R(t)$.

Latent variable analysis [42] is based on a suspicion that variables X, as obtained during stage (1), are not the best representation of independent inputs into the model, and there is a transformative mapping $X' = Q(X)$ that would result with a different, "hidden" list of true factors influencing the model.

Dimensionality reduction [43] is sometimes wrongly understood in medical literature as a form of feature selection, that is, using only the most significant variables to construct the model. It is more correct to understand it as any mapping $X' = Q(X)$ that also reduces the number of considered variables. This can be done to compress data (necessary when dealing with sparseness) or to reduce the number of degrees of freedom in the model thus allowing better fit on less available data. Principal Component Analysis [44] (PCA) is the most common technique.

A general term adaptive algorithm is used for methods when the mathematical representation F is allowed to iteratively change to achieve better fit with data. This includes techniques like LASSO, MARS [45] where inputs can be included into the model or rejected from it to improve performance; augmented polynomial regression methods where the polynomial basis is rewritten. The ultimate (although hard-to-tune) stage of development of the idea is the use of genetic algorithms where massively many parts of the mathematical representation are generated and then rejected if they degrade performance.

Bayesian (or belief) networks [46] are an example of graph representation of data. Under this representation, inputs X and outputs Y are arranged in a directed acyclic graph, with connected nodes representing conditional dependencies of variables on each other. These dependencies can be constant, gradually adjusted, or themselves designed as response models dependent on parameters.

Neural networks [47, 48] are an idea inspired by classical computer scientists attempting to imitate information processing in biological systems. They also model input-to-output relationships as a direct acyclic graph, with the weight of each edge gradually adapted until the entire network approximates an observed relationship between data and outcomes of an observed system. This approach represented a step away from rule-based modeling toward data driven paradigm; they can also be very effective at solving a class of large-scale optimization problems.

Stochastic, or random processes [49], with an important subclass of Gaussian processes [50] is a family of inference tools predicting multivariate distributions, rather than deterministic state vectors of multivariable systems. At each instance of the system evolution, statistics such as expected value, or covariance, are represented as functions dependent on parameters, schematically $Y \sim \mu(X), \mathrm{cov}(X, X)$ with $\mu(X) = F(X, P), \mathrm{cov}(X, X) = C(X, X, P)$.

Support vector machines are a relatively recent classification tool, originally set up as a high-dimensional binary classifier. It constructs a hyperplane separating data instances into two categories so that the gap between them is optimally wide. Modifications such as multiple application of the method, adaptive changes in data, and so on, allow multiclass classification and other forms of analysis.

Decision tree learning [51, 52] constructs a hierarchical graph from inputs to outputs, generally by multiple sampling of subset of data and establishing connections between objects, object classes, object superclasses, and so on, making each decision to include or exclude into a class based on a single variable at a time. Versions with adaptive and transformative variations exist.

Nearest neighbor method is a basic approach to proximity-based classification. It builds classes by including objects that are "nearby" a reference points, using some metric of distance and some stopping criteria (e.g., stop when a fixed number of points has been found, as in k-nearest neighbors, kNN [53]).

18.6 CASE STUDY: PREDICTION OF RARE EVENTS ON NONSPECIFIC DATA

The following section is somewhat technical, and assumes familiarity with mathematical methods listed in previous material. Here, we present selected material from one of our recent predictive analytics project: prediction of rare adverse events for general patient population, based on nonspecific, traditional clinical data.

This is not the most challenging, nor computationally demanding example, but it mentions multiple topics, and illustrates well the ideas of method-centric approach and equivalent/automatic inclusion of raw data into the predictive model. The work was performed without the use of external data science software; every step of the algorithm was designed and can be controlled by the investigators.

18.6.1 Problem Definition

In this study, we seek to develop a data-driven method of predicting risk of generic adverse events for a general population, based on nonspecific healthcare data collected over a fixed period of time. Our central example is mortality (from all causes). In clinical practice, having this prediction is useful for palliative care and end-of-life counseling. Details on a specific project are available in our academic publication [54].

In one such study, we use de-identified clinical and medical insurance records of 500,000 patients. For each patient, we use 6 months of data, time period defined backwards either the adverse event, or from the date of last transaction. The task is to predict a binary outcome (1 for rare event, 0 for normal state). This particular study is not longitudinal; we are inputting a collection of all medical events into the model, without timestamps.

This study is performed in the framework of personalized medicine: we want to automatically select only the most relevant training data for each patient, but at the same time don't want to restrict the study to any particular population group such as in-patients, or patients already recorded as having increased risk of an adverse event. This has led us to a decision to build an individual predictive model for each patient. To clarify: all such models will have very similar mathematical structure (2), a binary classifier using logistic regression. But each classifier will be constructed using a different subset of the available data (1).

The most relevant metrics of quality are precision (how many correct binary predictions?) and specificity (for patients that eventually do experience the adverse event, in how many cases is our prediction correct?).

18.6.2 Data Acquisition

The structure of the data is implied by the objectives of the study: we use routine, nonspecific healthcare information, aggregated with no assumptions, and without additional acquisition effort. This includes:

I. Standard biometric features (age, sex, ethnicity, weight, BMI), we expect about 10 features

II. Aggregated financial risk scores, we expect at most 10–15 features
III. Transactional clinical data

Some of the data will be missing or be declared unreliable by inspection; we recover missing entries by nearest-neighbor imputation [55] (the most similar entry is accessed for substitute value).

The dimensionality of transactional data is very high; we sort features by frequency and use a fixed amount of the most commonly occurring. Also, the last few days of an inpatient history may be characterized by an extremely dense activity (we call this "flurry of events"). For objectivity, we do not want to use this obvious clinical response to crisis as a predictive feature. We remove records from dates that fall into the upper 5% by frequency. After this, sparsity of transactional data will be very high, perhaps close to 99%; we will need to use dimensionality reduction to compress it into useable inputs.

18.6.3 Method Selection

Our personalized prognosis method consists of two parts: identify a reference group of patients "most similar" to a given one and use data for this reference group to construct an individual binary classifier. Accordingly, we break up data into two categories: information that is relevant for a long period of time is denoted as **static** (this includes biometrics and financial scores); the remaining data (transactions) is denoted as **dynamic**.

For each patient in the trial set, we form a reference group using an unsupervised clustering algorithm on static data. We use a weighted form k-nearest neighbors clustering [56]. The term "weighted" here refers to a form of adaptive selection of similarity metric; some variables are more important in determining similarity than others.

Data analysis in medicine is often motivated by rare events. For example, in databases available to us, patient death is recorded for only 3% of the observed population. For classification, it is helpful to rebalance the reference group, that is, to make sure that the proportion of binary outcomes in the training set is fixed. A set of dynamic data that corresponds to the reference group is extracted from raw data, cleaned up, and compressed using a weighted form of principal component analysis [57]. This produces a set of inputs X, now scrambled by compression so it cannot be annotated, and different for each patient.

We implemented the binary classifier F as an adaptive logistic regression model. We also attempted to use Gaussian Processes-based classifiers and found that they produce very similar results but at a much higher computational cost. The whole process: from setting up a reference group, to obtaining a patient-specific classifier model, to evaluating this model to obtain prediction for the individual patient, is visualized in Figure 18.7.

18.6.4 Partial Results and Discussion

The algorithm is validated as a standard binary classifier: comparison of the number of true positive, false positive, true negative and false negative predictions is performed on multiple sampled subgroups of patients from the available data. We report

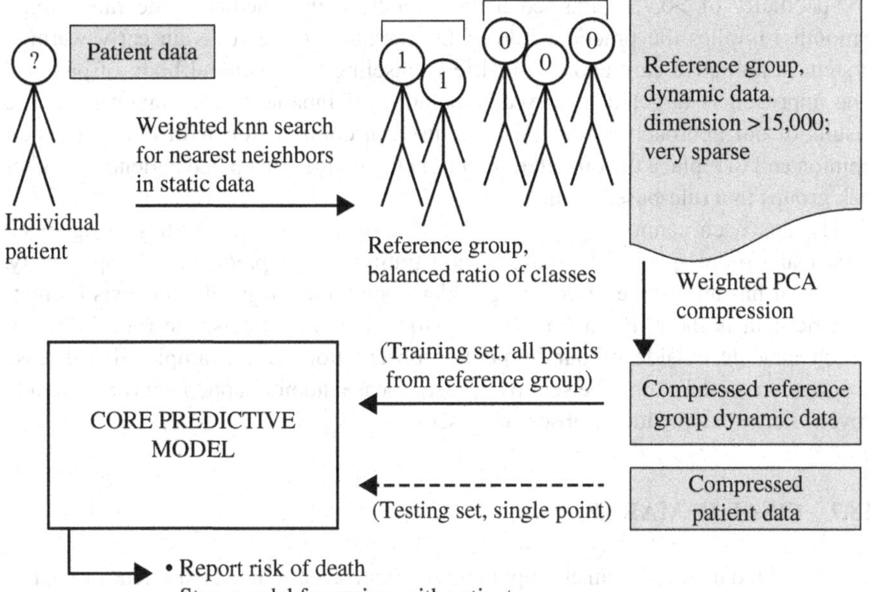

FIGURE 18.7 Schematic representation of predictive workflow. Reproduced with permission of Oleg Roderick.

TABLE 18.2 Performance of Binary Classifier for Mortality Prediction Based on Nonspecific Clinical Data

Model Setup	Correlation	Precision	Specificity	AUC
30 most frequent features, adaptive weighting	0.45	0.85	0.71	0.76
150 most frequent features, adaptive weighting	0.53	0.83	0.78	0.80
300 most frequent features, adaptive weighting	0.61	0.85	0.79	0.81
30 most frequent features, no weighting	0.65	0.81	0.61	0.70
300 most frequent features, no weighting	0.60	0.84	0.58	0.69

Reproduced with permission of Oleg Roderick.

correlation of a vector of predictions with a vector of true binary events, precision, specificity. We also report Area under the Receiver Operating Characteristic Curve [58] (AUC) metric for binary classifiers (in this case, for results of logistic regression).

In our central example, prediction of mortality performance of the method is affected by the number of "most commonly occurring" features were included in transactional data. It is also affected by details of model setup—in particular, by how well weighting was done (on every method used in core model construction (2) that was characterized by the term "weighted"). We compare the worst-case model selection (no weighting done at all) with the best performance of the model. Partial results are presented in Table 18.2.

Specificity of >0.75, obtained using data from intermediate time range (e.g., 6 months) implies the practical role of the method—to serve as an early warning system in palliative care and end-of-life counseling for a general body of patients. The approach is targeted at mixed population of inpatients and outpatients. The results of our approach are meant to be used in combination with medical expert opinion and to replace the outdated screening techniques that place patients in higher risk groups in a rule-based manner.

The approach cannot replace in-depth disease progression studies using much denser and more specialized data, demonstrating higher precision and specificity. Instead, it has advantage of covering wider population. Arguably, analysis on non-specific data is the only choice we have for predicting events, the root causes of which may lie outside of our immediate observations: for example, if a disease progression model cannot be set up because disease did not happen yet (or is already "over," i.e., went outside of observation scope).

18.7 FINAL REMARKS

The presented material is an attempt to disseminate accumulated experience of a data science group intensely involved in short-term and medium-term predictive analytics projects in support of a major healthcare system [30].

We tried to focus on management and ideological aspects of healthcare data science workflow. We believe that these are the parts of professional experience that cannot be reproduced outside of the medical environment, no matter how advanced was the training in mathematics and computer science.

The technical scope of the material was necessarily limited. We did not cover many of the exciting modern topics such as agent-based modeling, deep/representation learning, artificial intelligence, financial mathematics as applied to hospital environment, natural language processing, and automated sentiment analysis. We also focused on traditional clinical records as opposed to specialized genomics data. All these topics are also necessary for preparation of a modern researcher in the field of healthcare data science; we encourage independent exploration.

We purposefully did not overview the external, easy-to-use (or not so user friendly) computer packages for data analysis. We wanted to make this material educational objective, not an encouragement to buy solutions from a particular vendor. The market of healthcare informatics is being filled with database architecture solutions and models that are finely tuned to a narrow set of tasks. In this environment, fostering independent researchers is more important than employing personnel trained in the use of popular software tools.

For young career professionals choosing their first direction of development, as well as for experienced data scientists who started working in other fields, healthcare is a challenging, potentially very rewarding field. Industries such as marketing, financial sector, social media, and entertainment are full of both large companies and startups competing to provide essentially the same services. Engineering fields traditionally dealing with big data, such as astrophysics, or extreme-precision simulation, require advanced scientific degrees, and in respective subjects. In data-driven healthcare, on the other hand, many basic questions have not yet been answered—in fact, have not yet been asked. There is opportunity for impactful work on many levels of technical competence, and success depends comparatively less on luck and timing and more on creativity and dedication.

REFERENCES

[1] Hill, Thomas and Pawel Lewicki. *Statistics: Methods and applications: A comprehensive reference for science, industry, and data mining*. Tulsa: StatSoft, Inc., 2006.

[2] Toffler, Alvin. *The third wave*. New York: Bantam Books, 1981.

[3] Burke, Raymond R.. "The third wave of marketing intelligence." In: Krafft, Manfred and Murali K. Mantrala (eds.), *Retailing in the 21st century*. Berlin/Heidelberg: Springer, 2010. 159–171.

[4] Collins, Harry M. and Robert Evans. "The third wave of science studies: Studies of expertise and experience." *Social Studies of Science* **32**.2 (2002): 235–296.

[5] Berry, Michael J. and Gordon Linoff. *Data mining techniques: For marketing, sales, and customer support*. New York, NY: John Wiley & Sons, Inc., 1997.

[6] Brown, Brad, Michael Chui, and James Manyika. "Are you ready for the era of 'big data'." *McKinsey Quarterly* **4** (2011): 24–35.

[7] Groves, Peter, Basel Kayyali, David Knott, and Steve Van Kuiken. "The 'big data' revolution in healthcare." McKinsey Quarterly (2013).

[8] Davenport, Thomas H. and Dhanurjay Patil. "Data scientist." *Harvard Business Review* **90** (2012): 70–76.

[9] Dhar, Vasant. "Data science and prediction." *Communications of the ACM* **56**.12 (2013): 64–73.

[10] Hamburg, Margaret A. and Francis S. Collins. "The path to personalized medicine." *New England Journal of Medicine* **363**.4 (2010): 301–304.

[11] Khoury, Muin J., Marta L. Gwinn, Russell E. Glasgow, and Barnett S. Kramer. "A population approach to precision medicine." *American Journal of Preventive Medicine* **42**.6 (2012): 639–645.

[12] McClennan, Mark B., Michael J. McGinnis, Elizabeth G. Nabel, and LeighAnne M. Olsen. *Evidence-based medicine and the changing nature of health care: Meeting summary (IOM roundtable on evidence-based medicine)*. Washington, DC: National Academies Press, 2008.

[13] Murray, Michael F., David J. Carey, F. Daniel Davis, W. Andy Faucett, Monica A. Giovanni, Marc S. Williams, Christa L. Martin, and David H. Ledbetter. "The Geisinger-Regeneron Initiative: A New Kind of Genomics Research Collaboration." Report, Geisinger Health Systems, 2014.

[14] Conway, Drew. "The data science venn diagram." Dataists. Retrieved February 9, 2010: 2012. http://drewconway.com/zia/2013/3/26/the-data-science-venn-diagram, accessed September 13, 2016.

[15] "ODBMS Industry Watch Blog." http://www.odbms.org/blog/, accessed August 24, 2016 (2015).

[16] Costa, Fabricio F. "Big data in biomedicine." *Drug Discovery Today* **19**.4 (2014): 433–440.

[17] Tang, Paul C., Danielle Fafchamps, and Edward H. Shortliffe. "Traditional medical records as a source of clinical data in the outpatient setting." Proceedings of the Annual Symposium on Computer Application in Medical Care. American Medical Informatics Association, Washington DC, 1994.

[18] Safran, Charles, Meryl Bloomrosen, W. Edward Hammond, Steven Labkoff, Suzanne Markel-Fox, Paul C. Tang, and Don E. Detmer. "Toward a national framework for the secondary use of health data." *Journal of the American Medical Informatics Association* **14**.1 (2007): 1–9.

[19] Katsios, Christos and Dimitrios H. Roukos. "Individual genomes and personalized medicine: Life diversity and complexity." *Personalized Medicine* **7**.4 (2010): 347–350.

[20] Murdoch, Travis B. and Allan S. Detsky. "The inevitable application of big data to health care." *Journal of the American Medical Association (JAMA)* **309**.13 (2013): 1351–1352.

[21] Rodriguez, Jose A. and Giuseppe Giaccone. "Keynote comment: Are large-scale cancer-genomics projects ready to use?" *The Lancet Oncology* **7**.3 (2006): 190–191.

[22] Halamka, John D. "Early experiences with big data at an academic medical center." *Health Affairs* **33**.7 (2014): 1132–1138.

[23] O'Driscoll, Aisling, Jurate Daugelaite, and Roy D. Sleator. "'Big data', Hadoop and cloud computing in genomics." *Journal of Biomedical Informatics* **46**.5 (2013): 774–781.

[24] Dean, Jeffrey and Sanjay Ghemawat. "MapReduce: Simplified data processing on large clusters." *Communications of the ACM* **51**.1 (2008): 107–113.

[25] Borthakur, Dhruba. "The Hadoop distributed file system: Architecture and design." *Hadoop Project Website* **11** (2007): 21.

[26] Law, Maria Y.Y., and Hui-Kuang Huang. "Concept of a PACS and imaging informatics-based server for radiation therapy." *Computerized Medical Imaging and Graphics* **27**.1 (2003): 1–9.

[27] Florescu, Daniela and Donald Kossmann. "Storing and querying XML data using an RDMBS." *IEEE Data Engineering Bulletin* **22** (1999): 3.

[28] Dietterich, Thomas G. "Ensemble learning." In *The handbook of brain theory and neural networks*. Vol. **2**. Cambridge, MA: The MIT Press, 2002. 110–125.

[29] Achard, Frederic, Guy Vaysseix, and Emmanuel Barillot. "XML, bioinformatics and data integration." *Bioinformatics* **17**.2 (2001): 115–125.

[30] Paulus, Ronald A., Karen Davis, and Glenn D. Steele. "Continuous innovation in health care: Implications of the Geisinger experience." *Health Affairs* **27**.5 (2008): 1235–1245.

[31] Maeng, Daniel D., Jove Graham, Thomas R. Graf, Joshua N. Liberman, Nicholas B. Dermes, Janet Tomcavage, Duane E. Davis, Frederick J. Bloom, and Glenn D. Steele. "Reducing long-term cost by transforming primary care: Evidence from Geisinger's medical home model." *The American Journal of Managed Care* **18**.3 (2012): 149–155.

[32] Gardner, Elizabeth. "First steps for population health." *Health Data Management* **22**.3 (2014): 32–34.

[33] Benitez, Kathleen and Bradley Malin. "Evaluating re-identification risks with respect to the HIPAA privacy rule." *Journal of the American Medical Informatics Association* **17**.2 (2010): 169–177.

[34] Baumer, David, Julia Brande Earp, and Fay Cobb Payton. "Privacy of medical records: IT implications of HIPAA." *ACM SIGCAS Computers and Society* **30**.4 (2000): 40–47.

[35] Poggio, Tomaso and Christian R. Shelton. "On the mathematical foundations of learning." *American Mathematical Society* **39**.1 (2002): 1–49.

[36] Michalski, Ryszard S., Jaime G. Carbonell, and Tom M. Mitchell, eds. *Machine learning: An artificial intelligence approach*. Berlin, Heidelberg: Springer Science & Business Media, 2013.

[37] Seber, George A.F. and Alan J. Lee. *Linear regression analysis*. Vol. **936**. Hoboken, NJ: John Wiley & Sons, Inc., 2012.

[38] Dobson, Annette J. and Adrian Barnett. *An introduction to generalized linear models*. Boca Raton, FL: CRC Press, 2008.

[39] Hosmer Jr, David W. and Stanley Lemeshow. *Applied logistic regression*. New York: John Wiley & Sons, Inc., 2004.

[40] Lin, Danyu Y. and Lee-Jen Wei. "The robust inference for the Cox proportional hazards model." *Journal of the American Statistical Association* **84**.408 (1989): 1074–1078.

[41] Crestaux, Thierry, Olivier Le Maitre, and Jean-Marc Martinez. "Polynomial chaos expansion for sensitivity analysis." *Reliability Engineering and System Safety* **94**.7 (2009): 1161–1172.

[42] Muthén, Bengt. "Latent variable analysis." In Kaplan, David (ed.), *The Sage handbook of quantitative methodology for the social sciences*. Thousand Oaks, CA: Sage Publications, 2004: 345–368.

[43] van der Maaten, Laurens J.P., Eric O. Postma, and H. Jaap van den Herik. "Dimensionality reduction: A comparative review." *Journal of Machine Learning Research* **10**.1–41 (2009): 66–71.

[44] Ringnér, Markus. "What is principal component analysis?" *Nature Biotechnology* **26**.3 (2008): 303–304.

[45] Tibshirani, Robert. "Regression shrinkage and selection via the lasso." *Journal of the Royal Statistical Society. Series B (Methodological)* **58** (1996): 267–288.

[46] Heckerman, David. *A tutorial on learning with Bayesian networks*. Dordrecht: Springer, 1998.

[47] Tu, Jack V. "Advantages and disadvantages of using artificial neural networks versus logistic regression for predicting medical outcomes." *Journal of Clinical Epidemiology* **49**.11 (1996): 1225–1231.

[48] Baxt, William G. "Application of artificial neural networks to clinical medicine." *The Lancet* **346**.8983 (1995): 1135–1138.

[49] Chang, Chin Long. *Introduction to stochastic processes in biostatistics*. New York: John Wiley & Sons, Inc., 1968.

[50] Rasmussen, Carl E. "Gaussian processes in machine learning." In: Bousquet, Olivier, Ulrike von Luxburg, and Gunnar Ratsch (eds.), *Advanced lectures on machine learning*. Berlin Heidelberg: Springer, 2004. 63–71.

[51] Long, William J., John L. Griffith, Harry P. Selker, and Ralph B. D'Agostino. "A comparison of logistic regression to decision-tree induction in a medical domain." *Computers and Biomedical Research* **26**.1 (1993): 74–97.

[52] Quinlan, John R. "Induction of decision trees." *Machine Learning* **1**.1 (1986): 81–106.

[53] Tran, Thanh N., Ron Wehrens, and Lutgarde M.C. Buydens. "KNN-kernel density-based clustering for high-dimensional multivariate data." *Computational Statistics and Data Analysis* **51**.2 (2006): 513–525.

[54] Roderick, Oleg and Nicholas Marko. "Development of personalized models for mortality risk prediction on a general patient population," Geisinger Health Systems, 2015.

[55] Marlin, Benjamin M. Missing data problems in machine learning. Diss. University of Toronto, 2008.

[56] Yu, Cu, Beng C. Ooi, and Kian-Lee Tan. "Indexing the distance: An efficient method to KNN processing." *Proceedings of the International Conference on Very Large Data Bases* **27** 2001: 421.

[57] Daescu, Dacian and Michael I. Navon. "A dual-weighted approach to order reduction in 4DVAR data assimilation." *Monthly Weather Review* **136**.3 (2008): 1026–1041.

[58] Brown, Christopher D. and Herbert T. Davis. "Receiver operating characteristics curves and related decision measures: A tutorial." *Chemometrics and Intelligent Laboratory Systems* **80**.1 (2006): 24–38.

19

DATA ANALYTICS AND PREDICTIVE ANALYTICS IN THE ERA OF BIG DATA

AMY SHI-NASH[1] AND DAVID R. HARDOON[2]

[1]*Singtel, DataSpark, Singapore, Singapore*
[2]*Azendian, Singapore, Singapore*

19.1 DATA ANALYTICS AND PREDICTIVE ANALYTICS

19.1.1 Introduction

The raise of big data in recent years has opened up unprecedented opportunities, rapidly changing the landscape of analytics and technology. Data-driven decision making has become one of the most fundamental capabilities that not only results in strong revenue performance and superior customer experience but also drives innovation and strategic competitive advantage.

This chapter outlines the key principles of machine learning and predictive analytics. It explains the new fundamentals of big data and the evolving technology. It follows by the practical advice on how organizations can establish a new culture in order to truly transform their business in the new era.

19.1.2 A Brief History

The wave of data frenzy did not happen overnight. Rather, it is a crescendo of events happening since the early 1980s where the fields of business intelligence and predictive analytics were known as "data mining," a preexisting discipline with another closely related term known as Knowledge Discovery in Databases (KDD), which is the aim of performing data mining.

Internet of Things and Data Analytics Handbook, First Edition. Edited by Hwaiyu Geng.

Companion website: www.wiley.com/go/Geng/iot_data_analytics_handbook/

The first KDD workshop was held in 1989 in Detroit, MI, USA, during the International Joint Conferences on Artificial Intelligence (IJCAI-89) [1]. During the workshop and in those times, much of the emphasis was still on the development of expert systems. Expert systems were based on the notion that expert knowledge in any area was comprised of a set of rules. It was believed that by determining the rules an expert in a particular domain uses and replicating these rules in a system, the system would replicate expert behavior in that domain. These systems were and are still widely used in the biotechnology sector today to identify diseases in people. However, such systems were inflexible and brittle—they were unable to replicate human cognition across different case scenarios as they were limited to the preexisting set of defined rules, and whenever faced with a scenario which bent these rules, expert systems would fail.

The difficulties faced by expert systems highlighted the need for systems to be able to fall back on details of experience, detect similar cases, make decisions on which case was relevant, and apply existing knowledge. These systems would need to be able to identify similarities between recurring tough problems to create new cases and scenarios, as well as update their existing sets of rules. These systems needed to enable machines to "learn" from their experiences. This realization gave birth to the now widely used term of "machine learning"; a coming together of statistics, fuzzy logic, knowledge acquisition, artificial intelligence, databases, data mining, computer science, and neuroscience back in 1987 and 1989 in the respective fields of Neural Information Process Systems (or neuroscience) and KDD [2].

The benefits of embracing data mining and KDD can be seen clearly even in early adopters. A popular story is how Walmart employed machine learning in the form of an association rule learning to discover the relationships between the products it sold—it discovered relationships between products which no one would have thought of. For instance, Walmart found that diapers and beer were very likely to be purchased together [3]. In response, diaper and beer products were moved in closer proximity of one another and sales of both increased. Although we must remember that unlike domain expertise which is a top-down approach, data is bottom-up, that is, the insight uncovered at Walmart may not be applicable to other retailers as it is data driven from the behavior of Walmart's shoppers.

While association rule learning (also known as Basket Analysis) was one of the earlier methods of analysis which was popularized, especially in the retail sector, segmentation of customers has been performed for a much longer time. Market segmentation was increasingly used as a strategy to gain competitive advantage and increase sales since the 1970s. Businesses began to invest in understanding their customers, moved away from a one-size-fits-all approach, and began catering more diverse offerings to smaller groups within their customer base. This allowed businesses to not only better match customers' needs but also create new products and services, tapping new revenue streams and creating growth. Customer retention also improved as the business was aware of changes in needs, circumstances, and the lifestyle of customers and was able to adjust their offerings accordingly. Most importantly for Chief Marketing Officers, this allowed stronger branding and more effective communication to target segments. American Express was one of the early adopters of customer segmentation [2]. By introducing the Gold Card in 1966 and, subsequently, the Platinum Card in 1984 as a super-exclusive offering, the company reinforced its premium positioning and was able to derive more value from high income clients. Customer segmentation continues to be a core function of analytics today.

19.1.3 Descriptive, Diagnostic, Predictive, and Prescriptive Analytics

Analytics has a spectrum of methodologies, techniques, and approaches from descriptive, diagnostic, predictive and prescriptive analytics [4].

Descriptive analytics such as segmentation, clustering, and classification are generally performed at the initial stage of the analysis in order to get a good understanding of the shapes and the patterns of data. Descriptive analytics focus on finding the "what"—for example, what are our customers' behavior patterns? Who are our loyal customers? Descriptive analytics is widely used in consumer behavior analysis and marketing, providing a very effective way to understand the behavior of millions of consumers and creating targeted actions to each segment of customers instead of the masses. The descriptive analysis algorithms are the most natural in terms of accuracy and performance.

The next level of analytics is Diagnostic analytics which at times is coupled with descriptive analytics. This deals with the "correlations"—what is correlated to a customer choosing one product over another? Diagnostic analytics are generally more difficult to perform and produce more valuable insight. For example, marketers can sometimes interpret correlations based on business logic and take marketing actions accordingly. However, correlations cannot conclusively prove "causality" by machine learning algorithms [5], as the correlation is only true based on the limited learning data set used. Information produced from both Descriptive and Diagnostic analytics are generally considered hindsight as they largely concern things which have already happened in the past.

Predictive analytics is about using the information of the past to understand the likely occurrence of the future—for example, what is likely to happen next or what is a customer likely to buy next. Sophisticated models and machine learning are crucial in this area to perform the inductive reasoning needed. Finally, the most difficult type of analytics is Prescriptive analytics. Prescriptive analytics provides insight on what can be done to increase the probability of a desired outcome occurring—what can be done to make a customer more likely to choose product A over product B?

Predictive and prescriptive analytics are much more prevalent now than previously. There have been technological advancements in both software and hardware that enable more complex analytics to be performed. There are also many more sources of data, with high levels of digitization across the globe and with the Internet of Things taking shape. These factors, together with the exponential improvements in software and hardware, mean that a lot more data is available for analysis today than before. The extreme volumes of data created are now commonly referred to as "big data."

We use the term "Analytics" throughout the chapter to represent all descriptive, predictive, and prescriptive analytics.

19.1.4 Predictive Machine Learning Method: Supervised and Unsupervised

Machine learning algorithms are designed for specific types of problems and data; thus it is important to assure there is an alignment between the data, problem, algorithm to achieve maximum outcome. Machine learning algorithm largely fall into two groups, supervised and unsupervised learning. Supervised learning is used when the outcome (of target variable) can be clearly defined. Whereas in unsupervised learning, instead of a set of predefined examples, the algorithm present where data similarity is emergent.

For example, a supervised machine learning algorithm is being used to detect fraudulent transactions. Random subsamples of data are selected and manually classified as "fraudulent" or "nonfraudulent." These classified subsamples, also known as "training data," are then used to build models through the identification of patterns and trends. Thereafter, the algorithms will be able to independently classify new records as "fraudulent" or "nonfraudulent." Problems arise if there is too little training data or there are insufficient instances of fraud in the training data, as the requisite patterns and trends in the data will not be present.

In instances where unsupervised machine learning is similarly used for detecting fraudulent transactions, the nature of the issues faced by a lack of data is similar. In some instances of unsupervised machine learning, data is observed for anomalies in patterns and trends to highlight behavior which differs from the "norm." For example, Peer Group Analysis identifies accounts where behavior begins to differ significantly from other objects to which they used to behave in a similar manner. Another method, Break Point analysis, identifies when behavior on a single account begins to differ significantly from its previous behavior. Both approaches cannot function well if there is insufficient past data to establish relationships between accounts or determine what constitutes "normal" behavior for an account.

It is also important to note that in all machine learning processes, the insight produced is inductive in nature. The flagging of an object as "fraudulent" or as any other property only indicates that the object has a higher chance of being fraudulent than normal. It is not possible to determine for sure if an account is indeed fraudulent without further investigation.

Machine learning is focused on proving correlation not causation. Going back to the earlier example on diapers and beer, while it may be possible to prove that purchase of diapers and beer is correlated to one another, it is not possible to determine if diapers caused the purchase of the beer or vice versa without further investigation or the circumstances which led to such a correlation in the first place, without conducting further investigation such as a survey with customers. Therefore, while advanced machine learning algorithms have taken over much of analytics, methods such as surveys have not lost their relevance and still play an integral role in understanding customers.

One important element to consider is how to choose which of the different machine learning methodologies should be used as a solution, for example, in the question whether to use a decision tree or neural network structure in implementing a machine learning solution. Decision tree structures, as their name suggests, use decision trees as predictive models in deciding how to reach a conclusion on a particular set of data. Neural networks, on the other hand, mimic biological neural networks in estimating or approximating rules or functions which depend on a huge number of inputs and are generally unknown before discovery.

The choice of approach will be dependent not only on the degree of accuracy but also, for example, whether it is essential to understand what the model is doing for the model to be interpretable. The need for a more interpretable model will lend itself to decision tress, while potentially the more accurate models would be neural networks.

19.1.5 Cross Industry Standard Process for Data Mining

Most data mining projects today follow the Cross Industry Standard Process for Data Mining (CRISP-DM) which was conceived in 1996 [6]. The process breaks data mining into six stages:

1. The first, termed "business understanding," involves understanding how the project's objectives and requirements are related to business goals so that a data mining problem statement may be laid out.
2. The second, "data understanding," involves the initial stages of data collection and analysis to build familiarity with the data, identify quality issues, form preliminary hypotheses, and gain first insights on the data.
3. The third, "data preparation," involves determining and executing all activities which will transform raw data into the final data set—this includes selecting tables, records, and attributes, as well as transforming and cleaning the data so that it may be compatible with modeling tools.
4. In the next stage, "modeling," modeling techniques are selected, applied, and optimized.
5. By the "evaluation" stage, a completed model has been built, and it is evaluated on whether it achieves the business objectives that were identified in the very first stage.
6. Finally, "deployment" involves the organization and presentation of the knowledge generated by the model in a form that is easily understandable and useful to the end user. There are many forms of deployment—it could be as simple as generating a report or much more complex, such as implementing repeatable data scoring or data mining process.

It is important to note that the stages in CRISP-DM are not strictly sequential, and there is often iteration between the different stages in a project in an agile way. Traditional Software Development Life Cycle and PMP methodologies are most often *not* applicable in analytical projects.

19.2 BIG DATA AND IMPACT TO ANALYTICS

19.2.1 The Four Vs

The term "big data" was first mentioned by NASA researches Michael Cox and David Ellsworth in 1997 in the context of a problem statement—the volume of data was fast becoming too excessive for available computer systems to store and process [7]. Shortly after which, the world began to pay more attention to the growing volume and value of data due to the nature and exponential growth—humans were creating more data than they could handle.

The characteristics of big data are commonly described by the industry in terms of Volume, Variety, Velocity, and Veracity:

Volume: In 1999, the amount of information created in the world annually was comprehensively researched and quantified for the first time at 1.5 exabytes (one followed by 18 zeros) [8], and Enterprise Resource Planning (ERP) solutions and predictive analysis began to change how organizations did business [9]. Today, we create 2.5 exabytes of data every day [10]—so much that 90% of the data in the world today has been created in the last 2 years alone. The Internet traffic is predicted in a report by Cisco to reach 1.3 zettabytes (one followed by 21 zeros) by 2016 and 40 zettabytes by 2020 [11]. "There is no unambiguous way to measure the size of digital information" [12], but the trend and speed of the growth are exponential. The sheer volume of data presents the most immediate challenge to traditional IT architecture and data management systems.

Variety: The exponential growth of data is in both structured and unstructured forms. One of the largest sources of unstructured data is text and multimedia content—e-mails, documents, calls, instant messages, social media conversations, videos, audio, photos, webpages, and many more. This can be attributed partially to the widespread digitization and rapid growth of mobile device penetration across the globe. According to GSMA's The Mobile Economy 2015 [13], there are 3.6 billion unique mobile subscribers globally and 7.1 billion global SIM connections at the end of 2014 and a further 243 million machine-to-machine (M2M) connections. Vast amount of digital content is consumed. For example, in 2014 [14], 4 billion hours of videos are watched on YouTube every month; 400 million tweets are sent every day; an estimated 420 million wearables and wireless health monitors are contributing to the 150 exabytes of health data, in batch as well as real-time data streams. The variety of big data posts significant challenge to the conventional database or data warehousing solutions that require predetermined schemas.

Velocity: Data has always been generated, transferred, processed, and analyzed at different speeds. The "big" challenge presented by big data is the need to process data in (near) real time in order to maximize the value of data and make immediate decisions. High-velocity data analytics is particularly important in applications such as social media conversations, complex event management (e.g., traffic management), and IoT applications (medical monitoring, wearables, modern plane sensors, etc.). The latency of data processing and decision making generally has a spectrum. For example, a driverless underground train needs to control and monitor the precision of the movement within milliseconds; credit card fraud detection needs to make a decision within 30 seconds; real-time marketing offer might be reasonable within 10 minutes. Processing and analyzing data in motion (as supposed to data at rest) requires in-memory technologies and streaming machine learning techniques.

Veracity: With large volume, diverse formats and high speed, big data comes with more noise and higher uncertainty. The raising challenges go beyond data quality to aspects such as the ability to understand the data in the first place. In many organizations, particularly where data is gathered from new and legacy systems, data governance initiatives such as the data dictionary, metadata definition, data cleansing, and validation are sometimes the weakest link. Data stewardship is increasing a critical role in managing big data in large organizations.

Last but not least is Value: Organizations need to see clear use cases and the estimated potential return from the investment required in big data initiatives. One of the most challenging tasks to analytical leaders is for the short term to clearly articulate use cases and justify the approach and immediate investment and in the long term to gain cross-organizational commitment to build future transformational capabilities and innovation. A recent study by the Economist Intelligence Unit [15] showed that strong use cases include price optimization, new revenue generation, demand forecasting, and significant operational cost saving. Experimentation is one of the most practical approaches in developing a value-driven strategy [16].

Today big data is a myriad of synonymous and terms (as seen in the preceding text); thus as a rule of thumb, it is always good to clarify what one is *exactly* referring to when discussing big data in their context.

19.2.2 The Hadoop Wave and Big Data Analytics

Previously we explained the basics of contemporary analytics and the new characteristics of big data. Later in this chapter we will discuss the impact of big data on analytics, but first we need to understand the changing landscape of data management technologies.

Managing big data needs a new framework and principles. Although there are many new and too many confusing jargons, there are only a few key principles and frameworks. Once you understand these, you will be able to cut a path through the jargons and take advantage of the new emergent solutions.

These key principles are discussed in the following text as well as the combination of engineering and data science, which drives modern big data analytics and scalability.

First is the distributed data storage (as files or databases) and parallel data processing to deal with the large scale. If a file is too big for a single computer, then divide the big file into smaller files and distribute them in multiple servers. In distributed data storage, computer networks store information on multiple nodes, often in a replicated fashion [17]. Hadoop and Google File Systems are types of such distributed file systems that were developed by researchers at Yahoo and Google in 2003 and 2004 [18] for managing web search data. Today Hadoop Distributed File System (HDFS) is now widely used as the basic system for big data. Because multiple nodes can perform different tasks in parallel, this also massively increased the speed of data access.

Another type of distributed data store is called "peer network node data stores." A well-known peer network node data store is BitTorrent. In such data stores, users are able to reciprocate and allow other users to use their computer as a storage node. The blockchain is another peer network node data store, which maintains a constantly growing list of data records which are tamper and revision proof, even to operators of the data store's nodes.

The distributed approach can also be applied to create databases on top of HDFS in order to store and query data. These databases are often nonrelational (NoSQL, not only SQL) allowing unstructured data to be stored, proceeded, and queried over a large number of nodes. For example, Cassandra and HBase are used by Facebook, BigTable by Google, and Voldemort, a data store used by LinkedIn. In order to address and bring SQL-like capabilities to query-distributed databases, relational databases are also increasingly popular: Impala, Hive, Presto, Apache Drill, and EMC Pivotal HAWQ, to name a few. Many other BI tools like Tableau can also interface to data using SQL.

Then comes the processing of big data—since data is so large and moving data across the network is very slow, the framework of processing data is to push processing code to data in multiple nodes, instead of the other way around. This framework is called MapReduce that was first published by Google in 2014 [19] and further developed by the open-source community. Today, another faster and simpler processing framework called Apache Spark is replacing MapReduce in many organizations and big data projects. Spark's real-time capability and standard APIs for different languages win over many developers compared to the batch mode and complexity associated with MapReduce. Other more purposely built systems within Spark are also becoming very popular, for example, Kafka for real-time processing, Apache Solr for search, and GraphX for graph analysis.

The next step is to better make sense of big data. The abilities of existing machine learning algorithms fall short in handling real-time, unstructured, distributed, and large volumes of data. When dealing with different velocity and distributed data, associations and confidence levels are much more difficult to form. There have been new developments in algorithms that are designed to handle big data. For example, Deep Learning is one of the rapidly growing research topics particular for face recognition, bioinformatics, and natural language processing. It attempts to replace handcrafted features with algorithms that can learn from large-scale unlabelled data, for unsupervised or semisupervised feature learning and hierarchical feature extraction.

Increasingly industry application architecture design and software engineering best practice are integral components of analytics and not longer seen as two separate sequential steps. The new architecture design moves away from monolithic to microservice enabling a modular, reusable, and scalable structure across business. Similarly, this modular and parallel approach can also be taken to rewrite the machine learning algorithms, breaking a traditional complex algorithm into many smaller tasks that can be run in distributed and real-time environment. This necessitates data scientists to also have programming skills and understand concepts such as feature engineering, parallel processing, and performance optimization.

19.2.3 Current Status

Today, big data analytics, although still in its infancy, is one of the fastest growing fields driving new possibilities in data-driven decision making and business innovation across every industry. In Gartner's Hype Cycle Report [20], "Big Data" moved up the curve from the first stage of *Innovation Trigger in 2011* to *Peak of Inflated Expectations* in 2013. The exponential growth pushed Big Data into the third stage of *Through of Disillusionment* in 2014, indicating mainstream adoption in the next 5 years. Other enabling and application relevant technologies have also progressed forward, such as In-Memory Database, Streaming, Natural Language Processing, Hybrid Cloud computing, IoT platform, Wearable user interface, M2M communication, NFC, and so on 2015 has seen the strongest growth in business applications according to Gartner, led by Digital Marketing, Digital Business, and Autonomization such as virtual personal assistants and Autonomous Vehicles. However, core analytical topics such as Data science, Big data analytics for customer service, Software-defined analytics, and so on are still in their infancy.

The Hype cycle largely explains the question of why there is so much talk of big data yet a current lack of proven business cases linking investment in big data analytics with revenue returns. Early adopters are however emerging from a wide range of industries, from digital (Google, Facebook, Amazon) to Retail, FMCG and even Agriculture.

For example, Burberry uses radio-frequency identification tags to create customer profiles by keeping tabs on what they try on to better understand their customers and push the right products to them. CVS Health segments customers contact their call center into different behavioral groups and route them to the agents who interact best with them for shorter calls and higher issue resolution rates. Coca-Cola uses proprietary algorithms to ensure that its Minute Maid orange juice maintains one out of 600 identified flavor profiles. L'Oreal analyzes tweets, Facebook posts, product reviews, and news stories so it can leverage on and influence brand awareness and loyalty. Tom Farms LLC uses automated equipment which monitors tens of thousands of acres of farmland with sensors that produce a stream of real-time data to mobile apps for remote monitoring and control. Food Genius crawls the web for food demand-related material like restaurant menus to help companies like Kraft Foods and national chains like Arby's localize their offerings.

Another emerging trend is the cross-industry perspective. The potential of big data analytics opens up innovative business opportunities beyond the boundary of industries. For example, new data-driven smart solutions are starting to play a critical role in solving global challenges such as urban density, aging population, healthcare, mobility, and energy sustainability. Data gathered from smartphones, wearables, Apps, and other machines can be used to provide new solutions. One example of a data-driven urban mobility solution for city planning in Singapore, developed by DataSpark a wholly owned subsidiary of Singapore Telecommunications, uses telecommunication data to detect millions of public transportation trips everyday (underground trains and bus) [21, 22], monitors the congestion level of the transportation network, and recommends less crowded routes to regular commuters in real time (Figure 19.1) [23].

The technology advancement is only one side of a story. Mindset and culture change is more critical. The successful adoption of big data analytics is not just the acquisition of data streaming, high-capacity storage, and noSQL databases. Business leaders need to look deeper into the organizational enablers such as an experimentation approach, acquisition and retention of new talent, new operating and collaboration models, developing open ecosystems, and proactively managing data security, privacy, and trust. These topics are the key success factors in adopting new technology and building new competitive advantages. The following section will discuss these organizational enablers.

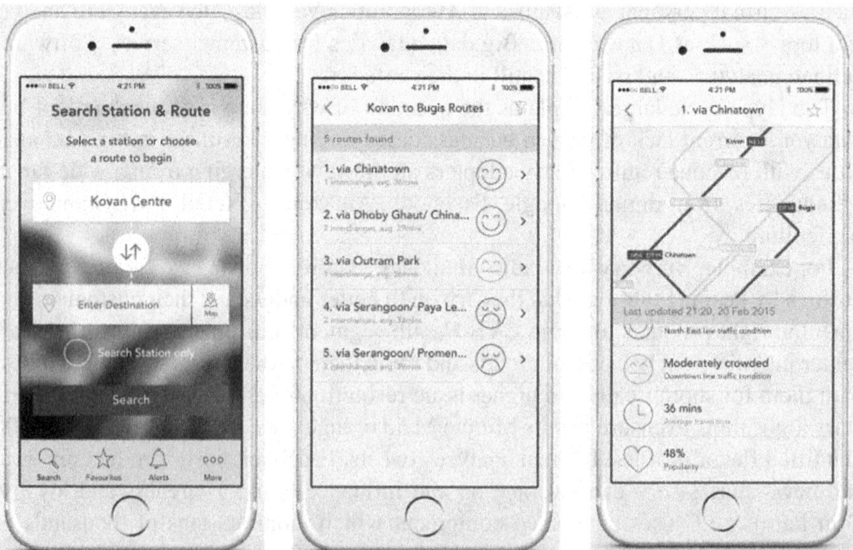

FIGURE 19.1 Screenshots of DataSpark's Mobile App that allows commuters to plan their route based on the average travel time to the destination and crowdedness of the boarding station in near real time.

19.2.4 Adopting Big Data Analytics at Corporate Level

The key question is where to start? With all the respect, typical IT-led projects often start from evaluating different vendors and solutions without a good understanding of business needs and use cases. This "starting with solution and figuring out the value later" approach often results in yet another expensive system with poor business adoption.

We recommend the opposite approach, starting from the end, testing the value with hypothesis-driven experimentation, shown in Figure 19.2.

The key to this approach is identifying a specific (small) business goal and a hypothesis to achieve the goal and implement a real-life experiment to prove or deny it. The experimentation approach creates basic prototype only but mimics all the critical components related to the goal. The process is to validate or identify flaws in the hypothesis, understand the complexity of the ecosystem, learn the know-how through doing, and build real business case in future target market based on the tangible result. It provides opportunity to identify risks during execution stage and demonstrates tangible benefit to business users. It takes skill and practice to design good and rapid experimentation but it is extremely powerful in order to gain internal stakeholders' buy-in and C-level commitment.

The experimentation also generates disruptive ideas. The hypothesis gives the starting point but many experimentation results in new ideas. It provides a vehicle for staff to ask the question "what if" and discover the answers in the process. This is a behavioral change, and the new mindset is fundamental for the organization to adopt big data analytics.

To summarize:

- DON't start from buying a big data solution (expensive mistakes!), or start from look at all of your existing data (data is not all equal!), without much thinking on what to do with it. (No, you cannot figure out later!)
- DO hypothesis-driven experimentation is a tangible way to valid disruptive ideas, build business cases, gain high-level buy-in, and learn what it takes to make it work.

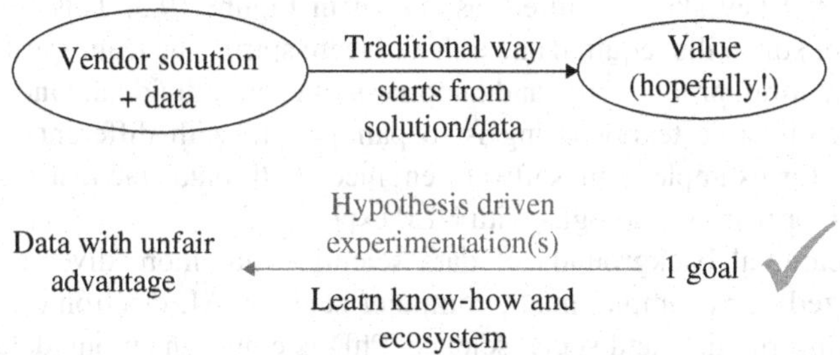

FIGURE 19.2 Hypothesis-driven experimentations are the smart way to start a big data initiative instead of the traditional obsolete approach of starting from your data or buying a big data solution.

- DO have a specific (often partial) goal—part of the value proposition that must be measured completely.
- DO have a hypothesis—which links to achieving the goal and can be measured tangibly. The outcome of the experiment is to confirm or deny the hypothesis.
- DO have a working system with a scientific framework—which mimics (faking is allowed) all the key components (internal and external) of the ecosystem involved in the value proposition and is able to be extrapolated to make a real business case in the target market.
- DO find/create the right data set that gives an unfair advantage (differential data, not all data). It doesn't matter if the right data set doesn't exist, or it is not possible to access, as the process is driven by the goal, not by the existing data. Differentiation is the key to answer "why us?"

19.2.5 Acquiring Talent and Skills

The article entitled "Data scientist: the sexiest job in the 21st century" published in *Harvard Business Review* in October 2012 is one of the early literatures about the role and job description of data scientist. Since then, the interest and demand for data scientist has grown exponentially. Shortage of this new breed of talent is becoming a serious constraint to many organizations. Acquiring the right talent is a no easy task.

So what are the key characteristics and skill sets of data scientists? Where to find them? How to motivate and retain them?

Good data scientists have strong curiosity and desire to solve real problems that are complex, ambiguous, and involving multiple sources of data. They are motivated by the discovery nature of the job, not afraid of challenging assumptions, and very intuitive with data-driven experimentation approach. Prototyping and visualization capability is a trademark of data scientist; programming skill is the one that separate data scientists from the traditional analysts. Java, Scala, Python are the most commonly used languages in data science community. Ability to work with embedded system and build hardware components such as sensors, mobile, and wearable are highly desirable today.

A good data science team is a multidisciplinary team, allowing a collective coverage of all the key skills required, as shown in Figure 19.3. This brings up the importance of soft skills required and a shared team spirit. The team spirit defines the culture, behavioral, productivity, and happiness of each individual. One of the practical tips on effective teamworking is to pair people with different strength and background, for example, pair software engineer with data scientist to build large graph model, or pair psychologist with NLP expert.

The educational backgrounds of data scientists are more diverse than many people realized, for example, math, computer science, AI, electronic engineering, physics, bioinformatics, and social science. PhD is common among data scientists, but young and talented under graduates also make great candidates. This is why many companies (particularly technology companies) have established internship programs and final year projects with reputable universities. There is also a group of

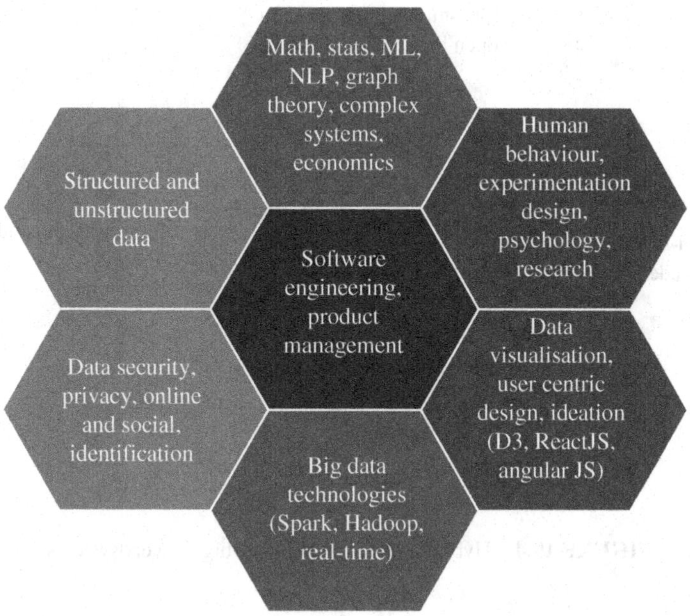

FIGURE 19.3 Multidisciplinary team is required to build big data analytics capabilities.

self-thought data scientists who are actually exceptional, and they are often very active in the open data science community. Taking part in the community, for example, via open projects or hackathon, allows the company to establish its name in the market and have direct dialog with a wider range of talents.

The talent market is highly competitive. Package, location, and company reputation are important deciding factors. Google used to be one of the coolest employers to work for, but now a Silicon Valley start-up with share option can be more attractive. Other determining factors are much softer but could have strong influence, for example, a strong and inspirational leader, real-life problems that worth solving, a degree of freedom to explore, high caliber team mates, and opportunity to make a difference.

19.2.6 Building an Open Ecosystem

Traditionally the analytics function often operates as a stand-alone department supporting the operation in areas such as marketing, customer service, and sales. However, due to the speed and complexity of the technology, in-house analytics functions are challenged to keep up. This had led to many organizations finding that the stand-alone Analytics operating model no longer works. Internal teams need to recognize that they are a part in a broader analytics ecosystem involving external sources of technology, innovation, and know-how. A buy or build strategy need to be considered carefully and strategically. Different strategies and mixes of internal and external partners can be applied at different stages of development, and the

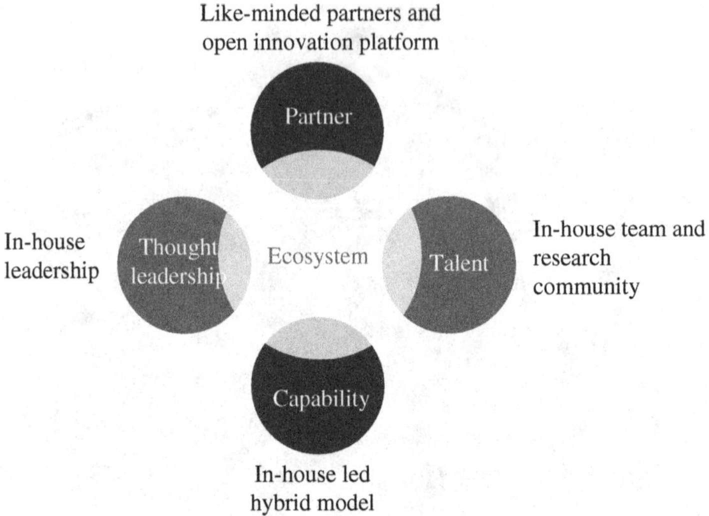

FIGURE 19.4 Hybrid strategy in building big data ecosystems.

strategy should therefore be fluid and responsive rather than fixed. Figure 19.4 shows one hybrid strategy shared in this chapter by the authors, based on their practical experience [16].

We strongly recommend that the "thought leadership" needs to come from the business leaders within the organization. External strategy or consulting firms can help but cannot take the driving seat. Thought leadership includes the vision and end state of big data analytics, the business case, priorities, and talent strategy. It is a key that the business leaders demonstrate both action and behavior that support the vision and strategy. On the other hand, capability development, such as user experience and interface design, visualization, software development, consulting and delivery, and so on, should take full advantage of the technology advancements available in the market whether through external providers or an in-house hybrid approach. Here, the key consideration is the architecture design that needs to be modularized and opened to allow for the introduction of future technology advancement without significant change of the infrastructure and rest of the technology choices.

The same strategy is true for talent. Since the global shortage, it is unrealistic for an organization to assume that it can acquire all the talent needed. The core IP should be kept in house for commercial reasons, but blue sky research and early prototyping can be executed through collaboration with consultancy firms, universities, research institutes, and so on. University graduates are a good source of talent acquisition, as they are the new generation whose first language is distributed file system, programming language, and NoSql. The open-source community is also becoming an important source for organizations to tap into, through hakathons, data challenges, or open projects. Government agencies, investors, and incubators can play an increasingly important role in the big data analytics ecosystem.

FIGURE 19.5 A framework to unlock the value of big data by protecting individual privacy and maintaining utility of the data.

Operating within an ecosystem requires in-house functions to work closely, comfortably, and efficiently with multiple players. This operating model itself can signify a culture change within the organizations, where developing trust and equal cocreation partnerships (not vendor management) can be new to many large organizations.

19.2.7 Actively Managing Privacy and Branding

Actively managing data privacy and branding is critical to any data-driven business. A transparent business attitude and technical best practice are required. A framework is proposed in the succeeding text by the authors, as shown in Figure 19.5. It requires a combination of user privacy protection (e.g., anonymization, encryption, measuring, and reducing the risk of reidentification [24]) and privacy-preserved methods that keep the utility (analytical) value of the data. The computational cost can be significant and should be considered when implementing both privacy protection and utility preservation methods.

19.3 CONCLUSION

The maturity level of data-driven decision making of an organization is often seen as the fundamental competency and true competitiveness of a business. Since the establishment of KDD in the late 1990s, the successful adoption of advanced data analytics and machine learning in business world has been a game changer on how a business makes decisions in their operations and how they communicate/engage with consumers.

In the most recent years, the raise of big data and rapid advancement of new technology has opened up unprecedented possibilities, triggering one of the biggest transformation waves in many industries. To most of the organizations, it is not a question of if, but when.

Experience has shown that an experimental approach is a smart and powerful way to get started. Define a tangible business goal, build a real working prototype, test hypotheses and validate assumptions, and learn the ecosystem and know-how. These specific steps can help organizations to create a new mindset and decision-making culture, from quantifying the business case to mitigating risks.

One of the biggest challenges that an organization will face is acquiring and retaining the right talent. Each organization will need to find its own way. However perhaps the most important success factor is strong leadership with vision and commitment to experiment and execute.

REFERENCES

[1] http://www.kdnuggets.com/meetings/kdd89/ (accessed January 11, 2016).

[2] http://www.kdnuggets.com/meetings/kdd89/kdd-89-report-aimag.html (accessed January 11, 2016).

[3] Wal-Mart: Sean Kelly's Data Warehousing: The Route to Mass Customization, a 1996 release from John Wiley & Sons, Inc., Hoboken, NJ.

[4] D. Hardoon and G. Shmueli, *Getting Started with Business Analytics: Insightful Decision Marking*, CRC Press, Boca Raton, FL, 2013.

[5] http://www.aaai.org/ojs/index.php/aimagazine/article/viewArticle/1230 (accessed January 11, 2016).

[6] http://www.taborcommunications.com/dsstar/99/0413/100687.html (accessed January 11, 2016).

[7] M. Cox and D. Ellsworth, Application-Controlled Demand Paging for Out-of-Core Visualization, NASA Ames Research Centre, Mountain View, CA, 1997.

[8] P. Lyman and H.R. Varian, *How Much Information*, University of California, Berkeley, CA, 2000.

[9] P. Preston, Choosing and Installing the Right ERP Solution, ComputerWeekly.com, 1999.

[10] http://www-01.ibm.com/software/data/bigdata/what-is-big-data.html (accessed January 11, 2016).

[11] http://www.pcmag.com/article2/0,2817,2405038,00.asp (accessed January 11, 2016).

[12] Gartner, Gartner Hype Cycle 2014 Report Puts Wearables Big Data in Their Place, 2014. https://www.gartner.com/doc/2814517/hype-cycle-big-data- (accessed September 13, 2016).

[13] http://www.gsmamobileeconomy.com/GSMA_Global_Mobile_Economy_Report_2015.pdf (accessed January 11, 2016).

[14] http://www.ibmbigdatahub.com/infographic/four-vs-big-data (accessed January 11, 2016).

[15] Economist Intelligent Unit, Competing Smarter with Advanced Data Analytics, Economist Intelligent Unit report, 2015. https://www.eiuperspectives.economist.com/sites/default/files/EIU_SAP_Competing%20smarter%20with%20advanced%20data%20analytics_ExecSummaryPDF.pdf (accessed September 8, 2016).

[16] IDC, Buyer Conversation: DataSpark's Journey in Establishing Analytics Business, IDC report, Framingham, MA, 2014.

[17] Y. Pessach, *Distributed Storage (Distributed Storage: Concepts, Algorithms, and Implementations ed.)*, 1 edition, CreateSpace Independent Publishing Platform, 2013. ISBN: 9781482561043.

[18] S. Ghemawat, H. Gobioff, and S.-T. Leung, The Google File System, 2013. http://static.googleusercontent.com/media/research.google.com/en//archive/gfs-sosp2003.pdf (accessed September 8, 2016).

[19] J. Dean and S. Ghemawat, MapReduce: Simplified Data Processing on Large Clusters, 2014.http://static.googleusercontent.com/media/research.google.com/en//archive/mapreduce-osdi04.pdf (accessed September 8, 2016).

[20] L. Columbus, 2014: The Year Big Data Adoption Goes Mainstream in the Enterprise, *Forbes*, New York, January 13, 2014.

[21] A. Shi-Nash, Invigorating the Telco Landscape: How Telcos Can Use Data Assets to Create New Applications, Strata+Hadoop World, Singapore, December 2015. http://conferences.oreilly.com/strata/big-data-conference-sg-2015/public/schedule/detail/45254 (accessed January 17, 2016).

[22] Z. Salim, Urban Mobility in the Smart City Age: DataSpark, ComputerWorld, Singapore, January 2015. http://www.mis-asia.com/resource/applications/urban-mobility-in-the-smart-city-age-dataspark/ (accessed January 17, 2016).

[23] T. Holleczek, D. The Anh, S. Yin, Y. Jin, S. Antonatos, H.L. Goh, S. Low, and A. Shi-Nash, Traffic Measurement and Route Recommendation System for Mass Rapid Transit (MRT), Knowledge Discovery in Databases, Sydney, Australia, 2015. https://www.researchgate.net/ (accessed January 11,2016).

[24] Y.-A. Montjoye, C. Hidalgo, M. Verleysen, and V.D. Blondel, Unique in the crowd, the privacy bounds of human mobility, *Nature*, 2013, doi:10.1038/srep01376. http://www.nature.com/articles/srep01376 (accessed February 17, 2016).

20

STRATEGY DEVELOPMENT AND BIG DATA ANALYTICS

NEIL FRASER

Macquarie University, Sydney, New South Wales, Australia

20.1 INTRODUCTION

One of the many challenges in strategic planning is engaging people on purpose, vision, and values of an organization. These can be articulated as three simple questions:

Why we exist?

What we aim to be?

How we act?

The interesting thing about asking simple questions is that they can be quite difficult to answer and also often lead to the most compelling and unusual answers.

The simple questions are also asked at specific times to help inform future direction in any organization. These moments of agency often come at a point of leadership change when strategy, and its development, is reviewed, renewed, or completely rewritten. The strategist's role in this process is to ensure that complexity is reduced where possible [1] and that the executives are asking the right questions and to make sure the vision and purpose are aligned with the promises made to customers. It is also a time to horizon scan for future events that might impact current or future strategy. All of these activities bring together a strategic understanding of a current situation at the moment of agency [2].

Internet of Things and Data Analytics Handbook, First Edition. Edited by Hwaiyu Geng.

Companion website: www.wiley.com/go/Geng/iot_data_analytics_handbook/

To actually build momentum around a single vision is no easy task in any organization. In Henry Mintzberg's famous five types of organizational structure, it is clear that to set and build unity around a single vision and purpose [3] has a different level of complexity depending on each type of organization. One observation is that more federated organization structures are harder to align around common purpose, vision, and values.

Further complexity to this engagement can also come when selecting which internal and external information will be used to set context. These information indicators intersect with strategic planning at various stages and with different levels of quality, comparability, and timeliness. To bring these quantitative information indicators into strategic planning requires consideration of many aspects of a community and its relationship to leadership. The risk of getting it wrong is to follow the riderless horse into nowhere.

This chapter discusses the essence of these agency moments with some case studies and tools of use. It covers two different analytical and engagement approaches to strategic planning which have been directly influenced by the onset of digital disruption and the age of data.

20.2 MAXIMIZING THE INFLUENCE OF INTERNAL INPUTS FOR STRATEGY DEVELOPMENT

It is common to use the quantitative results to help guide expectations on future performance and alignment to strategic outcomes set in the past. They are often reported in an organization's annual report or a competitor's annual report. These internal and external quantitative data inputs have always been part of the basic input as the first step in analytical approach to strategic planning. However the risks of just a quantitative approach to strategic planning are clearly written in history.

A classic industry example of this is the Blockbuster video chain where the CEO in the 1990s lacked the vision to see how the home video industry was shifting even though the evidence was clear: declining in store sales, new entrants, and a changing demographic. In fact Blockbuster was offered to buy Netflix for US$50 million in 2000. I am sure at the time there were many employees pointing out the external trends on the horizon to the CEO, but for some reason it never translated into a clear strategy and finally Blockbuster went bust in November 2013. So to take a transformational visionary step often requires listening to employees and a clear rationale based on quality information to avoid taking an organization on a strategic journey into nowhere on a riderless horse.

The acceptance of the need for a transformational strategy can be bridged by using a different approach to engagement and in both the quantitative and qualitative techniques used in strategic planning. This approach is also more applicable in the age of data where different forms of social engagement are found.

To move the strategic planning process from just pure knowledge at a senior management level to greater understanding of a current situation in the wider community requires a deeper form of engagement. Many of the best ideas about the future of an organization come from the general workforce, not just senior management, yet their genuine input into the strategic planning process is rarely sought in a deep and meaningful way. This is partly because of the time required to engage deeply and the time it takes to refine the gathered material into a strategic framework that encompasses many people's ideas. These qualitative inputs to strategy development are more challenging to collect and analyze. Nowadays new tools and techniques from the "big data" inventory can assist. There is also an interesting, but important, side effect of this style of engagement as it allows for a strategic planning consultation process to contribute to a cultural shift in thinking. The other benefit of seeking a broad community engagement phase is that you can genuinely stimulate people to think outside the box and come up with the transformational visionary aspects required in strategic planning that could be more acceptable and associated with a community.

The challenge with this approach is overlaying difficult qualitative text-based data sets captured from many groups of people and the time it takes. In the following sections we discuss communication and analytical frameworks that can be used to facilitate strategy development from deep community-based engagements and social networks.

20.2.1 Blue-Sky Thinking

To gather enough qualitative inputs for strategy development from a broad community, it is often important to get input from as many groups of people as possible. This can be done in a series of short workshops over a few months or alternatively by using the inherent social networks underlying the organization. Both options have their merits and can be adapted for "big data" analytics.

If a workshop approach is taken, it is important to quickly establish the purpose, outcome, and process so that people will willingly share their ideas in a short period of time. One technique is to visually depict where "transformational" thinking is placed on a spectrum of potential organization activity, as shown in Figure 20.1.

This simple graphic, when shown and described in a workshop process, encourages people to think about "transformation" as an outcome.

20.2.2 The Round Table

Another technique that can be used to get people to ideate about unity of purpose is to introduce "round table" communication into an engagement framework (see Figure 20.2). The metaphor of a "round table" is this instance when shown and described in a workshop encourages people to engage in a less formal way. The opposite to this is the "square table" metaphor which can be considered more formal and often used in performance assessment or a testing situation.

Round table thinking reenforces the "transformational" aspect that is being sought in the discussion, and any encourages more radical ideas to be brought to the table. It also allows people to feel less intimidated and to speak openly.

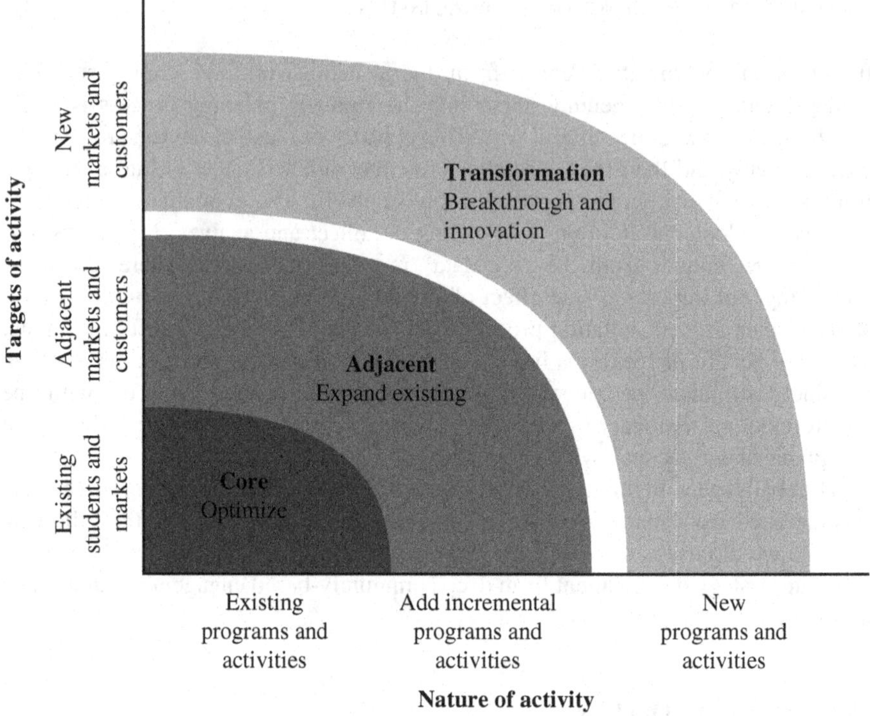

FIGURE 20.1 Transformational thinking technique. Reference [4] Nagji, B. and G. Tuff (2012). Adapted from H. Igor Ansoff.

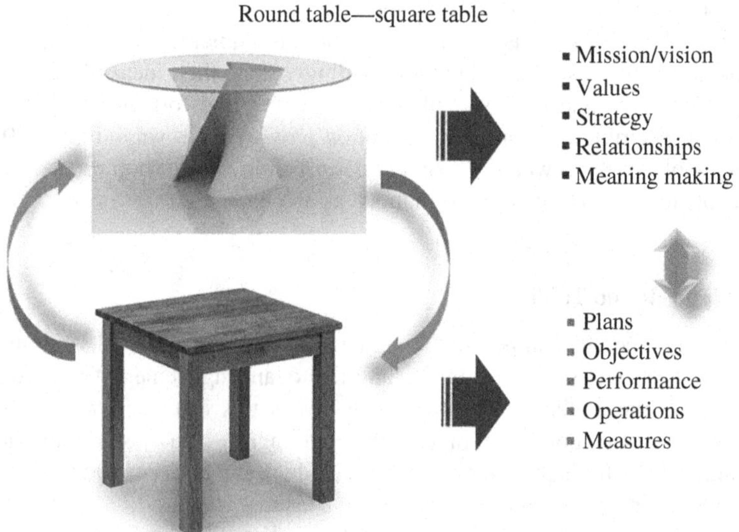

FIGURE 20.2 Round table versus square table strategic thinking. Reproduced with permission of Neil Fraser.

In the context of federated organization structures, such as that found in large universities where there are large faculties with vested interests, this broad and divergent engagement plays an important role in bridging the silos. Bridging silos raises the issue of organizational power flows and how they may interact with the aforementioned process.

20.2.3 Organizational Power Flows

A major consideration before embarking on a different analytical and process approach to strategic planning is the community structure. In federated community structures there are clear divides in culture within an organization by Mintzberg and Quinn [3]. An example depicted in Figure 20.3 is how this divide can appear inside a University:

1. *Power flow down*
   In the nonacademic part of a University, a command and control structure will often exist that is similar in characteristics to a commercial organization. Goals and objectives are cascaded from the top and embedded in teams and personal KPIs. Mintzberg called this the machine bureaucracy and is often found in the nonacademic staff at a University who run support areas in IT, finance, human resources, and property.
2. *Power flow up*
   In the academic part of the university, the power flow is more elusive and sometimes a former Head of Department (HoD) or Emeritus Professor (E) has more influence than the current HoD. Large research centers also tend to have a much larger influence in some faculties than others. The key bridging role between the two distinct parts of the organization is often a Faculty general manager (FGM).

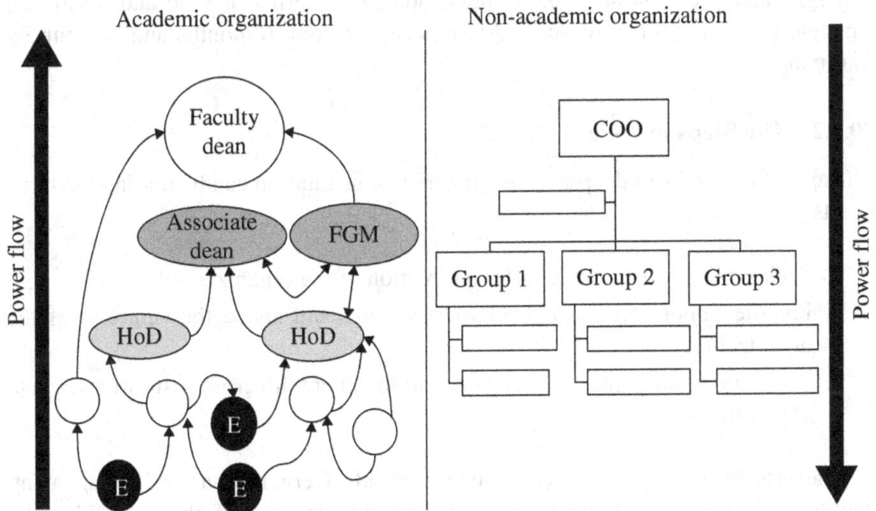

FIGURE 20.3 Organizational power flows in a University. Reproduced with permission of Neil Fraser.

A deep community-based engagement for strategic planning is helpful when these types of power flows coexist.

20.3 A HIGHER EDUCATION CASE STUDY

The following section covers a specific case study on deep community-based strategy engagement at a leading Australian University. The 6-month process allowed many people across the organization to come together in interactive sessions and frame different possible futures for the University.

The interesting thing about a University when asking questions about strategy is that it often leads to the even more interesting and unusual answers than in commercial organizations. This stems from a community difference in the individuals at the University. Academics often push for their own self-interest, particular in the area of their own research interests, when asked questions about University strategy. The nonacademic staff on the other hand often feel they have little to contribute to University strategy as they are in a support role to the main academic community but often have a better understanding of the functional weaknesses and strengths of the organization. So in this section we cover one way to run a representative strategic planning process—representative in the sense that it involves inclusiveness while remaining impartial and empirical throughout.

20.3.1 Time and Effort

The size and nature of the University are important to consider up front in the process design. In this example we had to bring together around 2,500 staff with diverse research and teaching. In addition the University services approximately 30,000 undergraduate students and 10,000 postgraduate. In terms of time and effort, the strategic planning process described in this section took 6 months and was run by four people.

20.3.2 The Steps to Take

The aim of a broad and deep strategic planning consultation can be outlined in three points:

1. Seek staff views on the long-term direction of the organization
2. Provide opportunities for staff to input into and shape the future(s) of the organization
3. Build collegiality through bringing staff together both cross-institutionally and within divisions

The process is simply to run through a broad divergent phase of engagement, capturing as many qualitative inputs as possible from staff through "blue-sky sessions," and then converge these views captured into a strategic framework.

This process is depicted in Figure 20.4.

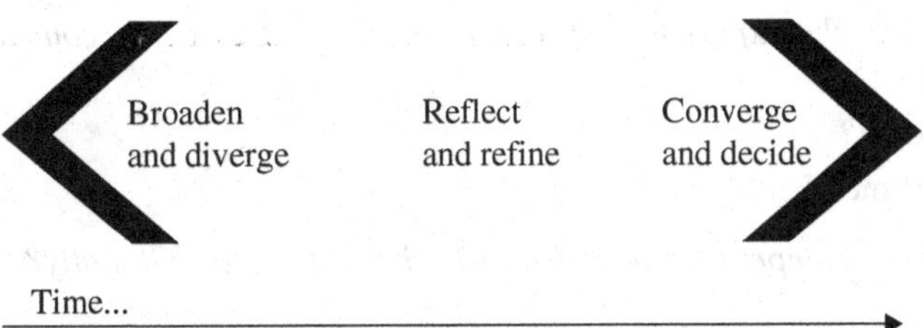

FIGURE 20.4 High-level planning process. Reproduced with permission of Neil Fraser.

20.3.3 Diverge and Broaden

The "diverge and broaden" phase was aimed at gathering insight from the university—both horizontally and vertically across the organization structure. Every staff member was able to voice their opinions about the university in writing or in open sessions in order to generate transformational ideas. In addition the key student bodies were engaged in the process in the same way.

In these "blue-sky sessions," different open-ended questions were asked to specific groups to help stimulate the conversation, and then the participants, working in small groups, were given time to synthesize their answers over a period of 1 hour before presenting this back to the group.

All participants were encouraged to give their ideas or suggestions at any time to the strategic planning team and email them to a central point for analysis. In all 45 cross-campus consultations, sessions occurred with over 800 staff involved from all faculties and divisions over a 3-month period. This generated over 3,000 comments for analysis and over 50 written submissions for further reading.

Following are some of the examples of the open-ended questions asked to the various groups at the University.

Academic Questions

1. *As you think about the far horizon for teaching and learning within the University*:
    - What will be the defining elements for a great educational experience for our students?
    - How will the students learn? What does this imply for how we organize our approach to (1) content (2) pedagogy?
    - [What does this imply for our Faculty and the University?]
2. *As you think about research in your Faculty and/or the University in the far future*:
    - What will differentiate the university that is sustainably successful in research?
    - How will research be different then compared with now?
    - [How well are we prepared to meet this opportunity?]

3. *What should the experience of a member of the University community be all about for:*
   - A student?
   - A staff member?
4. *Who will our competitors be in the far horizon? Locally? Globally?*

Professional Questions

1. When you think about what the successful university of the future will be like …
   - How will it be different from now? What will remain the same and why?
   - How will the teaching, learning, and research missions be best supported?
   - What will make the University an attractive place for you to work at?
2. Who will our competitors be in the future? Locally? Globally?
3. What in your view are the major three to five big issues and challenges that the planning process MUST address?

20.3.4 Reflect and Refine

The main challenge of this next phase of the strategy development is how to process information from the "blue-sky sessions" and interpret these findings in a meaningful and timely way for inclusion. Fortunately there are now analytical tools and techniques that overcome some of the strategic planning time constraints when dealing with large amounts of text that need to be processed.

The 3,000 answers captured in the sessions were collated into a single corpus of comments with metadata associated with question and group. Textual analytics was then used for answer analysis to identify themes and clustering of ideas. The methodology is described later in this chapter. In addition 50 separate long form submissions were sent through by academics which were read through and categorized into various themes that were emerging.

In order to gain further insight, the executive leadership group separately were sent all the comments, submissions, and the summary analysis. In total seven major themes emerged that reflected the major future needs and wants of the diverse University community based on what they had written in the "blue-sky sessions." The seven major themes were presented back to the community midway through the strategic planning process for further feedback as part of an interim green paper release.

Any direct strategic initiatives that had been brought up in consultation sessions (e.g., refresh the brand and website) were assessed against potential parameters [5].

The parameters for refinement and validation of strategic initiatives included:

- *Feasibility* (do they offer a path forward?)
- *Scope* (do they have practicality plus theoretical and empirical depth?)
- *Emergence* (are they capable of continuing reinterpretation in the context of the university's changing circumstances?)

- *Timefulness* (do they allow for measured contemplation and the long view, and not just focus on the short-term and fast-moving rhythms of academic and professional life?)
- *Locale* (are they sensitive not only to the university's global and local position but also to its universal calling?)

20.3.5 Qualitative Text Analysis in Strategic Planning

If enough answers to simple strategy questions are put to a community and answers are captured, then it might be feasible to use text analytics. In this case study 3,000 answers of semistructured text from sessions were transcribed by group and codified into a corpus. A broad answer category was also assigned based on the Ansoff criteria [4].

The codified input was then included in the analytical phase of the program as metadata at both a macro level and a micro level to get a closer association with individual group needs where possible.

20.3.5.1 Analytical Framework The analytical framework used on the session data required a four-stage text engineering process:

1. Basic word frequency analysis
2. Conceptual analysis
3. Correlational analysis of concepts
4. Theme analysis generated from the underlying concepts

The basic word frequency analysis picked up commonly occurring words in the 3,000 answers captured from the group sessions.

Conceptual content analysis provides a means of quantifying and displaying the conceptual structure of text and a means of using this information to explore interesting conceptual features.

The presence and frequency of concepts in the answers are the important variables measured in the analysis. Such concepts can be words or phrases, or more complex definitions, such as collections of words representing each concept. Visual concept maps were generated using overall presence and the co-occurrence of concepts in response to the same or similar questions.

The tools used had the capability for sentence splitting, keyword extraction, tokenizing, and key-concept finding. The analytical framework did not extend to extracting sentiment valence by topic or group.

The final step involves grouping the clustering concepts into broader "themes" and associating these with the appropriate response groups and/or questions.

A revalidation phase was also done manually to ensure nothing was missed which required a student to help with tagging comments by core subject areas.

20.3.5.2 Analytical Processing In Figure 20.5 the analytic process is shown. After initial text parsing a cluster analysis was undertaking to show concepts. Concepts could be visualized as circle where size and color represented influence.

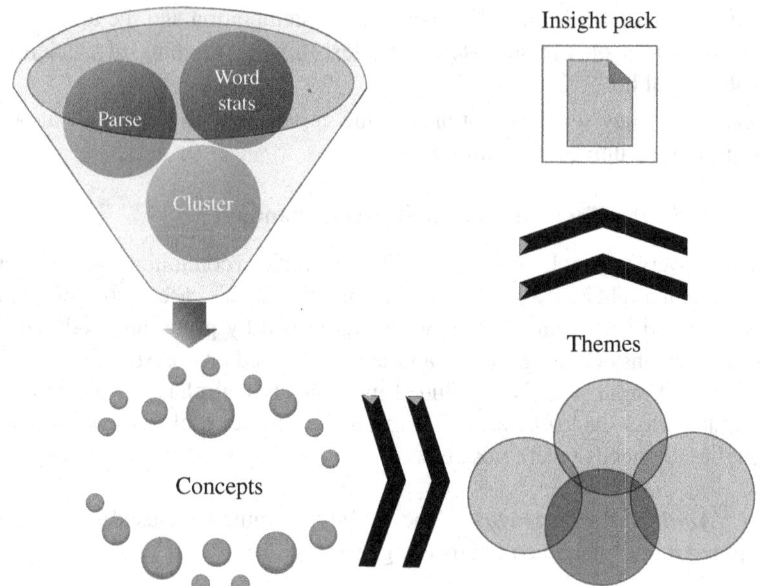

FIGURE 20.5 Textual analytic process. Reproduced with permission of Neil Fraser.

Clustering of the concepts was driven by representation of text similarity in answers. If same pieces of text attract appear, often then they tend to cluster together or attract one another strongly and so tend to settle near one another in the same visual map space. The concepts are then further clustered into themes, heat mapped, and are then analyzed for an insights pack.

20.3.6 Converge and Decide

To start to converge and decide on the legitimacy of the concepts and high-level themes identified in the analysis, a new round of consultations began with staff. An auditorium was used in these sessions to encourage large groups of people to attend and vote on the display boards for their favorite themes and initiatives identified from the analysis.

This allowed for a further level of convergence, and the votes and comments were again assessed and were used in the creation of a final strategic white paper. This final white paper contained seven strategic themes which were taken to the University governing body, approved and then finally released to the public in 2013.

20.4 MAXIMIZING THE INFLUENCE OF EXTERNAL INPUTS FOR STRATEGY DEVELOPMENT

In the commercial world the external inputs for strategy have grown into a flourishing online economy. There are now many data markets and analytics as a service supported online by the tools which underpin the "big data" movement and now form part of the business intelligence strategy suite. This massive online movement of data has led to rapid innovation that cuts through every sector, and it is simultaneously changing the skills required for strategists. To rise to this challenge, new more adaptive ways are needed to maximize the influence of external inputs for strategy development in order to give organizations an asymmetric information advantage over their rivals.

In strategic planning it has been relatively easy to find information from economic indicators, annual reports, or government statistics. The main challenge with these data sets from statutory reporting submissions is that they often lag considerably behind the current year. They also lack consistency in methodology across jurisdictions, making them difficult to compare or overlay with other external strategic data sets. This leaves gaps in quickly identifying what impacts current performance, and growth projections are anticipated for an organization or in a sector.

Another common feature is that these external statutory data sets are also often stuck in a proprietary nonmachine-readable format which makes it difficult to integrate, overlay, and ultimately analyze. There are some notable exceptions that have come from the open data movement, which some governments have participated in, which allow for open integration with new data standards. This fundamental aspect of openness in data will inform society at every level and should be encouraged as a future standard by regulatory authorities as long as it remains open. The power of the people to do great work with "open data" is greater than the people in power often realize. It is in effect a new age of data supported by new tools and techniques and fed by "open data" that is machine readable.

In this "age of data," companies that have embedded this data advantage into the "DNA" of their business model have transformed entire industries. A couple of well-known examples are the largest taxi firm is now Uber, the largest accommodation provider is Airbnb, the largest retailer is split between Amazon and Alibaba, and the largest phone company is split between Skype, WhatsApp, and WeChat. Most of these companies have generated their major growth since 2010 by gaining an asymmetric information advantage over their rivals. Strategists also need to change with the times and not choose to use old-fashioned ways to horizon scan or build strategy. This "age of data" can even be considered as an economic cycle in its own right which will follow its own dynamic boom-and-bust cycles.

20.4.1 Long-Term Cycles: The Age of Data

Historically long wave trends of prosperity and innovation cycles have been well described. One of the most commonly discussed in academia is the long wave Kondratieff cycles. The five main cycles in the Kondratieff long wave theory provide a strong visual perception tool to examine understanding on prosperity and innovation cycles over the last 200 years. The emergence and content of the next cycle are often debated in the literature.

Adams and Mouatt in 2010 [6] proposed that the fifth Kondratieff cycle should now be classed as the age of computing, while the sixth should be classed as the age of the information superhighway. Other authors have forecasted that the sixth Kondratieff cycle will be driven by resource efficiency, clean technology, nanotechnologies, and/or biomedical technologies [7].

A digital economic wave, such as "The age of data," may be different in nature and frequency compared to anything seen historically, so making direct comparison difficult. It is probably best hallmarked by the characteristics of the new entrants who have comparatively low overheads, limited fixed assets, global reach, and their main asset classed as data. They in turn are driving an economic cycle that has appeared relatively recently, approximately since 2010. This "age of data" also relates to the number of ways we can connect to each other wherever we go and ubiquity of fast data access provided by mobile technology.

It is abundantly clear that the number of companies that have come from nothing to achieve valuations of over US$10 billion since 2010, trading nothing but metadata and data, bare the hallmarks of a new wave of innovation and prosperity around data. This "age of data" also has an important technology differentiator in the form of distributed Hadoop-based computing and is commonly known by its mass media name: "big data."

So strategists need to look at the first of these digital economic cycles: "the age of data" understand its nature and make some predictions about relative frequency of these cycles. Thinking faster rather slower than the Kondratieff cycles may be a starting point and avoiding the strategy that leads to nowhere by following the riderless horse.

How to apply the information tools that come with the "age of data" into strategic planning is a relatively new emerging field, so the following section covers some underpinning concepts and a case study on using "big data" for strategy development.

20.4.2 The Role of Big Data in Strategy Development

The first step on the path to a better understanding of big data analytics is in the asymmetric information advantage it can provide. It is about looking at data in a different way with respect to the understanding of geography and the understanding of self. The power of the people rather than the people in power is a good starting point for strategists. In Figure 20.6 the data space now occupied by people in the social mobile world is represented.

This personal space people occupy digitally has changed with the launch of smartphones, high-speed mobile data networks, and high-quality mobile phone cameras. This social, local, and mobile phenomenon is driving the generation of data at an ever faster pace—hence the term "big data."

These sources of information are genuinely different to what has been seen before because a person's associations with locality are now tracked in multiple ways. It is also tracked down to a hyperlocal space (e.g., a supermarket aisle or on an escalator between floors in a train station). In addition the selfie-like obsession of the millennial child has led to a new cult of self-awareness and personal brand, generating even more data for analysis. Almost all of the social media communications of the millennial child are open to analysis, and they will become the most widely studied generation because of their openness to sharing information on a mass scale.

The implications of these changes for strategists are multifaceted. It changes the speed and frequency they can scan in the information horizon. It also changes our understanding of human behavior. Now you can study the formation of a new meme or put back together the daily reconstruction of events leading to an activity.

The concept of "Geography" in the big data world

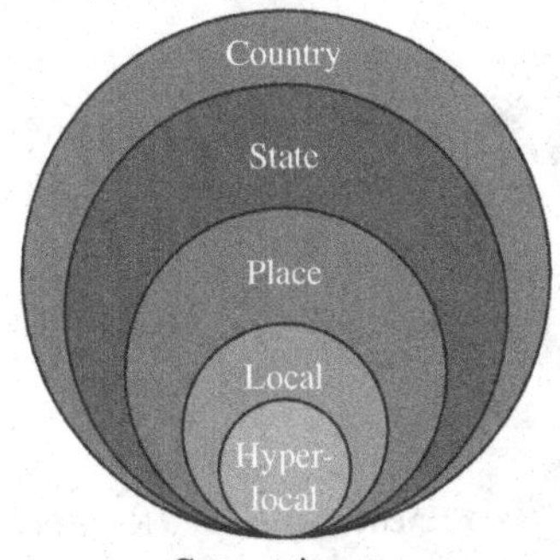

Geo-sentiment

The concept of "self" in the big data world

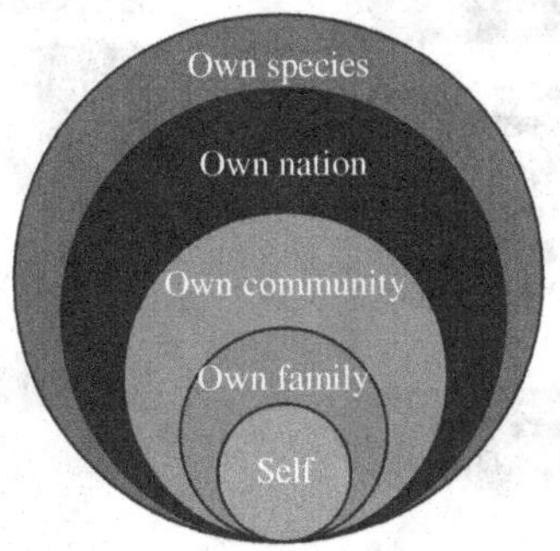

Socio-sentiment

FIGURE 20.6 Social mobile data space. Reproduced with permission of Neil Fraser.

At the core of this "big data" movement, you often see the natural language processing (NLP) techniques in use in conjunction with analytical engines to pick up sentiment, emotional content, instinct, and intent. Other forms of analysis have also emerged as the de facto standards, making "Big Data" analytics different and very applicable for use in strategy such as:

- *Pattern analysis*—Recognizing and matching voice, video, graphics, or other multistructured data types. Could be mining both structured and multistructured data sets.
- *Social network analysis*—analyzing nodes and links between persons. Possibly using call detail records and Web data (Facebook, Twitter, LinkedIn, and more).
- *Sentiment analysis*—Scanning text to reveal meaning as in when someone says, "I'd kill for that job"; do they really mean they would murder someone, or is this just a figure of speech?
- *Path analysis*—What are the most frequent steps, paths, and/or destinations by those predicted to be in danger?
- *Affinity analysis*—If person X is in a dangerous situation, how many others just like him/her are also in a similar predicament?
- *Empathy analysis*—Emotional classification of the users' comments.

A strategist will be able to know more than ever before about the interactions and communications of citizens, customers, leaders, social networks, and machines. Their open feelings about subject, product, service, brand, community are all new and important inputs for strategic planning. The next step in strategic questioning and analysis is whether the level of feeling is instinctive, what sentiment valence it has, what emotional label to attach along with arousal valence, and finally if there is an intent for action by either the observer or the agent.

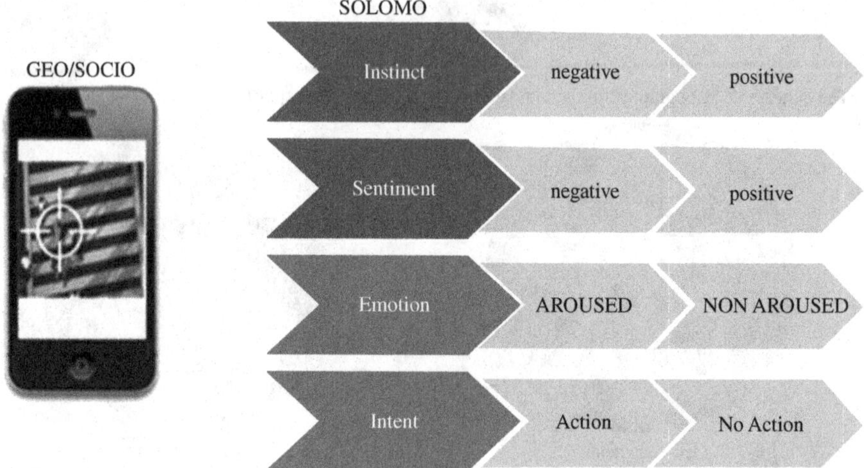

FIGURE 20.7 Social, local, and mobile emotional models. Reproduced with permission of Neil Fraser.

A strategist will also be able to know more than ever before about any machine entity and its communications. Machine entities provide valuable data about location, inventory, and proximity to other entities at ever-increasing rates. This is commonly known as the "Internet of Things." For example, if the scales can talk to the fridge, we may be locked out until we have reached our goal body weight; alternatively this machine collusion might drive you to eat out more.

Bringing together these two very different source data sets (machine generated and human generated) allows us to reconstruct a day in the life of somebody based on machine–entity interaction and/or people interaction. A strategist's questions of interest might be the following:

- What kinds of locations do they visit?
- How many people passed a place?
- Who are the people who pass by and where do they go next?
- What are their needs and wants at a particular time and place

Linking this to behavioral model such as SOLOMO (see Figure 20.7) brings the big data together into an analytical framework for strategists to use.

20.4.3 A Telecommunication Case Study

The moment it all changed for me personally was when mobile-only telecommunication networks began to carry more data than voice traffic. This signified the dawn of the mobile Internet for many mobile telecommunication professionals and had been anticipated since the early 2000s. In Australia this tipping point came in 2007 when mobile-only carriers, like "3" Mobile, celebrated the swing to greater data than voice traffic on the network, but this celebration did not last long. This tipping point toward data-centric mobile networks also led within the first few years to some of the most systemic major network failures as the capacity for carrying data over mobile networks was outstripped by the exponential demand in smartphones. It had taken a long time, approximately 7 years since 3G launched, for mobile telecommunication carriers to get to this point, but suddenly it became a harsh reality in 2007 when the iPhone launch drove mobile Internet adoption to unprecedented levels. Network planning teams had been scanning the horizon and predicting this tipping point event for many years but had clearly sometimes failed to strategically plan for the unexpected outcomes and customer backlash from any degraded network performance.

In response to this wave of change, a vast quantity of analysis was requested by telecommunication executives on network and traffic patterns. The managers and teams running the large-scale strategic data warehouses in the mobile carriers were tasked to capture and store all data traffic alongside all the voice call records. This satisfied the immediate needs for regulatory, capacity planning, and engineering. It also triggered a wave of investment into new data network probes and Hadoop-based disturbed technology to build larger and more linearly scalable computing solutions.

The information retrieved from these new platforms was so rich that it was possible to resolve the Internet better than ever before. It also became obvious that this new mobile Internet era was a holy grail for analytics with widespread applications in politics, policing, business, and the social sciences.

The rapid onset of the mobile Internet age also triggered the first major "big data" event in Australia when the Vodafone network suffered a catastrophic degradation in performance between December 2010 and March 2012. This led to around one million customers leaving the network and a class action law suit. Independent telecommunication analysts could easily observe the impact of the network failures from the outside by following the coverage in mass media, on social media, and on blog sites.

Many independent analytics teams at the time applied the latest tools and techniques on the customer commentary relating to the network performance issues. The open media semistructured data sets could be standardized for text analytics from websites and blogs and directly from social media datapipes. It was possible to detect hot spots of unrest across the country, across the Australian states and cities, and down to localities. Similarly a timeline of how customer's feelings toward different services and the company's response could be tracked. This gave rise to a full-blown case study on geosentiment [8] in Australia.

The concept of a geosentiment data mining tool based on social media, blogs, and websites was relatively new in 2011. The case study on the Australian Vodafone network meltdown involved taking a different analytical approach using NLP. This improved the accuracy and hit rate by a factor of 10 on geolocating hot spots. It was possible to quickly process what service people were complaining about, when they complained, and whether the geosentiment hot spot could be located from any gazetted placename mentioned by the customer in the online forums.

This NLP geosentiment approach is more sophisticated than simply using the pure geocoded data from Facebook or twitter based on the device location, which can be misleading. This case study brought together a number of telecommunication analysts in 2012 and members of the Macquarie University Analytics Department who were interested in developing geospatial intelligence as a strategic toolset to be used competitively. The team went on to win the Ventana Research Leadership Award in 2011 for geospatial business intelligence.

20.4.4 Competitive Geospatial Intelligence from Mobile Use

The problems facing geospatial intelligence from mobile devices we use today are twofold. Firstly, less than 2% of devices have their GPS tracking switched on at any time, so very few comments are able to be traced in real time back to their exact location. The reason for this is that users are more concerned about preserving battery life than knowing their location at all times, so tend to switch off their GPS tracking.

Secondly, even if the GPS tracking is on the location, it might be wrongly associated with what people are messaging about, and the geosentiment map would therefore be misleading. For example, a message sent from your mobile can be about an event in the past (e.g., last weekend) or even another person's experience of a service or product, so using the device location can give false positives—hence the need for NLP to improve the accuracy and precision.

A much better solution to these two problems, which gives a tenfold increase in geosentiment hit rate and accuracy, can be achieved using "big data" tools and techniques. This involves looking at the exact structure of the words and text to locate each tagged sentiment point against any associated gazetted placename. This augmentation of sentiment with location enables you to track opportunity and risk against return for any major event with greater certainty and location accuracy. This is an important input to strategy and operations.

To do this NLP style of analytics requires a "Gold Standard" for an industry (e.g., knowing the vocabulary of insurance or banking or telco). It allows you to see your competitor as well as yourself in a different but very strategic way. The second step is to run this domain-specific NLP engine on all sentence structures. Tagging each of these data types with a unique identifier allowed for the data to be replayed many times for different visualizations including heat maps and time series plots. The NLP engine was tuned to high performance so it could process and tag over 30,000 comments in less than 3 seconds in preparation for visualization. This data integration pipeline of text can pull in any website, document, or text to pinpoint hyperlocal sentiment data contained within the text. It goes deeper into the social media space and allows you to data mine commentary, email, and documents in your own organization or your competitors.

Strategists can use this technique to track product launches, brand activations, media events, network performance, or retail performance. All these events require independent verification tracking and comparison to competitive responses by location. This technique leaves strategists with more opportunity to act independently based on these open data sets and analyze competitor activity on the horizon in a very different way to what was possible before. Locating the place of origin for social sentiment, emotion, and intent helps in assessing risks and opportunities for any previous event and plan for future ones. It gives a unique and competitive insight into strategy, planning, marketing, sales, and enriching brand events.

So in the future, strategists may be asked for competitive geospatial intelligence to inform horizon planning.

20.5 CONCLUSION

Strategists have to look ahead and in the "age of data" they can get closer than ever to real-time information to scan for the emerging trends on the horizon. The tools and techniques described in this chapter have focused largely on semistructured textual data sets which are different but are now reachable within the relatively short time constraints imposed by strategic planning cycles. Capturing information from these new sources and turning it into new understanding is still the main new skill a strategic planning team needs to acquire.

Strategists in the "age of data" could consider these new approaches when building a strategic plan. One option to acquire these skills quickly is to bring in a specialist data scientist who could enable the strategic planning teams to analyze the answers to strategic questions that are asked on a mass scale to wider audiences or detect emerging issues which could lead to strategy failures if not picked up which is like tracking the riderless horse to nowhere.

Also in our case studies it is clear that the relationship of quantitative and qualitative data is vital in linking strategic planning to a deeper and more purposeful engagement to capture an organization's future purpose, vision, and values. What people say is as important as the hard quantitative data in a set of accounts or in benchmarked data. A balanced strategy development process will often use both sets of inputs at a point in time to set objective goals and devise frameworks for future investment. On this basis many organization will then make plans to disproportionally invest resources in line with a portfolio of initiatives to help implement a strategy.

In this chapter we have covered how strategic planning as an activity can be a transformational process for people and organizations. It builds from the communications, engagement, and other more subtle associations seen in complex structures with any organization. Ultimately though it is the people and customers that count the most in the implementation of any strategy and it is there voice that can now be heard in a more up-to-date way using new tools and techniques.

REFERENCES

[1] Montgomery, C., *How Strategists Lead.* McKinsey Quarterly, McKinsey & Company, New York, 2012.

[2] Bryan, L. L., *Just-in-Time Strategy for a Turbulent World.* McKinsey Quarterly, McKinsey & Company, New York, 2002.

[3] Mintzberg, H. and J. B. Quinn, *The Strategy Process: Concepts, Contexts, Cases.* Prentice Hall, Upper Saddle River, NJ, 1996.

[4] Nagji, B. and G. Tuff, Managing Your Innovation Portfolio. Harvard Business Review, Cambridge, May 2012.

[5] Barnett, R., *Imagining the University*. Routledge, New York, 2013.

[6] Adams, C. and S. Mouatt, The information revolution, information systems and 6th Kondratieff Cycle. The 5th Mediterranean Conference on Information Systems, MCIS 2010, Tel-Aviv-Yaffo Academic College, Tel Aviv, Israel, September 12–14, 2010.

[7] Moody, J. B. and B. Nogrady, *The Sixth Wave: How to Succeed in a Resource-Limited World*. Random House Australia Pvt. Ltd., North Sydney, 2010.

[8] http://www.geosentiment.com (accessed December 13, 2015).

FURTHER READING

Christensen, C. M. and M. E. Raynor, *The Innovator's Solution: Creating and Sustaining Successful Growth*. Harvard Business School, Cambridge, 2003.

Krishnan K., *Data Warehousing in the Age of Big Data*. Elsevier, Waltham, 2013.

Stubbs, E., *The Value of Business Analytics: Identifying the Path to Profitability Hardcover*. John Wiley & Sons, Inc./SAS Business Series, Hoboken, 2011.

Stubbs, E., *Big Data, Big Innovation: Enabling Competitive Differentiation through Business Analytics*. Wiley and SAS Business Series, Hoboken, 2014.

21

RISK MODELING AND DATA SCIENCE

JOSHUA FRANK
Intuit Inc., Woodland Hills, CA, USA

21.1 INTRODUCTION

Risk predictive modeling is an important and growing branch of applied data science. This chapter summarizes the state of risk modeling and discusses eight key lessons learned that are applicable to risk modeling as well as to other data science endeavors. The emphasis is on building a modeling system that stresses diversity and flexibility with a "modeling ecosystem" approach. The subject matter focus is fraud and financial risk modeling for payments and credit risk applications.

21.2 WHAT IS RISK MODELING

There are many forms of business risk and therefore many definitions of risk modeling. Some types of risks are common to any large enterprise. These can include first-party risks of hazard such as natural disasters that can damage plants and equipment, second-party risks of hazard such as worker injuries, third-party hazards such as liability from defective products, financial risks from sources such as foreign exchange rates and liquidity, operational risks such as the risk of labor relations issues, strategic risks from competitors and market demand, and strategic risks from regulatory/political issues as well as reputational risk [1]. There are also risks specific to the industry a company is in. For example, insurance companies have claims exposure. Lenders have risk of borrower default as well as interest rate risk for long-term fixed rate lending. Retail merchants and payments industry companies have risk for fraud. Investment companies must contend with a number of financial risks from their portfolio.

Internet of Things and Data Analytics Handbook, First Edition. Edited by Hwaiyu Geng.

Companion website: www.wiley.com/go/Geng/iot_data_analytics_handbook/

Risk modeling as a data science discipline is very well developed in a number of these areas. This is particularly true of the industry-specific risks such as lending risk modeling, payments fraud modeling, insurance risk modeling, and investment banking risk modeling. Many types of enterprise risk such as liability, regulatory, and reputational risk do not lend themselves as easily to quantitative analysis. However, this is not true of all types of general enterprise risk. For example, a large range of industry types now retain regulated personal data on consumers, and breach of these databases often results in severe costs for an enterprise. Advanced data science techniques are increasingly employed in preventing and detecting these breaches with a broad range of service providers available in the advanced security field.

21.2.1 The History of Risk Modeling

Risk modeling has used predictive modeling and large datasets since long before there was a term "data science." Previously the activities that are now often referred to as "data science" were referred to as statistical/predictive modeling, and the range of techniques employed in practice were more limited to traditional statistical techniques with some emphasis on the regression family of methods for predictive models (e.g., linear regression, nonlinear regression, logistic regression, etc.). Other activities were also similar to what is now included to in data science. Considerable time was spent collecting and cleansing data from a variety of sources and preparing variables (in the data science world commonly called "feature engineering"). However, the shift to "data science" terminology is appropriate since techniques have evolved. There are many more possibilities for types of data, more methods for manipulating data, and newer machine learning algorithms that go beyond traditional statistical methods.

While banks have long recognized the importance of credit risk modeling, there were several historical developments that pushed this technology further into the forefront. One was the development of standard credit risk models that are applicable across organization such as the FICO score developed by what was then Fair Isaacs. In the 1990s the credit card industry was further revolutionized by the growth to prominence of nontraditional financial organization that specialized in credit card issuance rather than being diversified across all areas of banking. These younger organizations tended to view data and analysis as a strategic advantage over the big banks and included MBNA, Capital One, Advanta, First USA, and Providian. The focus on data and modeling was both on the marketing/targeting side of the business and in the risk side. We are perhaps in another period of risk predictive modeling revolution driven once again by nontraditional financial institutions coming mostly from technology companies involved both in lending and payments processing.

21.3 THE ROLE OF DATA SCIENCE IN RISK MANAGEMENT

Limiting our scope now to lending and payments financial and fraud risk, data science/ statistical modeling plays a vital role in risk management. Small- and medium-sized loans to consumers are ubiquitous. Unlike a mortgage where the size of the loan and associated fees justify having a human underwriter spend significant time reviewing documents and making a personal decision regarding risk (putting aside the question for now of whether this human review leads to better decisions), credit cards and similar lending products are offered and underwritten on a massive scale, and data-based algorithmic decisions offer the most cost-effective method for risk decisions in the majority of cases. This applies even more in payments processing for financial institutions supporting both sides of the transaction (i.e., the card/payment instrument issuer and the merchant's payments processor) as well as for the payments processors. A single company may need to make millions of decisions on transactions in a single day (in some cases in near real time). The large quantity of decisions required in such a short period of time makes human decisions impossible and data science a necessary tool to prevent fraud and minimize other risks.

Aside from being necessary and cost efficient, data science tools are vital to risk management because they are also extremely effective. Fraud and risk are also very important to data science in that they are one of the more important in real-world usage and they have a longer development history than most other areas of predictive modeling.

21.4 HOW TO PREPARE AND VALIDATE RISK MODEL

21.4.1 Targeting

While unsupervised learning can be useful for some fraud problems (in particular anomaly detection), in most risk/fraud predictive modeling use cases, supervised learning will be the most useful type of modeling. Therefore, one of the first steps is to define the target variable. This is a common task in many data science projects, but there are some special issues that need attention for risk/fraud modeling.

One issue is time. Time complicates many steps of risk modeling, and targeting is one of these areas. First, whether the records being analyzed are loans, payments going through a network, or purchases at a retailer, the outcome will not be known immediately. For example, in the case of a credit card payment, it can take 90 days or longer in some cases to know that there is a "chargeback" (a dispute of the transaction by the cardholder). If the data scientist chooses to target actual losses rather than chargebacks, the process can be considerably longer. The amount of time needed to age a training/test observation is typically even longer for loans. A modeler may be concerned about the risk of mortgage or credit card defaults that occur years after the initial decision. For practical purposes, the data scientist analyzing a long-term loan will need to cut short the observation period and rely on only partially aged data. This is for three reasons: (1) while a default 10 years after being approved for a credit card has some relevance, it is far less relevant in terms of discounted value of cash flow than a default in the first year (in fact, a credit card account with a default in year 10 may even have a positive lifetime net present value in some credit card scenarios), (2) the predictive power of any model may weaken in the out years enough that including these late-year defaults as "bads" (using a 1/0 binary target) may add more noise to the model than helpful information, (3) the industries that would utilize these models tend to change rapidly creating a trade-off between full information and timely information—in most cases there will come a point where it is worth using more recent data that reflects current trends even if it is not fully aged. However, at the same time data scientists must remain very aware of the risks in using only relatively short-term data. One point of failure in the financial crisis was reliance on housing/mortgage market data that often only included a strong economic period with rapidly rising housing prices.

There is another aspect of time that often must be considered in targeting. Some analysis tasks will involve one-time decision points where the information set is collected once and remains static. For example, a decision to approve a credit card or mortgage application occurs once based on a fairly well-defined information set. However some decisions are made on a continual basis. For example, in addition to deciding whether to extend a line of credit, a credit card issuer decides whether to approve or decline each transaction made by the cardholder. At any point in time, the credit card information can be stolen by somebody submitting fraudulent transactions. The same applies for a bank processing payments on behalf of a merchant. The merchant's credentials could be stolen at any time by somebody committing fraud using the merchant's account information. Time complicates modeling of these ongoing decisions, and if time is ignored the model will have a bias. The solution to this bias is to consider the data as of a specific point in time. Inputs and outcomes are segregated based on this "snapshot" date/time. All inputs to the model must occur before the snapshot time, and all outcomes are measured after that snapshot time. The purpose is to look at the data as if the decision maker is standing at that particular point in time. The only input data known is what occurred before this point—this is what is used to predict outcomes. The only outcomes relevant to predict are those that happen after that point. So, for example, if an account is closed due to fraud the day before the snapshot date (Figure 21.1), it should be removed from the analysis because the outcome is already known as of the snapshot date.

A second issue that may come up with risk targeting is how/what to target. One question is whether to use a binary target (i.e., accounts that "go bad" such as fraud or default vs. those that do not) or another type of target. Binary targets are often the best in practice, but at times other targets are useful. For example, one may want to target the dollar losses experienced either due to fraud or defaults. Targeting dollar losses has some important uses, but it can create distributional problems for some types of modeling techniques if the data is skewed toward nonlosses as is often the case. For example, 99% of the records may be associated with a "good" outcome, 1% may be associated with a "bad" outcome, and the range of dollars lost in that 1% could be very broad. With this type of model target, a few large losses from outliers that may be unlikely to recur could distort the model results.

Another question that may come up related to what to target is how broadly to define your target. For example, a merchant processing bank may suffer losses both from merchants going out of business with orders unfilled and from fraud. Fraudulent merchants and merchants who go out of business show very different patterns. It is probably best to target these separately. One could go even deeper than this. For

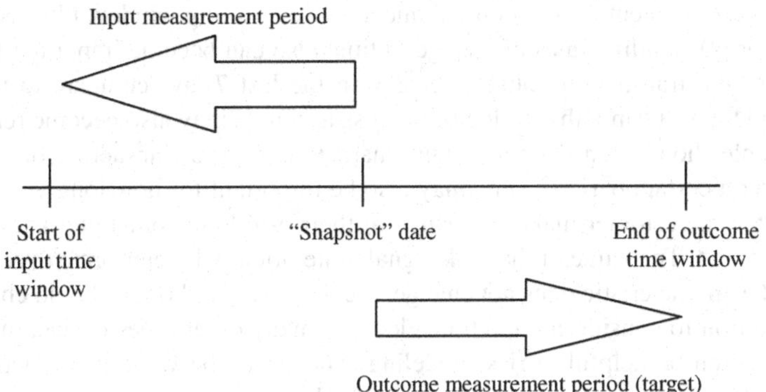

FIGURE 21.1 Using snapshot dates to predict fraud events. Copyright Joshua Frank.

example, there are many types of fraud in this industry. Some are up-front fraud at time of application (e.g., stealing a merchant's identity and applying for an account with that merchant information), and some are fraud on an existing account (e.g., stealing an existing merchant's credentials to take over the account). These can have very different characteristics. However, whether one wants to create separate models for each type of fraud depends on many things including how much data is available for each target category, the resources available to do the modeling, and how accurately can the modeler distinguish between the different categories of fraud used in the target.

Another relevant issue in targeting is dealing with targets that are not clearly defined in the data. While analyzing loan defaults in isolating the target is typically clear-cut (although aging can be an issue and sometimes a proxy might be used rather than waiting for definitive default data), fraud may not be. Also if the business use case requires multiple targets, distinguishing them may not be clear-cut. As an example, consider a financial institution processing payments on behalf of merchants. For various reasons, the merchant may voluntarily or involuntarily close the account before a final determination of fraud is made. So a final definitive determination of whether the merchant is fraudulent may never be made. In addition, if excessive chargebacks or a loss occurs and the merchant is closed for that reason, it may not always be clear whether this is due to merchant-initiated fraud or for other business reasons (financial risk). Therefore if multiple targets are used, there often will be ambiguity in target definitions. This can be addressed with simple rules of thumb to define the target, more complex algorithms, postmortem models where less ambiguous cases are used as training data to assign outcomes to more ambiguous cases. "Fuzzy" targets can also be used for ambiguous cases rather than discrete binary targets.

21.4.2 Feature Engineering

Time once again becomes important for feature engineering. If risk is being modeled in a dynamic environment, time will become a necessary element in the definition of some features. For example, if transaction volume is a relevant feature, then transaction volume over what time frame becomes an important question. The modeler may be interested in recent transaction volume and use a rolling window like last 3 days, 30 days, or 60 months. Rates of change in time also can become important features. How does the transaction volume per day in the last 7 days compare to the daily volume in the last 6 months? Velocities across accounts may also become relevant—for example, how often does a certain characteristic in a transaction or customer occur over a certain period? Time may also be important for how long you utilize a feature. Say a customer makes a change in their profile to something known to be risky for fraud. The value of that risk signal quite often will deteriorate over time, so even if the characteristic does not change, the way it is used often should change.

In addition to considering the time element, many other types of data manipulation will often be helpful in risk modeling. There may be value in extracting time elements (such as time of day). There may be benefit in extracting transaction elements and reaggregating in various ways. Elements of free text fields may be used by searching for specific patterns or characteristics of interest. Sometimes there may be value in comparing how elements of one text field relate to elements of another text field. Location information may also be most valuable when reaggregated or compared to other location information in various ways.

21.4.3 Model Selection

In selecting models for risk prediction, it is important to keep in mind some common characteristics of the modeling task:

- The target is typically a rare event.
- Depending on the use case, some models may need to be executed in production in near real time to assess the risk of individual events as they occur, while others can be executed in production as a batch job with timing being less critical.
- For many risk tasks, there are a large number of relevant variables.
- While there may be a moderately high number of customer records (such as millions), it may be necessary to go to subcustomer-level data (such as transactions), and there may be thousands of transactions per customer greatly increasing the number of records (quite possibly billions or more).
- The relationship between a feature and a target event's likelihood is typically nonlinear. For example, the chance of credit default may go up with utilization at time of application (i.e., the percent of credit line amounts with balances); however there may be less impact on risk in moving from 10 to 40% as there is in moving from 92 to 96%.
- There are often large interaction effects. For example, Feature A might be risky at high values only when binary Feature B is present.
- Relationships may change quickly and frequent retraining may be necessary. This is especially true for fraud.

Logistic regressions and Naive Bayesian Classifiers are common traditional statistical methods used. They have some advantages such as often being faster to train, especially with large datasets, and they are able to include large numbers of features. However, they are not very good at capturing complicated interactions between features, especially when the nature of those interactions may not be completely known up-front.

Decision trees are better with complex interactions but will not take advantage of the information in most variables when there are a large number of factors that are important to the risk model. Random forests have advantages over single trees, but they still suffer to a lesser extent from some of the same deficiencies. Some fraud detection models in production use neural nets (including deep learning), and this is particularly common in transaction monitoring for fraud by issuer banks (i.e., banks representing the cardholders); however my personal experience with them is that they typically do not outperform other modeling methods.

21.4.4 An Ecosystem Approach to Modeling

To retain many of the advantages of the various modeling methods while minimizing the disadvantages, my experience is that what I call an "ecosystem" approach works very well. An ecosystem approach could be a type of ensemble model or multilevel model. The difference between the ecosystem approach and a typical model ensemble is that the ecosystem approach stresses diversity in methodology. In addition, rather than randomly varying features or data inputs, the ecosystem approach intelligently combines features into subproblems. Each subproblem is made up of features that are thought to be likely to have some interaction. Each subproblem is solved with the optimal model (or set of models) for that particular subproblem (see Figure 21.2).

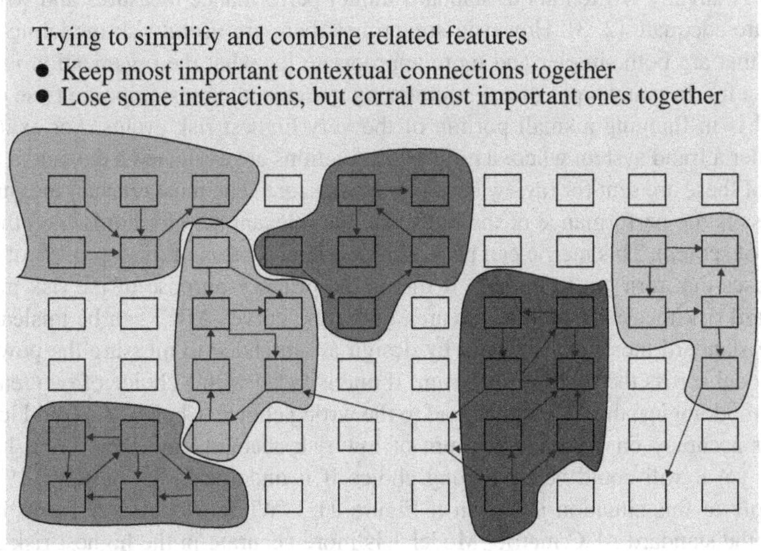

FIGURE 21.2 Simplifying feature space in submodels. Copyright Joshua Frank.

With the ecosystem approach, the top-level model is intended primarily as a method to aggregate all of the subcomponents, and it should be flexible and transparent. Therefore, to the extent that the data scientists want to add complexity to the modeling ecosystem, they should do so in the subproblems rather than at the top-level model which should be kept as simple and clean as possible.

There are several advantages to this ecosystem approach. First, dividing the modeling space into subproblems allows one to utilize the modeling methodology that is best suited for the particular set of features and concept in question. Second, to the extent that the modeler can very broadly anticipate the set of likely interactions, those interactions can be retained and will often be tuned better in the submodel because it is not diluted by the complexity of training the entire feature space at once. Perhaps the most important advantages are related to flexibility as well as functioning in a production environment. The fraud environment in particular can change at a very fast pace. Often, it will be useful to rapidly add a new feature or set of features that are found to be useful for the latest fraud attack. The ecosystem design allows submodels to be added in fairly flexible manner. This is also very helpful in general for real-world resource constraints and operational limitations. Unlike an artificial model competition, where the ultimate goal is single end-point optimal model, for most business environments models are always evolving works in progress. Resources are constrained, data is changing, and often something is needed now even if it is not yet at its end state (if there ever really is an end state). For this type of real-world scenario, a modeling structure that can evolve and grow with maximum flexibility is perhaps more important than a model that performs best on one particular day.

21.4.5 Measuring Performance

Much is already written about standard model performance measures and whether they are adequate [2, 3]. However, sometimes there are useful risk modeling measures that are both simpler and more appropriate for what the organization is truly solving for. In many types of risk monitoring systems, the most important use of the model is in flagging a small portion of the very highest risk events. For example, consider a fraud system where a million transactions are evaluated a day and a thousand of these are sent for review based on a risk score. The most relevant measure of success is the performance of the model on the relevant set of records (i.e., the top 0.1% of scores). This metric can be as simple as the portion of the top 0.1% of scoring cases that turn out to be bad. If this is the primary purpose of the risk model, standard risk measures such as the area under the curve (AUC) can be misleading. These standard measures typically by design are intended to measure the power of the model across the full risk spectrum. If one is faced with a choice of two features or methodologies, the AUC could lead to the wrong choice. Choice “A” could lead to higher accuracy on a broad spectrum of low-risk accounts and therefore a higher AUC, yet it still could be the wrong choice if it underperforms on the 0.1%. An example of this situation is shown if Figure 21.3. While Model 2 performs better using the standard AUC metric, Model 1 is more accurate in the highest risk score range. If these were fraud models and the ultimate purpose of the model when

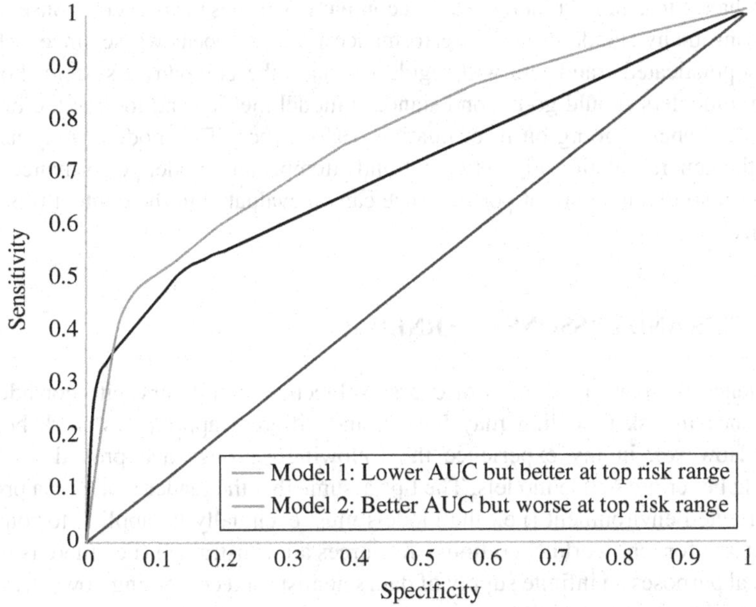

FIGURE 21.3 Choosing metrics for model accuracy. Copyright Joshua Frank.

operationalized is to be used as a criteria to alert on the top 1% riskiest events, Model 1 (with the lower AUC) would perform 2.5 times as well as Model 2 in catching the targeted events.

Though the performance numbers are not real here, this is not just a theoretical example. I have seen this dilemma occur more than once in practice. If repeated choices are made on individual features that maximize AUC, for example, rather than the performance of the model on the most risky accounts, the aggregate impact of these small choices can be significant and lead to a model that underperforms its potential for the primary production purpose.

In evaluating performance it is important to keep in mind that standard performance measures can often be misleading. In particular for fraud, performance on many measures often looks best when fraud is at its worst. Fraud attempts often come in waves. Many companies have a baseline fraud attempt level that is always present, but from time to time, a fraud ring may launch an organized and aggressive attack. Often there is a brute-force element to these concentrated attacks, where the goal is to overwhelm the defenses of the company and the fraud ring is betting that at least some of these attempts will be successful. Model performance metrics will often appear particularly strong when the fraud attempts are at their worst since there may be many cases of low-hanging fruit/obvious fraud that will boost performance statistics. But this should not be interpreted as a sign that the fraud systems are performing particularly well; instead it is just a consequence of a system facing heavy levels of fraud attempts.

By the same token, a company's fraud system that is performing extremely well in the long term may deter most fraud of moderate sophistication since they can easily change to a more vulnerable target company. This has the perverse consequence of making many standard model performance metrics appear worse since only the most sophisticated fraudsters will regularly attack the company's system. For this reason, modelers should go beyond standard model metrics and look at the broader context of what is going on in evaluating performance. The modeler may need to study the general relationship between fraud attempts and model performance in the long term so that the current performance can be evaluated in the context of similar historical periods.

21.5 TIPS AND LESSONS LEARNED

Experience on how to model is of course subjective, and others with considerable experience in risk modeling may have found different approaches work best for them. However, in my experience the following lessons have proved useful in building better predictive models. The tips assume that the reader works in a production-oriented environment (i.e., the models must eventually be applied to routinely predict risk for real-world operations), resources are constrained (i.e., there is not for practical purposes an infinite supply of data scientists and computing power), and the environment changes rapidly.

21.5.1 Everything Is Not a Nail

This comes from *the expression* "if all you have is a hammer, everything looks like a nail." There are many data scientists who are particularly knowledgeable in one modeling tool. They often view every problem they see through the lens of that tool and have a bias toward believing it is the best solution. Likewise, many fraud and risk shops develop considerable expertise in a particular methodology. This can lead to them inaccurately viewing that methodology as best for any given situation. Furthermore, even if a modeling team puts all risk data through various methodologies and has thoroughly and objectively determined that one particular technique is best when all data is considered at once in a single model, that does not mean that all of the pieces of the model work best using that technique. As discussed previously, a large problem may benefit from being broken down into smaller subproblems, and those subproblems likely will benefit most from using different methodologies.

The key point here is that every problem is not a nail, and modelers need to be open to trying different methodologies in different situations. Furthermore, even if using a different technique does not boost initial performance, it increases model diversity which can itself be beneficial as described in the next tip.

21.5.2 Diversity Matters

Diversity can be interpreted as applying at many different levels. For example, it can mean diversity in modeling methodologies. It can mean diversity in feature types. It can mean diversity in time frames, character string length, bins, or other specific parameters used to engineer features. But the same general rule applies for all of these levels: diversity improves model success.

Diversity has been known to improve the immediate performance of a model ensemble. A well-known example is the NetFlix prize where the winning model's performance improved substantially when different teams decided to combine their different methods into a diverse ensemble [4]. But even beyond improving initial performance of the model, there are other (perhaps more important) benefits gained from diversity. Diversity makes models more robust. They are more resilient in the face of environmental change. Often data science teams may find it difficult to keep up with the change in products, technology, and environmental context where that govern their predictions. Diverse models tend to stay relevant longer in these changing environments. This is somewhat offset by the complexity cost this diversity comes with. More complex models may be harder and more labor intensive to change. However, I believe the benefit of that diversity usually outweighs the complexity cost. In the case of fraud, there is an additional benefit. Fraudsters are actively trying to reverse engineer the company's fraud detection system to avoid detection. Diversity in a predictive modeling system makes that system less transparent to outsiders. This can be very important for long-term performance.

A real example from the payments industry of using diverse approaches that turned out to be very helpful in detecting fraud is the addition of a short-term fraud pattern model. Fraud can come from many sources and can take many forms, but often there will be large-scale fraud rings that will make intermittent but aggressive waves into fraud attempts on a system. We recognized that when we model fraud, there are really two distinct things we are trying to find. The first are general fraud "tendencies." These tendencies will remain true over time (at least using medium time horizons), and they are true across different fraud rings. The second thing is distinct "patterns" associated with a particular ring. Patterns remain true only as long as that particular ring is trying to enter your system. Trends and patterns tend to behave differently (see Table 21.1).

What we found is that rather than trying to fit these two very different things into a single model, we are better off modeling them separately. We have found that a set of decision trees and (to a lesser extent) neural nets worked best on the fraud ring patterns since nonlinear interaction effects were very important, while a variety of methods worked best for modeling general fraud tendencies. More important than method was the time frame and general process for modeling. For general fraud tendencies, we used a process that involved extensive hands-on analysis of individual features, and we build the feature library over time. However, for fraud ring patterns, we found that timeliness is more important than refining each feature; therefore we used a process that involved rebuilding the set of decision trees daily. These new models would be released into production daily without the intervention or supervision of a data scientist. This process worked very well to catch quickly evolving fraud patterns.

From the perspective of "diversity," the key point is that in detecting fraud, approaching the general problem from multiple angles using different methodologies and processes worked better than any single methodology.

TABLE 21.1 Distinguishing "Patterns" and "Tendencies" in Fraud

	Tendencies	Patterns
Overview	A variable correlated with fraud in general regardless of what specific fraud ring is hitting us	Factors associated with a specific fraud ring
Timing	Evolves w/products and technology but tends to be stable in the medium-term	Tend to be short-term
Identifiers tends to be…	Additive across variables	Not additive/interaction effects dominate
	Directionally consistent in a single variable	Tends to be a "sweet spot" within a variable
Examples	Free email accounts riskier Location mismatches riskier	Doctors with g-mail accounts from San Diego with specific browser settings all at once
Ideal Solution	Traditional Statistical Model or ML model Longer training time horizon	Pattern-focused algorithm like Decision Tree/Forest or Neural Net, with rapid retraining and a short-term time horizon

21.5.3 Flexibility Matters

The ability to quickly change model characteristics in the face of new fraud or general risk patterns is vital. In fact, it is often worthwhile to settle for slightly reduced model performance today in favor of a system built flexibly to allow quick changes tomorrow. Flexibility is important to be able to change as the environment changes, but it is also important for modeling teams that are learning and building as they go, even if the environment does not change. In other words, most real-world data science teams will be resource constrained and will be under pressure to deliver a product before they can add every feature they would like. So when a risk model is first built, there may be many promising ideas that are not yet tested or added. A flexible structure allows those features to be added easily later, which will shorten the model enhancement cycle and therefore improve long-term performance.

Figure 21.4 shows the general structure of a fraud modeling ecosystem. It involves a multilevel model which could be considered a special case of an ensemble. The aggregator (top-level) model should be one that adds many variables well and can be adjusted with maximum flexibility with a variety of lower level models feeding into the top level. One option for an aggregator model that has worked well in the past for me is a Naive Bayesian Classifier. Its advantages include flexibility and its transparency. For example, this type of model allowed us to keep a modular design when building out the system. We could add new pieces to the model while leaving the existing pieces intact. We did not need to retrain the entire model upon adding a new component nor even use a consistent testing/training period for all submodels since all components end up scaled as relative risk measures. The design of keeping the

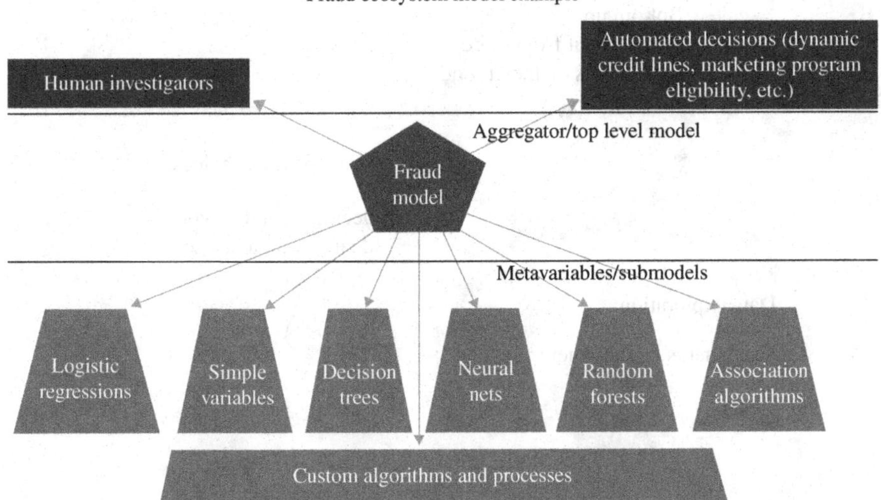

FIGURE 21.4 Fraud ecosystem model example. Copyright Joshua Frank.

complexity underneath and using a Bayesian Classifier on top allowed the system to remain flexible yet benefit from the advantages of the various modeling methods used underneath.

The modeling ecosystem also allowed us to have a highly diverse modeling system while retaining flexibility. The diversity is represented by the various arrows feeding into the naive Bayesian classifier. Besides standard modeling algorithms, past systems have also included some customized processes and algorithms.

21.5.4 Triangulate on Truth

Good insight can come by approaching the data from a number of directions (Figure 21.5).

Perhaps the first place to go when building a new risk model should be the staff with "boots on the ground." These are risk case investigators, manual underwriters, collection agents, and other staff who take a detailed look at individual risk cases and deal directly with the customer. These employees have a wealth of knowledge regarding signals, patterns, and trends that they see. They are a great place to start when trying to engineer features to assess risk. They should also not be forgotten as the model gets more mature and data scientists develop domain expertise. They are always a useful source for new insight.

Often we do not take full advantage of the information already available. Sometimes simple data exploration can yield great benefits. For example, there may be rich tables of transaction data, and exploring these tables can lead to new ideas and directions for risk signals to add. It is also useful to occasionally revisit these tables since exploring them with greater experience can lead to new ideas. One thing that is

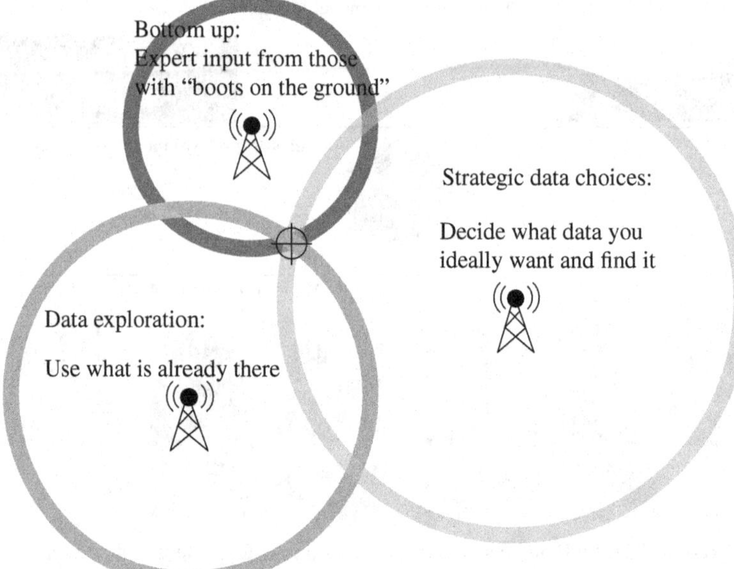

FIGURE 21.5 Triangulating on the right features. Copyright Joshua Frank.

not recommended is to rely on machine-based pattern/correlation searching alone to mine the data. If data possibilities are very large, even with reasonable thresholds and training/test protocols, spurious relationships will frequently come up due to over-mining. It is better to start with relationships that seem reasonable on a theoretical level and test those hypotheses than to brute-force mine the data for relationships that may not be real and even if real may not be sustained.

The third source of data that is important is strategic data choices from the data science team. These do not rely on what is already available but instead are intentional choices to seek out certain data deemed to be reasonably attainable and important. These strategic data choices can come from spontaneous ideas, but a good source for ideas is to do a detailed postmortem analysis on the "bads" perhaps starting with the worst cases (e.g., biggest losses). When looking at these cases, the question the modeler should ask themselves is whether there were any signals that could have been used to better identify the risk.

21.5.5 Be Wary of Black Boxes

Some modeling methodologies are transparent regarding what factors led to a high risk estimation while others are "black boxes." For example, the nature of neural nets and random forests make it difficult to gain intuition into why a particular case was deemed risky. While the accuracy gained may make it worthwhile to sometimes use these black box techniques, there are important downsides to this. Perhaps the most obvious one is that users (such as risk agents) do not understand what led to a high score which may make the case more difficult to investigate. However, even beyond that, black boxes can be hazardous because the data scientist is blind to potential problems with the model and will have more difficulty troubleshooting model problems that may develop over time. In my experience, it is not uncommon at all for the first run of a transparent model (such as a decision tree) to lead to results that reveal some kind of problem. For example, there may be a problem with the data for certain values or situations that was not obvious in advance. There may be a timing issue where it appears that the model has found an accurate early warning for risk, but it turns out that the information used is really only available after the risk is known (e.g., risk agents may take certain actions on fraudulent accounts before closing them and this fact may not be known to the modeler in advance), but those actions rather than predicting risk are really a sign that the investigator has already identified risk. There may be indirect but potentially illegal discrimination taking place unintentionally in the model. There may be product or system changes that the modeler should be aware of that get revealed by the output. All of this valuable insight depends on having visibility into what the model chooses to identify as risky. Without it, there is much more risk of creating a model that appears to work well but fails to perform some time after being put into production.

21.5.6 Automate What You Can

Risk modeling teams will often be faced with more things that would be helpful to do than they can possibly accomplish. Therefore automate anything that can reasonably be automated. Any task that is repeatedly done should be generalized into a procedure that can encompass all variations of that activity. For example, checking certain model or variable-level statistics may be done repeatedly and should be automated. Data health checks can be automated. Recalibration of certain types of model features can also be automated.

An example of a model feature general typology that may be useful to automate is calibration of categorical data where there are a large number of categories and data for each category may be sparse. For example, there may be categories for business types (if one is dealing with lending to or processing payments for businesses), sales channels, products, device categories, locations, transaction types, and so on. In some cases, there may be a single observation in a category, while another category may have thousands of data points. Also, the category universe may change over time (e.g., new sales channels may be added without marketing informing risk). The solution to this that we used was to create an automated process based on Bayesian probability to estimate the expected bad probability after adjusting for sample size for any category. This was built into a larger automated process to automatically recalculate short- and long-term sample size-adjusted risk for each category within each variable concept every day. This allowed us to have updated category risks without manual intervention.

21.5.7 Improving Stability

Risk models will often have some degree of instability in risk estimates that is artificially created by the model structure itself rather than caused by the true underlying risk. One common cause of this is the use of some kind of "windows" to define features.

For example, credit risk models will commonly score a risk event with a certain score as long as it falls within a particular time range (e.g., how many delinquencies exist within the last 6 months, or was there a foreclosure in the last 24 months), or a transaction velocity may be calculated over a specific period. The problem with these time windows is that they create a steep cliff in risk score (going up or down) at a specific point in time that is artificial and arbitrary. In reality the importance of a risk event typically changes gradually over time. Not only do these windows reduce accuracy to some extent, but they can also lead score users to lose confidence in the models if they see sudden shifts in the score that seem to be based on things that are inconsistent with common sense.

The issue can be addressed by shifting from using windows to a more nuanced method of handling change over time such as decay curves or fuzzy logic. This can complicate the model since it adds an extra dimension to many calculations, so it may turn out to be worthwhile to do this for only important drivers of model change.

21.5.8 Humans + Machines Work Best Together

While machine learning algorithms can perform repetitive calculations and many other tasks on a scale and with an accuracy beyond human capacity, there are many things humans still can do far better than machines. For example, human risk investigators can find connections between names of things that are beyond the capacity of state-of-the-art text analytics algorithms. Therefore a good risk system does not just use humans and does not just use algorithms but a combination of both that takes advantage of the best features of each.

It is also useful to try to find ways to take the human insight and feed it back into the machine learning risk systems. For example, we created a variable-level feedback mechanism that allows feature weights to be adjusted based on investigator insight on a particular case.

21.6 FUTURE TRENDS AND CONCLUSION

There are a number of trends that will impact how predictive risk modeling is done in the future. One is the well-known trend toward "big data." The sheer size of the growing "big data" available presents processing challenges to many organizations, and this is true of risk modeling along with other areas of data science. Fortunately, hardware and software innovations continue that attempt to allow organizations to keep pace with this growth. However it has also been pointed out that the challenge of "big data" is not just about size but also about velocity and variety [5]. The human resource challenges of keeping up with the velocity, and variety of data available are probably greater than the physical processing challenges the data represents. For this reason, organizations that find ways to generate insight more efficiently while being careful not to overmine will probably handle these challenges best.

In predictive risk modeling, there is also a move toward real-time and near-real-time decisions. This presents a challenge to many production systems. While timeliness is a very important goal, data science teams should be careful to not be overzealous in following this trend. Sometimes the cost of delaying certain decisions is not very high, while the benefit of waiting for additional information to become available may be very high in improving risk accuracy.

Machine learning algorithms continue to change over time. Deep learning techniques are a notable example of a method that is showing great promise. Aside from any accuracy gains, one advantage of deep learning is that it allows data scientists to rely on the algorithm to do more of the abstraction reducing the need for detailed feature engineering addressing part of the big data challenge. However, there are some important caveats regarding diversity, flexibility, and black boxes already noted that caution against overreliance on this method.

Credit data continues to evolve. Increasingly there are alternatives to traditional credit bureaus for assessing risk of customers with thin bureau data. These alternative credit risk data sources will become increasingly important in the future.

REFERENCES

[1] Enterprise Risk Management Committee (May 2003). "Overview of Enterprise Risk Management" (PDF). Casualty Actuarial Society. pp. 9–10. http://www.casact.org/area/erm/overview.pdf (Accessed August 19, 2016).

[2] Yun-Chun Wu and Wen-Chung Lee, "Alternative Performance Measures for Prediction Models", *PloS One* 2014, Vol 9, Issue 3, p. e91249.

[3] David J. Hand, "Measuring Classifier Performance: A Coherent Alternative to the Area under the ROC Curve", *Machine Learning* 2009, Vol 77, pp. 103–123.

[4] Eliot van Buskirk (September 22, 2009). "How the Netflix Prize Was Won". Wired Magazine. http://www.wired.com/2009/09/how-the-netflix-prize-was-won/ (Accessed August 19, 2016).

[5] "Challenges and Opportunities with Big Data: A Community White Paper Developed by Leading Researchers across the United States". http://www.purdue.edu/discoverypark/cyber/assets/pdfs/BigDataWhitePaper.pdf (Accessed August 19, 2016).

22

HADOOP TECHNOLOGY

SCOTT SHAW

Hortonworks, Inc., Santa Clara, CA, USA

22.1 INTRODUCTION

We are in the midst of a transformation in how we collect, process, analyze, and make use of data. Data is quickly becoming the new standard defining corporate success and competitive differentiation. Those companies who fail to make use of all their data risk falling behind to their competitors who gain from their data key metrics on customers and transactions. Furthermore, the new business models include data aggregators whose business exists because they have been able to collect data from a potentially unlimited number of sources and analyze the full spectrum of the data set to derive never-before-seen insights.

Of course, collecting, storing, and processing data comes at a steep cost as well as deep complexity. Since the advent of what we consider the modern computer, the means of storing data relationally or in data warehouses (essentially relational databases logically restructured to provide faster query access for unique analytic workloads) was the primary means to serve data to business analysts. What it took to begin shifting us away from the traditional approach was really big problems involving really big data of which only Internet could cause. Companies such as Google and Yahoo faced a change or die event, forcing them to completely rethink how they store and process data. Hadoop is the outcome of this event and, though it was built for a specific use case, it has emerged as a seminal technology driving similar changes in thousands of organizations.

E.F. Codd first presented his paper on relational database structure in 1970 [1] making the practical conceptualization, design, and implementation of relational systems nearly half a century old. Google came out with their GFS white paper in 2003 [2] making the public concept of Hadoop just over a decade old. Hadoop as a potentially

Internet of Things and Data Analytics Handbook, First Edition. Edited by Hwaiyu Geng.

Companion website: www.wiley.com/go/Geng/iot_data_analytics_handbook/

viable product began around 2008. This makes Hadoop as an Enterprise solution, outside of the Internet search companies, a meager 7 years old. The Hadoop journey is only beginning. The rate of adoption as well as innovation is mind boggling. Some companies struggle with understanding the landscape, while others have embraced it and are clearing new ground with incredible success. This chapter hopes to provide the reader with some clarity around core Hadoop technologies. Besides explaining the technology, special emphasis will be placed on explaining the why behind the technology. What is the value? Why build new skill sets or disrupt traditional processes to implement such an immature architecture. In the end it is value that drives adoption—not simply technology. This is truly the core benefit of Hadoop. It entertains the idea that we may have been doing it wrong all along.

22.2 WHAT IS HADOOP TECHNOLOGY AND APPLICATION?

I get asked all the time the question "what is Hadoop?" Technologists, architects, executives, and family members all ask me the same question. I have tried many different approaches, and some have had better success than others. Essentially, Hadoop is not a thing but a collection of things, but that definition tends to confuse the issue. If we really want to simplify things (and we do), Hadoop is storage. Hadoop allows you to take 1,000's of servers and make them look like a single directory. This definition works well for the nontechnical group, but the technical folks will beg the question that similar scale can be used in storage attached network (SAN) as well as pooled resources in virtual environments or distributed architectures in data appliances such as Microsoft Analytic Platform System (APS), Teradata, or Oracle Exadata.

Let us take a quick 50,000-ft view of what each of these architectures looks like. We will not go too deep into the benefits of each one but I will illustrate them for comparative purposes. A SAN is a collection of networked storage devices. This pool of storage is then carved out as logical unit number (LUN) and LUNs are mounted to servers as attached storage. LUNs can cross storage devices and a single disk can be carved (partitioned) into many LUNs. The following image (Figure 22.1) shows the high-level architecture.

The storage layer will be highly redundant via redundant switch paths as well as RAID configurations. Data is served to individual servers and is known as a *shared storage* architecture. For example, a database warehouse server could be using the same disk array as an OLTP ordering system. The primary purpose of a SAN is to provide flexible and managed storage to a large number of servers. Clients access the servers directly and utilize the individual servers for local storage and for memory and CPU resources.

SANs are storage solutions built to commoditize the storage layer and provide ease of storage management across a data center. But let's keep in mind storage is only half the equation when applying analytic solutions. Analytics is nothing without a compute framework. Traditional compute frameworks standardize on the assumption of a single system, or server, providing memory and CPU resources. This is a vertical scaling, or symmetric multiprocessing (SMP), architecture and proposes adding additional CPU or memory when compute performance no longer meets

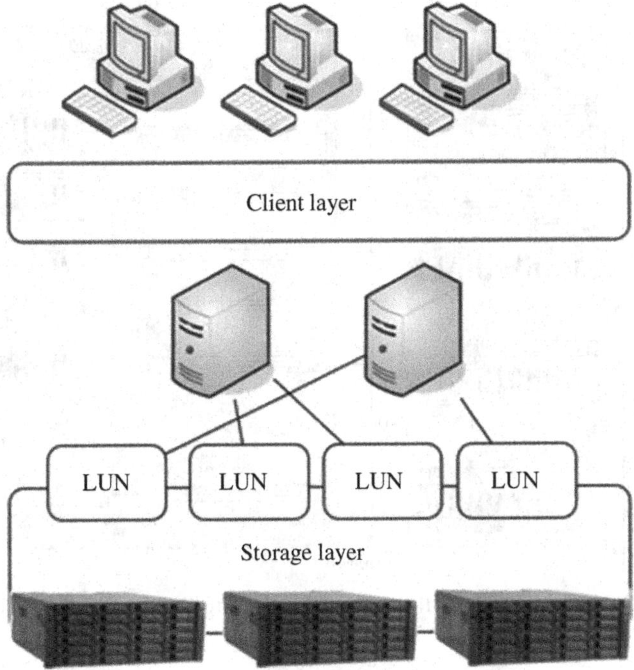

FIGURE 22.1 SAN architecture.

demand. Another option is to create an architecture that takes a given task and splits it up across compute nodes (servers) which then process each split in parallel. This is precisely what large appliances accomplish which is referred to as a "shared nothing" architecture.

Figure 22.2 shows a diagram of Microsoft's APS appliance. Oracle Exadata, Netezza, and Teradata have similar architectures. These are all referred to as Massively Parallel Processing (MPP) architectures. The infrastructure is split up between control nodes and compute nodes. Control nodes service data loading as well as user requests (queries) which are then executed in parallel across compute nodes along an internal dedicated InfiniBand network. Each server in the compute node is a complete install of the database system. Each one processes a subset of the total data set which is sent back to the control nodes, put together, and delivered to the client. This architecture is extremely redundant, extremely fast, and extremely expensive. It is capable of scaling much more than your traditional single-server architecture—there is only so much RAM and CPU you can add to a single server. In the case of APS, all you need to do is add more compute nodes, and you add more storage and compute capacity—APS can scale up to 6 petabytes (PB) [3]. Facebook stores upwards of 300 PB of data and processes 500 terabytes (TB) daily [4]. At this scale, it would be physically impossible through SMP architecture and prohibitively expensive with MPP architectures. In addition, when thinking of data warehouses, it is beneficial to have all your data in a single location for access. Data could all be

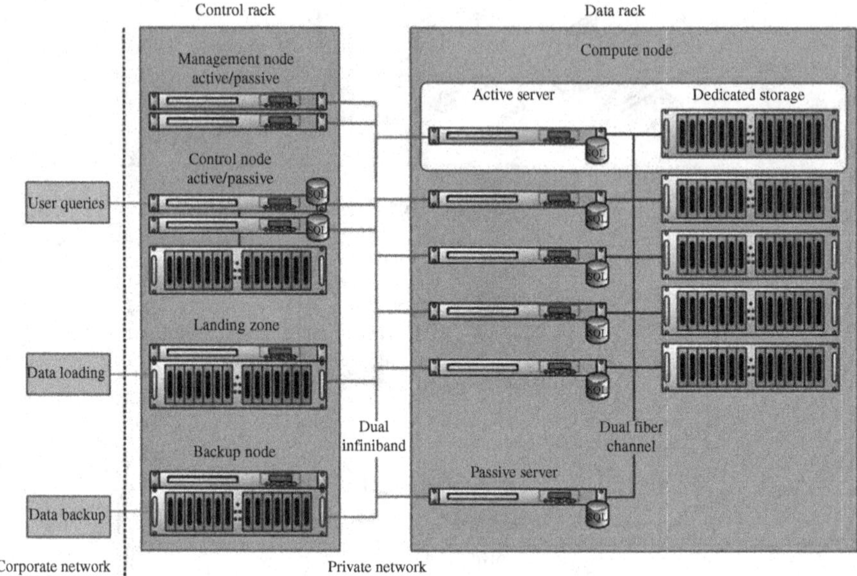

FIGURE 22.2 Microsoft's Analytic Platform System. http://www.pdwtutorial.com/2014/07/microsoft-parallel-data-warehouse_7.html.

stored on a SAN, but it would be difficult to service data at scale to clients because SANs do not provide an appropriate compute framework.

Hadoop tackles these SMP and MPP limitations in a unique fashion. Hadoop builds off the control/compute model but addresses the distributed computing in a much different way. Instead of leveraging an appliance-like design, Hadoop processes data across individual servers thereby using parallel processing across potentially hundreds or thousands of individual SMP servers. Figure 22.3 shows a high-level view from the client into a typical Hadoop cluster.

It is on the datanodes that the actual data resides. There can be many namenodes as well as thousands of datanodes in a single cluster. A "node" in this context is the same thing as a server. So, for example, you may have a hundred datanodes in your cluster, but the client will not know this. The client only sees a single directory such as \data\marketing\2015. In the case of Hadoop, both storage and compute are distributed across many servers. As will be discussed in the next chapter, this provides a number of key advantages and makes Hadoop a transformative technology for any company looking to leverage their data assets.

22.3 WHY HADOOP?

I get asked a lot the question of why use Hadoop especially when a company is already performing analytics using traditional data warehouses. The question underlies a common assumption. The assumption is that Hadoop, or Hadoop Distributed File System (HDFS) in particular, is as the same as a relational data warehouse.

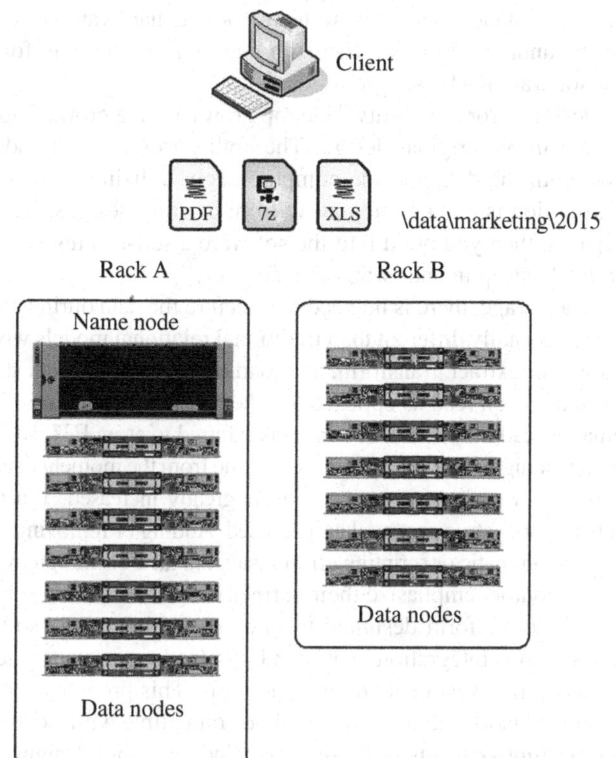

FIGURE 22.3 High-level view from the client into a typical Hadoop cluster.

The confusion is a by-product of the fact that a deployed Hadoop ecosystem can serve needs similar to a data warehouse. Hadoop is two things: storage and compute. As mentioned earlier, HDFS is a file system which assumes no logical or physical constraints on the ingested data. This is much different than how data warehouses ingest data. The compute is variable and can be any number of engines. The only assumption is that the engines understand parallel distributed processing.

Because HDFS is a file system and the compute engines can be any number of a solutions, the Hadoop ecosystem provides a few key advantages over traditional data warehouses, as well as traditional data storage solutions. Following is a short list of some of the major differences between Hadoop and traditional methods:

- Cost optimization
- Scalability
- Speed to analytics
- Flexibility

From purely a cost-per-TB view, you cannot get any cheaper than HDFS. In fact, the only thing cheaper than bare metal HDFS is cloud storage. How is this possible? Simply put, if you build hardware anticipating failure, you can build cheaper hardware. HDFS redundancy is a software feature so the hardware does not have to be designed for redundancy. You can focus on capacity and not pay for expensive interconnects or software RAID solutions.

Hadoop was designed for scalability. Hadoop grows as data grows, and this was a fundamental tenant in the original design. The ability to seamlessly add nodes as needed and have both the data and the compute begin utilizing the new resources without any distribution is a major step away from existing designs. Again, if node failure is anticipated, then you build into the software a self-healing solution which allows you to quickly swap in and swap out resources.

Since HDFS is a storage, there is no need to structure the data during the ingestion phase. This is fundamentally different than traditional relational models which require data transformation, or extract, transform, and load (ETL), prior to data storage. This is what is called schema on read as opposed to schema on write (Figure 22.4).

With a schema-on-read model, we have what is referred to as an ETL workflow. This is a much different paradigm than ETL. Your rate of time from the moment data is ingested to the moment of data visualization or reporting is greatly increased. You can quickly ingest all the data but only structure the data you need. Adding or removing data set features becomes a programmatic or scripting effort instead of an ETL design overhaul.

All Hadoop distributors emphasize their partner integrations. This is no accident. Hadoop is an analytic platform designed to integrate with analytic software solutions. Because Hadoop is integration centric, adding Hadoop to your enterprise will not negate your existing investments in analytic tools. This provides companies the flexibility to leverage Hadoop and know it will be compatible with existing software tools as well as existing skill sets in those tools. Hadoop is not designed to rip and replace, but to integrate. It does this by supporting common APIs such as ODBC and JDBC. It also supports common relational methods such as HCat for DDL and DML and HiveQL for SQL querying. Developers can leverage common Java APIs as well as other languages such as Python or Scala. Some people see Hadoop as complex, while others see it as providing options. Either way, Hadoop seeks to become the one platform all other solutions use for data management.

It helps to view Hadoop as a data platform. Ultimately, this will drive the "why Hadoop" debate. When we think of data platforms, we traditionally think of databases. The reality is that since the exponential growth in the variety of data, that is, data outside of the relational world, databases become simply another data application. The true data platform becomes the Hadoop system which serves up the data to the applications. We now start to think less about database administrators or database developers and more along the lines of data administrators and data developers. The "why Hadoop" becomes a product of what you want your organization to be. Do you want your organization to be driven by applications or by data?

22.4 HADOOP ARCHITECTURE

Hadoop is not just one thing. Hadoop has become a collection of many things—an ecosystem of open-source components. Traditionally, Hadoop is comprised of two components: storage and compute. This was back in the beginning when compute

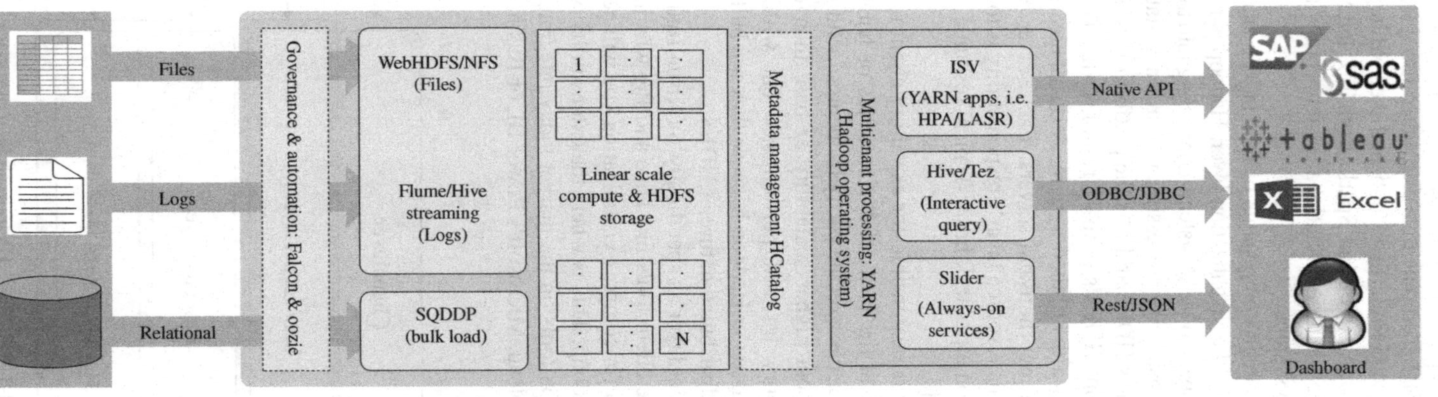

FIGURE 22.4 Hadoop as an ELT solution.

was only MapReduce (MR). Compute on a distributed file system such as HDFS has evolved significantly since Google first published their whitepapers. We now have processing frameworks such as Tez and Spark. Both Tez and Spark allow for faster data access comparable to traditional relational systems but with the ability to run at scale on a distributed file system.

Figure 22.5 shows the main components of the Hadoop stack. Keep in mind that Hadoop does not extend to the visualization layer. Hadoop is an integration platform which allows other developers to build systems connecting into the Hadoop framework. Hadoop is primarily responsible for processing, resource negotiation, and storage.

The data storage layer is HDFS and it is the platform for distributed data storage for the Hadoop ecosystem. HDFS's primary responsibility is to define how data is written to and read from a distributed system. A distributed system, or cluster, is composed of a few or a few thousand servers. Each server has resources such as memory, IO, and CPU. The resource layer's responsibility is to assign, schedule, and manage these resources on a per-task basis. Typical resource layer services include YARN [5] and Mesos [6].

The data access layer is responsible for executing client requests on the cluster. These requests could be in the form of SQL queries or languages such as Scala. Execution engines like Tez can execute partly in-memory with a few writes to HDFS for slower performance but higher scalability or base their execution like Spark almost entirely in-memory for fast performance but with reduced scalability. Execution engines are unique in Hadoop because they are built from the ground up to take advantage of the distributed nature of HDFS.

I have defined the client layer as specifically for visualization. The Hadoop ecosystem provides for some client tools like Hive and Pig which are included in most Hadoop distributions, but for complex visualization you will need to largely step out of what is considered Hadoop. This is where the extensibility of Hadoop shines, and most analytic visualization tools on the market today integrate with Hadoop. These include tools such as Tableau, Microsoft's Power BI, or BusinessObjects.

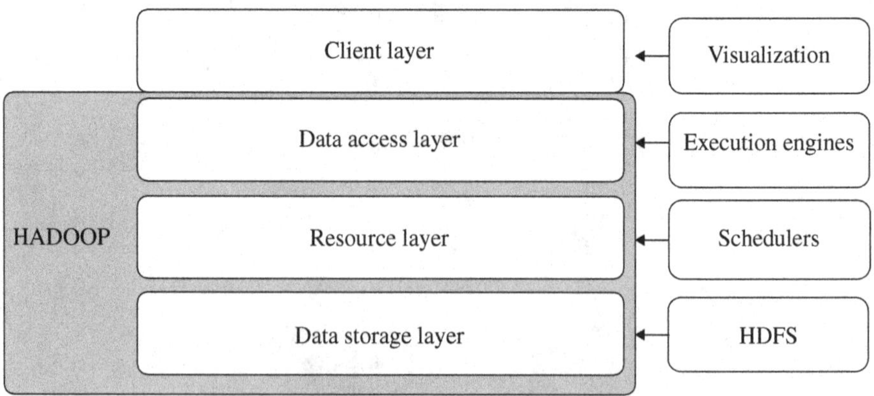

FIGURE 22.5 Hadoop architecture stack.

When thinking about the architecture of Hadoop, it helps to remember that Hadoop is a data platform. Hadoop is not an application solution any more than your data center is an application. Hadoop is designed to be a flexible platform capable of storing any type of data, as well as integration platform designed to work with existing tools. As you design your Hadoop solution, focus on data ingestion, data storage, and data access and/or movement. By focusing on how data will be ingested, stored, and accessed, you will develop a solid architecture capable of servicing any use case.

22.5 HDFS: WHAT AND HOW TO USE IT

I have talked briefly around the general Hadoop architecture as well as the two primary components of Hadoop: compute and storage. The HDFS is the storage component and along with resource negotiation via YARN, and administration via Ambari, storage represents a core Hadoop component. Conceptually HDFS is not difficult to understand. It is a private namespace living on top of the Linux file system and is designed to provide distributed, fault tolerant storage.

The original developers of HDFS had the following purpose around the platform: no data loss from hardware failure, capable of handling large data sets, can handle streaming data, simple coherency model, data locality, hardware, and software agnosticity [7]. Overall, to the end user it looks like any other file system. The end user is oblivious to the fact that when reading or writing to a HDFS directory, they are really accessing potentially hundreds or thousands of server nodes.

Operationally, HDFS acts like any other file system except HDFS requires additional commands to "see" the directories. From within a shell command, a user executes the "hdfs dfs" command prior to accessing the file system. Failure to execute the commands will result in a directory not found error. Figure 22.6 demonstrates these results:

```
[root@sandbox ~]# ls /user
ls: cannot access /user: No such file or directory
[root@sandbox ~]# hdfs dfs -ls /user/
Found 11 items
drwxrwx---   - ambari-qa hdfs          0 2015-10-27 12:39 /user/ambari-qa
drwxr-xr-x   - guest     guest         0 2015-10-27 12:55 /user/guest
drwxr-xr-x   - hcat      hdfs          0 2015-10-27 12:43 /user/hcat
drwx------   - hdfs      hdfs          0 2015-10-27 13:22 /user/hdfs
drwx------   - hive      hdfs          0 2015-10-27 12:43 /user/hive
drwxrwxrwx   - hue       hdfs          0 2015-10-27 12:55 /user/hue
drwxrwxr-x   - oozie     hdfs          0 2015-10-27 12:44 /user/oozie
drwxr-xr-x   - solr      hdfs          0 2015-10-27 12:48 /user/solr
drwxrwxr-x   - spark     hdfs          0 2015-10-27 12:41 /user/spark
drwxr-xr-x   - unit      hdfs          0 2015-10-27 12:46 /user/unit
drwxr-xr-x   - zeppelin  zeppelin      0 2015-10-27 13:19 /user/zeppelin
[root@sandbox ~]#
```

FIGURE 22.6 Accessing HDFS from the Command Line.

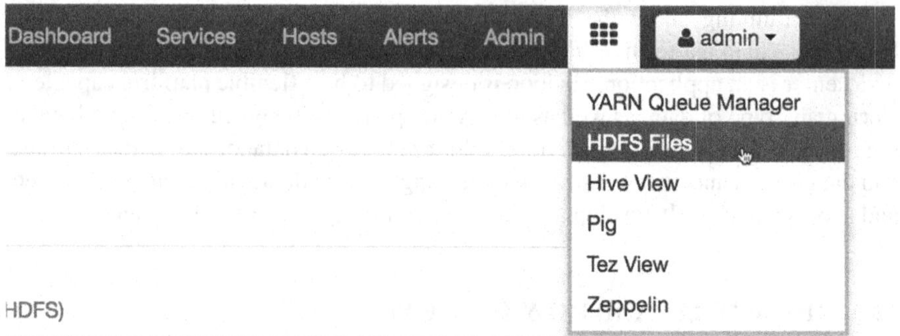

FIGURE 22.7 Ambari HDFS Files view.

HDFS is a *write-once-read-many (WORM)* [8] file system which means it is not fully POSIX compliant. There are advantages to this architecture. If the file system knows ahead of time that a file is immutable, then no locking mechanisms need to be in place. The lack of file locking significantly increases the rate of file writes and data ingestion. Beyond the WORM features, accessing files and directories on HDFS is similar to accessing files and directories on UNIX. Many of the same commands you use for UNIX are available on HDFS [9].

Another important concept to understand about HDFS is features such as system redundancy, block placement, and block recovery which are software enabled and not hardware enabled. By placing the onerous of redundancy and recoverability in the software, we greatly reduce hardware costs. Hardware represents a significant overall cost for any storage implementation, but this cost is greatly reduced because of the lack of need for redundant disk arrays or other redundant component features.

Though HDFS provides many file system features beyond the scope of this chapter, keep in mind that for the complexity and functionality HDFS provides, accessing HDFS is easy for anyone use to the standard UNIX file system. For those not comfortable with shell access, anyone can access HDFS through graphical tools such as Ambari HDFS Files View (Figure 22.7).

Because many of the critical business and distributed features of HDFS are abstracted away from the end user, companies can quickly and easily install HDFS and begin taking advantage of everything it has to offer without changing their existing process flows. In addition, there are hundreds of third-party software solutions which can read and write to HDFS. This allows for a full integration of HDFS into your existing software platform investments.

22.6 YARN: WHAT AND HOW TO USE IT

Any process or task running on a server requires resources. We can break resources into four categories: network, IO, CPU, and memory. Some tasks are considered CPU intensive, while others might be memory intensive. This simply means they require a specific resource to process more than another resource. Of course, not all resources are created equal. Task which process in-memory will perform much faster than task which requires significant IO. Resource may become saturated or become a bottleneck because you have too many tasks running on a server requiring the same resource.

When you implement a distributed system, the resources are also distributed. Hadoop is not virtualization, though it does have many of the same characteristic. From a resource perspective Hadoop does not create a pool of resources to be delegated to a particular task or, in the case of virtual machines, operating system. Hadoop excels at data locality or, in other words, bringing the processing to where the data resides as opposed to serving up the resources to the process.

When a job needs resources in a distributed platform like Hadoop the first step is to decide where to run the task. A task is a subset of a job. A job can have many tasks and each tasks can be run in parallel on the cluster. In a cluster consisting of a 1,000 servers and running millions of jobs and millions of tasks, deciding where to run can be complex. Once a task is assigned a server to run on, a negotiation must take place to assign that task the resources it needs to finish. Finally, with hundreds of tasks running for one job, someone needs to track when each task finishes and whether or not a task has failed. If and only if all the tasks successfully complete is the job done and the resources made available to other tasks.

The brains behind the scheduling of tasks and the assigning of cluster resources is YARN. YARN stands for "Yet Another Resource Negotiator", and since it was first proposed by Arun Murthy from Yahoo, the acronym may be a play of Yahoo which supposedly stands for "Yet Another Hierarchically Organized Oracle" [10]. Regardless of the origins of the name, the introduction of YARN into the Hadoop ecosystem was a major step in Hadoop becoming a true data platform.

Prior to Apache Hadoop YARN much of the resource governance work was performed by MR. MR was both a data processing framework and a resource negotiator and proved early on to inhibit the ability for MR to scale as needed [11]. Something is needed to be changed in order for processing to meet the needs of modern cluster sizes. In January 2008, Arun Murthy submitted the original JIRA for Apache Hadoop YARN [12]. The JIRA began the work to move resource governance out of MR and into YARN and further separated the resource management functionality from the job scheduling and monitoring process.

Hadoop experts view YARN as the operating system for Hadoop. YARN fences off resources between tasks and allows for multiple data access patterns including batch, interactive, and streaming. Following are YARN's primary components [13]:

- ResourceManager—responsible for allocating resources and accepting job submissions and is the primary master process
- NodeManager—a per-node process responsible for allocating containers and reporting back to the RM
- ApplicationMaster—a per-application process working with the NodeManager for monitoring and allocating container resources
- Containers—a single unit of network, IO, memory, and CPU resources (Figure 22.8)

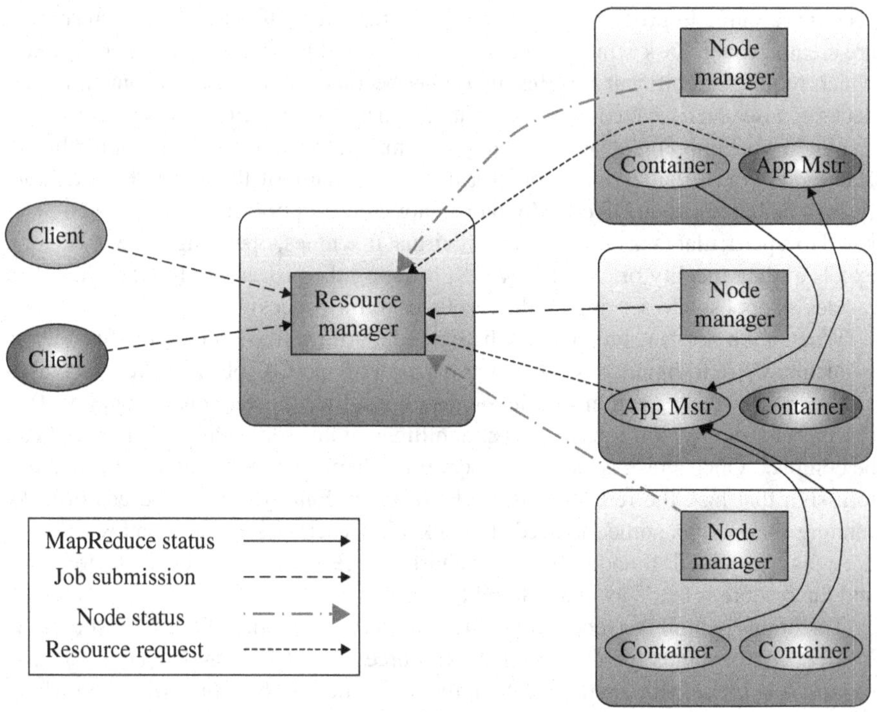

FIGURE 22.8 YARN framework.

YARN was built as a framework for applications which require cluster resources. APIs exist for YARN which allow applications to plug into the framework and abide by YARN's resource negotiation rules. This prevents any one application from utilizing all the cluster resources which were the case prior to YARN in MRv1 and traditional Hadoop. Developers certify most modern data applications on YARN. By certifying on YARN the application can become a key but safe player in the modern data lake. From an operation standpoint you may need to adjust YARN's container allocation settings as well as configure scheduling components like the Fair Scheduler or, more advisable, the Capacity Scheduler [14]. End users rarely have any direct need to understand YARN or even know that YARN exists. YARN is transparent from the end-user perspective so that the end user can focus on submitting jobs and feel confident those jobs will execute successfully and complete in a reasonable timeframe.

22.7 MAPREDUCE: WHAT AND HOW TO USE IT

Including a section on MR is useful as more of an example of distributed computing and its evolution than as a primer on how to use and execute MR applications. MR was key in the development of Hadoop as a data processing platform, but the majority of experts agree that MR as a useful data access method for modern data applications may be reaching the end of its life cycle [15].

Modern data architectures are complex systems used by enterprises for an emerging spectrum of use cases. These use cases range from real-time data flows, interactive and subsecond querying, ad hoc reporting, as well as OLAP. None of these patterns fit into the MR model. MR was specially adapted to executing parallel processing on a distributed system but inherent in that was slow and inefficient use of resources. Other data access solutions like Hive choose to use other, more efficient distributed engines such as Tez and Spark. As a Hadoop administrator or developer starting today, you may spend the rest of your career never troubleshooting an MR task or writing MR code.

We still should not underappreciate what MR did for the development of Hadoop as a core data platform. When in 2009 Facebook first developed SQL on Hadoop, MR was the primary engine for executing SQL queries on Hadoop [16]. At the time it was the only way and the existing Java map, and reduce functions worked perfectly for distributed processing. MR commoditized embarrassingly parallel processing and allowed the Hadoop ecosystem to begin its spread into more traditional, non-Silicon Valley organizations. Nothing else at the time was so accessible and yet able to perform calculations on such a large scale.

Look at MR today as the grandfather of subsequent distributed engines. It is a testimony to the popularity of the Hadoop space and the power of the open-source model that developers continue to refine and produce ever-improving data access engines. As with any great innovation, we would not be where we are today without MR though we could never get where we need if MR was all that we use.

22.8 APACHE: WHAT AND HOW TO USE IT

More and more companies are choosing to adopt an open-source software (OSS) strategy, but what does this mean and how does it affect day-to-day operations? OSS has a number of questions around both the viability of the open-source market as well as the risks companies assume when adopting OSS. What we are seeing in the field is the open-source framework and process, though certainly not perfect, providing companies significant cost improvements, technical innovation, and development agility.

A common misconception is that open source equates to nonrigorous development practice or that open source in some way compromises software integrity. Maybe the perception occurs because some open-source projects enter the Apache ecosystem as incubator [17] projects or in some beta form. While this is certainly true and is one reason for the rapid innovation of the open-source community, almost all open-source projects exist because they have been used and tested in one or many organizations, and open-source adopters need to be aware of the versioning process and the availability of general access (GA) releases as well as tags and branches. Documentation quality may vary from project to project as well as release cycles, and this can cause some organizations to view open sources as less rigorous than proprietary software solutions. For proprietary software, most of the software development dirty work (aka sausage making) occurs behind closed doors, while OSS makes this process transparent.

By definition OSS code can be modified by any developer, and the company can utilize the modified version. OSS empowers individual developers to enhance the software by filling functional gaps which may or may not prove useful to the greater community. The core tenet of open source is to encourage the crowdsourcing of these functional enhancements, whether they be in operations, integration, security, or usability, and submit them to the OSS process for inclusion into the main software trunk. It is important for companies to understand this process and realize all modified code, or enhancements are not automatically included into the product but, instead, the process dictates a somewhat democratic procedure between key committers who vote on the enhancements' benefit to the overall project.

Members fall primarily into two categories: contributors and committers. Contributors represent individuals who answer questions, provides documentation, provides financial support, or submits code [18]. A committer does everything a contributor does but also possesses the ability to vote on whether to submit code to the project and has a say in the project roadmap. The Apache projects are run as a meritocracy by which effort, quality of contribution, and involvement allows a developer to move up in rank and eventually become a committer.

OSS builds off the tenant of crowdsourcing. It challenges traditional methods of proprietary software development which encourages a behind-the-scenes development that hides innovation in the spirit of competitive differentiations. Alternatively, OSS thrives off the participation of the community who works together to provide the best product possible. Companies earn money off providing services such as training, consulting, and support in lieu of software licenses. More and more companies are opening up to the this model because it reduces costs and risks and prevents vendor lock-in. It yet to be seen if open source will be the primary software model in the future but, if the last few years are indicators, it will surely be a major disruptive factor for years to come.

22.9 FUTURE TREND AND CONCLUSION

The technology landscape is transforming. Data is becoming king as well as agile application deployment and development and the advent of deep analytics on a massive scale. OSS is rapidly innovating and changing the way we think about software. All of this rapid innovation comes at a cost. The cost is the inability of companies to keep up with the pace in terms of both their ability to adjust their corporate cultures to new methodologies and their ability to hire and retain resources capable of managing and maintaining the technology.

What we know is that both technology and corporations are good at adapting, and we see this in action as Hadoop begins to move away from the world of start-ups and dot-coms and into the mainstream of Fortune 500 companies. Hadoop and OSS in general are showing the same commitment to enterprise rigors like security, operations, and data governance as traditional, proprietary solutions. We are seeing a rapid adoption rate which allows for a whole new generation of technologists, engineers, and data scientists to get real-world, hands-on experience with the platform.

We saw similar movements as RDBMS became popular, as well as virtualization. Today containerization is the new virtualization, and Hadoop and its ecosystem is the new language of analytics. Look to the near future for a continuation of ebb and flow of acquisitions and attrition. It may be bumpy at times, but at the end we will be in an entirely new IT world which will serve our needs for a long time to come.

REFERENCES

[1] Codd, E., "A Relational Model of Data for Large Shared Data Banks", Communication of the ACM, Volume 13, Number 6, ACM, New York, NY, PP: 377–387, June 1970.

[2] Sanjay, G., Howard, G., Shun-Tak, L., "The Google File System", ACM, Bolton Landing, New York, October 19–22, 2003.

[3] http://www.microsoft.com/en-us/server-cloud/products/analytics-platform-system/ (Accessed September 9, 2015).

[4] http://www.adweek.com/socialtimes/orcfile/434041 (Accessed September 9, 2015).

[5] https://hadoop.apache.org/docs/current/hadoop-yarn/hadoop-yarn-site/YARN.html (Accessed December 15, 2015).

[6] http://mesos.apache.org/ (Accessed December 15, 2015).

[7] https://hadoop.apache.org/docs/r1.2.1/hdfs_design.html (Accessed January 28, 2016).

[8] https://en.wikipedia.org/wiki/Write_once_read_many (Accessed January 31, 2016).

[9] https://hadoop.apache.org/docs/stable/hadoop-project-dist/hadoop-common/FileSystemShell.html (Accessed January 31, 2016).

[10] http://archive.is/puqz (Accessed January 31, 2016).

[11] Murthy, A. C., Vavilapalli, V. K. (2014). *Apache HADOOP YARN*. Crawford, IN: Hortonworks.

[12] https://issues.apache.org/jira/browse/MAPREDUCE-279 (Accessed January 31, 2016).

[13] https://hadoop.apache.org/docs/current/hadoop-yarn/hadoop-yarn-site/YARN.html (Accessed January 31, 2016).

[14] http://docs.hortonworks.com/HDPDocuments/HDP2/HDP-2.1.2/bk_system-admin-guide/content/ch_capacity-scheduler.html (Accessed January 31, 2016).

[15] http://the-paper-trail.org/blog/the-elephant-was-a-trojan-horse-on-the-death-of-map-reduce-at-google/ (Accessed January 31, 2016).

[16] https://www.facebook.com/notes/facebook-engineering/hive-a-petabyte-scale-data-warehouse-using-hadoop/89508453919/ (Accessed January 31, 2016).

[17] http://incubator.apache.org/ (Accessed February 22, 2016).

[18] http://www.apache.org/dev/contributors (Accessed February 22, 2016).

23

SECURITY OF IoT DATA: CONTEXT, DEPTH, AND BREADTH ACROSS HADOOP

PRATIK VERMA
DB Research Inc., Hopkins, MN, USA

23.1 INTRODUCTION

The Internet is best understood as a globally distributed network of voluntarily interconnected devices that can communicate with each other over well-defined protocols to enable business operations. Until recently, the *devices* connected by the Internet have predominantly been computational (e.g., personal computers) or communicative (e.g., smartphones). In contrast, the Internet of Things (IoT) is a network of physical objects containing components that can collect data about the physical objects and exchange this data with other objects in the network. The physical objects communicate with each other in order to function as a collective unit. The electronic components embedded in these physical objects that collect and exchange data about these objects are typically designed for a single purpose based on the properties and interactions that are relevant to the physical objects; for example, the devices in a Smart light bulb need to be concerned with understanding and manipulating only a few attributes like hue and brightness of the light generated. Consequently, compared to multipurpose devices, the devices in IoT networks tend to require significantly less processing power and capacity, which also makes it possible to utilize these devices in significantly larger numbers. The number of devices in the IoT network is estimated to be an order of magnitude larger than the Internet; the total number of addressable devices (with IPv4 addresses) in the Internet was approximately 4.3 billion in 2011 [1],

Internet of Things and Data Analytics Handbook, First Edition. Edited by Hwaiyu Geng.

Companion website: www.wiley.com/go/Geng/iot_data_analytics_handbook/

whereas by one estimate IoT networks will contain more than 30 billion devices by 2020 [2]. For comparison, there are about 7.7 billion mobile phones in use among 4.7 billion unique subscribers as of 2016, according to GSMA Intelligence [3]. IoT technologies are being used in industrial (e.g., smart metering in utilities), enterprise (e.g., heating, ventilation, and air-conditioning management in buildings), and consumer markets (e.g., wearables for fitness tracking). The ability of devices to capture data associated with physical objects, the interconnectivity of devices with each other, and their connectivity with the rest of the information infrastructure define the main areas of security concern associated with the broad use of IoT technologies [4]. This paper briefly describes security concerns associated with the devices, the network, and the data associated with IoT technologies and then addresses data security within IoT platforms built using Hadoop technologies.

23.1.1 Devices

The common failures in security in IoT application attributable to device are primarily associated with vulnerabilities in the software associated with sensors and their logic of operation or incorrect data injected into the network that alters the operation of the device. The primary risks associated with the failure to secure the device include the malfunction of the device and potential compromise of the IoT network that affects the functionality of other devices. To achieve the low per unit cost and size, the power and capacity of individual devices in IoT networks is reduced. Significant progress has been made by device manufacturers in spite of this reduction in capacity by creating hardware-assisted security built into system on a chip [5]. These features are primarily focused on two areas: cryptography with secure access validation that prevents unintended access to data within the chip and secure remote firmware updates. In a recent study of attack surfaces of automotive Electronic Control Units (ECUs), authors were able to demonstrate malfunctioning of a car's brakes by injecting messages. The injected message required for malfunction was determined through reverse engineering of format of the message in a properly functioning ECU [6]. In some cases, the devices were built with security features that prevented an action from occurring on a message while the car was in motion. A full review of security related to operation of devices is outside the scope of this paper.

23.1.2 Network

The devices in the IoT network must communicate with each other in order to function as a collective unit and communicate with the rest of the infrastructure in order to be managed collectively. Based on the type of intercommunication, different protocols have organically emerged [7]: (1) Data Distribution Service (DDS) protocol addresses the movement of data among devices, (2) Message Queue Telemetry Transport (MQTT) and Extensible Messaging and Presence Protocol (XMPP) protocols address collecting device data and communicating it to platforms that use that data, and (3) Advanced Message Queuing Protocol (AMQP) addresses the movement of data among the servers of the platform that end up using the data. There are other protocols that address how devices communicate over wireless connections that include Radio Frequency Identification (RFID), Wi-Fi, or Bluetooth technologies. Any single IoT application relies on multiple protocols and technologies for its basic functionality. The common failures in security in IoT application attributable to network communication arise either from a lack of security functionality built into one of the many protocols used, from a lack of implementation of controls provided by the protocols, or from a lack of interoperability of the security functionality of different protocols. An example of the former includes a vulnerability in Bluetooth protocol that allows an unregistered service to connect to a device configured in discovery mode without authentication, which can be used to either gain access to the device or transmit virus through a file transfer from one device to another [8]. An example of second case includes lack of code in an IoT data analytics platform to verify credentials sent by a device per MQTT protocol before writing data received from the device to the storage platform. An example of the latter includes storing of credentials on multiple devices with the client identifiers that are used to authenticate devices (communicating over MQTT protocol) instead of using single sign on credentials management system within an IoT application. The primary risk associated with the failure to secure the network communication in the IoT platform is the potential compromise of the rest of the enterprise network connected to the servers receiving the data from these devices. For example, the initial intrusion in the 2013 data breach of retailer Target originated from loss of network credentials shared with an HVAC vendor that allowed attackers to upload malware to Target's point-of-sale devices [9]. A full review of the network security is outside the scope of this paper.

23.1.3 Data

The data generated by devices in IoT networks falls into three categories based on where and when this data is used primarily: (1) the data collected by the device is consumed by the device itself to operate at the time of collection, (2) the data collected by the device is aggregated to another location in real time where it is compared with data from other devices, or (3) the data collected by the device is aggregated to another location where the aggregated data can be analyzed across time or across devices at a later time. In practice, for the latter two cases, the devices must be individually addressable, and the data generated must contain that address and a time stamp for any real-time or aggregated analyses to be meaningful. In addition, due to the variety of devices in the IoT network and types of operations possible with each of these devices, the data collected tends to be semistructured in nature with a structure that may evolve over time. Unlike data exchange among devices in the Internet, in an IoT network, data exchange among devices or between devices and the analysis store tends to be much smaller (e.g., a single automated meter reading) and in many cases much more frequent (e.g., in case of a thermostat). Effective indexing, storing, and processing of the data aggregated from IoT devices requires a data management platform that is designed to accommodate semistructured data in large volumes that can be aggregated in real time and stored for the duration of analysis (typically min to months). Due to its low software and hardware

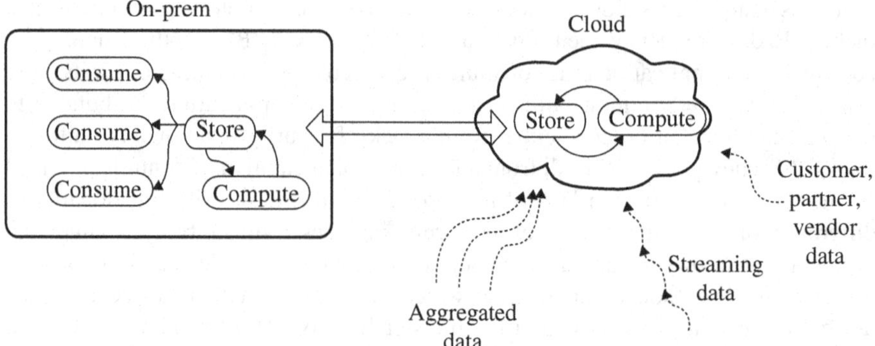

FIGURE 23.1 IoT data management built on Hadoop on-premises and in the cloud.

cost, large storage capacity, and wide availability of tools for analyzing unstructured data, Hadoop has become the storage and processing platform of choice for analyzing IoT data.

23.2 IoT DATA IN HADOOP

In many IoT use cases (e.g., automated meter reading), the devices generating data are distributed, and the resulting data must be aggregated in one location and preprocessed for deduplicating or cleansing prior to analysis. The availability of inexpensive and elastically scalable computation in the cloud makes cloud-hosted Hadoop an ideal choice for a data landing and cleansing area. Often one or more applications that end up using the data generated by the IoT devices are hosted on-premises as various mission critical applications. The data landed and aggregated in the cloud is often bulk ingested from the cloud and processed further on-premises. This paper addresses security and privacy of data as it is stored, processed, and consumed from Hadoop data platform when deployed on-premises or in the cloud (Figure 23.1).

23.3 SECURITY IN IoT PLATFORMS BUILT ON HADOOP

Data from across the IoT network is collected into one place, which in Hadoop platforms is HDFS storage. Using modeling, wrangling, and transformation tools, the raw data goes through a curation process before it is made available to business users. This curation and modeling is performed by data scientists or operationalized through computational models that provide business benefits like predicting market churn, detecting fraud, detecting defects, and so on. Often the discovered and rationalized data is put into relational stores so business users can use query tools like Excel on the curated data to draw insight from it (Figure 23.2).

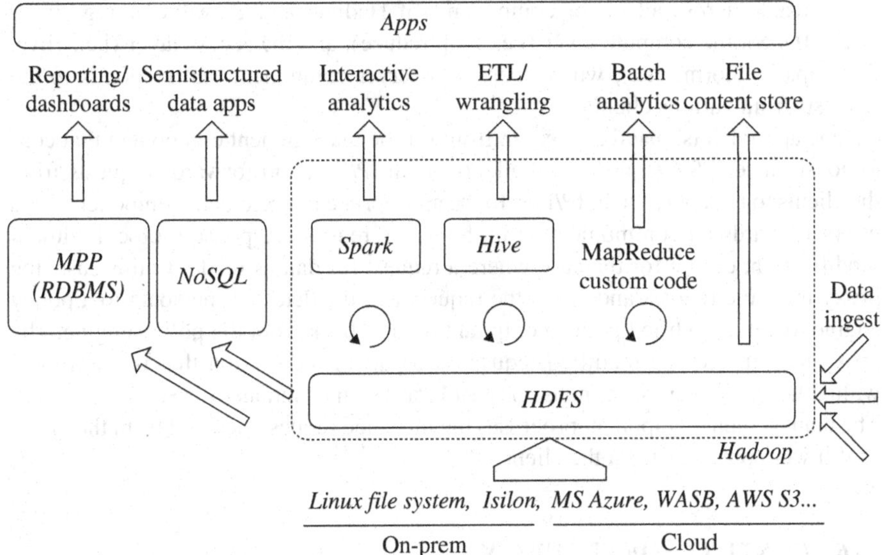

FIGURE 23.2 Hadoop and its relation to other platforms.

23.4 ARCHITECTURAL CONSIDERATIONS FOR IMPLEMENTING SECURITY IN HADOOP

One option is to build effective controls within Hive to manage end-user access to data in Hadoop. But access to the storage layer, HDFS, remains open to end users who utilize different access methods or to big data engineers who access the data directly at the HDFS level. Adding security to one or a few applications on Hadoop will lead to incomplete security given the proliferation of access methods within the Hadoop ecosystem. A better approach is to deploy security at the most common layer to all applications and access paths: the HDFS layer.

23.5 BREADTH OF CONTROL

In order to implement distributed computing, there are typically three main functional components, the client component, the management component, and the worker component. The client component deals with the input from the end user and converts it into requests that get sent to the management component. The management component gets the requests from the client component, splits a request into multiple subrequests, and sends the subrequests to the worker components for processing in parallel. The worker components each get a request to process the request by operating on the data. Typically, the management component serves as the first point of response serving many requests by many clients simultaneously and coordinating multiple worker components for any given request. This three-component paradigm is applicable across all major components of Hadoop including the storage layer (e.g., HDFS), the compute layer (i.e., map reduce), and the access layer (i.e., Hive, Solr, Spark, Storm, etc.), where there is a server component that responds to the requests from various clients.

One approach is to add a proxy in front for the management component that continuously listens for request from clients on an IP/Port and forwards requests from the clients to the appropriate IP/Port to the appropriate management component. This proxy forwards all communication in the native request–response protocol without modification, except for the case where a request for data is made. In this case, the proxy traps the request and sends the request and the detected metadata to a policy engine to get a modified policy compliant request back from the policy engine. The proxy then inserts the modified request into part of the session that is interrupted within the native request–response protocol format and continues the session forward. The management component processes the modified request received from the proxy as if it were obtained from the client.

23.6 CONTEXT FOR SECURITY

In Hadoop, the data elements being accessed in a given request may be present in the request itself (e.g., Hive query), in the file that the request operates on (e.g., in self-described files like Avro or Parquet), or in some other metastore that is populated prior to the request (e.g., HCatalog). The representation of the data elements can be different across or among these locations. For example, a file in HDFS may be named different from the table in Hive that uses the file, but both contain the same sensitive file like SSN. Or the same data values stored in different files or tables may be named differently as SSN or TIN. In either case the policies written once on a concept like SSN should be enforced equivalently whether the data is being queried via Hive, pulled from an HDFS file, or operated on in a MapReduce job. In order to achieve such result, policies must include the ability to map fields from different storage formats (file, table, collection, etc.) into a common logical concept and author policy on that concept. While the request is interrupted, the policy engine can add context from these logical concept mappings to make its policy decision. This allows a consistent policy to be enforced across different access methods.

23.7 SECURITY POLICIES AND RULES BASED ON PxP ARCHITECTURE

To manage the security and integrity of data, security admins create security policies to define which data each user is authorized to access using one or more rules. A *rule* is used to define an entitlement to a resource and is used to control the granularity of the access control policy. A *rule* specifies a *resource*, an optional *qualifier*, an *action*, and an *effect*. A *resource* is a data object or a service to be protected. A *data object resource* can be one of the following: database, schema, table, column, collection, column family, column qualifier, a folder, a file, a field within a file, and so on. A *service resource* can be one of the following: Hive, hue (application id), external, and so on. An *action* is one of the following: *read*, *write*, *use*, *connect*, or *execute*. The *actions* are dependent on the type of resource, for example, a data object or a service. An *effect* is either *allow* or *deny*. If a *qualifier* as a *record filter* is defined, then the *effect* is interpreted as *allow conditional* or *deny conditional* in the back end. A *qualifier* is a set of restraints that limit the *resource* entitlement either along the row dimension or within a cell dimension. This can be either a *record filter* or a *transformation*. If a *qualifier* is a *record filter*, then a condition must be defined in terms of variable = value conditions, where the variable must be a field within the resource and the value can be a static value, an in-line function, or a dynamically evaluated *attribute*. The rows, for which the condition is evaluated to be *true*, are affected by the *effect*. If a *qualifier* is a *transformation*, then an operation must be defined in terms of a static value, an in-line function, or a dynamically evaluated *attribute*. A *rule* or a *policy set* can be granted to a *user* or a *group*. A *user* or a *group* is a special type of *roles*. By *user* and *group*, we mean *dn* for an object of type person or group from LDAP.

A *policy set* is a collection of one or more *rules*. A collection of *rules* can either be grouped into a labeled *policy set* or be grouped by default into a *default policy set*. A labeled *policy set* is used to enforce purpose-based access control. For a given *user* or *group*, any *rule* that is not explicitly grouped into a labeled *policy set* gets assigned to the *default policy set* for that *user* or the *group*. For a given *user*, when a *policy set* is not specified at connection time, the policy applied is the (most allow or most deny) union of all the rules of the *default policy set* of the *user* and of all the *groups* that the *user* belongs to in the *user domain*. For a given *user*, when logging in through a client application, a list of all the labeled *policy sets* for that *user* or for the *groups* that the *user* belongs to is populated in the client side browser session for picking by the end user for the duration of the session. The access control is enforced only per the *rules* that are grouped into the *policy set* specified in the log-in. An audit entry is created in the policy enforcement log. For a given *user*, when a *policy set* is specified in a comment of the query, the specified *policy set* is compared against the list of all the labeled *policy sets* for that *user* or for the *groups* that the *user* belongs to. If the specified *policy set* is one of the allowed *policy sets*, then the access control is enforced only per the *rules* that are grouped into the specified *policy set*. Otherwise a deny-all policy is returned. In both cases, an audit entry is created in the policy enforcement log.

23.8 CONCLUSION

This paper briefly describes security concerns associated with the devices, the network, and the data associated with IoT technologies and then addresses data security within IoT platforms built using Hadoop technologies. By separating the management of the policies from the enforcement of these policies on the various repositories, data-centric security can be scaled across the enterprise, across a variety of data repositories while solving today's problem of siloed security policies. This approach also brings security closer to the data, directly above the database and the storage layer, preventing any user or application from bypassing controls.

REFERENCES

[1] "Free Pool of IPV4 Address Space Depleted." Smith, Lucie; Lipner, Ian. 2011.

[2] "How the 'Internet of Things' will impact consumers, business, governments in 2016 and beyond." Greenough, John. Business Insider. April 14, 2015.

[3] GSMA Intelligence. 2016. Current year-end data.

[4] "How to ensure security for Internet of Things devices". Vasudevan, Vinod. Economic Times. 2016.

[5] "ARM Brings TrustZone Security Technology to IoT Devices". Burt, Jeffery. eWeek. 2015.

[6] A Survey of Remote Automotive Attach Surfaces. Miller, Charlie; Valasek, Chris. 2014.

[7] "Understanding the protocols behind the Internet of Things." Schneider, Stan. Electronic Design. 2013.

[8] Security of Smart Phones. Mulliner, Collins Richard. MSc Thesis. University of California, Santa Barbara. 2006.

[9] "Target Hackers Broke in via HVAC Company." KrebsonSecurity.com. 2016.

PART IV

SMART EVERYTHING

24

CONNECTED VEHICLE

ADRIAN PEARMINE
DKS Associates, Portland, OR, USA

24.1 INTRODUCTION

One of the most intriguing market sectors in the Internet of Things (IoT) industry is the Connected Vehicle, as transportation affects virtually everyone, and the Connected Vehicle is the place where IoT and transportation converge. There are several different definitions of what a Connected Vehicle is, depending on your perspective in the automotive industry. Auto manufacturers, consumers, aftermarket retailers, traffic engineers, and regulators all have slightly different perspectives on what constitutes a Connected Vehicle, and there are many different reasons that the vehicle is connected in the first place.

At the broadest level, a Connected Vehicle may be defined as an automobile with a combination of an internal hardwired network between devices and one or more wireless connections to other systems, allowing access to other networks and services. This can be as basic as providing Internet connectivity utilizing commercial cellular carrier connections and providing wireless connectivity inside the car to other systems. It can also involve connection to other cars, to specific roadside infrastructure, and to third-party central systems.

Connected Vehicles manufactured after 2010 typically have in-dash displays that allow the driver or passenger to control and access different applications with a touch-screen monitor. These applications may include, but are not limited to, music, Internet, navigation, vehicle diagnostics, roadside service, parking applications, Wi-Fi connectivity to other devices in the vehicle, and a host of safe-driving applications.

Recently there have been significant advances in vehicle automation, including extensive testing of fully autonomous vehicles on the roadway. Autonomous vehicles can be considered a subset of Connected Vehicles, with the difference between Connected,

Internet of Things and Data Analytics Handbook, First Edition. Edited by Hwaiyu Geng.

Companion website: www.wiley.com/go/Geng/iot_data_analytics_handbook/

Automated, and Autonomous being the level of control and involvement of the driver. In the case of the fully Autonomous Vehicle, no human driver is involved.

This chapter will discuss Connected and Autonomous Vehicles, specifically their role in the IoT ecosystem. It will discuss the relationship between Connected and Autonomous and the various levels of vehicle automation, as well as other Connected Vehicle services and applications not related to vehicle automation.

Many experts identify the auto industry as one of the most significant market sectors for expansive growth of IoT technologies, but the numbers can be misleading or confusing depending on how they are portrayed. Regardless of whether one is looking at the number of wireless connections, number of individual connected sensors or devices, or the anticipated revenues associated with the implementation of these technologies, most analysts agree that the automotive industry is and will continue to be a major driving force in IoT deployment.

24.2 CONNECTED, AUTOMATED, AND AUTONOMOUS VEHICLE TECHNOLOGIES

Connected Vehicles and particularly Automated and Autonomous Vehicles utilize a number of onboard technologies to determine their location, monitor conditions around the vehicle, sense objects and obstructions, and communicate with other vehicles and roadside infrastructure. The configuration of devices varies from manufacturer to manufacturer and model to model, depending on the level of automation desired and a number of other factors. Figure 24.1 illustrates some examples of onboard systems that support vehicle connectivity and automation.

Vehicles can have automation functionality without being fully autonomous. In fact, features such as automatic transmission and cruise control have been around for many years, are widely adopted, and are now included in virtually all new vehicles. The National Highway Traffic Safety Administration (NHTSA) has designated five (5) levels of vehicle automation, as follows [1, 2]:

Level 0—No automation: The driver is in complete and sole control of the primary vehicle controls—brake, steering, throttle, and motive power—at all times.

Level 1—Function-specific automation: Automation at this level involves one or more specific control functions. Examples include electronic stability control or precharged brakes, where the vehicle automatically assists with braking to enable the driver to regain control of the vehicle or stop faster than possible by acting alone.

Level 2—Combined function automation: This level involves automation of at least two primary control functions designed to work in unison to relieve the driver of control of those functions. An example of combined functions enabling a Level 2 system is adaptive cruise control in combination with lane centering.

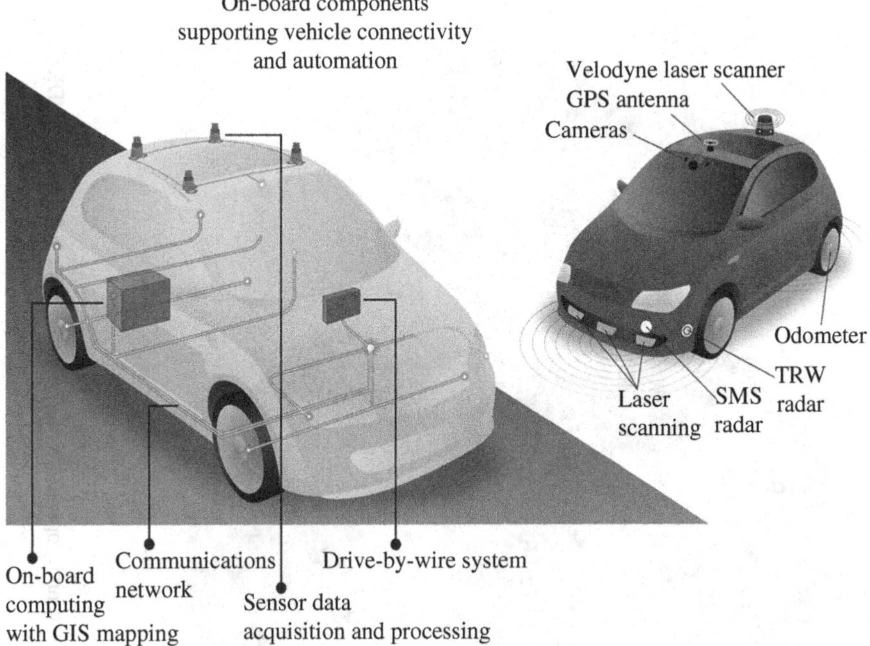

FIGURE 24.1 Example onboard components supporting vehicle connectivity and automation. www.ni.com. Reproduced with permission of DKS Associates.

Level 3—Limited self-driving automation: Vehicles at this level of automation enable the driver to cede full control of all safety-critical functions under certain traffic or environmental conditions and in those conditions to rely heavily on the vehicle to monitor for changes in those conditions requiring transition back to driver control. The driver is expected to be available for occasional control but with sufficiently comfortable transition time. The Google car is an example of limited self-driving automation.

Level 4—Full self-driving automation: The vehicle is designed to perform all safety-critical driving functions and monitor roadway conditions for an entire trip. Such a design anticipates that the driver will provide destination or navigation input but is not expected to be available for control at any time during the trip. This includes both occupied and unoccupied vehicles.

The complexity of the overall system and the number and configuration of sensors varies depending on the level of automation that a manufacturer is trying to achieve. Figure 24.2 illustrates some examples of the types of sensors that can be employed to support autonomous features. The combination of radar, video cameras, ultrasound and LIDAR (a sensing technology using a combination of light and radar) are used to augment the Geographic Information System (GIS) and Global Positioning System (GPS) technologies that track the vehicle's location in a highly precise and accurate manner.

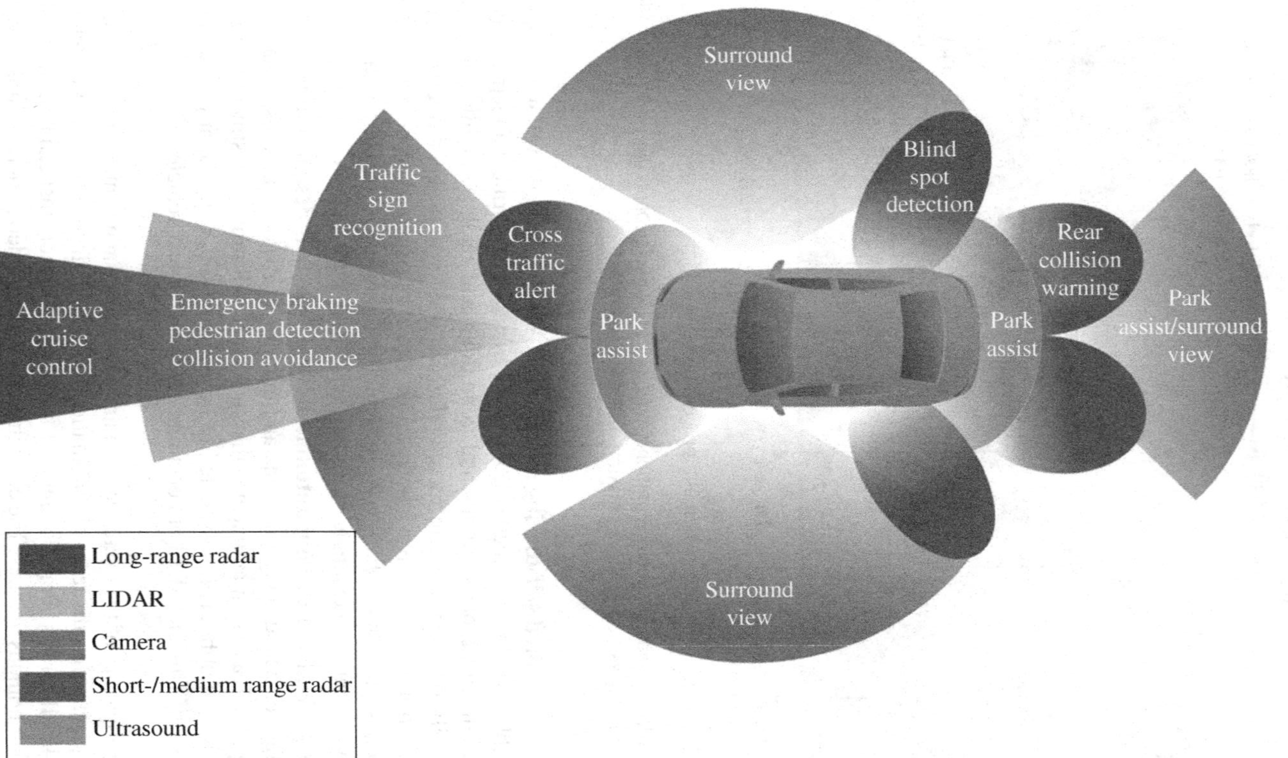

FIGURE 24.2 Example of sensors required to support autonomous vehicle operations. Reproduced with permission of DKS Associates.

These sensors can not only identify obstructions and obstacles but can also help with lane identification and support features like Adaptive Cruise Control and Blind Spot Detection. These automated functions can be supported even when other vehicles are not Connected Vehicles, which is critical as it will be decades before all vehicles are Connected Vehicles.

24.3 CONNECTED VEHICLES FROM THE DEPARTMENT OF TRANSPORTATION PERSPECTIVE

Our Federal, State, and local transportation agencies have a more specific definition of Connected Vehicles. Our departments of transportation (DOTs) define Connected Vehicles as using Vehicle to Vehicle (V2V) and Vehicle to Infrastructure (V2I) to allow cars to communicate with one another and with networks provided largely by the agencies themselves, through the infrastructure in V2I. In fact, V2I and V2V are considered the primary system components of an even larger ecosystem of V2X communications that includes pedestrians, bicyclists, buses, trucks, cell phones, and any other connected system or device.

Figure 24.3 illustrates the connected vehicle from the DOT perspective. In the DOT's definition, the primary form of communications for V2I and V2V is a specific technology called Dedicated Short-Range Communications (DSRC). According to

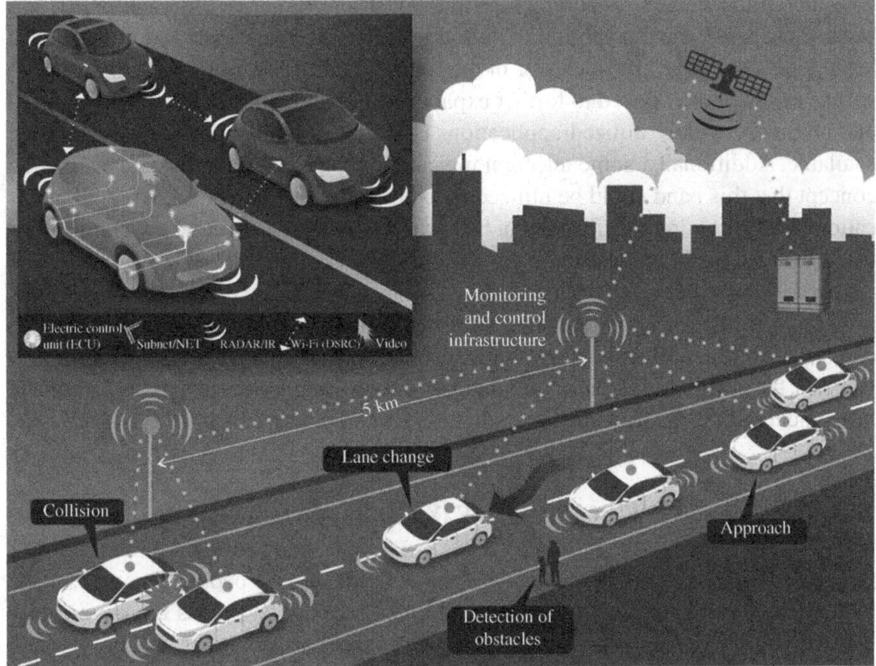

FIGURE 24.3 Examples of Vehicle to Vehicle (V2V) and Vehicle to Infrastructure (V2I) Communications. Source: Reproduced with permission of DKS Associates.

the US Department of Transportation (USDOT), DSRC "is a two-way short-to-medium-range wireless communications capability that permits very high data transmission critical in communications-based active safety applications. In Report and Order FCC-03-324, the Federal Communications Commission (FCC) allocated 75 MHz of spectrum in the 5.9 GHz band for use by Intelligent Transportations Systems (ITS) vehicle safety and mobility applications."[1]

DSRC is the preferred method of V2V and V2I communications because it provides very low-latency communications and very fast network acquisition capabilities, which are required for the safety applications envisioned in Connected Vehicle deployments. DSRC also provides an industry standard for radio frequency and network connectivity, which allows all auto manufacturers and local agencies to deploy standard technology. Finally, DSRC includes security and prioritization protocols in the standards, which are also critical requirements in a system that will support life safety applications.

24.4 POLICY ISSUES AROUND DSRC

DSRC is the DOT's preferred communication technology for V2V communications and for local V2I communications, but this preference is not without controversy. The 5.9 GHz band is considered "valuable real estate" in the radio spectrum world, as it is directly adjacent to the 5.8 GHz frequencies that comprise much of the unlicensed band covered under the IEEE 802.11a standards and commercially known as Wi-Fi.

Many in the IT and telecommunication industries argue that although DSRC has been set aside for the transportation industry for over 15 years, very few devices have been deployed either in vehicles or in the right of way. They argue that this valuable spectrum should be opened to further expand Wi-Fi bandwidth and that many beneficial and more widely utilized applications could be developed if this spectrum was available. Additionally, some telecommunication equipment providers are testing the concept that this band could be utilized by other systems when no V2V or V2I applications are attempting to utilize it, but that priority could be given to these applications when they are being requested.

At the time of authoring this book, decisions by the FCC and other regulatory bodies remain up in the air about the long-term dedication of DSRC to the transportation industry. While many automakers are beginning to implement DSRC radios in new vehicles, the traffic engineering community is still on the fence about making significant investment in DSRC infrastructure until these federal policy issues are resolved.

24.5 ALTERNATIVE FORMS OF V2X COMMUNICATIONS

Regardless of the long-term decisions about DSRC, most people involved in the industry recognize that there are many alternate means of vehicles communicating with one another, with infrastructure and with third-party systems. Wi-Fi, 4G, and

---

[1] See more at http://www.its.dot.gov/factsheets/dsrc_factsheet.htm#sthash.JkTd8dT9.dpuf.

5G cellular data communications and other evolving telecommunication technologies may all be utilized for V2X communications. In fact, most believe that it will take a combination of these technologies to support the different applications and use cases to ultimately support the Connected and Automated Vehicle industry.

Whatever the transport layers of communications ultimately become to support the deployment of Connected and Autonomous Vehicles, it is paramount that standards be agreed upon nationally and internationally for the applications, message sets, authentication, and security of these communications. Both the auto manufacturing industry and the traffic engineering industry need to have such standards in place to support the wide adoption of these technologies. Significant effort has occurred within the industry to develop these standards, and many have been—or are in the process of being—ratified both nationally and internationally. To support this effort, USDOT has partnered with the auto manufacturing industry, the American Association of State and Highway Officials (AASHTO), the Institute for Transportation Engineers (ITE), and ITS America to form the Vehicle to Infrastructure Deployment Coalition.[2]

24.6 DOT CONNECTED VEHICLE APPLICATIONS

As with other IoT market sectors, the purpose of various Connected Vehicle IoT subsystems and technologies is to provide data for the applications that they support. For example, Figure 24.4 illustrates the Pikalert Vehicle Data Translator, a specific Connected Vehicles Application currently being developed. With funding and support from the USDOT Research and Innovative Technology Administration (RITA) and direction from the Federal Highway Administration's (FHWA) Road Weather Management Program, the National Center for Atmospheric Research (NCAR) is conducting research to develop the Pikalert Vehicle Data Translator so that it combines vehicle-based measurements of the road and surrounding atmosphere with more traditional weather data sources, thereby creating road and atmospheric hazard products for a variety of users [3].

As illustrated in this figure, data from the foremost Connected Vehicle is collected from its onboard systems and delivered back to a data processing center. The center is also collecting detailed weather and road condition information from other systems, including national weather information providers and local roadside condition sensors. The center collects, analyzes, and processes the data. Then this data is transformed into information that is transmitted to traffic information dissemination systems and specifically sent as a warning to other approaching Connected Vehicles.

This application is just one example of dozens of Connected Vehicle applications defined by USDOT and currently in some stage of deployment or development. Figure 24.5 provides a list (as of the time that this article was authored) of the different Connected Vehicle applications. This list is divided into eight categories: V2I Safety, V2V Safety, Agency Data, Environment, Road Weather, Fee Payment,

[2] More information can be found at www.transportationops.org/V2I/V2I-overview.

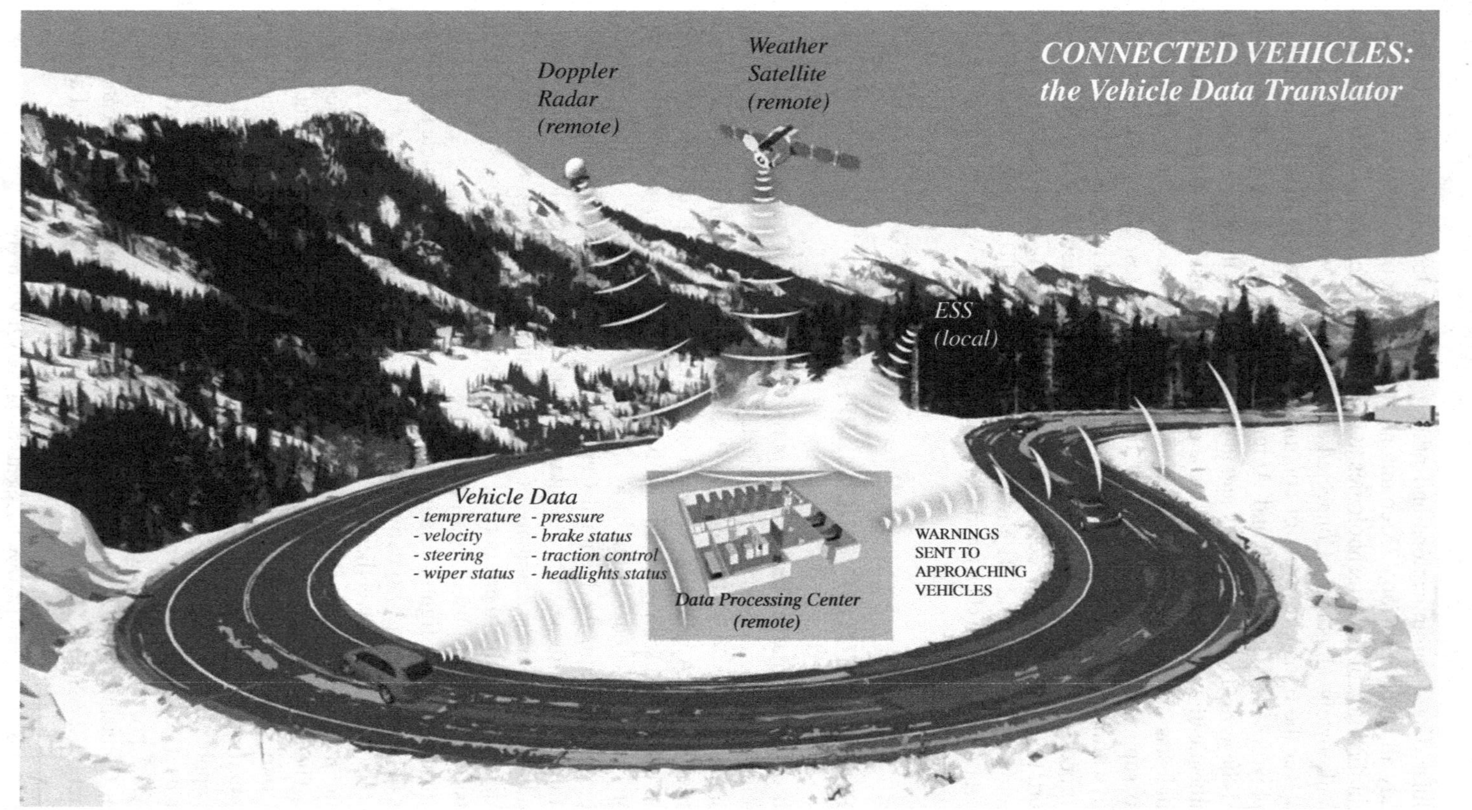

FIGURE 24.4 Pikalert Vehicle Data Translator—an example of Connected Vehicles Application. Reproduced with permission of University Corporation for Atmospheric Research.

Oregon Department of Transportation

CONNECTED VEHICLE APPLICATIONS

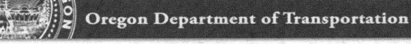

V21 Safety

Signal Phase & Timing (SPAT)
Red Light Violation/Driver Gap Warning
Curve Speed Warning
Stop Sign Violation/Gap Assist
Spot Weather Impact Warning
Pedestrian Warning
Railroad Crossing Warning
Disabled/Oversized Vehicle Warning

V2V Safety

Emergency Electronic Brake Lights (EEBL)
Forward Collision Warning (FCW)
Intersection Movement Assist (IMA)
Left Turn Assist (LTA)
Blind Spot/Lane Change warning(BSW/LCW)
Do Not Pass Warning (DNPW)
Vehicle Turning Right in Front of Bus Warning

Agency Data

Probe-based Pavement Maintenance
Probe-enabled Traffic Monitoring
Vehicle Classification Traffic Studies
CV-enabled Performance Measures
CV-enabled Turning/Intersection Analysis
CV-enabled O-D Studies
Work Zone Traveler Information

Environment

Eco-Approach/Departure Intersections
Eco-Traffic Signal Timing
Eco-Traffic Signal Priority
Connected Eco-Driving
Wireless Inductive/Resonance Charging
Eco-Lanes Management
Eco-Speed Harmonization
Eco-Cooperative Adaptive Cruise Control
Eco-Traveler Information
Eco-Ramp Metering
Low Emissions Zone Management
AFV Charging/Fueling Information
Eco-Smart Parking
Dynamic Eco-Routing
Eco-ICM Decision Support System
Dynamic Emissions Pricing

Road Weather

Motorist Advisories & Warnings (MAW)
Enhanced Maintenance Decision Support
Vehicle Data Translator
Weather Response Traffic Info (WxTINFO)

Fee Payment

Tolling
High Occupancy Toll Lanes
Congestion Pricing

Mobility

Advanced Traveler Information System (EnableATIS)
Multimodal Intelligent Traffic Signal (MMITSS)
Intelligent Traffic Signal System (1-SIG)
Signal Priority (Transit & Freight)
Mobile Accessible Pedestrian Signal (PED-SIG)
Emergency Vehicle Preemption (PREEMPT)
Intelligent Network Flow Optimization (INFLO)
Dynamic Speed Harmonization (SPD-HARM)
Queue Warning (Q-WARN)
Cooperative Adaptive Cruise Control (CACC)
Next Generation Ramp Metering (RAMP)
Response, Incident, Emergency (RESCUME)
Incident Guidance Emergency Responce (RESP-STG)
Incident Scene Work Zone Alerts (INC-ZONE)
Emergency Communications/Evacuation (EVAC)
Integrated Dynamic Transit Operations (IDTO)
Connection Protection (T-CONNECT)
Dynamic Transit Operations (T-DISP)
Dynamic Ridesharing (D-RIDE)
Freight Advanced Traveler Information (FRATIS)
Freight Dynamic Travel Planning & Performance
Drayage Optimization

Smart Roadside

Wireless Inspection
Smart Truck Parking

FIGURE 24.5 List of Connected Vehicle Applications from USDOT. Oregon Department of Transportation.

Mobility, and Smart Roadside. This list is modified and expanded frequently as new research continues and new applications are identified and defined.

As of early 2016, most of these applications were either still in prototype development and testing or in even earlier stages of concept definition. Auto manufacturers continue to work with local, state, and federal agencies on numerous pilot projects around the country to better define the applications. Together these partners are developing the standards, message sets, communication protocols, and security elements required for wider deployment of these applications.

24.7 OTHER CONNECTED VEHICLE APPLICATIONS

The Connected Vehicle and associated applications discussed in the preceding text actually make up a small percentage of the overall market space. Auto manufacturers, aftermarket providers, service providers, and insurance companies are offering new services including infotainment, remote vehicle diagnostics, navigation tools, Wi-Fi hotspots, and numerous other applications. The vast majority of these applications rely on cellular data communications, and many are a direct cell phone connection.

Hardware supporting these systems falls into two different categories. These technologies are either "built-in" or "brought-in," depending on the age of the vehicle. Newer vehicles often have cellular connectivity built into the onboard network, while older vehicles utilize hardware from aftermarket providers to make these connections. If vehicle diagnostics are required for the application, the onboard diagnostics (OBD) port of the car can be utilized to provide power to a device and access to the onboard computers. This is the same port used by mechanics on all newer vehicles to check and reset vehicle alarms and codes in those same onboard computers. Data on acceleration, braking, idling, fuel consumption, and many other onboard systems, as well as all system alarms can be accessed in real time from the OBD port by inserting a dongle as illustrated in Figure 24.6. The dongle usually includes computing

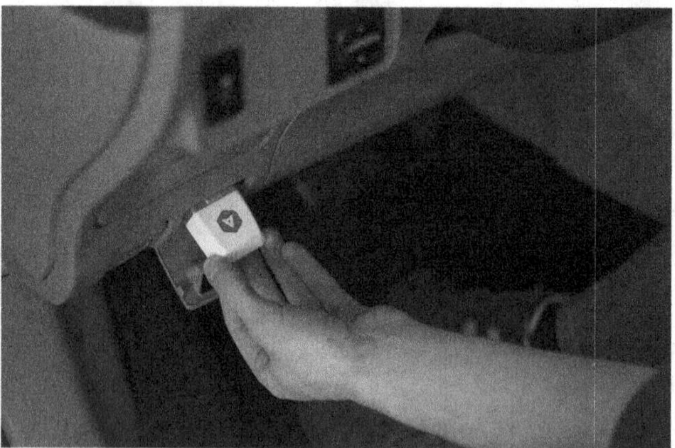

FIGURE 24.6 Onboard Diagnostics (OBD) Port and Dongle.

capabilities and a cellular communication radio, may include network routing functionality, and can be programmed to support many different types of applications.

24.8 MIGRATION PATH FROM CONNECTED AND AUTOMATED TO FULLY AUTONOMOUS VEHICLES

Two distinct paths have evolved in the migration path toward fully autonomous vehicles. Traditional auto manufacturers have been incrementally adding automated features to their cars for decades, starting as far back as the 1930s and 1940s with the introduction of automatic transmission, followed by the adoption of cruise control systems in the 1950s. The incorporation of additional automated driving functionalities has increased dramatically in the last several years. This includes automated parking, lane control, and adaptive cruise control, to name just a few examples.

The introduction of the disruptor community into the autonomous vehicle market has not only accelerated the drive toward fully autonomous vehicles but has actually created a second path to achieving this goal. When Google announced its intention to develop a "self-driving car," the company started with modified versions of a Lexus SUV but migrated quickly to building their own prototype of a fully autonomous vehicle. It is Google's stated intention that any cars that they produce for commercial availability or personal ownership not be incrementally upgraded from partial automation. Rather, these cars shall be fully autonomous from day one.

While Google has announced partnerships with major auto manufacturers for the development of mass-produced autonomous vehicles, the company has also started its own car company registered as Google Auto LLC. Thus far Google Auto is building only lightweight low-speed vehicles (LSVs), with a maximum speed of 25 mph [4]. This allows Google Auto to avoid onerous federal safety requirements and crash tests required by NHTSA and other regulatory bodies.

Other companies are following Google's path. The rideshare provider Uber is beginning to test its own fully autonomous vehicles in 2016. Like Google, Uber has stated that its intent in developing autonomous vehicles isn't to deploy autonomic driving features but instead to replace the driver altogether. It is becoming clear that whether there is incremental deployment or full automation from day one, we are rapidly heading toward a time when driverless vehicles are a reality.

24.9 AUTONOMOUS VEHICLE ADOPTION PREDICTIONS

Adoption rate predictions for autonomous vehicles vary widely from source to source, as there are many factors that impact how quickly autonomous vehicles will replace traditional cars on our roadways. Vehicles with partial autonomous functionality are already on our roads. Going beyond the examples mentioned earlier, we are now entering an era with more holistic vehicle autonomy. For example, in 2014, Tesla began equipping their Model S vehicles with cameras, sensors, and electric assist braking. In 2015 they then pushed out their "Autopilot" functionality via the

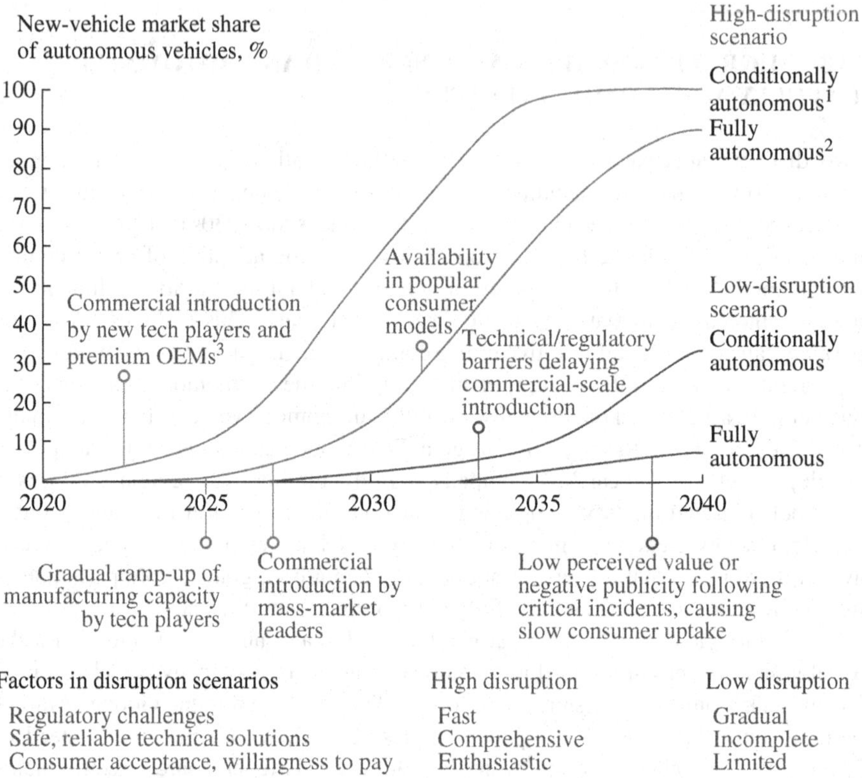

Factors in disruption scenarios	High disruption	Low disruption
Regulatory challenges	Fast	Gradual
Safe, reliable technical solutions	Comprehensive	Incomplete
Consumer acceptance, willingness to pay	Enthusiastic	Limited

[1] Conditionally autonomous car: the driver may take occasional control.

[2] Fully autonomous car: the vehicle is in full control.

[3] Original-equipment manufacturers.

FIGURE 24.7 Estimates of Autonomous Vehicle Market Penetration.

Version 7.0 software upgrade. Autopilot "allows the Model S to steer within a lane, change lanes with the simple tap of a turn signal, and manage speed by using active, traffic-aware cruise control. Digital control of motors, brakes, and steering helps avoid collisions from the front and sides, as well as preventing the car from wandering off the road" [5]. Many traditional auto manufacturers have also begun including various levels of autonomous functionality in their 2015 and 2016 fleets.

Figure 24.7 illustrates McKinsey & Company's [6] predictions about market penetration of autonomous vehicles over the next several decades. In this figure, McKinsey illustrates the wide range of adoption predictions, by showing both a "High-disruption scenario" and a "Low-disruption scenario." The figure shows the predicted percentage of new vehicle market share of both partial (conditional) and fully autonomous vehicles in both of these disruption scenarios.

Figure 24.8 projects fully autonomous implementation rates, based on previous vehicle technology deployment. In this report from the Victoria Transport Policy Institute [7], it is assumed that fully autonomous vehicles will be available for sale and legal to drive on public roads around 2020. The report also assumes that initial implementations of autonomous vehicles will be relatively expensive for individual consumers and that they will still have many technical and performance problems that are worked out over time.

Many other factors will impact adoption rates of fully autonomous vehicles. These include government mandates, that is, will different governments begin to mandate full automation for all new cars or eventually all cars on the road? Outside of government mandates, the broader development of new state and federal regulation and policy around autonomous vehicles will have a tremendous impact on the rate of adoption, helping to either speed up or slow down adoption, depending on what the regulations stipulate. It is also important to note that rates of adoption are also expected to vary widely between dense urban centers and more rural locations.

Another factor impacting overall adoption rates is the utilization of fully autonomous vehicles by mobility service providers or Transportation Network Companies (TNCs), such as Uber and Lyft, and car share companies, such as Car2go and Zipcar. If these mobility service providers implement autonomous vehicles as widely and as quickly as some predict, this could have the dual effect of helping reduce individual car ownership while increasing the sheer number of autonomous vehicles, and therefore the percentage of autonomous vehicles, that are on the road.

All of these factors combine to create a wide range of autonomous vehicle adoption estimates. Figure 24.8 illustrates this range as optimistic and pessimistic estimates. But outside of the predictions in this particular study, adoption estimates vary even more widely depending on the source.

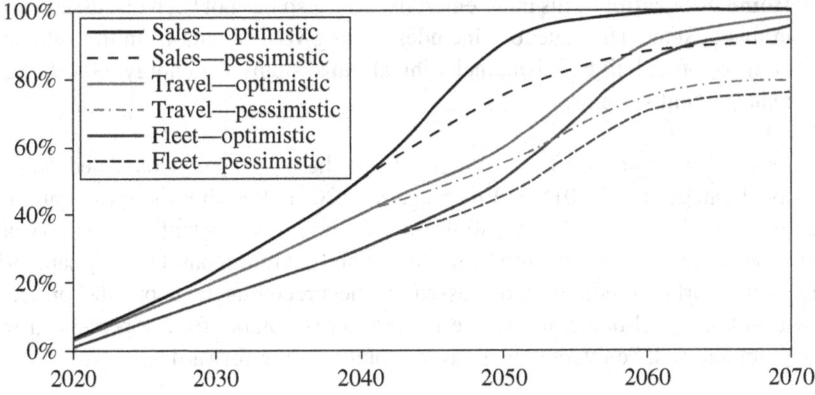

FIGURE 24.8 Autonomous Vehicle Sales, Fleet, and Travel Projections.

24.10 MARKET GROWTH FOR CONNECTED AND AUTONOMOUS VEHICLE TECHNOLOGY

The market potential for the Connected and Autonomous Vehicle is tremendous. In the 2015 edition of the annual Connected Car Study conducted by Strategy&, the strategy consulting team at PwC, the authors "foresee annual sales of connected car technologies tripling (from 2016 estimates) to €122.6 billion by 2021" [8]. This report identifies seven functional areas where auto manufacturers are focusing their Connected Vehicle development:

- **Autonomous driving**: The definition used for this category in this report includes all partially automated functionality, as defined in earlier sections of this chapter.[3]
- **Safety**: This category includes collision avoidance sensors, lane monitoring, emergency call functionality, and other safety systems.
- **Entertainment**: Includes functions that provide music and video to passengers and the driver, as well as smartphone integration, Wi-Fi, etc.
- **Well-being**: This emerging area has not been discussed earlier in this chapter but refers to sensors that are monitoring driver's health and competence. Examples include monitoring of heart rate and driver fatigue, which can then trigger audio or even physical alerts (e.g., vibrating seat or wheel) to the driver.
- **Vehicle management**: Includes the monitoring of vehicle diagnostic data and potential subscription service to notify mechanics or car dealerships for proactive maintenance.
- **Mobility Management**: Includes navigation information as discussed earlier, but expands on this to include guidance for faster, safer, more economical, and more environmentally friendly driving, based on historical and real-time data collected by the vehicle and from traffic sources.
- **Home integration**: This other emerging area also has not been discussed earlier in this chapter. This category includes linking vehicle data from the vehicle to home or other building data, including alarm systems and energy monitoring or management systems.

Figure 24.9 projects market potential of these various Connected Vehicle functional categories in 2016 and then again in 2021. Possibly the most important issue illustrated in this is the overall tripling of the market potential in just 5 years' time. The projected market growth in Safety and Autonomous Driving categories reflects the earlier predictions discussed in the preceding text, but the projected growth in both Well-being and Home integration is notable from a percent growth standpoint and will be interesting areas to monitor going forward.

---

[3] Note: This is an important distinction. In this chapter and in most discussion on this topic, the word autonomous is specific to no driver interaction at all. In this report, it is used for even partial autonomy.

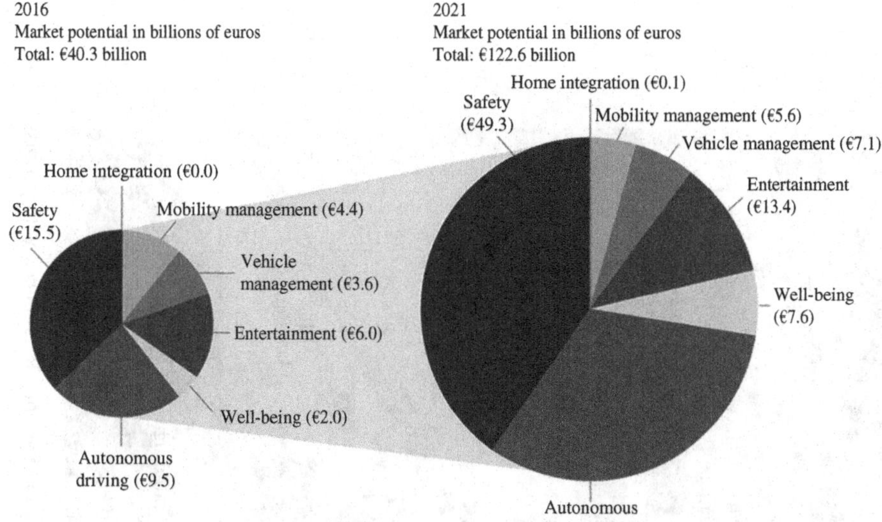

Note: Totals may not reflect sums due to rounding. Passenger vehicles only, excluding light commercial vehicles.
Source: Strategy&

FIGURE 24.9 Estimated Market for Connected Car Technologies, 2016–2021.

24.11 CONNECTED VEHICLES IN THE SMART CITY

Connected Vehicles play a crucial role in delivering on the promise of the Smart City. Many of the applications described earlier support the individual consumer and offer services for entertainment, comfort, safety, and efficiency. But the biggest societal benefits of the Connected Vehicle come when it is part of an overall Smart City system, including parking management, toll collection, traveler information and trip planning, and supporting intermodal and multimodal connections. The collection and analysis of the data from all of these systems and the information dissemination to citizens allow for more informed decisions by the agencies and by the general public.

Figure 24.10 illustrates how the European Telecommunications Standards Institute (ETSI) visualizes the Connected Vehicle participating in an overall Smart City environment. ETSI and ERTICO (the European ITS organization) refer to the Connected Vehicle and the broader regional ITS system as Cooperative Vehicle Infrastructure Systems (CVIS). This figure illustrates how the different systems fit together. Many of the other enabling technologies, Smart Cities elements, and architecture issues are discussed in other chapters of this book.

24.12 ISSUES NOT DISCUSSED IN THIS CHAPTER

There are many issues around Connected and Autonomous Vehicles that are not discussed in this chapter, and the issues that are discussed are only covered at a very high level. Indeed, a handbook focused solely on Connected and Autonomous Vehicles could easily create a volume the size of this IoT and Data Analytics Handbook.

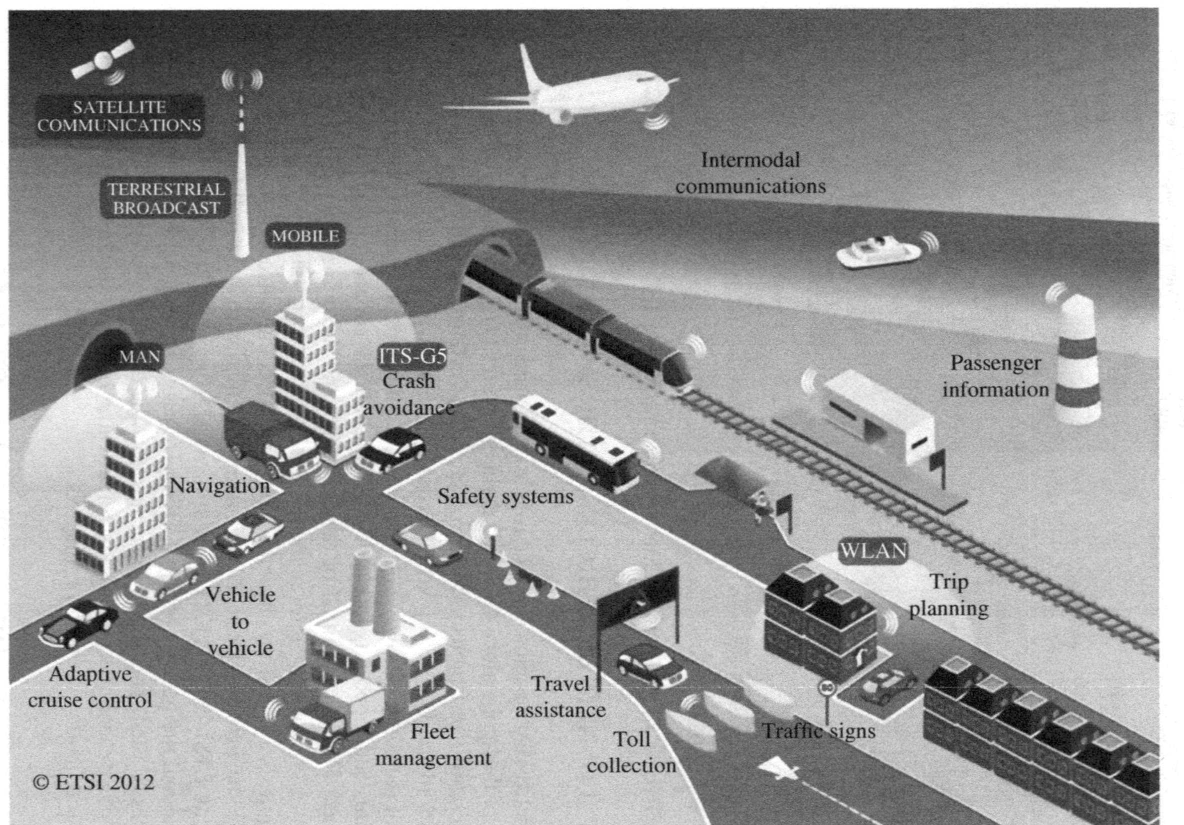

FIGURE 24.10 Smart Cities, Connected Vehicles, and Intelligent Transportation Systems. Reproduced with permission of European Telecommunication Standard Institute.

One of the main issues, not directly addressed in this chapter, is the impact that the Connected Vehicle, and particularly the Autonomous Vehicle, is expected to have on broader social issues such as land-use planning, equity, and mobility for the elderly and disabled. The combination of the technological advances discussed earlier, combined with the changing and emerging mobility business models (including the Share Economy), provides the potential for significant paradigm shift in access and mobility and therefore on these broader social issues. These are beginning to be researched, modeled, and discussed in many different forums.

Another key issue only briefly discussed earlier is network and system security. There have been many recent reports on the ease of hacking into Connected Vehicles. While it is mentioned in this chapter that security will be a factor on the implementation rate of Connected Vehicles, this is only part of the story. Acceptance of Autonomous Vehicles will require an extremely high level of confidence that the vehicles will be safe, and ensuring that they are secure will be critical to this. The promise of improved safety is one of the biggest selling points of Connected and Autonomous Vehicles from a societal standpoint, and again, ensuring that they are secure is a critical part of this.

24.13 CONCLUSION

Connected and Autonomous Vehicles offer some of the most intriguing examples of the IoT in action. The sheer number of individual sensors, the complexity of the onboard systems, the local (onboard) data processing, and decision making, combined with the remote data collection, analytics and information dissemination make the Connected Vehicle one of the best examples of IoT being used to solve everyday user needs. The Big Data Analytics elements are numerous and include both public and private sector examples. On the public sector side, the Connected Vehicle is one of the most critical elements in the transportation sector of the Smart City. The private sector is finding numerous ways to take Connected Vehicle data and offer users applications that impact their daily life. Navigation information, entertainment, vehicle health monitoring, and insurance pricing based on actual driver performance are all different existing applications that are being utilized at ever-increasing rates. New services and applications are being developed and deployed everyday that tie into smart homes, smart buildings, and smart health.

Automated vehicle functionality has been around for many years, but new functionality is advancing at a rapid pace by virtually all major auto manufacturers. Fully Autonomous Vehicles are arriving faster than many people understand. The industry is being driven not only by the traditional auto manufacturers but also by new emerging players including industry disrupters such as Google and Uber. Autonomous Vehicles have the potential to bring a significant paradigm shift in mobility and access to transportation and, along with that, promise to impact other broader social elements such as land-use planning, street design, and parking policies.

The market impact of Connected and Autonomous Vehicles is significant and is expected to continue to grow at staggering rates, at least for the near future. The overall impact of the IoT and Data Analytics to the transportation industry can't be overstated, and it is a very exciting time to be part of this industry.

REFERENCES

[1] http://www.nhtsa.gov/About+NHTSA/Press+Releases/U.S.+Department+of+Transportation+Releases+Policy+on+Automated+Vehicle+Development (Accessed August 19, 2016).

[2] NHTSA 2016: Federal Automated Vehicles Policy: Accelerating the Next Revolution In Roadway Safety, Sept. 2016. https://www.transportation.gov/sites/dot.gov/files/docs/AV%20policy%20guidance%20PDF.pdf (Accessed October 25, 2016).

[3] https://www.ral.ucar.edu/solutions/products/pikalert%C2%AE (Accessed August 19, 2016).

[4] "How Google quietly revved up its very own car company" The Guardian, August 2015 http://www.theguardian.com/technology/2015/aug/01/google-auto-car-making-company (Accessed August 19, 2016).

[5] "Your Autopilot has arrived" Blog: The Tesla Motors Team October 14, 2015. https://www.tesla.com/blog/your-autopilot-has-arrived (Accessed September 14, 2016).

[6] Disruptive trends that will transform the auto industry; McKinsey & Company, January 2016. http://www.mckinsey.com/industries/high-tech/our-insights/disruptive-trends-that-will-transform-the-auto-industry (Accessed September 14, 2016).

[7] Autonomous Vehicle Implementation Predictions Implications for Transport Planning December 2015; Todd Litman, Victoria Transport Policy Institute.

[8] Connected Car Study 2015: Racing ahead with autonomous cars and digital innovation, by R. Viereckl, D. Ahlemann, A. Koster, S. Jursch, September 16, 2015. http://www.strategyand.pwc.com/reports/connected-car-2015-study (Accessed September 14, 2016).

25

IN-VEHICLE HEALTH AND WELLNESS: AN INSIDER STORY

PRAMITA MITRA, CRAIG SIMONDS, YIFAN CHEN AND GARY STRUMOLO

Ford Research and Advanced Engineering, Dearborn, MI, USA

25.1 INTRODUCTION

The automotive industry is expanding beyond the traditional concerns of cars and transportation into personalized ownership experiences that would enhance the connected lifestyle of its customers. Helping them manage their health and well-being is fast becoming an important product development priority. Leveraging the intersection between automotive connectivity, connected healthcare, and proliferations of mobile apps, Health and Wellness (H&W) solutions are emerging as the newest set of automotive products and services in two ways: adding value to automotive connectivity and evolving notion of automotive safety.

Today we live in a connected world where the connected consumer expects or rather demands car manufacturers to make connected cars and provide cloud-connected services. As a result, cars are becoming one of the smartest consumer "electronics devices." In the automotive industry, there is an emerging shift of product development mindset from building cars to building transportation as a service where the unified digital ownership and driving experience is becoming a key differentiator. In-Vehicle Infotainment (IVI) systems have been the cornerstone technology to delivering such user experiences inside the vehicle. Launched in 2007, Ford SYNC® [1] has played a major role in driving the innovation for voice-activated in-car connectivity and infotainment systems. SYNC® allows users to make

Internet of Things and Data Analytics Handbook, First Edition. Edited by Hwaiyu Geng.

Companion website: www.wiley.com/go/Geng/iot_data_analytics_handbook/

hands-free phone calls, control music, and perform other functions with the use of voice commands. Furthermore, SYNC AppLink® is a suite of Application Program Interface (API) that enables third-party mobile developers to extend the command and control of a mobile app to the SYNC Human–Machine Interface (HMI). AppLink® transfers the normal functionality processed by touching a mobile device screen into the vehicle controls, such as voice commands, steering wheel and radio buttons, and capacitive in-dash touch screens. Additionally, new device integration features like Apple CarPlay® and Android Auto® are now added to SYNC, thus enabling iPhone users access to Maps, Messages, Phone, and Music through Siri® voice control or touch screen or Android users easier and safer access to Google Voice Search®, Google Maps®, Google Play® Music, and more via steering wheel controls and touch screen. As SYNC continues to offer enhanced capabilities and services to customers, we envision it will take on a larger purpose beyond catering to the infotainment needs of the customers and expanding into caring for their broader well-being. The US Department of Transportation has projected that the amount of time we spend behind the wheel is significantly on the rise [2]. As a result, vehicles are becoming a logical, private, and convenient place from which to manage our health while in motion.

The automotive industry's notion of safety has also evolved over the years, and changes were driven by a variety of factors. Originally, safety was all about crash-worthiness, that is, protecting the driver and passengers from harm in a crash. Gradually the definition of safety expanded into other factors as in-vehicle HMI became more comprehensive and complex in nature. Factors like usability, time on task, button size, brightness, and contrast ratio quickly became safety metrics, and standards for user interfaces were established through a variety of studies. These efforts provided insight into how people interact with machines and what other tasks could reasonably be expected of a human being while driving a car. As newer safety technologies (e.g., blind spot detection, front collision warnings, and crash mitigation) were developed, the definition of safety and the metrics that aid in quantifying it broadened. Recently, autonomous driving added yet another whole new dimension. It created the need for a new set of metrics and an even broader definition of safety. First, there is safety for people (and things) outside the vehicle while the vehicle is in autonomous mode. In this case, object detection and classification are key elements in avoiding crashes and minimizing the threat to pedestrians. In addition, the car must now understand if and when the driver is ready to take control, which requires a frequent if not continuous assessment of the driver's state. The car must also keep the driver informed of its intended path and of anomalies along the way. As the well-being of the automobile consumer beyond their survival has grown in importance, a shift from a one-size-fits-all safety mindset has occurred. We are now concerned about the particular health issues of the individual and how those affect safety and driving performance. For example, blood sugar imbalances can be detected before any outward manifestations that might affect driving actually occur. If a sensor detects the imbalance and notifies the vehicle, the vehicle's HMI can notify the driver so he or she can address the issue before a more serious situation arises.

Looking ahead, we can assume that the H&W of the driver will continue to grow in importance in both the sales and use of automobiles. From a sales perspective, this is similar to the way infotainment features were once seen as nice but not important. Now they are one of the key factors in a vehicle purchase decision. Likewise, the definition of H&W will continue to grow in breadth, and the vehicle features that are perceived to improve our H&W will become key factors among those seeking to buy a vehicle. They may also have a significant impact on how and when autonomous features are initiated.

The rest of the chapter is organized as follows. Section 25.2 provides a detailed overview of the in-vehicle technology enablers for H&W solutions. Section 25.3 outlines the top H&W features in the automotive industry, followed by a discussion in Section 25.4 on the technical challenges for implementing in-vehicle H&W products and services. Finally, Section 25.5 concludes the chapter.

25.2 HEALTH AND WELLNESS ENABLER TECHNOLOGIES INSIDE THE CAR

One of the major challenges for the automotive industry in the 1990s was the rapid growth of electronics features. Before that time, styling and drivability (ride, handling, and engine performance) were the key factors in consumer purchase decisions. As more and more electronics features became available, people began placing an ever-increasing emphasis on them when they looked to buy a new car. At the same time, engineers started to realize that only some of these desirable features could be completely built into a vehicle. Others required a built-in component plus interaction with the devices consumers carry into their vehicles. Some of those features also required information or services delivered wirelessly from off-board sources. Therefore, in this new paradigm, new features were often seen to have built-in, brought-in, and beamed-in components. That same categorization of components also applies to in-vehicle H&W solutions.

25.2.1 Built-In Technologies

Built-in technologies are enabled through hardware the automotive original equipment manufacturers (OEMs) built into their vehicles, such as the onboard spoken dialog system or the embedded modem that allows the vehicle to connect to the Internet directly. IVI has been the precursor and first enabler built-in technology for driver safety and H&W for a long time. In the 1990s, Ford developed a concept called the Vehicle Consumer Services Interface (VCSI) [3]. The VCSI (Figure 25.1) was designed to offer connectivity to brought-in devices as well as beamed-in services and offered an API that enabled software developers to create new features. The VCSI was the predecessor to Ford SYNC and similar automotive connectivity technologies.

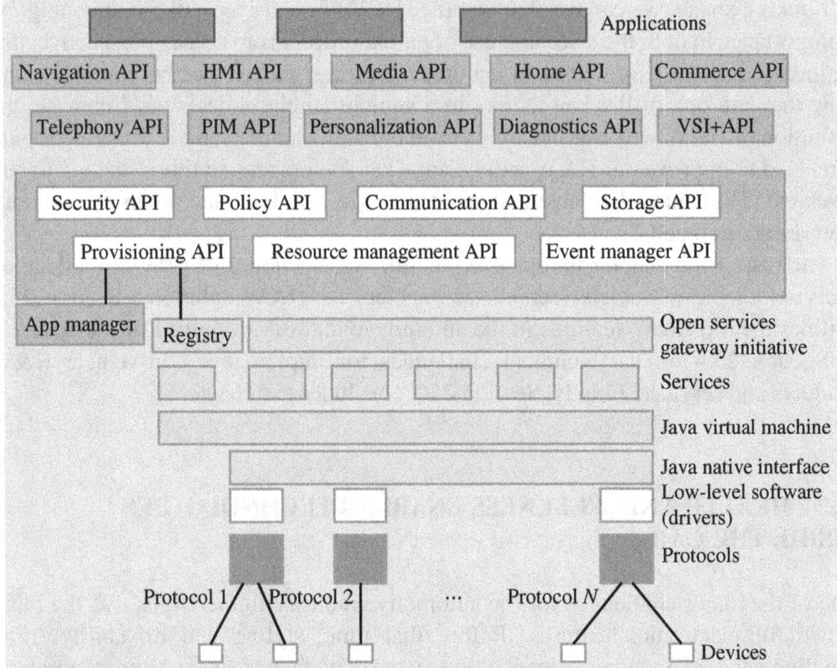

The figure shows an in-vehicle computing platform. Devices can be anything from vehicle sensors and actuators to radios, climate control modules, cell phones, or PDAs. The protocols layer indicates that most of these devices will be accessed through some sort of bus or network (such as MOST, CAN, LIN, etc. and so on). Above that are the drivers and interfaces between the hardware and the VCSI APIs. The applications layer sits on top, and the VCSI APIs provide a layer of abstraction from the hardware. Applications can thus be written once and used across many vehicle programs, and possibly over several model years.

FIGURE 25.1 Ford Vehicle Consumer Services Interface. Reproduced with permission from Ford Innovation and Research Center.

More recently, Ford partnered with a leading German university in 2012 to develop an in-vehicle heart monitoring seat that consists of six sensors embedded in the driver seat (Figure 25.2). A second experimental system [4] includes infrared sensors on the steering wheel to monitor the palms of a driver as well as her face looking for changes in temperature. A downward-looking infrared sensor under the steering column measures the cabin temperature to provide a baseline for comparing changes in the driver's temperature. There is also a sensor embedded in the seat belt to assess the driver's breathing rates. These sensors are able to provide real-time health data. One of the H&W solutions that leveraged this technology is a driver workload estimation that uses data collected from the sensors in combination with information about the driving environment and the driver's behavior. This is discussed in detail in Section 25.3.2.1.

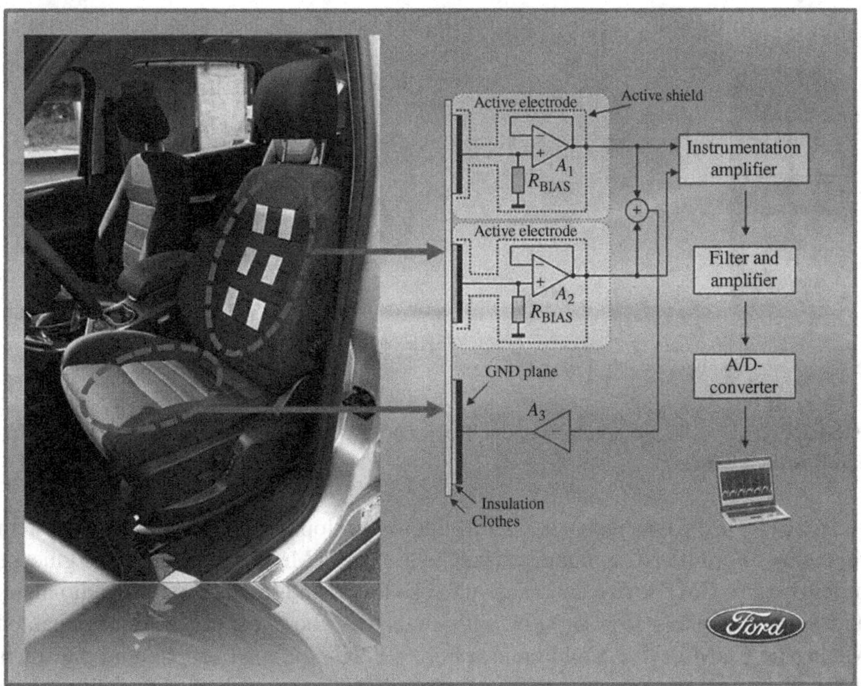

FIGURE 25.2 Heart rate monitoring seat. Reproduced with permission from Ford Innovation and Research Center.

25.2.2 Brought-In Technologies

In the late 1990s, consumers wanted to use their cell phones or MP3 music players in their vehicles. Unfortunately, at that point there were only a few, rather crude, after-market products that enabled this. Also, the introduction of digital music and devices like the iPod made the CD players that automobile OEMs built into vehicles seem outdated. Younger consumers had hundreds of songs on their digital devices, yet couldn't effectively or safely use them while driving. This pushed the industry to implement systems like Fiat Blue and Me® and Ford SYNC in the early 2000s, which connected to these devices through Bluetooth and made use of vehicle HMI systems to access and make use of content. Today, there are many portable and wearable devices in the hands of consumers, and most of them offer something useful to them while in their vehicles. This is why automobile OEMs have implemented Bluetooth and Wi-Fi connectivity along with the enhanced software and HMI solutions to make the best and safest use of these brought-in devices. Before smartwatches, most of these devices offered information or services that were accessed through the vehicle HMI. On the other hand, these new smartwatches offer their own HMI that can enhance what the vehicle already offers. Haptics, for example, can be very useful for safely alerting drivers about important issues or events. Likewise, biometric data from the wearable can give the vehicle insights into the driver's physical state and ability to respond to driving events.

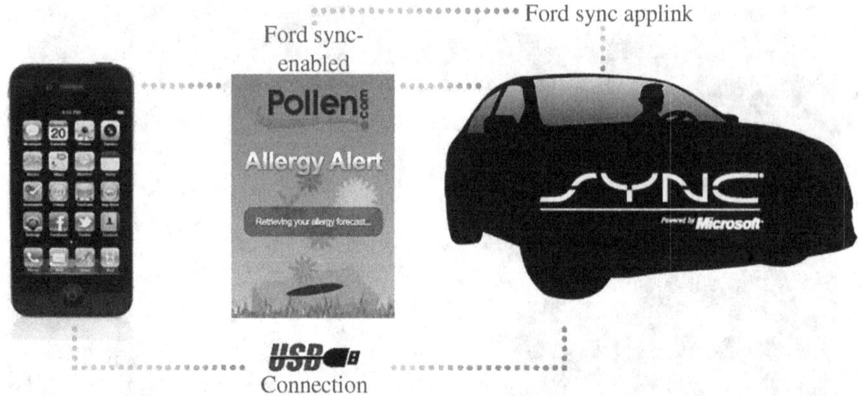

FIGURE 25.3 Allergy monitor system. Reproduced with permission from Ford Innovation and Research Center.

In 2011 Ford partnered with Medtronic to develop a research prototype system that allows Ford SYNC to connect via Bluetooth to a Medtronic continuous glucose monitoring (CGM) device and share glucose levels and trends through audio alerts and visual displays [5]. Drivers with diabetes who wear a Bluetooth-enabled Medtronic CGM device could enter a Ford SYNC-equipped vehicle and pair their device—as well as their smartphone—with SYNC, giving them the ability to use voice commands or steering wheel controls to receive audible alerts or center stack displays about deviations and trends related to their blood glucose levels. If a driver is going hypoglycemic, they may become confused or lightheaded, and their vision may become blurred, all of which could become a safety issue—hence being able to receive alerts related to their condition enables them to take corrective action in time. Ford also partnered with SDI Health to enable its Allergy Alert app to communicate via Ford SYNC using Ford SYNC AppLink®, giving users voice-controlled access to location-based day-by-day index levels for pollen, asthma, cold and cough, and ultraviolet sensitivity as well as 4-day forecasts (Figure 25.3). Easy in-car access to this type of information can help asthma and allergy sufferers plan healthier route choices and prepare for areas with high symptom triggers that could quickly lead to an attack.

In the future the car could respond even more intelligently to environmental conditions—automatically turning on the recirculation mode to keep pollen-loaded air from reaching passenger compartment—or even suggesting healthier driving route alternatives. The recent Ford partnership with Amazon in Echo® and Alexa™ (a cloud-based voice service system) [6] for combining the connected drive and home experiences (e.g., WeMo® [7], Hue® [8], and Wink® [9]) is an attractive path to extending home health monitoring into the car cabin.

To that end, this will help an automotive OEM become a more integral part of the Internet of Things (IoT) ecosystem without having to develop its own platform. With Ford SYNC and Amazon Echo, users can ask Alexa to start their car and check on the range of their electric vehicle from inside their home. While in the car, using a SYNC

AppLink enabled companion app, drivers can talk to Alexa and control their home environment. We envision that this platform will open up many opportunities for brought-in H&W solutions. For instance, the user can query the system while at home about the pollen condition on their planned route before starting a trip and precondition the vehicle cabin accordingly.

25.2.3 Beamed-In Technologies

Beamed-in products and services are enabled through harnessing the power of cloud computing and data storage by leveraging the connectivity brought in by consumer devices or by a secure embedded connectivity platform (e.g., a telematics control unit with embedded 4G LTE connectivity). Beamed-in H&W solutions have been fueled by the recent growth of the IoT. The networking and integration of things that were previously separated provide great opportunities. As we tie the devices and information we carry with us to our cars and homes, a world of data and services becomes available to us wherever we go. Also, our knowledge of healthy choices increases, and we participate in and benefit from solutions that come from big data analysis. We also have access to the knowledge of world's best diagnosticians through cloud-connected services.

Ford and WellDoc, a leader in the field of mHealth integrated services, collaborated in 2011 to extend Ford SYNC Services to WellDoc H&W solutions for people with chronic health conditions (e.g., diabetes and asthma) [5]. Ford SYNC Services leveraged connectivity to an off-board network called Ford Service Delivery Network (SDN)™ through brought-in connectivity (e.g., smartphone). SDN provided location-based traffic, directions, and information providers that drivers can access via their smartphone; and also made it possible for new H&W services such as WellDoc DiabetesManager™ and AsthmaManager™ to be added as in-car services for drivers. Another example of beamed-in H&W solution is the Ford–Microsoft–Healthrageous Mobile Connected Health Experience in 2011. This service promises *anytime–anywhere* health coaching at home (using a touch interface on Windows tablets) and in the car while driving (using the voice interface with Ford SYNC). Healthrageous specializes in connected and personalized health management and provides a solution that utilizes physiological data collected from blood pressure monitors, activity monitors, and glucose meters as well as behavioral data shared by the user, for identifying unhealthy habits and making healthier lifestyle suggestions. The integrated system architecture is presented in Figure 25.4. The system captures driver biometric and vehicle data as the basis for real-time H&W advice and monitoring. Additionally, the driver can provide voice inputs leveraging the Ford SYNC platform that allows them to detail aspects of their health routine (e.g., number of glasses of water consumed during the day) in a completely hands-free manner while driving. The data received from the driver is then uploaded into the Microsoft HealthVault® running on Windows Azure® Cloud Platform. The real-time health data is integrated with other health data (e.g., historical health data) and analyzed to create graphical reports the driver can access after having left the vehicle. This opened up a new horizon of future beamed-in H&W solutions that let people to monitor and improve their health wherever they go.

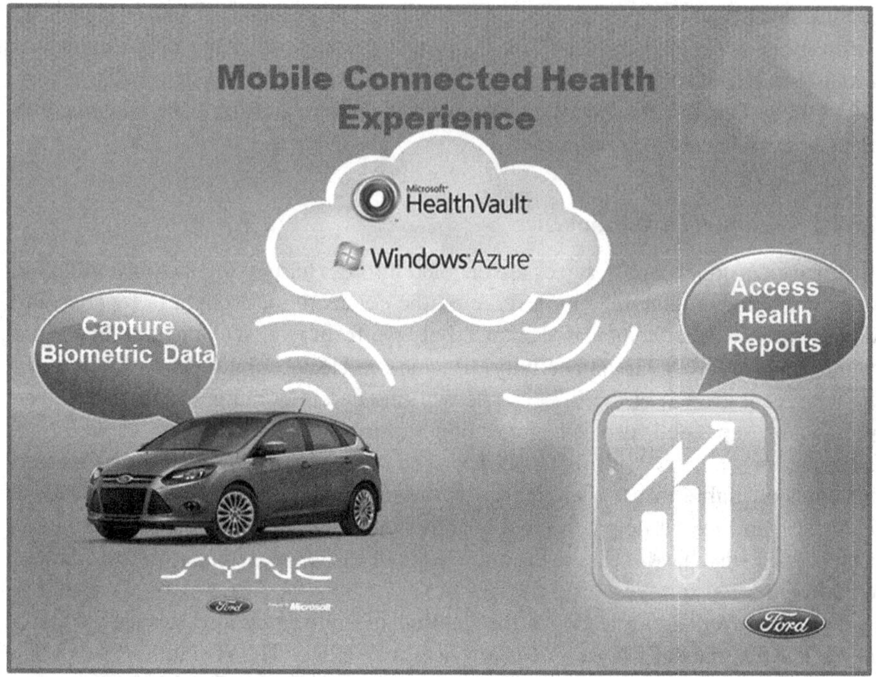

FIGURE 25.4 Mobile Connected Health Experience. Reproduced with permission from Ford Innovation and Research Center.

FordPass [10] is the newest wave in the category of beamed-in services from Ford. It is a cloud-based platform that would transform the relationship between automakers and the consumer in a similar fashion what iTunes™ did for music industry. FordPass features four smart mobility experiences to benefit members:

1. Marketplace includes mobility services that includes partnerships with ParkWhiz and Parkopedia to help people find and pay for parking more easily, and with FlightCar to borrow and share vehicles when they travel.
2. FordGuides help consumers move more conveniently by allowing them to speak directly to trusted and knowledgeable FordGuides using live chat launched from the app.
3. Appreciation where members are recognized for their loyalty—current program includes upcoming partnerships with McDonalds and 7-Eleven to reward FordPass members with merchandise and unique experience.
4. FordHubs where consumers can experience latest innovations at Ford. FordPass membership is free for everyone via online registration.

The appreciation experience of FordPass provides an attractive choice for the next wave of beamed-in H&W solutions. For instance, a connected nutritionist app could leverage the FordPass Appreciation column to build partnerships with healthcare providers in the future to provide drivers with safe and convenient search access (using voice commands or steering wheel radio buttons) to find healthy eating options nearby. Easy in-car access to this type of services can help members plan healthier eating choices while on the move. The users of the future would be able to (1) initiate personalized and multicriteria search for healthy dining options (based on preference, allergy, physical activity, etc.), (2) get turn-by-turn navigation directions to the selected restaurant, (3) pay using FordPay and get rewards, and (4) participate in a Pay-How-You-Eat (PHYE) health insurance program. The last feature can further provide a path to another powerful H&W solution for future users, that is, a holistic Usage-Based Insurance (UBI) for FordPass members. UBI programs are gaining fast popularity in the auto insurance industry (e.g., State Farm Drive Safe and Save Program [11] and Progressive Snapshot Program [12]) and can usher similar benefits in the health insurance industry by having policyholders' activity and eating behavior data reported via FordPass to the insurance carrier for possible premium reductions while promoting a healthy lifestyle.

25.3 HEALTH AND WELLNESS AS AUTOMOTIVE FEATURES

The ever-increasing demand for continuous innovation in connected automotive products and features has led automobile OEMs to work with noncompeting stakeholders (e.g., healthcare, IT, device manufacturers, insurance, etc.) to offer value-added services like H&W for vehicle occupants. According to recent study by Frost and Sullivan titled "Top Global Mega Trends to 2025 and Implications to Business, Society and Cultures," wearable healthcare revenue in 2020 is expected to reach $5.45 billion [13]. By leveraging the convergence of automotive connectivity, IoT, and Wearables, the automotive industry is pushing the technology envelope for enabling more H&W features that complement the infotainment and advanced safety features now in place. In the same Frost and Sullivan study, two classes of driver H&W features are predicted to be implemented as standard features by 2025, that is, driver physical state and environmental monitoring (i.e., allergen level, blood pressure, alcohol content, and fatigue) and luxury and comfort features (i.e., heating and massage seating). The current chapter focuses on the first classes of H&W features, that is, monitoring of driver and external conditions.

25.3.1 Monitoring of Chronic Conditions and Environmental Factors

There are a number of health conditions that can be monitored or detected through wearable devices. There are also a number of environmental factors the vehicle can detect or monitor through built-in systems. With the two working together, the H&W of the driver can be addressed reasonably well. For example, chronic conditions like allergies and asthma can be greatly affected by environmental conditions. Vehicles already sense atmospheric temperature and sun load. Powertrain control systems sense air density and sometimes atmospheric pressure. Other sensors can be added to sense things like humidity, CO_2, and allergens to aid in predicting the effect of these environmental conditions on vehicle occupants, especially those with known chronic conditions. With the onset of wearable/digital health devices came predictive algorithms that use sensed biometric data and sensed movement to derive various health conditions. For example, apps on smartwatches and fitness bands use heart rates and accelerometers to estimate activity and sleep quality, but some also sense skin PH and other parameters that can aid in deriving a broader understanding of the driver physical state.

One of the challenges that arise in causing an automobile to respond to a derived physical condition is the quality and accuracy of the sensor and predictive algorithms. Devices designed for use in medicine have been thoroughly tested and certified before doctors rely on them for diagnostic purposes. Consumer devices like smartwatches are not tested or certified in the same way. As a result, they may be more likely to lead to a misdiagnosis and therefore an inaccurate notification or unnecessary intervention by the vehicle. Timing is another significant factor affected by the quality of the sensor, the accuracy of predictive algorithms, and the time it takes to access off-board data and services. Some conditions that require a response from the vehicle happen slowly. For example, blood sugar issues can be sensed minutes before they affect the driver's performance. However, other health issues can cause an almost instant degradation in the driver's performance, and the time it takes to sense, calculate, communicate, and respond may be excessive. These issues force design changes for the H&W solution and most likely require preemptive steps, like predicting the likelihood of physical events based on driving scenarios and environmental conditions.

These H&W issues can be addressed by making the vehicle aware of road conditions and driving events that can increase the physical or emotional stress experienced by a driver. These road or environmental events can also aggravate a variety of conditions, including hypertension and breathing disorders. Traffic jams, road hazards, and weather can all add significantly to the stress experienced by drivers. If the vehicle knows ahead of time what is about to happen, it can then suggest alternate routes or a change in speed or take other steps to reduce the impact of such events. The vehicle can obtain prior knowledge of a driver's physical challenges or medical issues (e.g., sleep quality or allergies) in a variety of ways. The data may be stored in a smartphone or in the cloud and may even be tied to ongoing treatment that is being monitored and managed by a doctor. In the future, it is just as likely that a doctor will be notified as a result of something that happens while driving and information that comes from a vehicle may have a significant contribution to the diagnosis as well as any needed response.

25.3.2 Driver Monitoring

25.3.2.1 Driver Distraction Distracted driving, that is, the act of driving while engaged in other activities, is an ever-increasing concern across the automotive industry and government organization [14]. Activities like texting, talking on the phone or to other passengers, watching videos or reading on a personal device, or interacting with the in-vehicle systems, and so on take the driver's attention away from the road. Many automakers provide a "do not disturb" feature in their IVI systems to help the driver stay focused in busy situations by using the vehicle's own intelligence for managing incoming communications. Such a feature blocks incoming calls and text messages and can also be made to respond to text messages with an automated reply or to route calls to voicemail.

The "driver workload estimator" [15] is an algorithm that further helps the driver in high-workload driving situations using real-time data from existing vehicle sensors such as radar and cameras combined with driving behavior data such as inputs from the driver's use of the throttle, brakes, and steering wheel. This technology makes use of the intelligence of the car and provides smart management of in-vehicle communications and features tailored by the assessed workload of the driving situation and the value of the service to the driver. For instance, data from the sensing systems of driver-assist technologies are used to determine the amount of external demand and workload upon a driver at any given time including traffic and road conditions. The side-looking radar sensors used for monitoring blind spots and the forward-looking camera used for lane-keeping are on watch even when there is no active warning provided to the driver. These signals could indicate if there is a significant amount of traffic in the lane that the driver is merging into while entering a highway. If the driver is found to have increased throttle pedal pressure to speed up in this high traffic situation, the workload could be estimated high enough for an incoming phone call to ring inside the cabin (Figure 25.5).

In addition to using existing vehicle data to estimate demand on the driver, there is significant value in combining driver biometric or health data to get a holistic view of the driver workload and better tailor the experience when behind the wheel. The biometric sensors used for driver physical state monitoring could be vehicle-embedded (i.e., embedded in the steering wheel, seat, and seat belt) or brought-in (i.e., wearable devices). Using biometric data collected from a brought-in wearable device (as an alternative of using vehicle-embedded biometric sensors) for the driver workload estimator algorithm reduces the vehicle component engineering and integration costs and also mitigates challenges posed by the Food and Drug Administration (FDA) regulations. Many wearable devices (e.g., Intel Basis Peak®) provide additional insights into driver's physical state, such as heart rates, galvanic skin response, or last night's sleep score. Such data could be valuable in detecting high-risk scenarios (e.g., extreme heart rates or driver falling asleep on the wheel) that may impact the driver's ability to control the vehicle. With a holistic picture of the driver's H&W data blended with data from vehicle sensors, the car will have the intelligence to dynamically adjust the alerts provided to the driver and filter interruptions. For instance, the vehicle control system could increase the warning times for forward collision alerts and automatically engage the "do not disturb" feature helping the driver stay focused on the road.

25.3.2.2 Accident Reconstruction Monitoring the driver via access to integrated vehicle and driver health data could be of great interest for longitudinal data analysis applications such as accident reconstruction. For instance, auto insurance companies typically use the vehicle event data recorder (EDR) data as a source of reliable empirical evidence for accident reconstruction, evaluate liability, target claim costs and identify fraudulent claims, validate physical evidence and statements, and so on.

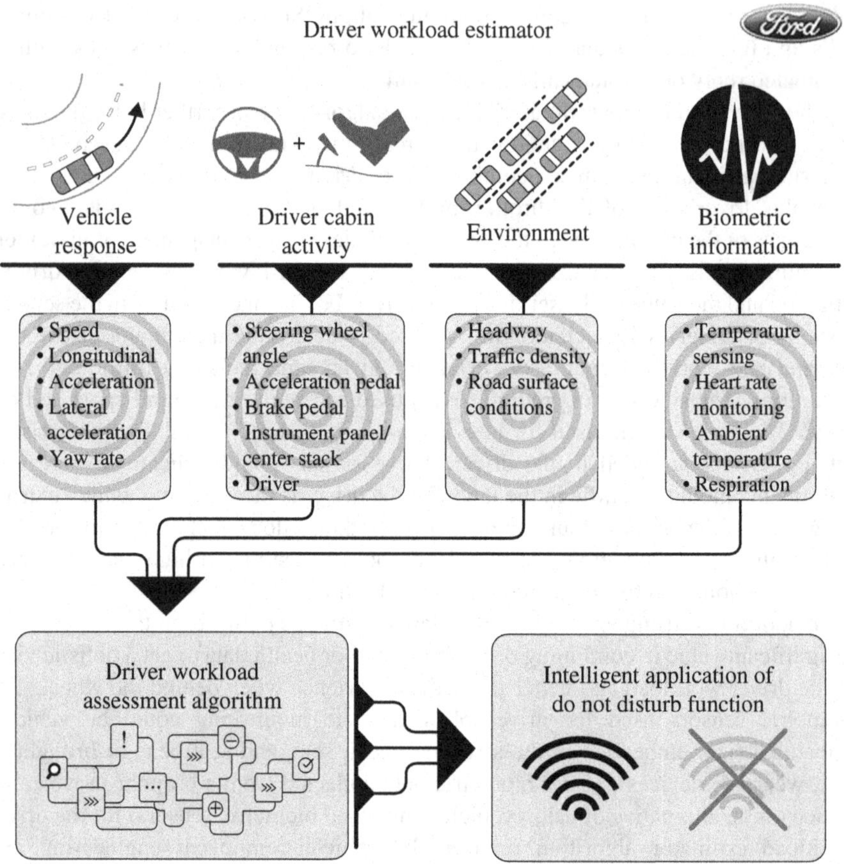

FIGURE 25.5 Driver distraction. Reproduced with permission from Ford Innovation and Research Center.

On the other hand, the state-of-the-art wearable devices can monitor precrash driver physical states (such as drowsiness, heart rates, blood glucose levels, etc.) and conditions (such as rheumatoid arthritis, injuries, etc.), which could have potentially impacted the driver's ability to drive and/or control the vehicle and ultimately led to the crash. It will be only more likely that in the future prosecutors and defense attorneys alike would use driver health data collected from wearable devices. Given the scenario, it will be of interest for advanced accident reconstruction systems and processes to have access to correlated precrash data from the vehicle as well as the driver. Driver health data collected from wearable devices may be pushed to cloud-based data fusion and analytics platform for performing real-time correlation of the vehicle data with wearable data. This will enable a comprehensive reconstruction of what happened inside the vehicle cabin prior to the impact. On the other hand, drivers could qualify for lower insurance (both auto and medical) premiums by sharing vehicle and health data as part of participating in "healthy living programs."

25.3.3 Driver Health and Wellness in Autonomous Vehicles

The ability to measure driver alertness and vital signs such as blood pressure and heart rates via wearable or vehicle-embedded technologies has potential applications for autonomous vehicles. The SAE Standard J3016[1] outlines six levels of driving automation: 0 (no automation), 1 (driver assistance), 2 (partial automation), 3 (conditional automation), 4 (high automation), and 5 (full automation).

There are many opportunities for improving the driver-assist technologies (level 1) using H&W features. In such situations, the driver biometric data can be utilized to adjust vehicle active safety system triggers based on certain medical condition. Say a driver is in rush-hour traffic, and the fitness tracker she wears picks up a spike in her heart rates. It could be an indication of increased driver stress level, and in that case, the adaptive cruise control system might automatically increase the distance between her vehicle and the driver in front of her. Another important application scenario is when a wearable device that monitors sleep levels could make a vehicle's lane-keeping assist function more sensitive if the device data shows the driver got a bad night's sleep—and therefore might be more likely to unwittingly drift out of his lane on the highway.

In a semiautonomous driving condition (levels 2–3), it will be critical to provide real-time alerts to the vehicle of the physical state of the driver if he/she is detected to be stressed, drowsy, or unwell. In this case the vehicle can present the driver with an option to relinquish control and navigate to a rest area or hospital or call for help. On the other hand, when a semiautonomous vehicle anticipates a need for the driver intervention or take over in certain situations (for instance, navigating around a construction zone or accident), it can alert the driver to take over via the wearable based on his location inside the vehicle and alertness to take control. If the driver is detected to be sleeping, a longer lead time is required to get him/her ready to take control of the vehicle. A wearable device monitoring driver physical states can detect this condition and help the vehicle manage driver reengagement appropriately. The same wearable can also provide a haptic notification to wake up the driver.

In the case of fully autonomous driving situations where driving is no longer the primary task of the traveler, there could be a number of opportunities in the area of connected healthcare. The rising prevalence of chronic diseases is forcing healthcare providers to find new ways to effectively manage patients across the entire continuum of care. Connected health solutions are gaining rapid adoption among providers to help monitor and improve patients' health outside of the doctor's office, thus surpassing the limitation posed by the few physical patient touch points per year. There could be a number of opportunities for utilizing the automobile and wearable devices as components providing an effective health and well-being program for customers and patients of all ages and conditions. Autonomous H&W solutions could be enabler for delivering healthcare inside a vehicle cabin. For instance, there could be value in delivering personalized and interactive clinician-directed surveys to vehicle occupants for addressing nonemergency health concerns and minimizing hospital

[1] http://standards.sae.org/j3016_201401/

readmission cost. At present we lack in a good handle on how diseases are spread. Healthcare providers are overwhelmed trying to deal with the panicking people during a potential pandemic situation. On the other hand, research organizations like Centers for Disease Control and Prevention (CDC) and universities have a hard time monitoring and tracking the migration of diseases. A connected health app for Ford SYNC could engage the vehicle occupant to run through an interactive (voice-assisted or touch screen-based) checklist of symptoms and also combine the response with live biometric data from the wearable to determine the risk level of the human. This data may also be stored in the cloud for monitoring the migration of disease for clinical research. If the risk level is above a certain threshold, the car will navigate to the nearest appropriate health facility. H&W solutions can also empower the vehicle occupants to manage and schedule their clinical appointments, get assistance to pre-check-in while arriving at a connected healthcare facility, send prescription refill request to desired pharmacy and schedule pickup, and so on.

25.4 TOP CHALLENGES FOR HEALTH AND WELLNESS

25.4.1 Off- and Onboard Communications

For H&W features to be more than gimmicks, they need to incorporate the data and features provided by a consumer's own devices, as well as the health data and services often provided through cloud services. There are too many potential health problems, and it is impractical for an automobile to become a medical diagnostic center. It is much more practical for the consumer to bring the needed sensors, information, and services with them through portable and wearable devices. The car can, however, offer a comprehensive set of user interfaces and the connectivity required to connect to devices and off-board services. For a while now industry experts have been debating whether or not connectivity to the cloud should be built in or brought in through a consumer device. There are reasonable arguments on both sides, but it appears the industry is moving toward utilizing both. The services that are vehicle or driving centric come through built-in connectivity, and the ones that the consumer already utilizes when away from their vehicle come through their own devices.

25.4.2 HMI Challenges

A number of organizations, like the Crash Avoidance Metrics Partnership (CAMP) and the Association of (international) Automobile Manufacturers (AAM), worked over a period of years to better understand automotive safety issues and establish guidelines for safe implementation of automotive features. Their work was intended to guide automotive manufacturers as they attempt to develop and implement new features, as well as to improve existing ones. They also shed light on the challenges of using automotive user interfaces while driving and helped quantify the risks caused by poor HMI design. The main issues they identified include too much complexity in user interfaces, difficulties in seeing visual components due to insufficient font/object size or contrast ratio, and too much time required to complete a step or task. With some infotainment and H&W features, part of the HMI solution is implemented on the consumer's device, either brought-in or wearable. This adds another layer of complexity in managing driver distraction and workload, but it also provides the opportunity of using haptics to alert drivers to appropriate issues. Not everyone notices when a warning light is illuminated in the instrument panel, but if that light is accompanied by an audible alert or a vibrating watch, it is much more likely to be noticed. With H&W one of the challenges is to appropriately communicate criticality. We don't want to startle the driver by making a simple informational notice seem like a critical alert. Neither do we want them to ignore factors that could grow into something much more critical if it not addressed in a timely fashion. It is not a good idea to make a diabetic think they are about to go into insulin shock when the issue is really that they may need to consume some protein soon to maintain their current stable state.

25.4.3 Lack of Standardizations

The overall advances in mobile and cloud computing and communications technologies and in particular in the automotive domain with connected cars, coupled with increasing use of wearables and other personal mobile devices (e.g., smartphones and tablets) result in the production of massive amount of sensed data from the vehicle as well as the wearable. While having access to this big volume of multi-source, multidimensional data could be highly beneficial for a number of automotive applications, the challenge of transforming the large-scale, raw, disparate data into useful, actionable intelligence in a timely manner is nontrivial. The various data sources, that is, vehicle, wearable, and smartphone, produce data in disparate formats (e.g., text, image, audio, video, etc.) at variable rates. In order to address this problem, the automotive industry should work toward generating standards for integration, processing, and analysis of multimodal data from the in-vehicle sensors and wearables, which would allow third-party developers to design and build new H&W features and use-cases on the fly, with absolutely no time required in implementing the underpinning IoT architecture for each specific stovepipe application. Without a common set of interoperable standards, consumers will quickly come face to face with a very real interoperable barrier preventing the exchange of their health data across disparate platforms or with provider healthcare IT systems.

25.4.4 Security and Privacy

One of the major concerns of vehicle electrical architecture design in a connected vehicle is security. A number of security experts have already demonstrated how "hackable" many connected production vehicles are. As new features with brought-in and beamed-in components are implemented, more core vehicle systems become exposed to hackers. Considering the fact that key political and military leaders are also driving production vehicles, the need for adequate security is critical.

Apart from security, data ownership and privacy preservation are two critical factors behind customer adoption. One of the challenges faced by the automotive OEMs in the early days of IVI was the fact that streaming data (music or video content) had to be acquired from the consumer personal devices and delivered to vehicle occupants without violating the intellectual property rights of the artists or other content owners. A similar argument exists today in the area of H&W about who owns the data that comes from consumer devices and the cloud and how their rights are protected or preserved as that data is utilized by automotive solutions. With the added security threat that comes from connected vehicles, this challenge may be even greater than the one presented a few years ago by infotainment solutions. In 2014, the Health Data Exploration Project at California Institute for Telecommunications and Information Technology performed consumer surveys [16] that revealed among individuals that agreed to make their personal health data available for research: 57% would only do so on the condition that their privacy would be protected and over 90% of respondents indicated the importance of anonymity of the data contributor. The consequence in the case of loss of privacy is expected to be disastrous, as witnessed by many business cases over the last decade. These surveys also showed that the vast preponderance of consumers insisted on sole ownership or at least shared ownership of the data with a mobile device OEM that collected it.

However, the highly multimodal nature of collected data and the need for multiparty computation pose significant challenge to implementing H&W features inside the car. In multiparty computations, multiple participants (e.g., automotive OEM, wearable OEM, healthcare provider, etc.) jointly compute on their respective private data without revealing privacy information to one another except for the agreed-upon computational outcome. Such computations require harnessing the power of the cloud computing given the enormous volume of data collected from various sources in various formats (e.g., vehicle, wearable device, highway sensors, etc.) and the task at hand of transforming it into useful, actionable intelligence in a timely manner. Hence, it is important to understand the privacy implication in the emerging IoT ecosystem and develop practical methods for achieving user privacy preservation.

Figure 25.6 provides an architecture overview of our vision for a unified vehicle and wearable data integration framework. The raw vehicle and wearable data are first uploaded to their respective OEM servers where they are semantically engineered and enhanced into the single common ontology format. This data is then ingested into the data integrator cloud upon encryption which produces time-aligned, correlated fusion data. Finally, this fusion data is exposed to the users as various types of services through a service API that allows for efficient and real-time indexing and querying. The service layer has a number of software modules for implementing various aspects of the provided service, for example, service governance, license agreement, interaction management, and so on. Note that encryption of each data item is very costly. Therefore, future research should focus on developing efficient and highly dynamic privacy protection schemes for secure interaction with third-party service (e.g., insurance firms, car sharing businesses, healthcare providers, clinical and transportation forums) for enabling connected and data-driven H&W apps and services in the upcoming era. Such schemes should utilize advanced

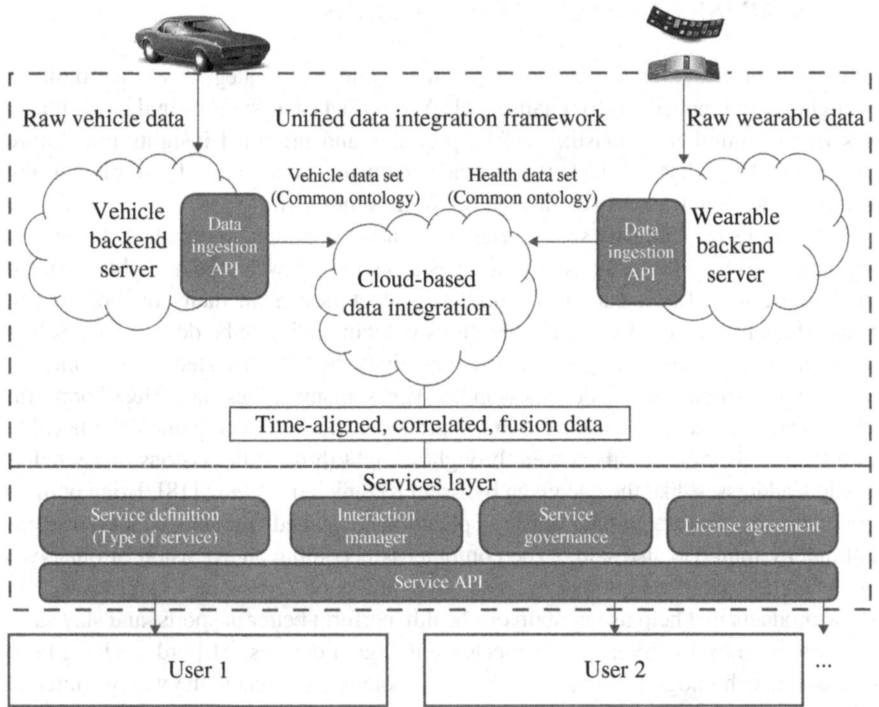

FIGURE 25.6 An architecture overview of the unified vehicle and wearable data integration framework. Reproduced with permission from Ford Innovation and Research Center.

cryptographic algorithms [17] that would allow the cloud to perform computations on the encrypted data and send encrypted results to the end-user devices and services. Thus the data would never be made available in an unprotected format to any entity other than the individual data owner (i.e., the end customer).

25.4.5 Government and Business Policies

With increasing interest in H&W solutions in the car, there is also an industry-wide concern for such solutions facing strict scrutiny from US FDA. Automakers want to steer clear from making the car into a clinical diagnostic and advice-dispensing device in order to avoid seeking FDA approval. Instead of trying to become an in-dash diagnostician, in-vehicle H&W solutions focus on providing enablers for the owners to access and manage their healthcare needs from behind the wheel. Another crucial challenge for H&W in the car is coming up with a strong business model. While brought-in and beamed-in solutions enable H&W features to be added to the car without increasing the cost of ownership, creating a business value proposition for customers across diverse partner industries (e.g., automotive, healthcare provider, insurance, etc.) is nontrivial, and so is the challenge of creating a revenue sharing model among multiple stakeholders.

25.5 SUMMARY AND FUTURE DIRECTIONS

This chapter provided a discussion of the technology categories (i.e., built-in, brought-in, or beamed-in) for enabling H&W products and features in the car. It also described a number of existing H&W products and provided insights into future trends. Finally, it discussed a few technical and business challenges for implementing H&W in the vehicle and also outlined possible solution strategies.

Many of us have already seen stories in the news of people in remote villages who are diagnosed by doctors thousands of miles away though wearable or highly portable medical devices. Local caregivers guided by doctors can, in many of these cases, respond appropriately to medical conditions without having to be doctors themselves through these new technologies and the connectivity the IoT ecosystem is providing. At the 2016 Consumer Electronics Show in Las Vegas, many of these new H&W opportunities were showcased. Some were in the products shown by companies like iHealth, Fitbit, and InBody, and others were brought to light through the visions presented in keynote addresses, like the one given by Brian Krzanich from Intel [18]. Brian pointed out that computers are gaining senses, a phenomenon called "the sensification of computing" by Intel. He also said, "The computer is becoming an extension of ourselves and is highly integrated into our everyday lives and our daily routines," and went on to show products that help us maintain our health, perform better in sports, and stay safer as we work through the use of connected, intelligent devices. At Ford we have been discussing technologies full of transformative potential to change the way we interact with our cars and creating personalized owner experiences that go well beyond the limits of infotainment. But there is one line we would never cross—it has never been our intention to turn the car into a medical device or robo-diagnostician. In this vision "the doctor in the car" is really the owner who is able to access information via SYNC that will help them take charge of their own H&W. Ford is making driver H&W a long-range strategic priority. In the upcoming era, we intend to pursue healthcare partnerships that will help develop smart mobile health solutions.

As an extension of the connected customer lifestyle, the "car that cares" is taking shape as an exciting new platform for connectivity, innovation, and transformation—every bit as much as smartphones or tablets. We believe it will fundamentally broaden the way we understand automotive safety—not so much accident avoidance as an integrated approach to ensuring the well-being of our customers.

REFERENCES

[1] "Ford SYNC", https://www.ford.com/technology/sync/, accessed August 10, 2016.

[2] "2013 Status of the Nation's Highways, Bridges, and Transit: Conditions & Performance", U.S. Department of Transportation, http://www.fhwa.dot.gov/policy/2013cpr/overviews.cfm#part1, accessed August 10, 2016.

[3] "An Embedded Architectural Framework for Interaction between Automobiles and Consumer Devices", E. C. Nelson, K. V. Prasad, V. Rasin, and C. J. Simonds, Proceedings of the 10th IEEE Real-Time and Embedded Technology and Applications Symposium (RTAS), Toronto, Canada, May 2004.

[4] "The Car Seat That Detects Heart Attacks: Ford Plans to Monitor Drivers' Pulses to Prevent Accidents", Daily Mail, http://www.dailymail.co.uk/sciencetech/article-2800101/car-seat-knows-heart-attack-ford-plans-monitor-heart-activity-cars-alert-authorities-necessary.html, accessed August 10, 2016.

[5] "Ford and Healthcare Experts Research SYNC Health and Wellness Connectivity Services Helping Manage Chronic Illness on the Go", Ford News Center, http://ophelia.sdsu.edu:8080/ford/12-20-2013/news-center/press-releases-detail/pr-ford-and-healthcare-experts-34627.html, accessed August 10, 2016.

[6] "Smart Cars Meet Smart Homes: Ford Exploring SYNC Integration with Amazon Echo and Alexa, Wink", The Ford Motor Company Media Center, https://media.ford.com/content/fordmedia/fna/us/en/news/2016/01/05/smart-cars-meet-smart-homes.html, accessed August 10, 2016.

[7] "WeMo Home Automation", http://www.belkin.com/us/Products/home-automation/c/wemo-home-automation/, accessed August 10, 2016.

[8] "Hue: Personal Wireless Lighting", http://www2.meethue.com/en-us/, accessed August 10, 2016.

[9] "Wink: A Simpler Way to a Smarter Home", http://www.wink.com/, accessed August 10, 2016.

[10] "FordPass: A Smarter Way to Move", https://www.myfordpass.com/, accessed August 10, 2016.

[11] "State Farm Drive Safe and Save", https://www.statefarm.com/insurance/auto/discounts/drive-safe-save, accessed August 10, 2016.

[12] "Progressive Snapshot", https://www.progressive.com/auto/snapshot/, accessed August 10, 2016.

[13] "Car as a Doctor—Automotive Industry Embracing In-car Health Monitoring and Wellbeing", Frost & Sullivan, Market Insight, August 2015. https://store.frost.com/car-as-a-doctor-automotive-industry-embracing-in-car-health-monitoring-and-wellbeing.html, accessed September 14, 2016.

[14] "Driver Distraction in Commercial Vehicle Operations", U.S. Department of Transportation, http://www.distraction.gov/downloads/pdfs/driver-distraction-commercial-vehicle-operations.pdf, accessed August 10, 2016.

[15] "Hybrid Intelligent System for Driver Workload Estimation for Tailored Vehicle-Driver Communication and Interaction", K. Prakah-Asante, D. Filev, and J. Lu, Proceedings of 2010 IEEE International Conference on Systems Man and Cybernetics (SMC), Istanbul, Turkey, October 2010.

[16] "The Health Data Exploration Project (2014) Personal Data for the Public Good", California Institute for Telecommunications and Information Technology, http://hdexplore.calit2.net, 2014, accessed August 10, 2016

[17] "Computing Arbitrary Functions of Encrypted Data", C. Gentry, *Communications of the ACM*, Volume 53, Number 3, Pages 97–105, 2013.

[18] "CES 2016 Report", C. Simonds, http://www.autotechinsider.com/products.html, accessed August 10, 2016.

26

INDUSTRIAL INTERNET

David Bartlett

General Electric, San Ramon, CA, USA

26.1 INTRODUCTION (HISTORY, WHY, AND BENEFITS)

The **Internet** fundamentally changed our lives, from the way we connect with each other to how we conduct business globally. It has ushered in a century where everyone is "a global citizen." It's a sweeping social disruption that not only brings with it new inventions and scientific advances but will also revolutionize both the methods of work and the workers themselves. The Internet Revolution is, according to some, the Industrial Revolution of our time. The Internet, however, is not the end of the transformational story. It is rather only the beginning and will continue to serve as a foundational construct for a number of new advances as well.

Perhaps one of the most impactful new advances is the **Industrial Internet** that is enabling a new transformation in industrial infrastructures at a global level. The Industrial Internet revolution is anticipated to be the most disruptive technology in the industry since the Internet revolution. An ever-increasing majority of surveyed industrial companies strongly believe that Big Data analytics (Figure 26.1) is a top priority for their organization. In fact, many believe that companies that lack an Industrial Internet adoption strategy will lose competitive edge and quickly fall behind their competitors. The Industrial Internet is just beginning, with its most important defining moments happening now. As we start to integrate the Internet with the industrial underpinnings of our world, from energy generation transportation, healthcare, and beyond, we are seeing radical improvements in productivity and efficiency that boost economic growth, benefit the environment, and make people's lives better.

Much of the initial focus is on using sensors, controls, and software applications to link machines, fleets, and networks with advanced analytics and with those who need

Internet of Things and Data Analytics Handbook, First Edition. Edited by Hwaiyu Geng.

Companion website: www.wiley.com/go/Geng/iot_data_analytics_handbook/

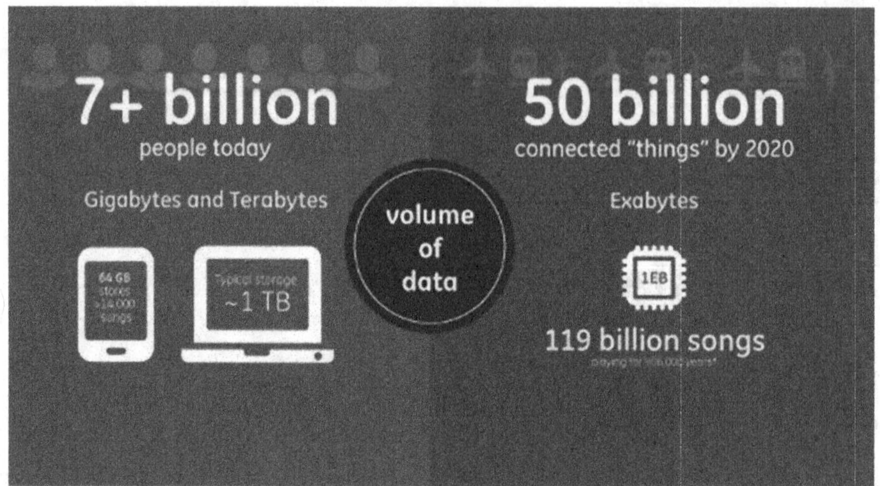

FIGURE 26.1 GE's Predix Cloud captures and analyzes the unique volume, velocity, and variety of machine data. Permission to use by GE.

better information to design, operate, maintain, and interact with that equipment. The Industrial Internet benefits us by the following:

1. Combines intelligent machines, advanced analytics, and people at work in ways that dramatically improve productivity and efficiency
2. Combines the global industrial system with the open computing and communication systems developed as part of the Internet revolution to rapidly advance analytics and insight based on "big data" generated by industries
3. Opens up new opportunity to accelerate productivity, reduce inefficiency and waste, and improve people's lives

The Internet has fundamentally changed the way people interact with each other—the Industrial Internet will fundamentally change the way we interact with industry and infrastructure. This integration of Internet-based technology and the global industrial system is just starting. Intelligent devices, intelligent systems, and automated decision making will increasingly blend the physical world of machines, facilities, fleets, and networks with the connectivity, big data, and analytics of the digital world.

26.2 DEFINITIONS OF COMPONENTS AND FUNDAMENTALS OF INDUSTRIAL INTERNET

What is the definition of Industrial Internet? Perhaps it is best to start with three basic questions that will help determine if your projects are best suited for Industrial Internet applications.

26.2.1 Do You Want to Optimize Industrial Assets and Operations?

This is a key value focus for applying the Industrial Internet. Where does one begin? It all starts with data. The world's industries have a plethora of data sources that are often neither connected nor integrated. The power of data integration is that it is foundational for system-level analytics and optimization of assets and industries.

Data capture and integration involve connecting to each source of data and monitoring events as they occur. This requires a time series database (TSDB), which is a software system that is optimized for handling time series data, arrays of numbers indexed by time. This enables you to understand industrial events as they happen—something called "real-time" or "near real-time" monitoring. Real-time logic evaluation in real-time systems performs event log analysis. This can be used to perform such activities as pattern matching, filtering of event occurrences, or aggregation of event occurrences. These types of activities can drive analytical processes that result in increased insight into certain industrial assets or operations that can then evoke actions to optimize those assets and/or operations.

This often requires a shift from traditional enterprise data warehouses to newer data lake constructs that do not fit data into predefined fields but rather preserves data more in its natural state and drives contextual analysis through the use of tagging or metadata which is basically data about the data. This enables a broader range of problems against which the data can be applied with more possibilities to drive value. There is also a trend for applying processing and analytics right in memory where data is stored to cut down on latency times and deliver results in the time frame required for real-time decision making.

Platforms are essential to enable reuse, scale, interconnectedness, and speed. Open platforms that leverage open source enable faster value creation and allow companies to focus on core value while not having to recreate foundational operating system and software application constructs. Open platforms should leverage cloud constructs as well. Cloud implementations maximize the effectiveness of the shared resources where resources are usually not only shared by multiple users but also dynamically reallocated per demand. GE's Predix® platform is an example of such a platform built for the Industrial Internet.

26.2.2 Does What You Wish to Perform Require Large-Scale Data and the Processing of Complex Analytics on That Data?

By large scale, I am referring to "Big Data" characterized today not so much in terabytes but in petabytes of data, especially in industrial applications. In 2012, Gartner defined big data with 3Vs: high Volume, high Velocity, and/or high Variety. Gartner and now much of the industry continue to use this "3Vs" model for describing big data. To extract value from big data, there is an evolving set of techniques and technologies that are able to quickly reveal complex patterns and valuable insight. A key focus area to extract value from big data is analytics. Analytics identifies and displays meaningful patterns in data. Analytics is key to understanding large complex sets of data by deploying such practices as statistics, computer programming, and data visualization techniques that make complex data sets easier to understand. There are many dimensions here to consider in addition to the use of mathematics and statistics. Data Science uses descriptive techniques and predictive models to gain valuable knowledge from data. These insights can drive automated actions or simply guide decision making in the contexts you wish to apply. Think about analytics as more of a comprehensive methodology as opposed to a discreet set of analyses.

26.2.3 Are You Concerned About the Cyber Security of Your Industrial Assets and Operations?

Industrial infrastructure security needs to be foundationally different from traditional IT detection systems. Securing connected machines has a unique set of complexities. The industrial systems and networks that are in place are key to life as we know. A disruption to just one of these systems, for example, rail, air, water transportation infrastructure, electricity, oil, gas, and nuclear can have dire consequences. Companies such as Wurldtech are emerging to provide cyber security services to assess, protect, and certify critical industrial infrastructure. Wurldtech, a GE company, protects critical infrastructure from cyber attack. The technology works with both device manufacturers and system operators to provide protection for critical infrastructure against the persistent and dynamic cyber threats. Such products and services reduce the risks and costs of a cyber attack while maximizing system uptime and a company's ability to meet compliance mandates.

26.3 APPLICATION IN HEALTHCARE

Healthcare touches everyone, and the value of improving the quality and speed of healthcare outcomes is readily apparent. For today's healthcare systems, big data is key to meeting the ever-increasing demands of doctors and patients. GE has been a leader in using the Industrial Internet to enable a new set of applications that are helping to transform the healthcare industry.

GE's Centricity 360® platform is a secure and reliable cloud-based platform that helps teams of physicians and caregivers work together in a clinical community—where they can quickly confer on patient cases, simultaneously access images and reports, and collaborate on diagnoses and treatment plans.

The Industrial Internet enables us to connect in real time to the medical equipment used by this team of physicians and caregivers. For example, many procedures now routinely use CT scans. CT uses X-rays, which are a form of radiation that patients who undergo scans are exposed to. While the dose of radiation is small from one scan, the radiation dose can be more substantial if the patient requires many scans over a small time period.

An Industrial Internet application on radiation dose has been developed to collect data directly from CT scanners. Doctors can then use this data to quantify the amount of practice-level radiation dose, and the information can be analyzed to tailor subsequent procedures, helping drive behavioral change to help ensure that clinicians are using the lowest possible doses. This is a good example of how the Industrial

Internet can integrate medical device, software, and cloud platform solutions to seamlessly work together to gain more accurate insights and most importantly deliver the best care to people.

This example is part of a broader Industrial Internet initiative in healthcare—a focus on technology, data, and analytics. This not only enables newer, smarter systems in a tech-enabled world but also connects caregivers with brilliant machines that provide more accurate information than ever before while enabling them to spend less time navigating the system and more time caring for patients.

Doctors are increasingly collaborating and combining their expertise to solve the most complex cases—what is increasingly referred to as leveraging the global brain to the maximum effect.

But how can people around the world more easily tap into that expertise? One way is leveraging a growing selection of wearables that can harvest real-time data from sensors located on your body. These wearable healthcare devices can tap into applications that are connected to that expertise. Such healthcare devices are on the rise and projected to overtake fitness wearables. The Industrial Internet makes this all possible, and it could not come at a better time as the generation of elderly patients begins to swell around the world and outpace the care facilities that are in place. Such connected devices and applications will enable the medical industry to better handle those in need of medical advice remotely. As in industrial equipment, the more predictive we can become with our physical beings by sensing any anomalies earlier, the more proactive we can be in reducing our own health risks and in some cases avoiding crisis situations entirely.

The findings of a recent report by the Health Research Institute recommended the following four actions:

1. Put diagnostic testing of basic conditions into the hands of patients: Close to 42% of physicians are comfortable relying on at-home test results to prescribe medication.
2. Increase patient–clinician interaction: Half of physicians said that e-visits could replace more than 10% of in-office patient visits, and nearly as many consumers indicated they would communicate with caregivers online.
3. Promote self-management of chronic disease using health apps: 28% of consumers said they have a healthcare, wellness, or medical app on their mobile device, up from 16% last year. Nearly 66% of physicians would prescribe an app to help patients manage chronic diseases such as diabetes.
4. Help caregivers work more as a team: 79% of physicians and close to 50% of consumers believe using mobile devices can help physicians better coordinate care.

26.4 APPLICATION IN ENERGY

Maximizing efficiency in power generation and distribution is essential not just to enable economic growth but also to do so in a way that improves environmental sustainability, especially as global growth raises standards of living—and desired energy consumption levels—across emerging markets. There are a number of new "as-a-service" offerings enabled by the Industrial Internet and Industrial Internet applications that are helping raise efficiency in this sector.

Energy-as-a-service is the newest entrant to the as-a-service business model, and it is redefining how businesses think about their relationship with energy. Today, energy is usually considered a cost beyond their control—a necessary expense required to run a business. In the new energy-as-a-service model, companies are changing their relationship with energy in two important ways:

1. Businesses are realizing their own ability to reduce, produce, and shift energy use by deploying financed on-site equipment to ensure immediate cash-flow benefits.
2. Companies are using sensors from intelligent lighting to align their energy spend to business metrics, such as dollars per square foot in a retail business.

They are able to do this with GE's newest Industrial Internet start-up: Current.

Current is a first-of-its kind energy company which integrates GE's LED, Solar, Energy Storage, and electric vehicle businesses with its industrial strength Predix platform to identify and deliver the most cost-effective, efficient energy solutions required by customers both today and in the future.

GE's Wind PowerUp Platform is a perfect example of the combined joint power of software and Hardware (Figure 26.2). The software analytics allow wind farm operators to optimize the performance of the turbines, based on environmental conditions. Raising the turbines' efficiency can increase the wind farm's annual energy output by up to 5%, which translates in a 20% increase in profitability. The Wind PowerUp Platform is already in operation on nearly 500 wind turbines, delivering an additional 86GWhr in annual power generation.

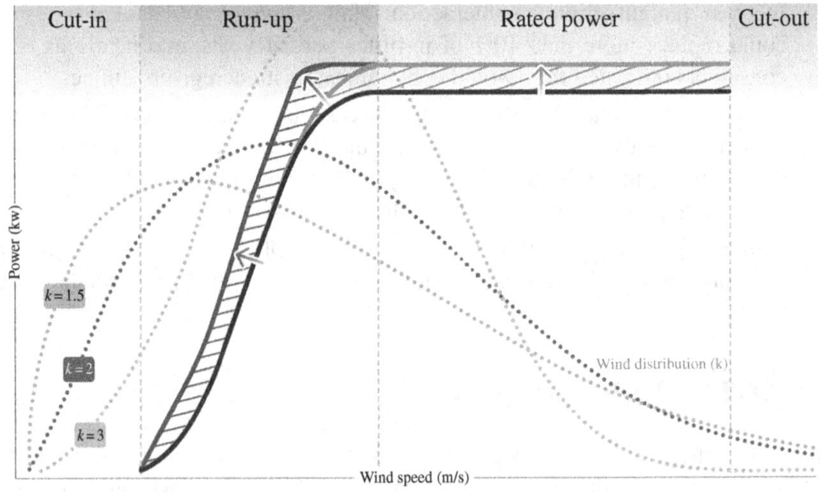

FIGURE 26.2 PowerUp performance curve. Permission to use by GE.

GE has developed a similar blend of software and hardware solutions for gas power generation: the FlexEfficiency Advantage Advanced Gas Path (AGP) solution. AGP combines sophisticated software with an upgraded hardware design enabled by the use of new materials. Based on 100 million hours of real-world operating data already, the flexibility of this system allows power generation plants to prioritize increased output, more efficient load responsiveness, reduced emissions, more robust start-ups, or lower turndown capability. Thanks to this flexibility, plant operators can quickly react to changing market conditions by maximizing output or reducing operating costs and to possible changes in emissions and power grid regulations. On one combined-cycle power generation plant with a net output of 525.2 MW, these solutions can reduce annual carbon dioxide emissions in an amount equivalent to that of 2,200 cars.

26.5 APPLICATION IN TRANSPORT/AVIATION AND OTHERS

The transportation network is the backbone of the economy, and inefficiencies in transportation translate to higher costs for business and ultimately in slower growth in output and incomes at the national level. In the railways sector, efficiency and productivity can be increased by raising the velocity at which the railway network operates and by reducing the "yard dwell time" that trains spend idle in a railway yard. Dwell time is closely and inversely linked to operating margins.

To improve performance, solutions built on the Industrial Internet can provide comprehensive support to rail mechanical and transportation departments on locomotive health, maintenance, and repairs. Solutions that collect and analyze performance data during locomotive operations can provide diagnostics and root cause analysis in real time. This would allow railway managers and technicians to schedule preventive, conditioned-based maintenance and repairs, maximizing reliability and availability.

Software solutions can guide an optimal reconfiguration of network and yard operations. Similarly, they can provide advanced planning of resources and materials. GE's Movement Planner and Yard Planner software have already achieved 10–15% improvement in velocity, a 5% reduction in dwell, and a 50% reduction in the need to change crews, substantially improving productivity. Similarly, GE's Trip Optimizer solution for railways has already delivered a 10% reduction in fuel consumption.

The top priorities in aviation are clear: maximizing safety and minimizing delays, limiting cancelations, and reducing fuel consumption.

The Industrial Internet provides the capability to monitor data collected from aircraft equipment and airline systems to predict, prevent, and recover from operational disruptions. Systems built in this way are capable of spotting and flagging issues not detected by traditional diagnostics, preventing operational disruption and lost revenue.

For an average US domestic airline (14 million passengers, 85,000 flights per year), there is an opportunity to prevent thousands of delayed departures and flight cancellations each year, helping more than 165,000 passengers get to their destinations on time.

Collecting real-time data generated by an aircraft and applying analytics can provide business intelligence and actionable insights to significantly improve an airline's overall efficiency in areas such as fuel management, flight analytics, navigation services, and fleet synchronization.

26.6 CONCLUSION AND FUTURE DEVELOPMENT

The convergence of the digital and the physical in the industrial world is a profound transformation that is far from fully appreciated. A connected device or machine becomes something entirely new and presents an unprecedented opportunity to drive higher value and efficiency.

The Industrial Internet is not only transforming individual machines and systems, but it is also changing the nature of work while transforming the economic landscape. Industrial companies that combine the digital and the physical open entirely new dimensions in the way they operate and in the value they can provide to customers and shareholders.

Connecting the digital world of research, design, engineering, manufacturing, and field performance enables a company to drastically reduce the time to introduce new products, leading to faster responses to customer needs and higher engineering productivity. Linking engineering, supply chain operations, and services data through the cloud means operators can optimize factories and products in real time and continuously improve them throughout their life cycle. As a result, machine uptime, throughput, and efficiency increase.

Combining deep expertise in both digital technology and industrial machines is not easy. Both fields require complex and sophisticated domain expertise and are experiencing fast-paced innovation. The new transformative digital–industrial company must focus on both and be able to merge them seamlessly in a way that maximizes value. Software development must be guided by the industrial machines' operating characteristics.

Combining software development skills with data science skills along with engineering population across industries is essential—both to maximize the joint value of digital and physical as well as to ensure the compatibility and adaptability of software solutions across industries.

Platforms are essential to enable and monetize the value of interconnectedness. Interconnectedness is all about communication, collaboration, and compatibility within the industrial ecosystem. It all starts with a platform, like GE's Predix platform, to enable this. The Predix platform is designed specifically to meet the requirements and characteristics of industrial systems: it guarantees data security as well as mobility, it is optimized for machine-to-machine communication, and it supports distributed computing and big data analytics. The Predix platform will support the rapid development of a growing number of applications for asset and operations optimization for a wide range of industrial sectors.

Such a platform can facilitate:

- Collecting and analyzing data from a larger set of different industrial assets, creating a deeper and more informative information set that delivers more effective insights
- Enabling the interoperability of a wider range of assets within an industrial operation or system, boosting operations optimization
- Allowing applications to be adapted and adopted across different industrial sectors
- Making it easier for developers, engineers, and data scientists to collaborate on a wider range of industrial solutions.

Industrial Internet solutions are now being developed and applied in a range of industrial sectors, including those that play a pivotal role in driving economic growth. In the future we will see applications evolving at greater pace that will solve the big system level problems in the industrial ecosystems of the world. Beyond asset optimization and operations optimization, we will see a new level of enterprise optimization that will be enabled by more interconnected data, a broader real-time sense of the operating environment in the context of the work being performed, and seamless cloud-based platforms that can deliver scalability and interoperability in a cost-effective and secure manner. This will ultimately translate to improving the quality of life—life that increasingly relies on the industrial infrastructure of today and tomorrow's world.

FURTHER READING

https://www.youtube.com/watch?v=zG1S3jIoqDI&list=PLenh213llmcZYClMpriSobSo3W-XyHmmj&index=4 (accessed August 22, 2016).

http://www.industrialinternetconsortium.org/ (accessed August 22, 2016).

https://www.ge.com/digital/industrial-internet (accessed August 22, 2016).

http://www.gereports.com/post/99494485070/everything-you-always-wanted-to-know-about-predix/ (accessed August 22, 2016).

27

SMART CITY ARCHITECTURE AND PLANNING: EVOLVING SYSTEMS THROUGH IoT

DOMINIQUE DAVISON[1] AND ASHLEY Z. HAND[2]
[1] *DRAW Architecture + Urban Design, Kansas City, MO, USA*
[2] *CityFi, Los Angeles, CA, USA*

27.1 INTRODUCTION

Architecture has long been understood as a system: a set of connected elements forming a complex whole that then functions as an affordance for inhabitants and pieces together to form towns and cities. The Bauhaus and especially one of its Directors, architect Mies van der Rohe, worked with systems of construction components to redefine modern architecture and expressions of materiality. Others, such as ecological architects, considered the multiple environmental subsystems in design of the built environment. William Whyte led the way in studying social interactions in urban environments. Yet, there is still considerably more work to be done to analyze and understand how the building itself is part of a larger ecology and environmental system—not just as a singular intervention on an urban site.

Ultimately, there are three scales to consider in architecture: the human level, the building level, and the urban or regional level. Architecture must consider not just the function and life cycle of the building itself but also its impact on the ecologies and communities around it once part of the urban network—throughout the entirety of its life cycle. And governance allows, and ideally empowers and educates, its communities to have an active voice in the manifestation of this evolution.

Internet of Things and Data Analytics Handbook, First Edition. Edited by Hwaiyu Geng.

Companion website: www.wiley.com/go/Geng/iot_data_analytics_handbook/

Buildings are largely static and yet dynamic objects that have broad implications, including the natural environment (air, water, energy), the built environment (other buildings, infrastructure such as streets, sidewalks, pipes, etc.), and social dynamics (public spaces—or lack thereof, amenities). A building relies on streets for access and public infrastructure for utilities in addition to demanding resources and creating waste as part of its operations. It is therefore impossible to isolate analysis at the building level without looking at the urban level when understanding impact. We will come back to the human level a little later.

The process of architectural design has evolved, albeit rather slowly, with the advent of new tools in recent years. Building integrated modeling (BIM) has reshaped how architects design the building as a whole and as a series of closely integrated systems with great dependencies and impact on the experience of the building occupant. Autodesk's Revit, Tekla BIMsight, Gehry Technologies, and Bentley are all example platforms that engage in BIM technology. This technology enables greater collaboration and information sharing throughout the design phase of project development and introduces the potential for better problem solving through more accurate three-dimensional modeling data and coordination of internal systems. Most importantly, it is the availability of this detailed data that has the potential to transform design decision making by giving architects, and their consultant and client partners, more opportunities to evaluate the impact of different design options—whether spatial, material, structural, or mechanical.

Equally exciting are new tools that have changed the way completed buildings can be monitored. These "smart buildings" can help save energy, water, money, and so on, and improve occupant experience by enabling operators to respond in real time to changing conditions, stresses, or failures in the building. These smart building technologies enable building operators to better understand the occupant use and its impact. Architects can benefit considerably from accessing this type of data to see how a building meets the design performance goals: are occupants responding and using the building as it was designed? Understanding how designs actually function could and should improve future designs. Even more impactful is the transparency available to occupants of real-time performance, which allows for a more responsive and efficient use of the building systems if monitored and managed.

Smart building technologies, however, are still limited to the extent of the site and do not help cities, communities, developers, owners, and so on understand the impact of this one building in relation to the larger urban scale. It will be essential to consider data at all scales: the human, the building, and the urban levels to truly adapt how we design buildings of the future to be part of a holistic, sustainable urban system. How, then, can we leverage and access the rapidly expanding sources of new data to fully understand and measure the relationship of development to our urban environments and move to a more data-driven approach to growing our cities? What if city planners and civic leaders could evaluate city plans and development informed with empirical evidence that pulls from the complexity of the systems defined and reshaped by the built environment to define policies for the future?

27.2 CITIES AND THE ADVENT OF OPEN DATA

Much has been written already about the importance of open data to government transparency and accountability. The general principle of open data is that this data is public and should therefore be readily available to the public to see, analyze, and use as appropriate. Cities such as Los Angeles, San Francisco, Kansas City, Missouri, Chicago, New York City, and Philadelphia have taken great steps forward to publish data sets online such as crime, 311 service requests, and permit information to give citizens access to information that they have a right to anyway per state open records legislation [1].

Open data policies and advocates continue to push for more data to be available, and communities of civic hackers have emerged across the country in a series of events and meet-ups to dive deeper into what this data means and how it shapes their city and to even identify new applications and tools that can run off that data to make cities more accessible for citizens and visitors. After all, government resources are limited, and these communities of civic hackers are creating new ways to look at how government operates and communicates with constituents and identifying problems that can be found by analyzing big data in ways that the government does not have the capacity to manage. Civic hacking has become a new tool for civic engagement. Code for America is a not-for-profit organization leading efforts nationally in this arena by organizing events and volunteer corps to engage with municipalities.

For a savvy architect and building owner, this open data can be a treasure trove of information to help in assessing the impact of a site's development on the larger context or urban level. Many open data portals can be tricky to navigate, however, and more work needs to be done to better present and visualize this public data for consumption. Additionally, much of the data available may not offer any insight to the environmental context of a site selected for a construction project, suggesting that even more data sets are necessary for effective analysis. For example, an architect and owner would benefit considerably from better access to public data to help in assessing the value a new development might bring to the neighborhood and city as planned.[1]

City government and other local officials would also benefit from better access and analysis of this data. Cities adopt plans and provide incentives for development of all scales, but it can be very difficult for a city to measure return on investment (ROI) from incentives. What if the ROI of development not only addressed first costs and land values but also assessed the impact on the local environment and broader community at a much more holistic level? What if a metric for a successful development included community input and feedback as to whether there was value to the neighborhood? The potential benefits of urban development, after all, do not need to be limited purely to financial returns.

As cities look to publishing more data online, administrators are looking at the sources of data and finding new ways to collaborate internally by using the data

[1] CartoDB OpenData.City, MySidewalk, and ESRI are all platforms that integrate data into mapping and data portals.

themselves. KC Stat, the City of KCMO's performance management system, is a monthly reporting on the data it collects on all of its operations as city government works toward meeting the priorities established by the City Council. This data is accessible online with a dynamic dashboard that enables the viewer to dive into the raw data or see a simple visualization as to whether the city is on track to meet its goals. Since many of these goals are linked directly to operational functions of local government, it is only a matter of time until smart city sensor data is fully integrated into this performance management program.

27.3 BUILDINGS IN SMARTER CITIES

With a growing emphasis on life cycle analysis or assessments at the building level, design decisions are often evaluated by the up-front impact of manufacturing through the full use of the material or component as a way to understand the impact of building design and to assess sustainability. Several measures have been created—such as the US Green Building Council (USGBC) Leadership in Energy and Environmental Design (LEED) program—to help architects and building owners evaluate their buildings. The LEED program continues to evolve with a stronger emphasis on building performance and credits for participating in demand response programs [2] but remains somewhat limited in the assignment of credits based on static measures. Even the Living Building Challenge or the American Institute of Architect's 2030 Challenge, which prioritizes on-site mitigation and carbon neutrality, focuses primarily on the site-generated data.

While to some extent these metrics provide a snapshot of the building's impact on the urban level with an emphasis on ongoing measurement and monitoring, the data is one-sided—coming entirely from the building itself and not its context or environment. There are also spreadsheets that do not integrate with the existing design and planning processes and so add a parallel workflow to the efforts of practitioners. It remains challenging to move beyond just measuring a single system for comparison on efficiency to understanding the entire building as a more complex system as a whole. How then might city planners and administrators address the potential impact of a building by factoring in a more complex set of data to enable the analysis of impact *before* anything is constructed?

Taking building technology and combining it with urban level data from smart cities technology could radically transform how architects, developers, and planners approach building design and scenario planning. The Internet of Things (IoT) or Internet of everything: the communication of a broad array of sensors in the built environment creates new opportunity for architectural design that will create truly smarter buildings and better, more sustainable and resilient cities.

While US cities have lagged behind European and Asian cities across the globe in adopting smart city technology, there are several success stories that illustrate the cost- and resource-saving potential of this technology at various scales. Cities such as New York, Boston, and Philadelphia have done a great deal of work to change the paradigm as there is still a lot to be proven in this space [3]. The majority of these applications, however, are specific to a single vertical: technology that addresses lighting and energy, parking, water, traffic and transit, or citizen experience. This requires a strategic investment in a specific solution that limits the agility of the municipality to evolve how it uses sensors to address systems that are often heavily integrated. More needs to be done to look holistically at how sensors and real-time data can inform integrated infrastructure management and planning to better leverage limited resources.

In June 2015, Kansas City, MO, launched a groundbreaking public–private partnership with Cisco Systems and Sprint to build a smart city and communications network around the downtown streetcar starter line. The goal is to create an open solution-agnostic platform to enable the City to install sensors that will help manage infrastructure more efficiently while creating a unique citizen experience in the urban core. Community information kiosks, for instance, will provide hyper local information about local businesses, events, and services that can dynamically change based on utilization, schedules, and so on, and create a direct feedback loop between the citizen and local government. This will also attract visitors and drive economic development as this platform will open up new possibilities for technology innovation while connecting visitors, residents, and businesses. These sensors will introduce unprecedented volumes of data to KCMO systems, which can contribute to a better situational awareness or perception of environment in real time. The goal is to make this data as open as possible once the system is online in 2016. How, then, do city officials and the private sector access this data to drive better decisions about how urban development should be planned and built for the generations to come?

27.4 THE TRIFECTA OF TECHNOLOGY

There have been three parallel movements that are major disrupters to how we might think about cities and the structures they contain. First is to make the information gathered by municipalities openly available to constituents—the rapidly growing open data movement, as mentioned earlier. Second is IoT and Smart City technology, which increases the three Vs of information (Volume, Velocity, and Variety) [4]. This, along with improved digital infrastructure to distribute the vast amounts of data that will be generated by IoT, is also being developed. Big Data can be proliferated across such gigabit fiber networks that have been installed in Kansas City, MO; San Francisco, CA; and Chattanooga, TN, to name a few. Though the most common and first-ranked example of sensor use (libelium.com) is of how to more easily find a parking space, the potential scale of problems that can be addressed through better data and, equally important, access to and analysis of that data far exceeds ideas of parking convenience. IoT holds the possibility of making our cities more resourceful and resilient places that can intelligently address major global challenges, such as Climate Departure. All three of these components are lining up in Kansas City, MO, and will allow this community to be a test bed of how comprehensive resource impact can assist in planning toward a truly resilient and sustainable future.

27.5 EMERGING SOLUTIONS: UNDERSTANDING SYSTEMS

This shift, in access to information that previously was very costly and sometimes time consuming to request and gain access to, has prompted the development of tools that would assist in making the architecture process more focused on resource use—to better understand how buildings affect their surrounding environment and, in turn, are affected by their environment.

Architecture professionals readily acknowledge that the available BIM or planning scenario tools to provide solid metrics on building performance and resource impact are arduous (steep learning curves), time consuming, or very expensive. Though interested in sustainable design, having a clear understanding of resource impact, such as energy use, storm water, access to transportation, water usage, and ROI, can mean referencing half a dozen disparate sources and relying heavily on engineers or building technology consultants (which charge an additional fee and incur more time) and sometimes inaccurate historical data, or, worse yet, rules of thumb. Project schedules and fees rarely allowed for this in-depth analysis, or when it was achievable, the results came too late in the design process, once the design was well established, and thus larger scale adjustments to siting and massing were no longer possible.[2]

A team at a March 2013 Google Hackathon in Kansas City, MO, tested the ability to pull a 3D immersive and ubiquitous modeling tool called SketchUp into Google Earth and cross-reference geo-specific data with national and local application programming interfaces (APIs). For example, rainfall data in a particular zip code as logged by NOAA used to calculate initial stormwater impacts, transportation scores from US census and Walkscore.com and local GIS data, and climate data. This tool, called PlanIT Impact, has the ability to cross-reference the Open Data gathered from Smart City sensors and benefits from being transmitted across Gigabit infrastructure (Figure 27.1) in order to provide a low-cost resource impact tool to answer the question of how city officials and the private sector could access data to drive better decisions about how urban development should be planned and built for the future.

An important aspect of the data analysis is not only transparent access to the base data but also provision of a metric by which performance and impact can be rated compared to code minimums or, alternately, what is required to achieve net zero. PlanIT Impact is correlating to existing rating systems established by the EPA to rate designs on a score from 1 to 100, with 100 being net zero across all data sets. The weighting of the various data sets requires input from the city stakeholders and provides a unique opportunity for community engagement to educate and align with a particular community's sustainability goals and priorities. Once the weighing of the scores is established however, it becomes a clear and objective benchmark against which projects can be compared to assess their impact on the community.

Other teams have developed dynamic analytics that also tap into the open data resources now available which could radically transform urban infrastructure management. Urban IQ, for example, is a team of professional data scientists who

[2] The Pareto rule outlined the importance of the first 20% of decisions within a project having 80% of the impact on resource use. Therefore early application of data is critical to optimize outcomes.

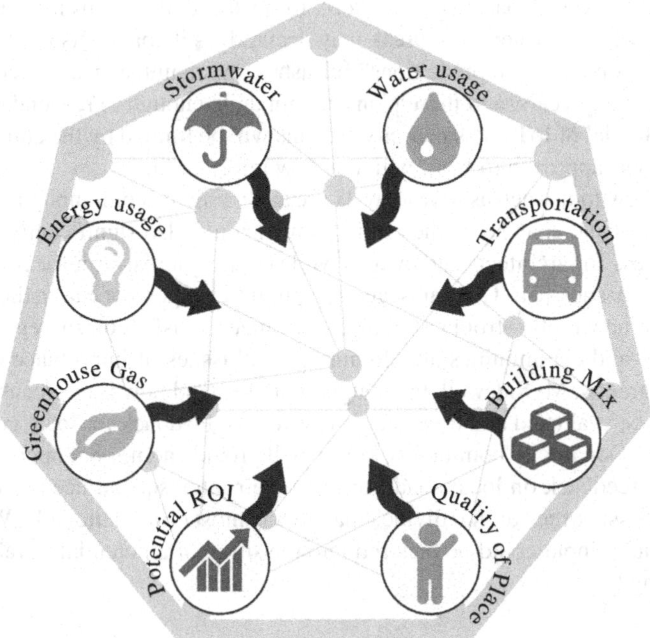

FIGURE 27.1 Courtesy of PlanIT Impact. Reproduced with permission of Dominique Davison.

volunteered their time to look more closely at available 311 data and reports of water main breaks. By layering additional sets of condition data such as weather and socioeconomic information, the team sought to identify a model that could better predict potential water main breaks by looking at trends in incident reporting and layering considerations of potential cultural impacts on physical infrastructure (i.e., is there a correlation between socioeconomics and culture on how we use our public infrastructure?).

Similar to PlanIt Impact, the Urban IQ team successfully created a prototype analytics tool over a weekend hackathon in Kansas City, MO, but relied on feedback and input from local government as a partner to truly evaluate the tool. The 311 data, for instance, required some additional interpretation by government officials to discern the differences in the types of incidents being reported (e.g., stormwater versus waste water issues) which highlights the need for government to continue to make its data more accessible and legible for potential use. The overall lack of data standardization and general omission of data dictionaries to outline what these data sets contain presents a barrier to truly opening up public data and making analytic tools and solutions more scalable.

An additional barrier is considering how these analytics will transform government processes. While analysis by tools such as PlanIt Impact and Urban IQ can radically change the knowledge government and the public have about infrastructure and the environment, there is more work to be done to ensure that the necessary policies and procedures are set up to enable civic leaders to apply that knowledge. If the Kansas

City Water Services Department has a tool, for instance, that helps it identify highly likely areas where a water main break may occur, does it change its service delivery model to respond to these predictions? Infrastructure maintenance and construction occurs over many years, and these plans are not agile enough to respond effectively. The full potential of IoT and smart city systems will be limited by the ability of cities to rethink their approach to work with this new information.

Finally, new tools such as MySidewalk are analyzing the community pulse around specific issues and changes in their cities, which could also significantly benefit the design process for architects and owners while creating a platform for conversations around civic issues [5]. The landscape for public engagement has radically transformed with new tools—from social media to citizen satisfaction surveys and online town halls—and communities are aligning around issues of importance rather than just by geography. MySidewalk has sought to understand how issue-focused conversations can be evaluated and translated into effective feedback for local government. Kansas City's smart city community kiosks will provide additional opportunities for community feedback on local conditions and their interests and needs and is a step toward addressing the "digital divide" that affects most cities nationally. What if this data and public input could be factored into the siting of a potential development or new building?

> A Smart City will combine public governance, people ownership and business collaboration. —Schneider Electric

27.6 CONCLUSION

The civic technology landscape is rapidly evolving and creating new opportunities for smarter and more sustainable and resilient cities in the United States. City leaders will have to push open data to new thresholds, however, to ensure this data is valuable and relevant to its citizens, and there is still considerable work to be done to standardize how this data is published. To be relevant, this data will need to be more accessible and sharable while creating a space to enable civic technologists to scale their solutions beyond a single city.

Improved resilience in urban design and architecture design processes require a deep understanding of both resource impact and quality of place. Resource impact assessment tools and tools that provide insight into human interaction with their environment, thereby addressing quality of place in design, are making important advancements toward shifting the design paradigm to a more informed and metric-based standard.[3] IoT provides the ability to more efficiently and quickly access this data, though analysis of the data and interpretation are still necessary. While these are promising advancements and international discussion is being initiated around Smart

[3] See the Lower Manhattan Smart Neighborhood Pilot where social interaction, and other data, is being charted through sensors to better understand what creates a more resilient neighborhood.

City tools and how they can benefit sustainability efforts, much work is needed around ensuring that inclusive civic engagement steers the application of Smart City IoT technology to ensure that maximum benefit is achieved for all and a metric is properly vetted and instituted.

REFERENCES

[1] Open Data Cities Index (2015). Retrieved from http://us-city.census.okfn.org (accessed August 24, 2007).

[2] From http://www.usgbc.org/leed (accessed August 24, 2007).

[3] Puentes, Robert and Adie, Tomer (April 2014) "Getting Smarter About Smart Cities," Brooking Institute. Retrieved from http://www.brookings.edu/research/papers/2014/04/23-smart-cities-puentes-tomer (accessed August 24, 2007).

[4] IEEE Trento Smart Cities Conference (December 2014) "Big Data and Open Data for a Smart City". Retrieved from http://smartcities.ieee.org/articles-publications/trento-white-papers.html (accessed August 24, 2007).

[5] MySidewalk. From http://app.mysidewalk.com/about (accessed August 24, 2007).

FURTHER READING

Open Data Cities Index (2015). Retrieved from http://us-city.census.okfn.org (accessed August 24, 2007).

GSMA-Connected Living (July 2014) "Understanding the Internet of Things". Retrieved from http://www.gsma.com/connectedliving/wp-content/uploads/2014/08/cl_iot_wp_07_14.pdf (accessed August 24, 2007).

International Electrotechnical Commission (2014) "Internet of Things: Wireless Sensor Networks". Discusses WSN uses for the smart grid, smart water, intelligent transportation systems, and smart home domains. Retrieved from www.IEC.com (accessed August 24, 2007).

People4SmarterCities.com—an IBM website created as a repository for ideas regarding applications of smart technology (accessed August 24, 2007).

Cisco, written by Clarke, Ruthbea Yesner (October 2013) "Smart Cities and the Internet of Everything: The Foundation for Delivering Next-Generation Citizen Services". Retrieved from http://www.cisco.com/web/strategy/docs/scc/ioe_citizen_svcs_white_paper_idc_2013.pdf (accessed August 24, 2007).

IEEE Trento Smart Cities Conference (December 2014) "Big Data and Open Data for a Smart City". Retrieved from http://smartcities.ieee.org/articles-publications/trento-white-papers.html (accessed August 24, 2007).

Townsend, Anthony (2014) "Smart Cities". WW Norton & Co., Inc., 2014.

From http://usmayors.org/publications/media/2013/04-water-localcosts.pdf (accessed August 24, 2007). Definitions: According to GSMA, The Internet of Things (IoT) refers to the use of intelligently connected devices and systems to leverage data gathered by embedded sensors and actuators in machines and other physical objects.

Climate departure: Climate departure, as described in a recent article in the journal *Nature*, is the point at which a city's coldest year will be hotter than any year on record before 2005. The Earth will pass climate departure in 2047, according to the study, but cities like Lagos, Nigeria, will pass this point in just 16 years. The International Energy Agency (IEA) estimates that urban areas contribute over 67% of global greenhouse gases, and this is expected to rise to 74% by 2030. It is estimated that 89% of the increase in CO_2 from energy use will be from developing countries (IEA 2008).

28

NONREVENUE WATER

KENNETH THOMPSON, BRIAN SKEENS AND JENNIFER LIGGETT
CH2M, Englewood, CO, USA

28.1 INTRODUCTION AND BACKGROUND

Nonrevenue water (NRW) is a critical issue facing water utilities around the world. A report by the World Bank defines NRW as "the difference between the volume of water put into a water distribution system and the volume that is billed to customers [1]." NRW may account for 25–50% of the total water supplied and has been found at higher percentages in extreme cases. However, many well-run and maintained utilities are able to achieve levels of NRW as low as 5%.

28.2 NRW ANATOMY

NRW includes three primary components: unbilled authorized consumption (metered and unmetered), apparent losses (e.g., water theft, meter inaccuracies, data handling errors), real losses (e.g., leakage in mains, service connections, overflows, main breaks). Unbilled authorized consumption includes authorized uses of water for operational purposes such as system flushing, firefighting, or water provided to users at no charge. Apparent losses are defined as water that reaches a user but is not properly measured and for which the utility is not paid. Real losses are defined as water that enters the water distribution system but doesn't reach a user. Every water system by nature will have some level of real losses.

Internet of Things and Data Analytics Handbook, First Edition. Edited by Hwaiyu Geng.

Companion website: www.wiley.com/go/Geng/iot_data_analytics_handbook/

28.3 ECONOMY AND CONSERVATION

As the demand on drinking water sources increases due to climate change and growing populations, NRW is an important consideration for water utilities. Water utilities are not only losing money but also treated water, which is an increasingly more valuable and scarce resource throughout the world. Excessive NRW can result in substantial lost revenue, severe environmental impacts, and increased risk to public health.

Properly quantifying NRW and investing in activities to reduce and manage NRW have numerous benefits:

- Efficient management of treated water can result in better service to customers and reduced disruptions, which in turn will increase customer satisfaction.
- NRW reduction can increase utility revenue by reducing apparent losses and lower operational costs by reducing real losses.
- NRW reduction activities may include and result in the potential for more accurate metering as well as improved data integrity.
- Increased knowledge of the distribution system gained by NRW reduction activities can decrease utility liability by reducing the risk of catastrophic main failures, safeguarding public health and property.
- NRW reduction activities create a potential for increased job opportunities to support such as replacement of aging infrastructure (Figure 28.1).

FIGURE 28.1 Replacement of aging infrastructure that caused water losses in Australia. Reproduced with permission of CH2M.

Major challenges of NRW reduction are funding limitations and few incentives or regulations. Water utilities are faced with meeting many costly and complex water quality regulations but little or no water quantity regulations. This is further complicated as access to safe, clean drinking water is taken for granted by many customers in the United States. Studies have shown that water is underpriced based on the value it provides and costs needed to operate and maintain a water system. Due to the limited budget of water systems and the prioritization of other issues, such as mandatory regulatory compliance for water quality, many water utilities struggle to address the funding and personnel needs required to properly address and control NRW and water losses. As stated, no national standards or regulations exist for water quantity management. However, some state regulatory agencies (Georgia, California, and Texas) have begun to require water systems to quantify, monitor, and address NRW through standardized water auditing.

28.4 BEST PRACTICE STANDARD WATER BALANCE

As an industry, a standardized approach was needed to determine NRW using an international water balance methodology and the associated terminology. The standardization by the International Water Association (IWA) shown in Figure 28.2 helped define and quantify the NRW components as many developing countries had not previously established a water balance [3]. Using this method, water utilities are better equipped to prepare baseline values establishing current levels of NRW and water losses through water auditing.

28.5 NRW CONTROL AND AUDIT

The initial process of the water audit method was developed by IWA and the American Water Works Association (AWWA) in 2000 [4]. Since then, it has been expanded and developed further. Currently, the water audit method includes both top-down and bottom-up approaches. The top-down approach requires the collection of existing information, records, data, and procedures from the water utility to complete the water audit. The bottom-up approach is used to validate the top-down results and increase confidence in the results of the water audit. Water audit validation may include various levels of detail and complexity depending on the required level of confidence and effort. The validation may include a component analysis process used to model leakage volumes by considering the nature and duration of the leaks, a detailed review of the billing system, a list of customer demographics, finished water meter testing, customer meter accuracy testing, and assessment of unauthorized consumption, among other activities.

The water audit and loss control practice and methodology continue to be developed and explained by AWWA through its Water Loss Control technical committee and the M36 Manual of Practice [2]. This is further facilitated through the use of a

Volume from own sources (corrected for known errors)	System input volume	Water Exported (corrected for known errors)	Billed water exported			Revenue water
		Water supplied	Authorized consumption	Billed authorized consumption	Billed metered consumption	Revenue water
					Billed unmetered consumption	
				Unbilled authorized consumption	Unbilled metered consumption	Non-revenue water
					Unbilled unmetered consumption	
			Water losses	Apparent losses	Customer metering inaccuracies	
					Unauthorized consumption	
					Systematic data handling errors	
				Real losses	Leakage on transmission and distribution mains	
					Leakage and overflows at utility's storage tanks	
Water imported (corrected for known errors)					Leakage on service connections up to the point of customer metering	

FIGURE 28.2 International Water Association best practices standard water balance [2]. Note: All data in volume for the period of reference, typically 1 year. Reproduced with permission of CH2M.

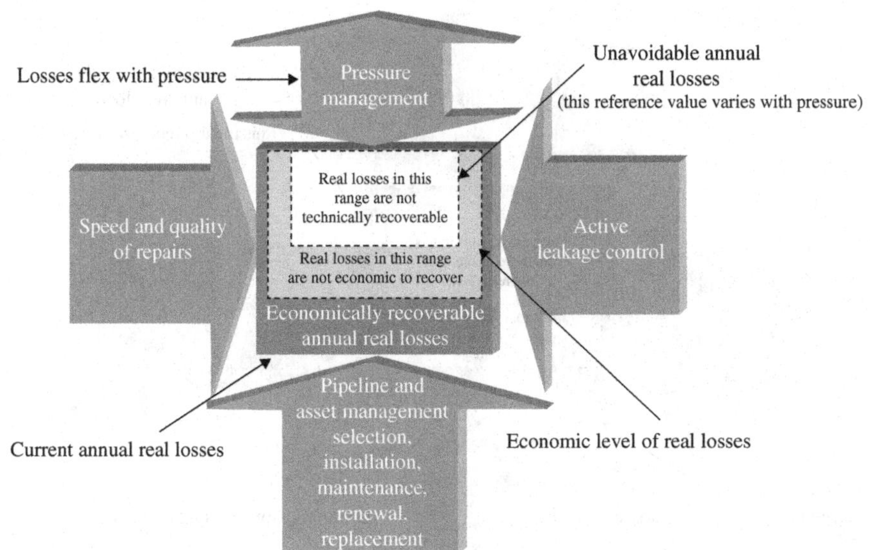

FIGURE 28.3 Four-pillar approach to control apparent losses [2]. Reproduced with permission of CH2M.

free water audit software tool [5]. The M36 manual describes a four-pillar approach to controlling real losses. The four pillars to control real loss include:

1. The current annual real losses (CARL)
2. The economic level of leakage (ELL)
3. Unavoidable annual real losses (UARL)
4. Four arrows representing how real losses occur (shown in Figure 28.3)

The CARL, ELL, and UARL are shown in the center box in Figure 28.3. The CARL is quantified through a water audit, and application of the four pillars of leakage control can reduce the CARL. The M36 Manual of Practice defines ELL as "the amount of leakage that can be economically avoided through leakage control measures whose costs are balanced against the savings of reducing ELL [2]." The UARL is a reference low level derived from data provided by water utilities able to achieve excellent leakage control and is for calculating the performance indicator called infrastructure leakage index (ILI). The calculation for UARL includes specific information about the water system attributes to make the determination.

The M36 manual describes a similar four-pillar conceptual approach to the control of apparent losses. The four pillars to control apparent loss include:

1. Current volume of apparent losses listed in the water audit
2. Utility-specific target for apparent losses or economic level of apparent losses (ELAL)
3. Unavoidable annual apparent loss (UAAL)
4. Four arrows representing how apparent losses occur (shown in Figure 28.4)

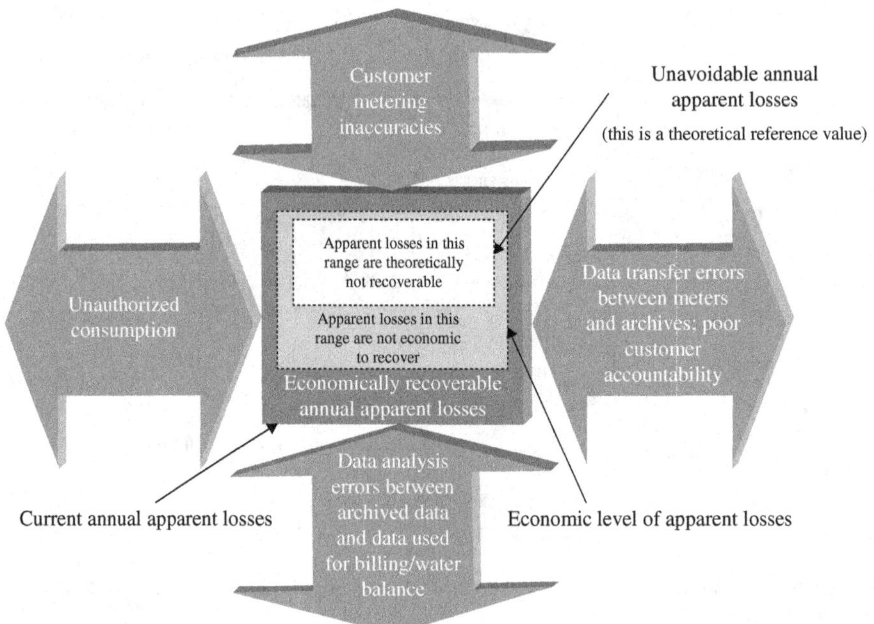

FIGURE 28.4 Four-pillar approach to control apparent losses [2]. Reproduced with permission of CH2M.

The current volume of apparent losses listed in the water audit, the ELAL and the UAAL, is shown in the center box of Figure 28.4. The M36 manual defines the ELAL as "the level at which the cost of the loss control efforts equals the savings garnered from the loss recovery," and the UAAL as "a conceptual level of loss representing the lowest level that could be attained if all possible loss controls could be exerted [2]." It should be noted that the UAAL is unavoidable and does not have an established reference value or formula. The four arrows surrounding the box are represented as dual arrows to indicate that control in each component of loss will result in a reduction in the total annual volume of apparent losses.

28.6 LESSONS LEARNED

Historically, water utilities have taken a reactionary approach and would only repair visible leaks within the distribution system using leak detection specialists to determine leak sources. Again, as many utilities lacked funding and staff, performance-based contracts were established to assist with NRW reduction. In response to legislation and the interest of NRW reduction, some countries have privatized water utilities through public–private partnerships. However, privatization is not always the most favorable option and can at times be viewed as controversial.

By using lessons learned from NRW reduction programs and existing technology coupled with the guidance in the M36 Manual of Practice and other documents, successful NRW reduction programs can be implemented and benefits realized. Advances in metering technologies and near real-time data applications have made NRW reduction a reality for a number of water utilities worldwide. Identification of NRW and water losses, improved management, and more reliable data, as well as the ability and willingness to invest, has driven the success of NRW reduction programs and control of water losses. The next section highlights the application of NRW programs worldwide through a series of case studies.

28.7 CASE STUDIES

28.7.1 United Kingdom

The UK approach to NRW management is based on achieving NRW reduction targets set by the regulator (Ofwat). Private water companies have incentives, such as increased returns to shareholders, related to achieving individual targets.

Ofwat repeatedly criticized Thames Water for the amount of water that leaked from its pipes and heavily fined the utility. In May 2006 the leakage was nearly 900 Ml/day, and Thames Water missed the leakage reduction target for the third year in a row. The Consumer Council for Water, a customers' group, accused Thames Water of missing the targets for the past 5 years. This accusation resulted in increased distrust of the utility by customers and the general public. In July 2006, instead of a fine, the company was required to spend approximately $180 million in repairs.

Under new ownership, Thames Water has met the Ofwat-agreed annual leakage reduction target for the past 8 years (2006–2014). In 2006–2007, the company stated that it had reduced its daily loss from leaks by 120 Ml/day to an average of 695 Ml/day. For 2009–2010, Ofwat-reported leakage was lowered to 668.9 Ml/day. In its price control determination for the period 2010–2015, Ofwat did not provide the funds needed to finance a significant further reduction in leakage and instead used the assumption that leakage would be 674 Ml/day in 2010–2011 and 673 Ml/day in 2011–2012. In 2011–2012, actual leakage was 637 Ml/day; in 2012–2013, actual leakage was relatively unchanged at 646 Ml/day; and in 2013–2014, actual leakage was 644 Ml/day.

The company achieved these reductions by:

- Implementing better pressure management of known problem sectors in its older water network
- Replacing 1,400 miles of worn-out Victorian pipes, mainly under London (Figure 28.5)

The recent successes in meeting leakage targets have mitigated the earlier failures. As a result, Thames Water leaks slightly less water than at privatization in 1989, having reduced leakage from its 19,300 miles network of water pipes by more than a third since its 2004 peak to its current lowest ever level. As of 2013 and with an older network profile, Thames Water leaked 25.8% of supply; as of 2014, Thames Water leaked 24.7% of supply.

FIGURE 28.5 Replacement of aged pipes.

Leak detection, tracking, and pipe replacement played a pivotal role in NRW reduction. During 2013–2014, over 45,000 repairs were completed in the network and a further 10,000 repairs on customers' pipework. Aging mains were replaced, including 11.8 miles of pipe in London, Reading, Slough, and Swindon, targeting those that leaked and burst the most, resulting in the replacement of over 280 miles of mains in the last 4 years. These repairs helped reduce not only leakage but also future disruption and inconvenience to customers. Furthermore, an extensive system of district metering area management was implemented, which included real-time data and controls to monitor and detect night flows, to improve pressure throughout the water network and minimize fluctuations and excessive pressures.

28.7.2 Bilbao, Spain

Udal Sareak, a water utility in the Basque region of Spain, serves more than 130,000 people with more than 38,500 connections using a distribution network of 1,273 miles. The utility is fully owned by Consorcio de Aguas del Gran Bilbao, a consortium of 80 municipalities in Bizkaia. The Udal Sareak water utility partnered with TaKaDu for NRW reduction. TaKaDu uses integrated water network management to help water utilities optimize operations and decision making. Integrated water network management also helps shorten repair cycles and improve customer service.

Udal Sareak has 257 flow meters and 53 pressure meters; field data were collected by data loggers and stored in a supervisory control and data acquisition (SCADA) system. The water utility struggled with data management and received large amounts of data, of which some was poor quality and therefore unusable. These issues,

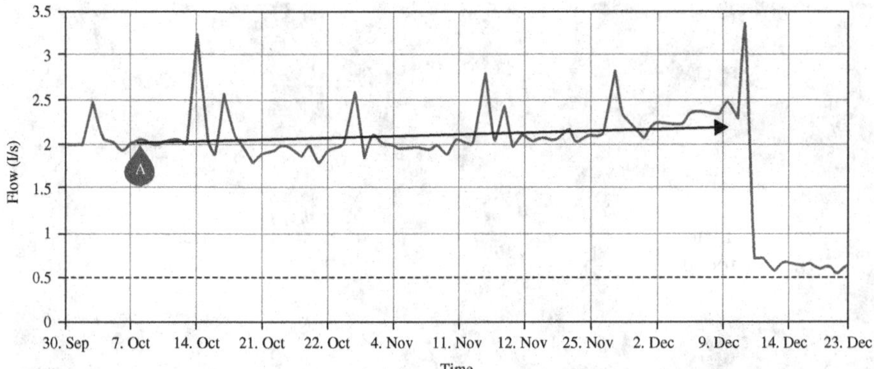

FIGURE 28.6 Identification of background leakage. The arrow indicates the flow increase trend detected by TaKaDu, which was not detected manually given the moderate change and background fluctuations. The statistical nature of the technology allows TaKaDu to identify small, invisible leaks.

coupled with infrequent data transmission rates, made it difficult to understand network patterns, trends, and specific event characteristics. Additionally, a low level of data integration further complicated the problem.

TaKaDu implemented an integrated water network management plan, with near real-time monitoring and advanced data analytics, reducing repair time and night flow by 50 and 15%, respectively. A 30% increase in locating invisible leaks was realized as shown in Figure 28.6.

The leak detected in Figure 28.6 would not have been detected using standard methods of nightline analysis or threshold-based alerts. In addition to improved leak detection, TaKaDu helped provide reliable repair verification. At an estimate of 7,000 work orders a year, repair verification was essential to reducing human error as well as the costs of repeat repairs, continuing leaks, and other inefficiencies. Overall, TaKaDu's solution helped Udal Sareak reduce water losses and improve customer confidence, further supporting the use of near real-time data in distribution systems.

28.7.3 South East Queensland, Australia

Unitywater supplies water and sewage services to approximately 765,000 individuals within a service area of 2,017 square miles in South East Queensland, Australia. The system comprises 297,266 water connections, 3,561 miles of water mains, 79 pump stations, 114 water reservoirs, 56,921 hydrants, and it delivered 52,896 Ml of water to homes and businesses from 2014 to 2015. With such a large distribution network, Unitywater had issues with both NRW and data management.

Unitywater struggled with data overload and 13% NRW with no single point of origin for NRW. Additionally, limited time, personnel, and resources to analyze resulted in limited tracking of leak events and a "best guess" targeting of areas with higher NRW. Unitywater needed automatic analysis of data with notification of potential issues within the distribution network as well as management of its large

FIGURE 28.7 Sites of leaks detected with TaKaDu's support.

amounts of data, including management of events and alarms from start to finish. Unitywater was also interested in a system that easily delivered financial and performance reports and was user friendly.

Unitywater partnered with TaKaDu to correct the items listed previously. TaKaDu established a System Loss Team, an NRW group, and provided operational processes review and updates. Furthermore, TaKaDu's integrated water network management system helped Unitywater avoid major water outages, found causes of dirty water complaints, found causes of leaks that ultimately prevented major events (Figure 28.7), identified pump faults, located breaches between areas, changed the management of the network, and tracked events from start to finish. TaKaDu also created a dashboard for use by Unitywater to prepare financial and performance reports.

Some of the additional benefits realized from Unitywater's partnership with TaKaDu included reduced chemical dosing costs, increased reporting abilities, fewer water quality events, reduced energy and personnel costs, and reduced response times by 60%. All of these items improved customer confidence and satisfaction. TaKaDu was also able to help Unitywater determine infrastructure costs through the application of real data to planning efforts to allow for changes in population-density planning.

28.7.4 Manaus, Brazil

Manaus, located in Northern Brazil, has a population of about 1.85 million with over 2,050 miles of pipes serving more than 374,000 customers. NRW was estimated at over 500 million liters per day or approximately 63% of the total volume supplied. Manaus Ambiental operates the water plants used to serve the area and became interested in NRW reduction. Manaus Ambiental partnered with CH2M to evaluate the existing conditions of the Ponta do Ismael service area through monitoring and performing a water balance of the water distribution system.

To successfully assess the water balance, the Ponta do Ismael service area was divided into two main areas and four macro districts. CH2M and Manaus Ambiental used a bottom-up water balance approach to conduct a minimum flow analysis of the recorded flow and consumption pattern. The minimum flow indicated the lowest water consumption level associated with the reduction of customer activity. The water demand during this time consists predominantly of actual consumption and water losses. Therefore, with a fair amount of certainty, it was assumed that minimum flow minus consumption equals the total water losses. The night consumption was estimated based on average night consumption data from "smart meters" placed in series with existing meters.

Continuous monitoring of night flows into water supply zones or district metered areas (DMAs) is an important operational tool for identifying water loss within the water distribution network. The project team conducted monitoring campaigns designed to monitor flow and pressure at each DMA in the system and the two main service areas. The NRW estimates for the two main areas were calculated using the minimum night flow analysis at 62.3 and 63.0% by volume. The NRW was responsible for an estimated $65 million in lost revenue annually.

Additionally, the project team assessed the distribution network condition by determining the various materials and lengths of water mains within the distribution system. The assessment indicated the predominant materials used were polyvinyl chloride, high-density polyethylene, and ductile iron. The assessment also identified the existing condition of the pipes to be in good condition with a report of main breaks or three breaks per 100 km/year for pipes 3 inches or larger in diameter. However, large main breaks were still an issue and caused significant water loss, costly repairs, substantial damage, and danger to the communities [6].

28.7.5 Asheville, North Carolina, USA

The city of Asheville is located in North Carolina. Asheville's Water Resources Department handles approximately 50,000 accounts for a city with a population of approximately 87,000. Water system facilities include three water treatment plants, approximately 1,674 miles of water lines, 64 pump stations, and 35 storage tanks. A NRW reduction program was established in partnership with Cavanaugh. Development and implementation of a 5-year water loss program began in 2012. A water loss reduction of 20% was achieved from 2012 to 2014 and was a direct result of the program implementation.

The NRW reduction program used by Asheville included:

- Validation of water supply volumes
- Design and implementation of a large meter-testing program
- Large meter right-sizing analysis
- Detailed analysis of consumption database to determine profiles, patterns, trends, and anomalies within customer consumption
- Detailed analysis of account database to identify billing, classification, and multiplier errors
- Development of DMAs in support of in-house leak detection survey work
- Establishment of a validated, reliable monthly data tracking system for key performance indicators
- Systematic improvements in data validity

In addition to the NRW program, Asheville developed a mobile phone application for customers to enter work orders to correct problems such as city water leaks. Ultimately, the mobile phone will be incorporated to further enhance the NRW program.

28.7.6 Southeast Pennsylvania, USA

In 2013, CH2M partnered with Aqua America to conduct a water audit and create an NRW loss control program for the southeast region of Pennsylvania. The system in southeast Pennsylvania provides over 100 million gallons of water per day to over 350,000 customers. It includes 114 municipalities, 4,400 miles of pipe, and eight surface water treatment plants.

The NRW project included interviews to determine the activities and to collect data, to analyze production metering and data, to review the billing system practices and processes, and to evaluate historical distribution activities. Initial review of the data revealed a 5% increase in NRW loss over a 10-year period (2003–2013). However, the data also indicated that breaks and leaks within the distribution system had decreased while the reported pressure within pipelines had increased. (An increase in pressure can lead to increases in pipe breaks and leaks.) Further review of the historical data found reporting errors and that the customer's meters had a low level of accuracy. When the data were adjusted to correct the errors, the increase in NRW was found to be 3% over the 10-year period. This audit helped provide a more accurate baseline of NRW and prevent reporting errors during implementation of the NRW loss program.

In 2015, CH2M assisted Aqua America with further investigations to refine and reduce NRW volumes, which included a review of leak detection practices. This review helped determine current leak detection practices and recommend improvements. Improved leak detection practices correctly identify leaks within the distribution system, fixing the leaks, and prevent future water losses. The improvement to the Aqua America system was in prioritizing the portions of the system that are more prone to leaks and standardizing the data collection results. Additional tasks performed by CH2M to help Aqua America included standardizing customer meter testing, selecting and designing DMAs in assessing production meter accuracy, developing theft reduction techniques, and conducting flow monitoring for system subzone water auditing.

28.8 THE FUTURE OF NONREVENUE WATER REDUCTION

The future of NRW reduction will rely heavily on cooperation, innovation, adoption of technology solutions, and support of changes within the water community. Water utilities should focus on increasing the efficiency of the distribution systems by promoting and increasing effective management, water auditing, pressure management, and automatic online monitoring. Demand management, demand-side efficiency, and water conservation can help extend the life of existing infrastructure and water supplies. The establishment of full-cost accounting and reasonable water rates coupled with improved policies and programs incentivizing conservation and efficiency could alleviate issues surrounding shortcomings in funding for NRW-related improvements.

Advanced Metering Infrastructure (AMI) is a growing component of Internet of Everything (IOE)-based NRW reduction programs, which have been increasing in number since 2010. IOE-based NRW reduction programs are part of a movement toward making city systems smarter and more readily available to detect and respond to threats (e.g., main bursts, pipe leakage) within the distribution system. AMI can help a utility reduce NRW; enhance its resilience to natural, accidental, and intentional incidents; improve the overall management of water resources; and optimize operations. The benefits of AMI not only include NRW reduction but also extend beyond the benefits of economic sustainability for utilities and cities worldwide.

28.9 CONCLUSION

Advances in technology and data management coupled with continuing education of those in the water industry are important factors for understanding NRW causes and developing reduction solutions. The availability of information on NRW from best practices, lessons learned, webinars, and other expertise is on the rise. The information is readily available through vendors, governments, and water organizations such as AWWA.

IOE, with an anticipated 50 billion smart devices worldwide by 2020 (CISCO), is creating a data revolution. New types of advanced sensing technologies are rapidly being developed for the industry, creating innovative disruptive approaches for NRW evaluation and reduction. One example of the new data-driven approach was demonstrated by Takadu's integrated management network–software application. Integration of multiple sensors in the distribution system (meters, pressure sensors, others), application of advanced data applications, and data visualization tools are quickly transforming a traditionally static approach for assessing NRW to a real-time continuous assessment. This is truly a transformational change for the water that addresses a global problem at a time where water shortage due to major climate changes and increased population growth has placed major stresses on available water resources.

REFERENCES

[1] Kingdom, W., Liemberger, R., and Marin, P.. *The Challenge of Reducing Non-Revenue Water (NRW) in Developing Countries—How the Private Sector Can Help: A Look at Performance-based Service Contracting.* Water Supply and Sanitation Sector Board Discussion Paper Series, Paper No. 8, The World Bank, Washington, DC, 2006, pp. 1–40. http://siteresources.worldbank.org/INTWSS/Resources/WSS8fin4.pdf (accessed August 24, 2016).

[2] American Water Works Association. *Water Audits and Loss Control Programs: M36*, Vol. 36, 4th Ed. American Water Works Association, Denver, CO, 2016.

[3] Frauendorfer, R., and Liemberger, R.. *The Issues and Challenges of Reducing Non-Revenue Water.* Asian Development Bank, Mandaluyong City, 2010. https://openaccess.adb.org/bitstream/handle/11540/1003/reducing-nonrevenue-water.pdf?sequence=1 (accessed August 24, 2016).

[4] Alegre, H., Hirner, W., Baptista, J., and Parena, R.. *Performance Indicators for Water Supply Services. Manual of Best Practice series.* IWA Publishing, London, 2000.

[5] American Water Works Association. Water Loss Control Committee Free Water Audit Software. http://www.awwa.org/resources-tools/water-knowledge/water-loss-control.aspx (accessed August 24, 2016).

[6] CH2M. *Water Efficiency Feasibility Study: Task 5 Evaluation of Existing Condition.* Georgia Institute of Technology, Atlanta, GA, 2013.

29

IoT AND SMART INFRASTRUCTURE

George Lu and Y.J. Yang
goodXense, Inc., Edison, NJ, USA

29.1 INTRODUCTION

There is an increasing recognition that our aging infrastructure needs to be monitored using the nascent Internet of Things (IoT) technology. By infrastructure we are talking about our bridges, roadways, railways, pipelines, radio and transmission towers, and so on. Infrastructure monitoring could mean monitoring the health of the structures or the many machines, such as motors and compressors, working alongside.

Let's say you have been asked to develop a system to monitor a critical asset, be it a structure or a machine. At its core, the task could be as simple as measuring and transmitting states or conditions of the asset to a remote host on Internet. This chapter walks you through the engineering decisions to realize such a system.

System design is an iterative process. It is common to start with a demonstration of the core functionality, possibly prototyped with readily available parts, to solicit support for development. It is essential to gather a comprehensive list of requirements so that they could be addressed early in the design process.

The scope of the requirements could cover aspects such as:

- Operating environment
- Power supply
- Connectivity options
- Data acquisition methodologies
- How to service/update after field deployment

Internet of Things and Data Analytics Handbook, First Edition. Edited by Hwaiyu Geng.

Companion website: www.wiley.com/go/Geng/iot_data_analytics_handbook/

- Using specific hardware or software component or interface to reduce production or operating cost
- Dates to achieve specific milestones

This is about designing the system to meet operational requirements and survive abnormal situations that are surprisingly common in real life, for example, flaky wireless connections and intermittent power sources. It is cost-prohibitive to overengineer a system to be able to handle every conceivable situation, so it comes down to the trade-offs you are willing to make. In the next section, we will go over the engineering decisions.

29.2 ENGINEERING DECISIONS

29.2.1 Power Supply

Will your device be powered by grid power source, battery, or some form of renewal energy? This is a surprisingly important decision because it would drive the key technological approach. A significant part of wireless sensor research and field work is based on the assumption that we will use very low-power wireless sensors for infrastructure monitoring. The very low-power consumption is usually achieved by very low-duty cycle, that is, the processors comes out of a low-power mode to gather and transmit data only for a very short period. These smart sensors could utilize batteries that last for months or years at a time or on rechargeable batteries recharged by solar or wind energy.

It is certainly convenient if the device has access to the main power from your local electrical utility, though you would still have to consider how your device could continue to function during power outages that are sure to happen. Chances are you will need an uninterruptible power supply.

The cost of renewal energy has declined phenomenally in recent years. Solar panels can now be acquired for less than US$1/W in commercial quantities. Where solar harvesting is not practical due to inadequate direct sunlight or foliage, small wind turbine often could be considered. With solar or wind energy harvesting, coupled with rechargeable battery of sufficient capacity, as with mains power, you could afford to install far more computation resource in the field that could handle sophisticated processing. You could also consider radio technologies that afford more bandwidth and range. As the next sections show, power supply could make a big difference in the options you have at your disposal to develop your monitoring service.

29.2.2 Monitoring Versus Control

Are you implementing monitoring and also providing supervisory control? In monitoring, we aim to capture the state or condition of the target asset. The captured data could be analyzed to assess current operation and predict how the state or condition might be in a future time, resulting in service activity to correct or prevent an undesirable asset state or condition. This broad category of applications is known as

Condition-Based Maintenance (CBM). The emphasis is on capturing the asset state or condition to provide timely maintenance as needed, as opposed to Schedule-Based Maintenance where maintenance tasks are performed based on a set schedule regardless of the asset state. This application scenario does not usually demand real-time response within a narrow time window.

In other application scenarios, you may need to provide supervisory control of equipment based on acquired data. Robotics, including Unmanned Aerial Vehicles (UAVs), are such cases. Many industrial and infrastructure processes depend on Supervisory Control and Data Acquisition (SCADA) systems continuously for their critical missions. It is important to know the amount of time given to provide the needed control; there may be consequences of failing to provide the needed control signal in time. Such stringent performance requirement could constrain your design decisions greatly.

29.2.3 Connectivity

The always connected nature is at the heart of a cyber–physical system (CPS). It allows us to integrate geographically distributed resources and deliver a solution quickly.

Networks have limited bandwidth and finite latency. Bandwidth is shared among a number of interface devices on the same network. There is a plethora of networking technologies to choose from. The ones affording longer distance and faster bandwidth tend to cost more in terms of power and carrier fees. You will need to decide what is appropriate for you given the network infrastructure that is available to you and the nature of your application. Here we try to consider from the attributes that are more relevant to the designers of a CPS. If wired Ethernet is available to your system, congratulations! Wired networking is simply not practical in vast majority of CPS scenarios.

29.2.3.1 Wireless Networking Infrastructure Mobile data services are offered through cell towers that cover wherever access terminals may be. With each new generation of technology, a base station (i.e., a cell tower) could handle more concurrent number of terminals, each sending and receiving higher volume of data, and possibly further away from the base station. In our homes and offices, Wi-Fi follows the same hub-and-spoke topology. If a terminal device is too far from the Wi-Fi access point (AP), it is well understood that a repeater could be used to extend the coverage.

If your system is expected to be deployed in the vicinity of a building in a not-too-remote area, chances are you could count on 2G/3G/4G mobile data services by your local telecom company, supplemented by Wi-Fi. Your systems could utilize Wi-Fi for communication among themselves at the site, either wired or wireless connectivity to Internet.

If you don't have the benefit of such infrastructure, you might need to build your own. Wi-Fi is most familiar to the developer and user communities and readily available. Most Wi-Fi AP/routers could support multi-AP network or mesh protocols.

Wi-Fi may not be a viable option to interconnect a number of smart sensors due to their power consumption. Soil moisture sensors across a large farm or strain sensors across the span of a bridge do not need the high bandwidth offered by Wi-Fi nor could they afford the power budget. IEEE 802.15.4 or Bluetooth Low Energy (BLE) could offer more appropriate connectivity options for these sensors to a nearby edge server. LoRa and Sigfox are new low-power options that could directly allow low-bandwidth smart sensors to reach a remote host.

Consider the case of sensors along the span of a bridge that is more than 1 km long. You could require every sensor to transmit data directly to a cloud server via 3G interface, at substantial subscription cost and power consumption. You could place a Wi-Fi AP near the midpoint and use high-gain antennae to cover the 500 m distance for furthest devices. You could use ad hoc mesh network where each node could act as a relay for each other; each node therefore only need enough power to reach the closest neighbor. Such mesh networks often have the added benefit of fault tolerance, since there are usually more than one path to reach a destination. 6LoWPAN is a mesh networking protocol that has been implemented on IEEE 802.15.4 and BLE.

You need to think of network connectivity as intermittent and unreliable. Wireless connectivity is subject to radio interference, especially when using crowded spectrum. Having redundant network interfaces to different networks would be nice, but it is a rare luxury. For some applications, mesh networks that could provide multiple paths between edge nodes and network gateways could greatly increase system robustness. It is essential to utilize messaging protocols that ensure reliable message delivery. That means keeping a local archive until it is certain that the data has been successfully delivered to the remote data repository. Persistent data storage is expensive, especially on a remote device in the field that might not have the luxury of grid power. The size of the local archive and the storage technology used is a decision you would have to make.

Your edge node should be able to function autonomously; by that we mean the system in the field should not have to rely on remote services for its essential tasks. This suggestion may contradict your desire to utilize a remote Web service or API, which is a common practice in modern information systems. Remote services will not be reachable at all, or suffer unpredictable latency, due to less than ideal network connectivity. Strong reliance on remote service means network connectivity issues could render edge devices useless. Ideally, your edge node could function normally even in the total absence of network connectivity, accurately time-stamping gathered data into local storage to be transmitted when network connectivity is restored. That is, your application on the edge should not utilize remote service synchronously but could utilize remote services asynchronously so that absence of network connectivity would not hinder essential operation.

29.2.3.2 Data Telemetry Versus Event Notification A system could start with having data acquisition module forwarding acquired data to another host to analyze the data and detect abnormal events. This makes perfect sense in a lab setting when we want to use readily available software tools for analysis and the data acquisition module does not have the resource to run that software tool. Besides, you may want data acquisition module to focus on its job of acquiring data at precise sampling intervals. In this case, the network connectivity is just enabling telemetry.

When deployed in the field, the cost of telemetry could change the thinking. Are you sending data to a remote host to detect meaningful events? Is there a lot of data to sift through to determine if there is an abnormal event to raise attention? Can the event detection only be computed on a remote host with more resource? It may now make sense to process the raw data on the edge device to drastically reduce the amount of raw data to be transmitted to a remote server. This could effectively improve the scalability of your solution. You will now need to design your deployed system to be able to process data and detect abnormal events, in addition to data acquisition and telemetry. Since local computation is often cheaper than transmitting data to remote server for processing, this strategy could be preferred to yield a more responsive and lower energy-consuming system. It is conceivable then to use a much lower-cost wireless interface to send regular heart beat and occasional event notification and use a more costly interface only when it is necessary to send supporting data.

For example, let's say you are monitoring the vibration at many points along a civil structure to detect and localize damage. Transmitting all the raw data, possibly sampled from 3D accelerometers at multiple points on the structure at 100 Hz or higher, to a server for processing would require a lot of bandwidth. It would be more efficient for the edge devices to have sufficient resource to analyze vibration data. This is a very valid assumption given the low cost of microcontrollers (MCUs) and microprocessors (MPUs) today. Secondly, the communication pattern to localize structural damage is inherently among adjacent nodes that have detected abnormal vibration. It would make sense to utilize mesh network along a pipeline infrastructure to facilitate such peer-to-peer communication without going through a distant central hub. The solution now utilizes far less bandwidth to transmit the likely location of the detected damage, resulting in lower operating cost.

29.2.4 Protocol Considerations

There is frequent talk of the need to standardize protocol for IoT. This is a topic that could elicit strong feelings. There may be many reasons to support a particular protocol. Technically, you would look at it from the perspective of efficiency, reliability, and ease of implementation. From business perspective, there may be strong motivation to use same protocol as current products or be part of a community to benefit from the ecosystem of users, suppliers, and developers that work together to increase its value.

There are many low-level protocols used primarily for interfacing to sensors and actuators. Modbus is a well-known serial communication protocol for connecting industrial electronic devices. Controller area network, or CAN bus, is a message-based protocol that allows MCUs in a vehicle to communicate with each other. SDI-12 is a serial communication protocol for sensors that monitor environment data. Some buses may even require dedicated interface hardware.

There are many high-level protocols used primarily for data exchange between hosts. We consider only Internet Protocol (IP). One could have edge sensors write data directly to a central database server at a public IP address, serving on a specific port. This straightforward approach may not be efficient over a relatively slow and unreliable connection. A better way is to utilize an asynchronous message queue to deliver messages reliably between data publishers and data subscribers. More broadly, one could identify sensors as the data publishers and the analytic programs as data subscribers. Many tools cater specifically to such *pub–sub* communication pattern common in IoT applications. It is possible that you are asked to use one of the more popular tools, for example, MQTT.

You should also decide how to service a deployed system in the field. Your solution would not scale well if it requires site visit by qualified technical staff in order to perform software or firmware updates. Everyone who has used a smartphone already comes to expect that embedded software could be updated over the air reliably. From the very start, consider how to service deployed units from the comfort of your office. Could you ssh into each unit? Would it make sense for each field device to connect into a Virtual Private Network? How would you provide for remote service yet still offer strong protection against hackers?

While a standard protocol for all IoT devices might sound appealing, the protocol that is appropriate for a full-blown computer with operating system (OS) may not make sense for a smart sensor system that is built on an MCU with low-power RF interface. While it is possible to run a Web server on an MCU, it would be far more appropriate to run a Web server on a larger MPU that can support large number of concurrent connections, secured by adequate encryption. Modern web frameworks allow one to quickly develop and deploy Web services associated with sensing and control, which could be easily accessed by people using web browsers or utilized by other information systems in the form of RESTful API. Use of such tool stack could greatly increase developer productivity. It could be a key factor in your ability to add feature improvements efficiency.

The choices of protocols to support should follow directly from the requirements. The choices would very strongly influence architectural choices that we will discuss next.

29.2.5 Architectural Choice

Many engineers are already familiar with using MCUs for sensor interface, data acquisition, and control. It would be fair to assume that all readers have worked with computers running an OS, most commonly Windows, Linux, or Mac. The smartphones of today are effectively a miniature computer running an OS, constantly communicating with services on remote servers to exchange messages. Technically they are all MPUs. For better distinction we describe this latter approach as an MPU with OS. Either or both architectures are viable options for your design. It should be noted that the distinction between these approaches is not absolute. It is possible to run OS on some MCUs and possible to run applications on MPUs without an OS. This chapter aims to discuss the relative fit of these architectural approaches for your application.

29.2.5.1 Microcontroller (MCU) MCUs are embedded inside all sorts of appliances around us: electronic coffee makers, rice cookers, water sprinkler controllers, and so on. A new car today is likely to have more than a hundred MCUs. MCUs have many built-in peripherals such as UART, I2C, SPI, USB to interface with other electronic devices. MCUs have many hardware counters or timers to facilitate counting or executing a task at precise time or intervals. MCUs usually have analog-to-digital conversion (ADC) ports to sample analog voltages into a digital value. MCUs offer varying number of general-purpose input/output (GPIO) pins to facilitate digital input and output.

In the distant past it might be necessary to struggle through assembly code to fit the firmware inside a very limited EEPROM of an 8-bit MCU, using preciously little SRAM (possibly under 1 kb). Today there are many 32-bit MCUs, with clock speeds over 100 MHz, megabytes of flash memory for program and data storage, and hundreds of kilobytes of SRAM. They are usually programmed in C. It is now possible to use high-level scripting languages such as Lua and JavaScript on some MCUs as well.

MCUs could draw very little power relative to the conventional computer sitting on your desk. Even at full operation, it draws only tens of milliamps, usually at 3.3 or 5 V. Most notably, many MCUs support low-power modes that could reduce power consumption down to sub-microamp levels. In most cases, the firmware could put the MCU into a low-power mode, to be waken upon an event (such as interrupt due to an external trigger or an internal timer), perform certain actions, and then return to low-power mode again. For a well-designed application, the MCU could be active less than 1% of the time. The percentage of time an MCU spends in active tasks could be described as the duty cycle. Such a low-duty-cycle strategy makes it possible for battery-powered devices to operate for months or even years. This extends well to radio-frequency communication using BLE and IEEE 802.15.4 radios. This is essential for consumer applications.

Low-latency I/O is a big reason to use MCU. This means your firmware application could detect input or control an output within a very short time. This could be as short as one machine cycle. On a 100 MHz MCU, a 1-cycle instruction would take 10 ns to execute. That is not the typical latency that you could count on though. Most likely your firmware application is in some kind of low-power waiting state that would have to be awaken before it could act on the event. Latency then is highly dependent on interrupt latency of the MCU. It could be hard to determine the true interrupt latency because an interrupt service routine could be further interrupted by one or more interrupts of higher priority. It could also depend on the level of low-power mode the MCU is in when interrupted. It is possible to experience interrupt latencies in microseconds, a very long time for a modern MCU that runs in tens of megahertz or higher clock frequencies. In most cases this is still preferred over the even longer latencies for an MPU-running OS.

Kim et al. have detailed discussion of an MCU-based wireless sensor network for health monitoring of civil infrastructures, using Golden Gate Bridge as the case study [1].

MCU summary:

- Pros
  - Directly operate on the hardware, that is, "bare metal," not through layers of abstraction and OS.
  - More deterministic; if an instruction takes one machine cycle to execute, you can be sure that it will take 10 ns on an MCU that is clocked at 100 MHz.
  - Low-power modes afford long operating time on battery power.
- Cons
  - Developing directly on bare metal MCU means mapping your application directly to the hardware resources, usually orders of magnitude less than an MPU with OS; the operational efficiency comes at a cost of higher learning curve and dependency to the target platform (i.e., tied to the specific component supplier and tool chain for development), resulting in much reduced portability.
  - Firmware application may be harder to update after field deployment, highly dependent on the boot loader provided by the MCU manufacturer.
  - The firmware application is Inherently single-threaded operation. You will need a Real-Time Operating System (RTOS), essentially a task manager, to help manage multiple parallel threads, trading off latency.

29.2.5.2 Microprocessor (MPU) There have always been companies that offer smaller versions of desktop or notebook computers for embedded applications. Using the familiar desktop OS and peripherals for storage and display means a mature component supply chain to rely on. The advent of modern smartphones in recent years has greatly increased our expectation of embedded computers. A burgeoning ecosystem of single-board computers (SBCs) has sprung up, leveraging on open-source OS (usually Linux) and development tools. While often catering to the *maker movement*, mostly referring to amateur developers and students, this ecosystem is immensely beneficial to professional developers alike. Many monitoring and control systems have been developed on Raspberry Pi and BeagleBone. It is now very reasonable to consider having for your own embedded system a multicore 32-bit or 64-bit MPU, gigabytes of DRAM, tens of gigabytes of solid-state flash storage, high-resolution touch-screen display, built-in support for Wi-Fi and Bluetooth, and constantly reading data from an assortment of sensors.

An MPU offers high-throughput processing capability that is simply not available on an MCU. While some MCU supports DSP operations, and even hardware floating point unit, it would not be a practical platform to implement a persistent database on an MCU. Many MPUs come with one or more Graphic Processing Unit (GPU) cores that could be utilized for certain types of computation. Image processing is a class of application that would require relatively high bandwidth and latency to transmit raw data to remote servers for processing, which could be processed locally, possibly utilizing multiple MCU and GPU cores, at much lower cost.

A big part of the benefit of this ecosystem is the wealth of software tools for OS, development, deployment, and operational support. While Windows and Mac remain the dominant desktop OS, Unix, and Unix-like OS (including Linux and BSD variants) run majority of the servers on public Internet [2]. The same OS is embedded in most of the network disk servers found in many homes and offices. The Linux OS that comes with most SBCs already supports the many protocols and development tools used for these servers. It makes sense for an infrastructure monitoring system to support the very same protocols and development tools that are already embraced by the developer community.

Latency is an important issue that we had first mentioned in earlier MCU discussion. On an MPU system latency is largely a feature of the OS. Many developers come to realize that Linux is not an RTOS when their 1.6 GHz quad-core MPU could not toggle GPIO pins as fast, nor as precisely, as an Arduino, whose 8-bit MCU is clocked at a mere 16 MHz. There are OS variants such as RTLinux and Xenomai that would make Linux more appropriate for real-time applications. It is usually not as simple as booting up Xenomai to enjoy lower latency.

The drivers you might require for your hardware devices may or may not work well with an RT kernel. There are certainly commercial RTOS options if your requirement calls for it and your budget allows for it, though the license cost and learning curve would have to be justified against the benefits.

Someone would point out that an SBC such as Raspberry Pi or BeagleBone could not meet the operating temperature requirement for most outdoor deployments for infrastructure monitoring applications. Several SBCs have open-source designs where you could substitute parts to achieve higher operating temperature range. You could talk to a PCB assembly supplier, who is more than happy to work with you on your custom modifications.

MPU summary:

- Pros
  - Ease of development in many high-level programming languages.
  - OS and library afford rich development environment, interface protocols, and operational support.
  - MPU system more likely has the resource to do on-the-edge event detection and data compression.
  - Likely easier to update after field deployment.
  - Many SBCs to choose from, leveraging on MPU designed for consumer tablets and smartphones.
- Cons
  - Higher resource requirement (in terms of power, bandwidth).
  - Higher latency.
    - Linux is not "real-time" OS; you can use RTOS available for your MPU to reduce the latency, possibly to sub-microsecond level.

    - RTOS options exist but cost and familiarity concerns.
- Flexibility comes with higher system complexity.
- OS and library may require frequent update (bug fixes).
- Security concerns (bug in a common packages may likely impact your systems).
- Rapid product cycle of key components may mean the same part may not be available in 10 years; that could be a concern for products that are subject to regulatory compliance.

29.2.5.3 MPU–MCU MPU and MCU are not mutually exclusive. A way to get the best of both worlds is to combine MPU and MCU; utilize MPU for the mature OS, high-throughput computation, and wealth of development tools and network services; and use MCU for the latency-sensitive data acquisition and control. This approach has already been utilized in a number of smartphones (e.g., the M8 in Apple iPhone 6 and 6 Plus or the Atmel AVR MCU in Samsung Galaxy S4). The MCU in this MPU–MCU combination is called a sensor hub [3].

A relatively modest MCU that costs a few dollars could satisfy this need competently. One would need simple firmware on the MCUs to sample data in response to some internal or external trigger and transmit the data to MPU. This approach could scale to a number of MCUs working in conjunction.

The larger MPU could indeed go into low-power mode, waiting for MCU to wake it up to do the heavy lifting only when necessary. This could result in much lower overall power consumption.

For applications with stringent latency requirement, some designers resort to using FPGA to achieve deterministic I/O. FPGA could also be used for hardware-accelerated data processing. FPGA vendors now offer large number of gates integrated with MPU cores; these programmable gates could be thought of as highly customizable sensor interfaces or soft MCU or DSP cores. Conventionally the benefit of FPGA comes at cost of complexity and power requirement though new product offerings could be very competitive for those that have overcome the steep learning curve. In the context of our discussion, we will consider FPGA as equivalent to a dedicated MCU. XMOS is a novel multicore MCU architecture that also offers deterministic I/O (to 10 ns cycle time) without suffering variable interrupt latencies.

Some semiconductor vendors offer MPU–MCU combinations directly in a single package. Texas Instruments (TI)'s Sitara AM335x series combines an ARM Cortex-A8 MPU at 1 GHz with two ARM Cortex-M3 MCU at 200 MHz. TI calls these two MCUs as Programmable Real-time Units to emphasize their particular utility in low-latency sensing and control applications [4]. Intel has an MPU–MCU offering in its Atom family. At the center of Intel Edison is an Intel Atom 22 nm SoC with a dual-core CPU at 500 MHz and an MCU at 100 MHz. This allows the product to collect and preprocess data via the MCU in a low-power state and hand the filtered data off to the CPU for analytics [5]. NXP has a VFxxx controller family that integrates ARM Cortex-A5 MPU and Cortex-M4 MCU cores in a solution [6].

Besides having the integrated MCUs, these vendors also provided shared memory access and interprocess communication API. This could be more efficient than developing one's own protocol.

We anticipate more such offerings from semiconductor vendors as they are good fit for IoT applications.

MPU–MCU summary:

- Pros
  - Use the dedicated MCU as real-time sensor hub to handle the latency-sensitive data acquisition and control, where MCU is a good fit.
  - Use MPU with Linux OS for network interfaces; leverage the many protocols and services available: servers (Web services), data processing (event detection), secure access (TLS, SSL), data compression/storage, date–time synchronization (Network Time Protocol (NTP)). Use MPU for more involved data processing of the gathered data for quick local control, or reduce bandwidth requirement to a centralized service.
  - Combination could reduce overall latency and power consumption.
- Cons
  - Increased system complexity and cost.
  - Reduced solution portability to different component suppliers.

29.2.5.4 MPU–MCU over a Network The MPU and MCU combination could be loosely coupled and interconnected over a network. A smartphone with a number of BLE sensors is one such implementation. Similarly, IEEE 802.15.4 smart sensors can be connected through MCUs with low-power RF interface to an MPU. The MPU acts as a hub for a number of smart sensors. The MPU typically runs an OS and protocol stack, acting as sensor aggregator and router. This type of application is rapidly growing in homes and offices with wearable smart devices.

29.2.6 Data Acquisition

Now that we have discussed the architectural choices, it is appropriate to revisit data acquisition in more detail. As mentioned previously, data acquisition involves utilizing the many peripherals on MCU or MPU to interface to sensors; UART, SPI, I2C are the most likely peripherals. You might need to capture analog sensor output into digital values. ADC ports are relatively scarce on MPUs. If you need more than a trivial amount of ADC, chances are you have to utilize MCU in your design.

Infrastructure monitoring encompasses sampling of many static and dynamic physical attributes across the structure. Temperature of a structure or a machine is a relatively slow varying physical attribute. It is sufficient to sample such slow varying attributes once every few seconds or even minutes. Dynamic attributes such as vibration are more challenging because of the larger amount of data acquired and the need for precise timing. There are three aspects of precise timing: precise sampling period, precise trigger, and synchronization across network of devices. If an accelerometer is sampled at, say, 100 Hz, you would want the time between successive samples to be at exactly the same sampling period, 10 ms, not subject to variable latencies due to task schedulers or interrupt handling. The sampling should commence uniformly after the trigger, be it a timer interrupt or an external trigger on an input pin. This is a key consequence of your architectural choice. Some applications require data sampling to be synchronized across multiple devices so that the relative phase of the data could be compared. Monitoring electrical grid and vibration modes of civil structures are examples where such synchronization may be needed. Embedded electronics rely on oscillators or crystals as local time sources, possibly with a battery-backed real-time clock. Such self-contained time sources could not maintain sufficient synchronization in the field because of temperature-dependent changes and varying rate of crystal aging over time. One would have to utilize some sort of synchronization protocol among each other or against a reference time server through Internet or GPS. The NTP is a protocol designed to synchronize the clocks of computers over a network. NTP would utilize time servers on Internet, in conjunction with GPS, to act as a good reference time source. Experienced readers would point out the risk of relying on external services for critical time synchronization, as network connectivity and reliable GPS signal could not be taken for granted. These risks could be mitigated by a local GPS Disciplined Oscillator, at a cost of size, complexity, and power consumption. You have to decide if your application requires this level of effort for synchronization.

How much data to acquire is likely a question to be answered by the asset owner. For long-term structural monitoring, it is often sufficient to archive the representative state of the structure. This goal could be achieved by periodic sampling covering different times of the day, through different seasons, each lasting a relative short period. For most civil structures, 10 seconds of continuous sampling repeated every hour should be more than sufficient to meet its long-term monitoring requirements. That may not be adequate during forecast events when strong wind or water flow could be expected in advance, and most certainly unanticipated events will be missed all together. Only continuous monitoring could capture unanticipated events that could cause structural damage. If you have the luxury of power, bandwidth, and storage, you may choose to archive all raw data. Even for a modest number of sensors, one could end up with gigabytes of raw data a day that does not seem to vary from 1 day to the next. It may be sufficient to only archive periodic samples for long-term analysis and events worthy of detailed analysis. Lee and Yen discuss their field results of using continuous monitoring for bridge monitoring in [7, 8].

29.3 CONCLUSION

Infrastructure monitoring is still in its relative infancy, though benefiting from rapid technological innovation in MEMS sensors, fiber optic sensors, wireless communication, embedded computing, and energy harvesting. Interested readers are encouraged to peruse the references at the end of this chapter.

REFERENCES

[1] S. Kim, S. Pakzad, D. Culler, J. Demmel, G. Fenves, S. Glaser, and M. Turon, Health Monitoring of Civil Infrastructures Using Wireless Sensor Networks, Proceedings of the 6th International Conference on Information Processing in Sensor Networks (IPSN '07), Cambridge, MA, April 2007, ACM Press, pp. 254–263.

[2] Usage Share of Operating Systems: https://en.wikipedia.org/wiki/Usage_share_of_operating_systems (accessed October 26, 2015).

[3] Sensor Hub: https://en.wikipedia.org/wiki/Sensor_hub (accessed August 24, 2016).

[4] Texas Instrument Programmable Realtime Unit Subsystem: http://processors.wiki.ti.com/index.php/Programmable_Realtime_Unit_Subsystem (accessed August 24, 2016).

[5] Intel Edison: http://www.intel.com/content/www/us/en/do-it-yourself/edison.html (accessed August 24, 2016).

[6] NXP VFxxx Controller Solutions Based on ARM Cortex A5 and M4 Cores: http://www.nxp.com/products/microcontrollers-and-processors/arm-processors/vfxxx-controller:VYBRID?tid=vanVYBRID (accessed August 24, 2016).

[7] W. F. Lee, T. T. Cheng, C. K. Huang, C. I. Yen, and H. T. Mei, Performance of a Highway Bridge under Extreme Natural Hazards: Case Study on Bridge Performance during the 2009 Typhoon Morakot, *Journal of Performance of Constructed Facilities*, **28**, 1, 49–60, 2014. (ASCE, Reston, VA).

[8] C.-I. Yen, C.-K. Huang, W. F. Lee, C.-H. Chen, M.-C. Chen, and Y.-C. Lin, Application of a High-Tech Bridge Safety System in Monitoring the Performance of Xibin Bridge, *Proceedings of the Institution of Civil Engineers: Forensic Engineering*, **167**, 1, 38–52, 2014. (ASCE, Reston, VA).

FURTHER READING

http://www.goodxense.com/iotsmartinfrastructure (accessed September 7, 2016).

Industrial Internet Consortium, Industrial Internet Reference Architecture, Needham, MA: http://iiconsortium.org/IIRA.htm (accessed August 24, 2016).

Health Monitoring of Civil Infrastructures Using Wireless Sensor Networks: https://www2.eecs.berkeley.edu/Pubs/TechRpts/2006/EECS-2006-121.pdf (accessed September 13, 2016).

Y.-L. Xu and Y. Xia, *Structural Health Monitoring of Long-Span Suspension Bridges*. CRC Press, Boca Raton, 2011.

30

INTERNET OF THINGS AND SMART GRID STANDARDIZATION

GIRISH GHATIKAR[1,2]

[1] *Greenlots, San Francisco, CA, USA*
[2] *Lawrence Berkeley National Laboratory, Berkeley, CA, USA*

30.1 INTRODUCTION AND BACKGROUND

Smart Grid and digital advancements for energy are in midst of a revolution. On one side, next-generation decarbonized grid and significant investments in its modernization are driven by the challenges from decentralized generation and high penetration of variable on nondispatchable renewable generation. This drive is fueling a significant growth in unmanaged distributed energy resources (DER), new loads (electric vehicles (EVs) and battery storage), and distributed generation (DG) at various domains of the Smart Grid, particularly the distribution, and at the grid-edge domains where Smart Grid interfaces the electricity consumers (or users). On the other side, the digital energy revolution, largely driven by the profuse Internet of Things (IoT) technologies, is also undergoing transformative changes with an increase in number of connected devices or systems that can communicate and intelligently act upon information to support the dynamism in the Smart Grid and support the modernization efforts at the state and national levels.

To enable the grid of the future requires collective efforts to address some key challenges:

- Lack of dynamic demand-side flexibility to encourage electricity consumers' response to changing generation conditions for economic, societal, and environmental benefits

Internet of Things and Data Analytics Handbook, First Edition. Edited by Hwaiyu Geng.

Companion website: www.wiley.com/go/Geng/iot_data_analytics_handbook/

- Inability to integrate large-scale renewables and DER to balance supply and demand to facilitate efficient, resilient, and reliable clean energy grid
- Insufficient adoption of standards-based digital network that can sense, collect and transmit data, and provide interoperability and low-cost integration for fragmented systems and services from multiple electricity operators, providers, and vendors
- No provision of cyber security and user privacy by design for systems integration and control, in particular, for distribution systems and grid edge or the last mile for demand

To address these challenges and provide seamless interfaces between the next-generation grid and the digital energy revolution, this chapter focuses on the following key objectives:

1. Describe digital energy in the contextual framework of the IoT and Smart Grid applications with emphasis on data communication interoperability.
2. Describe Smart Grid power systems integration and standards framework that are unique drivers of the IoT framework and technology applications.
3. Identify the IoTs and standards' role for electric grid, energy systems integration, and DER for energy customers' (or users') benefits.[1]
4. Identify IoT and standards application areas with examples to transform and cost-effectively integrate Smart Grid systems at different domains and new markets.

The focus of this chapter is to provide a basic and contextual representation of the IoT and Smart Grid systems integration requirements and why the IoTs must leverage data standards to effectively address the unique challenges faced by the current and next-generation grid. The chapter addresses on "what" and "why" part of these objectives and not describe on "how" it is achieved, albeit specific examples of IoT and Smart Grid standards applications allude to it.

Smart Grid technical standards enable ubiquity for communication networks and applications through interoperability. Interoperability enables seamless integration and interfaces of the IoT devices and systems, thus increasing the potential for their successful scaling and cost-effective deployments within the existing industry fragmentation. The chapter considers *application domain* Smart Grid standards for IoT data communications using constructs such as eXtensible Markup Language (XML). The chapter does not cover *physical domain standards*, which by default use Internet Protocol (IP) via Ethernet, Wi-Fi, and ZigBee technologies. The chapter does not cover *network and transport domain standards* either, which are supported by means such as transmission control protocol (TCP), hypertext transfer protocol (HTTP), REpresentational State Transfer (REST). Standards are useful when adopted by the industry, or they address a key gap in Smart Grid interoperability and cyber security.

[1] IoTs, as a plural, are indicative of many IoTs that can be collectively benefited from the qualifying description.

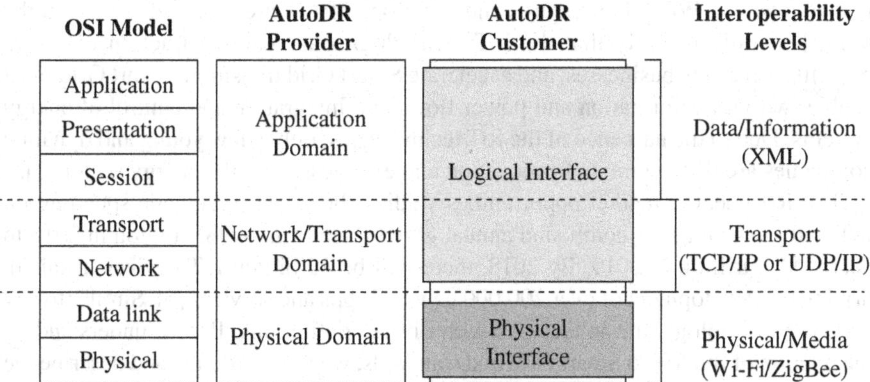

FIGURE 30.1 Smart Grid standards interoperability at different levels for AutoDR signals.

Figure 30.1 shows interoperability standards adopted by the industry for electric grid transactions to leverage demand flexibility through a mechanism called automated demand response (AutoDR) [1]. Interoperability is represented using the seven-layer Open Systems Interconnection (OSI) model, which all IP-networked devices support. Smart Grid interoperability of AutoDR transactions is defined by the application-level standards that are agnostic of any transport or physical mechanisms that industry or market forces adopt.

The IoTs that support Smart Grid must achieve end-to-end interoperability and security. When systems with diverse characteristics interact within the Smart Grid, the IoTs must possess basic ability to process application-level messages. Previous studies have covered specific applications for Smart Grid and customer transactions and are not included in this chapter [2, 3].

30.1.1 Chapter Contents and Structure

- Section 30.2 defines and describes IoT-applicable Smart Grid and interoperability standards.
- Section 30.3 describes Smart Grid standards and DER integration links for IoT applications.
- Section 30.4 describes the Smart Grid digital energy, as described by the IoT framework.
- Section 30.5 provides conclusions and key recommendations.

30.2 DIGITAL ENERGY ACCELERATED BY THE INTERNET OF THINGS

Digital technologies are accelerated by the Internet growth in every facet of our lives—health, environment, energy, transportation, and manufacturing. In this context, the term IoT seems to have originated circa 1999 from MIT Media Lab's innovations [4]. While the features and functions of IoT are more relevant than the term, it is only recently that the IoT technologies are gaining traction to derive benefits, drive new businesses, and accelerate Smart Grid transition. Smart Grid must enable two-way information and power flows and integration and control of energy systems. Due to the nascence of the IoT technologies, only a few young and dynamic companies are driving the IoT innovation and even lesser for the energy sector [5].

The IoT-specific market opportunities studies show that worldwide spending on IoT will grow at a 17% compound annual growth rate from $698.6 billion in 2015 to nearly $1.3 trillion in 2019. By 2018, there will be 22 billion IoT devices installed, driving the development of over 200,000 new IoT apps and services [6]. Similar to any evolving technology, the IoT comes with diverse definitions. Before understanding the relation of the IoT to Smart Grid and standards, we need to define and describe the IoT and its emerging applications. Many organizations have made efforts to define the IoT. The definitions from key organizations and activities are included in the succeeding text [7]. The intention of this exercise is to derive a contextual definition of the IoT that considers the basic features and differentiates it for Smart Grid applications:

National Institute of Standards and Technology (NIST): NIST references IoT to Cyber-Physical Systems (CPS) and describes it as "Cyber-physical systems (CPS)—sometimes referred to as the Internet of Things (IoT)—involves connecting smart devices and systems in diverse sectors like transportation, energy, manufacturing and healthcare in fundamentally new ways. Smart Cities/ Communities are increasingly adopting CPS/IoT technologies to enhance the efficiency and sustainability of their operation and improve the quality of life."

Internet Engineering Task Force (IETF): The institution that is primarily responsible for making the "Internet work better" and has the largest network international community defines IoT as "The basic idea is that IoT will connect objects around us (electronic, electrical, non-electrical) to provide seamless communication and contextual services provided by them." While not defined here, the IETF also provides definitions for the *Internet* and *Thing* [8].

International Telecommunications Union (ITU): Defines IoT as a network that is *available anywhere, anytime, by anything and anyone*.

Organization for the Advancement of Structured Information Standards (OASIS): A reputed open protocol and standards development organization (SDO) describes IoT as "System where the Internet is connected to the physical world via ubiquitous sensors."

Institute of Electrical and Electronics Engineers (IEEE): The IoT activities within IEEE are part of P2413 project and working groups. The IEEE defines IoT as "A network of items—each embedded with sensors—which are connected to the Internet."

European Telecommunications Standards Institute (ETSI): The IoT is described in the context of conceptual technologies as "machine-to-machine (M2M) communication" and defines it as "the communication between two or more entities that do not necessarily need any direct human intervention. M2M services intend to automate decision and communication processes."

These definitions reveal that there is no common understanding of the features and functions of an IoT and it is nebulous—likely for a good reason. Deciphering the key terms, in relation to Smart Grid, the IoT is referred to as objects, connected smart devices and systems, ubiquitous sensors, network of items with embedded sensors, automated decisions, and seamless communications and services all having a common denominator of connectivity through the Internet. In the context of this chapter's Smart Grid focus, the IoT features are distinguished from its functions or applications that it can support (e.g., intelligence for automated decisions). This is important not only for physical and logical separation of the IoT but also for deriving unique business models and value IoT-based technologies can provide. **Smart Grid IoT here is defined as an object (device or system) of secure Internet-based connected and addressable technologies with three key features: (1) communications, (2) sensing, and (3) controls (also termed Actuators).**

In the definition, it should be noted that IP communications, addressability, and cyber security are inherent features of the IoT. Consumer privacy, as applicable, is an important consideration for the IoTs. The Internet-enabled (referenced as "Internet of") network of connected devices (referenced as "things") supports Smart Grid applications to either improve existing practices or offer emerging technologies and services for the Smart Grid. Interoperability standards are key for "seamless communications" and "ubiquity" of IoT objects. Equipped with these features, the IoTs can enable innovation for billions of Smart Grid-connected devices and consumers with technologies and services for functional applications such as[2] supply and demand coordination, customer engagement, advanced intelligence and data analytics, end-to-end automation, device management, and service-based platforms. This framework is critical to support innovation and new business models through the proliferation of the IoT technologies for Smart Grid.

Figure 30.2 shows the framework for these IoT features and applications in the Smart Grid context. The key feature communications enables IoT to transmit and

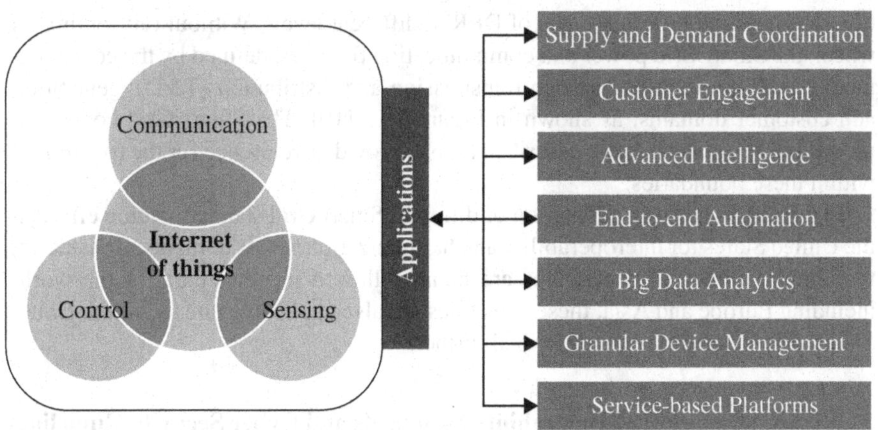

FIGURE 30.2 Framework for the Internet of Things and applications for Smart Grid.

[2] Connected devices for Smart Grid includes the following, but not limited to: energy systems, distributed energy resources, electric vehicles, phones, appliances, thermostats, and smart meters.

receive the information with cyber security and appropriate user privacy incorporated by design. The sensing is a key feature of an IoT and includes information such as measurement (e.g., energy, power) and detection (e.g., occupancy, temperature). The control feature is important for Smart Grid application and can include local intelligence to execute energy-related decision management objectives. The IoT must support communications and sensing and can include control feature to provide any functional value to its users. For example, customers managing EV charging or reading the thermostat temperature set points using smartphones.

Innovation is driven by ideas and enhanced by the intrinsic process of diffusion. Without the diffusion the innovation would have minimal social and economical impact [9]. Innovation is also driven by the field results. The two-way arrow between the IoT and the enabling applications in Figure 30.2 represents this state, which drives innovations and identifies new IoT technology inventions. For example, to address variability from high penetration of renewables is driving intelligent consumer devices and controls that can automatically receive and respond to generation conditions based on consumer choice. These applications enable Smart Grid users, which can be business-to-consumers, business-to-businesses, or M2M interactions, to leverage the IoT benefits. To understand the value of IoTs to the Smart Grid, it is important to dissect this problem further in the context of power systems interoperation using Smart Grid standards and enable high penetration of renewable generation, DER, and balancing of supply and demand.

30.3 SMART GRID POWER SYSTEMS AND STANDARDS

Similar to IoT, many definitions for Smart Grid exist. In the context of this chapter, Smart Grid refers to a decarbonized grid with high amount of renewable generation, distributed generation or DG, demand flexibility, two-way power, and communication flows among different types of DER at different levels. Without reinventing the wheel, the Smart Grid power and communication flows are defined by the conceptual model to be between and among transmission and distribution (T&D), generation, and customer domains, as shown in Figure 30.3 [10]. The Smart Grid conceptual model and key standards are described here to provide a reference for the role of IoTs within these boundaries.

Figure 30.3 shows an adaptation of different Smart Grid standardization efforts in the United States for interoperability and harmonization across domains. Considering that similar Smart Grid practices are being followed in other parts of the world, including Europe and Asia, these principles are also applicable outside of the United States. These standards are described further.

30.3.1 Smart Grid Interoperability Standards and Cyber Security Guidelines

Standards are in references to power, data, and protocol interoperability among or with the IoT systems. Interoperability references the capability of an IoT to interface and work with other IoTs or disparate products or systems without any implementation-specific data translations. This chapter focuses on Smart Grid interoperability standards for data communications that support the IP-based protocols for Internet network connectivity.

Table 30.1 describes key Smart Grid standards supported by different standards organizations, in reference to the Smart Grid conceptual model. These standards

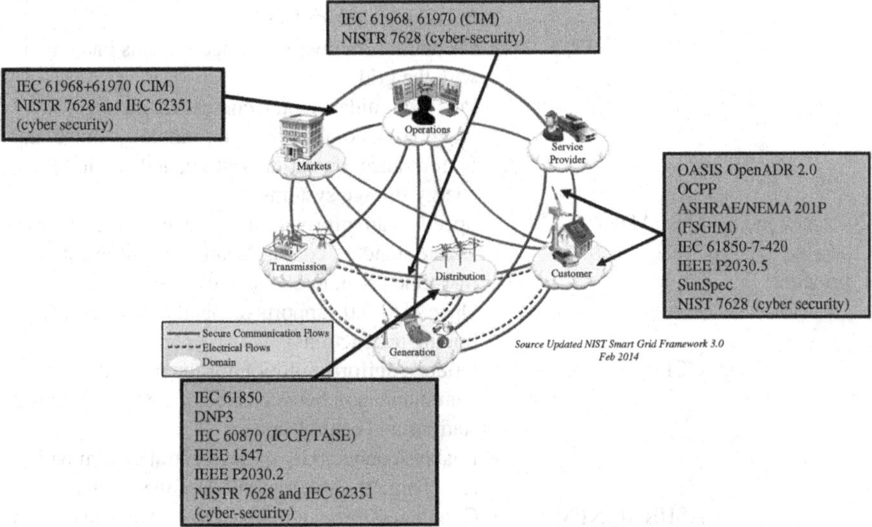

FIGURE 30.3 Conceptual Smart Grid architecture and interoperability standards interfaces. Figure adapted from New York State Department of Public Service Technology Working Group.

TABLE 30.1 Smart Grid Standards for Distribution and Customer Domains and Cyber Security

Domain	Standard	Description
Markets and operations	IEC CIM (61968, 61970, 62325)	• Deployed and increasingly used within the utility/system operator's enterprise systems • Originally, these provided a common definition for power system components for use in the Energy Management System (EMS) Application Programming Interface (API)
Distribution systems	IEC 61850 suite	• Standard for the design of electrical substation automation • IEC 61850-7-420 focuses on distributed energy resources (DER) communication systems
	DNP3	• Used by Supervisory Control and Data Acquisition System master stations, Remote Terminal Units (RTUs), and Intelligent Electronic Devices (IEDs) • Primarily used for communications between a master station and RTUs or IEDs
	IEC 60870 (ICCP/TASE)	• Developed to allow two or more utilities to exchange real-time data, schedule, and control commands

(*Continued*)

TABLE 30.1 (Continued)

Domain	Standard	Description
	IEEE 1547	• Standard for interconnecting DER with electric power systems • Provides a set of criteria and requirements for the interconnection of DG
	IEEE P2030.2 (in dev.)	• Guidelines for energy storage systems integrated with the grid • Provides guidance on technical characteristics, integration, compatibility, terminology, functional performance, operation, testing, and application of energy storage systems
Customer and service provider	OpenADR	• Specifies an information model and messages to enable standard communication of DR events, real-time price, market contexts • OpenADR 2.0 supports demand response and pricing transactions
	OCPP	• Offers a uniform solution for the method of communication between electric vehicle charging stations and central system • Enables connectivity of any central system with any charge station, regardless of the vendor
	ASHRAE/NEMA 201P (FSGIM)	• Common information model to enable management of residential commercial and industrial DER in response to communication with a smart electrical grid and electrical service providers
	IEEE 2030.5 (SEP 2.0)	• Specifies a standards-based application profile for use in Smart Grid home area networks (HANs), following a RESTful architecture on an Internet Protocol (IP) stack
	SunSpec	• Distributed energy industry information specifications to enable “plug & play” system interoperability for inverter-based DER
Crosscutting cyber security guidelines	NISTIR 7628	• Describes guidelines for cyber security for the Smart Grid • Considers North American Electricity Reliability Corporation (NERC) critical infrastructure protection standards
	IEC 62351	• Defines security requirements for power system management and information exchange, including communications network and system security issues

ASHRAE (American Society of Heating, Refrigerating, and Air-Conditioning Engineers), CIM (Common Information Model), DNP (Distributed Network Protocol), FSGIM (Facility Smart Grid Information Model), ICCP (Inter-Control Center Communication Protocol), IEC (International Electrotechnical Commission), IEEE (Institute of Electrical and Electronics Engineers), NEMA (National Electrical Manufacturers Association), NISTR (National Institute of Standards and Technology Report), OCPP (Open Charge Point Protocol), OpenADR (Open Automated Demand Response), SEP (Smart Energy Profile), TASE (Telecontrol Application Service Element).

define functions for the key domains of electricity markets, operations, distribution systems, service providers, and customers, including their interfaces. The generation and transmission domains are not addressed because the majority of interoperability and harmonization requirements for IoT-based innovations and new business models shall have the maximum potential in these domains.

IoTs can enable the dynamic and real-time management of this complex next-generation grid through communications, sensors, and controllers or actuators. Such dynamic grid infrastructure can better balance supply-side variability, DER system integration, and electricity demand resulting from high penetration of renewables and decentralized or DG.

30.4 LEVERAGING IoTs AND SMART GRID STANDARDS

Smart Grid challenges from the widespread adoption of DER are (1) high costs associated with enabling grid connectivity for energy systems, (2) high integration complexity and cost, (3) proprietary technologies and custom-engineered design, (4) more and diverse data sources to measure and verify value streams, (5) difficult to access data where and when it is needed, and (6) rigid and single-function enablement. Most of utility implementation is silo oriented and single function that raises costs and limits the flexibility of the solution for new problems, creates vendor lock-in, and suboptimizes the impacts, including future requirements such as DER integration at different Smart Grid domains and enabling new concepts such as economic-based transactive energy (TE). Taking this to a higher level, the US Department of Energy (DOE) in its Grid Modernization Plan highlights the following three relevant Smart Grid challenges that must be addressed for its transformation [11]:

1. Changing mix of types and characteristics of electric generation
2. Supply- and demand-side opportunities for customers to participate in electricity markets
3. Emergence of interconnected electricity information and control systems

These challenges relate directly to the definition of the Smart Grid and provide opportunities and need for interoperability standards and facilitation of role of and value from the IoT technologies. In fact, the DOE Plan also mentions, "interoperability standards define technical requirements for defining the capability of two or more networks, systems, devices, applications, or components to externally exchange and readily use information securely and effectively."

To better understand the roles of IoTs for Smart Grid transformation with high penetration of renewable generation and DER integration and interoperability standards, Figure 30.4 represents a framework of interconnected grid with features and functions that establishes the value of the IoTs. In this specific case, an EV, as a "roaming" DER, can be likened to stationary storage sited in different Smart Grid domains. The IoTs can manage intermittency from renewable generation (as shown by

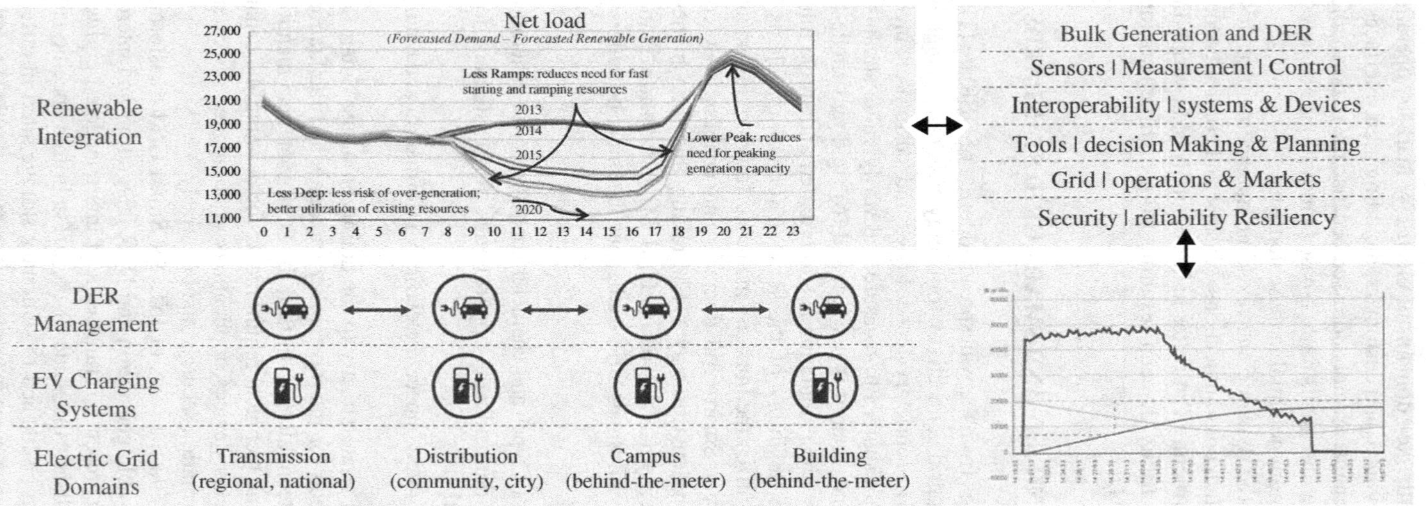

FIGURE 30.4 Example framework for interconnected grid, DER integration, and the IoT role.

the California Independent System Operator's net load model with 33% renewable energy mix by 2020 [12]) and integrate them for cost-effective demand flexibility using sensing, measurement, and control features. Interoperable devices and systems provide grid operators and service providers with cost-effective tools to integrate power systems and enable decision making, planning, and monitoring that drive efficient grid operations and electricity market design that transform Smart Grid to be secure, reliable, and resilient.

In terms of electricity systems and how the aforementioned framework can be used by the standards-enabled IoTs to support energy systems integration and enablement of user participation in the electricity markets, we should dive further to understand the different layers and functions.

30.4.1 Power Systems Integration for Smart Grid

The integration and interoperability goals that the standards can support using IoTs can be best understood using the national efforts by the DOE's Office of Energy Efficiency and Renewable Energy (EERE) grid integration initiative [13]. The energy systems integration (ESI) activity that resulted from this initiative describes different layers that must interoperate with each other as an integrated system to serve different Smart Grid objectives. This presents a key opportunity for optimal utilization of newly developed technologies [14].

Figure 30.5 from the EERE's grid integration initiative shows five essential layers that each Smart Grid domain (building/campus, distribution, and regional/transmission) must interface with:

Device layer: Consists of the physical energy devices and networks that produce, consume, store, or transport energy.

Local control layer: Consists of the electromechanical-, electronic- or software-based modules necessary to control a single device (in the Device Layer) in a stand-alone manner. This layer includes necessary device sensors, power electronics controller stages, and softwares.

Cyber layer: This, including communications, information, and computation platforms, is necessary to support control applications at the system level.

System control layer: This, including system monitoring, system state estimation, energy network security assessment, and so on, is responsible for the system-level concerns of security and reliability of a collection of connected devices.

Market layer: Addresses economic, regulatory, financial, and policy aspects of the system.

The ESI requirements can be a fundamental building block for IoT and interoperability standards, as interfaces at every power systems layer and within the Smart Grid domains to drive value of IoT technologies and facilitate the evolution of new business models.

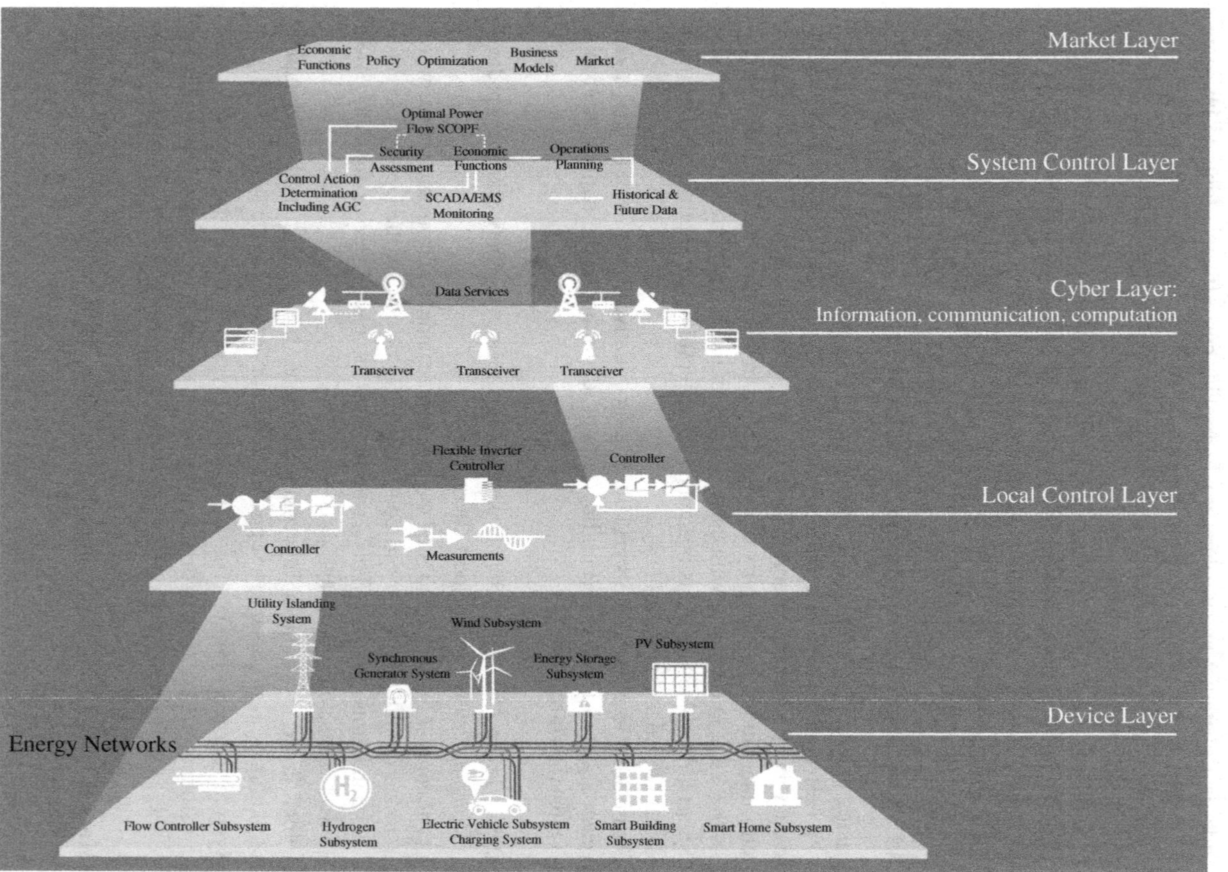

FIGURE 30.5 Five layers of integration for Smart Grid domain interfaces. DOE and NREL.

30.4.2 Smart Grid Interoperability Standards Interfaces and IoT Applications

The IoT technologies that are supported by interoperability standards are key enablers for open interoperable platforms and encourage innovations. Thus standards become a key part to not only address the key Smart Grid challenges but also provide a use case for widespread adoption of IoT technologies and its resulting services.

From previous sections we know that EVs, as a DER, require standards-based integration with all three power flow domains—systems operators, electricity service providers, and customers. The benefits of interoperability standards for customer-side transactions such as AutoDR are well studied and not included here [15]. Figure 30.6 describes these interfaces and open standards that are adopted by the industry for the distribution, markets, operations, and transmission domain interoperability [16]. Open standards are in reference to publicly available standards developed with a consensus process and have no intellectual property or have consumer-friendly terms. Such standards lower the barrier for adoption and improve cost-effectiveness of IoT technologies for Smart Grid applications. Open interoperability standards that also enable open system architecture that "extends across multiple smart grid resources that would allow optimized and reliable control of the resource both autonomously and as part of a system" are lacking [17]. Examples of enabling open Smart Grid standards supporting EV and DER adoption are Open Charge Point Protocol (OCPP), Open Automated Demand Response (OpenADR), and Smart Energy Profile (SEP). This list does not include power flow and inverter interconnection standards. While these standards support distribution network electricity providers' need for information exchange with prosumers, the International Electrotechnical Commission (IEC) standards, common information model (CIM), and IEC 61850 play a dominant role for power systems' interfaces within and with transmission network system operators. Secure Internet-based protocols and automated metering infrastructure (AMI) are widely used transport mechanisms between customer and distribution network. The higher cyber security needs for transmission network communications with bulk power plants and perceived lack of IP-based protocols supporting it, distributed network protocol 3 (DNP3) and inter-control center communication protocol (ICCP), are widely used.

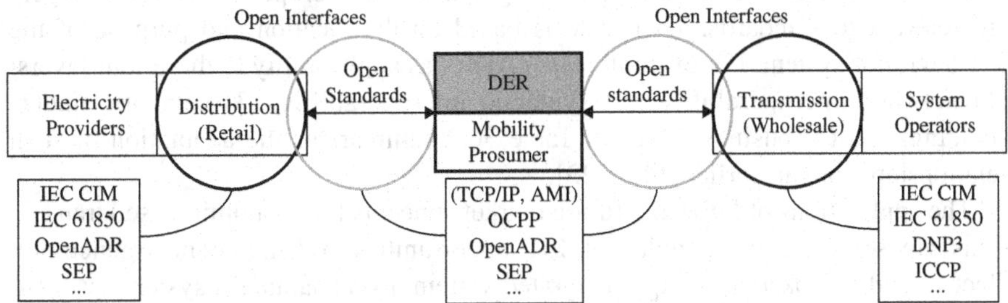

FIGURE 30.6 Open standards and interfaces for DER at transmission and distribution domains.

The IoT-enabled integrated and connected devices and power systems that can seamlessly interoperate with the electric grid system create opportunities for different clean energy generation types and characteristics to easily integrate with the electric grid systems and enable new electricity market opportunities and applications for users.

30.4.3 IoT Applications for DER Value and Transformative Electricity Markets

Many benefits can be envisioned from an integrated Smart Grid that is secure and interoperable with abundant clean energy generation. Without going into details on the existing services such as demand flexibility, price-responsive loads, or different systems and tools that enhance the Smart Grid operation, two specific IoT applications within the evolving DER systems (EVs and battery storage) and enablement of TE markets are discussed.

30.4.3.1 Smart Grid Value of Electric Vehicle Battery Storage Numerous studies by the US federal and state agencies, in partnership with research institutes, electric mobility industry, and electric grid participants, have shown the impacts of electric mobility for grid services [18–20]. The results of five relevant case studies of EV battery storage value for electric grid services using various forms of IoT are summarized in Table 30.2 [16].

30.4.3.2 Transactive Energy and Networks Emerging concepts such as TE markets and supporting IoT-based networks represent a unique opportunity for IoT technologies to provide value to the users and address the reliability and large-scale integration of renewable generation. While the seven-layer OSI model describes the scope of interoperability standards and its relationship to different network layers, the GridWise™ Architecture Council (GWAC)'s interoperability context-setting framework (or "GWAC stack") is an important link to the emerging TE concepts [21]. The GWAC, supported by the US DOE, coordinates an effort to define a common framework for TE and defines it as "A system of economic and control mechanisms that allows the dynamic balance of supply and demand across the entire electrical infrastructure using value as a key operational parameter" [22].

Although the GWAC stack describes eight layers of interoperability, not all layers are relevant to standards. Their role is based on the location and purpose of the IoT device or system. For example, OpenADR has applicability to these four layers: (1) network interoperability, (2) syntactic interoperability, (3) semantic understanding, and (4) business context. Table 30.3 summarizes the adaptation of such relationship from an earlier study [23].

The applications of DER and future market concepts for economic-based transactions are some basic examples of IoT opportunities. When the next-generation electric grid is looked as a large-integrated system, its disparate subsystems or components, business and ownership models, and regulatory framework represent even larger opportunity for the IoT applications.

TABLE 30.2 Electric Vehicle and Battery Storage DER Value for Grid Services

Southern California Edison	One of the largest workplace DR charging study project leverages the networked EV charging station management services to extend it for DR. Using OpenADR and OCPP, the project successfully demonstrated the value of standards for utility power system interoperability for demand flexibility with successful network management and smart charging of EV charging stations
Sacramento Municipal Utility District	The project by Sacramento Municipal Utility District (SMUD) focused on residential charging station infrastructure and evaluation of technical performance and grid impacts using time-of-use and dynamic rate. The project results showed high customer satisfaction and issues with driver behavior and meter-to-charging station interoperability
University of Delaware	One of the first projects for two-way EV battery storage power flow services has led to commercialization of the technology with industry partners. The project-enabled aggregated market participation of the EVs to show driver and grid operator benefits for grid stability resources
Ft. Carson	Study focused on two-way EV battery storage and microgrid simulation has identified significant potential for EVs. The project captured 45% of the total connected power capability of two EVs and demonstrated the potential of and improvements in the system and EVs to follow electricity market regulation signals
Open Vehicle—Grid Integration Platform	The project is a collaboration among automotive manufactures, utilities, research institutes, and technology providers to develop standards-based communication platform to support BEVs for grid integration. The project demonstrated cloud-based platform for interoperable interfaces using OpenADR, SEP, and AMI. The next phase of the project is considering onboard vehicle IoT telematics option for direct business-to-BEV driver interactions.

TABLE 30.3 GWAC Data Models, Transport Mechanisms, and Security Applications for OpenADR

Standards Organization	Specification
Business context	Standards must define the business context under which it operates within the Smart Grid (e.g., standards for price and DR communications)
Semantic understanding	The publicly available schema and specification describe the semantic meaning of data contained in XML data structures
Syntactic interoperability	The schema defined in the XML standard defines the data models and message payloads
Network interoperability	Supported by transport mechanisms where key IP-based transports and interactions between the participants are defined

30.5 CONCLUSIONS AND RECOMMENDATIONS

The IoT-based technologies with a minimum feature of communications and sensing and paired with control features can provide support to many objectives of the current and next-generation Smart Grid and represent a significant growth opportunity. The consideration for interoperability, device- and system-level integration, and seamless interfaces across different Smart Grid domains can be supported by different standards that are specific to grid power system and user needs. In particular, the application domain standards for IoT data communications have the potential to address the diversity in DER, DG, distribution systems, and supporting markets and operations for electricity consumers. The Smart Grid standards in the markets and operations, distribution systems, and customer and service provider domains are primarily served by SDOs such as the IEC, OASIS, IEEE, ASHRAE, and NEMA.

The technical and market design of electricity power systems and its integration at different Smart Grid domains enhances the value of IoT technologies and necessitates the interoperable standards interfaces at the edges of the domains (interdomain) and at different layers of the power systems within the domains (intradomain). For Smart Grid to gain value from the IoT for intra- and interdomain Smart Grid applications, the technologies need to be developed for unique contextual features and functions of the Smart Grid, prevailing standards, and benefit(s) it can offer to its users. Such value of innovative technologies and services can jump-start the grid modernization investments and benefits the businesses from new business models such standards-supported IoT technologies can offer.

The new DER technologies such as EVs and battery storage represent unique opportunities that can use the IoTs for grid services. With increasing variable generation mix in the electric grid and its decentralized operation, IoT technologies and strategies that support demand flexibility will provide unique value to the grid (e.g., avoided generation, grid balancing). Many pilots such as two-way power flows and market design concepts such as TE are evolving to address the next-generation grid. The evolutions of IoT technologies that are cost-effective and ubiquitous and seamlessly integrate with users have the potential to provide temporal and spatial visibility and control that are yet to be envisioned.

Cyber security and privacy are often cited as major issues for widespread adoption of the IoTs. Compliance to Smart Grid cyber security requirements for IoT is critically important for success and must be considered as a design principle. User data privacy must be considered, wherever applicable. With any exponential increase in the use of IoT devices for Smart Grid, customer education on its benefits, energy management of IoTs when not connected to the electric grid, and other features such as unique IPv6 addressability for IoTs shall require a closer look.

REFERENCES

[1] Ghatikar G., E.H.Y. Sung, and M.A. Piette, Diffusion of Automated Grid Transactions Through Energy Efficiency Codes, ECEEE Summer Study on Energy Efficiency, June 2015, France. LBNL-6995E.

[2] Ghatikar G. and R. Bienert, Smart Grid Standards and Systems Interoperability: A Precedent with OpenADR, Grid-Interop Conference, December 2011. LBNL-5273E. doi: 10.13140/2.1.4163.4081.

[3] Ghatikar G. and E. Koch, Deploying Systems Interoperability and Customer Choice within Smart Grid, Grid-Interop Conference, December 2012. LBNL-6016E. doi: 10.13140/2.1.5162.8320.

[4] Forbes, A Very Short History of The Internet of Things, http://onforb.es/STpSh7 (accessed on February 16, 2016).

[5] Gartner, Startups and Small Vendors Are Driving Innovation in The Internet of Things. http://www.forbes.com/sites/gartnergroup/2015/11/24/startups-and-small-vendors-are-driving-innovation-in-the-internet-of-things/ (accessed on August 2016).

[6] International Data Corporation (IDC): Internet of Things Spending Forecast to Reach Nearly $1.3 Trillion in 2019 Led by Widespread Initiatives and Outlays Across Asia/ Pacific. http://www.businesswire.com/news/home/20151210005089/en/Internet-Spending-Forecast-Reach-1.3-Trillion-2019 (accessed on August 2016).

[7] Institute of Electrical and Electronics Engineers (IEEE), Towards a Definition of the Internet of Things (IoT), Revision 1, May 2015.

[8] Internet Engineering Task Force (IETF), The Internet of Things: Concept and Problem Statement, October 2010. https://tools.ietf.org/id/draft-lee-iot-problem-statement-02.txt (accessed on September 10, 2016).

[9] Hall B.H., Innovation and Diffusion, National Bureau of Economic Research, Working Paper Series, January 2004.

[10] National Institute of Standards and Technology (NIST), NIST Framework and Roadmap for Smart Grid Interoperability Standards, Release 3.0, September 2014.

[11] U.S. Department of Energy (DOE), Grid Modernization Multi-Year Program Plan, November 2015. http://www.energy.gov/sites/prod/files/2016/01/f28/Grid%20 Modernization%20Multi-Year%20Program%20Plan.pdf (accessed on September 10, 2016).

[12] California Independent Systems Operator (CAISO), Demand Response and Energy Efficiency: Maximizing Preferred Resources, December 2013. https://www.caiso.com/ documents/dr-eeroadmap.pdf (accessed on September 10, 2016).

[13] U.S. Department of Energy (DOE), Office of Energy Efficiency and Renewable Energy (EERE), Grid Integration Multi-Year Program Plan, February 2014. http://iiesi.org/ assets/pdfs/iiesi_lynn.pdf (accessed on September 10, 2016).

[14] Ruth M.F. and B. Kroposki, Energy Systems Integration: An Evolving Energy Paradigm, *The Electricity Journal*, Volume 27, Issue 6, 36–47, 2014.

[15] Ghatikar G., J. Zuber, E. Koch, and R. Bienert, Smart Grid and Customer Transactions: The Unrealized Benefits of Conformance, IEEE Green Energy and Systems Conference (IGESC), Long Beach, CA, November 2014.

[16] Ghatikar G., Decoding Power Systems' Integration for Clean Transportation and Decarbonized Electric Grid, Proceedings of the India Smart Grid Week, March 2016, New Delhi, India. 10.13140/RG.2.1.3555.4960.

[17] U.S. Department of Energy (DOE), Office of Electricity Delivery and Energy Reliability (OE), Smart Grid Research and Development: Multi-Year Program Plan 2010-14, September 2012. http://energy.gov/sites/prod/files/SG_MYPP_2012%20Update.pdf (accessed on September 10, 2016).

[18] U.S. Department of Energy (DOE), Evaluating Electric Vehicle Charging Impacts and Customer Charging Behaviors—Experiences from Six Smart Grid Investment Grant Projects, Smart Grid Investment Grant Program, December 2014. https://www.smartgrid.gov/files/B3_revised_master-12-17-2014_report.pdf (accessed on September 10, 2016).

[19] Charged: Electric Vehicle Magazine, Workplace Charging with Smart Grid and Demand Response, Issue 20, July/August 2015.

[20] National Renewable Energy Laboratory (NREL), Multi-lab EV Smart Grid Integration Requirements Study, Technical Report NREL/TP-5400-63963, May 2015.

[21] The Gridwise® Architecture Council (GWAC), GridWise Interoperability Context-Setting Framework, March 2008.

[22] The Gridwise® Architecture Council (GWAC), GridWise Transactive Energy Framework: Version 1.0, January 2015.

[23] QualityLogic, Inc., Transactive Control and OpenADR Roles and Relationships, Revision 1.07, May 2014. https://www.qualitylogic.com/tuneup/uploads/docfiles/TC_OpenADR_Roles_and_Relationshipsv107.pdf (accessed on September 10, 2016).

31

IoT REVOLUTION IN OIL AND GAS INDUSTRY

Satyam Priyadarshy
HALLIBURTON Landmark, Houston, TX, USA

31.1 INTRODUCTION

Internet of Things (IoTs) is a popular term in the context of the emerging technology revolution that has created a highly connected world. IoT can be viewed as a concept that promises to connect everything, everyone, and everywhere forever. One of the most promising applications of this ubiquitous connectivity will be Augmented reality—the new way of looking at the connected world. Any application that leverages ubiquitous connectivity will be a use case of IoT irrespective of the industry.

As the Internet was gaining popularity and usage in connecting people, another revolution in terms of radio frequency identification (RFID) was also gaining momentum for tracking and monitoring of business assets and inventory. In that context, the first use of the term Internet of Things was proposed by Kevin Ashton in reference of optimizing supply chain [1, 2], but "The Internet of Things" as a concept was first coined and used by MIT Auto-ID Center and linked to RFID and electronic product code (EPC). In brief, IoT refers to the *complex network of software, physical, and virtual entities* which are embedded or implemented in *devices* that have software elements to performing activities that are predetermined or by leveraging computing paradigm. Here *devices* include sensors, smartphones, tablets, computers, electronic products, RFID tags, wearables, and many others. These *devices or entities* are referred as things in the IoTs. In the IoT paradigm, these entities are believed and expected to be capable of communicating and working together with similar entities

Internet of Things and Data Analytics Handbook, First Edition. Edited by Hwaiyu Geng.

Companion website: www.wiley.com/go/Geng/iot_data_analytics_handbook/

through communication channels like Internet or, otherwise, to perform the functions, more efficiently and effectively, which are assigned to these entities.

One of key enablers for IoT revolution is the evolution of the Microelectromechanical (MEMS) technologies that helped miniaturization of hardware components like sensors, actuators, microcontrollers, wireless transceivers, and so on. As a result a large amount of new innovation in software for embedded devices has taken place at various levels including protocols, operating systems, architectures, and so on. These enablers allow one to optimize Machine-to-Machine (M2M) interactions (Machine refers to *device*) and Human-to-Machine (H2M) interactions, which are essential for autonomous capabilities in the connected world.

Even though IoT concept's adoption is moving with full speed, some open-ended challenges are yet to be address by the industry. One of the key challenges relates to the choice of communication protocol. Many proprietary and domain-specific protocols are being used by the industry. As a result of these diverse protocols, the need for sophisticated application gateways to connect devices using different proprietary protocols keeps growing, thus increasing the complexity of architecture and application development (Figure 31.1). There is however push to adopt Open standard communications based on Internet and Web technologies for generating long-term value from IoT [3, 4].

Irrespective of these open challenges, in business terms, IoT is going to play a significant role in many industries. A 2015 IDC report suggests that IoT market could grow to $7.1 trillion by 2020. The markets would be in many industry verticals including healthcare, retail, e-commerce, intelligent homes and business buildings, transportation, agriculture, manufacturing, energy, military application, and so on. In

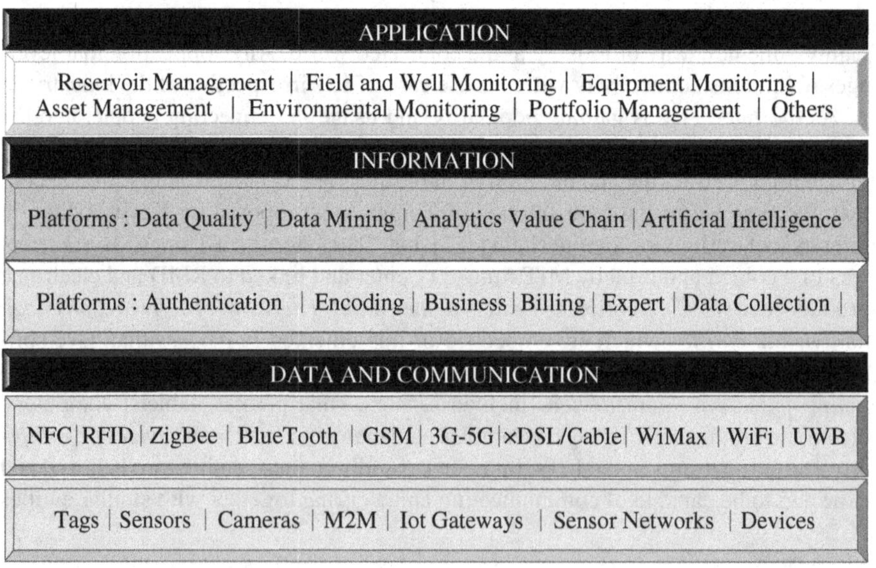

FIGURE 31.1 Schematic IoT technical architecture. © Satyam Priyadarshy. Use with permission.

fact a report by US Intelligence Council suggested that even quotidian things like furniture, food packages, basic home apparels, and so on could become Internet nodes or *entities* by 2025. They refer IoT as one of the most disruptive civil technologies [5, 6]. A Pew Research report predicts that "The Internet of Things will thrive by 2025" and suggests that "the opportunities and challenges from amplified connectivity will influence nearly everything, nearly everyone, nearly everywhere [7]."

The oil and gas (O&G) industry cannot remain immune to the impact of IoT. O&G industry has one of the most complex operations of any industry. It has components of all the industries at much higher scale but with two major differentiations that do not exist in other industries—one being risk and other being regulations. O&G industry is one of the highly regulated industries ranging from scrutiny from local authorities to global regulatory bodies. A simple error in operations could be fatal in terms of human life; hence the industry focuses a lot of health, safety, and environment (HSE) metrics, and the higher safety score of HSE derives the business strategy and then the revenue and profits.

31.2 WHAT IS IoT REVOLUTION IN OIL AND GAS INDUSTRY?

IoT revolution is just beginning. In 2010, it was estimated that 50 million out of 50 billion machines were connected; that's just one 1% with a potential of 99% that is open to IoT market [8]. To address this market gap, major Information and Communications technology (ICT) companies are addressing IoT concept using terms like *Industrial Internet* (GE), *Smarter Planet* (IBM), Internet *of Everything* (CISCO), Smart + *Connected Manufacturing* (CISCO) [9–11]. In 2015, these ICT companies and other 80 companies came together to form the Industrial Internet Consortium (IIC) to speed the adoption of latest IoT technologies in the industrial applications.

As described earlier IoT is about ever connectedness of devices and people, so what are the requirements for IoT revolutions? Here are some of major requirements:

- Connectivity is paramount.
- Integration and collaboration of heterogeneous devices that are connected by disparate technologies in wired networks and wireless networks and running a gamut of application-specific software.
- Scalability beyond millions of devices, where centralized solutions become bottlenecks.
- Agility in terms of adapting to continuous transitions in device configuration, new service deployment, and environment is critical.
- Security issues.
- Computational ability.
- Fault tolerance.
- Power consumption.
- Smart sensing technology.

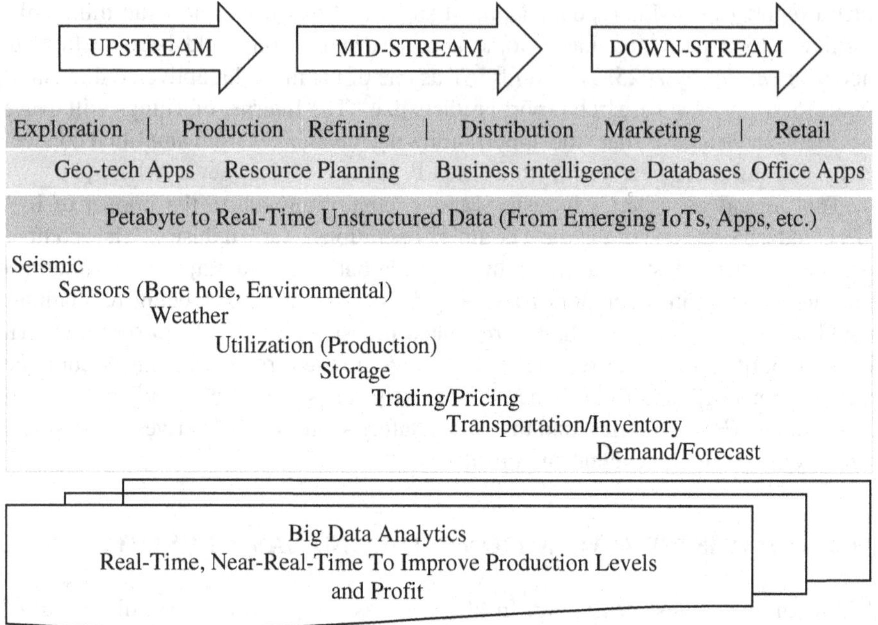

FIGURE 31.2 An oil and gas life cycle at a high level. Throughout the life cycle IoT and Big Data analytics can be used. © Satyam Priyadarshy. Use with permission.

Oil Industry faces extreme challenges whether be conventional resources or unconventional resources. In Figure 31.2, it shows the O&G life cycle at a high level. Irrespective of the phase of oil life cycle, there are many challenges that the industry faces [12]. In the following sections, the use of IoTs will be discussed in three use cases for the O&G industry. In this chapter, the focus will be upstream O&G industry, also referred as exploration and production (E&P) industry.

31.3 CASE STUDY

For the case of O&G industry, we focus on three key technologies:

1. Sensing technologies that obtain asset information simultaneously and accurately through the use of sensors, RFID, product code like QR codes, and other equipment ID methods.
2. Data and Information transfer technology for delivering near real-time, real-time, and bulk data asynchronously for appropriate actions at the right time, through a deep integration and modular platform that connects variety of communication networks and the Internet.
3. Big Data processing technology for analyzing and creating actionable insights from large volumes, and wide variety of data that is being delivered at different velocities, with varying degree of value associated with it. Depending upon the location of data source, the connectivity between data sets could be over virtual

data sets instead of data being in the same data lake or silo. Also, the Big Data processing technology will vary on the context of oil life cycle. In summary, the Big Data technology platform needs to adhere to the Priyadarshy's seven pillars of Big Data [13, 14].

These key technologies will play a critical role in following cases:

Case 31.A The Automated and Self-correcting Remote Operations for Drilling and Completion of Wells

The increasing complexities of drilling wells are very evident for industry professionals, as industry moves to tap reservoirs in unconventional resources. At the same time, the industry is facing a shortage of skilled field experts and workers due to a massive retirement of skilled workforce and lack of trained next-generation workforce in this domain. IoT will alleviate some of these concerns by enabling automate remote operations. Effective IoT implementation will allow to field metrics in real time, be able to connect to ongoing captured tacit and explicit knowledge base (hosted on back office), allow for deploying machine learning and artificial intelligence-based models to act on events, and so on (Figure 31.3).

Thus IoT will increase safety and reduce the time a person spends on these operation sites, thus reducing the cost of operations. Leveraging the Virtual pillar of Big Data through the extended connectivity paradigm, the predictive maintenance and the feedback on-site professional will be available in much shorter time than currently possible.

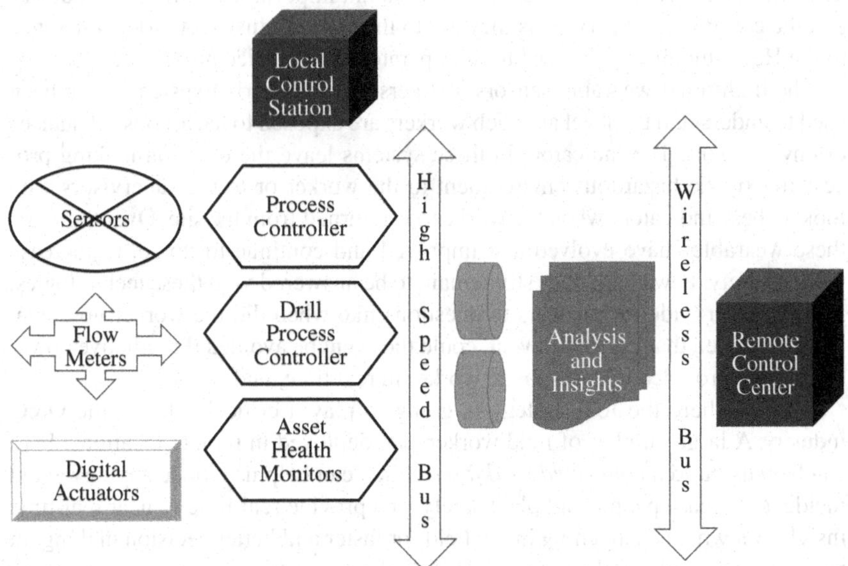

FIGURE 31.3 Several critical areas for IoT Deployment in Oil and Gas Industry. © Satyam Priyadarshy. Use with permission.

Case 31.B SCADA and IoT

Supervisory control and data acquisition (SCADA) is the name for remote monitoring and controlling machines using coded signals over communication channels. It is a type of industrial control system (ICS). ICS by definition are computer-based systems that monitor and control industrial processes that exist in industrial sites. The traditional SCADA systems leverage extensively the Human–Machine interface (HCI) so that human operators can take appropriate actions needed to run smooth operations.

SCADA systems have evolved from being largely monolithic to being distributed to being networked. In the networked SCADA design, the system may be spread across geographically over one or more LAN networks. With the growth of IoT revolution, the SCADA systems are moving to the next generation, where it can take advantage of cloud technology and become increasingly real time. IoT enables the implementation of complex control algorithms at the edge now, which could not be done with traditional programmable logic controls easily. The use of open network protocols inherent in the IoT technology provides better and reliable mechanism to manage decentralized SCADA implementations. As more complexity of IoTs becomes manageable, so will SCADA systems become more efficient and reliable.

Case 31.C The Health, Safety and Environment (HSE) Guidelines

The three issues related to HSE are of paramount importance to the E&P industry and the community it serves, as they are to the O&G industry. A strict adherence to the HSE guidelines is a must for all operators and service providers.

The traditional wearable sensors and personal monitoring systems have been used to understand the level at which workers are exposed to hazardous substances or environment. The indicators in these systems leave the decision-making process to exit the hazardous environment to the worker or to the supervisors who look at these indicators when the worker has returned from the site. Over the years these wearables have evolved and improved and continue to do so, to increase worker safety. Few challenges still remain to be answered with these technologies. It does not provide the answers to questions like when did the worker go down, what happened that led to the event, could the event be avoided if predictive power of analysis provided insights to the worker in real time, and so on.

This is where the IoT strategy is going to play a critical role for the O&G industry. A large number of field workers are deployed in remote locations; IoTs can help us build a *connected workforce* which can help us reduce any unwanted incident, increase people and plant safety, and provide real-time or near real-time insights on what is happening in the field for faster and better decision making, in the interest of keeping the score for HSE high.

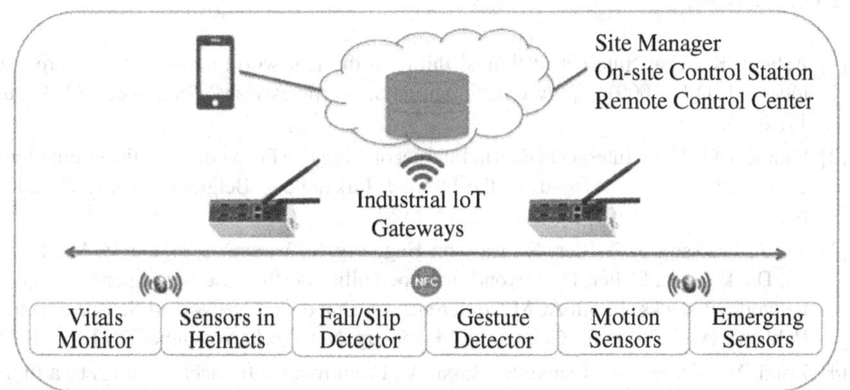

FIGURE 31.4 A simple framework for IoT-based Health, Safety and Environment at Oil and Gas.

IoTs will enable companies to collect data from a variety of sensors and, using the distributed architecture for remote and on-site data hubs, can process the data locally and remotely to make informed decisions in shortest possible time. The increased storage capacity and processing speeds of chips could enable one to build wearable *worker black box* such that a large degree of metrics could be stored during operations, providing a mechanism to continuously monitor activities and safety through a remote operation center and providing safety guidelines should it become necessary in unwanted situations. IoT devices that include a heart rate monitor, a toxic gas monitor, a self-contained breathing apparatus, a nonverbal gesture monitor, and motion active sensors will improve operational efficiency and productivity. The data and information collected from such IoT devices through the *connected workforce* network will become an important part of the E&P industry in the future (Figure 31.4).

31.4 CONCLUSION

IoT is certainly a next-generation technology that if leveraged could save billions of dollars in cost for O&G industry.

There remains significant number of challenges associated in realizing the full potential of IoTs in O&G industry. These challenges are also opportunities for cost-effective operations for O&G. Some of these challenges can be summarized as the availability of low-cost communication networks at the fields and between fields and back offices; the physical and cyber security issues associated with sensors, devices, and networks; the development of low-cost smart sensing systems and network; the battery technology; the deployment of computational algorithms at the edge; the scalability; the standardization of protocols, and the acceptability among the O&G industry. The emergence of new computational technology, the scalable distributed data and cloud frameworks, the Big Data revolution, and the mobility adoption provide hope for solving these challenges and wider adoption of IoTs in O&G industry.

REFERENCES

[1] Ashton, K., That 'internet of things' thing, in the real world things matter more than ideas. RFID J. (2009) http://www.rfidjournal.com/articles/view?4986. Accessed: August 17, 2016.

[2] Santucci, G., From Internet of data to Internet of things. In Proceedings of the International Conference on Future Trends of the Internet, Luxemburg, Belgium, January 28, 2009; pp. 29–47.

[3] Ko, J., Eriksson, J., Tsiftes, N., Dawson-Haggerty, S., Vasseur, J. P., Durvy, M., Terzis, A., Dunkels, A., Culler, D., Beyond interoperability pushing the performance of sensor network IP stacks. In 9th ACM Conference on Embedded Networked Sensor Systems, Publisher Association for Computing Machinery, New York, November 2011; pp. 1–11.

[4] Zorzi, M., Gluhak, A., Lange, S., Bassi, A., From today's Intranet of things to a future Internet of things: a wireless- and mobility-related view. IEEE Wirel. Commun. 17(6), 44–51 (2010).

[5] Council, N. I., Disruptive civil technologies: six technologies with potential impacts on us interests out to 2025. Conference Report CR 2008–2007, April (2008)

[6] Atzori, L., Iera, A, Morabito, G., The Internet of things: a survey. Int. J. Comput. Telecommun. Netw. 54, 15 (2010).

[7] Anderson, J., Rainie, L., The internet of things will thrive by 2025 (2014) http://www.pewinternet.org/2014/05/14/internet-of-things/. Accessed: August 17, 2016.

[8] ETSI, European Telecommunications Standards Institute, Annual Report (2010).

[9] GE Industrial Internet GE Intelligent Platforms, http://www.ge-ip.com/industrial-internet. Accessed: June 8, 2014.

[10] CISCO Internet of Everything http://internetofeverything.cisco.com/ Accessed: June 8, 2014.

[11] IBM – Smarter Planet – United States http://www.ibm.com/smarterplanet/us/en/overview/ideas/?re=spf. Accessed: June 8, 2014.

[12] Velda Addison Internet of Things could transform the oil patch, E&P Magazine (2015) http://www.epmag.com/internet-things-could-transform-oil-patch-815016#p=full. Accessed: August 17, 2016.

[13] Priyadarshy, S., Big Data – the 7V's that are must for describing it. http://chiefknowledge guru.com/2012/10. Accessed: August 17, 2016.

[14] Davis, B., The 7 pillars of Big Data. https://www.energyinst.org/information-centre/ei-publications/petroleum-review/petroleum-review-january-2015. Accessed: August 17, 2016.

32

MODERNIZING THE MINING INDUSTRY WITH THE INTERNET OF THINGS

Rafael Laskier
Vale, Rio de Janeiro, Brazil

32.1 INTRODUCTION

Disruptive technologies will have an enormous impact on the global economy over the next years and decades. Many believe that this new wave will have a stronger impact than the industrial revolution. It is going to affect the way people live and companies operate and entire business models.

The economic impact of the main disruptive technologies will be in the range of trillions of dollars. According to *McKinsey Global Institute*, just Internet of Things (IoT) alone will have an economic impact between three and six trillion dollars by 2025, much more than most of largest global GDPs in 2014 (Figure 32.1).

The IoT technologies are enablers for innovation, growth, and differentiation. Smart, connected products will change how value is created for customers, how companies compete, and the boundaries of competition itself.

IoT technologies will disrupt entire value chains and force companies to rethink nearly everything they do, from how they develop products to how they manufacture and operate.

Internet of Things and Data Analytics Handbook, First Edition. Edited by Hwaiyu Geng.

Companion website: www.wiley.com/go/Geng/iot_data_analytics_handbook/

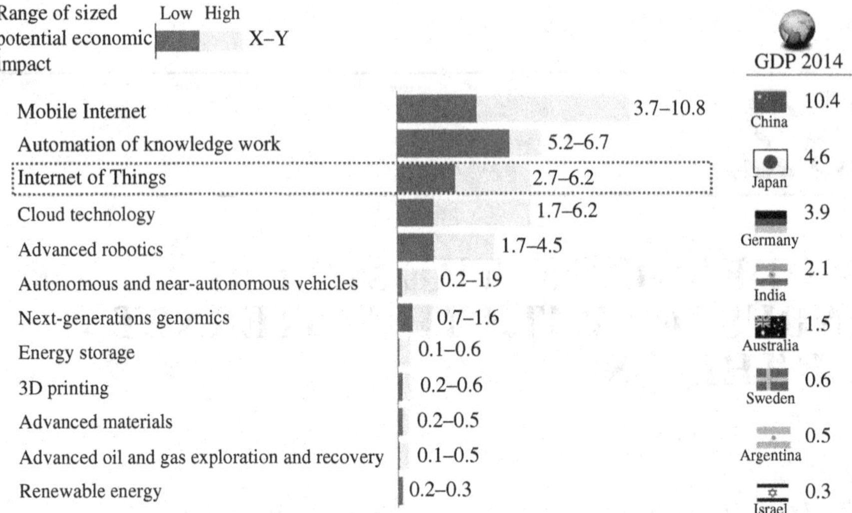

FIGURE 32.1 Economic impact of the 12 most significant disruptive technologies by 2025, in US$ trillions per year and countries' current GDP in 2014, in US$ trillions. Permission to use by McKinsey.

According to Professor Michael Porter, 2015 (HBR Edition), the capabilities of smart, connected equipment in large operations could be mainly grouped into four areas—Monitoring, Controlling, Optimizing, and Autonomy:

1. Monitoring: Equipment with sensors and connected with external data sources enable the monitoring of the conditions and environment at each stage of the value chain as well as monitoring equipment usage and respective impact to overall production.
2. Controlling: Software embedded in the equipment enables control of equipment functions and customization for specific aspects of each operation location when needed.
3. Optimizing: Monitoring and controlling capabilities enable algorithms that optimize equipment and the value chain operation, enhancing equipment and value chain performance and allowing predictive diagnostics and maintenance.
4. Autonomy: Combining monitoring, controlling, and optimizing enables autonomous equipment operation and self-coordination of equipment and systems.

Industrial IoT will dramatically impact all sectors, from Agriculture to Mining and Metals. While the world struggles to surpass a recent past of weak productivity growth and fragile employment, the industrial IoT offers an opportunity to redefine many sectors and accelerate economic growth.

However, many companies are not ready to benefit from industrial IoT. According to a survey made by *Accenture* (2015), 73% of a group of 1,400 C-Level decision makers have said that their companies have yet to make any concrete progress, and only 7% have developed a comprehensive strategy to take advantage of IoT and other digital technologies.

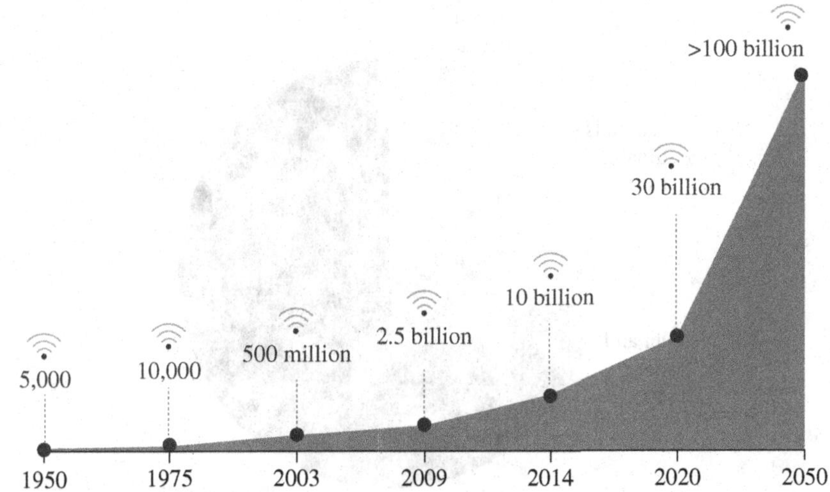

FIGURE 32.2 Historical and forecasted number of connected devices in the world. Permission to use by IBM.

We are in the middle of the largest revolution ever that has already reached a point of no return. According to *IBM*, there will be around 30 billion connected devices globally by 2020 (see Figure 32.2).

Companies can just choose between being prepared or not and being disrupt or disrupted.

32.2 HOW IoT WILL IMPACT THE MINING INDUSTRY

Experts agree that IoT is set to change the world, with up to 30 billion connected "things" likely to be in use by 2020. But how could this digital innovation wave revolutionize the Mining Industry landscape?

Looking into the iron ore industry cost structure and using BHP Billiton as an example (see Figure 32.3), it is quickly noticed how intense in people the mining industry is. The labor and contractors and consultants categories dominate the cost base representing 65% of total costs. As such, IoT technologies which increase labor productivity would appear to be appropriate (i.e., automation, remote monitoring, and other applications).

Other key drivers for mining industry costs are fuel and energy which are mainly consumed by trucks, equipment, and buildings at the mine sites as well as by the ore product transportation (e.g., railway, port, and shipping). Those costs could represent from 10 to 15% of total operating costs. Raw materials and consumables costs are also relevant and often as high as fuel and energy costs. This points to applying technologies, IoT related, which could minimize consumption of these items and/or using the items in a more productive manner (i.e., control systems, energy monitoring systems, advanced engine control systems, and other examples).

Real-time smart remote monitoring of equipment, for example, would enable that parts are changed out as required rather than according to a schedule. This could yield benefits in terms of either lower expenditures on spare parts (as the useful life

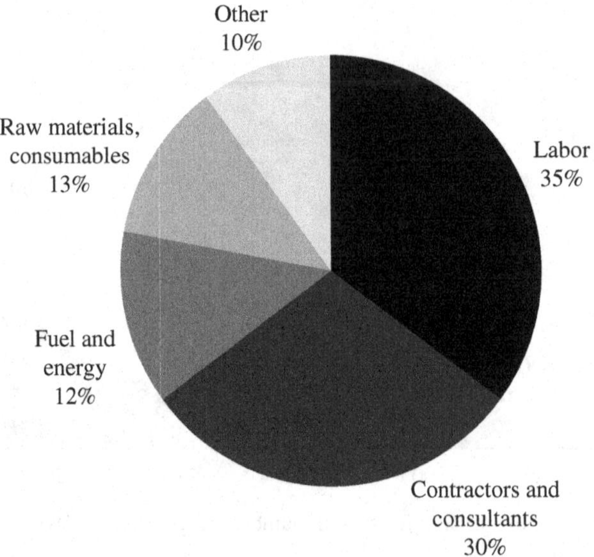

FIGURE 32.3 Cash cost breakdown in BHP Billiton's Western Australian iron ore assets, FY 2014.

of parts is extended) or improved equipment availability due to a reduction in unplanned outages (because a part fails before it is changed).

Automation systems could increase productivity, as each unit of labor become more productive ultimately allowing for lower spending on labor or increased production with the same labor input.

Automation and remote monitoring may also yield benefits in terms of improved worker safety. Moreover, remote monitoring, automation, and advanced engine control units could (has) lead to lower fuel and energy consumption by equipment like haul trucks and mills.

Safety is another area where Mining should be importantly improved by IoT technologies. Safety is, and always will be, the first priority in mining.

Despite the efforts and positive results in accident rates at some of the largest global miners, the fact is that workers are still getting injured at the mine sites across the globe. This is not acceptable for any serious mining company. Improvements in the safety area should come with automation, wearables, location tracking, and other technologies that also focus on productivity.

The Mining Industry should be one of the sectors that will be impacted the most by IoT because of two main factors: Scale and industry context.

32.2.1 Scale

From the enormous size of the value chain and equipment to the large amount of volumes produced every year, everything in the mining industry is huge. Off-road trucks, vessels, railways, port facilities, and several other mining equipment are already using thousands of sensors that generate a massive amount of data on a daily basis.

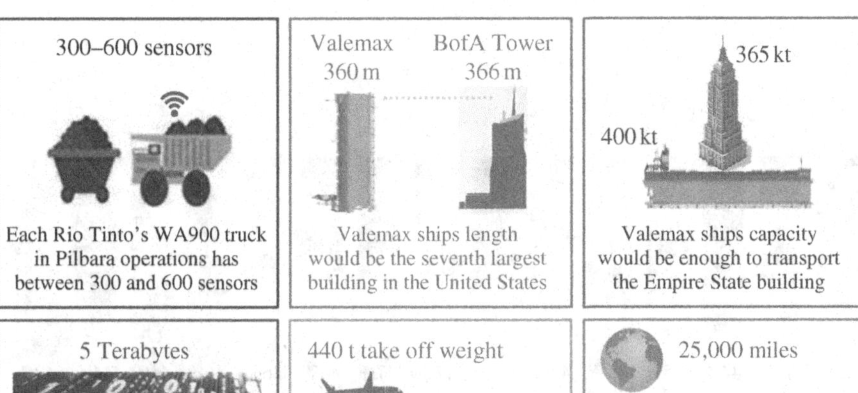

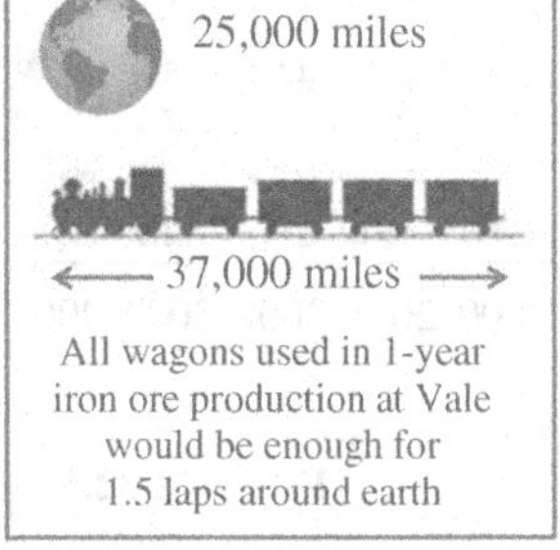

FIGURE 32.4 Mining industry facts.

Each Rio Tinto's WA900 truck in Pilbara region (Western Australia), for example, could have up to 600 sensors. All WA900 trucks together, which are not the biggest portion of Rio Tinto's fleet, daily generate five terabytes of data (see Figure 32.4 for other mining industry scale facts).

The large scale of mining operations could enable technology developers to transform their disruptive innovations into a large business. Total world's iron ore production, for example, was above two billion tons in 2014. Every dollar per ton of cost reduction that a technology developer delivers could mean an economic impact of hundreds of millions of dollars.

However, the aforementioned massive amount of data is usually not yet fully used for making the fastest and smartest possible decisions in real time at the mine sites. There is still a big room for IoT technology developers to help miners to improve how they use data to make decisions.

32.2.2 Industry Context: Sustainable Productivity Imperative

32.2.2.1 A Decade of Growing Volumes at Any Cost (2001–2011) The first decade of this millennium was remarked by a boom of commodity demand which is commonly called in the mining industry as the *super cycle*. During this period there was a big rise in most of the mining commodity prices driven by an above than expected demand boom. Supply side was not able to respond to the demand surge on time because, on average, it takes 7–10 years from the deposit discovery to the operation start-up and ramp-up of mining projects.

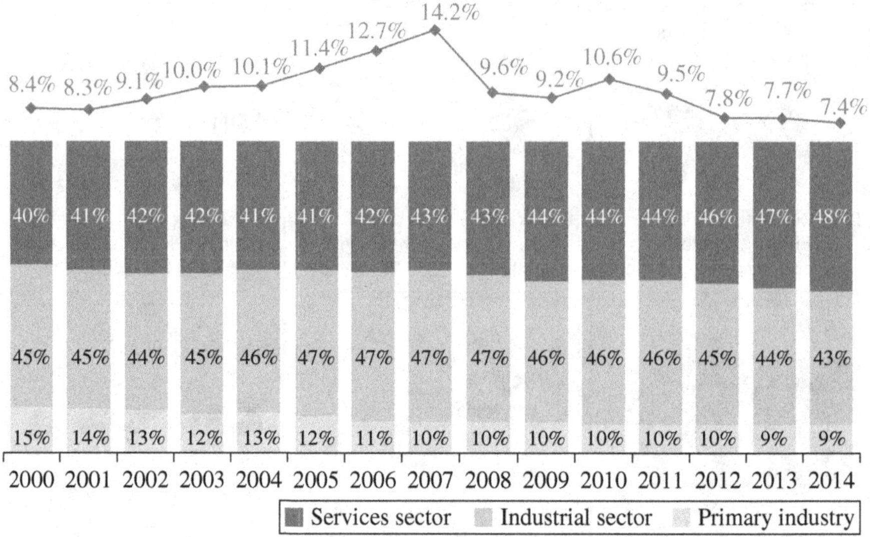

FIGURE 32.5 Chinese GDP growth and Chinese GDP by sector.

The demand surge was largely due to the rising demand from emerging markets such as the BRIC countries, as well as the result of concerns over long-term supply availability. By far, the Chinese economic growth was the main driver for the mining product demand spike during this period.

Chinese GDP has grown around 10% per year or more during the 2000s decade, and the share of industrial sector in the GDP grew from 45% in 2000 to 47% in 2008 (Figure 32.5).

The Chinese economic growth during the super cycle was driven by massive investments in infrastructure and urbanization which are highly intensive in mining and metal products like steel, iron ore, metallurgical coal, copper, nickel, and others. Chinese iron ore imports, for example, increased from around 70 million tons in 2000 to more than 900 million tons in 2015.

Infrastructure development was a top priority for the Chinese government over the past decade. They recognized that a modern economy runs on reliable roads and rails, electricity, and telecommunications. The Chinese government goal was to bring to the whole nation an urban infrastructure up to the level of infrastructure in a middle-income country while using increasingly efficient transport logistics to tie the country together.

From the late 1990s to 2005, around 100 million Chinese benefited from power and telecom upgrades. Between 2001 and 2004, investment in roads from rural areas grew by a massive 51% annually. And in recent years, the government has used substantial infrastructure spending to hedge against flagging economic growth.

This way, China has overtaken the United States and the European Union to become the world's largest investor in infrastructure by investing 8.5% of 1992–2011 GDP into infrastructure (Figure 32.6) and, consequently, boosting the metal commodities demand.

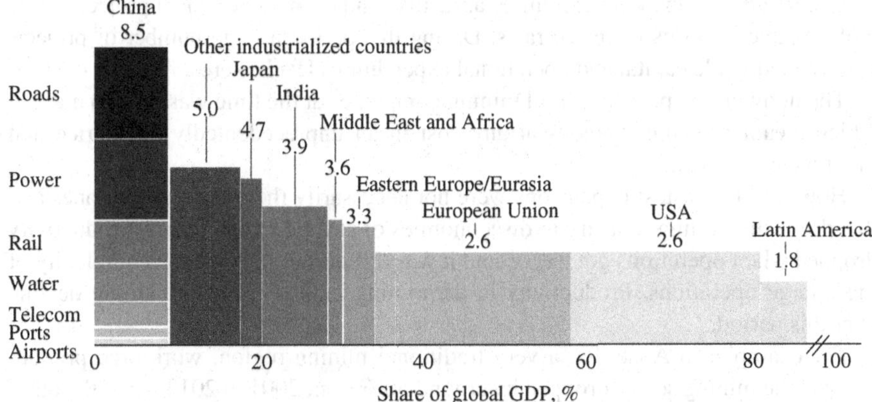

FIGURE 32.6 Investments in infrastructure, 1992–2011, weighted average, % of country's GDP. Permission to use by McKinsey.

As a consequence of the massive construction and infrastructure investments, the Chinese urbanization rates, measured by the percentage of the total population living in urbanized areas, climbed from 25% during the 90s to levels above 50% in 2015. The Chinese urbanization boom also have doubled the country's housing sector, which has risen from a total floor space completed of approximately 500 million square meters in 2000 to levels above 1 billion square meters in 2015.

In such a positive business environment, there was a big race among the top global mining producers to meet the demand surge and gain market share. For example, in the iron ore sector, Vale, the largest global iron ore producer, increased its iron ore sales from 138 million tons in 2000 to 319 million tons in 2014. BHP Billiton and Rio Tinto have sold in 2014 close to four times their iron ore sales volumes in 2000, and a large new entrant came in, FMG, which increased iron ore seaborne supply by 148 million tons (see Figure 32.7).

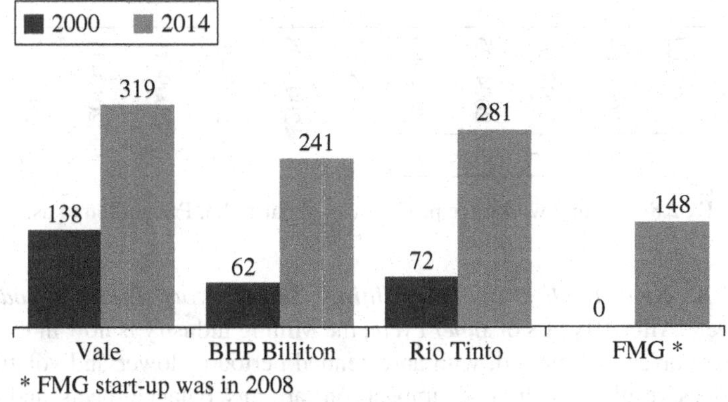

FIGURE 32.7 Iron ore sales per company, in Mt. Author analysis using companies' annual report.

Prices and margins were extremely attractive, and even less efficient projects could look feasible in terms of return rates. During this period a large number of projects were started while capital and operational expenditures levels were still under control.

The number one priority for all mining companies at the time was time to market, which meant growing volumes at any cost in an unprecedentedly high price and margin environment.

However, the fastest expansions were not necessarily the most efficient ones and that drove the mining industry to diseconomies of scale. In many cases, productivity dropped when operations got larger, and it was difficult to manage the complexity of these large operations. Productivity in the mining industry was in a strong decline over this period.

For example, in Australia, a very traditional mining region, workforce productivity in the mining sector dropped by around 50% from 2001 to 2013 while the other sectors in the same country slightly improved (see Figure 32.8).

> **We were a little more entrepreneurial and innovative** in what we did. The last decade has taken some of that out of us.
>
> (Senior Executive in Mining Industry in a survey coordinated by EY)

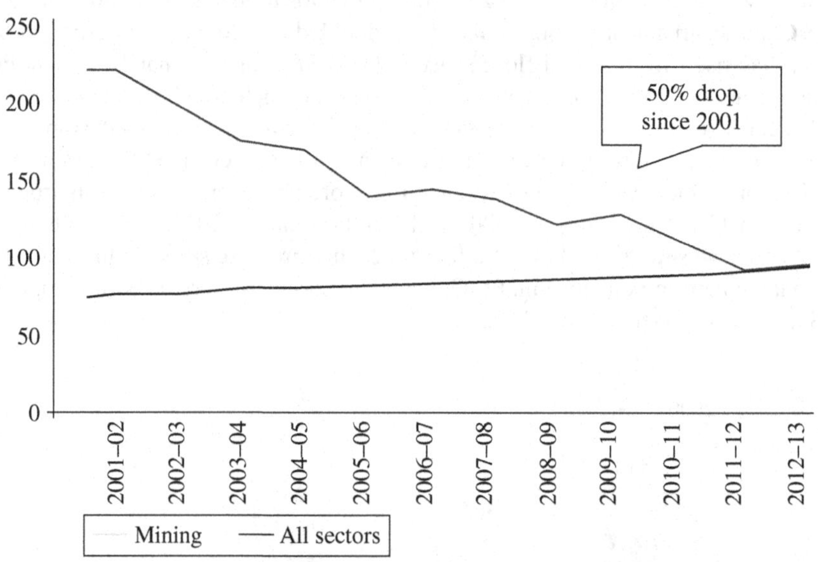

FIGURE 32.8 Mining workforce productivity in Australia. Permission to use by EY.

32.2.2.2 A New Reality in the Mining Sector: Sustainable Productivity Imperative After 10 years of *super cycle*, the Mining industry is now in the middle of a "super correction" period, with an extended period of lower and volatile commodity prices, resulting in dramatic impacts on earnings, balance sheets, and investor perceptions of the sector. As a result, miners became strongly focused on margin, cash flow, and capital returns.

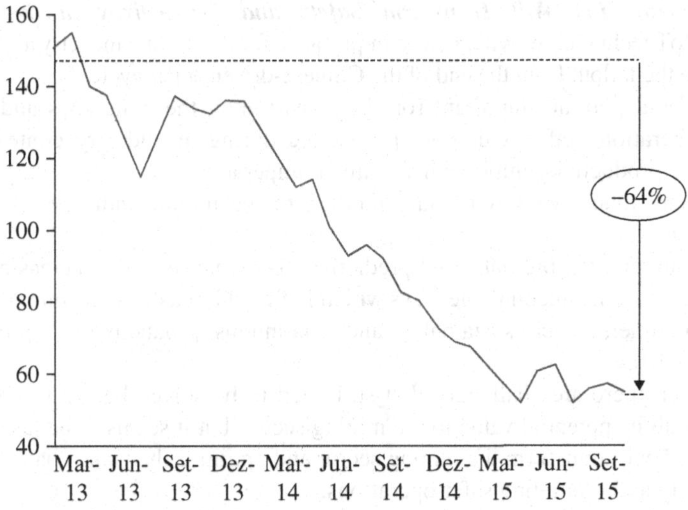

FIGURE 32.9 Platts IODEX 62% iron ore CFR China, in US$/dmt.

In the current lower commodity price environment, operational margins are squeezed, and instead of growing volumes mining companies now need to focus on value. For example, iron ore prices have dramatically dropped by more than 60% since January 2013 (Figure 32.9).

Many initiatives labeled by the miners as "productivity" initiatives up to date have been focused on short-term cost cutting—both Capex and workforce cuts—which have led to some results over the last few years. However, the mining industry is not passing through a 1- or 2-year crisis, and the current environment must be faced as the mining sector new reality for many years ahead.

The short-term measures will come to an end and are not going to be enough. The miners should move beyond those point solutions and adopt an end-to-end solution to transform business and management models.

Current industry context increases the need for reaching sustainable productivity improvements, and innovation has become imperative to the mining industry sector. Innovation is the only possible way miners could continuously and consistently keep improving productivity along the years.

> We have a marathon coming up—we need a lean company. The **low ferrous and non-ferrous prices** and the low oil and gas prices are a **marathon**, and we need to be lean and efficient with regards to **cost and productivity**.
>
> Murilo Ferreira, CEO, Vale

> Who would have thought today we would be using **Big Data** to analyze mines in Mongolia? These **developments allow us to be much more efficient**
>
> Sam Walsh, CEO, Rio Tinto

> …If we don't **start to bring innovation back**… The major miners will be subsidiaries of General Electric …
>
> Mark Cutifani, CEO, Anglo American

32.2.2.3 How IoT Will Transform Safety and Productivity in the Mining Sector IoT technologies will grow in importance for the mining industry as the sector deals with the fallout from the end of the Chinese-driven super cycle.

IoT should play an important role in transforming the mine sites and making mining operations safer and more productive during an industry context where sustainable productivity improvements are an imperative.

There are many ways IoT could modernize the mining industry all over the value chain.

From automation, big data, and predictive maintenance to the increasing use of wearables, there are uncountable ways which IoT could transform the mining activity and help miners reach sustainable and continuous productivity improvements (Figure 32.10).

However, there are challenges that still need to be solved before IoT can fully aggregate all its potential value to the mining sector, but it seems to be just a matter of time. IoT will transform the mining sector and optimize the value chain, boosting productivity and promoting safer operations.

32.2.3 Creating a Model of Success for Innovation in the Mining Industry

Typically in the mining sector, companies are hierarchical and command-control based with a strong top-down culture. The new context of the industry will require a change on mindsets and behaviors, including the way how they understand innovation.

A model of success for innovation should include four key pillars: (1) high level of collaboration, (2) a culture of innovation, (3) expanded scope, and (4) appropriate structures.

32.2.3.1 Pillar 1: High Level of Collaboration The mining companies need to realize that innovation is not about their R&D budget size. For example, the Global Innovation 1000 study by Strategy&—one of the most traditional global innovation rankings—has demonstrated, for the 10th consecutive year, that there is no significant correlation between how much you spend and how well you perform over the long term (Figure 32.11). You can't buy your way to the top!

Actually, innovation is much more about interaction than how much money is spent. It requires transformational changes from organizational culture to the way people are organized. A collaborative approach is a must if any miner wants to be successful on innovation, including both internal and external collaborations.

Disruptive innovations have more chances to happen when a company has a culture of exposing and solving problems together. Completely different areas like Operations (mine, railway, port, shipping), IT, Engineering, Procurement, Exploration, Technology, HR, and Communication will need to efficiently collaborate. Making so many different departments (and different bosses) to work well together into very hierarchical companies is the biggest challenge. Finding models that can break the internal silos and boost collaboration is decisive to make it work.

Furthermore, companies from many sectors are increasingly adopting open innovation models to boost collaboration externally (Figure 32.12). Expanding the

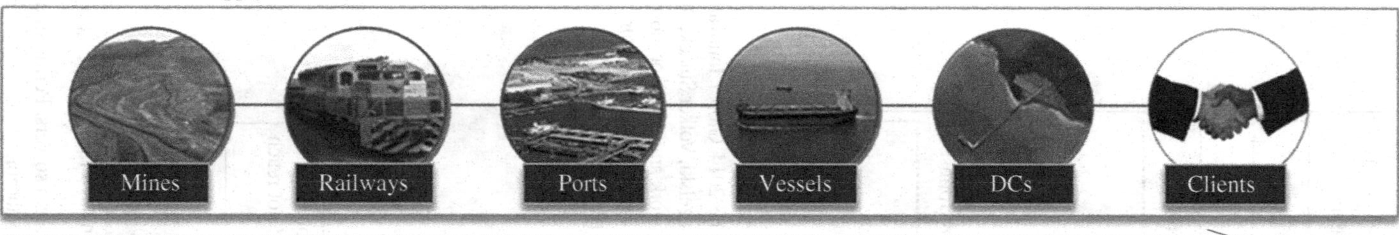

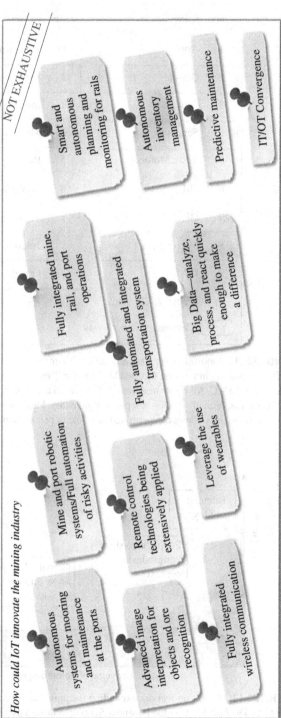

FIGURE 32.10 How could IoT transform mining operations? From Rafael Laskier's presentation at IoT World 2015, San Francisco, USA.

10 Most innovation companies 2014		The top 10 R&D spenders 2014	
Rank	**Company**	**Rank**	**Company**
1	Apple	1	Volkswagen
2	Google [in both lists]	2	Samsung [in both lists]
3	Amazon	3	Intel
4	Samsung [in both lists]	4	Microsoft [in both lists]
5	Tesla Motors	5	Roche
6	3M	6	Novartis
7	GE	7	Toyota
8	Microsoft [in both lists]	8	Johnson & Johnson
9	IBM	9	Google [in both lists]
10	Procter & Gamble	10	Merck

[dashed box] Companies in both lists

FIGURE 32.11 Innovation versus R&D ranks. Adapted from "The 2014 Global Innovation 1000: Proven paths of innovation success (Media report)" by Barry Jaruzelski, Volker Staack, Brad Goehle.

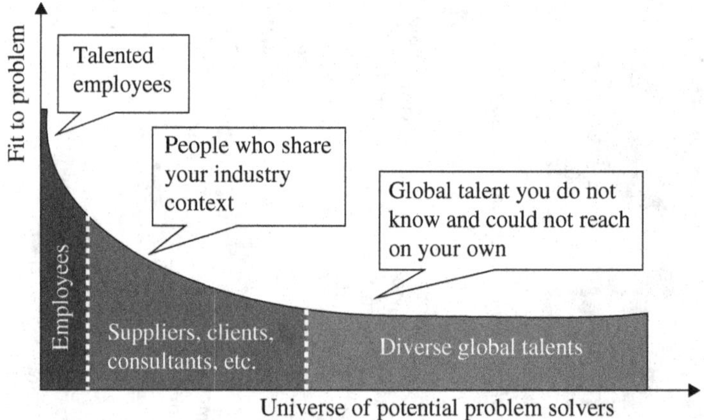

FIGURE 32.12 Open innovation: expanding the universe of problem solvers. Permission to use by Gary Dushnitsky, London Business School; Adjusted by the author.

universe of potential problem solvers increases the chances of success. Miners could develop improved models where they could work better with suppliers, clients, and, why not, competitors.

32.2.3.2 Pillar 2: Culture of Innovation The organizational culture required to successfully operate is different from the culture to successfully innovate. However there is no right and wrong or better and worst type of culture for mining. Both are needed, and the real challenge is to find a way to make so different types of culture efficiently coexist inside the same company (Figure 32.13).

Any miner will require a culture of discipline management to successfully run huge-sized operations. Mining operations can have thousands of workers, hundreds of enormous equipment, and hundreds of millions of tons of materials yearly moved. The only way to keep running such a complex operation efficiently and safely is through focus and specialization, routine management, and a failure safe process (a mine site is a "no failure zone;" any failure could cost a life or cause injury).

On the other hand, to increase chances of success on the innovation side, it will require diversity of knowledge and experiences instead of specialization. It will also require fast learning processes where the company can quickly repeat cycles of prototyping seeking for disruptive changes. An appropriate culture of innovation would

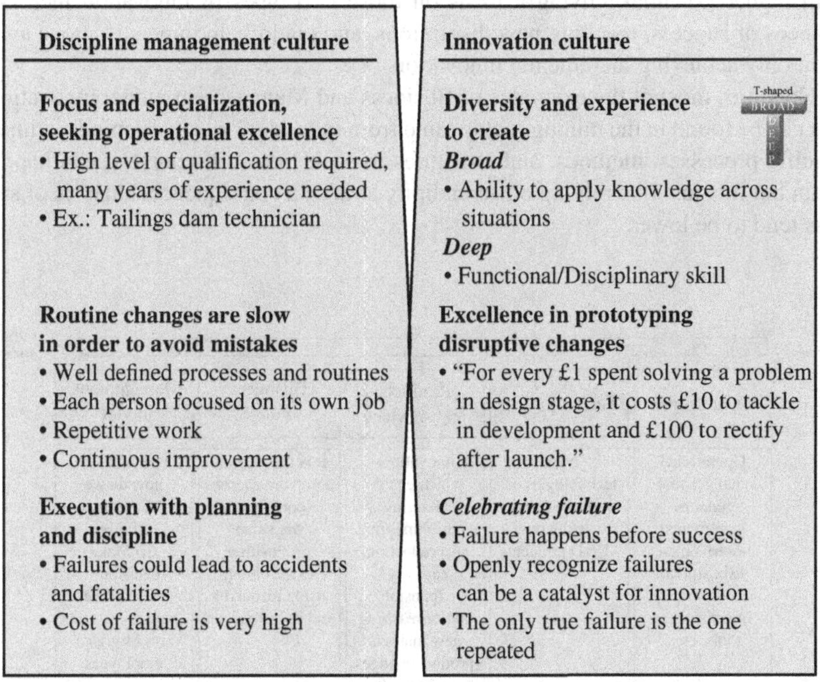

Discipline management culture	Innovation culture
Focus and specialization, seeking operational excellence • High level of qualification required, many years of experience needed • Ex.: Tailings dam technician	**Diversity and experience to create** (T-shaped: BROAD / DEEP) ***Broad*** • Ability to apply knowledge across situations ***Deep*** • Functional/Disciplinary skill
Routine changes are slow in order to avoid mistakes • Well defined processes and routines • Each person focused on its own job • Repetitive work • Continuous improvement	**Excellence in prototyping disruptive changes** • "For every £1 spent solving a problem in design stage, it costs £10 to tackle in development and £100 to rectify after launch."
Execution with planning and discipline • Failures could lead to accidents and fatalities • Cost of failure is very high	***Celebrating failure*** • Failure happens before success • Openly recognize failures can be a catalyst for innovation • The only true failure is the one repeated

FIGURE 32.13 Discipline management versus innovation. From Rafael Laskier's presentation at the Chief Innovation Officer Summit 2015, New York, USA.

also celebrate failures as steps for reaching success instead of keeping a "no failure zone" environment.

Organizational culture in mining companies should incorporate both discipline management and innovation cultures, and workers will need to know when to exactly adopt each of them.

32.2.3.3 ***Pillar 3: Expanded Scope of Innovation*** Realizing the five types of innovation (Figure 32.14) and expanding the internal scope to support each type are key to successfully reach disruptive innovations in the mining sector. It requires having available tools, processes, and organizational structures that can support each of the five types.

Typically, at most of the mining companies, processes, structures, methodologies, and tools to support two types of innovation can be found: (1) Operational (Six Sigma, Lean, quality tools, continuous improvement) and (2) Technological (R&D).

As a consequence, the mining sector has seen only incremental innovation, which is important but doesn't move the needle to the miner's cash flow challenges.

The last important disruption in the mining sector was the heap leaching process for copper during the 70s, which enabled the exploration of new deposits that were previously not feasible. In order to seek disruptive innovation, the mining companies should focus on the Management model and Business model innovation types.

Management model and Business model innovation are harder to reach but also with higher potential. Giving a focus on those two types of innovation increases chances of success, reaching new disruptions, and could help miners to move away from only achieving incremental innovation.

However, most of the examples of Business and Management model innovations that can be found in the mining sector came from individual visionary minds, without specific processes, methods, and structures to seek them. Of course it can happen again but it is just like relying on serendipity and, as a consequence, chances of success tend to be lower.

Easier/ incremental → *Harder/ disruptive*

Operational	Technological	Products and services	Business model	Management model
Operational changes that generate incremental improvement (Six sigma, continuous improvement tools, etc.)	New technologies or improvement in existing ones (R&D projects)	Innovations which enable charging a premium price, cost reduction, capacity expansion, and/or reaching new markets (product changes, new products)	It is the way the company chooses to comply with the value proposition to the client, using resources and key processes	Principles of how do we (i) Coordinate activities, (ii) Make decisions, (iii) Define objectives, and (iv) Motivate employees

FIGURE 32.14 Types of innovation.

32.2.3.4 Pillar 4: Appropriate Structures As expected, hierarchical companies like miners traditionally have more rigid structures with a triangle-shaped way of organization. This kind of structure provides focus on the status quo and works well for managing mining operations when combined to the previously mentioned Discipline Management culture.

However, the same kind of structure cannot efficiently support the type of interaction and culture required to foster innovation. Boosting innovation requires new ways to organize. It is not about creating more boxes in the traditional organizational chart. Actually, it is about having parallel ways of organization where the same people from the traditional hierarchized structure can collaborate without silos and away from the top-down mindset (Figure 32.15).

32.3 CASE STUDY

32.3.1 Kennecott Mine: Rio Tinto

The Rio Tinto's Kennecott Utah Copper Mine is fully integrated, from extraction and concentrating to smelting and refining. This mine, located in the United States, at over 100 years old, has produced more copper than any other mine in the world. It is currently more than 2.7 miles across and roughly 0.75 miles deep, making it the world's largest human excavation on earth.

In every large mine like this one, operating 24×7×365, uptime is critical, and any unplanned downtime reduces the facility's potential throughput. Operational optimization will always continue to be the focus for asset-intensive companies like this.

In 2014, the Rio Tinto's Kennecott operations had a programmed shutdown of its copper smelter. That was a US$170 million initiative which would involve thousands of workers from contractors and the company being on site all over the shutdown period.

Then a collaborative initiative between Rio Tinto and Accenture took place for creating an IoT innovation using pervasive Wi-Fi, RFID tags, and an overall digital shutdown solution designed to improve safety, productivity, and efficiency.

Rio Tinto wanted to prove the concept of integrating real-time location data with other enterprise data to provide benefits of improved safety, support reconciliation for bills, and increase worker's productivity. If they could do it for a big shutdown like this and see what the savings were and collect the data, they could perhaps prove the concept to not only use it for other shutdowns elsewhere within Rio Tinto businesses but also adopt it by the contractor management team for normal operations.

Commercially Rio Tinto saw some benefits around being able to reconcile work hours as well as the productivity of time that people spent actually doing tasks versus doing administrative work.

This case would make everybody involved to realize what a huge opportunity a smelter shutdown is in terms of gathering data and trialing the system because they have roughly 2,000 contractor workers on site per day for the smelter shutdown, so it was a great opportunity to capture data, trial the system, and learn as much as they could (see Figure 32.16 for additional smelter shutdown demographics and statistics).

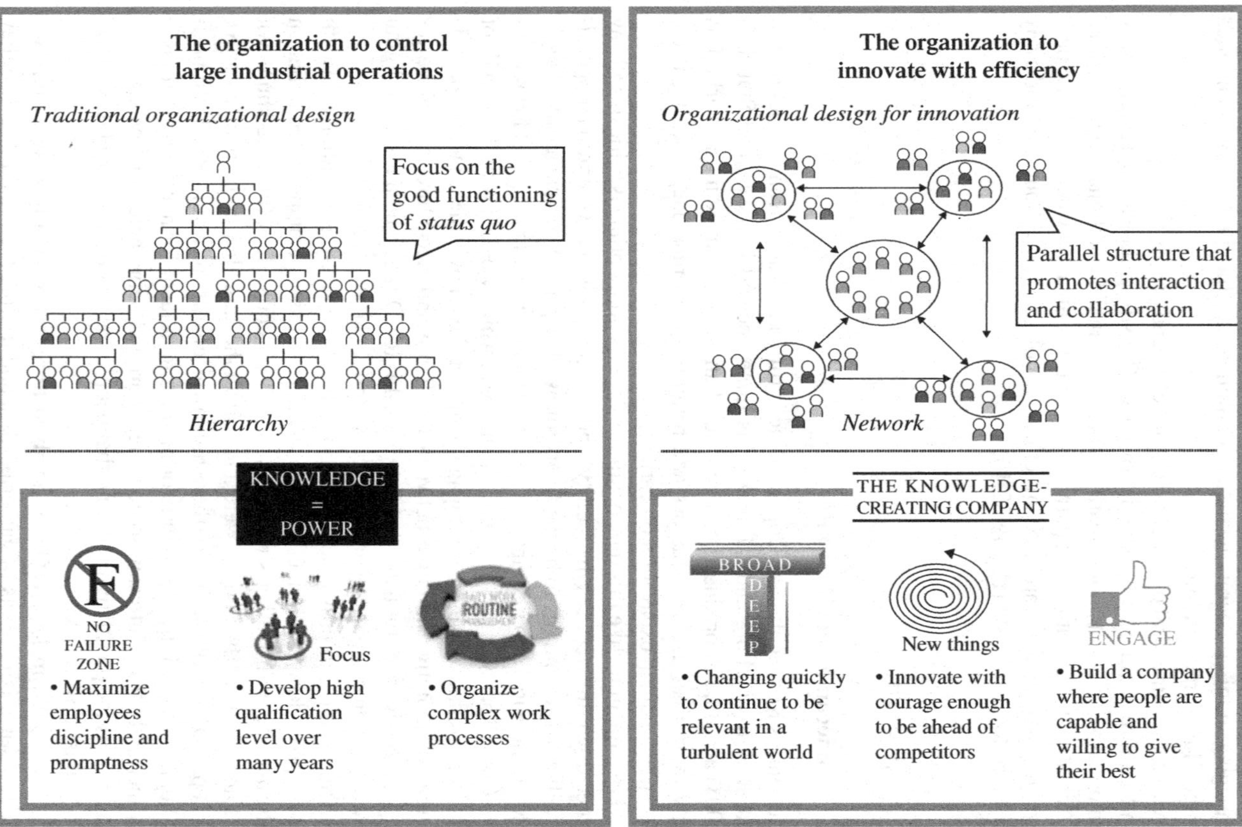

FIGURE 32.15 Organizational structures: traditional mining versus innovation. From Rafael Laskier's presentation at the Chief Innovation Officer Summit 2015, New York, USA.

2014 Smelter Shutdown Demography

Exposed hours	925,000
Duration of plant outage	75 days
Budget at completion	20% under budget
AIFR	0.43

Mobility PoC statistics

Tracked personnel	1,885
Tracked equipment	146
Wi-Fi access points	98

FIGURE 32.16 Smelter shutdown demographics and mobility PoC statistics. Permission to use by Accenture.

What Rio Tinto wanted to do through this proof of concept was using existing technology to allow them to track individual movements of contractors through the plant. By doing that and integrating with other data sources, Rio Tinto created this data-rich environment where they were able to improve safety, productivity, and commissioning.

The components of this end-to-end digital solution for safety and operational improvements in plant shutdowns were (1) wireless network across process area, (2) RFID tags used to triangulate location for people, equipment, and vehicles and visualize on a real-time tracking system, (3) Analytics dashboard to review real-time data as well as evaluate trends, (4) Mobile apps for contractors to sign their workers onto the jobs at the correct job code, and (5) Integration with existing systems.

As previously said in this chapter, a high level of collaboration and diversity is key for a successful model to foster disruptive innovation. In the Kennecott case, everything achieved was only possible due to an intense collaborative work between many Kennecott plant areas such as Smelter, contractor management group, engineering service team, process safety, operational services, and other areas as well as the vendors and contractors involved in the shutdown and Accenture's digital turnaround team. This collaborative approach allowed Rio Tinto to have everything quickly implemented, in approximately 70 days, on time enough before the beginning of the shutdown program.

The purpose of having the tags in place was to be able to track where people were at any given time. In the past, at the Kennecott mine, they could know who was coming on site and who was leaving at the end of the day, but they never knew where they were at any given time. So the tracking actually allowed Rio Tinto to know where people would be and then be able to tie that back with the projects that they were working on.

That digital solution for contractor tracking allowed Rio Tinto to know, for example, once workers came to sight and how long could it take to go from their initial entering the gate to their lead-up areas and then getting out to the productive areas. The company could also find out how much workers would have to walk back and forth to tool rooms to pick up tools or how much time they would actually spend in the assigned productive work areas as well as to get an understanding of what that division of travel time versus productive time was.

One of the things that this innovation let Rio Tinto do was actually find equipment. Occasionally it happens during a shutdown that one contractor or someone may borrow equipment from another contractor. Rio Tinto was able to use the tags in order to be able to go quickly find where that was. They also have the issue of sometimes light vehicles being at certain locations blocking access, and they were able to very quickly find the people that own the vehicles and get them moved out, so it didn't have the interference on the work being done.

The data generated from the tags are integrated with existing data such as gate pass and others, in order to allow Rio Tinto to plan shutdown more safely, more productively, and less costly in the future. Additionally, this innovation is also something to be applied in the operational area as well and not just for large shutdowns.

Main insights and findings from this IoT proof of concept during the smelter shutdown were the following:

Safety

- Monitor worker fatigue status.
- Warn of unqualified personnel in high-risk areas.
- Capture data to prevent incidents.
- Evaluate correlation between incidents and fatigue (time of day, day in shift, time in shift, etc.).
- Support incident investigation and hot spot identification.

Effectiveness

- Provide real-time project management.
- Provide real-time root cause analysis (RCA) capability.
- Maximize workforce productivity.
- Accelerate planner feedback.

Efficiency

- Reduce contractor billing errors.
- Support informed planning to lower overtime charges.
- Improve equipment utilization.

Quality

- Ensure only qualified contractors conduct work.
- Enhance supervisor visibility to project work.
- Support more effective and guided project management activities.

In summary, the result of this proof of concept provided visibility to drive value across:

(i) *Vendor Management*: Labor reconciliation, equipment/vehicle reconciliation
(ii) *Productivity*: Real-time location visibility, resource availability, increased time on tools
(iii) *Worker Safety*: Fatigue management, incident risk factor analysis

In the future Rio Tinto hope they can be able to use these tags both in the areas of safety, to be able to understand better where people are and how their interactions are with equipment, and supporting productivity, to be able to better allocate where equipment needs to be and in general have an improvement at both in terms of shutdown and day-to-day operations.

That proof of concept also enabled Accenture to develop an end-to-end digital solution for plant turnarounds, shutdowns, or outages called Digital Turnaround as a Service (dTAaaS). With this model, according to Accenture, productivity could increase from 5 to 15%, billing accuracy could be optimized from 3 to 5%, outages, and cycles could be potentially shorter, and it also enhances safety through fatigue management, risk factor analysis, site security, and gas monitoring. According to Scott Tvaroh, Accenture's Global Lead for this solution, Accenture has implemented a newer version of this solution at two different mines in Canada with similar productivity results in spring and fall of 2015.

32.3.2 Other Examples

Rio Tinto's Mine of the Future and Anglo American's FutureSmart initiatives are other examples which make clear that the IoT is the direction the industry is headed.

Anglo American's FutureSmart initiative is an approach which blends together a number of initiatives for both innovative technologies and open collaboration. Initial investments of FutureSmart will be around US$30 million per year, and in mid-2015 more than 50 initiatives and key partnerships were under way to increase mining efficiency and eliminate potential hazards.

Some of the main components of FutureSmart initiative are using lasers as cutting tools (for the drill and blast process), utilizing 3D technology to provide richer data for faster and better decision making, and automation and robotics (for underground mining and fleet management).

Automation has long been a part of the mining industry priorities.

Rio Tinto's Mine of the Future, unveiled in 2008, focuses on two key themes: (1) Achieving efficiency in surface mining through autonomy and (2) Improving recovery practices. The program initiatives involve next-generation technology and big data utilization. Main examples of step-change technologies from the program are a centralized operations center for remotely managing operations (launched in 2009) and autonomous truck haulage, trains, and drilling.

Rio Tinto currently is the largest owner and operator of autonomous trucks in the world and mid-2015, they already had 57 trucks operating across three sites in the Pilbara region, Australia.

One of Rio Tinto's most recent innovations is the AutoHaul technology, which could deliver the world's first fully autonomous heavy haul, long-distance railway system. It has the potential to eliminate the need for approximately 70,000 km of remote area driving each week to get train drivers in place to start or finish their shift, as well as provide flexibility in scheduling and eliminate driver changeovers.

Rio Tinto also uses an Autonomous Drilling System (ADS) at its West Angelas mine in the Pilbara region. This is now the world's first full-time autonomous drill mine with seven rigs in operation having drilled almost two million meters by June 2015. See Figure 32.17 for a more complete summary of Rio Tinto's initiatives for automation.

According to Rio Tinto, the company already reached attractive results so far, such as:

- Drills: 10% increase in equipment utilization in West Angelas, Pilbara
- Grade recovery: 2% increase in high grade recovery at West Angelas
- Load utilization rates: 14% higher for automated trucks
- Operating costs: 13% lower at Hope Downs 4 site

Key initiatives	Value delivered
Productivity programmes Several sites of the group	• Improved asset reliability • Reduced operating risk
Equipment automation Pilbara and Hunter Valley	• Improved safety • +14% truck utilization • –13% load and haul operating cost • 3× improved drill labor productivity • Real-time ore body data
Mine automation systems/big data Pilbara, Hunter Valley, and Kennecott	• RTV is platform: 3D modelling tool that allows to visualize in real-time data generated from sensors located on trucks, drills, and other mine equipments. • +2% high grade ore at West Angelas
Operations and Processing Excellence Centres Perth, Brisbane, and Pune	• Perth Operation Centre • Brisbane Processing Excellence Centre (+US$ 80 million cash flow) • Pune Analytics Excellence Centre

FIGURE 32.17 Summary of Rio Tinto's automation efforts.

Vale, the largest iron ore global producer, is also running its own "future of mining" initiatives. Vale's portfolio of IoT initiatives embraces several new technologies with focus on productivity gains, cost reduction, and safety improvements. Some of the main initiatives focus on autonomous mine (trucks, trains, and drills), wearables, and location tracking.

One of the most promising initiatives for location tracking took place at Vale's nickel operations in Canada. Vale has been using RFID in Canada since 2005 to track containers, rail cars, and other assets. In 2008, the company deployed an RFID solution at its Stobie mine in Ontario to monitor the ore grade (mineral concentration) as it is mined in real time.

The Totten mine case is the most recent proof of concept implemented at Vale's nickel operations in Canada, and it is an important one to understand RFID's future in the mining industry. At Totten mine, Vale is assessing how IoT technologies could optimize the underground mine operation, improving safety and reducing costs.

The company installed a wireless network enabling people and equipment to be locatable underground. Supervisors could be able to use mobile phones and tablet computers while working underground. A real-time location system was tested to track underground equipment and personnel using active Wi-Fi RFID tags.

A centrally controlled ventilation system that could adjust itself according to local operations was also tested at Totten mine. Using the location data of people and equipment from the tracking system, Vale could identify what kind of piece of equipment is in a certain area and adjust the ventilation to accommodate it. This ventilation on demand system is more efficient because it only sends fresh air to areas of the mine in use, and it can adjust as necessary when larger pieces of equipment are brought into specific areas. The system can also help to control temperatures underground because running ventilation fans increases underground temperatures, so minimizing their usage is a significant accomplishment.

The RFID tags also increase safety by allowing the control room to know instantly where employees and equipment are located in the event of an emergency. Production and maintenance data from mobile equipment such as scoops, trucks, and drills can also be captured via the wireless network. That information could be used for predictive maintenance analysis on equipment, and real-time production data can be captured for immediate reporting.

Equipment manufacturers are also quickly jumping onboard. Caterpillar Inc., for example, has already integrated its MineStar technology to manage everything from material tracking to sophisticated real-time fleet management, machine health systems, autonomous equipment systems, and other applications. Many of the largest service providers like IBM, Cisco, Hitachi, GE, Komatsu, and others are also going in the same direction.

32.4 CONCLUSION

There is no doubt that IoT will continuously progress in the mining industry and transform how mine sites will be operated in the future. It is just a matter of time. The current industry context and the need of sustainably improve productivity will increasingly move miners on that direction because it is a matter of survival to them.

The large scale of the mining industry also makes it highly attractive to IoT technology developers which could find in the mining industry landscape the opportunity to transform their innovations into hundreds of million dollar-sized businesses.

Forecasted economic impact of IoT globally could be compared to some of the largest GDPs in the world like Japan or Germany, and the mining sector tends to be a relevant share of it. According to Gartner Inc., a technology research consultant firm, the number of connected products in the mining industry is expected to rise from 24 million in 2014 to 90 million in 2020. The 25% annual growth rate in mining connectivity is one of the fastest among the industry sectors tracked by Gartner.

The IoT revolution will dramatically impact productivity, costs, and safety at the mine sites and processing plants. However, the mining industry environment is still a bit away from what is required to fully build and capture the benefits that IoT and other disruptive technologies could bring.

Mining companies are usually extremely hierarchical and organized into silos. In such a nonfavorable environment, information doesn't flow, and it is not difficult to find several and duplicated systems along a mining value chain that becomes a barrier to integration. Collaboration (and innovation) becomes harder to happen.

In order to fully take advantage of the IoT revolution, miners will have to learn how to work better in terms of collaboration, both internally (breaking the silos, changing culture and mindsets, and finding new ways of organization) and externally (working better together with clients, suppliers, industry experts, and competitors). It will require mining companies to shift from a culture of "right and wrong" to a culture of "solving problems together."

Mining suppliers and service providers in general also need to change their mindsets and the way they interact with the miners. Many of them are massively investing in IoT solutions for the mining industry, and some promising new technologies are emerging.

However, in general, they are still with a stronger focus on their solutions rather than the real problems miners face. Putting tools and products instead of being effectively problem-centered could make the IoT revolution to come slower for the mining industry. Shifting to a real problem-centered approach would increase everybody's chances of success.

The IoT represents a huge opportunity to the mining industry, but there is still a lot of work (and change) to be done.

FURTHER READING

"*Creative Confidence – Unleashing the Creative Potential Within Us All*", Kelley, T. & Kelley, D., Crown Business New York, USA, 2013.

"*Digital Solutions for Contractor Safety Tracking*", Tvaroh, S. & Sjogren, J., Solutions 2.0 Conference, USA, 2015.

"*Industrial Internet of Things: Unleashing the Potential of Connected Products and Services*", World Economic Forum in collaboration with Accenture, USA, 2015.

"*Industry 4.0: The Future of Productivity and Growth in Manufacturing Industries*", The Boston Consulting Group, Germany, 2015.

"*The Key Elements Required to Make Innovation a Success*", Laskier, R., Chief Innovation Officer Summit, USA, 2015.

"*Making Innovation Happen in the Mining Industry*", Laskier, R., 6th Mine Tech Forum, Chile, 2015.

"*Modernizing the Mining Industry with the Internet of Things*", Laskier, R., Internet of Things World 2015 Conference, USA, 2015.

"*Productivity in Mining: Now Comes the Hard Part - A Global Survey*", Ernst & Young, UK, 2015.

"*The Case for Innovation in the Mining Industry*", Clareo^{SM}, USA, 2015.

"*The Industrial Internet: Six Critical Questions for Equipment Suppliers*", The Boston Consulting Group, USA, 2014.

"*The Internet of Things: Mapping the Value Beyond the Hype*", McKinsey Global Institute, USA, 2015.

"*The Simple Rules of Disciplined Innovation*", McKinsey Quarterly, USA, 2015.

"*Winning with the Industrial Internet of Things*", Accenture, USA, 2015.

33

INTERNET OF THINGS (IoT)-BASED CYBER–PHYSICAL FRAMEWORKS FOR ADVANCED MANUFACTURING AND MEDICINE

J. CECIL

Computer Science Department, Center for Cyber Physical Systems, Oklahoma State University, Stillwater, OK, USA

33.1 INTRODUCTION

The term Internet of Things (IoT) is becoming popular in the context of the ongoing IT revolution which has created a greater awareness of emerging and smart technologies as well as phenomenal interest in IT-based products in the world community. IoT can be described as the network of physical objects or "things" embedded with electronics, software, sensors, and connectivity to enable it to achieve greater value and service by exchanging data with the manufacturer, operator, and/or other connected devices [1]. In a nutshell, IoT refers to the *complex network of software and physical entities* which are embedded or implemented within sensors, smartphones, tables, computers, electronic products as well as other devices which have software elements to perform computing or noncomputing activities [2, 3]. These *entities* are the "things" referred to in the term "Internet of Things" which are expected to be capable of collaborating with other similar entities as part of the Internet and other cyber infrastructure at various levels of abstraction and network connectivity. The underlying assumption is that by interacting with each other, a large range of services can be provided using this network of collaborations. This subsequently enables these entities to provide greater value to customers and collaborating organizations [4].

Internet of Things and Data Analytics Handbook, First Edition. Edited by Hwaiyu Geng.

Companion website: www.wiley.com/go/Geng/iot_data_analytics_handbook/

Example of *things* are weather monitoring sensors, heart monitors, monitoring cameras in a manufacturing work cell in a factory, safety devices in a chemical processing plant, advanced home cooking equipment, etc. [5, 6]. There are many risks and benefits to embracing such a vision [7, 8]. The benefits lie in being able to form collaborative partnerships to respond in a more agile manner to changing customer preferences while being able to seamless exchange data, information, and knowledge at various levels of abstraction. Such an emphasis on adoption of IoT principles can also set in motion the realization of advanced next-generation Cyber–Physical relationships and frameworks which can enable software tools to control and accomplish various mundane as well as advanced physical activities [9]. Two of the dominant application contexts are advanced manufacturing and medical fields. While they are two distinctly different fields, the adoption of such IoT-based cyber frameworks has similarities in the design of the overall cloud-based frameworks which enable distributed users to share and use collaborative users for their respective applications. In the manufacturing context, engineers can access a cloud of resources using various IoT devices and sensors which can exchange and share data/information from remote locations including interacting with physical manufacturing resources [10]. In the medical context, surgeons can interact with surgical residents using such shared collaborative frameworks implemented on cloud-based platforms; such a framework also holds the potential to control physical surgical equipment for surgical procedures using cameras and other sensory feedback which are critical to the success of such activities. The role of the Next Internet under development in the United States, the European Union (EU), and other countries ushers in the next revolution in such innovative cyber activities where the physical location of experts and users as well as software and physical resources is no longer an issue.

In the IoT frameworks discussed in this chapter, we have explored the use of cloud principles to support information and data exchange among IoT devices and software modules. Cloud-based technologies are becoming popular and have become the focus of many industrial implementations [11–15]. In Ref. [16], a computing and service-oriented model for cloud-based manufacturing is outlined. Three categories of users are described including providers, the operators, and consumers. Some of the benefits for cloud-based frameworks which are identified in Ref. [15] include reducing up-front investments, reduced infrastructure costs, and reduced maintenance and upgrade costs. In Ref. [14], the potential of cloud computing is underscored in transforming the traditional manufacturing business model and creating intelligent factory networks that support collaboration. A service-oriented system based on cloud computing principles is also outlined for manufacturing contexts [14, 15, 17].

33.2 MANUFACTURING AND MEDICAL APPLICATION CONTEXTS

In an IoT context, one of the core benefits is from the cyber–physical interactions which help facilitate changes in the physical world. The plethora of smart devices emerging in the market serve as a catalyst for this next revolution which will greatly impact manufacturing and manufacturing practices globally. Imagine being able to design, simulate, and build a customized product from a location hundreds or thousands of kilometers away from engineering and software resources, manufacturing facilities, or an engineering organization. Today, using cloud technologies and thin clients such as smartphones and smart watches, the potential of using such IoT principles and technologies for a range of domains including advanced manufacturing and medicine is very high [18].

In a manufacturing context, such cyber–physical approaches also support an agile strategy which can enable organizations functioning as Virtual Enterprise (VE) partners to respond to changing customer requirements and produce a range of manufactured goods. The concept of a VE has been of interest to the manufacturing communities for the past three decades [19]. The need to rapidly form such VEs and enable the sharing of partner resources in different locations (irrespective of the heterogeneous nature of the platforms and tools) continues to be a driving theme today. With the help of advanced computer networks, such cyber (or software) resources and tools can be integrated with physical resources including manufacturing equipment. Thin clients, sensors, cameras, tablets, and cell phones can be linked to computers, networks, and a much larger set of resources which can collaborate in an integrated manner to accomplish engineering and manufacturing activities. When customer requirements change, such an approach can also help in interfacing and integrating with a variety of distributed physical equipment whose capabilities can meet the engineering requirements based on the changing product design. Against this backdrop, it is important to also underscore recent efforts to develop the next generation of Internets.

In the context of medical applications, the adoption of IoT-based cyber–physical frameworks and approaches will enable doctors to be better trained as well as take advantage of distributed resources including simulation-based training environments as well as engage in telemedicine activities to serve a wide range of patients who are in remote geographical locations.

In the United States, the Global Environment for Network Innovations (GENI) [20] is a National Science Foundation-led initiative that focuses on the design of the next generation of Internets including the deployment of software-defined networking (SDN) and cloud-based technologies. GENI can also be viewed as a virtual laboratory at the frontiers of network science and engineering for exploring Future Internet architectures and applications at scale. In the context of advanced manufacturing (such as microassembly), such networks will enable distributed VE partners to exchange high-bandwidth graphic-rich data (such as the simulation of assembly alternatives, design of process layouts, analysis of potential assembly problems, as well as monitoring the physical accomplishment of target assembly plans). In the EU and Japan (as well as other countries), similar initiatives have also been initiated; in the EU, the Future Internet Research and Experimentation (FIRE) Initiative is investigating and experimentally validating highly innovative ideas for new networking and service paradigms (http://www.ict-fire.eu/home.html). Another important initiative is the US Ignite [21] which seeks to foster the creation of next-generation Internet applications that provide transformative public benefit using ultrafast high gigabit networks. Manufacturing is among the six US national priority areas (others include transportation and education). Both these initiatives herald the emergence of the next-generation computing frameworks which in turn have set in motion the next Information-Centric revolution in a wide range of industrial domains from engineering to public transport. These applications along with the cyber technologies are expected to impact global practices in a phenomenal manner.

GENI and FIRE Next-Generation technologies adopt SDN principles, which not only reduce the complexity seen in today's networks but also help Cloud service providers host millions of virtual networks without the need for common separation isolation methods such as VLAN [22]. SDN also enables the management of network services from a central management tool by virtualizing physical network connectivity into logical network connectivity [22]. As research in the design of the next-generation Internets evolves, such cyber–physical frameworks will become more commonplace. Initiatives such as GENI and US Ignite are beginning to focus on such next-generation computer networking technologies which hold the potential to radically change the face of advanced manufacturing and medicine (among other domains).

33.3 OVERVIEW OF IoT-BASED CYBER–PHYSICAL FRAMEWORK

IoT entities and devices will greatly benefit from the evolution of Cyber–Physical approaches, systems, and technologies. The term "cyber" can refer to a software entity embedded in a thin client or smart device. A cyber–physical system (CPS) can be viewed as an advanced collaborative collection of both software and physical entities which share data, information, and knowledge to achieve a function (which can be technical, service, or social in nature). In a process engineering context, such CPS can be viewed as an emerging trend where software tools can interface or interact with physical devices to accomplish a variety of activities ranging from sensing and monitoring to advanced assembly and manufacturing. In today's network-oriented technology context, such software tools and resources can interact with physical components through local area networks or through the ubiquitous Word Wide Web (or the Internet). With the advent of the Next-Generation Internet(s), the potential of adopting cyber–physical technologies and frameworks for a range of process has increased phenomenally.

33.4 CASE STUDIES IN MANUFACTURING AND MEDICINE

33.4.1 Advanced Manufacturing

In the context of manufacturing, collaborations within an IoT context can be realized using various networking technologies including cloud-based computing [23, 24]. According to the National Institute of Science and Technology (NIST), Cloud computing can be viewed as a model for enabling ubiquitous, convenient, on-demand network access to a shared pool of configurable computing resources (including networks, storage, services, and servers) [25, 26]; the computing resources can be rapidly provisioned with reduced or minimal management effort or interaction with service providers. Some of the benefits for cloud-based manufacturing include reduced infrastructure costs, reducing up-front investments and lower entry cost (for small businesses) [27–29].

The IoT-based framework outlined in this section is being implemented as part of a US Ignite project dealing with advanced manufacturing and cyber–physical frameworks [30]. The manufacturing domain of interest is the assembly of micron-sized devices. This is one of the first projects involving Digital Manufacturing, cyber–physical frameworks, and the emerging Next-Generation Internet (being built as part of the GENI initiatives). It is also the first implementation of an IoT framework in the manufacturing domain [2, 31].

The preliminary implementation of this IoT framework has been completed in the form of a collaborative Cyber–Physical Test Bed (CPTB) whose long-term goal is to support cyber–physical collaborations for a complex set of life cycle activities including *design analysis, assembly planning, simulation, and finally assembly of microdevices*. The presence of ultrafast high gigabit networks enables the exchange of high-definition graphics (in the Virtual Reality (VR)-based simulation environments) and the camera monitoring data (of the various complex micromanipulation and assembly tasks by advanced robots and controllers). Engineers from different locations interact more effectively when using such Virtual Assembly Analysis environments and comparing assembly and gripping alternatives prior to physical assembly.

The resources of the CPS using cloud technologies are illustrated in Figure 33.1. The users in different locations can access the CPS modules/resources through thin clients and IoT devices including tablets, cell phones, work cells, computers, and other thin clients. Thin clients refer to devices with less processing power that relies on the server to perform the data processing [32]. In this IoT-based cyber–physical

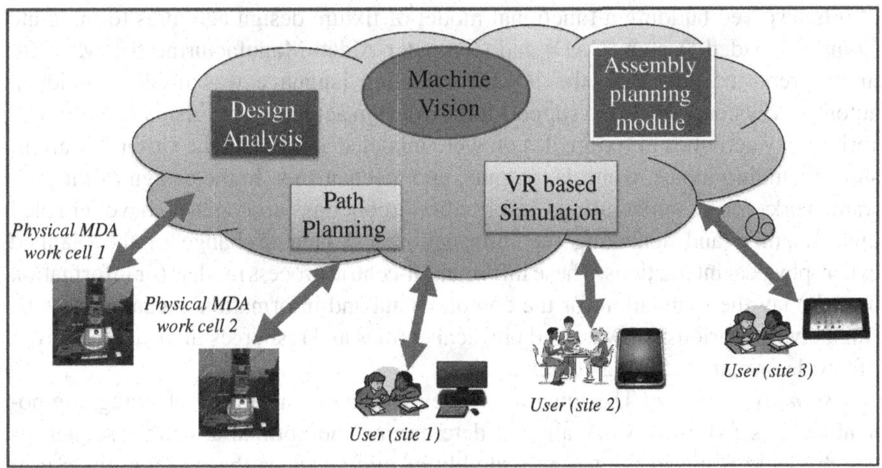

FIGURE 33.1 The IoT-based cyber–physical test bed. Courtesy of J. Cecil.

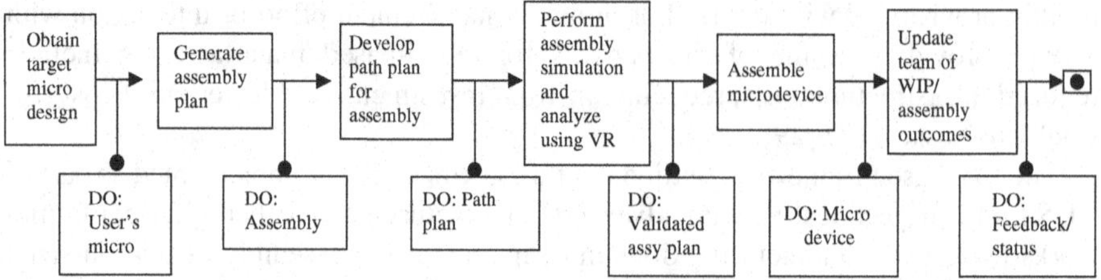

FIGURE 33.2 Overview of the main cyber–physical tasks in the collaborative framework. Courtesy of J. Cecil.

frameworks, engineers can collaborate from geographically distributed locations and share resources as part of an agile collaboration process; they can interact with CPS resources using computers and/or thin client such as smartphones and tablets. The main cyber–physical tasks were modeled using the engineering Enterprise Modeling Language (eEML) [33] shown in Figure 33.2. The term DO refers to Decision Outcomes as a consequence of completing various cyber–physical tasks in this collaborative process. The overall "mini" life cycle activities in this cyber–physical collaboration include obtaining target microdesign, generating assembly plan, developing path plan for assembly, performing assembly simulation and analyzing using VR, assembling microdevice, and updating WIP/assembly outcomes. The resources in this Cyber–Physical implementation include the following (Figure 33.1): assembly/path planning modules, VR-based simulation environments (to compare, analyze, and validate candidate assembly and path plans), assembly command generators, machine vision-based sensors/cameras (for guiding, monitoring physical assembly), and physical microassembly equipment (to assemble the target microdesigns). An overview of some of these resources is provided in the following sections.

The creation of the information models lays the foundation for designing complex software systems in manufacturing and other fields [34–36]. One of the earliest efforts involved building a functional model of fixture design activities to integrate Computer-Aided Design (CAD) and Computer-Aided Manufacturing (CAM) [36]; in this reported approach, the IDEF0 modeling language was used to build an automated system [34, 35] to support fixture design activities for prismatic parts. The various key activities in fixture design were modeled along with the various relationships including inputs, controls, outputs, and mechanisms. In the design of this IoT framework for manufacturing, information modeling approaches have enabled understanding and analyzing the information and data exchange of the planned cyber–physical interactions. These information-centric process models (or information models) lay the foundation for the complex data and information exchanges occurring between various software and physical entities and resources involved in a given life cycle or manufacturing context.

Assembly Planning: The function of the cyber-based assembly planning components in this IoT framework aims at determining the optimal assembly sequences which would result in the target assembly of microdevices (based on a given user microdesign). The outcome of the assembly plan generated is used to study various

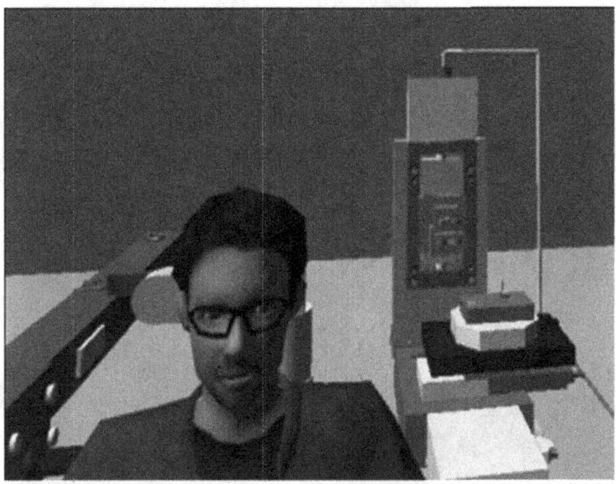

FIGURE 33.3 The virtual assembly environment. Courtesy of J. Cecil.

process issues using a VR-based assembly simulation environment. Several algorithms have been studied for use in developing a near-optimal assembly plan. For example, a Greedy Algorithm (GRA)-based approach has been employed to generate near-optimal assembly sequences. Another assembly planning approach involves the use of Genetic Algorithms in generating assembly sequences for given assembly scenarios. The underlying context for such multiple planning approaches is to mimic a real-world scenario where multiple organizations have such planning capabilities which become part of the larger set of cyber–physical resources using IoT-based frameworks.

Virtual Assembly Analysis Environments: The VR-based simulation environments are part of this IoT framework; in a real-world context, such simulation environments can be some of the resources made available by one or more VE partners using such IoT frameworks. To demonstrate this feasibility, several simulation environments were created using Coin3D, Unity, and other graphical tools. Such VR-based environments enable distributed teams to compare proposed assembly plans, change the layouts of physical resources, identify potential problems, and make needed modifications to the assembly plans [37]. As part of a demonstration, the distributed users and engineers used the Next Internet (being developed under our involvement of the GENI initiative) to propose, compare, and modify assembly/path plans rapidly. The high gigabit data relating to these VR images were transmitted using this Next-Generation Internet technology. Figure 33.3 shows an example showing a view of the virtual environment (built using Unity 3D engine) to help engineers and users interact with it. Through such cyber interactions, the most feasible assembly plan can be identified. Subsequently, based on the assembly plan chosen or developed, the corresponding robot assembly commands can be generated and then communicated to the appropriate work cells; this results in the physical assembly of a target microdesign.

Physical Resources: Two physical work cells were part of this collaborative CPTB. One of the Work cells has cameras and an advanced microgripper mounted on

FIGURE 33.4 Assembly tasks. Courtesy of J. Cecil.

FIGURE 33.5 A physical cell. Courtesy of J. Cecil.

micropositioners for manipulation and assembly. The assembly plate rests on a support fixture which has two linear degrees of freedom and a rotational degree of freedom. This work cell is capable of assembling micron-sized parts rapidly. Another work cell also part of this CPTB has a shape memory alloy-based gripper. Figure 33.4 shows a view of the assembly tasks and Figure 33.5 provides an image of one of the physical assembly cells used in this test bed.

33.4.2 Medical Surgery and Telemedicine

The second case study deals with a VR environment designed to train medical residents who are specializing in orthopedic surgery. A VR environment can be used as a simulation-based training tool which can provide training and learning experiences for budding medical residents pursuing surgery. Typically, Medical residents learn

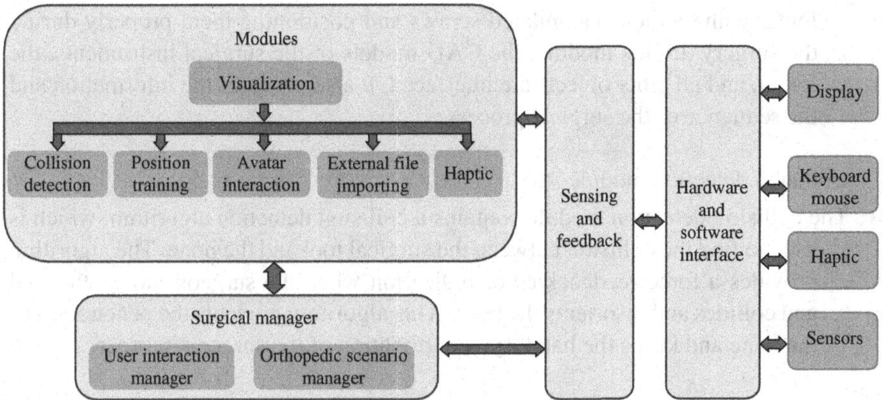

FIGURE 33.6 Architecture of virtual environment for orthopedic surgery (VEOS). Courtesy of J. Cecil.

surgery by observing the more experienced surgeon performing it and then eventually assisting them, before moving on to performing their own surgical procedures on patients. A VR environment will equip the residents with a more varied and detailed set of experiences and knowledge that is critical in determining alternative ways to perform surgery while addressing various medical conditions. Other traditional ways to learn surgical skills is to practice surgical procedures on cadavers and lab animals. The use of cadavers comes with certain risks of infection, while use of lab animals is diminishing substantially for various societal reasons. In this context, the design of virtual surgical environments holds the potential to develop a technology-based alternative that provides unique learning capabilities which can be extended and modified based on various factors while increasing the training opportunities (as they are simulation and computer based).

Using an IoT-based cyber–physical framework, orthopedic residents can be trained by a master surgeon from a different location using VR-based environment; the Virtual Environment for Orthopedic Surgery (VEOS) has been created for such an innovative application (shown in Figure 33.6). The scope of the simulation environment is currently restricted to an orthopedic surgical process known as Less Invasive Stabilization System (LISS) surgery. The primary knowledge source for the LISS surgical process was Dr. Miguel A. Pirela-Cruz (Head of Orthopedic Surgery and Rehabilitation, Texas Tech University Health Sciences Center (TTUHSC)). The VEOS was designed and developed on a PC-based platform. Prior to building this virtual environment, the complex orthopedic surgical process was modeled using the eEML [33].

The overall architecture of VEOS is shown in the previous figure, and functional responsibilities of the various modules are discussed in this section. An overview of the key modules follows:

Visualization module

The purpose of the visualization module is to display different stages of the simulation from the initial position of LISS parts to the fracture reduction of the femur using various cannulated screws and positioning them properly during the surgery. In this module, the CAD models of the surgical instruments, the femur, and all other objects are interfaced. It also provides the information and the sequence of the surgical process.

Collision detection module

The collision detection module contains a collision detection algorithm, which is used to find the collision between the surgical tool and the bone. The algorithm provides a force feedback/color indication when the surgeon moves the tool and collides and disorients the bone. This algorithm prevents the penetration of the bone and keeps the hand–eye coordination of the surgeon in order.

Position training module

The position training module helps residents to predict the exact location of the surgical tools in the nonimmersive environment (shown in Figure 33.7). The position training module measures the distance between surgical implants and the bone and also the lower and the upper proximity of the LISS plate. In this module, the indicator ensures the exact place of the implant by turning on the red color. In case the surgeon placed the implant in the wrong direction, it indicates the changes in the form of green and yellow light. Thus, the surgical resident can virtually practice and improve his/her skills in the placement of the implant using the virtual environment. This is a very useful tool that the experienced surgeons can utilize to show the residents the proper directional placing of the LISS plate. The simulation environment incorporates a color indicator module and provides the feedback to the user by changing to a specific color depending on the orientation of the place. This feedback is mainly focused on the upper and lower proximity of the femur. Two different methods were developed in the module to indicate the position of the plates at the upper and lower ends of the femur. Each indication method has the same features to show the changes in the orientation of the plate. These methods have red, yellow, and green indicator lights to indicate the changes in the environment whenever there is a misplacement of the plate. In order to maintain accuracy in the system, each indicator in the module is designed with varying dimensions. These dimensions are the primary resource for the color indicators to provide positional feedback to the residents. The color changes are used as an alert system for the surgeon to understand the complexity, difficulties, and the importance of the LISS surgery process. The only allowable dimension is 0.75 cm and is exhibited by a circular indicator in the color of red. If a yellow light is indicated, the resident will know that he or she is close to reaching the ideal dimension. If the red light comes on, it indicates that the positioning is far from the allowable dimensional limits. The developed position training module provides quick response and feedback through the indicators and helps the residents learn quickly and efficiently.

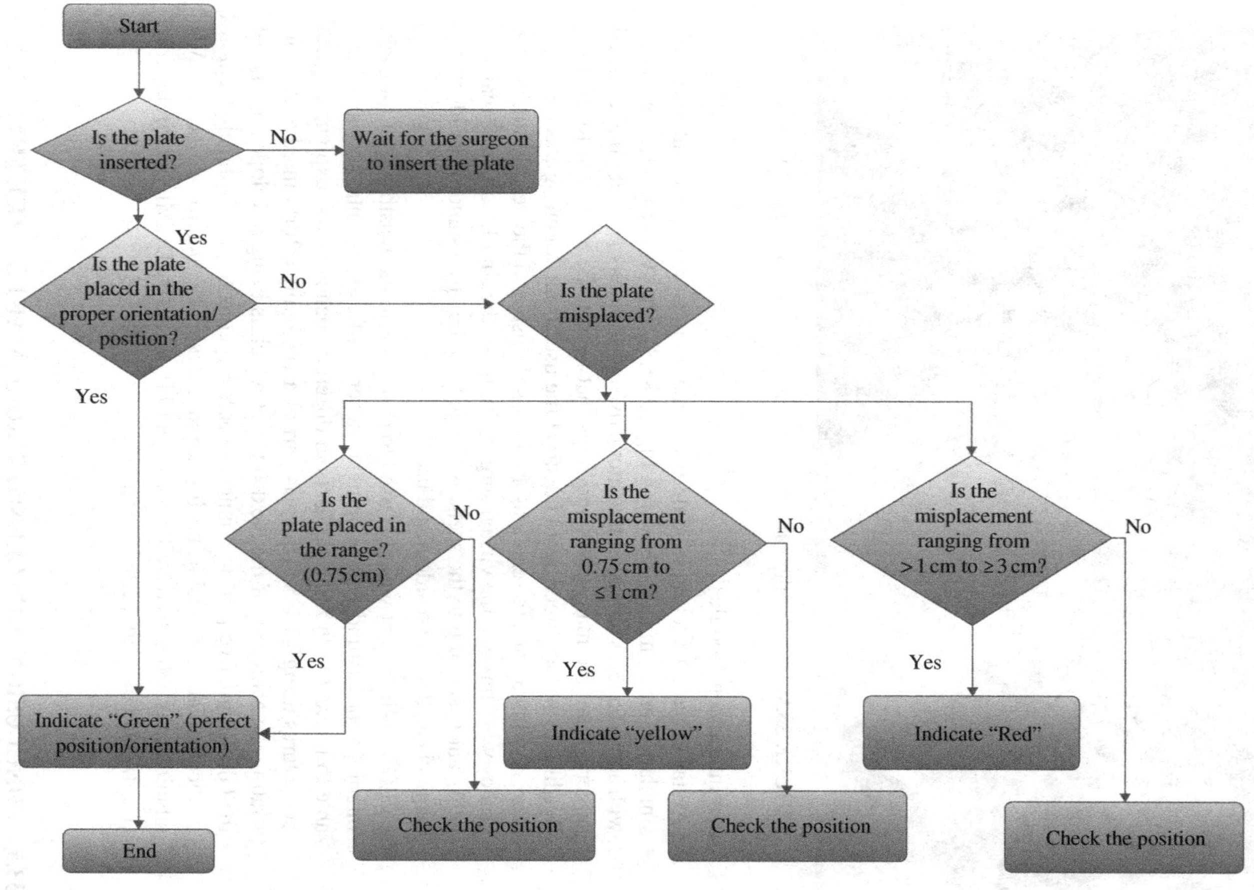

FIGURE 33.7 Flowchart for position training. Courtesy of J. Cecil.

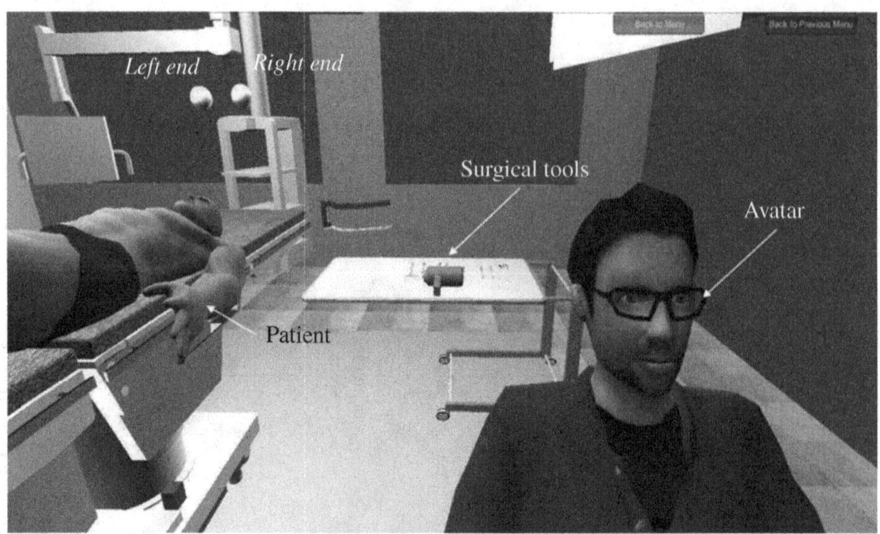

FIGURE 33.8 View of the virtual surgical environment. Courtesy of J. Cecil.

Avatar interaction module

An Avatar-based interaction module is created to assist the user to interact with simulation environment (shown in Figure 33.8). The Avatar model developed with audio features enhances the interactive capabilities of VEOS. In the virtual surgery, there are many parts and steps that need to be explained to the user. Reading from the screen would distract the user from watching a certain step or examining a part; instead a user can simply listen to the description or step as they work through the virtual surgery. To make the simulation more realistic, an Avatar is added into the scene which aids in the understanding of identified surgical steps to the resident and user.

The VEOS environment was also used to demonstrate the feasibility of the Next Internet which is under development. In series of experiments, surgeons were able to interact with multiple users in different locations and explain surgical procedures using VEOS. This was part of a US Ignite effort aimed at demonstrating the evolution of the Next Internet which is being developed as part of the GENI initiative [20] to connect expert surgeons to geographically remote medical trainees, reducing healthcare and medical education costs. The VR simulation is being used by TTUHSC in El Paso, led by Dr. Miguel Pirela-Cruz (Dept. Head of Orthopedic Surgery).

33.5 CONCLUSION: CHALLENGES, ROAD MAP FOR THE FUTURE

Based on the two case studies, a number of issues have been identified for industrial adoption of such IoT-based cyber–physical frameworks.

For the manufacturing case study, the cyber–physical resources collaborated using the IoT cloud framework discussed in previous sections; user inputs were given through the Web; subsequently, assembly plans were generated which were then compared and modified using the VR-based simulation environments; finally, the validated plan was assembled using physical work cells. Several target micro- and meso assembly designs (Figure 33.4) were assembled using the implemented cyber–physical framework; meso/microcomposite part designs were built to study the capabilities of the two work cells; while there is no universally accepted definition of *Meso assembly*, we use the term meso scale to include part sizes greater than 1 mm, with accuracies greater than 25 μm.

Two rounds of validation were conducted. In the first set of collaborative activities (within the United States), users at multiple locations (three locations including Stillwater, Tulsa, and Washington, DC) were able to interact and collaborate on the assembly planning, path planning, and gripping approach activities through the cloud-based framework; they were able to propose, compare, and modify assembly plans which could be visualized and studied using the VR environments. Subsequently, during the physical assembly activities, the monitoring cameras were able to share the progress of the assembly tasks through the cloud-based framework. In the second round of validations, we tested the robustness of the overall cloud-based approach with users in France (ENSAM Aix En Provence) interacting with engineers in Stillwater (United States). This intercontinental demonstration involved supporting collaborative activities including proposing/modifying assembly plans and studying the alternatives using simulation-based environments. This was a milestone achieved as it highlighted the capabilities of the Next Internet across continents.

Orthopedic surgical residents participated in several rounds of validation activities aimed at studying the impact of using Virtual Reality-based simulators for education and training. The procedures performed by the residents were assessed by the senior surgeon using pretest and posttest methods. The performance rating scale was developed in the assessment of the skills demonstrated by residents/surgeons during the training. During the first round of validation (fall 2015), a majority of the residents scored higher in their posttests regarding their knowledge and understanding of the surgical process after using the simulation environment. Modifications to the environment were incorporated including additional modules as well as a haptic interface to the virtual environment. In the later validation activities, all the medical residents scored higher in their posttests compared to pretests. These results have indicated that the VEOS environment has positively impacted the learning of LISS plating surgical procedures. Additional modifications to the environment are under progress based on feedback received from the surgeons and residents involved in this case study.

The IoT-based framework outlined in this chapter is a step toward ubiquitous computing where engineers and users will not be required to have computing resources to accomplish engineering tasks; instead, they will be able to access and use resources in a "cloud" through thin clients to conduct engineering activities. The approach developed uses a cloud-based approach which seeks to make it easier for engineers who may be geographically distributed (and collaborating from different parts of the world) to be able to conduct simulation and physical engineering activities using next-generation Internet technologies.

Cyber–physical frameworks hold significant potential in support agile collaborations in medicine and manufacturing; when customer requirements change, adoption of such cloud-based frameworks enables engineers and manufacturing partner organizations to exchange and interact using thin clients, computers, and other devices that are part of the IoT landscape. The emerging Next Internet being developed as part of the GENI initiative enabled the sharing of data and information among the distributed teams using the IoT frameworks outlined in this chapter.

This chapter outlined IoT-based frameworks to support collaborations among distributed partners in medical and manufacturing contexts. Such IoT-based frameworks are needed to support collaborative engineering and medical activities. Cloud-based approaches can facilitate the collaboration of the various cyber–physical components using advanced cyber infrastructure (related to the Next Internet as part of the GENI initiative). The role of the Next Internet under development in the United States, the EU, and other countries needs to be underscored for the realization of such next-generation smart technology-based approaches for both engineering and medical applications. The major game changer is the availability of ultrafast high gigabit networks which is becoming more available in various cities. For example, Google has adopted several cities for laying out their high gigabit networks (Google Fiber) which will enable a variety of industries and healthcare organizations to benefit from. In the medical context, for example, surgeons can interact with other surgeons and residents and guide them through surgical procedures using such ultrafast networks which will allow the high-definition camera views and VR data to be exchanged faster with less latency; in manufacturing applications, engineers can propose new ideas and study them collaboratively in a concurrent engineering context using VR-based engineering environments. This will enable earlier and effective detection of downstream problems, reducing cost and time in the development of new products and processes. With the advent of this Next Internet, the stage is set for the next revolution in distributed cyber activities where the physical location of experts, users as well as software and physical resources is no longer an issue.

The cyber–physical framework and approaches discussed in this chapter emphasize how such sharing of engineering resources using next-generation Internet technologies can be realized. Such frameworks facilitate the realization of global VEs where collaboration between distributed partners is possible especially when responding quickly to customer requirements or healthcare needs whether it is in manufacturing or medicine. As IoT devices become ubiquitous, such "smart" interfaces and thin clients are expected to play an important role in facilitating collaborations in various fields beyond medicine and manufacturing. The use of cloud technologies also is vital to the success of such IoT-based frameworks; with the popularity of such technologies in industry today, the next wave of manufacturing and medical collaboration is underway which will further enable national and global collaborations.

ACKNOWLEDGMENTS

Funding for the research activities outlined in this chapter was obtained through grants from the National Science Foundation (NSF Grant 0423907, 0965153, 1256431, 1257803, 1359297, and 1447237), Sandia National Laboratories, Los Alamos National Laboratory, the Mozilla Foundation, and Oklahoma State University.

REFERENCES

[1] http://en.wikipedia.org/wiki/Internet_of_Things (accessed August 29, 2016).

[2] Lu, Y. and Cecil, J., An Internet of Things (IoT)-based collaborative framework for advanced manufacturing. *The International Journal of Advanced Manufacturing Technology*, **84**, 5, 2016, 1141–1152.

[3] Atzori, L., Iera, A., and Morabito, G., The Internet of things: a survey. *Computer Networks*, **54**, 2010, 2787–2805.

[4] Tan, L. and Wang, N., 2010, Future Internet: The Internet of Things. In Proceedings of the 3rd International Conference on Advanced Computer Theory and Engineering (ICACTE 2010), Chengdu, China, August 20–22, (Vol. **5**, pp. V5-376). IEEE.

[5] Ashton, K., That "internet of things" thing. *RFID Journal*, **22**, 7, 2009, 97–114.

[6] Houyou, A.M., Huth, H.P., Trsek, H., Kloukinas, C., and Rotondi, D., 2012, Agile Manufacturing: General Challenges and an IoT@ Work Perspective. In Proceedings of 2012 IEEE 17th International Conference on Emerging Technologies and Factory Automation (ETFA 2012), Krakow, Poland, September 17–21, pp. 1–7. IEEE.

[7] Sundmaeker, H., Guillemin, P., Friess, P., and Woelfflé, S., 2010, Vision and Challenges for Realising the Internet of Things, Cluster of European Research Projects on the Internet of Things—CERP IoT. http://www.internet-of-things-research.eu/pdf/IoT_Clusterbook_March_2010.pdf (accessed September 10, 2016).

[8] Gubbi, J., Buyya, R., Marusic, S., and Palaniswami, M., Internet of Things (IoT): a vision, architectural elements, and future directions. *Future Generation Computer Systems*, **29**, 7, 2013, 1645–1660.

[9] Buyya, R., Yeo, C.S., Venugopal, S., Broberg, J., and Brandic, I., Cloud computing and emerging IT platforms: vision, hype, and reality for delivering computing as the 5th utility. *Future Generation Computer Systems*, **25**, 2009, 599–616.

[10] Buckley, J., (ed.), 2006, *The Internet of Things: From RFID to the Next Generation Pervasive Networked Systems*, Auerbach Publications, New York.

[11] Cecil, J., Ramanathan, P., Prakash, A., and Pirela-Cruz, M., 2013, Collaborative Virtual Environments for Orthopedic Surgery. In Proceedings of the 9th Annual IEEE International Conference on Automation Science and Engineering (IEEE CASE 2013), Madison, WI, August 17–21.

[12] Berryman, A., Calyam P., Cecil, J., Adams, G., and Comer, D., 2013, Advanced Manufacturing Use Cases and Early Results in GENI Infrastructure. In Proceedings of the Second GENI Research and Educational Experiment Workshop (GREE), 16th Global Environment for Network Innovations (GENI) Engineering Conference, Salt Lake City, March 19–21.

[13] Wu, D., Thames, L., Rosen, D., and Schaefer, D., 2012, Towards a Cloud-Based Design and Manufacturing Paradigm: Looking Backward, Looking Forward, Vol. **3**, 32nd Computers and Information in Engineering Conference, Parts A and B, Chicago, IL, August 12–15.

[14] Tao, F., Zhang, L., Venkatesh, V.C., Luo, Y., and Cheng, Y., Cloud manufacturing: a computing and service-oriented manufacturing model. *Proceedings of the Institution of Mechanical Engineers, Part B: Journal of Engineering Manufacture*, **225**, 10, 2011, 1969–1976.

[15] Xu, X., 2012, From cloud computing to cloud manufacturing, *Robotics and Computer-Integrated Manufacturing*, **28**, 1, 75–86.

[16] Cecil, J., 2013, Information Centric Engineering (ICE) Frameworks for Advanced Manufacturing Enterprises, Proceedings of Industry Applications and Standard Initiatives for Cooperative Information Systems for Interoperable Infrastructure, OnTheMove (OTM) Conferences, Graz, Austria, September 10, pp. 47–56.

[17] Wang, X. and Xu, X., 2013, ICMS: A Cloud-Based Manufacturing System, *Cloud Manufacturing*, Springer Series in Advanced Manufacturing, Li, W. and Mehnen, J., (eds.), pp. 1–22, Springer Verlag, London.

[18] Deboosere, L., Vankeirsbilck, B., Simoens, P., De Turck, F., Dhoedt, B., and Demeester, P., Cloud-based desktop services for thin clients. *IEEE Internet Computing*, **16**, 6, 2012, 60–67.

[19] Cecil, J., Virtual Enterprises, refereed paper in a book, The Internet Encyclopedia, 2003, Volume **3**, pp. 567–578. http://www.wiley.com/WileyCDA/Section/id-101393.html (accessed September 10, 2016).

[20] Global Environments for Network Innovation (GENI), http://www.geni.net (accessed August 29, 2016).

[21] https://www.us-ignite.org/about/what-is-us-ignite/ (accessed August 29, 2016).

[22] http://www.serverwatch.com/server-tutorials/eight-big-benefits-of-software-defined-networking.html (accessed August 29, 2016).

[23] Liu, J. and Yu, J., 2013, Research on the Framework of Internet of Things in Manufacturing for Aircraft Large Components Assembly Site. In IEEE International Conference on Green Computing and Communications (GreenCom) and IEEE and Internet of Things (iThings/CPSCom) and IEEE Cyber, Physical and Social Computing, Beijing, China, August 20–23, pp. 1192–1196. IEEE.

[24] Bi, Z., Da Xu, L., and Wang, C. Internet of Things for enterprise systems of modern manufacturing. *IEEE Transactions on Industrial Informatics*, **10**, 2, 2014, 1537–1546.

[25] http://csrc.nist.gov/publications/nistpubs/800-145/SP800-145.pdf (accessed August 29, 2016).

[26] Cecil, J. and Chandler, D., Cyber Physical Systems and Technologies for Next Generation e-Learning Activities, Innovations 2014, W. Aung, (eds.), iNEER, Potomac, MD, 2014.

[27] Benefits of Cloud Computing, http://www.mbtmag.com/articles/2013/05/how-manufacturers-can-benefit-cloud-computing (accessed August 29, 2016).

[28] Tao, F., Cheng, Y., Xu, L.D., Zhang, L., and Li, B.H., CCIoT-CMfg: cloud computing and Internet of Things based cloud manufacturing service system. *IEEE Transactions on Industrial Informatics*, **10**, 2, 2014, 1435–1442.

[29] Tao, F., Zuo, Y., Da Xu, L., and Zhang, L., IoT-based intelligent perception and access of manufacturing resource toward cloud manufacturing. *IEEE Transactions on Industrial Informatics*, **10**(2), 2014, 1547–1557.

[30] Cecil, J., https://vrice.okstate.edu/content/gigabit-network-and-cyber-physical-framework (accessed August 29, 2016).

[31] Lu, Y. and Cecil, J., An Internet of Things (IoT) Based Cyber Physical Framework for Advanced Manufacturing. *On the Move to Meaningful Internet Systems*, Lecture Notes in Computer Science **9416**, I. Ciuciu, H. Panetto, C. Debruyne, A. Aubry, P. Bollen, R. Valencia-García, A. Mishra, A. Fensel, F. Ferri, (eds.), Springer Series, New York, pp. 66–74.

[32] http://www.devonit.com/thin-client-education (accessed August 29, 2016).

[33] Cecil, J. and Xavier, B., 2001, Design of an Engineering Enterprise Modeling Language (eEML), Technical Report. Virtual Enterprise Technologies, Inc. (VETI), Las Cruces.

[34] Cecil, J., TAMIL: an integrated fixture design system for prismatic parts. *International Journal of Computer Integrated Manufacturing*, **17**, 5, 2004, 421–435.

[35] Cecil, J., A clamping design approach for automated fixture design. *International Journal of Advanced Manufacturing Technology*, **18**, 11, 2001, 784–789.

[36] Cecil, J., A functional model of fixture design to aid in the design and development of automated fixture design systems'. *Journwal of Manufacturing Systems*, **21**, 1, 2002, 58–72.

[37] Cecil, J and Jones, J., An advanced virtual environment for micro assembly. *International Journal of Advanced Manufacturing Technology*, **72**, 1, 2014, 47–56.

PART V

IoT/DATA ANALYTICS CASE STUDIES

34

DEFRAGMENTING INTELLIGENT TRANSPORTATION: A PRACTICAL CASE STUDY

ALAN CARLTON[1], RAFAEL CEPEDA[1] AND TIM GAMMONS[2]

[1] *InterDigital Europe Ltd, London, UK*
[2] *ARUP, London, UK*

The authors acknowledge the support of Innovate UK by providing funding for the oneTRANSPORT project and the oneTRANSPORT partners for their input and insightful discussions.

34.1 INTRODUCTION

The Transport industry is at a crossroads: embrace the future or try to preserve a status quo. Today, many industries are on a journey from closed data silos to the open possibilities of the Internet of Things (IoT). This legacy of silos has resulted from many decades of highly custom projects, protocols, and integration efforts, which have created a world of isolated islands of functionality. This approach is not sustainable in a world where increasingly profits and perceived value are moving to services and data.

The Transport industry has made good strides in the right direction. However, transport sector initiatives are not directed to opening data. Instead, to date it has been about defining protocols and technology to support the management of closed transport data and *control center* applications. Likewise, opening up existing data without addressing opening up the systems and the market will limit the use of that data in a geographically scalable and sustainable fashion. The value of the IoT lies in moving away from this single-purpose-asset paradigm to a multiuse disruptive one. Furthermore, it is not good enough if just one party opens up data for their own

Internet of Things and Data Analytics Handbook, First Edition. Edited by Hwaiyu Geng.

Companion website: www.wiley.com/go/Geng/iot_data_analytics_handbook/

benefit and ends up creating yet another silo. Instead, what is needed is a truly *open marketplace for data*, where a multitude of stakeholders can play, innovate, and benefit.

34.2 THE TRANSPORT INDUSTRY AND SOME LESSONS FROM THE PAST

The Transport industry of today parallels the telecommunication industry 30 years ago. In the 1980s, the telecom industry was characterized by national monopolies (or oligopolies) of local champions and manufacturers. This strangled innovation and growth due to separate national standards, small markets, poor economies of scale, and high prices. This did not ebb until after progressive acts of deregulation and the advent of standardized global technologies—like Global System for Mobile Communications (GSM) digital mobile phones—and prior to the consolidation, and massive growth, of the telecom industry.

Two key factors triggered change in the telecom industry: one of technological and one of regulatory nature. Technological change allowed the advent of affordable mobile phones, providing new capabilities that found a latent mass-market demand. Later on, many of the traditional telecom manufacturers struggled in accepting the need for change when Apple and Google introduced their new operating systems, iOS and Android, and their App stores. Long-standing industry leaders, such as Nokia and Motorola, blindsided and convinced of their traditional long-held views could not compete. Both companies are no more.

Regarding regulation, European integration and the emergence of the Single Market in 1992 were the key drivers for the GSM digital mobile phone standard. Europe-wide market opportunity encouraged the industry investment that drove the necessary technological advances. Countries that embraced this change by deregulating its telecom industry, 5 years or more ahead of other nations, emerged as the winners. With the United Kingdom as the change leader in Europe, the combination of early deregulation and support of this technology change resulted in many major telecom companies from Japan and the United States, making very substantial research and development (R&D) and manufacturing investments in the United Kingdom in the 1990s.

A key difference between the transport management and the telecom industry is that while both started as state-owned regulated industries, one has seen privatization and deregulation across the world and the other has remained state owned and managed. This, in part, reflects the slow and growing importance of transport management as an essential public service and the historic delegation of this to multiple disparate Local Authorities. As a consequence, and as summarized in Table 34.1, there is a stark contrast between transport and telecom industries today.

The transport sector is also facing technological and regulatory issues that today are leading to multiple proprietary solutions to common needs. For instance, a local authority transport manager, seeking a solution for parking, is faced with a choice of over a dozen different solutions. All these solutions offer similar capabilities, but are

TABLE 34.1 Comparison of Transport and Telecom Industries Today

Issue	Transport	Telecom
Market nature	Geographically fragmented (regional, not even national)	Global
Overall market size	Low volume	High volume, massive
New service development	Slow, costly	Very fast, third-party developer markets
Product types	Proprietary, bespoke, niche solutions	Standardized solutions
Pricing	High price	Low price, economies of scale
Competition	Low competition	High competition

Courtesy of InterDigital. Use with Permission.

implemented in different ways, with different benefits, costs, and, perhaps most importantly, incompatible systems across vendors and even across same-company products.

With this in mind, the next logical step is to explore the evolution of the transport industry and apply lessons from the telecom industry to revolutionize the transport experience for its users. A novel and innovative solution using the emerging global IoT standard oneM2M [1] is later presented in this chapter as the tangible creation of an open marketplace for data to benefit citizens, cities, and stakeholders.

34.3 THE TRANSPORT INDUSTRY: A LONG ROAD TRAVELED

The earliest forms of automated traffic systems were deployed in the 1920s in New York and Detroit. These were isolated traffic signals deployed at intersections to improve road safety. Over the years, these isolated traffic control systems were enhanced with the use of microprocessors in the 1950s and early system integration in 1970s. These emerging coordinated traffic control systems were known as Urban Traffic Control (UTC) in the northern hemisphere and Area Traffic Control (ATC) in Asia and the southern hemisphere. In the 1980s, the Transport and Road Research Laboratory (TRRL), then part of UK Government, independently developed traffic responsive coordinated systems in cooperation with other governmental and industry players. The main objective of this system was adjusting the traffic signal settings (cycles, green splits, and offsets) by real-time optimization of an objective function, such as minimizing travel time or stops, based upon estimates of traffic conditions. This system later morphed into two competing proprietary solutions known as Split Cycle Offset Optimization Technique (SCOOT) in the United Kingdom and Sydney Coordinated Adaptive Traffic System (SCATS) in Australia.

SCOOT and SCATS have been successfully deployed in many cities around the world. The SCATS was developed to deal with corridor management and capacity problems along routes radiating out from a central core—a feature typical of many Asia-Pacific road networks. SCOOT, on the other hand, is a real-time modeling-based system, which manages a network as a whole and distributes "green time" to optimize overall queue lengths and congestion—a feature typical of European cities, especially in the United Kingdom. Both systems, however, can be adapted to function in different environments.

Also in the 1980s, Japan developed a coordinated traffic system with responsive capability. Although the United States had a number of R&D projects, it was generally accepted by the US traffic engineers that fixed-time plan systems were adequate. Thus, transport control products were developed by Japanese and US firms independently using national technology norms with very little global cooperation or standardization.

Through the 1990s, only a handful of traffic system suppliers survived worldwide due to the limited market size and the high up-front cost for new entrants. By the mid-1990s, there were only about 400 traffic systems deployed worldwide. With a life cycle of 10–15 years, the demand for new systems was also limited. Meanwhile, SCOOT and SCATS systems continued to be highly proprietary and fiercely defended by suppliers (e.g., Siemens) and government organizations (e.g., Roads and Maritime Services, NSW). This proprietary model was also common with Japanese suppliers, while in the United States some transport authorities had more success integrating products from different suppliers. The Chinese traffic industry followed the lead of international manufacturers' products, which enabled them to circumvent some integration issues. However, many of the more sophisticated features of international products did not fully function with the Chinese equipment.

By the mid-1990s, the urban traffic systems market was in a "technical cul-de-sac." The technology was outdated and traffic data was not accessible outside of the proprietary systems. Having invested over many years, the major suppliers saw little benefit in opening their systems and were therefore reluctant to embrace change. At around this time, the urban traffic management and control (UTMC) concept was developed in the United Kingdom. With sponsorship from the Department for Transport (DfT), UTMC aimed to provide a way forward for the UTC market to meet current needs and to provide a platform for future competitive development.

The development of UTMC brought about a much needed change and moved the traffic industry forward with benefits such as the introduction of IP communications, rich data mapping, geographic information system (GIS) interfaces, and interoperability. This also led to some competition among peripheral device manufacturers such as those for CCTV cameras. Other suppliers also developed new functionality such as journey time measurement and interfaces with interurban systems. Nowadays, however, UTMC remains a capital-intensive facility funded entirely by Transport Authorities forced to purchase additional enabling features from a closed set of traffic equipment suppliers.

DATEX II has also been developed in Europe to offer a framework that defines an open and standardized data format and a model for transport data exchange between traffic centers, traffic operators, and media partners. Most UTMC systems deployed today support DATEX II adaptors.

While transport systems have moved to some open interfaces, it is still the Traffic Authority that remains referred to as the "end user" in deployed systems. Hence, they determine what the consumer or "citizen end user" actually requires. In contrast, other industries such as the mobile communication sector have seen tremendous growth through the globalization of standards, opening up of data, and the enablement of market forces (e.g., application stores) in driving the definition of end-user needs.

The intelligent transportation systems (ITS) industry is similarly trying to respond and adapt to the emerging market for open transport data. Yet, the proposed mechanisms for the release of data tend to follow traditional models (e.g., a licensed one-way data stream and selected data sets rather than full release). Suppliers are also understandably protective of their systems since they charge for the ongoing support, product enhancements, and other services.

While traditional industry suppliers had viewed their own customers, the Local Authorities, as the end user, the early 2000s saw other players outside of the traditional traffic industry recognize the opportunity to empower the traveling public—the era of user-driven transport applications had begun. The challenge that now needs to be addressed is a wider integration across the whole transport sector for the benefit of the complete citizen end-user experience.

The first steps in this journey came with the arrival of commercial satellite navigation devices. This was fundamentally enabled by the US government decision to allow access to the higher-precision GPS satellite codes previously restricted to military use only. The potential of a mobile phone to evolve into a handheld mobile Internet device, or smartphone, was pioneered by Nokia with the incorporation of GPS chips in their high-end phones along with map functionality. Unfortunately, small screens and poor user interfaces rendered them a poor competitor for GPS devices. Shortly after though, Google recognized that mobile Internet access could enable a new level of local search capabilities, serving a real customer need, while also enhancing the value of their existing product offering. This marked the advent of real-time mapping services. The subsequent introduction of the iPhone presented a much easier-to-use interface replacing the traditional layered menus used by Nokia and other phone manufacturers at the time. The second-generation iPhone, launched in 2008 and complemented with the Apple's App Store, enabled easily downloadable applications starting a full revolution in the telecom industry. This chronologic timeline is illustrated in Figure 34.1.

The new App store paradigm opened the door for empowerment of end users. Early applications were simple downloadable city guides, transport timetables, and navigation. Over time, transport applications became more sophisticated evolving from the use of static to dynamic data. Within the United Kingdom, government support for the opening up of publicly owned data has been an important factor in facilitating this movement. The launch of data.gov.uk in 2010, which included access to the National Public Transport Access Nodes (NaPTAN) database, was perhaps one of the earliest steps. Availability of such data enabled public transport apps to quickly evolve from simple timetables to delivery of real-time service updates (e.g., location) to users for specific buses and trains. Combination of multiple data sources enabled real-time comparison of different mixed transport modalities, for example, Citymapper. Mass usage navigation apps, such as Google Maps, have also evolved to

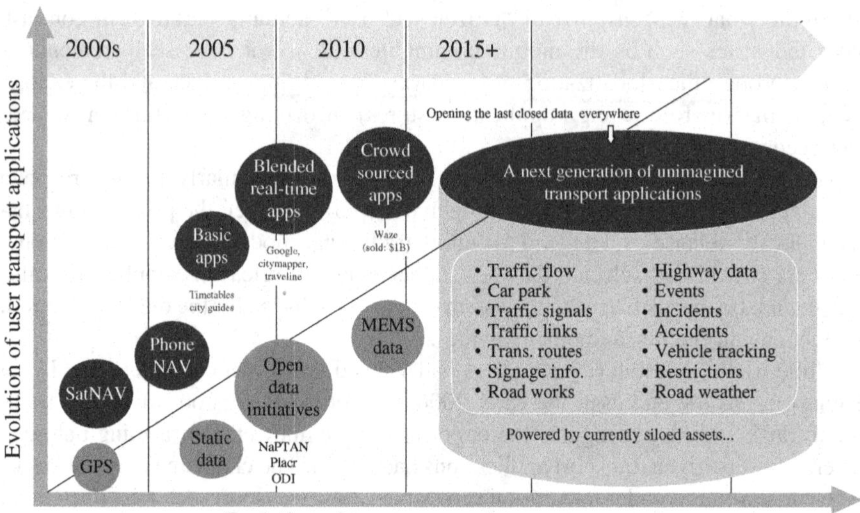

FIGURE 34.1 Evolution of transport applications and transport data assets. Courtesy of InterDigital.

incorporate such capabilities. Another major advance was the launch of the London Datastore (LDS) later in 2010, making much of the new transport data available free of charge to developers for commercial use. This was followed by the creation of the Open Data Institute (ODI) in 2012.

In parallel, with the emergence of Open Data and new applications, smartphone hardware continued to evolve. This included more sophisticated sensor capabilities—compass, gyroscopes, accelerometers, magnetometers, multiaxis inertial measurement units, and others. This evolution enabled devices to measure location, movement, and other factors more precisely, which effectively provided GPS-like performance even when outside of the GPS coverage range. Crowdsourcing of location data, speed, and traffic incidents has leveraged such capabilities notably led by the Israeli start-up company Waze, acquired by Google in 2013.

34.4 THE TRANSPOPRT INDUSTRY: CURRENT STATUS AND OUTLOOK

While much of the transport data has been unlocked in the UK's largest cities, much more still remains closed and inaccessible. The LDS, for example, offers access to a wide range of historical data as well as live feeds of traffic cameras, traffic disruptions, etc. Live access to detailed raw data on individual traffic signals (e.g., SCOOT data), and many other UTMC input data, remains problematic in large part due to closed system design. Moves toward a more open and integrated approach are starting to take off. However, without an integrated approach, there remains a large risk of maintaining the status quo, where individual Transport Authorities adopt different technology paths, an approach that puts dominant suppliers in a strong position to exploit this fragmentation. Currently, given trends in Local Authority funding, such approaches are just not sustainable in most countries across the globe.

In summary, there are two clear stories emerging:

1. The first driven by geographically fragmented Local Authorities' legislated responsibility to manage traffic—which has resulted in an industry cul-de-sac of high-priced, customized systems in an era of diminishing financial budgets
2. The other driven by the telecommunication and IT industries and by the Government-sponsored trend toward Open Data, which has resulted in a mass market for smartphone applications that end users want and for which many are willing to pay

At first glance, one might envision that these two stories perfectly complement each other, with the second being the solution to the problem of the first. While in one sense this is true, real challenges remain in large part due to conflicting cultural perspectives, experiences, and goals of different industries and communities. Some important transport-specific and time-specific requirements need to be highlighted:

- *Local responsibility and accountability* for Transport Management needs to be retained in order to account for the specific local needs when undertaking transport planning. However, the activities within adjacent counties and authorities and mutual impacts are not to be ignored. Furthermore, other local nontransport-related activities are becoming increasingly relevant. For example, the emergence of the Smart Cities concept and the opportunity to harness data across different conventional silos to deliver local benefits and improvements.
- *Local authority trends* related to funding in all areas including transport. Together these have contributed to a constrained market and deployment of ITS products in the United Kingdom and other countries.
- *The role of data* in creating new economic value and the subsequent adoption of a proactive policy in regard to Open Data are important. Local Authorities are being encouraged, at least in the United Kingdom, to make data publicly available [2]. Thus, Authorities are increasingly recognizing the market potential of the data assets they have in their possession.
- *The challenge of opening up data* as some authorities have been approached by major Web players wishing to acquire access to their transport data. For a Local Authority this is both an opportunity and a challenge. It represents an opportunity to secure new income, but it is a major challenge as well for at least three reasons:
    - Valuing the data—Quantifying the value of data is difficult as in many cases the true value is not yet known by a seller or purchaser.[1]

[1] A similar situation arose when Technology Transfer Offices (TTOs) were introduced at Universities to increase research commercialization. Many TTOs initially hugely overvalued IPR, with very limited University income generated as a result—the opposite of the intention of their establishment.

- ◦ Skilled staff—Local Authorities have limited staff resources capable of drafting appropriate contracts, as exemplified in Partridge Report [3], which revealed how poorly managed licensing contracts had resulted in both inappropriate control and release of personal health data through lack of understanding.
- ◦ Commercial staff—Multiple users are emerging for Local Authority data. Even if the skills were there, the time taken to negotiate multiple low-value contracts will require more commercial resource than are available.

These issues will be even further magnified when considering that Local Authorities face similar issues regarding their data assets from other public sectors.

"Open Once, License to Many" inherently offers a simple solution to these issues. The platform model in the real use case discussed in the next section is one of Publish/Subscribe, whereby standard contracts can be developed with suppliers and with consumers of data to enable significant savings both in staff resources and proper control of data assets. This model will also offer benefits to larger Web players who also prefer to negotiate a single contract with platform providers for all the available data. For example, this was the case in the deal between Google and ITO World [4].

It is clear that time is running out for the existing outdated business models. The reduction in public funding and the desire for independent Application Developers to solve the problems in the transport industry are two strong factors in driving this forward. The traditional assumption that governments have no alternative but to pour more money into the management of transport solutions may continue to hold true to a degree, but this is simply a recipe for survival, not for growth—which is what is needed to address the challenges of more increasing population densities and increased traffic. A novel and innovative solution that encourages such growth is discussed next.

34.5 USE CASE: oneTRANSPORT—A SOLUTION TO TODAY'S TRANSPORT FRAGMENTATION

oneTRANSPORT is an Innovate UK-supported [5] project initiative in the area of "Integrated Transport–In-Field Solutions." The project addresses both immediate and anticipated future challenges facing the transport industry (e.g., congestion, shrinking Local Authority budgets, end of subsidies, etc.). These challenges are complex and interdependent. Today, oneTRANSPORT involves 11 cross-sector partners including five Transport Authorities (data owners and use case providers), a technology platform provider, a transport industry specialist, and four transport sensor/device manufacturers and transport analytics providers.

The oneTRANSPORT solution is based on oneM2M, an international standard to truly enable the IoT revolution (Figure 34.2). This standard was created with the singular intention of breaking down industry and data and technology silos and enabling the possibilities of the IoT. oneM2M is a *Service Enablement Layer* standard whose core functionality is to provide a set of service capabilities that enable manageable data sharing for new services and application development from multiple parties.

FIGURE 34.2 oneM2M service enablement layer international standard. Courtesy of InterDigital. Use with Permission.

oneM2M is based on a "store and share" paradigm. Hence, interested applications are notified about resources and relevant changes by means of subscription. Access rights, usage, and privacy are ensured through the defined service capabilities.

oneTRANSPORT is unique and complementary to existing transport technologies (e.g., UTMC, DATEX II). It enables new layers of functionality to address challenges from the transport industry by opening up the data, on a multicity scale, that many Local Authorities have locked behind existing systems. The data is made accessible to the innovative power of the application development community in a fair and equitable manner for all stakeholders. oneTRANSPORT will enable new transport applications to be highly transferable, delivering both local and interregional impact. The challenge here is doing this in an economically attractive manner to Local Authorities. To this end, oneTRANSPORT defines an innovative cloud-based model of brokers that enable an "open once, sell to many" vision.

oneTRANSPORT, by virtue of its approach, offers the opportunity to trigger a transition in the transport management industry to a standard-based ITS market with associated economies of scale. The oneM2M standard defines a set of RESTful[2] APIs[3] and protocol interface spanning devices (e.g., traffic sensors, actuators), gateways, oneM2M servers, and application servers. As a service enablement standard, oneM2M is agnostic to the underlying connectivity and can operate over wired or wireless systems, standards-based or proprietary, as would be found in different geographical regions. The oneM2M standard has been in development for several years. There are currently more than 300 participating partners and member organizations from all around the world. Much of oneM2M core definition was derived from previous work in European Telecommunications Standards Institute (ETSI). In 2012, the standard transitioned into a joint[4] activity involving all of ETSI's regional

[2] Representational state transfer (REST) is a standardized way to create, read, update, or delete information on a server using simple HTTP calls (the foundation of data communication for the World Wide Web).

[3] APIs = Application Programming Interfaces, standardized ways in which software entities interact/communicate.

[4] The bodies involved in standardizing oneM2M are ETSI (EU), ATIS and TIA (America), ARIB and TTC (Japan), CCSA (China), and TTA (South Korea). These same bodies collaborated to create the mobile broadband 3G and 4G standards, which today have greater than two billion users, 30% of the global population (cf. 22% in 2012), and a 75% penetration in the developed world.

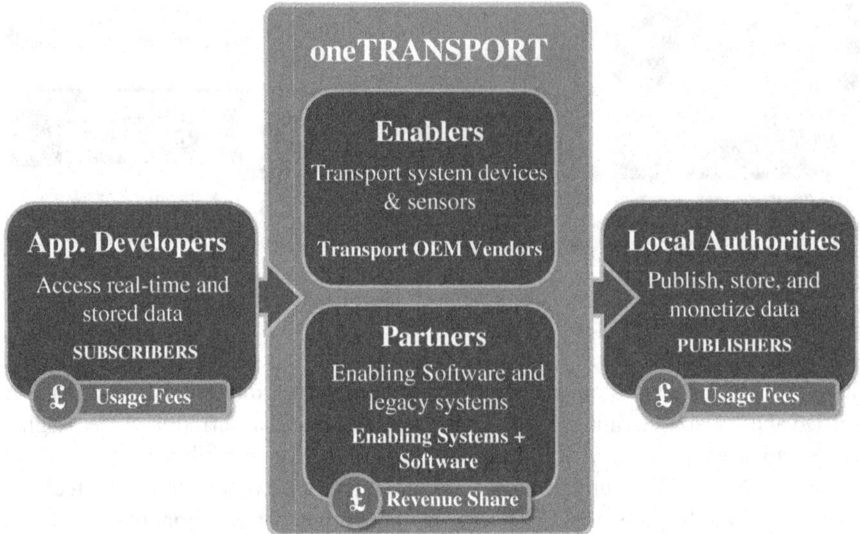

FIGURE 34.3 oneTRANSPORT platform vision. Courtesy of InterDigital. Use with Permission.

counterparts across the United States, Japan, China, and South Korea. The first formal release of the standard was in February 2015.

The oneTRANSPORT platform vision is illustrated in Figure 34.3. This vision is simple: to create an open marketplace for data, analytics, and software enablers through the power of the oneM2M open interfacing. oneTRANSPORT defines a platform that allows Local Authorities as well as third-party software vendors to monetize the array of data assets that their technologies provide. The open interface approach implicitly avoids vendor lock-in issues—hence allowing the proliferation of an open ecosystem that in time is expected to expand well beyond transport.

The *oneM2M Service Layer* provides the "core system services" of oneTRANSPORT and full-featured M2M Service Delivery Platform (SDP) functionality. This includes interworking of transport data assets from other systems, event management, gateway services, device management and discovery, configurable charging, filtering and semantic services, and so on. All data that is made available to higher-layer entities flows through, and is exposed by, the oneM2M API. This layer supports a RESTful architecture, allowing higher-layer entities to register for event notifications and other supported services. *oneTRANSPORT* offers new "premium data service capabilities" to Application Developers. There may be multiple embodiments of software services, from different vendors, designed for the same or similar purposes in this layer. Two categories of premium service capabilities are currently anticipated:

1. *Transport data analytics* that are predictive and/or simple analytics functions. There is no inherent limit to the number of analytics engines and providers that can be hosted by oneTRANSPORT.

2. *Transport application enablers* that provide core application enablement software and API support to higher-layer applications. These software entities typically encompass functionality to more than one higher-layer application.

oneTRANSPORT may be used to enhance and expand the value of current applications such as Citymapper or Waze through integration at the oneTRANSPORT API. It should be noted that today's applications such as Waze are *smartphone data-centric* applications (i.e., relatively simple). oneTRANSPORT will open up a new world of *machine data available everywhere*, enabling a whole new category of possibilities and expansions.

The following use case exemplifies the "predictive" experience that will be possible when this world of fluid data and competitive analytics is realized. Below is an excerpt of a future day in the life of the oneTRANSPORT end user:

> *The Travel Avatar Experience*—Joe woke to the usual sound of his smartphone alarm. After grabbing his phone from his bedside table, he smiled at the message that greeted him. "Turn over Joe, the way things look right now you might be better leaving about 15-minutes later." Joe really like this new application—it really worked! Over the years, he had crowded his phone with countless travel apps that never really delivered. This app did and it made a difference. It brought everything together in a uniquely personal experience: buses, trains, cars, incident delays, car shares, and even parking information near his usual destinations. He had challenged it a few times, but had long since learned not to waste his time because it was always right. Most days he would take his car and the app would guide him to the cheapest (and available!) parking spots near his office. Other times, it would recommend that he take the bus or a train. As long as he never forgot his phone, the app just seemed to know what he was doing and the awards system provided him free trips on regular occasions. Furthermore, the longer he used this app, the adverts in the background just got better and more useful. He had actually seized on some of the ideas suggested to him for lunches, coffees and even a new gym that had opened close to his work.

34.6 oneTRANSPORT: BUSINESS MODEL

Today, there are numerous physical and virtual data sources that might be leveraged in any number of ways to drive intelligent transport decisions. However, the highly fragmented nature of these data sources makes it nearly impossible for an open application ecosystem, with all its benefits, to emerge in any meaningful way. oneM2M provides a solution to this problem in its core value proposition by defining a standardized horizontal service layer. This common service layer allows the exposure of data sources via a standardized API, which enables easier application creation, ecosystem proliferation, and high scalability.

By introducing a new "layer" of functionality to complement traditional and existent technologies, oneTRANSPORT will open up transport data assets in a monetizable manner going beyond simply Open Data. The impact of oneTRANSPORT will be nothing less than the completion of the value chain up to the "citizen end user."

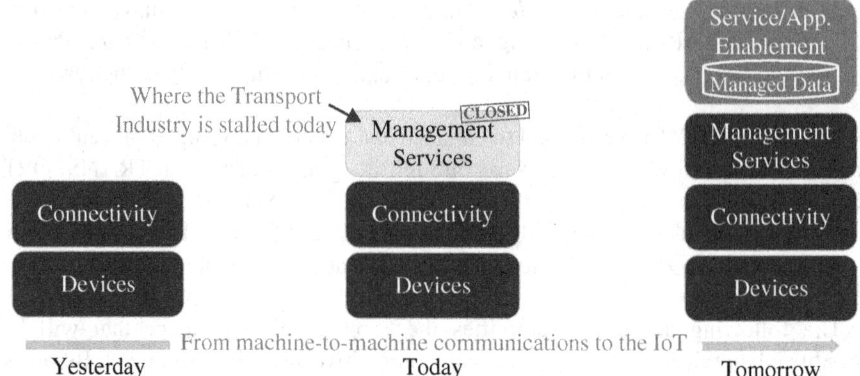

FIGURE 34.4 oneTRANSPORT opens data assets once to many users. Courtesy of InterDigital. Use with Permission.

This will open up a new level of possibilities in transport application experience following the structure in Figure 34.4.

The evaluation of oneTRANSPORT as a business proposition requires a thorough analysis of what costs are actually involved and what strategies can be used to deliver revenues and long-term profitability. While some technology propositions may have the potential to deliver integrated transport and reduce congestion, if they cannot deliver this in a practical and profitable way for Transport Authorities, it has no value.

The proposed business model of oneTRANSPORT is envisioned in three evolutionary phases. These phases have a central player, a *oneTRANSPORT Broker*, orchestrating the services offered by the solution. The oneTRANSPORT broker role can be occupied by anybody: a Local Authority, a technological provider of the platform, or an appointed third-party manager as a business expert in the area. This central player may change from phase to phase depending on exploitation agreements and the competitive environment that develops. Figure 34.5 outlines the evolution from today's status quo, with Local Authorities solely supporting all their transport solutions, to a oneTRANSPORT-based system with multiple data brokers, IoT platform providers, app and analytic vendors, and data owners. This transition is enabled by the three phases described as follows:

Phase 1: In-field trials—This initial phase is a 2-year trial starting in late 2015 and characterized by Local Authorities opening up their data assets to the software development community through a oneTRANSPORT broker. This phase is also characterized by the integration of data assets, installation of new sensors, development and testing of mechanisms for usage and charging, development of embedded analytics, creation of an exemplary software application, and the deployment of initial data filtering tools. Since the revenue stream is expected to be low during this phase, the initial system development will be covered by funding from a private–public partnership.

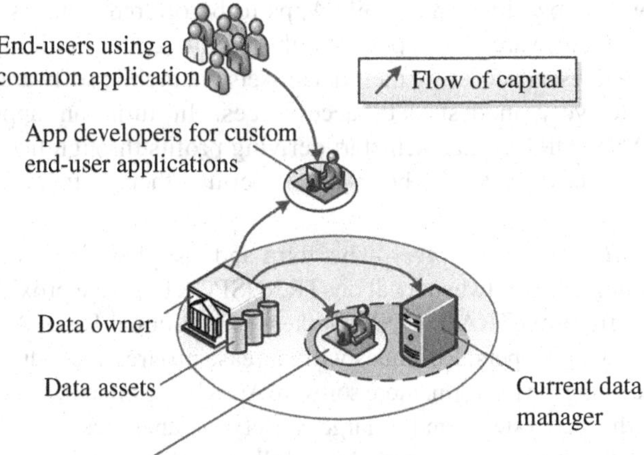

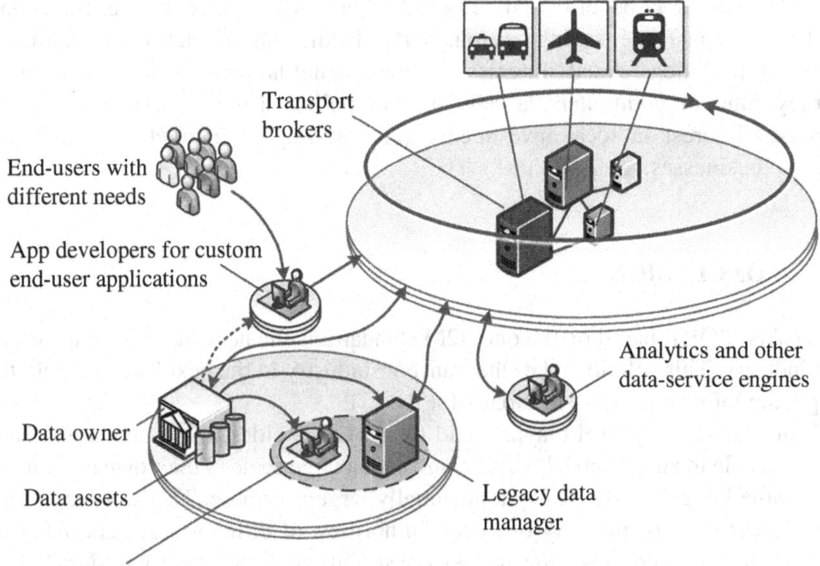

FIGURE 34.5 Business model evolution of oneTRANSPORT. Courtesy of InterDigital. Use with Permission.

Phase 2: Initial commercial stage—During the second phase, or a 1-year initial commercial rollout, a more comprehensive Data-as-a-Service (DaaS) scheme will be offered to Application Developers with both enhanced analytics and premium data offerings. This phase will create an open marketplace for data, which will allow different mobile Apps to be offered to users from a large number of software developers. At this point, Local Authorities will see profits from users accessing their data assets via mobile software applications subject to very modest data access fees. In turn, an appointed *local* oneTRANSPORT broker will start deriving profits through licensing fees. A portion of these fees will be used to recruit other partners and grow the ecosystem.

Phase 3: Full commercial stage—This third and last phase is characterized by a tighter integration between local oneTRANSPORT service proxies and service offerings from oneTRANSPORT brokers to additional Local Authorities and application developers. Revenues will increase progressively due to more data users accessing the platform, more software Apps being offered, external data users joining the ecosystem, and a large variety of analytics and premium data services. This phase is envisioned as a fully scalable business solution with an extensive rollout as a result of low expected investment from counties and a competitive market.

Additional direct benefits to the Local Authorities will increase over time coming from, for instance, savings in transport subsidies and reduction in health services for pollution-related illnesses. Indirect benefits will result from reduced journey time of commuters, a reduction in pollution-related expenditure, and increased interest in local investments and thriving job market for small and medium businesses.

34.7 CONCLUSION

oneTRANSPORT, based on the oneM2M standard, is the next step-up for the transport industry. This solution takes the transport industry to the next level and into the deep realm of possibilities envisioned for the IoT.

Transport is not a level playing field today. Large cities, like London and New York, are able to autonomously drive change on a large scale in their own systems by using large budgets and offering a potentially large user base. This story is not the same across the vast majority of Local Authorities, of different sizes and budgets, who are facing ever-increasing budget constraints and "do-more-with-less" challenges. Even within the larger city economies, there are limits to what can be achieved in a truly scalable fashion. Time and time again, history has shown us that in order to achieve great change and impact, scale and standards are required.

The Transport industry needs to change in order to address the range of challenges that it currently faces in a holistic manner. The benefits of such changes could be enormous and may include: much needed new revenue streams for Local Authorities, significant socioenvironmental improvements, and the realization of a truly internationally integrated transport system enabled by a fair and ubiquitous open data environment.

To realize these benefits, oneTRANSPORT is delivering the next generation of integrated transport solutions building on the following principles to ensure readiness (and a winning position) for the inevitable IoT future:

- *Avoid proprietary systems*—to reduce the risk of downstream vendor lock-in.
- *Use globally standardized technology*—to open up the global market.
- *Support multiple existing infrastructures, rather than being "clean slate" solutions*—to ensure operation across diverse systems and multiple geographies.
- *Involve multiple Transport Authorities and ecosystem players*—to achieve and demonstrate the preceding points.
- *Deliver new sources of revenue into the transport ecosystem*—so that Local and Transport Authorities find the solution (at a minimum) cash neutral and hence create rapid global adoption.
- *Encourage solutions that facilitate new, open, competitive markets*—to create opportunities for both the existing transport industry players and new entrants.
- *Encourage Authorities to open up their transport data*—by creating new possibilities and revenue streams for them.
- *Employ an open, standardized, interface for new transport sensors*—to support and make use of data from future IoT sensors, which will be deployed outside of transport (future proofing), and to encourage interface standardization (lower costs) among transport sensor and system suppliers.
- *Support services and applications from other emerging systems*—to maximize value and interoperability of solutions.
- *Provide measures of the modal shift effectiveness of services*—to allow their comparison.
- *Enable Local Authority services beyond transport*—to support the next moves to smart counties and to the future "smart world."

oneTRANSPORT is not another industry, data, or technology silo. A fundamental premise of this solution is the integration and discoverability of multiple data assets, platforms, vendors, and systems. This integration process will be established through multiple open events to share knowledge and resources among interested stakeholders, which include Transport Authorities in any country, other platform providers, analytics and device vendors, software developers, universities, citizens, and so on. This integrative project is thus a real opportunity to unleash the true value of the IoT.

ACKNOWLEDGMENT

With special thanks to Vanja Subotic, InterDigital, for her leadership and many useful editorial inputs.

REFERENCES

[1] oneM2M is a trademark of the Partner Type 1 of oneM2M. See http://www.oneM2M.org (accessed August 31, 2016).

[2] “Release of Data Fund”—see https://www.gov.uk/government/news/15-million-funding-to-open-up-public-data (accessed August 31, 2016).

[3] “Review of data releases by the NHS Information Centre.” Sir Nick Partridge, June 17, 2014. http://www.hscic.gov.uk/datareview (accessed August 31, 2016).

[4] http://www.bbc.co.uk/news/technology-27394691 (accessed August 31, 2016).

[5] connect.innovateuk.org (accessed August 31, 2016).

35

CONNECTED AND AUTONOMOUS VEHICLES

LEVENT GUVENC, BILIN AKSUN GUVENC AND MUMIN TOLGA EMIRLER

Ohio State University, Columbus, OH, USA

35.1 BRIEF HISTORY OF AUTOMATED AND CONNECTED DRIVING

Early work on autonomous vehicles can be traced back to work on autonomous operation of mobile robots motivated by the need to use such devices in unmanned space exploration. An autonomous mobile robot is a completely independent unit with its own power source, actuation system, an array of sensors to perceive the environment, and a computing system used for decision making. Research on mobile robotics has, thus, been inspirational for autonomous road vehicles and automated driving. However, it is not possible to directly jump from autonomous mobile robot technology to automated vehicles operating on roads. There are several reasons for this. First of all, a mobile robotic platform is not designed to operate on roads, especially on highways. Any vehicle that is operated on roads has to pass rigorous certification tests called homologation which no mobile robotic platform will be able to satisfy. Secondly, while mobile robotic platforms can stop and wait to assess the current situation, this is not possible in the case of an automated vehicle which cannot arbitrarily stop in the middle of the road, thus obstructing traffic. Early researchers working in the area of autonomous road vehicles have concentrated on low-speed driving in closed test areas for this reason. This is a very significant issue which has been a differentiating factor in successful implementations. Those research groups that were able to implement their automated driving algorithms at speeds up to highway driving speeds have been successful, while others who were not able to do

Internet of Things and Data Analytics Handbook, First Edition. Edited by Hwaiyu Geng.

Companion website: www.wiley.com/go/Geng/iot_data_analytics_handbook/

this have failed. Unlike mobile robotics, automated driving vehicles are constrained to follow and share the road and obey all rules of traffic.

There have also been early research efforts that were concurrent with the work in autonomous mobile robots. One significant example is the pioneering research work on autonomous vehicles at the Ohio State University during the 1960s and 1970s led by Prof. Fenton of its Electrical Engineering Department [1]. The cars used in this pioneering work had a sensor in front that sensed the current inside a wire carefully taped to the road and served as the path to be followed. This work had to stop in 1980s due to lack of funding. The California Partners for Advanced Transportation Technology (PATH) program was one of the well-known succeeding programs on automated highway systems in the United States originating in 1986 [2]. The California PATH program was funded by the State of California and also used guided wires embedded in the road initially, followed later by the use of magnetic markers embedded in the road for autonomous trajectory following.

The well-known and well-publicized Defense Advanced Research Projects Agency (DARPA) Grand Challenge autonomous vehicle races of 2004 [3] and 2005 [4] and the subsequent DARPA Urban Challenge of 2007 [5] were more recent demonstrations of the feasibility of autonomous vehicles. While the feasibility of autonomous driving was demonstrated in the DARPA challenges, the solution approach of bulky actuators, an excessive amount of redundant sensors surrounding the vehicle and the large computers used with insufficient computing power based on today's standards, meant that the developed technology was nowhere close to implementation in a series production road vehicle. In those days, consumers also did not demand cars with automated driving capability as the technology was considered to be expensive and not able to cope with the large variety of highly demanding real-world situations.

Some drastic changes took place in recent years in both the automated driving and the connected vehicle (CV) areas that changed both the consumer and federal and state government acceptance of these new technologies. In the automated driving area, the success of the Google self-driving cars [6] that were actually built based on DARPA Challenge experience of winning teams for obtaining lidar-based maps in the United States and around the world was a big game changer. The public opinion changed to a much more favorable view of autonomous vehicles and their prospect of being available soon. A similar event took place in vehicle connectivity around 2009–2010 as modems based on the new IEEE 802.11p communication standard for vehicle-to-vehicle (V2V) and vehicle-to-infrastructure (V2I) communication became available. A wide variety of alternatives had been tried in the past for V2V and V2I communication without much success. With the new standard in place and with the newly available modems, automotive Original Equipment Manufacturer (OEMs) and suppliers were immediately interested in the possibilities offered by CV technologies. The presence of major automotive OEMs and research labs backing most of the teams that took part in the Grand Cooperative Driving Challenge (GCDC 2011) [7] was a clear indication of this interest. In the GCDC 2011, connected automated driving based on the new communication standard was demonstrated by multiple runs by two platoons of vehicles from different vendors in a highway, at regular highway speeds. Research in the CV area has progressed tremendously since then, as is evidenced by the highly successful CV Pilots program of the US Department of Transportation (DOT) [8]. As a result of this success, automotive OEMs will soon be asked to manufacture their vehicles with onboard units that have modems with V2V, V2I, and vehicle-to-everything (V2X) communication capability.

While not taking a direct interest in the autonomous vehicle results of the DARPA challenges, automotive OEMs and suppliers took a different approach of building vehicle automation piece by piece, starting with active safety systems like the Electronic Stability Control (ESC) [9] system and continuing with Advanced Driver Assistance Systems (ADAS) like Adaptive Cruise Control (ACC) [10], Lane Departure Warning and Keeping [11], Collision Risk Warning and Avoidance [12] using emergency braking, and so on. Their approach was to build automated driving functionality incrementally and first offer the new technology in higher-end models. As a result, it is possible to have technologies like automatic car following (ACC), lane centering control, and automatic parking even in the lower-priced models today. Based on this success and on public acceptance and also based on nontraditional competition from the likes of Google and Apple, automotive OEMs are preparing to make automated driving available in their series production vehicles as early as 2020. These automated vehicles will also be connected to other vehicles, the road infrastructure, and to other road users like pedestrians and bicyclists through smartphone apps.

35.2 AUTOMATED DRIVING TECHNOLOGY

Automated driving has been categorized into six categories [13] displayed in Figure 35.1 by the Society of Automotive Engineers (SAE). Currently available automated driving technology falls under level 2 and level 3 which are partial and conditional automation, respectively. Level 2 partial automation is available in series production vehicles with lane centering control for steering automation and ACC and collision avoidance for longitudinal direction automation. Partial automation is characterized by all driving actuators being automated and the presence of a driver who can intervene when necessary. Recently introduced autopilot systems for cars are examples of conditional automation where the car takes care of driving in some driving modes (like highway driving), but the human operator is always in the driver seat to take over control if necessary. In level 4 highly automated driving, the human driver is available to intervene, but the driving tasks are taken care of autonomously even if no intervention takes place. There are a lot of research vehicles with level 4 automation. The goal, of course, is to reach the highest of them all, that is, level 5 automation, which is characterized by all driving modes being executed autonomously without the need for someone in the driver seat. This means that all driving tasks except for the entry of the final destination should be planned and executed autonomously by the car. While some example applications exist in the case of predefined missions, no vehicle with that level of autonomy currently exists.

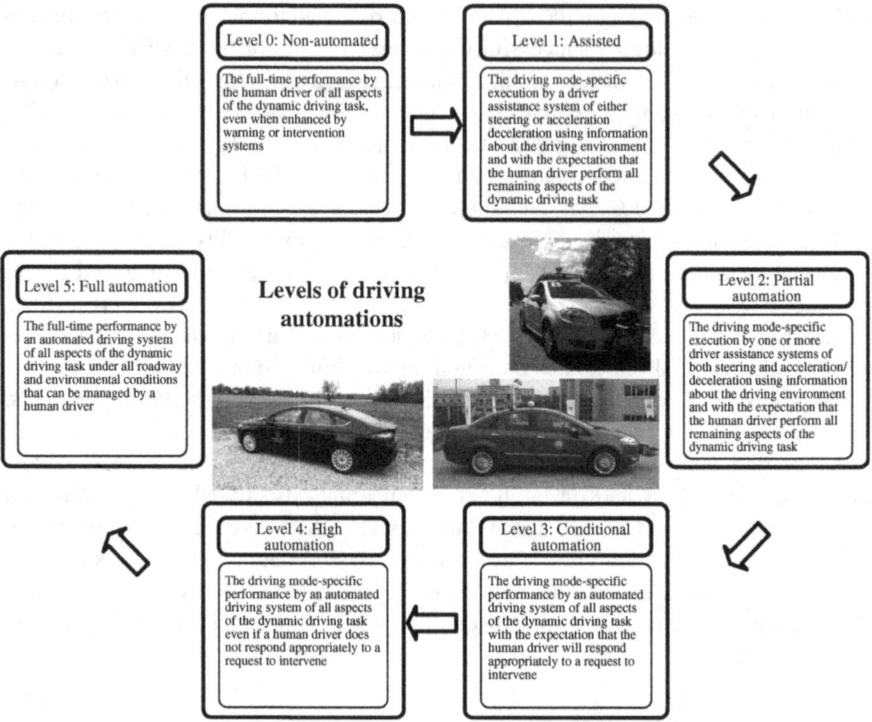

FIGURE 35.1 Categories of automated driving as defined in SAE J3016 are illustrated noting that level 5 full automation is the final goal. © Copyright by Levent Guvenc. Use with permission.

Regardless of the level of automation desired, all automated driving systems need to possess similar underlying features. Firstly, the actuators of the vehicle in the longitudinal and lateral directions have to be automated. These include the throttle, brake, and gear shifting in the longitudinal direction and steering actuation control in the lateral direction. Current road vehicles use electronic throttle control (ETC), so it is relatively straightforward to automate throttle commands. Use of electronic commands for automated braking is possible in road vehicles that have electronic brake-force distribution (EBD). Even though EBD intervention is the preferred method, braking automation can also be achieved by interfacing to the brake pedal sensor in a hybrid electric vehicle. The electronic transmission controller needs to be accessible by software for automating gear shifting. As the automatic transmission will take care of power shifting automatically, it needs to be accessed by the automated driving controller for higher performance. The automated driving controller in level 4 and level 5 automation will also need to shift to reverse gear autonomously. Finally, the electric power-assisted steering (EPAS) electric motor can be used for autonomous steering in current road vehicles through its torque overlay function.

Access to all these actuators is available on the controller area network (CAN) bus in current road vehicles, meaning that implementing the actuators needed for driving

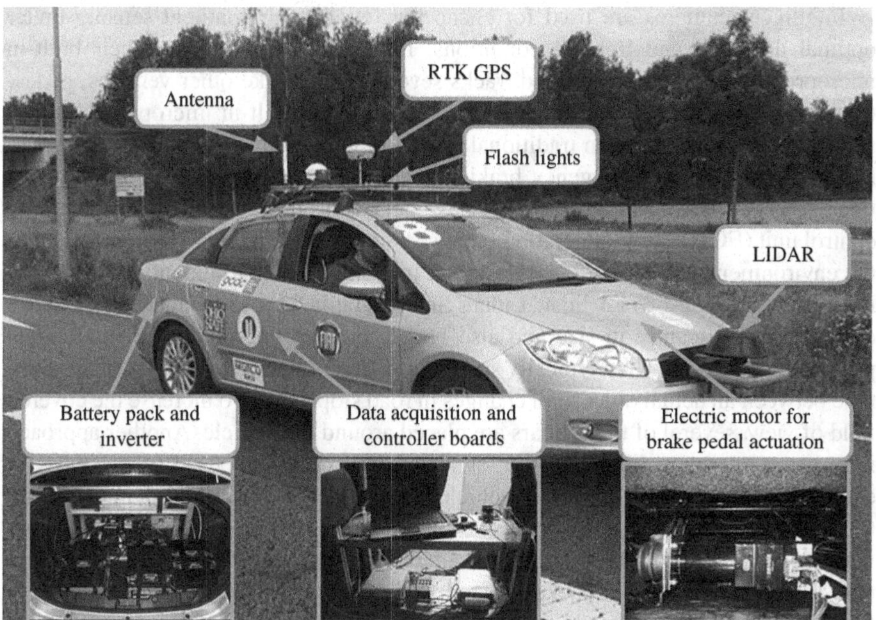

FIGURE 35.2 An interesting example of a level 2 partially automated driving vehicle from 2011 is illustrated here. The vehicle automation hardware is completely add-on with no additional holes drilled in the car. All the add-on hardware was easily removed completely in about 2 hours by two people. This platform was extended in 2012 to a level 3 conditional automation of the same brand of vehicle and demonstrated in autonomous driving. © Copyright by Levent Guvenc. Use with permission.

automation is a matter of interfacing and sending the correct CAN bus signals. The actuators needed for vehicle automation already exist in the car for an automotive OEM. This should be compared with the bulky actuators that were used in the past as is shown in Figure 35.2. A state-of-the-art series production road vehicle also has several sensors like the speed, longitudinal and lateral accelerations, yaw rate, steering wheel position/velocity, and wheel speed sensors that can be used as part of the sensor suite needed for vehicle automation.

One of the most important sensors needed in automated driving is the Global Positioning System (GPS) sensor which is needed for positioning. The GPS sensor is actually just a receiver of GPS satellite signals. A differential antenna real-time kinematic (RTK) GPS sensor and an inertial measurement unit (IMU) with at least a 10 Hz update rate and 1–2 cm of accuracy is currently needed for level 3–5 automated driving. Inertial navigation system calculations are used with the IMU readings to obtain correct position information during short GPS outages [14]. IMU readings occur at much higher frequency of 100 Hz or more and can be combined with the less frequent GPS signals in a Kalman filter to obtain smooth, higher-frequency position updates.

Intelligent cameras are used for cheap but reliable environment sensing under optimal dry road and lighting conditions. Intelligent cameras have their built-in microprocessor which detects and tracks several objects like other vehicles, pedestrians, and bicyclists in real time. Radar sensors with built-in microprocessor and signal processing have been traditionally used in the automotive industry as part of ADAS like ACC and emergency braking. These radar sensors can also detect and track several objects and have a CAN bus interface to the corresponding electronic control unit (ECU). New-generation radar sensors can also be used to do mapping of the environment. When one is interested in mapping the environment, however, the most powerful sensor is a lidar. Lidars that are built for use in automobiles have decreased significantly in size and also have a built-in microprocessor and signal processing software. They typically scan four different planes to be able to differentiate between an actual object and changes in road slope [15]. To increase the covered field of view, several of these lidars are placed around the vehicle. Another approach is to use a scanning lidar like the 64 plane ones that were used in the DARPA challenges. The Google driverless vehicles also use such scanning lidars placed on top of the vehicle for environmental sensing. Much smaller 32 and 16 plane scanning versions of these lidars are currently available at much lower cost. However, these types of lidars are still not packaged compactly enough for use in series production vehicles. Furthermore, they do not have a CAN interface or a built-in microprocessor with signal processing. They only provide raw data and, therefore, require the presence of a powerful, real-time operating computer on board the vehicle. Successful series production-level automated driving will require all automation actuators and sensors to have CAN bus interfaces and all sensors to have built-in intelligence. All of these vehicles will necessarily have a navigation system which will have a high-resolution digital map providing electronic horizon information in real time. This map which will have very accurate representations of lane positions will function like an extra sensor (map as sensor) in the vehicle. The GPS and V2V modem will be integrated as part of the navigation system. As an illustrative example, a level 4 automated driving vehicle under preparation is shown in Figure 35.3.

Success of future series production of automated driving vehicles will require either considerable reductions in the cost of high-performance sensors used in automated driving or improved sensor signal processing, fusion, and control methods to be used with low-cost, low-performance sensors. The currently used approach of different suppliers providing different ADAS functions has resulted in separate ECUs for each main function like ACC or lane centering control. A new hardware architecture comprising of a smaller number of more powerful ECUs has to be adapted and utilized in the future. Different suppliers should be provided space in one of these smaller numbers of ECUs for their functions.

As compared to the big automotive OEMs, the smaller high-technology companies that currently want to concentrate on low-speed automated driving are facing significant cost-related challenges as they cannot use the benefits of series production. Down the road toward widespread acceptance and production of automated driving vehicles, a major breakthrough in approach similar to the PC revolution or similar to the one that partially took part in the robotics area is expected to take place.

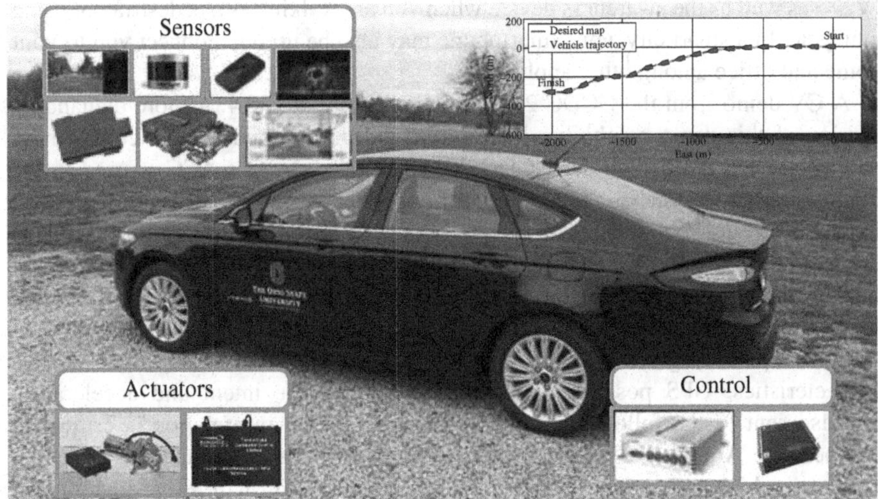

FIGURE 35.3 A level 4 highly automated driving vehicle under construction is displayed here. The actuators, sensors, and electronic control units used for automation have a very low footprint in this type of automation where driving is taken care of by the autonomous system with the driver taking control only if there is a need. © Copyright by Levent Guvenc. Use with permission.

First of all, the highest benefit from automated driving will be achieved when using a fully electric vehicle. This fully electric vehicle will have electromechanical brake-by-wire and steer-by-wire actuation. Its hardware and software architecture will be open source, modular, and plug and play. This will initiate new actors which will be small start-up companies to easily enter this market with their innovative, alternate solutions. New nontraditional manufacturers will be able to easily assemble and configure these vehicles just like assembling and configuring a PC. Pretested and certified vehicle chassis frame and body combinations will be readily available. Highly accurate digital modeling tools will be used for highly realistic virtual testing of the car before it is manufactured.

35.3 CONNECTED VEHICLE TECHNOLOGY AND THE CV PILOTS

It is expected that CV technologies will be present in series production cars before higher levels of automated driving. Cars will be equipped with a modem capable of V2V, V2I, and V2X connectivity. This modem is just an extra sensor with endless possibilities for application development and enhancement from the perspective of automated driving. For safety critical applications that are typical of automated driving, these modems use the wireless access in vehicular environments (WAVE) protocol based on the IEEE 802.11p standard and use the intelligent transportation systems (ITS) band of 5.9 GHz. In the future, with faster communication rates, your smartphone may actually be used as the main communication device in your car

(V2X) as well as the awareness device when you are walking as a pedestrian or using a bicycle. In a smart city, your smartphone may also be used to connect you to your home and office among other applications.

A CV demo from the GCDC 2011 is shown in Figure 35.4. A more demanding version of this demo combining both automated driving and vehicle connectivity will take place in 2016 as the GCDC 2016 [16]. Cooperative ACC is used in the longitudinal direction to reduce the time gap between vehicles in a platoon. Infrastructure-to-Vehicle (I2V) modems broadcast speed limits and traffic-related events to the vehicles. Vehicles on the same lane send requests to each other before forming a convoy. Vehicles in the convoy receive a recommended speed profile which they all follow. The lead vehicle broadcasts this speed information which is broadcast from each preceding vehicle to the one following it. Other information like vehicle characteristics, GPS position and velocity, and vehicle intent and acceleration are also sent to the following vehicle. The acceleration information is critical in Cooperative Adaptive Cruise Control (CACC) as it helps reduce the time gap between vehicles. A properly tuned and executed CACC will allow platoons of vehicles to damp shockwaves that they will be faced with in traffic, thereby improving mobility. A coordinated combination of CACC with lateral control and decision making results in the most basic form of cooperative automated driving in highways.

FIGURE 35.4 An example of connected driving is seen here with two platoons of connected vehicles following a speed profile communicated to them by the lead vehicle. The vehicles use CACC in the longitudinal direction. Traffic light SPAT information and speed limits are communicated by roadside units to the vehicles. © Copyright by Levent Guvenc. Use with permission.

A low-cost longitudinal cooperative driving system that uses communicated position/velocity/acceleration information with a low-cost GPS has been implemented and tested successfully [17] in the GCDC 2011 and later and is a good starting point for low-cost highway driving automation when used together with an intelligent camera for lane keeping and collision avoidance.

Along with improved mobility in the form of avoiding congestion during traffic shockwaves, CV applications also have safety and environmental benefits. Intersection safety is the most important use of CV safety applications as CV technology allows nonvisible vehicles to be seen through the communication link. Other safety applications include collision warning and curve speed warning. Vehicles are also warned of pedestrians and bicyclists who are nearby. Environmental benefits are obtained indirectly through the use of optimal speeds for fuel minimization and through avoiding congested traffic.

Research on CV has advanced significantly in Europe, Japan, and the United States. The current approach in the United States is the most advanced due to the fact that real deployments of CV technology are already taking place. The DOT in the United States has been funding CV test sites, CV application development efforts, and now large-scale CV deployments on real roads. The CV applications and hardware and software architecture have been standardized. Three large-scale CV deployments were recently funded. Along with the incorporation of roadside units, weather information sensors, and traffic lights with signal phase and timing (SPAT) information, large traffic processing centers that collect information from the equipped vehicles and use it dynamically for ecorouting purposes are also used in the deployments. The deployments use tens of thousands of vehicles equipped with V2V modems or simpler awareness systems. The ecospeed regulation, ecolanes management, and Eco-Cooperative Adaptive Cruise Control (eco-CACC) are the latest CV applications that automated driving vehicles will definitely benefit from.

35.4 AUTOMATED TRUCK CONVOYS

User demand dictates which new technology ends up in series production vehicles. Automotive OEMs will not develop new technological features unless there is a demand from the buyers or a mandate by the regulating authorities. Interestingly, the biggest need and demand for driving automation comes from the logistics sector due to many reasons. The logistics sector wants their trucks to be able to drive fully or semiautonomously as this will reduce the driver workload considerably. It is very difficult to find expert truck drivers who need to spend long hours of driving which is a very tiring and boring task. Drivers are also mandated by law to rest after a certain period of driving. Some driving tasks like those in ports or loading/unloading situations are very slow paced, with the driver being forced to waiting for hours in the truck without traveling a significant distance. These types of situations make automated driving very attractive for trucks. This is the reason why truck manufacturers and suppliers are working on highway automated driving systems for trucks. These systems are currently very basic in nature and use a camera and radar as the

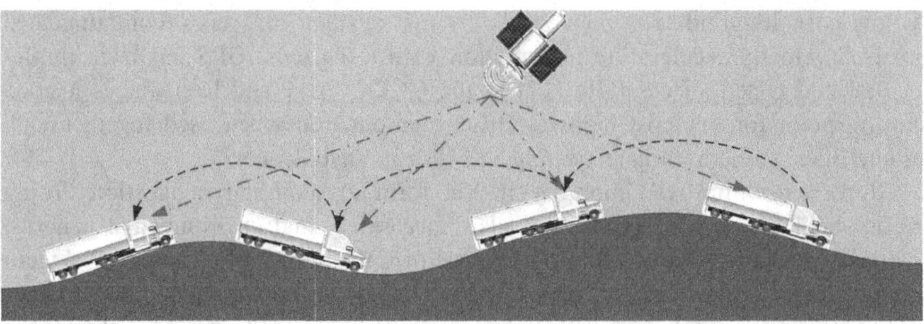

FIGURE 35.5 Predictive cruise control of a platoon of trucks is shown. Predictive cruise control systems for trucks are currently available from manufacturers. Current research on predictive cruise control focuses on incorporating traffic speed and traffic light information to the cruise speed profile optimization. The highest fuel economy benefits will be obtained if the trucks are also equipped with regenerative braking. © Copyright by Levent Guvenc. Use with permission.

main sensors. The camera is used for lane keeping control and the radar is used for longitudinal control. A currently available technology for trucks that is illustrated in Figure 35.5 is predictive cruise control where the slope data in the electronic horizon coming from a digital map is used to find the optimal speed profile for fuel efficiency. It is expected that these predictive cruise control systems will find widespread use in the truck industry and will form an integral part of the automated driving truck. Work on truck platooning has been going on for decades with a large number of successful research implementations. All trucks in a truck platoon will follow the optimal speed trajectory. The following trucks will also benefit from a reduced aerodynamic drag due to the presence of the truck in front of them, resulting in further fuel savings. We should soon be expecting to see automated convoys of trucks on the roads with possibly only one driver operating the leading truck.

35.5 ON-DEMAND AUTOMATED SHUTTLES FOR A SMART CITY

In recent years, the notion of a smart city has gained popularity around the globe. We are actually already living in relatively smart cities. Important aspects of smart cities are that they are green, connected, and automated. Automated driving technologies are, thus, a very important part of smart cities. Small, fully electric automated vehicles are expected to solve mobility problems in smart cities on an on-demand basis [17]. This on-demand usage will reduce the total number of vehicles. The smaller fleet of vehicles will move both people and goods and will be connected to all other aspects of the smart city. While their main function will be to pick people up and transport them to their desired destination, it will also be possible for them to deliver online orders to your doorsteps. These automated shuttles will also be integrated with public transportation systems and solve the first mile, last mile problem. They will also solve the mobility problems of an aging population.

35.6 A UNIFIED DESIGN APPROACH

Research and development work on automated driving trucks and automated driving cars currently concentrates on high-speed highway driving as that is where the biggest benefits lie. These vehicles can currently use relatively expensive sensors and actuators due to the final cost reductions made possible by series production. On the contrary, on-demand automated shuttles usually use low-speed vehicles and do not benefit from series production as their production volumes are relatively small. Widespread use of such automated low-speed shuttles will only be possible upon the use of low-cost sensors, actuators, and ECUs. Currently, the design of high-speed and low-speed driving automation uses two different approaches, with low-speed automation being similar to mobile robotics. In the long run, however, both the high-speed and low-speed automated driving worlds have to converge as we reach full level 5 automation of driving. The converged design procedure will look like the V-diagram shown in Figure 35.6. A highly realistic and validated model of the vehicle will be used in model-in-the-loop (MIL) simulations first. This will be followed by software-in-the-loop (SIL) simulations. Actuators, sensors, controllers, and modems of the vehicles will be networked together in a hardware-in-the-loop (HIL) simulation setting for final lab testing with real hardware. The rest of the vehicles and traffic situations will be virtually generated using real-time capable simulation computers. The last step will be road testing with the actual vehicle in a proving ground. It is possible to do vehicle-in-the-loop (VIL) simulation at this stage. One interesting example of VIL testing is platooning with a convoy of virtual vehicles [18].

It is clear that a unified design approach that handles both low- and high-speed driving automation will be necessary in the future. This unified design approach will benefit greatly from the future possibility of an open-source, modular, and plug-and-play software and hardware architecture for automated driving vehicles.

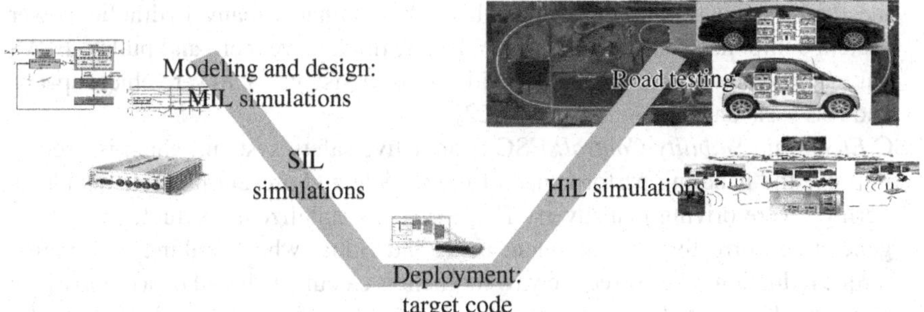

FIGURE 35.6 The V-diagram of a unified approach for developing connected and automated driving systems starting with model-in-the-loop simulations and ending in road testing. © Copyright by Levent Guvenc. Use with permission.

35.7 ACRONYM AND DESCRIPTION

ACC *Adaptive Cruise Control:* Cruise control systems maintain a constant speed set by the driver, with the driver being required to apply the brakes when a slower preceding vehicle is encountered. In contrast, a vehicle equipped with ACC adjusts its speed automatically, to safely follow other vehicles with possibly varying speeds. In doing so, the ACC system uses a sensor to detect the presence of the vehicle being followed while measuring relative distance and relative speed. Using this data in its control system, a vehicle with ACC adapts to changes in the speed of the vehicle that precedes it [10].

ADAS *Advanced Driver Assistance Systems:* The general name of driver support systems in order to increase the safety and comfort of the driver and the passengers such as ACC, automatic parking system, collision avoidance systems, lane departure warning system, lane keeping assistance, hill descent control, etc.

CACC *Cooperative Adaptive Cruise Control:* CACC system extends the capability of ACC system using the information of the state of the preceding vehicle obtained via wireless communication. Using this extra information, the time gap in standard ACC systems can be reduced [18].

CAN *Controller Area Network:* CAN is a serial bus system, developed to be used in vehicles to exchange information between different electronic components. The CAN protocol is standardized by the International Standards Organization [19].

EBD *Electronic Brakeforce Distribution:* EBD system is a braking technology that automatically varies the amount of force applied to each wheel based on road conditions, speed, loading, etc. in order to maximize stopping power while maintaining vehicle control [20].

Eco-CACC *Eco-Cooperative Adaptive Cruise Control:* Eco-CACC is a kind of CACC. Eco-CACC includes longitudinal automated vehicle control while considering ecodriving strategies (minimization of fuel consumption and emissions). The Eco-CACC application incorporates information such as road grade, roadway geometry, and road weather information in order to determine the most environmentally efficient speed trajectory for the following vehicle [21].

EPAS *Electric Power-Assisted Steering:* EPAS supplements the required torque that the driver applies to the steering wheel. It eliminates many hydraulic power system components such as the pump, hoses, fluid, drive, belt, and pulley. EPAS has variable power assist, which provides more assistance at lower vehicle speeds and less assistance at higher speeds [22].

ESC *Electronic Stability Control:* ESC is an active safety system, which is used to improve the handling performance of the vehicle and prevent possible accidents during severe driving maneuvers. These systems stabilize the vehicle motion by generating corrective yaw moment using individual wheel braking and engine torque reduction. The corrective yaw moment is calculated based on a comparison between the desired and measured vehicle yaw rate and/or vehicle sideslip angle [23].

ETC *Electronic Throttle Control:* ETC establishes the essential connection between the acceleration pedal and the throttle valve using electronic signals instead of a mechanical link. The desired throttle position is calculated based on the information from both the acceleration pedal and other systems such as engine controller, ESC, etc. The control unit compares the desired throttle position with the actual throttle position and sends the appropriate signal to the motor to drive the throttle to the desired position [24].

GPS *Global Positioning System:* The GPS is a space-based navigation system that provides location and time information in all weather conditions, anywhere on or near the Earth where there is an unobstructed line of sight to at least three or more GPS satellites [25]. In practice, however, information from at least five satellites should be used. High-end GPS systems can use information from hundreds of satellites.

HIL *Hardware-in-the-Loop:* The actual control hardware is tested in a real-time simulation environment, which contains the plant model as real-time software. It is useful for testing interactions between the control hardware and the plant [26]. Since the plant exists as software, effects of changes in the plant and disturbances are very easily tested in a lab setting.

IMU *Inertial Measurement Unit:* The IMU is a sensor to measure acceleration and angular velocities along a body's three main axes. It consists of accelerometers and gyroscopes. An internal navigation system is the combination of the IMU sensor with mechanization equations. These equations convert IMU data to vehicle position, velocity, and rotations [27].

MIL *Model-in-the-Loop:* The controller and the plant are represented entirely as models in a simulation environment. This is also called offline simulation. MIL simulation is useful for initial controller development [26].

RTK *Real-Time Kinematics:* The RTK satellite navigation is a technique used to enhance the precision of position data derived from satellite-based positioning systems such as GPS in real time, that is, with low computational latency such that it can be used instantaneously in a moving vehicle. It uses measurements of the phase of the signal's carrier wave, rather than the information content of the signal, and relies on a single Earth-fixed reference station(s) or interpolated virtual station to provide real-time corrections, providing up to centimeter-level accuracy [28].

SIL *Software-in-the-Loop:* The control algorithm is represented as executable code, and the plant is represented as a model in a simulation environment. It is useful for testing the executable control code for possible errors before controller implementation [26].

SPAT *Signal Phase and Timing:* SPAT is a message type that describes the current state of a traffic light signal system and its phases and relates this to the specific lanes (and therefore to movements and approaches) in the intersection [29]. The signal phase is green, yellow, or red and the timing is the time to change phase.

V2I *Vehicle-to-Infrastructure:* V2I communication is the wireless exchange of critical safety and operational data between vehicles and the roadway infrastructure, intended primarily to avoid motor vehicle crashes [30]. V2I communication is also used to collect data from vehicles passing by. This data is sent to a traffic control center for further processing.

V2V *Vehicle-to-Vehicle:* It is a dedicated wireless communication system handling high data rate, low latency, low probability of error, line-of-sight communications between vehicles [31]. V2V communication is necessary for CV platooning and emergency braking.

VIL *Vehicle-in-the-Loop:* The VIL test setup combines the test vehicle with a synthetic test environment such as a traffic simulation. The actual vehicle is made to believe that it is in traffic by supplying it with artificially generated sensor signals. Therefore, repeatable and safe tests are realized using the real hardware and dynamics of the vehicle. This setup combines the advantages of driving simulators and a real test vehicle [32]. The vehicle may be stationary or moving.

REFERENCES

[1] http://researchnews.osu.edu/archive/naefenton.htm (accessed December 25, 2015).

[2] http://www.path.berkeley.edu/about (accessed December 25, 2015).

[3] http://archive.darpa.mil/grandchallenge04/ (accessed December 25, 2015).

[4] http://archive.darpa.mil/grandchallenge05/ (accessed December 25, 2015).

[5] http://archive.darpa.mil/grandchallenge/ (accessed December 25, 2015).

[6] https://www.google.com/selfdrivingcar/ (accessed December 25, 2015).

[7] http://ieeexplore.ieee.org/stamp/stamp.jsp?arnumber=6266747 (accessed December 25, 2015).

[8] http://www.its.dot.gov/safety_pilot/ (accessed December 25, 2015).

[9] Aksun Güvenç, B., Güvenç, L., Karaman, S., Robust Yaw Stability Controller Design and Hardware in the Loop Testing for a Road Vehicle. *IEEE Transactions on Vehicular Technology*, **58**: 555–571, 2009.

[10] Aksun Güvenç, B., Kural, E., Adaptive Cruise Control Simulator, a Low-Cost Multiple-Driver-in-the-Loop Simulator. *IEEE Control Systems Magazine*, **26**: 42–55, 2006.

[11] Emirler, M.T., Wang, H., Aksun Güvenç, B., Güvenç, L., Automated Robust Path Following Control based on Calculation of Lateral Deviation and Yaw Angle Error. *Dynamic Systems and Control Conference*, ASME, Columbus, OH, October 28–30, 2015.

[12] Ararat, Ö., Aksun Güvenç, B., Development of a Collision Avoidance Algorithm Using Elastic Band Theory. *17th IFAC World Congress*, IFAC, Seoul, July 6–11, 2008.

[13] http://www.sae.org/misc/pdfs/automated_driving.pdf (accessed December 25, 2015).

[14] Altay, I., Aksun Güvenç, B., Güvenç, L., A Simulation Study of GPS/INS Integration for Use in ACC/CACC and HAD. *9th Asian Control Conference*, Istanbul, June 23–26, 2013.

[15] Sankaranarayanan, V., Güvenç, L., Laser Scanners for Driver-Assistance Systems in Intelligent Vehicles. *IEEE Control Systems Magazine*, **29**: 17–19, 2009.

[16] http://www.gcdc.net (accessed December 25, 2015).

[17] https://engineering.osu.edu/news/2015/08/smart-ride-ohio-state-researchers-develop-automated-demand-transportation (accessed December 25, 2015).

[18] Güvenç, L., Uygan, I.M.C., Kahraman, K., Karaahmetoğlu, R., Altay, I., Şentürk, M., Emirler, M.T., Hartavi Karcı, A.E., Aksun Güvenç, B., Altuğ, E., Turan, M.C., Taş, Ö.Ş., Bozkurt, E., Özgüner, Ü., Redmill, K., Kurt, A., Efendioğlu, B., Cooperative Adaptive

Cruise Control Implementation of Team Mekar at the Grand Cooperative Driving Challenge. *IEEE Transactions on Intelligent Transportation Systems*, **13**: 1062–1074, 2012.

[19] http://www.can-wiki.info/doku.php (accessed January 1, 2016).

[20] https://en.wikipedia.org/wiki/Electronic_brakeforce_distribution (accessed January 1, 2016).

[21] http://www.iteris.com/cvria/html/applications/app101.html (accessed January 1, 2016).

[22] http://www.cvel.clemson.edu/auto/systems/ep_steering.html (accessed January 1, 2016).

[23] http://www.cvel.clemson.edu/auto/systems/stability_control.html (accessed January 1, 2016).

[24] http://www.cvel.clemson.edu/auto/systems/throttle_control_demo.html (accessed January 1, 2016).

[25] https://en.wikipedia.org/wiki/Global_Positioning_System (accessed January 1, 2016).

[26] https://www.youtube.com/watch?v=dzgMYGWXpUc (accessed January 1, 2016).

[27] Emirler, M.T., Uygan, I.M.C., Altay, I., Acar, O.U., Keles, T., Aksun Guvenc, B., Guvenc, L., A Cooperating Autonomous Road Vehicle Platform. *Workshop on Advances in Control and Automation Theory for Transportation Applications*, IFAC, Istanbul, September 16–17, 2013.

[28] https://en.wikipedia.org/wiki/Real_Time_Kinematic (accessed January 1, 2016).

[29] http://www.iteris.com/cvria/html/glossary/glossary-s.html (accessed January 1, 2016).

[30] http://www.its.dot.gov/factsheets/v2isafety_factsheet.htm (accessed January 1, 2016).

[31] http://www.iteris.com/cvria/html/glossary/glossary-v.html (accessed January 1, 2016).

[32] Bock, T., Maurer, M., Farber, G., Validation of the Vehicle-in-the-Loop—A Milestone for the Simulation of Driver Assistance Systems. *Proceedings of the IEEE Intelligent Vehicles Symposium*, Istanbul, Turkey, June 13–15, 2007, pp. 612–617.

36

TRANSIT HUB: A SMART DECISION SUPPORT SYSTEM FOR PUBLIC TRANSIT OPERATIONS

SHASHANK SHEKHAR[1], FANGZHOU SUN[1], ABHISHEK DUBEY[1], ANIRUDDHA GOKHALE[1], HIMANSHU NEEMA[1], MARTIN LEHOFER[2] AND DAN FREUDBERG[3]

[1] *Institute for Software Integrated Systems, Vanderbilt University, Nashville, TN, USA*
[2] *Siemens Corporate Technology, Princeton, NJ, USA*
[3] *Nashville Metropolitan Transport Authority, Nashville, TN, USA*

36.1 INTRODUCTION

The allure of smart city technologies lies in its promise to enrich the lives of residents by empowering the stakeholders to make efficient and informed decisions and help alleviate complex issues. One such issue that is universally faced by large cities of the world is the problem of commuter traffic. Solutions are difficult, partly because so much of our physical infrastructure was designed for a different era—when no one could imagine just how many cars would hit the streets each day—and partly because options like mass transit and adopting car shares, biking, and walking are seen by many as "trading down" making their commute less convenient. The irony is that very often, a short walk or bus ride would be far faster, easier, and cheaper than driving. Unfortunately, there is a general lack of awareness among people about the effectiveness of these options.

For example, consider the city of Nashville, TN. The traffic congestion in the city has nearly doubled over the last decade and is expected to grow at an even faster rate in future—a recent study found it to be the top 25 congested cities in

Internet of Things and Data Analytics Handbook, First Edition. Edited by Hwaiyu Geng.

Companion website: www.wiley.com/go/Geng/iot_data_analytics_handbook/

the United States [1]. Therefore, there is an urgent need to act now. Given that adding infrastructure is difficult and slow, we are focusing on increasing the use of shared mobility options like public transit. Nashville ranks as 74 out of 100 in the number of passengers per capita [2]. Improving transit services and increasing their efficiency and usage are strategic priorities of the Nashville Metropolitan Transport Authority (MTA). Key to attracting new users, in addition to providing services that are competitive with personal automobiles and that can meet basic service needs, is to make public transit easier to use and to improve the image of the services. For those who seldom use public transportation, it can be difficult to figure out how to interpret schedules and how to pay the fare, for example. Potential new customers have to be convinced that riding transit will be a pleasant experience, and people like themselves use transit. However, without the ability to model, it is almost impossible to efficiently design, configure, and deploy such a system.

This lack of understanding is an impediment to improving and deploying of smart city technologies at large scale. To that end, researchers from the Institute for Software Integrated Systems at Vanderbilt University have teamed up with the Nashville MTA and Siemens Corporate Technology to work on the Transit Hub [3] project (Figure 36.1). This project aims to put accurate, real-time information about potential travel options into citizens' hands as soon as they choose their desired destination from their current or a specified location.

Static schedule information for transit has been available in Nashville for a number of years through tools such as Google Transit. However, vehicles do not always exactly follow the schedule, making real-time vehicle locations and bus stop ETAs valuable pieces of information for the customer. In order to provide this type of real-time information, Nashville MTA has partnered with Trapeze Group to design and install a fleet-wide vehicle tracking system called Automated Vehicle Location (AVL). Transit Hub makes use of transit schedules in combination with real-time vehicle location and service alert information from the AVL system, as well as data from rider smartphones and dozens of other sources. Among other analysis, anyone with our free app[1] can compare travel times using all available options—biking, walking, public transit, driving, and more.

As part of the larger smart city movement, we are implementing the Transit Hub project as a large-scale distributed human cyber–physical system (CPS) [4], wherein the human and the sensors in the physical environment provide the information, which is analyzed by computational services to provide contextual decision support to the commuters. The data collected by the system also helps us create predictive analytical models that help the MTA better understand the transit network bottlenecks and areas for future improvements. This article focuses on the technical underpinnings of the Transit Hub application and the backend data services.

[1] The application is currently undergoing tests and is not yet available for wider dissemination.

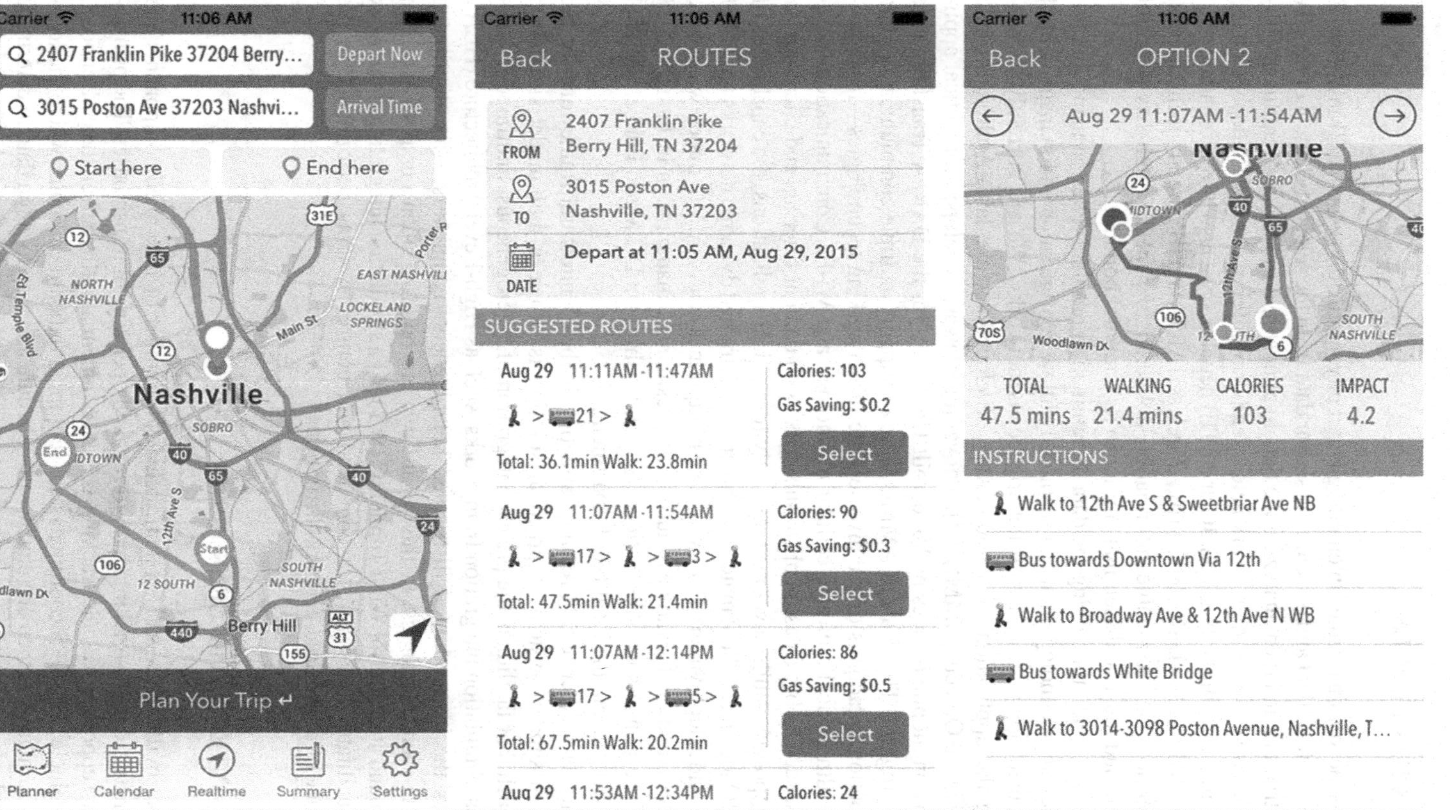

FIGURE 36.1 Trip planner, route comparison view, and details view. Reproduced with permission from Abhishek Dubey.

36.2 CHALLENGES

One of the key challenges in enabling smart city applications such as Transit Hub is the ability to collect and disseminate sensor data from the selected element of the city infrastructure and then analyzing the data in almost real time. However, this is difficult because of scale and network heterogeneity. While we can mostly depend on the availability of cellular (e.g., 3G/4G and LTE) networks, we also have to contend with Wi-Fi LAN (e.g., DSRC/802.11p), WirelessHART, and Bluetooth technologies. Furthermore, all actionable information delivered to a nearby hub has a useful lifetime and must be disseminated prior to its temporal bound.

Our approach for solving the data dissemination problem relies on using modern middleware technology that can seamlessly allow configuration and management of distributed application across heterogeneous networks. We are currently investigating the application of a middleware developed by our lab—Android Mobile Middleware Objects (AMMO) [5], a novel middleware developed for tactical applications on mobile devices by Vanderbilt University for the DARPA Transformative Apps program. The goals of AMMO were to support multiple communication/data distribution paradigms over a diverse suite of networking technologies—from low-bandwidth tactical radios to cellular technology and Wi-Fi. Communications paradigms, such as broadcast publish–subscribe with content filtering, and client–server are supported. While developed primarily for tactical applications, the capabilities of AMMO are an excellent match for resource-constrained dynamic networks typical of a city-wide deployment. For example, smartphones with AMMO middleware were used for situational awareness during the presidential inauguration in January 2013 by first responders with the National Guard, Washington DC Metro Police, National Park Service, and the local Fire Department.

Another challenge that requires attention is the architectural framework for the decision support system. A practical solution is to use a multimodel simulation approach that facilitates the precise integration of heterogeneous, multimodel simulations. Simulation integration frameworks, such as High-Level Architecture (HLA), address the integration of distributed heterogeneous simulators using distributed discrete-event semantics. In this regard, we are building upon our prior work on Command and Control Wind Tunnel (C2WT) [6] a simulation integration framework that facilitates assessment of command and control systems performance in the presence of adversarial network disruptions.

36.3 INTEGRATED SENSORS

The Nashville MTA AVL system includes a number of integrated components and sensors to aid in the tracking, monitoring, and operating of transit buses. The primary sensor supporting vehicle tracking is a GPS antenna and receiver located onboard each vehicle. The GPS coordinates are sent to an onboard computer where the location information is then correlated with schedule and stop location data to determine where the bus is on the current route and trip. This location information is also used to calculate schedule adherence (minutes ahead of or behind schedule) for the bus.

In addition to the GPS receiver, there are additional integrated sensors to aid in the tracking of vehicles. For example, vehicle odometer readings are used to supplement GPS information in areas where signals are weak or unreliable. At Music City Center (MCC), the downtown transit hub for MTA, radio-frequency identification (RFID) transmitters have been installed at each bus bay. Every vehicle has an RFID tag that reads the signal from the MCC bay tags to determine when the bus arrives and departs the transit center. This equipment also allows transit supervisors to monitor the location of buses within the facility down to the individual bay level.

Other in-vehicle components are also integrated with the AVL system. The farebox receives route and location information from the AVL, enabling the correlation of fare payment activity by bus stop. Engine alarms are sent directly to the AVL by the Engine Controller Module to enable real-time vehicle health monitoring and historical data analysis. Infrared-based Automated Passenger Counter (APC) sensors installed at each bus door provide passenger boarding and alighting activity data to the AVL which then correlates the data with schedule information to determine passenger activity by stop, trip, route, and direction. Even the vehicle destination sign is integrated with the AVL to enable the sign to change automatically and without driver intervention when switching between routes.

36.4 TRANSIT HUB SYSTEM WITH MOBILE APPS AND SMART KIOSKS

The Transit Hub system is accessed by a mobile application that can be deployed on individual user's smartphones. It features smart trip planning, service-alert integration, personal transit schedule management and notification, and real-time transit tracking and navigation. The Trip Planner utilizes origin and destination address and departure and arrival time to search for the future transit trips that meet the user's requirements.

The Trip Planner offers a user-friendly interface (Figure 36.1) for users to enter the information for route searching. For people who are not familiar with the Nashville area, for example, first-time visitors, they can enter the start and end address in the search bar. For the local residents, they can just drag the map view to pinpoint the start and end locations.

The app's real-time view (Figure 36.2) displays a map of the scheduled trip with lines indicating the walking/bus route and markers indicating the bus stops to transfer as well as original and destination locations. Users can tap the Go button at the very bottom to start real-time navigation. If real-time data is available for the trip, the exact bus that the user is supposed to get on/off for the next step will be shown in the map, with time label indicating the remaining time. If there is no real-time data for the scheduled bus, time left to get on/off the next bus will be calculated and displayed based on the static schedule time.

The application not only aggregates the real-time transit data that is already available from local transit authorities, but it also provides a crowd-sourced alternative to official transit tracking feed. When a user is using the mobile application, they can choose to provide their anonymous location and the data about when they are on the bus to the backend server. The server uses anonymous data to protect user's privacy.

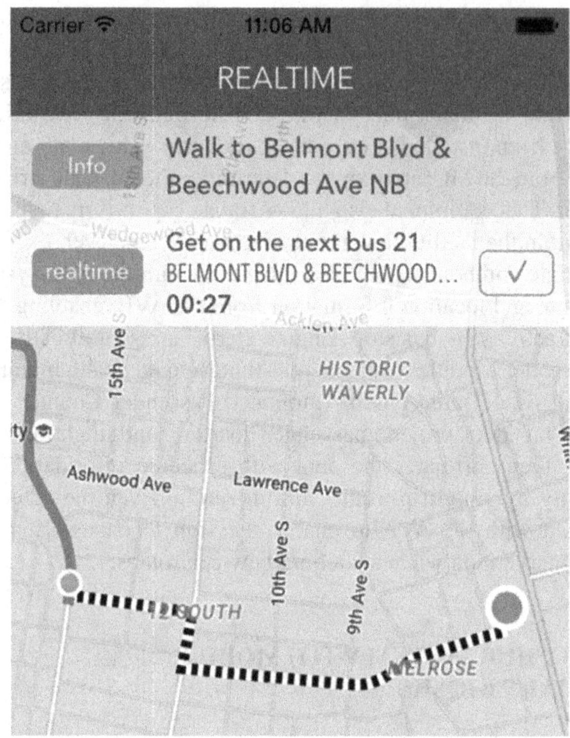

FIGURE 36.2 Real-time view. Reproduced with permission from Abhishek Dubey.

As the data accumulates over time, techniques like transit simulation and machine learning can be utilized to process the huge database of collected data, predict transit delay with more accuracy, and make the transit service more reliable by adjusting transit schedules. Furthermore, it could even provide potential bus stops to add and remove to better meet the evolving demand patterns.

Notification is usually an important part of any transit application. Without proper notifications, users can still miss the buses if they forget to keep an eye on the departure time after searching the suggested routes and planning the ideal trips to destinations. In the Transit Hub application, users can enable push notifications easily from the app and customize how long they want to leave alerts to be pushed ahead of time. When a user schedules a trip into the future and adds it to the Calendar, the notification is set to be pushed to the user's device when it is time to leave to catch the scheduled bus. Moreover, if the backend server predicts that the scheduled trip might be delayed for some reason, the user is notified accordingly.

36.4.1 Transit Hub Information Architecture

Figure 36.3 illustrates the technical design of the overall Transit Hub decision support system which enables the smartphone application. At the center lies the hub middleware responsible for coordinating all the Transit Hub activities. Its roles include

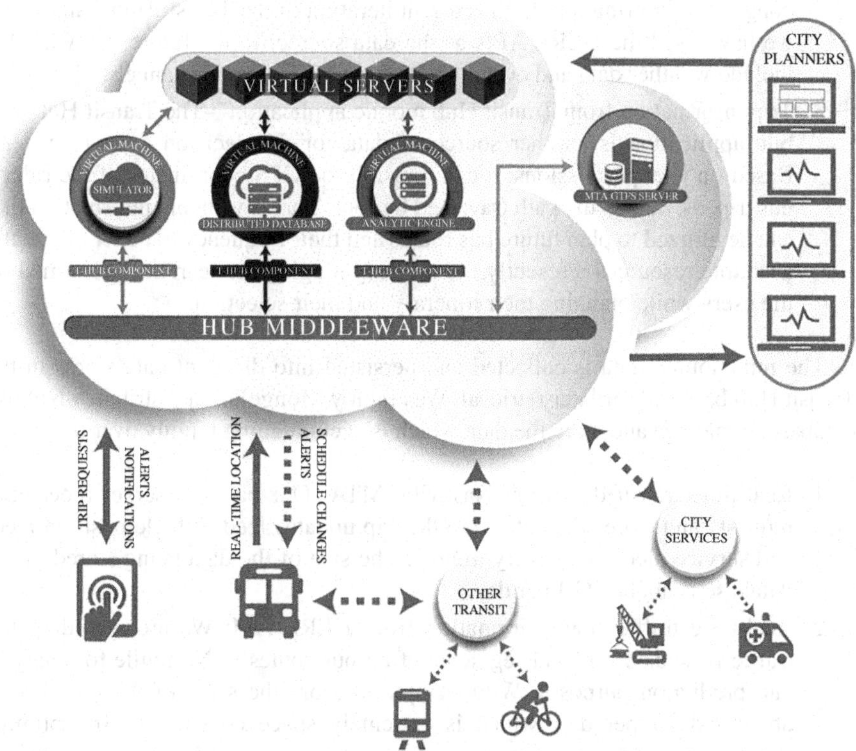

FIGURE 36.3 Transit Hub design. Reproduced with permission from Abhishek Dubey.

running data collection service, analyzing the collected feed, running simulations to provide a decision support framework in response to client requests. The data collection service is responsible for collating data from different sources and persisting into the Transit Hub's distributed database for consumption by the decision support system and its clients. The service adheres to the timeliness and data quality guarantees needed by the decision support system. The service is also replicated across multiple servers in master–slave architecture such that one of the slaves take over the data collection role in case the master service fails. There are three types of data being collected:

1. Real-time feeds from Nashville MTA—Google has defined the General Transit Feed Specification (GTFS) [7] for transit authorities to release transit data feeds. The Nashville MTA publishes real-time feeds using the GTFS format. Transit Hub collects this data for supporting mobile applications and performing historical analysis for the city-level decision support system.
2. Traffic-related feeds—There are several sources of traffic-related information in Nashville city. The Tennessee Department of Transportation (TDOT)

publishes traffic-related feeds. HERE API [8] is another source for traffic congestion information. In the current iteration of the Transit Hub framework, we have used the HERE APIs as the data source. In the future, we will also include weather data and event details from social media channels.

3. Trip information from Transit Hub mobile application—The Transit Hub mobile application is another source of data for the decision support system. Based on user permissions, it can transmit the GPS coordinates of the rider's bus trip, as well as the path traversed by foot. This is valuable information that can be utilized to plan future bus routes and their frequency to better utilize the available resources. Presently, we are only logging the searches performed by the users while planning their itinerary and their selection.

The multisource data is collected and persisted into different data stores in the Transit Hub backend for later retrieval. We employ MongoDB, a distributed NoSQL database to manage and store the data which is accumulating rapidly over time:

1. Real-time transit data from Nashville MTA: Our backend server repeatedly requests and stores the data from the trip updates feed, vehicle position feed, and service alerts feed every minute. The size of the data being stored in the database is about 3 GB per day.
2. Real-time traffic flow information from HERE API: We are recording the traffic flow data for road segments of all bus routes in Nashville for analysis and prediction purposes. Without optimization, the scale of the raw data is about 2.8 GB per day, which is extremely space consuming. To optimize performance and save storage space, we remove the static fields from original data such as road segment, physical layout information, and speed limits. Furthermore, since the traffic condition typically remains the same in a short term, we adopt a time series format that only stores the traffic condition which changes since last update. The optimized traffic data for storing is reduced to nearly 10% of its original size, about 0.27 GB per day.
3. Static bus schedule dataset: This dataset is updated only when Nashville MTA releases new bus schedules to the public.
4. Crowd-sourced data from Transit Hub app: Our backend server collects these data anonymously upon user permissions when users plan for bus routes or track their trips in the app. The size of the data depends on the quantity of mobile app users and how frequently they use the app.

36.4.2 Decision Support Framework for Transit Hub

The Transit Hub mobile application helps users in trip planning and real-time tracking. It relies on the Transit Hub middleware for supplying optimal trip plan options and better trip tracking than what is provided by the MTA feeds. To achieve this, Transit Hub performs analysis and provides decisions at two levels. At the global level, it integrates historically collected data with simulations to provide augmented feeds to the mobile application and trip recommendations that helps MTA to optimize the load (future work). At the user level, the user's travel history and current information, such as planned trip, location, time, and cost constraints are utilized to provide best recommendations for the user—this is part of the ongoing work.

36.4.2.1 Analyzing the Collected Data Feed In the current version of Transit Hub, we provide augmented feeds to external entities and the Transit Hub mobile applications containing predicted time for different bus routes. The analytical engine of Transit Hub consists of a simulation-based predictive model, a data-driven statistical model, and a real-time prediction model.

The simulation-based model works with the real-time feed and current traffic delay information to simulate bus movement on various routes and predict delays. This model uses the Simulation of Urban MObility (SUMO) [9] microscopic simulator for simulating city traffic. We use an OpenStreetMap (OSM) [10] of Nashville and convert it in the format that SUMO understands. The static routes for buses from MTA GTFS feeds are mapped to SUMO format for simulating. We maintain a pool of virtual machines to run simulations for different buses from their current locations. The real-time traffic conditions are obtained using HERE APIs. Going forward, we will also take into account the historical data from traffic congestion at different times of the day. The traffic congestion information contains a "jam factor" that provides information about congestion level at all road segments within the queried region. Based on the jam factor, we configure the simulator for lane speeds and periodically run multiple simulations to collect the result to augment the feeds with delay results produced in the simulation.

The statistical model applies long-term analytics methods on historical transit data to explore persistent delay patterns in route segments and bus stops. This model utilizes K-means clustering algorithm to group the historical delay data into different clusters according to the time in the day and performs normality test on each clusters. The analytics results, including the mean value with confidence interval of each cluster, are then provided to the analytics dashboard and Transit Hub mobile app. The model also helps to identify the outliers in real time. By investigating the bus trips with severe delay, we can understand better what factors cause bus delay and how to optimize the system. Our backend server runs the statistical methods for all the route segments at the end of each month when a new monthly historical dataset is ready.

The real-time prediction model utilizes the real-time transit feeds such as trip updates and vehicle locations to provide short-term delay prediction. Since the data rate of the original real-time vehicle location feed is not stable and there may be errors and delay in hardware and communication, we developed a Kalman filter to reduce the noise when estimating the arrival time at each bus stop. To predict the delay in a route segment, we utilize not only the predicted route's data but also the data from multiple other routes that share the same route segments. Another Kalman filter is applied on the preceding trips' delay data to predict the delay for the requested route. We will integrate more data sources into the system, such as weather conditions, traffic congestions, special events, and so on.

36.4.2.2 Dashboard and Recommendation Engine for City Planners The Transit Hub decision support system will also cater to the needs of MTA engineers and city planners. The current implementation of the Transit Hub analytics dashboard is shown in Figure 36.4. Choosing the options from different routes, directions, weekdays, and time period in the day, the analysts from MTA and the city can utilize the information provided by the dashboard to check the historical delay patterns and outliers in each route segment, identify which parts are the bottle necks in the routes, and come up with solutions to modify the bus schedules and optimize the performance of the transit system.

It will enable what-if analysis based on simulations and historical traffic and demand data and help in answering questions like how do the riders get affected if the number of buses are reduced or increased, buses get rerouted or a traffic accident occurs? It will also assist in simulating cascaded delays due to congestion. Designing incentives for passengers to take buses from particular stops and routes based on simulated results is another goal.

36.4.2.3 Infrastructure Requirements for Transit Hub The Transit Hub has been designed to form the backbone of Nashville MTA's scheduling and planning services. This imposes strict requirements on the system that needs to be fault tolerant and provide timeliness guarantees. The users of Transit Hub application also expect availability and timeliness bases service level agreements (SLAs). The scale of the system brings its own challenges.

The real-time feeds accumulated by the data collection service are at the scale of several gigabytes per day. This number will keep increasing as MTA expands its services and we introduce new data sources. This requires us to design infrastructure that can efficiently persist and query data at terabyte scale and handle petabyte scale data in the longer run. We also need data replication to withstand failures. To fulfill these requirements, we use a distributed NoSQL database, MongoDB residing on a cluster of servers. However, going forward, we need a database which can support even larger datasets.

Another resource-intensive component of Transit Hub is the Analytical Engine that periodically or on demand runs simulations to predict delays on different routes. However, maintaining a huge pool of virtual machines to perform simulations is expensive. We need efficient resource management algorithms so that the service is viable. To that end, we are developing algorithms utilizing Linux container-based virtualization technology [11].

36.4.3 Kiosk Systems for Human–CPS Interaction

Today's platforms have evolved from being simply technical innovations to also focus on realizing powerful, socioeconomic platforms affecting the daily lives of millions of users. These socioeconomic platforms enable a new class of applications and services to emerge at an unprecedented speed, quality, and cost. These trends can help cities and urban areas to increase their pace of innovation, while the effort expended in planning and implementation can be reduced from time spans ranging

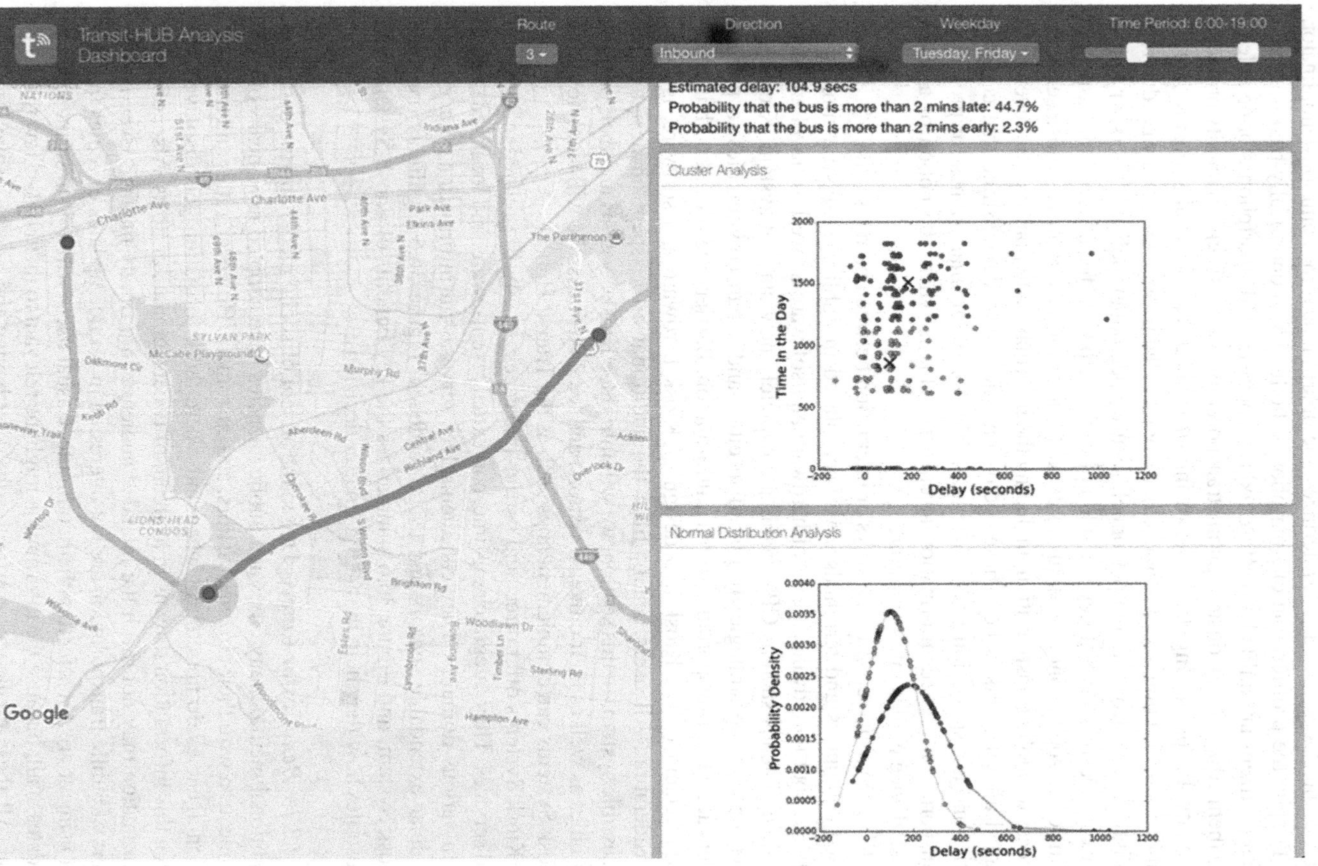

FIGURE 36.4 Transit Hub analytics dashboard. Reproduced with permission from Abhishek Dubey.

from years or even a decade to just months. As an example, consider the state of art for user interfaces or wearable interfaces: the smartphone is the common denominator.

Despite the proliferation of smartphones, both socioeconomically backward and senior citizens are unlikely to afford today's sophisticated smartphones or be able to use them effectively. Consider public transportation as a basic civic amenity provided by a city. People having difficulties using smartphones are often the most reliant on public transportation, for example, elderly or disabled people. Often they need to rely on information from station personnel, security personnel, police officers, or fellow citizens. Despite several new mobile apps that help plan for transportation, people with lower socioeconomic status often lack the resources to access those services. Access to public transportation, including the information to plan trips has even been named a civil rights issue. These socioeconomic issues demonstrate that communities and transportation providers cannot rely solely on smartphones for providing transportation-related information to citizens. This is where the notion of a smart kiosk device comes into the picture.

Kiosk Systems enable new ways of interaction between humans and CPS. Touch-based computers and terminals have been placed in public settings for more than 20 years, but the advancements of hardware and software as well as novel architectural approaches such as CPS enable a completely new class of system. Kiosks are envisioned to be strategically placed in cities and urban communities to provide riders access to transportation-related information. Barrier-free access, both in terms of placement of the Kiosk and design of Kiosk hardware and software, is a key consideration. All users benefit from the additional effort spent on designing an accessible system. In this section, we will introduce Kiosk Systems and their benefits for cities as well as their citizens, visitors, and local business. We will also show how Kiosk Systems can complement apps such as the Transit Hub.

Kiosk Systems offer direct and barrier-free access to city-life-related information and services. They integrate real-time, contextual, event-aware, and localized content from various domains. They can themselves serve as an information source using modular extendable sensors and usage feedback. Figure 36.5 highlights one such Kiosk System, and Figure 36.6 summarizes the key benefits of Kiosk Systems and illustrates how they help in making smarter decisions.

36.4.3.1 Benefits for Cities Due to the advent of social media and the decline of traditional media, many cities and public agencies have adopted new means of communication with their citizens, visitors, and customers. However, cities and transportation agencies do not control when and where the information gets displayed. So, often this information drowns in a stream of tweets, emails, or posts. Also, often the information may not reach the intended audience in a timely manner. Even regular users of public transportation are not interested in service updates of routes they are not using or are not affecting their commute. Casual users will not subscribe to those updates at all, because the information might be relevant to them only a few times per year. All these challenges can be resolved by having the kiosks display the right information at the right time.

FIGURE 36.5 Example of a Kiosk System—Siemens smart city hub. Reproduced with permission from Abhishek Dubey.

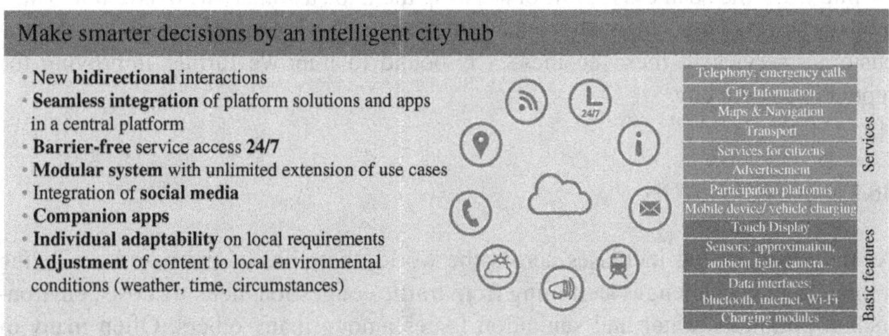

FIGURE 36.6 Decision support by a Kiosk System. Reproduced with permission from Abhishek Dubey.

Additional benefits accrue from the ability of the kiosks to provide instant feedback to city officials because the kiosks can track and report on the likes and dislikes of citizens who utilize the services of the kiosks thereby enabling the officials to refine their offering. A prominent capability of the kiosk is the fact that it can be customized for the needs of the city. This capability stems from the understanding that cities have different needs, they face different challenges, they have different geographical features, and they may have different historical past. Having a kiosk be tailored to the needs of the city is a significant benefit for the city.

36.4.3.2 Benefits for Citizens Kiosk Systems enable quick access to services such as public transport, bike sharing, parking, games, and navigation integrated into one interactive platform. The reason why this is more attractive is because this approach does not incur the same impediments as those incurred in traditional communication mechanisms, such as mass mailings or local newspapers. These latter approaches cannot convey real-time information and updates. The Kiosk System enables information on and booking of such services. The hardware platform itself facilitates easy information consumption by using large multitouch screens. Free (wireless) Internet access and charging facilities for smartphones or tablets are additional benefits for citizens. Charging facilities can also be extended to electric bicycles and cars. Many more capabilities can be included in a kiosk. For example, it can be equipped with medical equipment ranging from simple first-aid kits all the way to lifesaving defibrillators. They can also be used to track localized crime, such as an assault, and help dispatch security personnel to the scene in a timely manner.

36.4.3.3 Benefits for Entrepreneurs and Local Businesses Kiosk System enable winning new customers by means of advertising that are not considered as such. If a traveler is looking for a small, independent café that is not located at a nearby location, a Kiosk can provide him or her directions to the café. Incentives such as coupons can also be targeted effectively, for example, limiting them to a certain period of time when the business is slow or showing them to customers who have not visited a business yet. Thus, kiosks have the potential to boost local economy. In turn, the customer service in these businesses is bound to improve further improving the reputation of the city.

36.5 CONCLUSION

As the urban sprawl increases across the world, the cities of the world are facing unprecedented challenges stemming from traffic congestion, housing costs, environmental pollution, water and sanitation issues among many others. Often many of these challenges cannot be effectively addressed due to the bureaucratic structuring of the city government and the lack of interoperability across various departments. Beyond these limitations, however, many of these challenges remain unresolved also because technology has not been harnessed to its fullest potential despite significant advances in wireless and mobile connectivity, proliferation of smart end devices such as smartphones and other IoT technologies, and powerful computational platforms such as cloud data centers.

This chapter described our efforts in harnessing these technological advances in the context of smart cities. Specifically, we have focused our efforts on the Transit Hub project being designed and deployed for the city of Nashville, Tennessee, USA. The Transit Hub is a multitier architecture comprising the Nashville MTA's buses which act as the sensors of our IoT architecture and providing their location information, a trip planning mobile app running on smartphones that enables travelers to plan their trips, and a cloud-based decision support system that analyzes real-time traffic and bus location data to serve the trip planning requests made by smartphone users. Complementing the smartphone-based app is also the Smart Kiosk system, which can be deployed at strategic locations within the city to guide visitors and also city residents, particularly those from socioeconomically backward strata who cannot afford smartphones or the elderly who cannot operate the smartphones, in guiding them to their destinations.

Preliminary ideas and working artifacts from our Transit Hub project were demonstrated at the 2015 Global City Teams Challenge program [12] hosted by the US National Institute for Standards and Technology (NIST). Our team is now partnering with the Nashville city government addressing multiple additional challenges beyond just the bus trip planning. We aim to make further progress in the future and demonstrate our ideas at venues such as GCTC 2016, as well as evaluate the efficacy of our technology and its benefits to Nashville. Based on these outcomes, we plan to reach out to other cities providing them insights gained from our work and make our technology available to them.

ACKNOWLEDGMENTS

This project uses derivatives from work funded in part by the US National Science Foundation and Siemens Corporate Technology.

REFERENCES

[1] Report: Nashville's traffic congestion 33rd worst in the Western Hemisphere. http://www.bizjournals.com/nashville/blog/2014/06/report-nashvilles-traffic-congestion-33rd-worst-in.html (Accessed September 28, 2015).

[2] Comparison of Nashville MTA with Peers. http://www.nashville.gov/portals/0/SiteContent/MTA/docs/StrategicTransitMasterPlan/06Ch4PeerReview.pdf (Accessed September 28, 2015).

[3] TRANSIT-HUB. https://thub.isis.vanderbilt.edu (Accessed September 28, 2015).

[4] Cyber Physical Systems. http://www.nsf.gov/news/special_reports/cyber-physical/ and http://www.nist.gov/cps/ (Accessed September 28, 2015).

[5] AMMO – Android Mobile Middleware Objects. http://www.isis.vanderbilt.edu/sites/default/files/u352/Sandeep-ammo_poster.pdf (Accessed September 28, 2015).

[6] Graham Hemingway, Himanshu Neema, Harmon Nine, Janos Sztipanovits, and Gabor Karsai. 2012. Rapid synthesis of high-level architecture-based heterogeneous simulation: a model-based integration approach. *Simulation*, **88**, 2, 217–232 (Sage Publications).

[7] General Transit Feed Specification Reference. https://developers.google.com/transit/gtfs/reference?hl=en (Accessed September 28, 2015).

[8] Here: Maps for Developers. https://developer.here.com/ (Accessed September 28, 2015).

[9] "Simulation of Urban MObility" (SUMO). http://sumo.dlr.de/wiki/Main_Page (Accessed September 28, 2015).

[10] OpenStreetMap. https://www.openstreetmap.org/ (Accessed September 28, 2015).

[11] Shashank Shekhar, Hamzah Abdel-Aziz, Michael Walker, Faruk Caglar, Aniruddha Gokhale, and Xenofon Koutsoukos. 2016. A simulation as a service cloud middleware. *Annals of Telecommunications*, **71**, 3, 93–108 (Springer).

[12] Global City Teams Challenge. https://www.us-ignite.org/globalcityteams/ (Accessed September 28, 2015).

37

SMART HOME SERVICES USING THE INTERNET OF THINGS

GENE WANG[1] AND DANIELLE SONG[2]

[1] *People Power Company, Redwood City, CA, USA*
[2] *University of California, Berkeley, CA, USA*

37.1 INTRODUCTION

Smartphones are the most popular consumer electronics device on the planet. We carry little supercomputers around in our pockets or purses every day. And we use them for all kinds of purposes, not just making calls. In fact, when you talk to your friends about something you're trying to do, they're likely to say, "Well, there's an app for that." At People Power, we have been researching how smartphone control can extend to the Internet of Things (IoT). There are infinite possibilities, but we are targeting the areas in your life where you'll get the most out of connected devices. We are building smart home services running on smartphones that connect users to the people and things they love most.

The service starts free. You can download our app Presence to turn your old smartphone into a free webcam or security camera with motion detection and video alerts. In addition to this, you can scale up and begin forming an ecosystem of connected and cooperating devices in your home.

37.2 WHAT MATTERS?

Homes are expensive, and people want to protect them.

Internet of Things and Data Analytics Handbook, First Edition. Edited by Hwaiyu Geng.

Companion website: www.wiley.com/go/Geng/iot_data_analytics_handbook/

As IoT has taken off, people have come up with all kinds of new innovations, from remote-controlled home appliances and smart baby monitors to smart wristbands that track your activity level. The tech world as we know it is expanding to reach every part of our lives, and each added function brings another level of control and possibility.

Although IoT is multidirectionally expanding by leaps and bounds, the innovations that will be most impactful relate to what people care about the most: family, close friends, pets, and valuable belongings. The value is clear; the IoT services that have caught on and become part of people's daily lives are the ones that deal with security, healthcare, and saving money. IoT devices that do things like help you fix your posture or improve your golf swing will surely bring a positive change to your life, but to make the most positive impact and do the most good, our innovations must be driven by what people prioritize.

The home is the locus of these values. Not only is a home expensive, but it is also invaluable for the things it protects: family, pets, and your valued belongings. People's lives are centered around their home, which is why smart home services will be adopted into our daily patterns and lifestyles. So you can spend less time worrying and more time living and doing the things that are important to you with the ones who are important to you.

37.3 IoT FOR THE MASSES

How do you bring IoT to the masses? There are a number of technologies through which you could attempt to build the most far-reaching solution, but the ideal solution is an end user sticky app. We have already discussed that IoT services will be the most impactful when they relate to what people care about the most—but they will also catch on the best when the services are easily accessible and able to support all of an end user's needs. Smartphones have replaced computers as the main hub from which people connect to IoT, so they will be the hub through which we reach the masses.

We need a solution that can support many different kinds of devices and protocols so that you can do things like reduce your energy usage, secure your home, and check up on your loved ones from the palm of your hand. To support such a system requires powerful analytics, and to empower users to build a secure community, there must be a social networking aspect.

An IoT solution for the masses must allow large-scale deployments of IoT networks to be supported. In order to achieve this, the software must be able to weave together disparate platforms, devices, and protocols to create an ecosystem of smart devices in your home.

The great thing about smartphones is that not only can you use it as a hub, but every smartphone is already equipped with a number of sensors. Most people do not return or donate their old smartphones when they upgrade, so those unused phones can be used as free IoT sensors.

37.4 LIFESTYLE SECURITY EXAMPLES

Lifestyle security is the term that acknowledges the varied and dynamic nature of your unique lifestyle and offers the solution of a home security system that matches your every need. It's not your grandfather's security system. Gone are the days of installation fees and expensive hardware for a security system that leaves much to be desired and has rigid constraints. Lifestyle security becomes integrated into your daily routine and is optimized for your smartphone to meet your daily needs.

37.4.1 Smartphones as Internet of Things Devices

The smartphone is not only a controller but also a sensor. Utilizing the sensors in the phone as IoT end points, you can repurpose any old, unused smartphone into an IoT device. Here is a collection of case studies using a repurposed smartphone as a motion-detecting security camera.

Our first case study involves catching burglars. One family had set up their motion-detecting security camera to keep an eye on their kitchen area while they were on vacation. When their dog sitter came into the kitchen, opened their cabinets, and began stuffing the family's prescription medication into her shirt, the motion-detecting webcam was prompted to capture the act and notify the family with a video alert. As it turns out, this particular dog walker had done this many times in the past. She was arrested for stealing type 2 narcotics.

When the Ryu family was out of town, they had a house sitter care for their house and make sure their cat was fed and doing well. The resourceful house sitter wanted to keep track of the cat and set up a camera to watch for any irregular activity but ended up catching armed burglars on video stealing the family's computer.

The security webcam has been used to supplement regular security systems as well. A user—Chris—cleverly installed a security camera in a vending machine. There was already a regular security system in place, but the camera placed inside the vending machine captured the three people breaking into the machine from a clearer angle, joking around as they use a small knife to work at the machine, with faces in plain sight. The video was blasted on local news to aid in finding the people.

We know many parents who work long days for the sake of their families and so aren't able to be at home when their kids finish a day at school. A camera installed in the entranceway as a front door camera can be used to check whether your kid has come home (or not!) from school. It'll give you that extra bit of security being able to see for yourself that Johnny has made it home unscathed from his third grade classroom.

These are also great for watching pets. People have used them to make sure their sick or lonely dog is doing okay when they are left at home alone. These have also been used for sneakier purposes like checking to see which of the neighbor's dogs is pooping on your lawn. Another use has been veterinary care. If you bring your dog to the veterinarian for surgery, you can watch your dog in its cage as it receives aftercare. You don't have to change your work schedule or leave your other responsibilities behind to be assured that your pet is recovering safely.

Video recordings also open up a whole new world in being able to observe wildlife that you would normally miss. It has been used to observe owls at zoos, and people have caught giant herds of deer or a couple of skunks going through their yard. Users who care for chickens have also used the motion-detecting cameras inside their chicken coops to make sure no suspicious characters, be they burglars or foxes, were making off with eggs. This way, you can also be aware of all of your home's visitors; maybe it was a bird—and not the mailman—that was tampering with the contents of your mailbox!

37.4.2 Entry Sensors as Internet of Things Devices

The door sensor can be affixed to your front doors and windows. Now you can know when your teenager comes home at night, and when you are away from home or on vacation, you can be alerted when someone opens your front door or your child's bedroom window.

37.4.3 Water Sensors as Internet of Things Devices

If your house floods due to old water pipes breaking, that's thousands of dollars of damage. I've had neighbors who have had old pipes burst in their house—twice! But what if you were able to catch the leak sooner? This goes for people who have vacation homes as well. Keep your mountain cabin secure so that you are alerted right away when pipes freeze, thaw, and burst. That's a lot of damage and expense avoided.

37.4.4 Touch Sensors as Internet of Things Devices

The touch sensor alerts you if it moves. This allows you to see if your babysitter has been stealing from your liquor cabinet or if your children have been reaching into the cookie jar. Simply attach the touch sensor to the lid of the jar or your bottle of expensive whiskey and it will go off if someone moves it or lifts the lid.

37.4.5 Motion Sensors as Internet of Things Devices

The motion sensor lets you detect motion—day or night. It is infrared so it can see in the dark, meaning you can have coverage around the clock. You can mount the motion sensor onto a wall or near the ceiling so that it can look down into the room you want to monitor, and it will be able to tell if anyone is moving.

37.4.6 Smart Plugs as Internet of Things Devices

Smart plugs can be installed at critical points throughout the house so that energy use is shut down after everyone is asleep. With commands that repeat themselves daily, you can be sure your TV and entertainment station are turned off every night so that you save money every night you go to bed and do your part in conserving energy.

37.4.7 Robots as Internet of Things Devices

Robots can help give you more control. You can mount your motion-detecting iPhone onto a main motorized iPhone mount, and then from your smartphone, remotely control the direction and angle of your camera's field of vision.

37.4.8 Ecosystem of Devices

One of the true benefits to having a smart home system is having an ecosystem of devices working in concert with one another (Figure 37.1). A natural language rules engine lets your connected devices cooperate for you. Not only can we connect to any device easily, but we can also make your devices work together with simple "if, then" commands. For example, if I detect someone opening the front door and it's after 9 pm and before 8 am, then I will beep the alarm, take a picture, and turn on the light.

You can also switch your entire entertainment system off every night after you finish using it or set your system to turn the lights on every evening when you are away on vacation to deter burglars from targeting your home… the possibilities are there at your fingertips!

37.5 MARKET SIZE

The market for smart home services is growing rapidly as it becomes clear that the future for home security must be accessible, easy to install, and affordable. By 2020, five billion people will depend on 50 billion connected devices surrounding them. By

FIGURE 37.1 This screenshot is an example from the app Presence. It depicts the screen of an end user coordinating connected devices to work together for their benefit.

2022, an affluent home in the developed world will have over 500 smart devices of all types. By 2025, 50% of our homes in the United States will be "smart" homes, mostly populated by Do-It-Yourself (DIY) systems that the end users install themselves.

37.5.1 Homes and Home Security

To give you an idea of the scale of this market, we will first look at the housing market. The United States has 133 million housing units, and China had almost 456 million households in 2012. The United States and China are among the biggest markets for home security, with 61% of the market share of home security concentrated in the Americas and around 21% in Asia and Pacific.

The global market is growing. According to a report from MarketsandMarkets, the global home security market was valued at $28.3 billion in 2014 and is predicted to reach $47.5 billion by 2020.

37.5.2 Home Security Market Challenges

The home security market has plateaued at 25% penetration and traditionally consists of middle- and upper-income groups. This leaves lower-income populations completely unprotected by home security firms. This is paradoxical as lower-income people tend to live in neighborhoods with higher rates of crime and thus have a higher need for security measures.

Those who live in multifamily buildings and rent their homes are also disadvantaged by a system that targets wealthy homeowners. According to the chief analyst at NextMarket Insights, one in three US households rent their homes, and younger consumers, who tend to be more mobile, are less likely to consider purchasing a traditional home security system because they are not easily transferable to new homes. There is a need for a more flexible security service that targets those who are priced out of expensive traditional security alternatives.

37.5.3 The IoT Solution and IoT Market

The demand for flexible and low-cost home security is being met through the smart home and IoT, which includes all internet-connected devices that can be controlled remotely through a smartphone or remote.

The IoT market is blooming. Business Insider says that IoT will become the largest device market in the world by 2019, double the size of the smartphone, PC, tablet, connected car, and wearable markets combined.

This coincides with a trend in home security toward DIY solutions as an emerging alternative to traditional security solutions. According to a report from NextMarket Insights, the DIY self-installed home security solution will account for $1.5 billion in equipment and services by 2020.

37.5.4 IoT Challenges

Despite the huge potential for growth, there remains a general lack of adoption of IoT solutions. Sixteen percent of households currently have at least one smart home device, including things like smart sensors, lights, and door locks. Of those households that do not have a security system, 20–40% of US households express willingness to adopt home security with smart aspects. The lack of awareness of smart home solutions is to be expected, given that the market for smart home solutions is still relatively new. In order to take advantage of the market, social media and word of mouth will be important factors in spreading awareness.

37.5.5 A Future of Success

There are opportunities for implementation in other nations like China, which has a huge market with more people. China has recently become the greatest iOS app-downloading country. iOS downloads grew 30% from Q1 2014 to Q1 2015, surpassing the United States and making China now the largest market for quarterly iOS downloads. China will be a large market for IoT home security, with a growing tech sphere and a prevalence of apartment homes, which will benefit from a cheaper security system.

In addition to the national growth, it appears that apps in China related to lifestyle are growing at a faster rate than apps for entertainment and photo and video. This indicates that people are downloading an increasing number of apps that concern their lifestyle relative to fun apps or photo and video apps. As we have mentioned before, the apps and technology that impact the areas people care about most will succeed—and it seems that lifestyle apps are the types of apps that are being integrated into people's daily lives.

37.6 CHARACTERISTICS OF AN IDEAL SYSTEM

37.6.1 Characteristics for End Users

The ideal system will fit your mobile lifestyle, be managed from the palm of your hand, set up swiftly and easily, have the potential to expand to cover whatever you need, and connect you to your community.

37.6.1.1 Mobile Lifestyle The ideal system suits you and fits with your mobile lifestyle. In the increasingly mobile lifestyle of the twenty-first century, it is necessary to find new and innovative ways to integrate the things you value into your mobile device. This is what we call lifestyle security. This could mean anything from being able to ensure that your house is secured while you are away on vacation to checking whether your child came home from school while you are at work or at the grocery store.

The ideal smart home security system would target the safety of the house. It combines industry-leading sensors and motion detectors with the mobile app to monitor doors and windows, detect motion, and receive updates on temperature and water detection through your smartphone. This is the ultimate use of your smartphone as the hub and "remote control" from which you have the power to ensure your home's security and your freedom from worry.

The ideal energy management system on the other hand would deliver real-time, whole-home energy monitoring and smart plug control, giving you the ability to manage electricity use from a smartphone or tablet. This is perfect for residential and small business users to save money every month with just an app and some plugs.

37.6.1.2 Managed from the Palm of Your Hand The ideal system can be managed from the palm of your hand. This means you should be able to do everything from one app instead of needing multiple apps to control multiple devices (Figure 37.2). Whether you are managing your energy or keeping track of your home security, both tools should be controlled and monitored from the same app.

37.6.1.3 Out-of-Box Experience An ideal system sets up in minutes and doesn't require advanced knowledge of technology or electronics in order to begin using it expertly. Out-of-Box Experience is a key component of product design and should be easy enough for end users to do without help.

The ideal out-of-box experience is as easy as placing the sensors where you want them and deciding on rules and actions in your app. It would be a simple, clean process that has the system up and running in minutes, with an in-app video to tell you what to do so that you don't need to rely on technicians. The ideal Out-of-Box Experience allows you to set up your own smart home system independently (Figure 37.3).

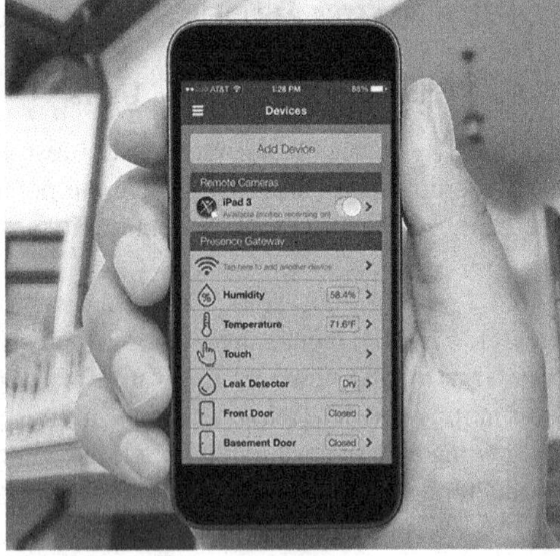

FIGURE 37.2 We are increasingly utilizing mobile devices to keep up with our increasingly mobile lifestyles.

FIGURE 37.3 This is an example of a security pack with a clean, elegant design and a vast array of sensors so that users don't need to go out of their way to purchase their own sensors.

37.6.1.4 Framework of Innovation The ultimate component of a solution that adapts to your life is one that expands as your needs grow (Figure 37.4). The ideal solution evolves as you and your ideas evolve, expanding to add more sensors or outlets under your command or downsizing as you see fit. You should be able to custom-make new commands and actions as you require from within your app, and the system should expand or collapse with just a few changes in the app, switching the desired sensor off.

37.6.1.5 Social Networking The ideal system is about more than just you and your belongings. A social aspect would work with your family and friends, creating a connective community aspect both on social media and in the neighborhood.

An ideal system lets you share with your friends on social media and involve your family and friends in your community or neighborhood who you trust to help you watch over your house. This proves useful if you are leaving for vacation. Instead of relying on a call center, you can have someone receive alerts from your home for you so that they can check up on your home or help as needed. This is all about integrating the system with your life, as part of the connections you already have with people (Figure 37.5).

FIGURE 37.4 This screenshot shows an example of an app that is customizable to the end user's needs. The user can add or delete devices at the touch of a button.

37.6.2 Characteristics for Partners

For partners, there are other factors on the back end to ensure that the solution is ideal for each service provider or manufacturer. An ideal system is customizable, able to be translated into different languages, and offers a framework for innovation.

37.6.2.1 A Customizable System Customizable systems would allow partners to allow partners to easily add new capabilities, applications, and analytics that allow them to build new systems in the future that they haven't even imagined today. This makes for a flexible solution that allows the company to concentrate on what they think is most important now for their customers while allowing the service to evolve into the future.

The ideal software stack offers our partners the chance to become part of the growing IoT world without having to build the infrastructure on their own. Regardless of the suite of programs the company wants to build, a good software stack will

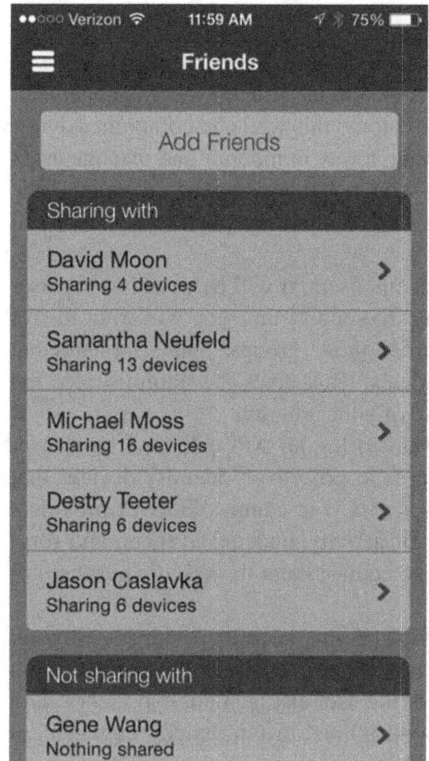

FIGURE 37.5 This figure displays a friend's list on the first screen and on the second screen, an example of how you can have your friend David Moon receive alerts about your "expensive whiskey" for you.

weave in easily with what products a company already has, or become the platform on its own. This will allow companies to offer life-changing services in areas like home security, energy management, and elderly care with ease.

37.6.2.2 Localizable for Other Languages A system that can be localizable for other languages allows us to come out with a French or Chinese version of an app, creating more opportunities in the global market for expansion.

37.6.2.3 Framework for Innovation The ideal system offers a framework for innovation. It grows and evolves to support new connected devices and powerful new analytics, providing a constant innovative outlook for future services and use cases and allowing your company's visions and dreams to be seen to completion. Not every company has the time or resources to scale up to match and beat their competitors. An ideal system helps take care of that job for them.

37.6.3 Administrative Console

One great use of IoT has been for service providers to utilize IoT software to have an administrative role-based web console for device management, user management, user communication, and engagement. In order to do this well and comprehensively, you would need a dashboard for basic reporting, a way to monitor and manage users and groups, a communication method to send messages, a way to engage the community via challenges, and points and rewards to reinforce positive behavior and form habits.

In order to give great technical support to end users, it will be useful to log user activity and statistics and diagnose problems. To support this, you will want to sort users into groups, track various metrics from those groups, and send targeted, customized messages to groups and individuals. This gives the administrator the knowledge and power to manage every phase of your program.

For diagnostics and support, you will want to log all API calls and track user activity and statistics. There should be features to proactively identify devices that have issues so that you can manage devices across your campus. Service providers will want to manage users, take notes on their activity, understand them, and form Groups of users, manually or automatically, to control them for specific purposes or better manage your actions.

Communication must be enabled in various forms such as in-app messaging, e-mail, and text message. Another way to engage your user base is through challenges and games. Challenges are part of the user engagement experience, and you can reward habit-forming behaviors with points, driving sticky engagement programs.

37.7 IoT TECHNOLOGY

In order to carry out all of the requirements of an ideal system as discussed earlier, you need a fully developed software stack. The example we will use is People Power Company's answer to the demand for IoT technology.

People Power's IoT Suite includes Presto, which connects IoT devices to the Cloud for free (Figure 37.6); Symphony, which provides social engagement, data analytics, and a blazingly fast mobile rules engine; Virtuoso, which allows telcos and utilities to offer compelling apps for home security, energy, healthcare, and more, under their own brand; and Maestro, which enables service providers to technically support end users; engage them with challenges, points, and rewards; and upsell high-potential users to new devices and services.

37.7.1 Connectivity Layer

The fastest and most cost-effective way to cloud-enable any product Presto is free for products with up to 2048-bit SSL, optional bidirectional authentication, up to 12 measurements and status updates on average per hour, and near-instantaneous command delivery.

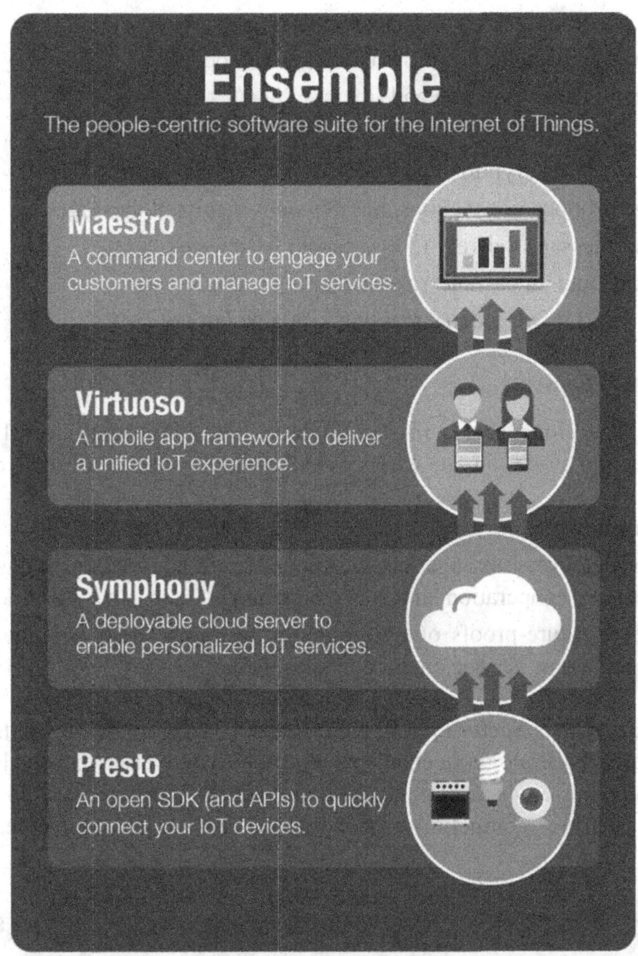

FIGURE 37.6 IoT software architecture.

Presto is the fastest and most open connectivity layer available today for IoT. It is also free for manufacturers to Internet-enable any device, and with open APIs and an open developer console, it's extremely easy to get connected. It also connects to other clouds through cloud-to-cloud integrations and OAuth 2.0 and can even connect with devices that are already shipping in market by teaching Presto how to speak that device's language. Once connected, manufacturers can take advantage of the upper layers in IoT stack, including Composer, Harmony, Virtuoso, and Maestro.

Presto's Key Features:

- HTTP GET and POST to send measurements and receive commands.
- Both JSON and XML support, it's the developer's choice.
- Bidirectional data streams, usually less than 50 bytes per minute.
- NAT and firewall penetration.

- Synchronization in 0.25–0.50 second latency
- Online and offline recognition of devices and people
- Automatic data correction and filters
- Vector measurements for maximum database efficiency
- 2048-bit SSL encryption and optional bidirectional authentication
- Manage dropped connections, reliably deliver missed messages
- Marshal message delivery to low-capability devices
- Broadcast and unicast messaging, sharing a single pipe
- RTMPS and RTSP video streaming available but not free

37.7.2 Cloud Server

A carrier-grade cloud server includes family and community social networks, e-commerce, and a rules engine which can all be deployed into any data center.

Why is Symphony important?

- Harmony delivers an engaging customer experience that encourages sharing, competition, cooperation, and social commerce.
- Composer future-proofs organizations by enabling smart learning, data analytics, and a new service creation environment.

37.7.2.1 Harmony: Social Engagement Layer Harmony implements the social and behavioral psychology mechanisms required to deliver an extremely powerful and engaging customer experience (Figure 37.7). Harmony enables customer interaction, sharing, competition, cooperation, and social commerce. Harmony

FIGURE 37.7 Harmony social engagement layer.

enables Trusted Social Networks and Community Social Networks and shifts the user experience from individual smart homes to collective smart communities. Uniquely, Harmony also bridges IoT with popular social networks, such as Twitter and Facebook (Figure 37.7).

Trusted Social Network is like a neighborhood watch program, enabling users to share events from their home from specific devices with people they trust, typically while they're away or on vacation. Another form of a Trusted Social Network enables caregivers to all participate in taking care of a loved one and receive alerts if that person falls or hasn't taken their medicine.

Community Social Networks are for communities and neighborhoods. We connect people who may be physically nearby each other but may not necessarily know each other directly. There are many benefits to doing so, including the powerful behavior-modification strategies that can be employed across a population. For example, an energy engagement program employs the following psychology strategies based on the formation of a Community Social Network:

Social proof—People want to do what others like them are doing. Harmony enables groups of users to see what others are doing, as well as share their own experiences.

Motivational bridges—Motivational bridges reward good habit-forming behaviors and word-of-mouth advertising with points, which translates into meaningful rewards.

Goal setting and public commitment—Setting attainable goals that are worthy enough is proven to motivate people to make lifestyle changes. Making goals public, rather than private, increases the likelihood they will be achieved. Accountability is key and a fun way to merge community and goal setting together.

Incentives and competition—Winning a competition is a prize unto itself—status—and competitions inherently incorporate aspects of social proof and goal setting. Competitions can range from living healthier lives or being in touch with your home environment to being an active promoter of smart home service to others.

37.7.2.2 Composer: Smart Learning Analytics and Automation Composer is a real-time big data and smart learning services engine. Simply put, this is the foundation for an application development and soon app stores for IoT—which will add new value and features for residential and business customers. The power of Composer lies in its ability to enable data scientists and developers to creatively develop apps on top of any Internet-connected device. Unlike a mobile app, apps for IoT run 24/7 in the background of user's lives. The app store of the future will not be in your face on a mobile screen but operate silently in the background to automate tasks, notify users of important events, generate monthly reports, learn a user's or appliance's patterns and behaviors, and connect service providers with their customer base without being overalerted.

Today, Composer enables developers to create apps for connected devices in the very popular scripting language Python. With appropriate user permissions, apps have the ability to listen to data streams from devices, control devices, generate alerts and emails, and learn. In the near future, apps will be able to ask users questions to gain additional context around the data.

To ensure app quality, our Composer SDK and app store have been designed to enable users to rate and provide feedback on apps, just like the Apple App Store.

Analytics services add new value and features and/or open new markets for residential and business customers.

Features can be added to connected products beyond what the manufacturer intended, enabling clients to differentiate themselves from other smart home services.

Composer enables developers to create and customize apps and allows end users to provide immediate app feedback and rankings.

37.7.3 Mobile App Framework

An app framework for iOS, Android, and Web enable the rapid deployment of compelling IoT services featuring your brand.

37.7.3.1 Why Virtuoso? It's all about the user experience—their connection and their trust. Powerful cloud servers, intelligent analytics, and the smartest devices in the world are all useless if the interface isn't simple and easy to use.

37.7.3.2 How Does It Work? We've done the heavy lifting for you. Thanks to Virtuoso, you can offer mobile-first services that easily expand with add-on products and features. Specifically built to support IoT, Virtuoso will jump-start your interface to your users by allowing easy integration with third-party devices, cloud services, and apps.

Businesses large and small are struggling with how to quickly roll out IoT services. But end users don't want to have to use multiple apps to manage their devices. In a mobile world, Virtuoso solves these issues by combining heterogeneous device services into a single, unified app framework, empowering enterprises to deliver mobile app services that truly matter.

37.7.3.3 What Is It? Virtuoso is a customizable app framework for iOS, Android, and Web that allows enterprises and others to get a head start on new revenue streams leveraging IoT. The framework enables the control of connected devices and develops solutions such as security, care, and energy management. People Power's own award-winning Presence app is built with Virtuoso and is the first and only "freemium" model for IoT:

The Virtuoso's key features include:

- A white-label framework with proven user appeal
- Offers a time-to-market advantage
- Ease of deployment and integration with third-party devices

- Incorporates a behavioral-design approach to the user experience
- Provides for better engagement and more satisfied customers
- Enables higher average revenue per user (ARPU) and greater retention

37.7.4 Command Center

A command center enables device management and services across a large user base. Sign up users, support their needs, issue challenges, and offer rewards to keep them engaged.

37.7.4.1 Why Maestro? When deployed in large scale, IoT can be daunting. Maestro takes the anxiety out of the equation by providing you with the administrative tools you need to tend your flock of users and devices.

37.7.4.2 How Does It Work? Monitor millions of devices and control them remotely. You can onboard new users with ease and run customer engagement programs to keep them connected. Diagnostics, dashboards, and predictive analytics simplify your life.

The social engagement tools within Maestro help service providers connect with users and ensure they leverage IoT services effectively. Gamification rewards higher user engagement, leading to higher ARPU for the service provider.

Maestro can facilitate large-scale population behavior change, realizing the true value of IoT across homes, businesses, and communities alike.

37.7.4.3 What Is It? Maestro is a unique management platform for businesses to easily administer their IoT devices, services, and customer programs remotely. Our web-based management tool includes onboarding, monitoring, messaging, engagement, analytics, and reporting. Think of it as Customer Relationship Management (CRM) for your devices.

Maestro collects user, device, and program data for key insights into device monitoring and customer activity and lower administrative costs. It enables the management of multiple deployments and sites from an easy-to-use interface with big-data reporting that provides the actionable intelligence needed by service providers to keep things running smoothly and their customers engaged.

Maestro's Key Features Include:

- IoT service recruiting and device fulfillment
- Device status monitoring and troubleshooting
- In-app messaging to groups or individuals
- Gamification with Points and Rewards

37.7.5 Geofencing

A lot of people take their smartphone around with them wherever they go, so we can use the location of your smartphone to register whether you are at home or away. Geofencing lets you create a virtual fence around your home so that when you leave the range and go into "away" mode, any protocols you have set to activate when you are "away" will be triggered. This becomes useful when you are in a rush, because all you have to do is leave home and when your phone registers as "away" from home, all of your lights turn off. Geofencing is also useful for when you are on vacation. If you live in Silicon Valley but your smartphone is now in the East Coast, the system can infer that you are traveling and continue to turn your lights on and off to make it appear that you are at home and deter burglars from targeting your home.

37.7.6 IoT Services for Anybody with a Smartphone

The Presence app from People Power is leading a whole new model for IoT. For both our hardware and the total cost of our service, there is little to no cost. This puts us in a strong market position to reach those who would be deterred by expensive hardware and high recurring fees. As Tom Kerber, the Director of Research at Parks Associates said, "as consumer awareness of smart home products and services increases, the smart home market will shift to lower cost channels. People Power Presence is the lowest cost entry point for the smart home."

Our freemium model allows those who were previously priced out of traditional security alternatives to get the benefits of an award-winning app that can give them a sense of security over their lives.

37.8 CONCLUSION

In order to bring IoT to the masses, solutions must be developed that improve lifestyles in the areas people care about the most, in a way that people can easily access and use, at an affordable price.

An end user app solution empowers users to take their home security and energy management into their own hands. Powerful software that allows platforms, devices, and protocols to be woven together allows manufacturers and service providers to coordinate and manage a comprehensive ecosystem of smart devices. A powerful user management console will let you do anything from find and diagnose problems to engage communities in games that will change their behavior and begin real social change across an entire building or campus.

The Internet of Things offers endless possibilities for innovation, and the market is growing rapidly for smart home solutions on smartphones.

38

EMOTIONAL INSIGHTS VIA WEARABLES

Gawain Morrison
Sensum, Belfast, UK

38.1 INTRODUCTION

Technology can read our emotions—a subject matter that stirs deep questions about ourselves, our societies, our methods of communication, our future civilization.

There are many science fiction books and films that vary from harmonious to dystopian futures depending on how paranoid the author or the characters are. Who is to know what way the future will play out, but we have the opportunity to shape that direction by grabbing hold of the devices, the data, and the opportunities to do better for humanity right now.

And this handbook will provide a comprehensive view of the array of hardware and software technologies being created to record data of our place in space, the parameters of the world and universe around us, and what is biometrically happening in our bodies. And that's a lot of data!

While we are right to be cautious of how the data may be used and by whom, we should take this opportunity to open up the discussions for this new age of data capture and transmission—we call it the Age of the Digital Self. If we treat the digital form of ourselves with the same rights as we would treat the human form of ourselves, we have a bright future in achieving harmony with emotional technology, machines, industries, medicine, and beyond.

This chapter will give you an introduction to how you can look at the measurement of emotions with these technological advances and why anyone would bother.

Internet of Things and Data Analytics Handbook, First Edition. Edited by Hwaiyu Geng.

Companion website: www.wiley.com/go/Geng/iot_data_analytics_handbook/

This will be largely in the domain of market research, branding, and entertainment, but the same principles could be applied to any use.

38.2 MEASURING EMOTIONS: WHAT ARE THEY?

Emotions are complex. Obviously.

But it's largely understood that they're a naturally occurring response to a situation or stimulus. They are a response to changes in environment or self. They are different to moods that can last for hours, days, or weeks. They are really to evoke a reaction. A survival reaction. Anger is to make you fight back. Disgust is to make you get it out of your system or stay away from it. Fear is to make you run. Laughing or crying is to make you empathize with your fellow "flock."

Emotions research from the mid-1800s to now has shaped today's understanding of the measurement of human emotions, and even after that time theorists disagree when we get beyond the core emotions.

Most psychologists agree that our core emotions are evolutionary survival tools that most species have.

And then as humans have evolved with morals and self awareness, these have expanded.

There are largely two schools of thought to emotions—one is that your body generates the physiological change and that we then feel it, and the other is that we judge the emotion, and then the physiology changes accordingly.

But the main thing to note is that there are physiological changes that are aligned with your emotions, and these can be measured.

Some statistics would have us believe that up to 95% of decisions are made instinctively and emotionally before we even know it. The nonconscious.

But traditional market research, and therefore the decision making based on those insights, only looks at the deliberative and logical processes, which are carried out via surveys and focus groups which are open to massive bias. The conscious.

Companies then spend millions of dollars based on this one-sided consciously expressed emotional information.

It's important that we look at both conscious and nonconscious processes to gain a holistic view of emotional response.

Even more importantly is to be able to contextualize what the triggers of that emotional response and expression are.

38.3 MEASURING EMOTIONS: HOW DOES IT WORK?

We express emotional responses in one of three ways:

1. *Externally* through our face, body, and voice:
   - How we sit, carry ourselves, or express ourselves can be both conscious and nonconscious across our face, through our voice, and through our body movements. By analyzing the muscle movements in our face, the tones of our voice, and the way we sit toward or away from someone, we can express a lot about the emotions we feel.

- Paul Ekman's Facial Action Coding System (FACS) basic emotions model has allowed us to measure basic emotions, panculturally, and since you can do this kind of measurement via any camera, it is unintrusive, while some argue it impacts on nonanonymity, although this is down to how the images are captured, processed, and used.
- The main benefit of voice and face analysis, using microphones and cameras that are inbuilt to most mobile and computing devices, is that they don't interfere with the test stimulus response, since they are nonintrusive.
- Sometimes both voice and facial responses can be consciously controlled by a respondent, so not indicating their true emotion.
- Additionally mild emotions and mixed emotions, beyond the basic emotions, can be difficult to assess.
- Some expertise is still required to be able to interpret what the signals mean.

2. *Internally* with physiological changes in "arousal," which can be measured via changes in heart rate, skin conductance, skin temperature, breathing among others:
   - These operate at the nonconscious level and cannot be controlled by a person thus providing an objective point of view on how that person emotionally responds to the stimulus or experience. These responses are also outside of cultural and social variables so they can be measured across cultures and groups.
   - While these signals are great for identifying when a stimulus has occurred, it is still not possible to associate an emotion to that stimulus without secondary data to assist. Also it is important to gather as much contextual data as possible to qualify whether the response is an emotional stimulus or driven by an alternative physical response, for example, exercise.
   - Some expertise is still required to be able to interpret what the signals mean.
3. *Consciously* expressing how we think we feel:
   - Through self-reporting how we feel, be that a survey, a focus group, or a conversation, we can offer up a wide range of emotions describing how we feel, from positive to negative.
   - A wide number of tools exist for asking questions to express how we feel and also in displaying them, requiring little expertise in designing or interpreting them.
   - The problem with conscious measurement of emotions is that they can be biased based on what people think you want to hear, or people can find it difficult to provide the precise emotion.

38.4 LEADERS IN EMOTIONAL UNDERSTANDING

There are a wide and varied number of models and theories when you look into this space of emotions, with thought leaders specializing in psychology, medicine, biometrics, neuroscience, and behavioral economics, among the disciplines.

I'll lightly touch on three individuals here: Paul Ekman, Dr. Daniel Kahneman, and Robert Plutchik, but there are many in the space of emotions research, and specifically as the interest in understanding emotions and behaviors grows, this list grows.

FIGURE 38.1 Use of FACS within a PR campaign. Courtesy of Sensum.

0 sec	0.1	0.2	0.3	0.4	0.5	0.6	0.7	0.8	0.9	1 sec	1.1
Observe			React			Express					
System one—implicit Non-conscious/emotional/fast						System two—explicit Conscious/logical/slow					

FIGURE 38.2 System 1 and System 2. Courtesy of Sensum.

Paul Ekman has been one of the main proponents of understanding emotional responses using facial coding analysis.

His research has been used to look at nonverbal emotional detection and pancultural emotional understanding and to develop the FACS (Figure 38.1) with the aim of generating a taxonomy for every human facial expression.

In 2011 Economics Nobel winner *Dr. Daniel Kahneman* released a book, *Thinking, Fast and Slow*, where he coined these terms for our two modes of thought (Figure 38.2):

"System 1" is fast, instinctive, and emotional:

- The nonconscious. The under-the-surface gut responses
- Biometric, Neurometric, Psychometric types of research

"System 2" is slower, more deliberative, and more logical:

- The conscious. The calculated processed response
- Surveys, Focus Groups, Ethnography research

Robert Plutchik was professor emeritus at the Albert Einstein College of Medicine, and his theories were based on psychoevolutionary classification of emotions that basic emotions were evolutionary.

His was an integrative approach taking a number of overlapping theories on evolutionary principles resulting in the development of the "Plutchik Wheel," illustrating in 2D and 3D a wheel of polar emotions.

Created to illustrate variations in human affect and the relationship among emotions, he looked at how emotions pair, and this kind of model is being used in quite a lot of robotics work and sentiment analysis.

38.5 THE PHYSIOLOGY OF EMOTION

Let's start with facial analysis—our body's outward facing tool to express emotion.

Paul Ekman is largely seen as the godfather of emotions research. While not the first person to research emotions, he pioneered the study of emotions in facial expressions.

His research in the 1970s identified basic emotions across cultures including anger, disgust, fear, happiness, and sadness and that the only real differentiator was in the "display rules," where a culture may conceal certain effects of the expression.

As his research has progressed, he has added a range of positive and negative emotions.

Since then there have been moves to bring that down to four since anger and disgust have similar facial muscle movement and so do fear and surprise.

Alongside this research an array of biometric and neurometric research has grown, with an ever-expanding set of tools to research with, and the area of emotions research has dramatically increased in psychology circles and beyond.

While the face expresses your emotion to your fellow humans, your body is going through continuous fluxes and changes in physiology as your emotions change, from heart rate, to sweat response, to pupil dilation, to breathing, to blood pressure.

And as such there are many ways to be able to measure those emotional changes. And this has far reaching implications for understanding emotions, from health and well-being through customer behavior to behavioral economics.

With the increasing number of wearables and sensors coming onto the market, produced from a number of sources from university spinouts, to crowdsourced prototypes, to high-end medical grade products, there is a wide disparity in the quality of the data capture.

This causes some concerns when applied to the world of medicine. Clinicians must make decisions based on the quality of the data they have access to, and if they are to integrate consumer grade data into this process, they need reliability in what they're seeing. On the flip side if it's to act as a controller for entertainment or an additional data feed into a nonmedical app or game, then the reliability of the data is nowhere near as important—you just need to know if it's going up or down and assign the parameters as necessary.

38.6 WHY BOTHER MEASURING EMOTIONS?

You can consciously describe the emotion when you've had an argument, fallen in love, or lost a family member, but it's very challenging when it comes to a glass of orange juice or packaging concept or latest commercial, and you can certainly forget about being able to quantify that emotion. Until now. With these new methods of data capture and analysis, we are getting close to being able to do that.

That's all well and good, but why bother? How does this apply to brands and customers?

In the case of advertising, it has been proven that emotive advertising campaigns perform better on *every* business metric:

A *30% increase in sales* when your ads engage emotionally with your customers

As much as a *three-fold increase in brand loyalty* and motivation to purchase

These metrics were researched and published by Les Binet, Head of Effectiveness at the London-based agency Adam&EveDDB who runs DDB Matrix, the network's econometrics consultancy. Adam&EveDDB is the producer of commercials for UK department store John Lewis, widely recognized as being some of the most emotional commercials created.

And understanding emotions isn't just important for making better advertising but for better products, services, communications, media, and experiences.

Depending on the experimental design for your research, whether it's a large quantitative study or a deep qualitative study and whether it's in lab or in the field, there are a selection of methodologies for capturing emotions, behaviors, and system 1 and 2 responses. This could be eye tracking, ethnography, biometric responses, or implicit response testing.

The key is to establish the most appropriate tools for the study, from wearables to mobile devices to webcams, and then upload the aggregated data for analysis and reporting.

Every emotion has a physiological and psychometric response, increased heart rate, muscle movement, or response time, and once consolidated deep insight is the result. And it's all about the insights.

38.7 USE CASE 1

38.7.1 "Unsound": The World's First Emotional Response Horror Film

"Unsound" was a collaborative project that brought together the disciplines of film production, music composition, environmental art, technology, and engineering to research "future cinema" and the ever-increasing demand for audience interactivity and immersion in the audiovisual experience.

Beginning as a conversation about creating films that helped the audience feel more involved and more immersed in the experience, Gawain Morrison (Sensum CEO and cofounder) and Dr. Miguel Ortiz Pérez (Sonic Arts Research Centre, Queen's

FIGURE 38.3 Image from "Unsound" screened in SXSW 2011. Courtesy of Sensum.

University, Belfast) discussed a number of techniques before deciding that tapping into emotions of a film audience could be really interesting. The aim was to create a film that was unique for every audience that watched it based on their emotional response.

Horror feature-film writer Spencer Wright scripted the film, and film director Nigel (N.G.) Bristow directed it. The film was 15 minutes in length, and a number of permutations could be viewed or heard depending on how the audience felt as they moved from scene to scene.

Small attachments to the audience member's hands pick up electrocardiogram (ECG) signals, measuring and recording the electrical activity of the heart, and electrodermal activity (EDA) which measured the change in conductance of a person's skin, which is highly sensitive to emotion arousal in people.

The world premiere of the film was screened in SXSW 2011 (Figure 38.3) and attracted interest from Disney Research to Coca-Cola, resulting in an article in the New Scientist and the creation of the Sensum platform for measuring emotional insights (Figure 38.4).

38.8 USE CASE 2

38.8.1 Thrill-Seeking Seniors? Identifying the Pension Personal

In a first-of-its-kind experiment, Skipton Building Society, the United Kingdom's fourth largest building society, was keen to gain a true understanding of people's retirement wishes, hooking up the nation's preretirees to scientific probes, revealing their conscious and subconscious reactions to images of life after work (Figure 38.5).

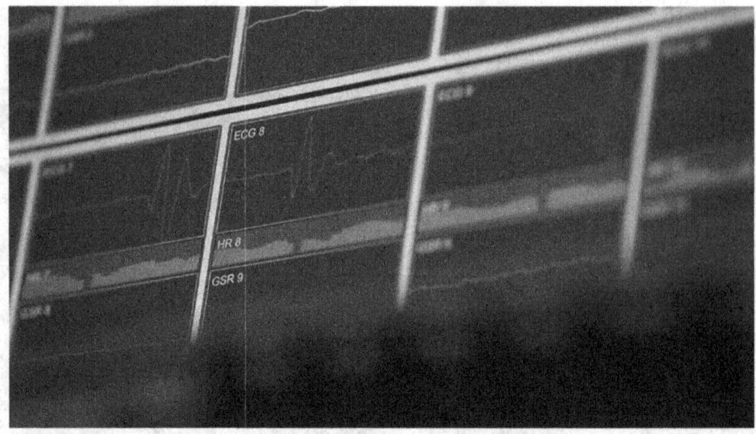

FIGURE 38.4 Sensum platform for measuring emotional insights. Courtesy of Sensum.

FIGURE 38.5 Conscious (dotted line) and nonconscious (solid line) reactions to images of life after work. Courtesy of Sensum.

The starkest finding was their dramatic physical and emotional rejection of traditional views of retirement. This included increased perspiration and goose bumps when shown key words and images associated with it, ranging from it being "the end of a chapter" to the start of their "golden years." The study also found that today's preretirees are bored by traditional "pipe and slippers" images of life beyond work.

Dr. Jack Lewis, a published neuroscience consultant and author of Sort Your Brain Out, said "Skipton has broken new ground by using physiological and sensory research, together with traditional methods. By applying this cutting edge new technology, the Society has been able to dig deep into its respondents' true feelings and combine this with qualitative and quantitative findings, to give the most comprehensive insight yet into what really makes individual people tick when it comes to retirement."

The key to these findings was combining conventional qualitative and quantitative fact-finding techniques with a scientific twist. Portable skin sensors provided by

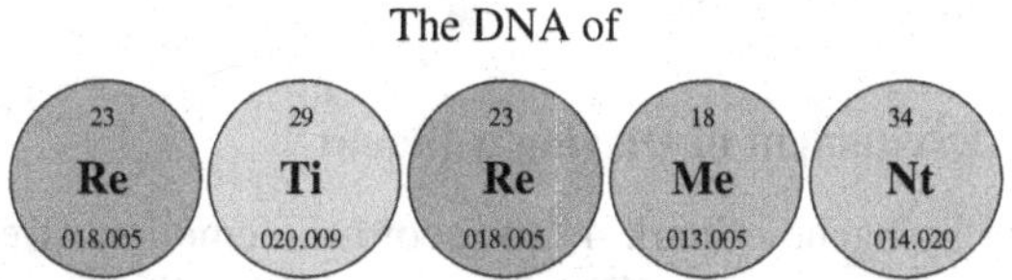

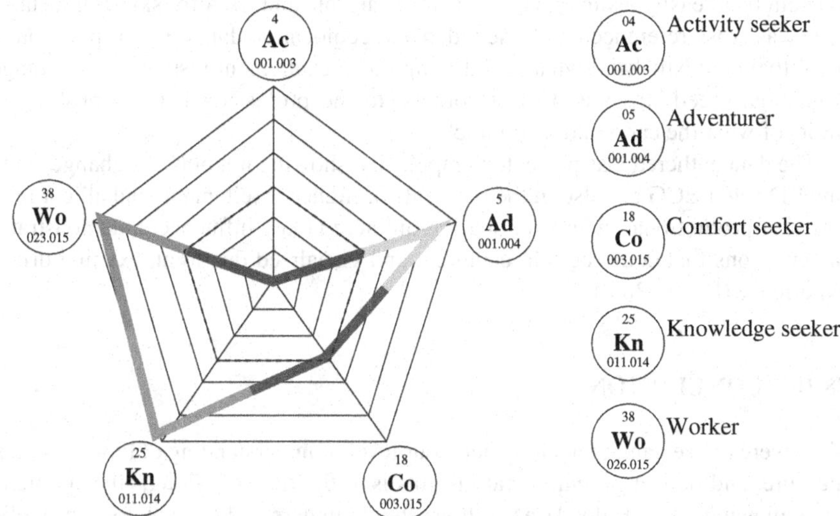

FIGURE 38.6 Retirement DNA. Courtesy of Sensum.

research technology firm Sensum were used in focus group and interview settings to help their agency Jaywing to track latent responses alongside people's mindful, verbal reactions.

- Five retirement personas discovered—research pinpoints individuals' most and least dominant traits to give profound individual fingerprint.
- People physically rejected stereotypical ideas of retirement, while welcoming suggestions of exciting new beginnings.
- In 64% of participants there was, however, a telling difference between their conscious and subconscious visions of their retired selves.
- Encouragingly, most people are aspirational about their retirement, and 51% are looking forward to it.

Armed with this research, Skipton and its customers are now better placed to understand their specific individual preferences and retirement ambitions, as well as creating a mobile app for their staff to use whenever identifying these key personas, the retirement DNA (Figure 38.6).

38.9 USE CASE 3

38.9.1 Measuring the Excitement of Driving a Jaguar

Sensum was approached by agency Spark 44 to record the emotion of excitement while drivers raced Jaguar's new car, the XE, and generate visualizations to overlay video footage captured of each driver (Figure 38.7).

Since excitement is a high arousal emotion, we were able to establish the increases in excitement easily using ECG signals for heart rate and EDA for skin conductance. This was cross-referenced with the video and geolocation data for their place on the track that contextualized what was driving those excitement responses; that ranged from high-speed straights, to fast corners, to the professional driver pushing the limits of what the car could do on track.

The data gathered was presented graphically, showing not only the changes in the raw EDA and ECG but also in the moments of highest excitement, and all cut into a series of promotional videos for sharing online, via key influencers, demonstrating the emotions that could be felt, captured, and visualized due to an exciting driving experience (Figure 38.8).

38.10 CONCLUSION

If we were to use gaming analogy here, our level of understanding of how to capture, measure, and deliver on emotional insights is at the time of "Pong," the simple bat and ball game, NOT Halo. As a result we should understand that any tools providing

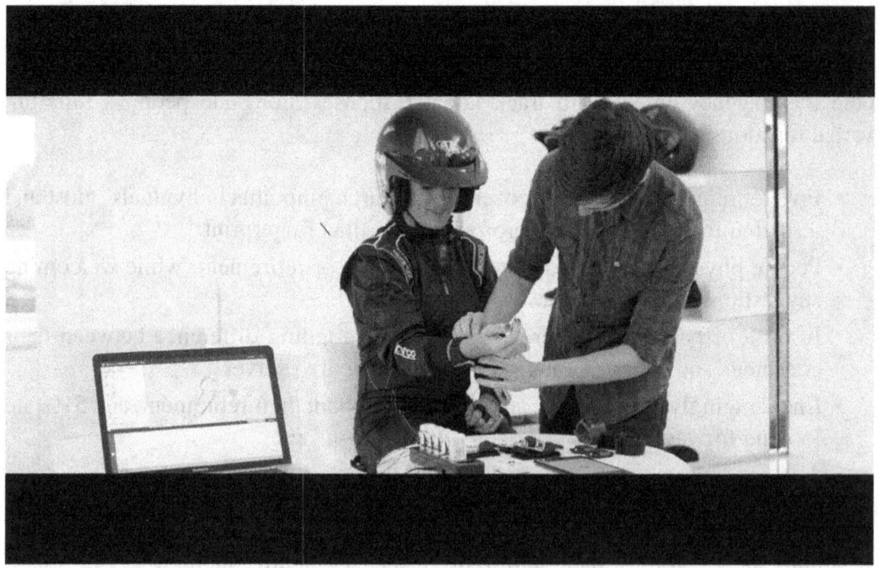

FIGURE 38.7 Measuring the excitement of driving a Jaguar. Courtesy of Sensum.

FIGURE 38.8 Captured and visualized due to an exciting driving experience. Courtesy of Sensum.

insight into emotions are presently at the foundation stage, but very quickly the data will grow, both in terms of scale and in terms of context, and before we know it we will be at the level of Halo, so the conversations on how it should be used need to be had now, and we'll create an amazing world for all people—human and digital.

The Internet of Things. Wearable technology. Smart devices. Quantified self.

A world where everything around you, including yourselves, can be measured, visualized, and reacted to, providing a hyperpersonalized understanding of the self. From health to education to entertainment.

It's all happening around us, through the devices that we see, we carry on our being, we work on, we have in our homes, and that's just the devices that we will notice. And before we are aware enough of what kind of data can be captured and used for decision making, at a personal, corporate, or governmental level, it may be too late to be able to take control of that data. But it's important to talk about it now and have an opinion.

Trust and transparency in this age will be the making of those who succeed.

FURTHER READING

Averill, J. R. (1980). A constructivist view of emotion. In R. Plutchik & H. Kellerman (Eds.), *Emotion: Theory, research, and experience* (pp. 305–339). New York: Academic Press.

Bridger, D. (2015). *Decoding the irrational consumer*. London: Kogan Page.

Damasio, A. R. (1994). *Descartes' error: Emotion, reason, and the human brain*. New York: G. P. Putnam.

Ekman, P. (1977). Biological and cultural contributions to body and facial movement. In J. Blacking (Ed.), *The anthropology of the body* (pp. 39–84). London: Academic Press.

Ekman, P. (1992). An argument for basic emotions. In *Cognition and emotion*, Vol. 6 (pp. 169–200). London: Routledge. https://www.paulekman.com/wp-content/uploads/2013/07/An-Argument-For-Basic-Emotions.pdf (accessed September 10, 2016).

Lewis, M., Haviland-Jones, J. M., & Barrett, L. F. (Eds.). (2008). *Handbook of emotions* (3rd ed.). New York: Guilford Press.

Roseman, I. J. & Smith, C. A. (2001). Appraisal theory: Overview, assumptions, varieties, controversies. In K. R. Scherer, A. Schorr, & T. Johnstone (Eds.), *Appraisal processes in emotion: Theory, methods, research* (pp. 3–19). New York: Oxford University Press.

39

A SINGLE PLATFORM APPROACH FOR THE MANAGEMENT OF EMERGENCY IN COMPLEX ENVIRONMENTS SUCH AS LARGE EVENTS, DIGITAL CITIES, AND NETWORKED REGIONS

FRANCESCO VALDEVIES
Selex ES Company, Genova, Italy

39.1 INTRODUCTION

The world today has become increasingly digital: at work and at home, and while traveling, people are constantly connected through the Internet via a variety of devices and networks. However, thanks to the Internet connection, the majority of the future eight billion people on Earth will have the possibility to access a wider range of information and use it in a different way to improve the quality of life.

Moreover, the "Internet of Things," connectivity in a wider sense, is enabling a global interconnected continuum of devices, objects, and people, normally referred to as "Digital City."

The Digital City expression encompasses a new concept for urban context that aims at increasing the sustainability, livability, and quality of life standards in forward-looking cities. These cities incorporate distributed intelligence at all levels of their constituting parts to improve sustainability, reliability, security, and efficiency.

Internet of Things and Data Analytics Handbook, First Edition. Edited by Hwaiyu Geng.

Companion website: www.wiley.com/go/Geng/iot_data_analytics_handbook/

All these improvements are possible thanks to an intelligent coordination of several activities that take place in urban environments.

Numerous heterogeneous entities are involved in this city vision: the citizens and their interactions with social, economic, and public organizations; infrastructures for public services delivery; and innovative technologies that provide the possibility to enrich traditional passive objects with elaboration and communication capabilities, as the Internet of Things paradigm testifies.

Instead of meeting on a single reference model, the Digital City concept has been developing through a range of distinctive aspects, depending on city characteristics, as well as on:

- Different contingent city needs, such as energy savings and urban mobility improvements
- The need to provide social inclusion to citizens, in order to make them service and information users as well as providers
- Geographical and cultural diversity
- Technological context, which is continuously evolving and strictly related to legacy systems
- Differences in proposition and funding models, which are related to the innovation level required and to the heterogeneity in city components and economic scenarios
- A systemic approach that tends to combine information managed within different vertical domains, to provide a broader view of the monitored situation

A fundamental trait that lies behind all of these Digital City's distinctive aspects is the heterogeneity and interconnections in needs and solutions provided. This aspect on one side brings benefits and opportunities, but on the other it introduces also vulnerabilities and risks; in fact malfunction and disruptions to a single component, both intentional or accidental (terroristic attacks or natural disasters), can have a strong impact on the others with high economic, societal, and even politic costs.

In this scenario the capability to reduce these risks and mitigate the impact of possible natural disasters or human attacks is a key element for a "safe" and digital city.

"Resilience is the capability to prepare for, respond to, and recover from significant multi-hazard threats with minimum damage to public safety and health, the economy, and security of a given urban area" [1].

For example, in the urban context resilience intends to:

- Prevent disasters and service interruptions with a careful monitoring
- Reduce impact of natural disasters caused by climate change
- Improve the response to critical security events, such as terrorist attacks
- Coordinate the intervention of different forces involved in the emergency management

Summarizing a Digital City connects many domains; therefore an attack to one component has a strong impact on the others: to prevent this situation it is necessary to strengthen the safety and security of systems and people and assure the resilience of the city with a holistic approach.

To achieve these ambitious goals, it is imperative to use digital technology: the Selex ES's City Operating System is a significant example of a solution aimed at managing a digital city, also capable of being tailored for major events and territory control. It is an integration and management framework, designed to add intelligence and resilience and to support safety and security processes for complex infrastructures.

Selex ES has successfully delivered security solutions in many complex environments (e.g., NATO, Glasgow 2014, Milan EXPO 2015) addressing security needs through a single platform both for the domain of standard operations and emergency management, capable of covering growing areas, from sites to cities and territories, through the targeted application of dual-use capabilities.

In the following pages we develop these aspects in more details.

39.2 RESILIENT CITY: SELEX ES SAFETY AND SECURITY APPROACH

According to the ARUP and Rockefeller Foundation study "City Resilience Framework" dated April 2014 (Figure 39.1), the City Resilience Framework provides a lens through which the complexity of cities and the numerous factors that contribute to a city's resilience can be understood.

It comprises 12 key indicators that describe the fundamental attributes of a resilient city, grouped into four main categories: *people* (the health and well-being of

Category	Indicator
People	Minimal human vulnerability
	Diverse livelihoods and employment
	Adequate safeguards to human life and health
Place	Reliable communications and mobility
	Continuity of critical services
	Reduced physical exposure and vulnerability
Organization	Availability of financial resources and contingency funds
	Social stability and security
	Collective identity and mutual support
Knowledge	Effective leadership and management
	Empowered stakeholders
	Integrated development planning

FIGURE 39.1 ARUP categories and indicators synthesis. Reproduced with permission from Francesco Valdevies.

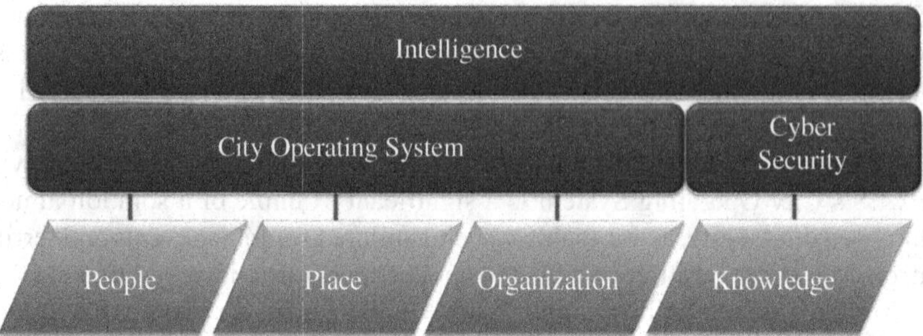

FIGURE 39.2 Resilience framework categories and Selex ES-integrated security solutions. Reproduced with permission from Francesco Valdevies.

individuals), *place* (infrastructure and environment), *organization* (economy and society), and *knowledge* (leadership and strategy).

Selex ES approach to Resilience is in line with the ARUP analysis and addresses in particular security (physical and logical) and safety needs arising from the four categories identified.

Selex ES solution is focused on physical and logical security domains and based on Observe, Orient, Decide, Act (OODA) methodology, supporting a wide range of proactive or reactive security activities to achieve strategic or tactical outcomes in an integrated security vision.

The concept of integrated security is based on Figure 39.2:

- Integration of Information from different domains (logical and physical)
- Creation of an overall view for a wider situation awareness
- Elaboration of acquired data to provide enriched information to support critical events prevention and to identify trend detection
- Coordination among forces during operations

The following picture represents the Resilience framework categories and the corresponding Selex ES-integrated security solutions components.

39.3 CITY OPERATING SYSTEM: PEOPLE, PLACE, AND ORGANIZATION PROTECTION

A digital city improves the quality of life of its citizens and the local economy through individual and joined-up programs across transportation, energy, communications, public services, infrastructure, and security.

The City Operating System is Selex ES's solution for managing a digital city. It can also be tailored for major events and territory control. It is an integration and management solution, providing a set of connectors and modules designed to add intelligence and resilience to complex infrastructures. The City Operating System allows city managers, administrators, and agencies to effectively respond to the challenges associated with sustainable development at a local level.

Selex ES solution supports different stakeholders not only in their daily operations accomplishment but also in emergency situations management.

In case of critical events, situation awareness, provided by the system, allows the user to follow the evolution of the incidents and support the decision making processes. Besides, during the intervention phase it is possible to communicate with the involved field forces thanks to the integrated Communications Service Platform.

The solution is a valid support not only in physical security domain but also in the prevention of cyber attacks in the information security context, as described later.

39.3.1 A Modular Urban Infrastructure Platform

Selex ES's modular solution for Digital City comprises a number of specialized city subsystems:

- **Urban security** for increasing public security and citizens' safety through zone control
- **Intelligent transport system** for providing information about road networks, promoting the use of public transportation, and regulating access to urban areas
- **Infrastructure for secure communications and emergency** for providing professional radio systems based on Terrestrial Trunked Radio (TETRA) technology for voice communications and data, messaging and interoperability with existing heterogeneous networks (Communications Service Platform —PERSEUS®)
- **Cyber security** to guarantee information integrity, systems resilience, and continuity within city governments and supervisory centers

All of these can be monitored via Selex ES's City Operating System, which integrates all city subsystems in a single dashboard providing views, alerts, and workflows for governance and control of the city environment.

39.3.2 A Comprehensive View for Taking Informed Decisions

The City Operating System gives users all the tools they need to manage the day-to-day operations of the city and the ability to intervene effectively during an emergency. The system integrates and organizes data generated by the sensors and other information-gathering systems in the area, including human communications, giving operators a complete strategic overview of the city's status and allowing inclusive communication with citizens.

39.3.3 Key Points

The City Operating System:

- Integrates both new and legacy systems through connectors or ad hoc interfaces
- Gathers information and data from a multitude of existing systems
- Correlates data in order to generate new, richer information

- Provides a control room platform for event governance, providing a customizable workflow for user support
- Includes a communication service platform able to connect separate communication networks and to assure interoperability with other control rooms, aiding territory security and quality of life (Police, Fire Brigade, Civil protection, First Aid, etc.)
- Integrates a range of add-on tools including a crowd management module for detecting people's movements and behavior
- Provides 2D and 3D views, with mapping and 3D rendering service platforms
- Supports secure interaction (Open Access) with users, partners, or third parties with applications for citizens, companies, and organizations

The City Operating System is a Service-Oriented Architecture (SOA) platform based on web services (Figure 39.3). Using a publisher–subscriber method, information in the system is published in a number of classes which receivers can subscribe to. The platform consists of three functional layers:

1. **Field layer:** Includes all subsystems and sensors that acquire information directly in the field. At this level, the information obtained may already be subjected to a first elaboration according to the business logic of its domain.

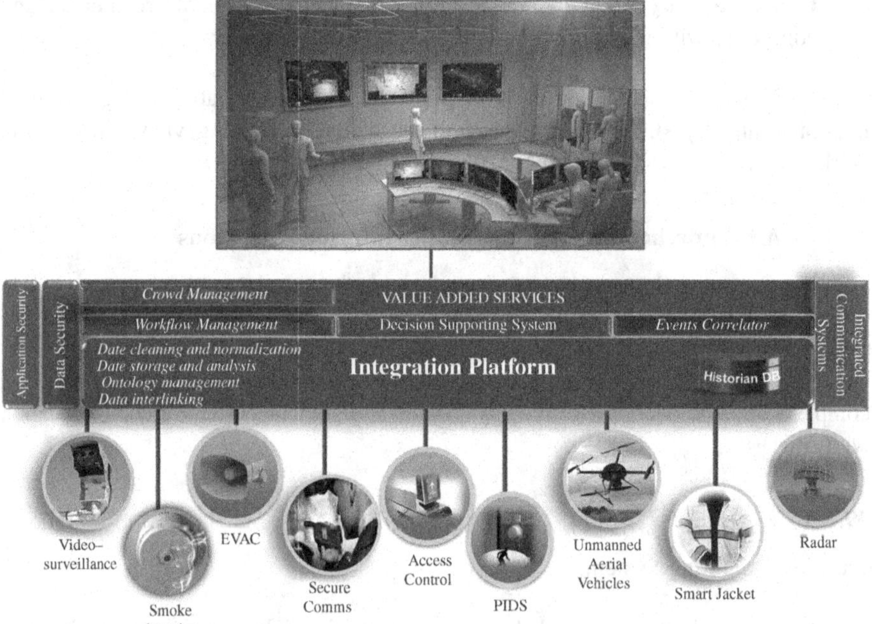

FIGURE 39.3 Selex ES City Operating System. Reproduced with permission from Francesco Valdevies.

For example, in the **physical security domain**, the main specialized subsystems and devices integrated in the framework are the following:

- Perimeter Control: Acoustic Sensors; Cables, fiber, microwaves, infrared; Mid-/Long-range sensors
- Access Control: Biometric Sensors, Mobile Access Control Systems (Patrol Support System (PSS)), Automatic Car Plate Reading, Preferred Access Lane
- Intelligent Video Surveillance: Video Acquisition from a number of Third Parties, ONVIF Compliant, Video Analytics
- Advanced Sensors: Electro-Optic Vision, Radars
- Mini UAV and Micro UAV: Drako 5, Falco
- Network Communication Systems

2. **Skill layer:** It is the business logic level and it is the actual kernel of City Operating System. Here, the data and events coming from the field layer are collected through a software infrastructure based on a bus and made available to the various processing engines, in order to "apply intelligence" to the system according to rules and algorithms oriented to specific domains. This approach allows the framework of City Operating System not only to be used in multiple contexts such as cities and major events, first of all, but also to be tailored for usage in airports, buildings, critical infrastructures, and so on. Different business logics, different processes, and different event correlation rules will be implemented for each of them within the same application framework.
3. **Presentation layer:** Information built in the Skill Layer is presented in dedicated dashboards for each user profile, with a three-dimensional view of specific areas of interest where applicable (3D rendering) and through various interfaces: web, mobile, messaging, voice, haptic, and so on. Each specific user function can access the system through a customized layout.

39.3.4 Main Capabilities

- **Integration**—System Integration Facility (SIF) allows the acquisition and distribution of all information to and from all layers of the system (field, skill, presentation). The events are normalized and stored in a persistent database.
- **Correlation**—Event Correlator is a rule-based engine, aimed at searching for relationships between different, even apparently unconnected events, from different subsystems, in order to generate alerts and alarms, identify false positives, and generally provide smarter information;
- **Management support**—Workflow Engine is a software engine that configures and executes automatic or semiautomatic procedures, consisting of sequences of actions and reactions which can be triggered by a predefined event. It has to be equipped with a graphic designer for easily programming new workflows.

- **Resources management**—The City Operating System allows operators in the control room to manage staff, vehicles, equipment, and facilities out in the field. This service is particularly useful for operations during major events and more generally for the management of security events.
- **Planning**—The Planner is a module that allows for the comprehensive planning and management of activities relating to scheduled events.
- **Localization and representation**—GIS and Cartography services are application for browsing and processing georeferenced data. The service provides visualization and query in 2D/3D maps and allows for the management of metadata associated with map coordinates on multiple levels (multilayer). The service also allows for the indexing of temporal layers and manages features like geocoding and reverse geocoding.
- The UI can be structured as a Web Portal (**City Cockpit**) or be specialized for simplified/vertical interfaces (e.g., mobile, messaging). An operator, depending on their role and access rights, can monitor an integrated view of the city (**Situation Awareness**) or can be fed individual pieces of data or messages and can be guided step by step in taking the right decisions or implementing the correct action.

39.4 CYBER SECURITY: KNOWLEDGE PROTECTION

Selex ES provides real-time services and technological solutions, designed to prevent, protect, deter, detect, and respond to the most advanced and persistent cyber attacks—protecting and managing the data and information that constitute the core of any modern infrastructure and organization.

Selex ES thus supports public and private organizations and institutions in guaranteeing the safety and security of citizens and city assets against cyber threats, safeguarding privacy and preserving the resilience of systems, required to guarantee the continuity of services in a digital city.

Selex ES considers security as an iterative process, starting with the scenario and risk analysis and leading to the implementation of a flexible solution, tailored to the characteristics of the existing infrastructures. This process is iterative since, as the threats evolve, so does the technology.

Consequently, the security processes must ensure a continuous review of the security procedures and systems adopted in order to manage the threat and deploy appropriate countermeasures (Figure 39.4).

For this reason a reliable and effective security system must support and facilitate continuous evolution, not only for processes and systems but also for people.

Selex ES demonstrates its deep understanding of this iterative process by supporting some of the largest and most complex security environments, such as the design and delivery of the NATO Computer Incident Response Capability (NCIRC), Cyber Security programs serving 22,000 users in 28 nations, encryption systems, critical data exchange, intelligence, and investigation for domestic and international clients.

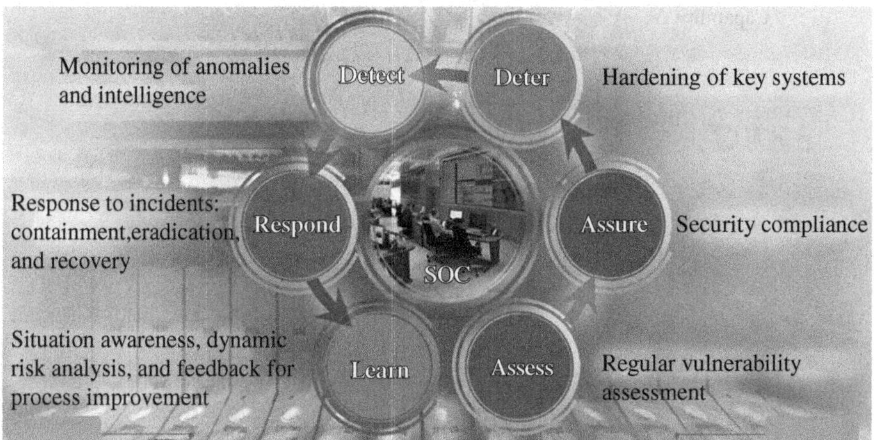

FIGURE 39.4 Selex ES Cyber security services. Reproduced with permission from Francesco Valdevies.

39.5 INTELLIGENCE

Across the Internet people are talking. Mostly harmless, often very specific to them and the world around them. But there are also less innocent and more targeted conversations. They may refer to your organization, how to exploit it or how to damage it.

Selex ES is targeting this traffic. Using one of the most powerful computers in the world, Selex ES is able to deep mine specific content that is matched to specific requirements.

Using semantic engines and parsing functions, Selex ES is able to crawl through the Internet, including the less frequently visited areas and the so-called dark web. In this way, references and threats to your business and your organization, even related to specific functions or people, can be identified.

39.5.1 Semantic Engines

With the huge computing power available, the systems are able to scour for information relating to specific sets of queries. Because it is semantic the query is able to differentiate meanings. If required, the intelligence gathered can be then extrapolated to use in other contexts, such as resilience, emergency planning, or disaster prevention, to protect citizens and city assets from external and internal threats.

39.5.2 Intelligence Platform

Selex ES XASMOS platform is based on a High-Performance Computer, specialized in high-speed processing of massive information on the Internet:

- Defines problem-oriented analysis to support the intelligence process
- Applies different mathematical algorithms and vertical applications to generate actionable intelligence
- Provides timely reports and suggests immediate actions for remediation

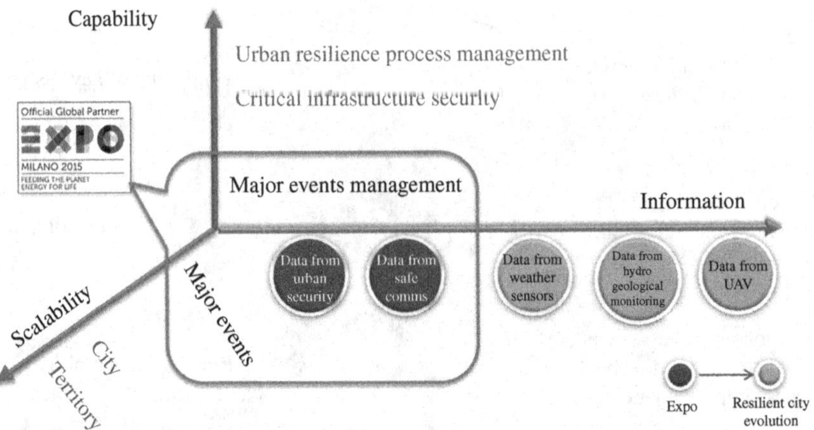

FIGURE 39.5 Selex ES solution from City Operating System to Resilient City. Reproduced with permission from Francesco Valdevies.

39.6 A SCALABLE SOLUTION FOR LARGE EVENTS, DIGITAL CITIES, AND NETWORKED REGIONS

Selex ES solution is scalable and can evolve along three dimensions (Figure 39.5):

- **Information:** Data acquisition and elaboration progressively growing through the integration of new sensors and systems
- **Capabilities:** Continuous enrichment of supporting processes and of intelligence analysis and algorithms
- **Scalability:** Application to different and more complex contexts (from neighborhoods to city to metropolitan area to region).

The solution described can be applied to respond to the pressing challenges posed by the hydrogeological instability; indeed it can be an effective support for territory and environmental monitoring activities, as well as for managing emergencies caused by extreme weather events.

39.7 SELEX ES RELEVANT EXPERIENCES IN SECURITY AND SAFETY MANAGEMENT IN COMPLEX SITUATIONS

Some of the most relevant experiences that contributed to the growth of the approach and of the City OS platform are described in the following:

39.7.1 Turin 2006: XX Olympic Winter Games

- For the Winter Olympic Games held in Turin, Italy, Selex ES designed and implemented the integrated security system for **21 Olympic sites** scattered across Piedmont.

- The project involved the installation of **700 video cameras**, 100 km of optical fiber, and 50 km of power cable.
- Training courses were also organized for operators and, during competition events, constant technical assistance was provided by on-site personnel.

39.7.2 G8 from La Maddalena to L'Aquila

- As part of the company's continued support to the Department of Civil Protection in Italy, Selex ES managed the safeguarding of the G8 Summit in 2009.
- SES was committed not only for design, integration, and implementation of the Coordination Center and all subsystems but also for the relocation of the entire integrated security solution and command centers from La Maddalena, Sardinia to L'Aquila, Abruzzo just a few weeks prior to the event.
- Quoted at the time, the head of the civil protection department commented that the combination of technology and services delivered by Selex ES *provides exceptional support for the activities that are crucial for the success of the G8 meeting in L'Aquila.*

39.7.3 Sochi 2014: XXII Olympic and Paralympics Winter Games

- The event required to support reliable, instantaneous call establishment among **10.492 users**, organized in **2.752 call groups**, with **175.263 memberships**.
- This was possible thanks to a TETRA network, supplied by Selex ES, operated by MS-Spets telecom and with radio terminals provided by Sepura.
- The system supported a very complex service organization:
  - Network Infrastructure Technical Support Service
    - 100 engineers for Network and backbone Operation and Management
  - Radio terminals distribution and management
    - 100 engineers on 20 distribution points
    - 8 distribution points active H24
  - Service Management
    - 10 senior engineers

39.7.4 Glasgow 2014 Commonwealth Games

- On December 4, 2013, Selex ES was announced as the Official Protective Perimeter Security Provider for the Glasgow 2014 Commonwealth Games.
- Under this agreement, Selex ES provided a wide range of physical perimeter security measures to secure over 20 Games venues, including the Athletes' Village.
- The program consisted of security fencing, CCTV, and security lighting as well as security management systems.

- The numbers:
  - 19 concurrent installations in less than 4 months
  - Physical Security Systems for 17,000 m fencing perimeter
  - Securing the Athletes' Village site Size of a city with a population of 7,000
  - Networking over 500 remote sensors
  - Sharing data over 20,000 m of fiber optic cable
  - Averaging 50 daily, on-schedule deliveries to 20 venues
  - 80% equipment reused from previous events
  - Protecting the perimeter of 20 unique venues

39.7.5 Milan Expo 2015

Selex ES represents Finmeccanica Group as "*Expo 2015 Official Global Partner*" providing the "Safe City and Main Operation Center" platform, including three main subsystems:

1. *Smart City Main Operation and Security Centre (SC2–Selex ES' City OS Platform)*

   The Operations Centre for the ordinary and extraordinary event management, consisting of a center for integrated management of information and activities that can provide a picture of the event situation and of the Exhibition Site in real time, including emergency and crisis situations. It will interface external key Systems, dedicated to security and safety aspects of major events (Ministry of Interior, Police, Civil Defense, Fire Brigade, etc.)
2. *Infrastructure for Secure Communications and Emergency*

   Professional radio systems based on TETRA technology accompanied by the new Long-Term Evolution (LTE) technology for voice communications/data, messaging, and interoperability with PS, 118, fire fighters, and so on. Such infrastructure will be able to interact with existing heterogeneous networks, in order to ensure mission critical encrypted communications, as well as messaging and user localization systems. The use of the Communications Service Platform (PERSEUS) will make possible the integration of the different networks.
3. *Special equipment*
   - CCTV system with anti-intrusion
   - Fire and smoke detection system
   - Sound system of Emergency Voice Alarm Communication (EVAC)
   - Security Kits of Pavilions and Clusters

39.7.6 Genova Marassi Living Lab: Harmonise Resilience Project

HARMONISE—EU FP7 Security ongoing project (from June 1, 2013, to May 31, 2016). The central aim of A Holistic Approach to Resilience and Systematic Actions to Make Large Scale Urban Built Infrastructure Secure (HARMONISE) is to develop a comprehensive concept for the enhanced security, resilience, and sustainability of large-scale urban-built infrastructure and development.

HARMONISE Main Objectives are the following:

- Improve **knowledge**, **comprehension,** and **planning** of the **urban** infrastructures **resilience** against existing and emerging threats.
- Support local administration and public security in the **security management** and **territory control.**
- Integrate **heterogeneous sensors and legacy systems** in a single Operation Room to support the **urban resilience.**
- **Selex ES** provides **City OS** with two integrated tools: *Flow Analysis* and *Crowd Monitoring* for addressing mitigation, preparedness, and response phases of the Resilience Cycle.

Living Lab Genoa area is Marassi, a residential district and crucial traffic node, with:

- A football stadium
- Two shopping centers
- Two schools
- A prison
- An outdoor market twice a week
- A river

39.7.7 Operation Control Rooms for Italian Civil Protection Department

Selex ES developed the main and local Operation Control Rooms for Italian Civil Protection Department as part of National System for Civil Protection (Figure 39.6).

The purpose of the system is to provide support during the whole emergency life cycle including:

- Predisaster (monitoring, forecasting, mitigation, and preparedness activities to reduce communities vulnerability and minimize impact of disasters)
- Response (information collection, decision making, rescue, and assistance)
- Postdisaster (short- and long-term restoring activities)

39.7.8 SNIPC

Sistema Nazionale Integrato Protezione Civile (SNIPC) Project, developing the Italia Control Room and the Information System for Civil Protection. The solution provides to operators all information useful for their activity improving interoperability among field forces increasing their national and local situation awareness, also thanks to the interaction with Regional Control Rooms.

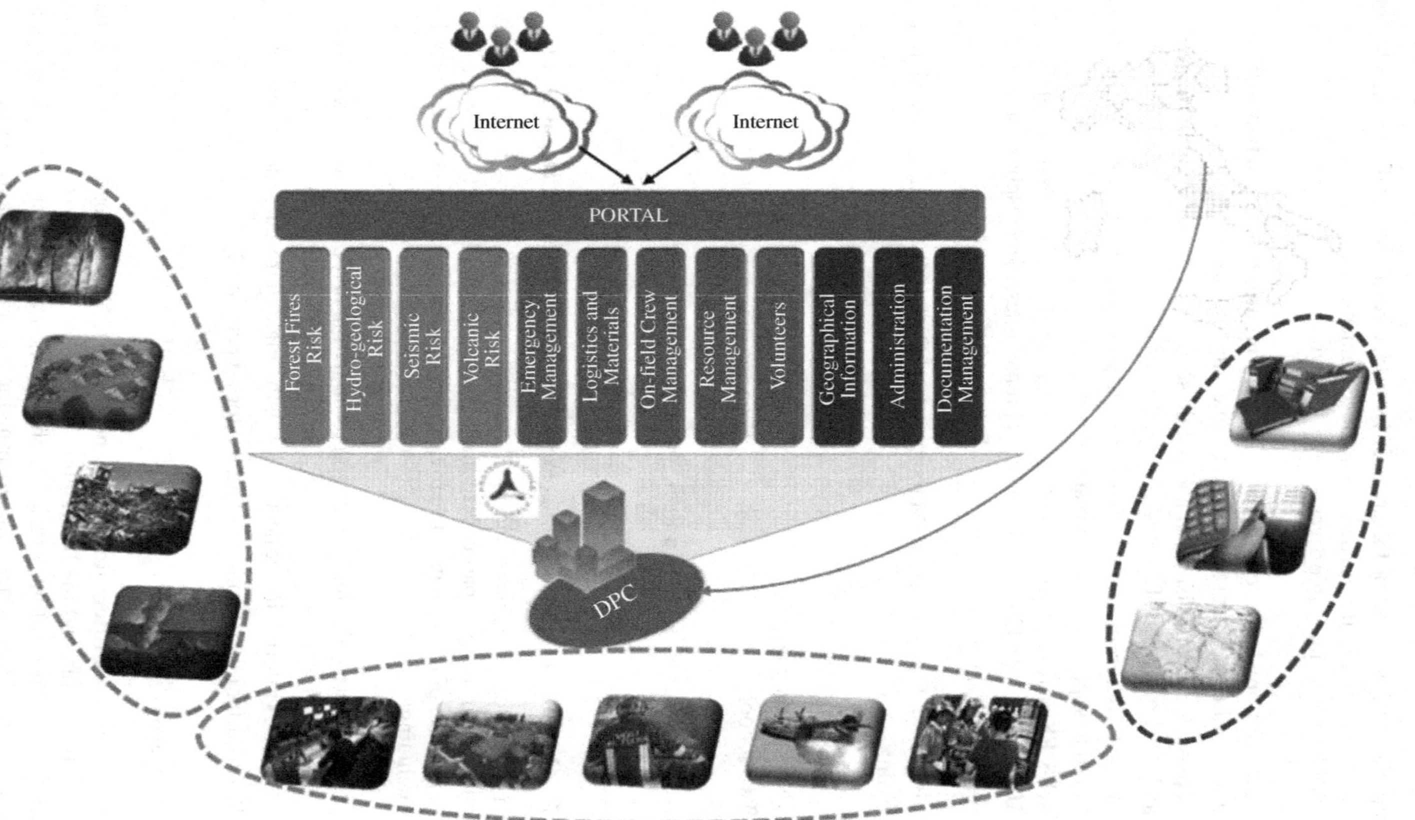

FIGURE 39.6 Operation control rooms for Italian civil protection department.

39.7.9 NCIRC

The Finmeccanica Cyber Solutions team has provided the NATO Computer Incident Response Capability (NCIRC)–Full Operating Capability (FOC) requirements.

NCIRC FOC provides a highly adaptive and responsive system to help protect NATO from cyber attacks against mobile and static Communications and Information Systems. This world-class team led by Finmeccanica (FNC IM, SIFI.MI), comprising its companies Selex Elsag, Selex Systems Integration, and VEGA, together with its partner Northrop Grumman (NYSE:NOC), is leveraging its wealth of experience in addressing complex cyber defense requirements.

39.7.10 VTS

The national Vessel Traffic System (VTS) has been developed by Selex ES and cofunded by the Ministry of Infrastructure and Transport. In addition to the monitoring of maritime traffic, security at sea, and environmental safety, the system also provides for the electronic management of activities connected with the maritime/port transportation cycle, the monitoring of dangerous loads, and the management of security and alarms in a port platform environment and information, info mobility, and booking services for the coastal navigation of ships/goods and passengers.

39.8 CONCLUSION

We have described an approach to the management of everyday city operations and of emergency events based on a single platform, capable of integrating many heterogeneous systems and infrastructures, of using relevant information built from open source and social intelligence, and of connecting professional users and citizens, doing this through interfaces and standard procedures tailored to the capabilities of the specific user.

This approach did not appear by chance or overnight but is the result of a long-standing experience that Selex ES has gained in security management systems, in Command and Control for Defense and in control rooms for civil and military use, and that it decided to condense in a single platform, designed having in mind the needs of future digital cities and citizens.

APPENDIX 39.A HOW BUILD THE PROPOSITION

To understand customer requirements and processes, during 2014 Selex ES begins to apply LIVING LAB activities with public administration, in particular with Genoa municipality: this experience showed that a private and public partnership with the participation of university and SME is the correct and sustainable model to apply technological solutions in an urban context.

Besides some international collaborations in 2014 showed that City OS framework can be adopted in foreign countries, and LIVING LAB model could be certified by international organizations as ENoLL.

CANVAS model has been applied in LIVING LAB Genoa located in Marassi district with particular attention to resilience aspect; the following figure shows the results obtained.

Hyogo protocol defines a checklist of ten points to enhance urban resilience.

Ten-Point Checklist—Essentials for Making Cities Resilient

Essential 1: Put in place **organization and coordination** to understand and reduce disaster risk based on participation of citizen groups and civil society. Build local alliances. Ensure that all departments understand their role to disaster risk reduction and preparedness.

Essential 2: Assign a budget for disaster risk reduction and provide incentives for homeowners, low-income families, communities, businesses, and public sector to invest in reducing the risks they face.

Essential 3: Maintain up-to-date data on hazards and vulnerabilities, prepare risk assessments and use these as the basis for urban development plans and decisions. Ensure that this information and the plans for your city's resilience are readily available to the public and fully discussed with them.

Essential 4: Invest in and maintain critical infrastructure that reduces risk, such as flood drainage, adjusted where needed to cope with climate change.

Essential 5: Assess the safety of all schools and health facilities and upgrade these as necessary.

Essential 6: Apply and enforce realistic, risk compliant building regulations and land use planning principles. Identify safe land for low-income citizens and develop upgrading of informal settlements, wherever feasible.

Essential 7: Ensure **education programs and training** on disaster risk reduction are in place in schools and local communities.

Essential 8: Protect ecosystems and natural buffers to mitigate floods, storm surges, and other hazards to which your city may be vulnerable. Adapt to climate change by building on good risk reduction practices.

Essential 9: Install early warning systems and emergency management capacities in your city and hold regular public preparedness drills.

Essential 10: After any disaster, ensure that the needs of the survivors are placed at the center of reconstruction with support for them and their community organizations to design and help implement responses, including rebuilding homes and livelihoods.

APPENDIX 39.B DETAILS ABOUT REVISION OF THE INITIATIVE

39.B.1 Introduction

Analyzing activities carried on during 2014 with customers of local and national Public Administration and examining emerging technological and innovative programs, it seems that the Smart City Model is well known, but in general it is adopted more in principle than effectively. Certainly some exceptions exist due to cities that have started projects and best practices in this field years ago but in the majority are non-Italian cities.

To face an evolution toward the Smart City Model is important, but it requires a well-structured road map which has to be supported with specific business models, which are often innovative and organized on a mid-/long-term basis.

On the other side, the interest about the Smart City Model for what concerns urban security and urban resilience is becoming more and more pressing. In this domain, also because of recent climatic events, attention is absolutely high. More in general, Public Administration is perfectly aware that risks and criticalities are increasing in complexity and numbers, and they require a new vision of the city in order to achieve a more effective governance.

Starting with these assumptions, SC2 platform developed by Selex ES and enriched also with innovative technologies, deriving from collaborations with SME and from cooperation with Public Administration, represents an effective opportunity to match demand and solutions aimed to solve problems related to critical situation governance.

City Administration demands, especially during a financial crisis period, for solution on how to manage predictable and unpredictable critical situations, using limited resources. It seems therefore that it is possible to find effective solutions thanks to innovation and creativity in the technological field, as well as in the methodological one. Engaging public and private stakeholders, for example, in a cooperative partnership is an effective way to allow to Public Administration to locate and increase resources to support projects sustainability.

Coherently with what is described earlier, an in-depth analysis and enrichment of capabilities developed within FP7 HARMONISE project, already addressed in 2014 Business Plan, whose main objectives are about resilience, has been considered particularly important. In particular, the scouting activity of market innovative technological solutions has been carried on with the main aim of identifying best of breed sensors and systems to be integrated in SC2 to enrich its capabilities in the resilience field.

39.B.2 Selex Es Living Lab

Selex ES, a Finmeccanica company, is an international leader in electronic and information technologies for defense systems, aerospace, data, infrastructures, land security and protection, and sustainable "smart" solutions.

As **technology partner** for **EXPO Milan 2015** (Universal Smart Cities Exhibition), Selex ES developed the Safe City and Main Operation Center, a support framework to manage the safety and security, crisis, and emergency situations for major events and urban environments.

The *Smart City Security Center* (SC2) collects and correlates data from sensors located across multiple sites, providing operators with a real-time 24 hour overview for surveillance, threat mitigation, and emergency response. An advanced graphic interface offers a realistic view of the environment to be monitored, assisting operators in planning interventions.

SC2 is a modular open architecture solution that **can be tailored to specific urban needs** through the **integration** of legacy systems and specialized solutions. It offers an infrastructure which enables planning, monitoring, and governance of the sustainable, resilient, and participating city.

In addition to delivering real-world smart city solutions and proven safety and security technologies, Selex ES is an active participant in two current EU-funded (FP7) Smart City demonstrators.

39.B.2.1 Looking for Partners for Collaborative Innovation Our **approach is to use our** "state-of-the-art **technology" to work in partnership** with City Councils, Public administrations, Local Enterprise Partnerships, and Universities **to deliver** safe and **smart city solutions**.

We use *Living Labs* as an engagement model (Figure 39.7), working with partners to drive rapid innovation and real-world solutions in the cross-disciplinary area of smart cities, starting from the safety and security aspects already addressed by our technology.

Living labs are **real-world areas** or "functional regions" where cross-sectorial stakeholders such as industries, public agencies, universities, institutes, and individuals collaborate toward the **creation, prototyping, validating, and testing of new services, products, and systems in a real-life context**.

Living labs could be city regions, villages, or industrial plants. For example, Selex ES is developing a living lab in Genoa's district Marassi, an area with high building density and social activities (see attached sheet for details).

Selex ES is looking to invest in the United Kingdom, using the unique open, extendable architecture developed for the SC2 platform **for the creation of a Living Lab** in partnership with a British University/Municipality or other appropriate entity.

In such a Living Lab, we aim to further develop a highly flexible interface to effectively manage information and events by integrating our own and our partners' heterogeneous technologies or legacy systems and sensors. Furthermore the opportunity exists to create new and more efficient processes for the management of ordinary operations and emergencies in the city.

With our partners we will **test and deliver original and innovative smart city solutions**.

39.B.3 Selex Es–Smart City Security Center Architecture

39.B.3.1 Living Lab in Genoa Selex ES will develop a living lab in Genoa's district Marassi, an area with high building density and social activities (Figure 39.8).

Main characteristics of the Living Lab are the following:

- Two crossing municipalities involved (northern and southern Bisagno valley).
- Our lab activities are to integrate in SC2 framework the following systems: Hydrogeologic Monitoring, Sporting event Monitoring, Traffic, and Mobility Monitoring.

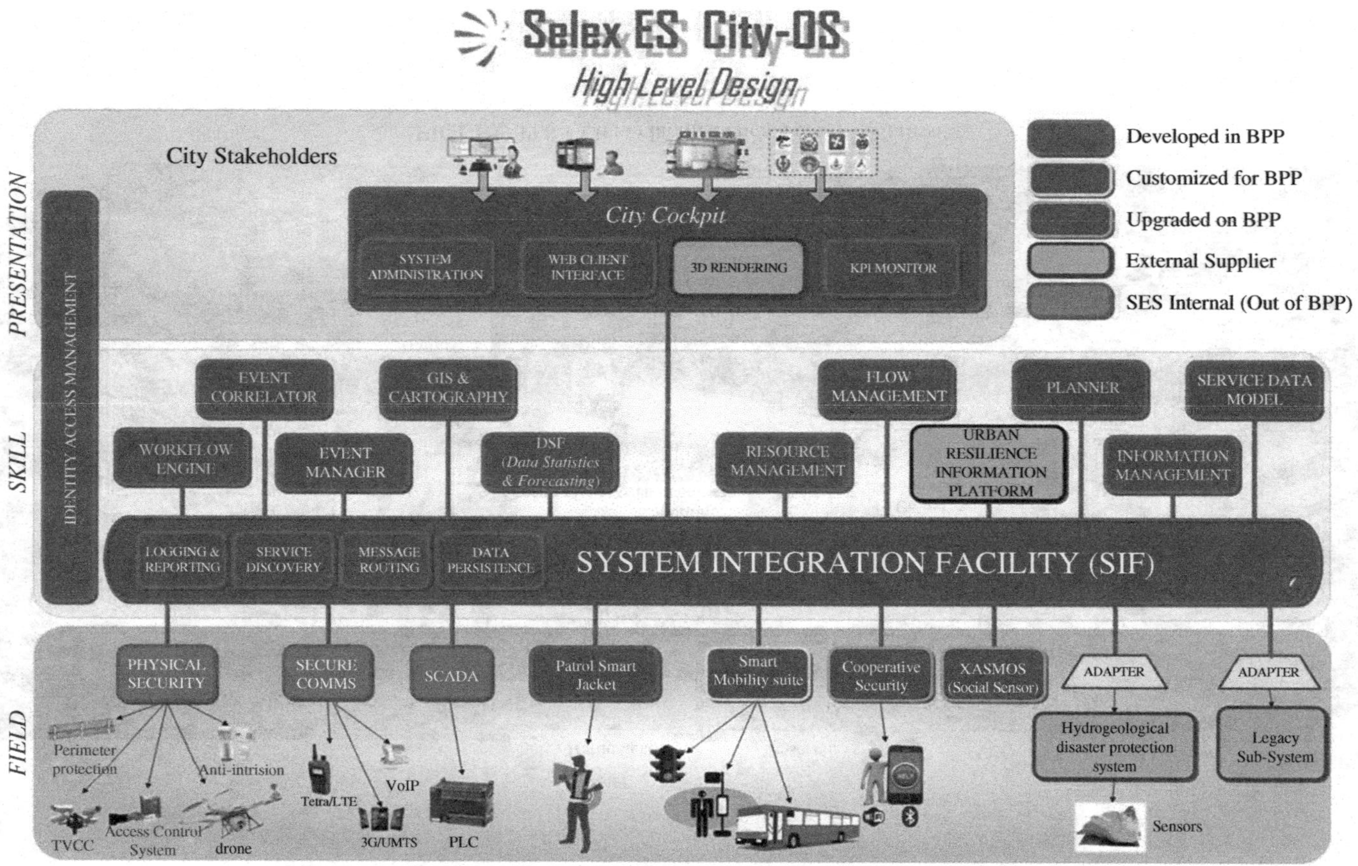

FIGURE 39.7 Living lab engagement model.

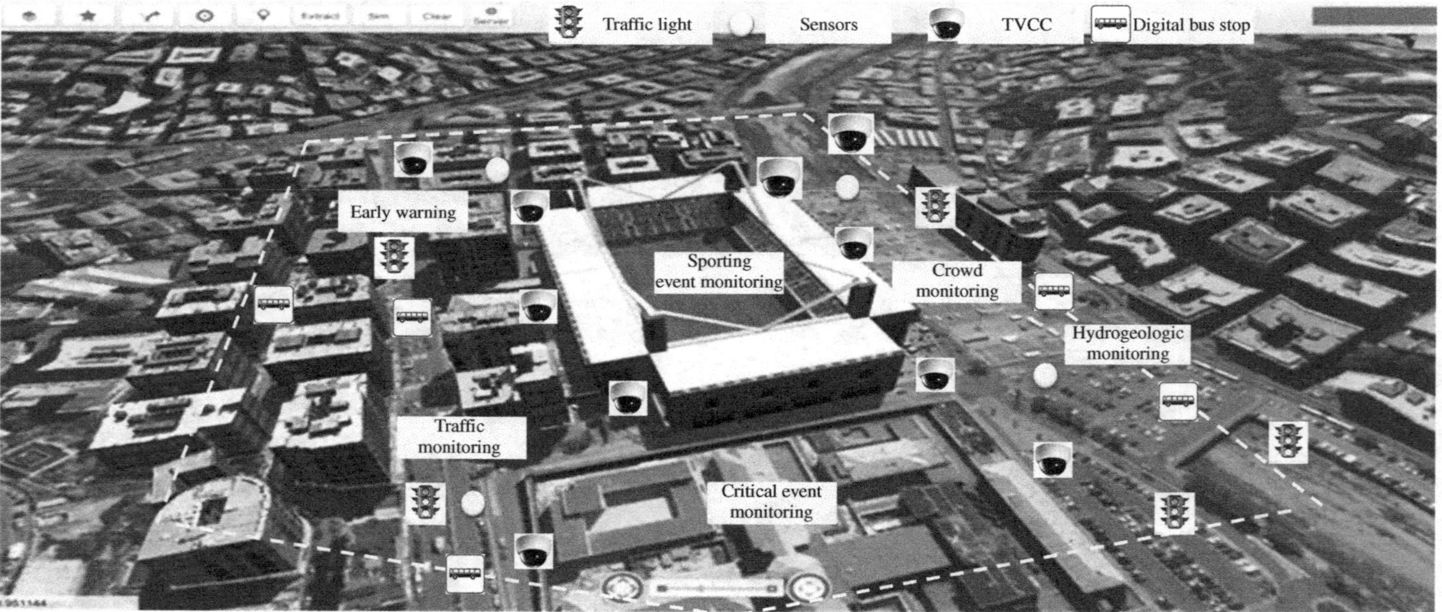

FIGURE 39.8 Living lab in Genoa's district Marassi.

- European Projects HARMONISE (Resilience) and Plug In (Mobility) are in the same location.
- The solution could integrate Energy Saving System and Weather conditions Monitoring.
- Deadline: HARMONISE case study by September, live by December 2014.

REFERENCE

[1] Wilbanks, T. (2007). "The Research Component of the Community and Regional Resilience Initiative (CARRI)". *Presentation at the Natural Hazards Center, University of Colorado-Boulder, Boulder.*

40

STRUCTURAL HEALTH MONITORING

George Lu and Y.J. Yang
goodXense, Inc., Edison, NJ, USA

40.1 INTRODUCTION

Structures are all around us. Structures are ubiquitous in the physical world we live in. The structures we live or work in need to shield us from the weather elements, withstand some degree of degradation over time, and stay strong over a time scale that is comparatively longer than one's lifetime. The vehicles we travel in need to offer comfort and safety through the modes of transportation and be able to protect us through collisions. We expect likewise of the transportation infrastructure we travel on: the railroads, bridges, ramps, etc.

Structures don't last forever; to build and operate structures within a reasonable budget, they are designed to support a specific load over a limited lifetime and withstand environmental changes to a limited degree. Structural health degrades over time as intended. It is important to make sure that a structure continues to offer the required level of service as it ages. Structural failures are very expensive and often cost lives. Structural health monitoring (SHM) is essential to inspect a structure's health over time to ensure it continues to meet operational requirement. SHM is essential to ensure public safety and efficiency.

SHM has been done mostly manually and periodically. The interval between inspections varies, depending on the operating organizations' practice. It is not rare to hear of annual or even biannual inspections. Modern SHM systems take advantage of the rapid advance in low-power processors, sensors, wireless communication, and data storage to automate the inspection process. Such SHM systems started to deploy

Internet of Things and Data Analytics Handbook, First Edition. Edited by Hwaiyu Geng.

Companion website: www.wiley.com/go/Geng/iot_data_analytics_handbook/

within the past decade [1]. With the rapid decline in the cost of enabling technologies, it is becoming practical to consider real-time continuous monitoring broadly.

This chapter presents the authors' firsthand experience at designing and implementing a SHM solution for bridges. It is intended as a case of practicing the engineering approach described in the earlier chapter on IoT and smart infrastructure.

40.2 REQUIREMENT

The SHM system was designed based on requirement from a team at National Taiwan University of Science and Technology, led by Prof. Chung-I Yen and Prof. Wei F. Lee, that has expertise in applying modern SHM for bridge monitoring [2, 3]. From here onward we will refer to each edge sensor device as a Bridge Monitoring Unit (BMU).

40.2.1 Operating Environment

The BMU will be deployed directly on bridges. The system needs to be able to withstand direct exposure to all sorts of weather conditions.

40.2.2 Power Supply

At some target sites there could already be grid power available from prior installation of environmental monitoring equipment. In most cases there is not prior power supply installation. It is strongly preferred if the BMU could operate without the expensive installation of grid power. Whether solar or grid power is used, it is required to have backup battery that could keep the system operating for 5 days after losing primary power source.

40.2.3 Monitoring Only

The BMU is only required to perform passive monitoring. No supervisory control is required.

40.2.4 Connectivity

The BMU is required to support both wired and wireless connectivity. Occasionally, some sites already have wired connectivity from prior monitoring equipment installation. Nevertheless, wireless connectivity should be the norm in most cases.

The BMU is expected to utilize local wireless service provider's 3G service for uplink to server. On structures where multiple monitoring units are deployed, it would be cost efficient to share uplink.

Connectivity, wired or wireless, could be unreliable. Connectivity could be lost during critical event when the data is most important. This would be the case during an intense typhoon or earthquake that damages a local cell tower. It is therefore required for the BMU to retain up to 20 days of data. When data could not reach the server, it could be retrieved from the BMU eventually. In the event of major structural failure, such data could be extremely valuable for forensic analysis.

40.2.5 Data Acquisition

The BMU is required to monitor both dynamic and static parameters. Dynamic parameters include acceleration in three axes and inclination in two axes. Static parameters include water level, water velocity, and structure temperature. Sampling rate of dynamic parameters should be configurable, from 50 to 200 Hz.

As specific choice of sensors may change over time, it would be ideal if the BMU could accommodate different sensor interfaces.

The data samples should be grouped into a file every 30 seconds and timestamped.

Notably, all sampled data needs to be transmitted to server for analysis and long-term archive.

40.2.6 Robustness

The BMU should be able to tolerate intermittent wireless connectivity to server, as 3G connections are known to drop often. In the case of unreliable connectivity, data transmission to server should resume when connectivity is restored.

The BMU should be able to operate in the absence of GPS reception and still meet all previous requirements over short duration.

40.3 ENGINEERING DECISIONS

40.3.1 Power Supply

We understand that grid power may not be available in most sites. Even when grid power is available, its constant availability could not be counted on. We decided to utilize a rechargeable battery as the primary power source, which is recharged by grid power or solar panels as applicable.

We chose lithium iron phosphate ($LiFePO_4$) chemistry over the more common lithium ion or lithium polymer because it better retains capacity over longer cycles, ensuring adequate capacity over years of deployment. Lithium iron phosphate has lower energy density, resulting in larger and heavier battery. This is less of a concern for SHM application since it would be static once deployed at a site. Capacity retention over large number of cycles is more important.

Due to obstruction by the structure and foliage, it is possible that the site could afford direct sunlight on the solar panel only during certain time each day. For such sites one should choose solar panel of higher power rating so that it could adequately recharge the battery in the limited hours of direct sunlight.

40.3.2 Connectivity

The requirement that all sampled data needs to be transmitted to server for analysis and long-term archive is a tall one. It rules out low-power mesh networking options like 802.15.4 which could not support the needed bandwidth.

Wi-Fi could provide the needed bandwidth. Wi-Fi is common enough that we could easily get parts that support the long-distance transmission. Typical bridges span hundreds of meters. Some bridges could span kilometers. This one-dimensional topology is not a good fit for the typical Wi-Fi installation where one Access Point serves edge devices over a circular area. We decided that Wi-Fi mesh would be a practical way to link multiple systems deployed on one bridge to an edge router, using intermediate BMUs to relay the traffic from BMUs further away. An edge router would act as the gateway to remote server. For sites with only a single system, it could be simpler to support 3G uplink directly in a BMU.

With either 3G or Wi-Fi, to reduce bandwidth and power consumption, we would compress data before transmission. Each 30 seconds window of data is saved into a file, whose name includes the current date–time and unique ID of the BMU. The file is then compressed and saved to flash storage before queueing it for transmission. The compression effectively reduces size of data to be stored and transmitted by 70%.

There are many Wi-Fi modules designed for IoT-type application. These modules typically could interface to a microcontroller (MCU) via UART or SPI. These Wi-Fi modules tend to have their own MCU to run their protocol stack. We are not aware of any such modules that could support mesh protocol. They also tend to have limited RF power (under 20 dBm) that is not suitable for the kind of long-range outdoor application we have in mind. We found success with Wi-Fi USB dongles that support 27–30 dBm output power, with Linux kernel drivers that support mesh networking. Even with omnidirectional antenna, we were able to achieve close to 400 m range between adjacent BMUs.

For 3G connectivity we found many 3G USB dongles designed for consumer or industrial uses.

These connectivity device choices would necessitate use of a Linux single-board computer (SBC) due to the driver and network protocol stack that is required for operation.

40.3.3 Protocol Considerations

While not part of the requirement from customer, we wanted to be able to *ssh* over Virtual Private Network (VPN) into the BMU for miscellaneous services, mainly to update application and MCU firmware as needed. This is not mere convenience as the deployment sites are thousands of miles away from us physically. Requiring on-site work by field engineer that is familiar with embedded computer is inefficient and costly. This meant we would want the field devices to support TCP/IP networking on Linux. Linux would also afford us ample choice of open-source tools for application development and deployment.

We had planned simple Web server on each BMU so that we could directly monitor status of BMU's operation during development and deployment. It could also evolve to support a RESTful API.

We used an asynchronous message queue to transmit compressed data files from BMUs in the field to server. The asynchronous message queue provides us the required robustness against connectivity disruptions.

40.3.4 Architectural Choice

We wanted a SBC-based on ARM microprocessor. ARM-based SBCs are known to have low power consumption. They are available from several suppliers, with Linux OS. Most importantly, growing number of open-source SBC was starting to be available at low prices. When we were making the SBC decision, BeagleBoard had already been introduced for a few years and BeagleBone was just released. We liked the active developer community and were very attracted by the large number of peripherals exposed on the 92-pin headers.

Linux is not a real-time operating system and is not an ideal platform for low-latency tasks such as data acquisition. We said "not an ideal platform" because there existed patched versions of the Linux kernel that could make it more suited for real-time applications. Nevertheless, we needed many ADC ports that are usually available in MCUs but not in Linux SBCs. While we could use external ADC IC that interfaces to Linux SBC via SPI or I2C, MCU could offer far more flexibility in addition to cost and performance advantages. We decided it would be easier to use a combination of MPU and MCU.

To accommodate large variety of possible sensor interfaces, we decided to use two ARM Cortex M3 class MCUs as dedicated data acquisition processors, DAQ1 and DAQ2. Their function could be implemented in firmware. MCU firmware could be updated in the field from the Linux SBC, offering us a convenient way to fix bugs or add new functionality after field deployment. Two UART channels could be utilized to interface with these two DAQ processors.

40.3.5 Data Acquisition

The critical aspect of data acquisition is the timing for sampling dynamic parameters. The sensor interfaces to be utilized and sampling rate, set by customer for the particular site, would be transmitted from a software application on Linux SBC to DAQ processors. Our MCU firmware would set a hardware timer based on the specified sampling rate. On timer overflow, data would be sampled from the specified interfaces and passed to SBC.

To ensure correct timestamping of sampled data, we use ntpd and GPS to set time on each BMU [4, 5].

40.4 IMPLEMENTATION

We designed and assembled a printed circuit board for our BMU design. It contains:

1. Two ARM Cortex M3 MCUs (with the necessary passive parts and clock sources)

2. GPS module
3. Terminal blocks to connect:
   a. Temperature sensor
   b. 3D accelerometer
   c. Two inclinometers
   d. Power supply
   e. SDI-12 bus for water velocity and water level sensors
4. Headers to mount the BeagleBone SBC
5. Many extra signal pins for future expansion

The USB host port on SBC provides the wireless interface, either 3G or Wi-Fi.

The BeagleBone SBC was mounted on the PCB and loaded with Debian Distribution, ntpd, VPN client, openSSH server, and application software that

FIGURE 40.1 BMU mounted on a pilot site in Taiwan.

FIGURE 40.2 Yuan Shan Bridge in Taipei where two BMUs have been operating since 2014.

manages the DAQ processors as well as data file compression–storage transmissions.

The PCB is mounted on a rigid stainless steel plate inside an IP67 enclosure, with flanges to facilitate mounting to the target structure. Field deployment usually takes only a couple of hours (Figures 40.1, 40.2, 40.3, and 40.4).

Although the requirement is to transmit sampled data to server for analysis and archive, we have demonstrated doing FFT on the data immediately after sampling. We could identify frequency peaks corresponding to vibration nodes and track their deviation over time.

40.5 CONCLUSION

The underlying technology for SHM is still evolving rapidly. New protocols are still being proposed to help achieve standardization and accelerate wider adoption. We recommend developing your SHM system with loosely coupled functional modules, using open standards as much as possible, so that individual modules could be upgraded without reengineering the whole solution.

We believe it makes sense to utilize computation on the edge to analyze data for anomaly before transmitting it to servers. Vast majority of the sampled data in SHM applications will not indicate a significant event and thus is not worth transmitting and archiving. Long-term analysis using snippets of periodical data is sufficient to monitor a structure's health over time. Event detection on the edge sensor could

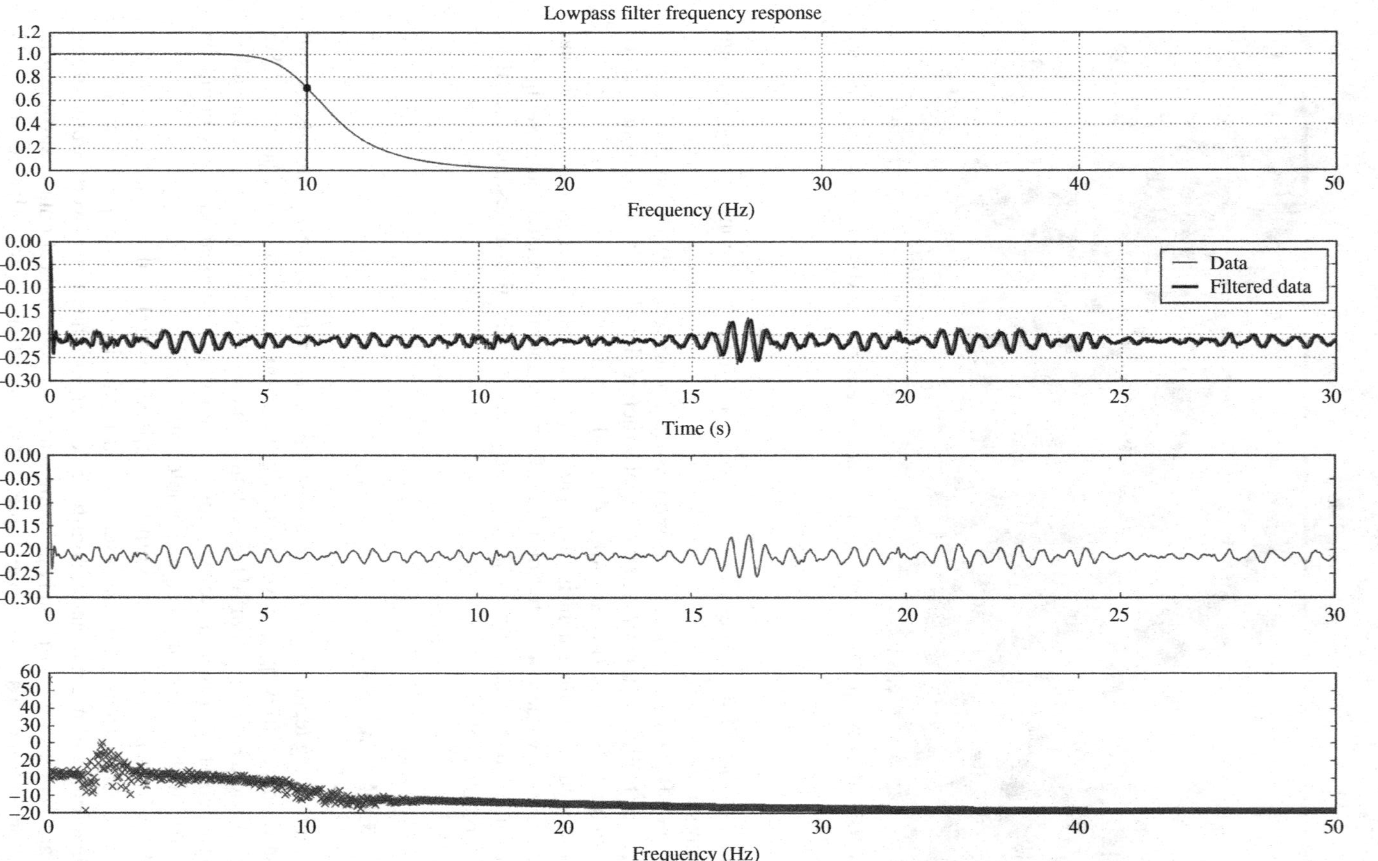

FIGURE 40.3 Data from a dynamic inclinometer processed to show clearly identifiable peak on the frequency spectrum. Shifts of frequency peaks are important indicators in SHM.

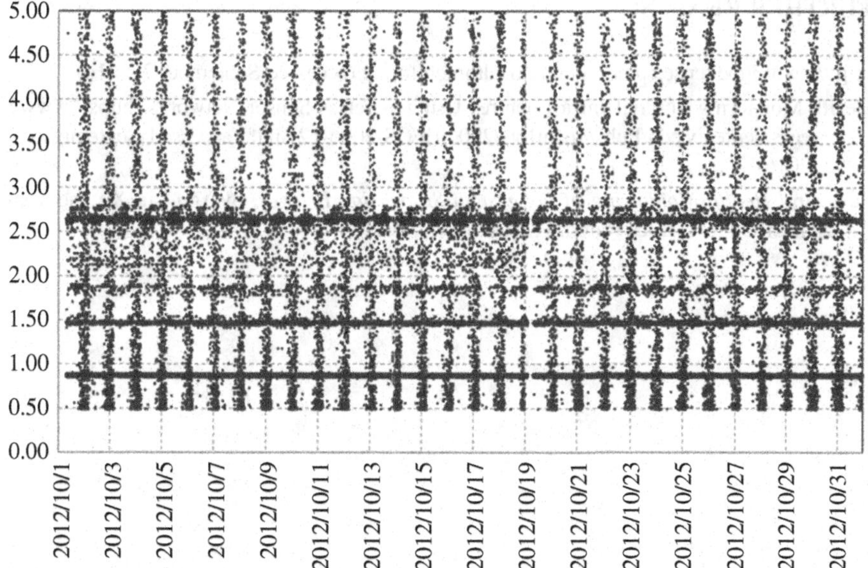

FIGURE 40.4 Spectrum of one of the dynamic inclinometers over time. The darker bands indicate the vibration nodes of the structure.

significantly reduce latency to event detection and reduce the bandwidth to transmit data indicating routine conditions. Such bandwidth reduction could make it practical to take advantage of nascent low-power wide-area networking technologies, such as LoRaWAN, for efficient connectivity for SHM applications.

REFERENCES

[1] Health Monitoring of Civil Infrastructures Using Wireless Sensor Networks, Sukun Kim, Shamim Pakzad, David Culler, James Demmel, Gregory Fenves, Steven Glaser, and Martin Turon. In *the Proceedings of the 6th International Conference on Information Processing in Sensor Networks (IPSN'07), Cambridge, MA*, April 2007, ACM Press, New York, pp. 254–263.

[2] Performance of a Highway Bridge under Extreme Natural Hazards: Case Study on Bridge Performance during the 2009 Typhoon Morakot, W. F. Lee, T. T. Cheng, C. K. Huang, C. I. Yen, and H. T. Mei. *Journal of Performance of Constructed Facilities*, Vol. 28, No. 1, February 1, 2014, pp. 49–60 (ASCE, Reston, VA).

[3] Application of a High-Tech Bridge Safety System in Monitoring the Performance of Xibin Bridge, C.-I. Yen, C.-K. Huang, W. F. Lee, C. H. Chen, M. C. Chen, and Y. C. Lin. *Proceedings of the Institution of Civil Engineers—Forensic Engineering*, Vol. 167, No. FE1, 2014, pp. 38–52 (ASCE, Reston, VA).

[4] https://en.wikipedia.org/wiki/Network_Time_Protocol (accessed August 13, 2016).

[5] https://en.wikipedia.org/wiki/Synchronization_in_telecommunications (accessed August 13, 2016).

FURTHER READING

http://www.goodxense.com/structuralhealthmonitoring (accessed September 7, 2016).

Health Monitoring of Civil Infrastructures Using Wireless Sensor Networks: https://www2.eecs.berkeley.edu/Pubs/TechRpts/2006/EECS-2006-121.pdf (accessed September 13, 2016).

Structural Health Monitoring of Long-Span Suspension Bridges, by You-Lin Xu and Yong Xia, CRC Press, Boca Raton, FL (October 7, 2011).

41

HOME HEALTHCARE AND REMOTE PATIENT MONITORING

KARTHI JEYABALAN
University of Utah, Salt Lake City, UT, USA

41.1 INTRODUCTION

The *Internet of Healthcare Things* is the theme for Connected Health Symposium 2015 for *Internet of Things (IoT) in Healthcare*. As we are all in the evolution phase for the devices that communicate through wireless technology, there are a lot of lessons to be learned with real-world Healthcare usage scenarios.

The application of IoT in the Healthcare (Figure 41.1) can be classified as follows:

- Critical care
- Follow-up care
- Preemptive care
- Monitoring
- Diagnosing
- Treating

At the time of writing this chapter, we were still evaluating the reliability of IoT in Healthcare, and still Healthcare industry is evaluating the use of IoT Device to treating patient in real world. The Clinical usability sometimes does not fulfill the Healthcare services outcome, for example, why would I need a wireless device that is not reliable in the hospital? I can use an existing wired Medical device that is reliable. Assumption is that it is dangerous to put an IoT device to treat Critical care

Internet of Things and Data Analytics Handbook, First Edition. Edited by Hwaiyu Geng.

Companion website: www.wiley.com/go/Geng/iot_data_analytics_handbook/

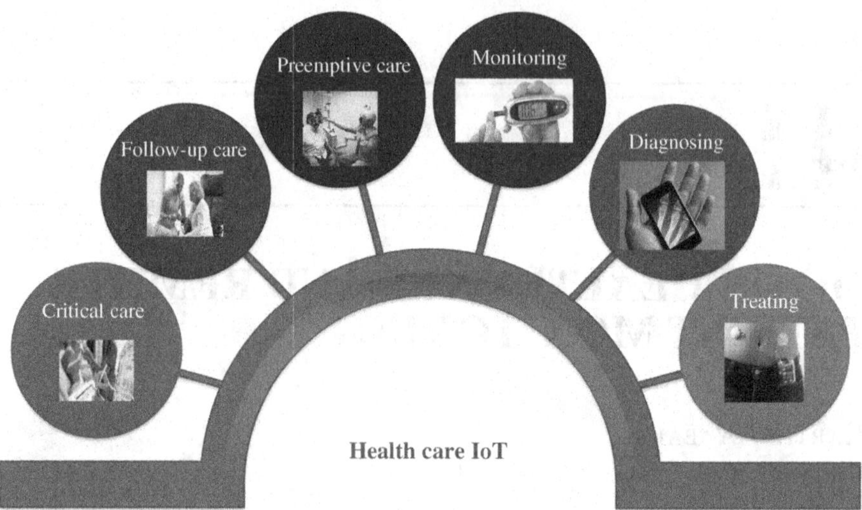

FIGURE 41.1 Application of IoT in healthcare. © University of Utah TeleHealth Services.

patients. We can send home IoT devices for follow-up care for patients by getting their biometric Data. Using IoT medical device, getting biometric data can be used as Preemptive Care approach for and create change in patient's Life Style. We have to identify the class of IoT application and prove the use case. Certain usability can be only performed through wireless IoT Devices for tracking a moving person or object. In this case we cannot tether device with a wire attached to a person or an object to monitor. The application of IoT for different class has to be based on the Use Case. The Use Case has to have a basic need or critical need. Technical possibilities and hurdles can be evaluated. When setting up an IoT device in a Hospital environment, it has a lot of challenges. The Hospital may have some or all of the ambulatory division, inpatient, Emergency, specialty, pharmacy, rehabilitation, Laboratory, and so on. The IoT application varies from monitoring patient, to biotest sample tracking, to nurse practitioner spending time with patients.

41.2 WHAT THE CASE STUDY IS ABOUT

CASE STUDY 1

Out of these classifications we chose to start with simple and less intrusive case study like tracking patient's movement in University of Utah Hospital physical medicine and rehabilitation (PM&R) (Figure 41.2). There are different kinds of patients who come to PM&R facility for treatment such as patients who are mentally not capable but physically strong enough to walk within the hospital premises, and there are some occurrences that patients wander off to the street and even to freeway. It is not viable to lock the patients undergoing therapeutic treatment

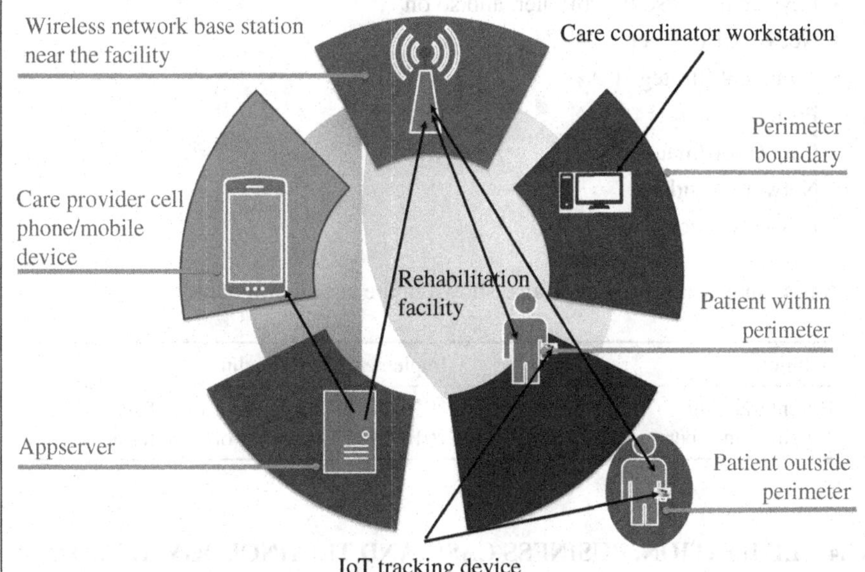

FIGURE 41.2 IoT patient tracking system diagram. © University of Utah TeleHealth Services.

into a confined room or area. So we need to track a patient with a lightweight device, and network should be capable to pick up the device chirps (emitting message) to care provider and alert them through their phone or other devices. Using this approach patients can move around freely within the facility but with constantly monitored performance.

41.3 WHO ARE THE PARTIES IN THE CASE STUDY

The following are members who participate in the case study. An organizational chart is shown in the succeeding table with their responsibilities.

Role	Specific Role Name	Responsibilities
Care Provider	Physician	Diagnosing and treating patient
Care Provider	Nurse practitioner	Patient care
Technical strategist	IT person	Coordinating, implementing, and assisting in planning
Project manager	Program manager	Manages overall project development
Care coordinator	Nurse manager	Oversee any patient notification
Network engineer		Set up perimeter and security with network
Network vendor		Provides networking hardware tools
Device vendors		Any vendor provides I/O device

- Physician, Nurse Practitioner, and so on
- Network Engineer
- Technical Strategist
- Project Manager
- Care Coordinator
- Network Vendor
- Device vendor

A road map with timeline is also shown in the succeeding table.

Solution	Implementation Timeline
Patient tracking	2016–2017 University of Utah fiscal year
Nursing time spent on tracking	2016–2017 University of Utah fiscal year

41.4 LIMITATION, BUSINESS CASE, AND TECHNOLOGY APPROACH

Currently, there is no mechanism to track a patient; moreover, there is limitation in Ethernet Wi-Fi and Bluetooth in the Hospital perimeter. The device cannot emit its presence after certain distance due to limitation in network capability or some dead Spots. So the simple tracking device sends tiny packets of data. So we need Ultra Narrow bandwidth and dedicated network for IoT devices. A very lightweight IoT device wearable in patient's Wrist.

41.5 SETUP AND WORKFLOW PLAN

Equipment

1. Base station which transmits and receives IoT device Messages should be installed on the Roof of the nearby facility so that it has better coverage.
2. IoT device is put on the Patient's wrist (Figure 41.3).

The IoT sensor devices are worn on patient's wrist during the therapy period. The devices will constantly emit message to base station about the location of patients. The Network operator-provided Base station will establish an Ultra Narrow Bandwidth that is dedicated for IoT devices. So the patient wearing IoT devices on boot-up will communicate with network (Ultra Narrow Bandwidth base station) pre-programmed frequency and message. This is publish/subscribe model, so the devices registered to the network, and only it can send a message by communication token established by Business Support System.

All the Device to Antenna are Radio Links. There are very few network operators in this unlicensed Bandwidth. Assume the security (tampering, resending messages,

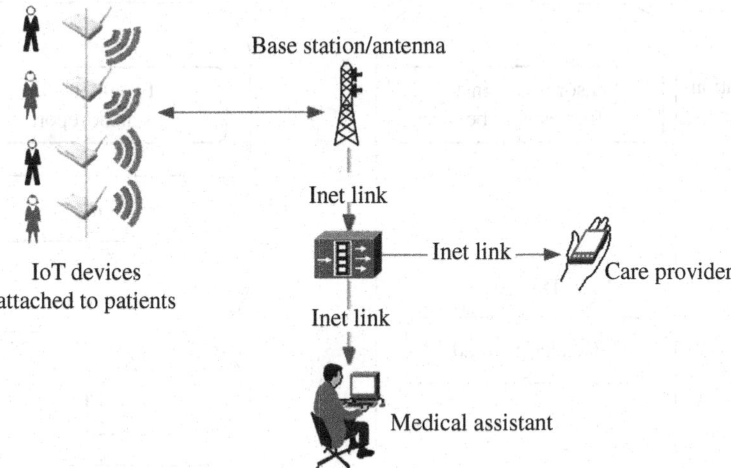

FIGURE 41.3 Patient tracking system using IoT sensor devices. © University of Utah TeleHealth Services.

integrity of messages, no open-inbound port listening, authentication, authorization) of the IoT devices is taken care of.

Classification is crucial to manage the ecosystem of Healthcare IoT devices (Figure 41.4). For example:

- Device Type = Tracker, Temperature, and so on
- Device Group = Medical
- Device Type Related to Business Support System = To track Patient movement
- Device taking a communication token in Business Support System Order = Communication Token for Device Group and Device Type

41.6 WHAT ARE THE SUCCESS STORIES IN THE CASE STUDY

1. The prepilot exercise points out the person's movement from room to room or location to location, and we can see that the status is Web portal dashboard. This feature can help the care coordinator to actively locate the patient movement.
2. There is no issue in network proxy settings for the Bluetooth listeners.
3. Using commercially available consumer-friendly wearable devices will be welcomed by Healthcare community.

Note: Due to Hospital-wide Network Wi-Fi is enabled. This project is reevaluated to use Wi-Fi 5G bandwidth replacing all Bluetooth frequencies.

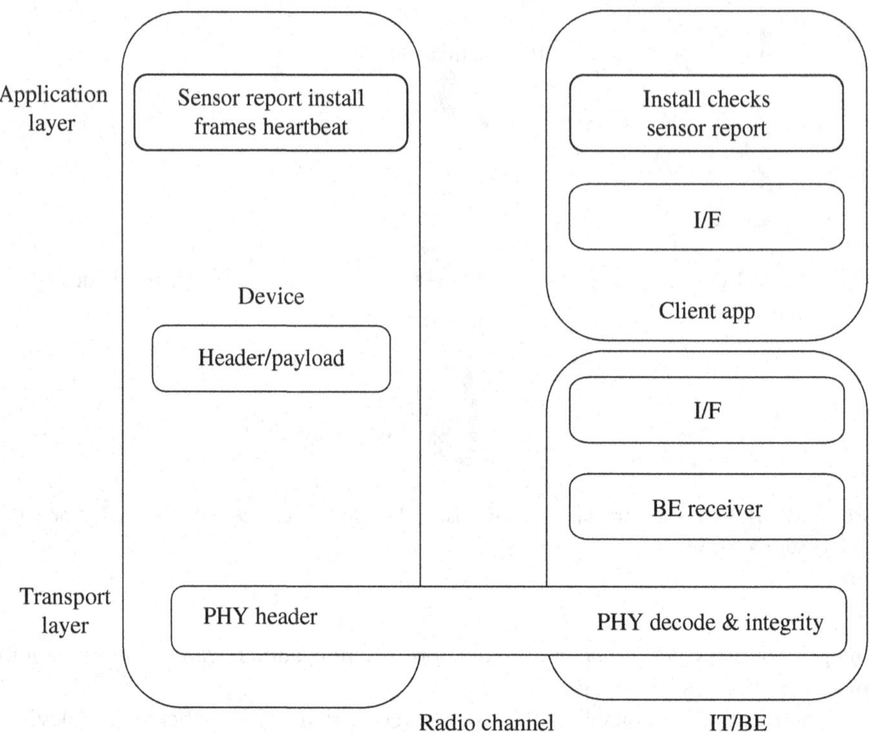

FIGURE 41.4 Healthcare IoT devices and classification. © University of Utah TeleHealth Services.

Serendipity Finding 1

When this demo of the solutions caught the attention of our orthopedics clinic operations director, we found that this can be applied to operational efficiency. He mentioned the typical current workflow of the department.

1. Patient checks in the front desk.
2. Patient is given a tablet for survey.
3. Patient is taken to X-ray by technician.
4. Nurse practitioner takes vitals.
5. Doctor checks the patient details.
6. Doctor meets the patient face to face.

The current workflow process has delays between the next steps. This is due to lack of patient location awareness. They perform action only when they see the patient, and it involves actual person monitoring and notifying. Alert and notification is missing in real time. So this patient tracking solution can be used to alert the care coordinator to move the patient to the next step and avoid waiting time. So the patient will be able to see the physician ASAP. This is a good candidate for patient satisfaction. So this is another potential pilot with IoT.

Serendipity Finding 2

Second doctors want to find how much time the MA/Nurse spends with Spinal cord-injured patients and Quadriplegic patients. This information helps the department to better allocate resource and capacity management.

Currently, IoT devices have potential opportunities for the following:

1. Send home Bluetooth-enabled medical devices with Gateway. This gets instant biometric readings from patients and update in the patient and nurse portal.
2. In patient setup will also benefit from continuous reading of vitals using IoT devices without any wire tangled.
3. IoT Compression suite having electromyography sensors (EMG) will provide real-time and recorded-time muscular activities, heart rate, and respiration.
4. IoT Shoe Insole provides real-time Gait, Symmetry, Jump details of Rehabilitation patient. These details give care providers information to adjust the physical therapy exercise for patients.
5. Ingestible pills through IoT technologies provide medical adherence, tracking, and so on.

41.7 WHAT LESSONS LEARNED TO BE IMPROVED

- The latest IoT devices in Healthcare and evolution are ongoing. Currently, Skin patch biometric devices which are also noninvasive are going to be the future within couple of years (current year 2015). So adoption of these devices is essential, and interface with the mainstream Healthcare application is crucial.
- When IoT Healthcare devices come to the market (like consumer-related step monitor, sleep monitor, and heart rate monitor), it is necessary to communicate with the provider (Physician, Nurse, etc.) for multifactor diagnosis. They are the sole authority on the decision-making process. Since the patient empowerment is necessary from home perspective, physician empowerment on clinical setup is crucial. Data generated with IoT medical devices/sensors fed to the smart applications can provide rapid preliminary diagnosis and provide prescription conjunction with provider.
- With this explosion of IoT devices, there should be very effective implementation governance for secure and solid integrity of data delivery to be taken care of for successful Healthcare.

Figure 41.5 shows IoT components in a Digital Health Ecosystem. The Digital health ecosystem starts with IoT Devices at the bottom. Those devices provide a wide range of information on biometrics, laboratory sample tracking system, health and

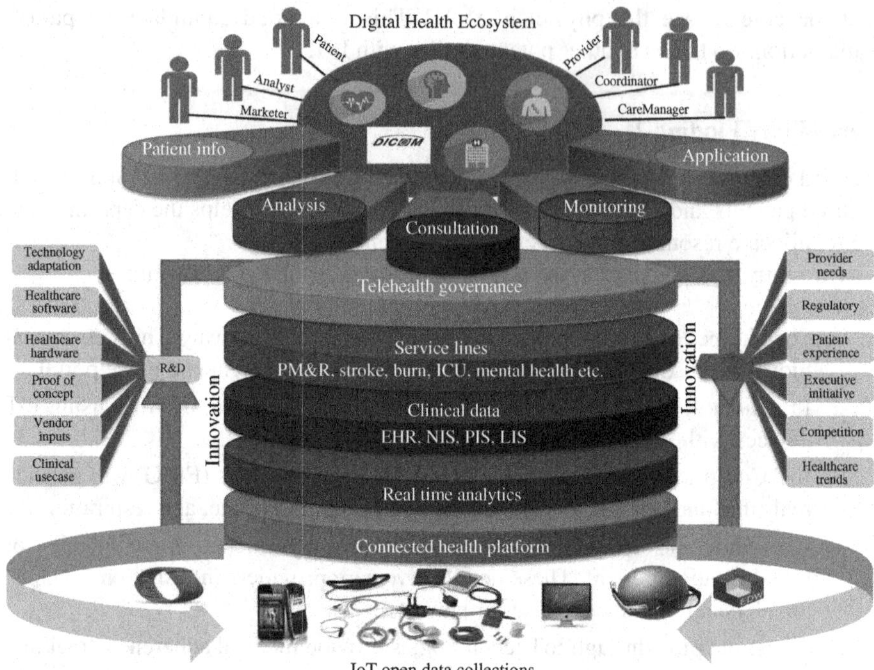

FIGURE 41.5 Digital health ecosystem with IoT components. © University of Utah TeleHealth Services.

wellness devices (wearables), and so on. They are connected to a health platform and move upward by applying analytics and workflow. The information that reaches to clinical data EHR, NIS, and so on will be more actionable data or decision-making data. This actionable data can be available to different service lines like PM&R, Stroke, and so on. The whole deployment will be notified and tracked by Telehealth team to facilitate Telehealth calls. The whole Telehealth system's goal is to automate the process and notify the critical actions to take place. This process will be constantly influenced by R&D on the one hand and innovation on the other hand. The top figures consist of care provider, patient, and other actors. These actors use the appropriate action tools (DICOM for seeing patient's MRI) to provide care. For example, the action tools and IoT data at the bottom can provide shared decision making on certain ailments. The patient's insole sensors provide data on patient rehabilitation exercise adherence, and the MRI provides patient improvement after the surgical procedure.

FURTHER READING

http://mobihealthnews.com/ (Accessed January 19, 2016).
http://www.telehealth.com/ (Accessed January 19, 2016).
https://digitalhealthsummit.com/ (Accessed January 19, 2016).
https://www.wearable-technologies.com/ (Accessed January 19, 2016).

PART VI

CLOUD, LEGAL, INNOVATION, AND BUSINESS MODELS

42

INTERNET OF THINGS AND CLOUD COMPUTING

JAMES OSBORNE
Microsoft, Redmond, WA, USA

42.1 INTRODUCTION

Both "machine-to-machine (M2M) systems" and "Internet of Things (IoT)" are broad terms that are sometimes used interchangeably, and there is little point in debating where one ends and the other begins. But it would be a mistake to think of IoT as simply the latest buzzword. It's more useful to think of IoT as an evolution of M2M made possible by disruptive forces in three key areas of technology (Figure. 42.1):

1. As observed by Moore's Law [1], integrated circuits have been getting reliably denser (and therefore more powerful) over time. The disruption for IoT was the "jump" of digital technology from fixed devices such as desktop PCs to battery-operated devices like mobile phones and sensors.
2. Koomey's Law [2] is less well known, but it's just as important for IoT because it relates to power consumption. It observes that the number of computations per joule of energy has been doubling about every 1.57 years. Many IoT scenarios involve placing sensors in locations that are not attached to a power grid. In some cases an IoT device may be expected to run on a single battery for 5 years or longer. Because of these types of constraints, IoT developers as a rule must assume that power is *not* a limitless resource; with Koomey's Law we can predict that IoT devices will become more and more capable per unit of energy over time and therefore continue to present opportunities for innovation.

Internet of Things and Data Analytics Handbook, First Edition. Edited by Hwaiyu Geng.

Companion website: www.wiley.com/go/Geng/iot_data_analytics_handbook/

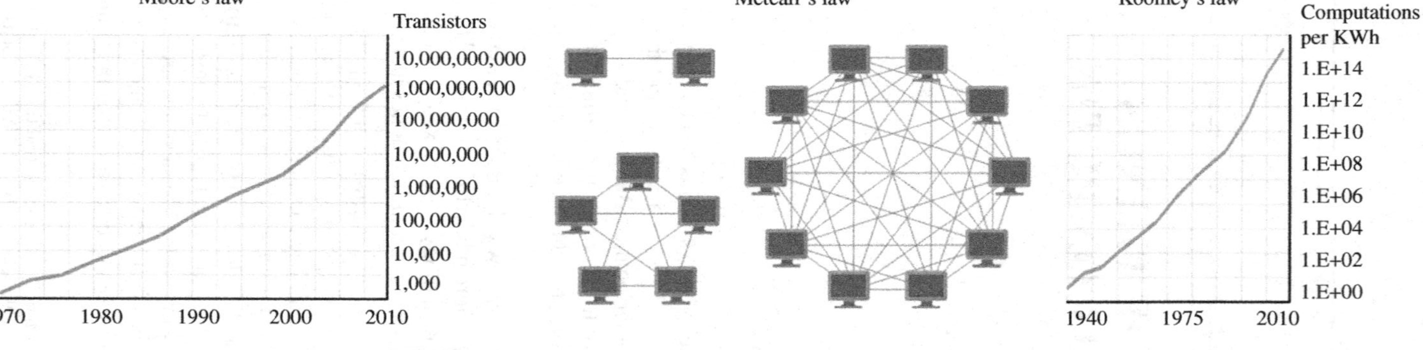

FIGURE 42.1 Disruptive forces in IoT. Courtesy of Microsoft.

3. Finally, Metcalfe's Law [3] deals with rapid increase in the value of a network relative to the number of nodes in the network. This is pretty intuitive for social networks like Facebook, but it also has powerful implications for IoT scenarios, especially when considering business intelligence capabilities. It's one thing to know sensor details about a single "thing," like a car. It's quite another to know details about 10,000 or 10 million cars.

These three forces—compute performance, power efficiency, and networking capability—work together to enable powerful new scenarios in digital electronics, from wearable biometric devices [4] to high-efficiency agricultural "plant factories" [5] to Smart Cities [6]. However, none of these scenarios would be possible with just the devices themselves. The true power of IoT comes when you give sensors and other devices the ability to *send* data ("telemetry") to servers over the Internet and *receive* instructions sent from an operator or another device.

42.2 WHAT IS CLOUD COMPUTING?

An IoT solution provider could stand up a server or set of servers to receive telemetry and perform other duties like device management and data analysis, but it's challenging to scale custom server deployments to match the load presented by large-scale IoT scenarios. Cloud-based services (i.e., "Cloud Computing") offer a more attractive alternative.

Cloud Computing (or more simply "the Cloud") refers to a class of on-demand compute services available over the Internet. These services include foundational offerings like computation and data storage (called "Infrastructure as a Service" (*IaaS*)), as well as more specialized services like machine learning and parallel data set processing (called "Platform as a Service" (*PaaS*)). The Cloud offers many appealing benefits for customers, including scalability, affordability, and manageability. It has become a critical part of IT infrastructure for companies large and small.

Cloud providers use economies of scale to offer compelling Cloud services at affordable prices. In so doing they take on the role of "system operators to the world." This frees up technology innovators to focus on their key differentiating products and features because they can effectively outsource much of their IT needs. As with any technology platform, though, care must be taken to use it wisely. Cloud-based products and solutions must be properly designed to account for several factors such as latency, compatibility, network infrastructure, security, and data privacy.

One of the ways Cloud service providers help to address these factors is by providing different deployment models of Cloud services. There are three main deployment patterns of Cloud services—public, private, and hybrid.

A *public* Cloud offers services for use by any registered customer. Services are accessed over the open Internet and are powered by computers and other infrastructure located in the data centers owned and operated by the Cloud providers. Unless customers pay for higher quality of service, Cloud providers will often run services for many different customers on a single hardware server. This can sometimes lead to the "noisy neighbor effect"—this occurs when multiple services run on a single hardware instance or virtual machine in the Cloud. Because any of the services make exhibit "bursty" behavior—times when their resource utilization peaks far beyond normal—other services running on the same hardware or VM may experience intermittent periods of diminished performance [7].

A *private* Cloud is operated for the benefit of a single organization and often deployed on dedicated infrastructure that eliminates the noisy neighbor problem. Whether operated by the organization itself or by a third party, a *private* Cloud's main value is the high level of privacy and isolation from the open Internet. This pattern has become popular as Cloud security and privacy concerns have become more and more important to businesses worldwide.

Hybrid Cloud deployments generally include aspects of both *public* and *private* Cloud implementations. The exact dividing line between the public and private services varies case by case and is usually custom-tailored to meet the individual requirements of enterprise customers.

42.3 CLOUD COMPUTING AND IoT

IoT solutions generally comprise three or four parts (Figure 42.2):

1. Sensors that monitor the physical environment or attached machinery.
2. A CPU connected either directly to the sensors or by a wireless network.
3. A remote server in the Cloud that receives telemetry from devices and often provides a command and control capability for the devices.
4. (Optional) A "field gateway" device that facilitates communication between one or more IoT devices and the Cloud. For example, a field gateway may communicate with devices on a local network like Bluetooth or 6LoWPAN and then relay the device messages to the Cloud over IP. Field Gateways can also do protocol translation (e.g., MQTT to AMQP) and even offload some processing and analytics from the Cloud (e.g., preprocessing video frames).

The Cloud offers features and benefits that make it particularly well suited for IoT. For example:

- Cloud services are designed to "scale as you go." That is, you can start out with a low-cost, low-capacity deployment and then easily move up to more powerful server hardware when your solution grows (in terms of connected devices and/or backend analysis). As with a household utility like water or electricity, you only pay for what you use. Contrast this to a self-hosted scenario where you are responsible for selecting and procuring server hardware to meet the demands of your IoT solution. You must deploy infrastructure to handle not just today's peak demand but also for your projected peak demand in the near future.

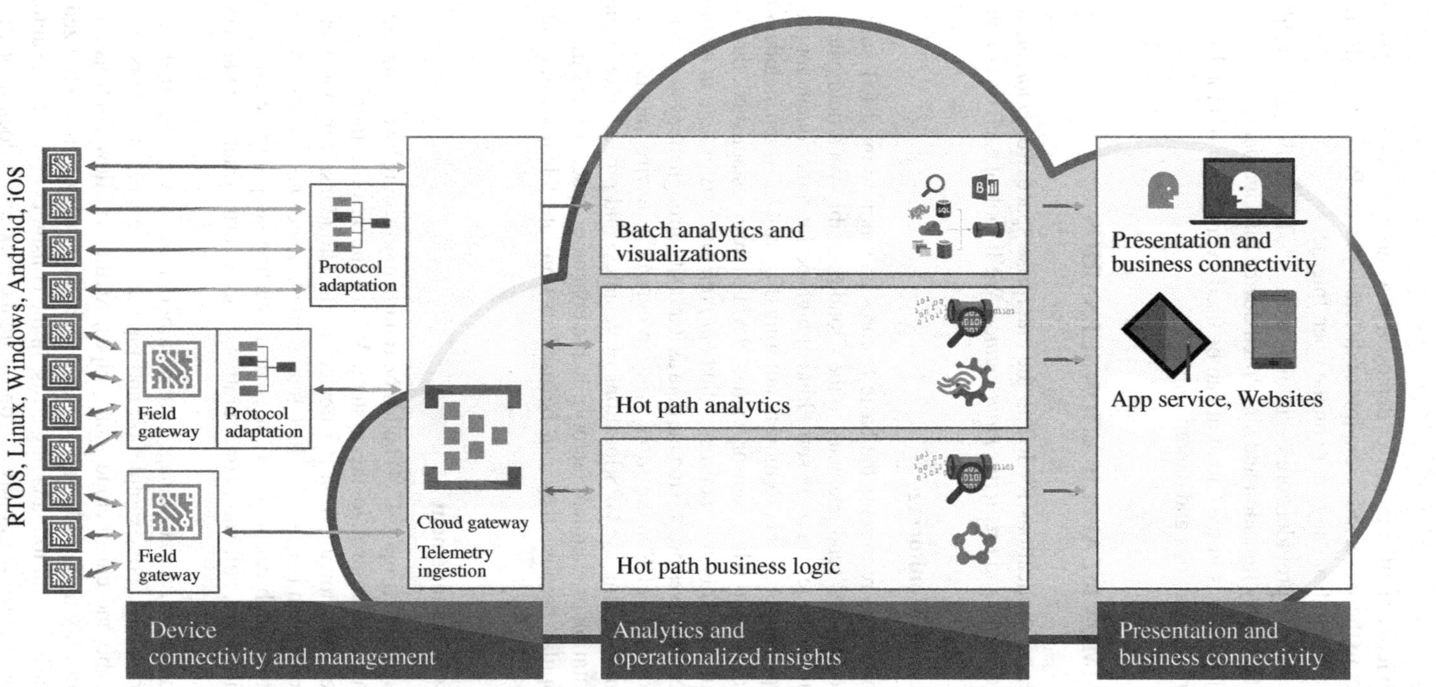

FIGURE 42.2 Example IoT device and cloud pattern. Courtesy of Microsoft.

- IoT devices are often mobile or deployed in a variety of locations. They need to connect to the server side from lots of different places. Public Clouds are generally reachable from anywhere on the open Internet.
- Cloud service providers are beginning to offer IoT-specific services like high-speed telemetry ingestion and simplified device management.
- Once IoT data is in the Cloud, it can be used by a variety of cloud-based tools like stream analytics and data mining.

42.4 COMMON IoT APPLICATION SCENARIOS

Three common applications for IoT in the enterprise space are *remote monitoring*, *asset management*, and *predictive maintenance*. We'll look at each one in turn.

42.4.1 Remote Monitoring

Remote Monitoring represents the basic features common to most IoT solutions. In this scenario, devices send telemetry to the Cloud (possibly via a field gateway), and the Cloud stores the telemetry in some kind of high-performance queue for processing by the enterprise application. Remote monitoring (Figure 42.3) often includes *anomaly detection* (detecting when sensor values deviate from established thresholds or "normal" values). This is sometimes called *hot path analysis* or *data in motion analysis* because it involves processing the data as it transits through the system in real time.

Some examples of Remote Monitoring in enterprise environments include fleet management (tracking the GPS locations of vehicles) and process monitoring (e.g., in a power station or chemical factory). In the consumer space, fitness trackers have become popular because they can be used to translate telemetry into attainable health goals.

42.4.2 Asset Management

Asset Management (Figure 42.4) builds on remote monitoring by adding capabilities of command and control to the system. Where the remote monitoring solution encompasses data flowing *from* IoT devices to the Cloud, asset management adds a new communication path where the Cloud communicates back *to* IoT devices. This communication path can be used for control (starting, stopping, resetting devices), updating (firmware, software, configuration), and security (managing access rights, ownership).

In the enterprise, Asset Management is important for companies that have assets in distant locations. For example, oil and pipeline companies have operations that span the globe and can include tanks, pipes, regulators, flow controls, etc. Their ability to not only monitor but also control their machinery from centralized operations centers is critical to managing costs. In the consumer space, home automation systems now include the ability to control lights, lock and unlock doors, and even start cooking dinner all with the tap of a button on a smartphone.

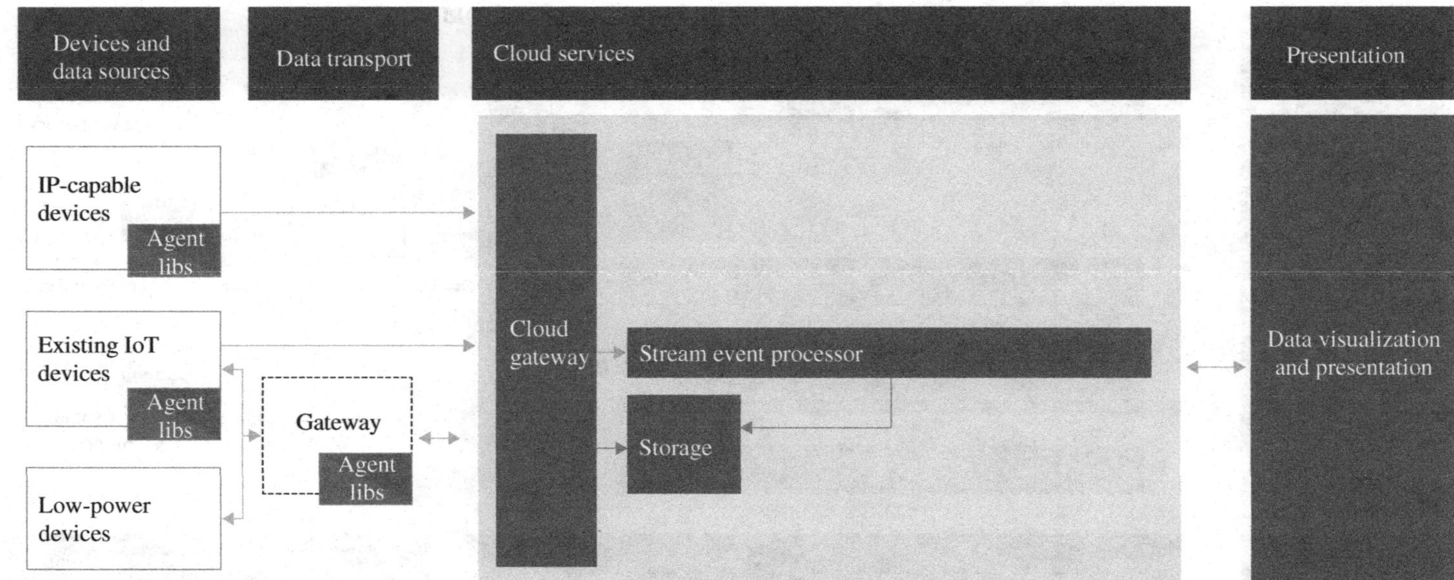

FIGURE 42.3 Remote monitoring. Courtesy of Microsoft.

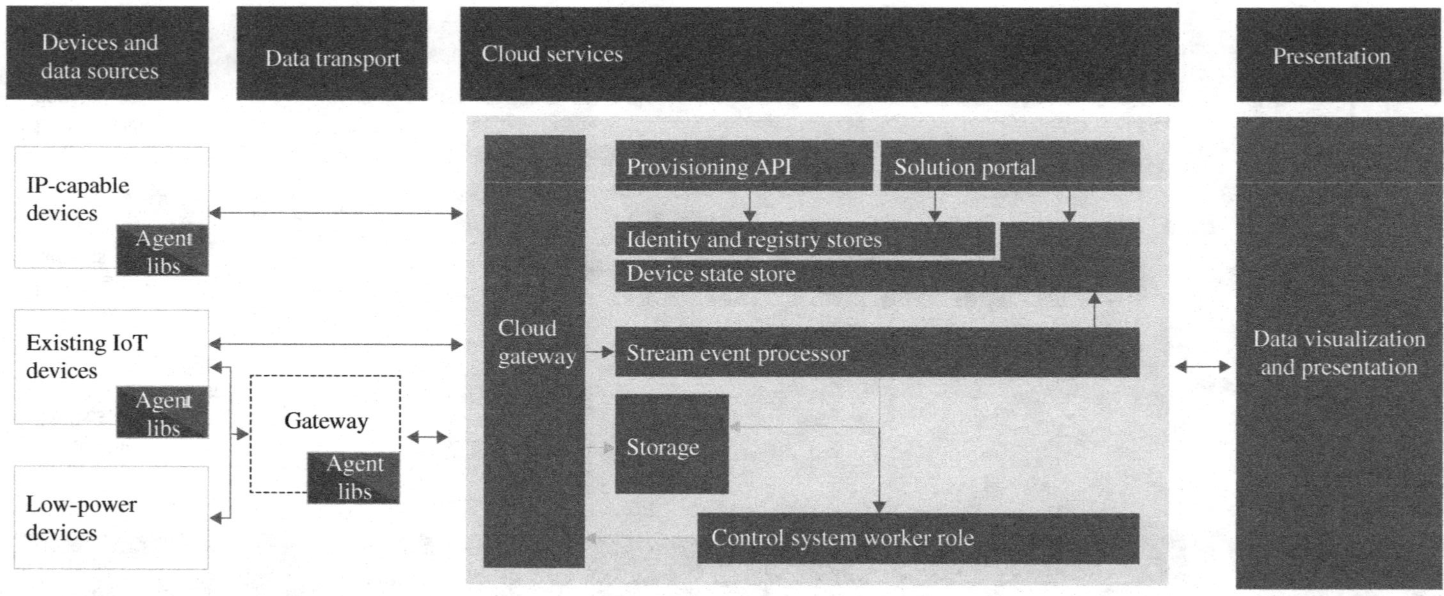

FIGURE 42.4 Asset management. Courtesy of Microsoft.

42.4.3 Predictive Maintenance

In Predictive Maintenance (Figure 42.5) scenarios, we bring in additional Cloud services like machine learning and distributed analysis tools like Apache Hadoop. These services operate on structured and unstructured data to mine for insights not immediately apparent in the telemetry. This activity is sometime called *cold path analysis* or *data at rest analysis* because the analysis is generally performed on data that has been placed into a data lake [8] or other storage container, where the data is "resting" rather than flowing through the system.

Vending machines provide an excellent example of the benefits of predictive maintenance. Vending machine failure can result in multiple losses—loss of business because a machine is unable to vend product, loss of product when items in a machine must be kept hot or cold, and loss of customer trust when customers "give up" on using a machine because it has failed them in the past. Vending machines are commonly maintained by field technicians who make rounds, stocking machines, and inspecting each machine in their area one at a time. If a problem is found, a repair technician is summoned to address the issue. While this process works, it can be improved by changing the dynamic. Instead of "polling" each machine to see what product and maintenance it requires (if any), IoT-enabled machines can "notify" when they require restocking. With remote monitoring and machine learning, failure patterns can be recognized and addressed before actual failure happens—this is predictive maintenance. By enabling vending machines to request restocking on their own and by watching their telemetry for indicators of impending failure, the entire vending machine business can be transformed.

42.5 CLOUD SECURITY AND IoT

While it's beyond the scope of this paper to fully cover information security as it pertains to IoT, there are some key points to consider:

The stakes are high. Before IoT, the biggest impact of a computer security breach was usually financial. Now with IoT, security breaches can result in physical damage in the real world or even loss of human life. It's therefore imperative to take security very seriously.

Encrypting communications between client and server is just the beginning. Encrypting data assures that it cannot be observed or altered in transit, but it does nothing to guarantee that the data can be trusted. For trust, both the sender and receiver need to provide each other with irrefutable identity credentials.

Physical tampering is IoT's "analog hole" [9]. Even with secure communications and credentials, IoT sensors can be tampered with to provide misleading data. Consider a thermal sensor that is manipulated with a flame from a match or lighter to report misleading values. There is no single solution for this kind of tampering, but some approaches include redundant sensors, sensors within physically secured enclosures, and sensors that monitor other sensors.

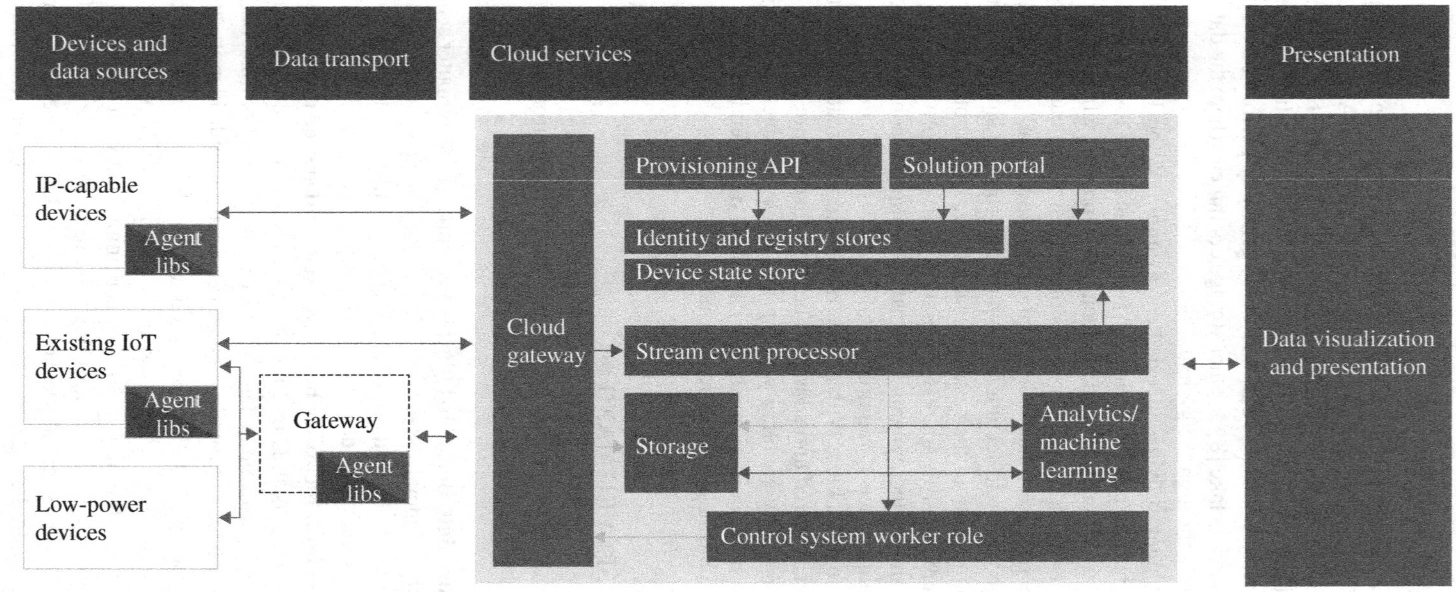

FIGURE 42.5 Predictive maintenance. Courtesy of Microsoft.

You cannot secure software that you cannot update. We have seen again and again that all software is vulnerable to exploits. It's just a matter of time and effort on the part of hackers until the exploits are discovered. Software updates are the primary defense against exploits—once an exploit is discovered, software makers can correct the defect in the code and then send out an updated version with the fix. However, if there is no mechanism to update the software running on a device, then it can't be fixed and the exploit remains forever.

42.6 CLOUD COMPUTING AND MAKERS

While much of the discussion so far has addressed enterprise IoT, Makers can also benefit greatly from Cloud services. In fact, one could argue that Makers and small companies benefit from Cloud Computing even more than big companies for three main reasons:

1. Cloud services and in particular the latest generation of IoT-focused Cloud services do much of the "heavy lifting" for IoT systems, leaving the Maker free to focus his or her limited time on what makes his or her solution special and different. This is particularly important in the self-funded start-up model.
2. Cloud providers have strong motivation to create open standard-based systems that interoperate freely with other platforms. This is great for Makers as it enables them to create solutions that avoid vendor lock-in, where hardware or software vendors offer products that do not readily interoperate with offerings from competing businesses [10].
3. Makers only pay a modest fee to benefit from all the enterprise-quality expertise and engineering that is built into every Cloud platform. Even if the Maker has the time to build their own server infrastructure from scratch, it's unlikely that any single individual has expertise in all the areas requiring attention—reliability, scalability, security, redundancy, etc.

Even a Maker who has a very targeted IoT project with narrow scope can benefit from connecting to Cloud services. Consider for a moment an example Maker project where an IoT device is attached to a pet to monitor the animal's location. While such a device could be built to connect to a single server that provides a Web interface (e.g., with a map and locator), connecting it to a Cloud service would offer the same capability while also providing options for new features such as data sharing, lost and found, and trend analysis. By connecting to the Cloud from the start, they afford themselves more options and opportunities than they would have with a closed system.

42.7 AN EXAMPLE SCENARIO

Let's walk through a concrete example of a Cloud-enabled IoT scenario (Figure 42.6). In our scenario, we will have a Maker device like a Raspberry Pi and an attached temperature sensor. Our goal is to have the device signal an alarm if the temperature goes above a specified value—in this case 35°C:

1. Code running locally on the board samples the value of the temperature sensor.
2. Board periodically sends temperature reading to an IoT telemetry ingestion service running in the Cloud. This assumes the board is connected to the Internet (over Wi-Fi, Ethernet, cellular, or other transport mechanism). Frequency of telemetry messages is determined by the user/developer and should be based on the needs of the application. An obvious side effect of sending too frequently is network congestion and potential per-message fees/charges from the Cloud provider.
3. Telemetry received at the IoT Gateway is passed on to another service for analysis. Note that the IoT Gateway usually has no understanding of the contents or meaning of data inside the telemetry messages. The developer is responsible for sending data that can be interpreted by the Stream Analytics module on the Cloud side. For example, a telemetry message from the Board may be sent as a JSON-encoded [11] structure as follows:

```
{
    "deviceId" : "1"
    "temperatureCelcius": "50"
}
```

In the Stream Analytics component, a developer-created SQL query to find temperatures greater than 35°C might look like:

```
SELECT deviceId FROM <stream> WHERE temperatureCelcius
> 35
```

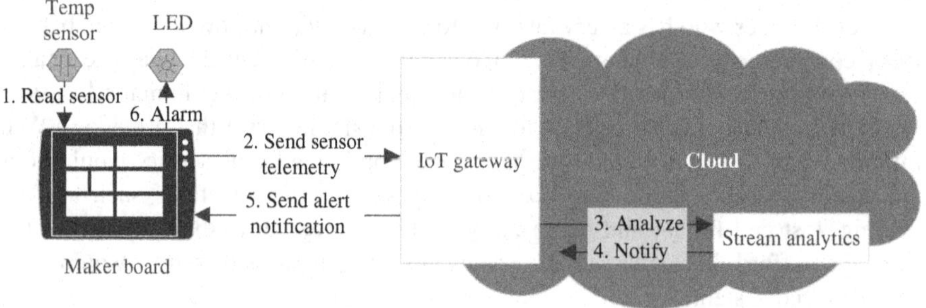

FIGURE 42.6 Example cloud-enabled IoT scenario. Courtesy of Microsoft.

4. When a message in the stream matches the query, a delegate action can be taken. It could be sending the data to another Cloud service, logging a message to a file, or, as in this case, sending a message back to the client device. In this scenario, the message is sent back to the client using the IoT Gateway, but that is a mechanism that may vary from one Cloud service provider to another.
5. A notification message is sent back to the device (usually, but not necessarily, over the same connection as the incoming telemetry).
6. Code running locally on the board receives the notification from the Cloud and is programmed to react by flashing a light on the board for a few seconds.

The same net effect of flashing a light when a sensor exceeds a threshold value could have been programmed into the device to run completely independent of the Cloud, but the result would be a "dumb" device that only knows how to do a single thing a single way. It would not be able to contribute data for aggregation, analysis, or machine learning. It could not be part of a remote monitoring scenario, nor could it contribute to predictive maintenance. In other words, it could do what it does (flash a light when a temperature gets too high), but no one outside the immediate area would know about it. That may not be an issue in this contrived scenario, but it could be a very big deal if we're talking about measuring temperature in a medical refrigerator, a power plant, or even a human body.

42.8 CONCLUSION

We are entering the golden age of data. Where compute power dominated the last 50 years of technology innovation and drove tremendous economic growth, *data*—specifically data generated by sensors of all kinds—will drive the next wave of technology innovation. We are already beginning to see this in the form of fitness trackers, industrial automation advances, smart cities, and more. As we continue to improve our capabilities with streaming analytics, machine learning, and artificial intelligence, we will invent new and surprising ways to improve both our factories and our individual lives. The key to all of this is the raw material—the data. When the data exists in a silo, or is discarded, it does not provide the fuel necessary to power our analytics engines. To get the most out of our IoT data, it must be delivered to the place where the magic happens. It must go to the cloud.

REFERENCES

[1] Moore, G.E., "Cramming More Components onto Integrated Circuits," *Proceedings of the IEEE*, vol.86, no.1, pp.82, 84, January 1998.

[2] Koomey, J.G.; Berard, S.; Sanchez, M.; Wong, H., "Implications of Historical Trends in the Electrical Efficiency of Computing," *Annals of the History of Computing, IEEE*, vol.33, no.3, pp.46, 54, March 2011.

[3] Shapiro, C.; Varian, H.R., *Information Rules*, Harvard Business Press, Cambridge, MA, 1999.

[4] www.fitbit.com (accessed August 3, 2015).

[5] http://www.fujitsu.com/global/about/resources/news/press-releases/2015/0412-01.html (accessed August 3, 2015).

[6] http://smartcitiescouncil.com/article/cities-hack-their-way-livability-gains (accessed August 3, 2015).

[7] http://searchcloudcomputing.techtarget.com/definition/noisy-neighbor-cloud-computing-performance (accessed August 30, 2015).

[8] https://en.wikipedia.org/wiki/Data_lake (accessed August 30, 2015).

[9] https://en.wikipedia.org/wiki/Analog_hole (accessed August 30, 2015).

[10] https://en.wikipedia.org/wiki/Vendor_lock-in (accessed August 30, 2015).

[11] https://en.wikipedia.org/wiki/JSON (accessed August 30, 2015).

43

PRIVACY AND SECURITY LEGAL ISSUES

FRANCOISE GILBERT*

Greenberg Traurig LLP, Silicon Valley, East Palo Alto, CA, USA

43.1 UNIQUE CHARACTERISTICS

43.1.1 Significant Volume of Data

Connected devices and Data Analytics (DA) projects can collect a wide range of data about a specific individual. This data may include, for example, data related to a person's habits (e.g., sleep patterns or distance walked), the person's precise geographic location (including presence at or absence from home, a bar, or a theater), financial account numbers, or health data (such as weight or blood pressure). Since data storage is cheap and given that even smallest IoT devices have the capability to store more data than an entire computer system occupying hundreds of square feet used to be able to store, there is usually little concern about the amount, volume, or nature of the data collected.

* Francoise Gilbert focuses her practice on US and global privacy and security in a wide variety of markets, including, among others, cloud computing, big data, connected devices, robots, and other emerging technologies. She is the author of the leading two-volume treatise "Global Privacy and Security Law" (Aspen Wolters Kluwer Law & Business), which covers in depth the privacy and data protection laws of 68 countries. She is a founding member of the Cloud Security Alliance, cochair of the PLI's Annual Privacy and Security Institute, and the recipient of numerous awards and accolades for her work in the privacy and security field. More information on her practice may be found at http://www.gtlaw.com/People/Francoise-Gilbert.

Internet of Things and Data Analytics Handbook, First Edition. Edited by Hwaiyu Geng.

Companion website: www.wiley.com/go/Geng/iot_data_analytics_handbook/

43.1.2 Data Collected over Time

In addition, this data is generally collected regularly and over long periods. For example, devices that collect weight or sleep patterns are likely to collect an individual's records on a daily or weekly basis for an entire year or longer. Devices collecting information about electricity consumption, such as a smart meter, measure both the volume and amount of electricity used in a dwelling all day, every day over several years.

43.1.3 Eavesdropping Ability

Given the nature, variety, volume, and periodicity of the collection, transfer, use, processing, and sharing of the data, there is a risk that a data scientist, device manufacturer, or an intruder might be to eavesdrop remotely into a person's private life.

In the utility example earlier, not only does the data collected by the smart meter allow the utility company to gather the information necessary to invoice the customer, but it also can be used to determine whether an individual is present at home or not and what type of appliances he uses, for example, a dishwasher or a television (TV) set, at what time of the day and how frequently.

The interception and analysis of unencrypted data transmitted from a smart meter device could result in the identification of specific activities occurring in a person's home, such as whether the home is occupied, whether a dishwasher or a TV is being used, or even, perhaps, what TV show an individual is watching. Further, combined over an extend period of time, the data may provide information regarding the times when the individual is present at home or not, as well as reveal personal habits and routines, for example, "watches TV on weekends until 2 or 3 am," thereby allowing a service to eavesdrop into its customer's habits.

43.1.4 Risks and Vulnerabilities

The collection of data everywhere, anywhere, and at any time allows the creation of large databases that can be connected to myriad other databases, devices, sensors, and, at times, third parties.

The massive volume of granular data by itself causes concern because this raw data could be used for unintended purposes. For example, information that a refrigerator is empty could conveniently trigger the repeated automatic ordering of groceries, which are then left on the doorstep of the homeowner. In reality, the empty refrigerator might only signify that the dwellers are away from home and might be simply vacationing. However, the fact that groceries are accumulating on the doorstep signals the home is unattended, which might mean a bounty for a burglar.

43.1.5 Research and Profiling

Big data analytics techniques and significant computing capability, applied to a compilation of a large volume of data, could allow data scientists to build individual profiles, with demographics such as gender, marital status, job status, or age. They might be able to infer very intimate information about an individual, such as a person's overall well-being, the progression of certain diseases, sleep patterns, frequency of exercise, or even perhaps the level of happiness.

What happens if the conclusions drawn are flawed or the inference is inaccurate because the data is insufficient—for example, the individual exercises everyday but forget to wear his health band most of the time?

What if the inference is inaccurate? If the smart meter is able to detect that two neighbors open their garage door at the same time every Sunday morning, does it not mean that they are doing the same thing? One could be ready to go to his weekly tennis game and the other to his weekly service at the local church. Flawed analysis could infer that they are both very athletic, very religious, or that they have a personal relationship.

43.1.6 Errors: Intentional Modifications

What if the data has been maliciously modified by a prankster or a disgruntled employee? Will the flawed modified patient data cause a change in treatment resulting in the patient's death?

The complex—and at times, uncontrolled—ecosystem surrounding many IoT devices makes the data that they collect vulnerable to numerous accidental uses, misuses, unintended uses, unanticipated uses, or unauthorized uses, which are likely to have adverse consequences on the individuals whose data has been collected through these devices. These risks include, among others, violation of privacy, breach of security, decisions based on flawed or inaccurate profiles, and product liability.

43.2 PRIVACY ISSUES

43.2.1 Protection of Personal Information

IoT devices and DA projects are within the continuum of the evolution of computer and sensor technologies. IoT devices combine many of the preexisting techniques and capabilities that were developed in the nonconnected world and take advantage of the capabilities of the Internet. Ultimately, their primary function is to facilitate the collection and processing of data for a variety of purposes.

From a privacy perspective, the ability to collect a constant and regular flow of personal information may cause significant privacy risks to users. For example, the collected information may reveal the habits, locations, or physical conditions of an individual over time. This information, if shared without proper restrictions, might be reused for purposes unknown by the customers, shared with unscrupulous third parties, or retained even when it is no longer relevant and the individual's circumstances have changed.

In the United States, the general rules that apply to the protection of personal information are rooted in the long-standing Fair Information Practice Principles (FIPPs). The principles that form the FIPPs include:

- Notice
- Choice

- Access
- Accuracy
- Data minimization
- Security
- Accountability

The FIPPs were first articulated in 1973 in a report by the US Department of Health, Education, and Welfare. The Privacy Guidelines, adopted in 1980 by the Organization for Economic Cooperation and Development (OECD), contain principles that are similar to the FIPPs [1]. The 1995 European Union (EU) Directive on the Protection of Personal Data [2] and the EU General Data Protection Regulation (GDPR)—which will replace the 1995 EU Directive on the Protection of Personal Data on May 25, 2018—also contain principles similar to FIPPs [3].

The FIPPs are also present in numerous seminal documents that encompass the US views on the protection of personal information. For example, the FIPPs are incorporated in the Health Insurance Portability and Accountability Act (HIPAA) [4], the White House Consumer Privacy Bill of Rights [5], the Federal Trade Commission (FTC) Privacy Framework set forth in the 2012 Privacy Report of the FTC [6], the 2000 FTC report Privacy Online: Fair Information Practices in the Electronic Marketplace: A Report to Congress [7], as well as numerous self-regulatory guidelines [8].

43.2.2 Notice and Choice

One of the basic FIPPs is that customers should be informed that personal data is being collected. Additionally, the customer should be given the opportunity to choose whether she wants her data to be used in the manner described by the vendor or service provider.

IoT devices and DA projects provide the opportunity to collect a large amount of data that may have significant implications for the privacy and/or security of individuals. Therefore, it is crucial to provide customers with the adequate level of information, especially when the collection of data may lead to the identification of sensitive information, such as medical information or an individual's personal habits (e.g., the routes taken by a jogger or the times when a person is away from home).

43.2.3 Practical Difficulty in Interaction with Customer

It is frequently difficult to provide notice and choice especially when there is no discernible consumer interface. Many IoT devices have no screen or suitable interfaces to communicate with the consumer, thereby making notice on or through the device itself difficult. This is the case, for example, in the case of many home appliances and medical devices. When a device does have a screen, it is frequently smaller and less customer friendly than those of mobile devices where providing notice is already a challenge [9]. Further, even if a device has screens, the sensors may collect data at times when the consumer may not be able to read a notice, for example, while driving.

In addition, some IoT products and DA services may not be consumer facing. This is the case, for instance, for a sensor network that monitors electricity use in a building. In this case, it is even more difficult to find a practical and effective way to provide notice.

However, the FTC has indicated that it expects these notices and mechanisms for consumers to exercise their right to make choices to be provided in addition to the typical lengthy privacy policies and terms of use [10].

To address the notice and choice principles, developers of IoT or DA solutions should devise and distribute their products and services in a manner that allows the consumers to receive, and have easy access to, clear notices of the capabilities and intended uses of the products or services, as well as the types of data collected and the use to which the data will be put.

Further, to the extent that choices are offered, the customers should be provided with a clear description of their options and a simple means to exercise their choices with respect to the collection, use, or sharing of their data.

43.2.4 Consumer's Reasonable Expectation

In its 2012 Privacy Report [6], which sets forth recommended best practices, the FTC recognized that it is not necessary to provide choices for every instance of data collection in order to protect privacy.

The FTC also explained that some data activities are within a consumer's expectations and that others are outside this "common use" range. For example, while consumers know that a fitness band is collecting data about their physical activity, they do not expect this information to be shared with data brokers or marketing firms. If the fitness band vendor expects to do so, then it should provide the customer with clear and simple notice of the proposed uses of his data along with a way to consent to the sharing of his data.

The FTC indicated that, in the case of practices that are consistent with the context of a transaction or the company's relationship with the consumer, a business should not be compelled to provide choice before collecting and using consumer data, because these data uses are generally consistent with consumers' reasonable expectations. In this case, the benefit of providing notice and choice is likely outweighed by the cost of doing so.

On the other hand, for uses that would be inconsistent with the context of the interaction (i.e., unexpected), companies should offer clear and conspicuous choices. Where the collection or use is inconsistent with the context, it is necessary to provide consumers with the requisite information and a clear statement of the customer's options.

In addition, if a company collects a consumer's data and deidentifies that data immediately and effectively, the FTC has stated that the company does not have to offer choices to consumers about this collection [6].

43.2.5 Choice

The FTC has indicated that it believes that providing consumers with the ability to make informed choices remains practicable. It also indicated that not every instance requires offering a choice to a customer [6]. For some practices, the benefits of providing choice are reduced, either because consent can be inferred or because public policy makes choice unnecessary. When data is used in a manner that is generally consistent with the reasonable expectations of the customers, the cost to consumers and businesses of providing notice and choice likely outweighs the benefits.

For example, if a consumer purchases a smart oven from ABC, and the oven is connected to an ABC app that allows the consumer to remotely turn the oven on to a specified setting, and ABC decides to use information about the use of the oven by the customer to improve features of the oven or to recommend another of its products to the consumer, it need not offer the consumer a choice for data collected for these uses, because they are consistent with its relationship with the consumer. However, if the oven manufacturer shares the data collected from the customer with a data broker or an ad network, such sharing would be inconsistent with the context of the consumer's relationship with the manufacturer, and the company should give the consumer a choice.

43.2.6 How to Communicate Privacy Choices to the Customer

Whatever approach a company decides to take, the FTC recommends that the privacy choices it offers should be clear and prominent and not buried within lengthy documents.

Different companies have adopted different methods. These include:

- Providing consumers with opt-in choices at the time of purchase
- Offering video tutorials
- Affixing QR codes on devices
- Providing choices at the point of sale
- Incorporating notices and providing choice as part of set-up wizards
- Incorporating privacy notices in a dashboard
- Providing a management portal including menus for privacy settings that consumers can configure and revisit, possibly from a separate device such as a smartphone
- Use of icons
- Communicating with the customer outside the device, for example, through email

43.3 DATA MINIMIZATION

Data minimization refers to the concept whereby companies should limit the data that they collect and retain and dispose it securely once they no longer need it. This principle can be found throughout the numerous privacy frameworks addressing the protection of personal data around the world. It has also been included in several policy initiatives. See, for example, including the 1980 OECD Privacy Guidelines, the 2004 Asia-Pacific Economic Cooperation (APEC) Privacy Principles, and the 2012 White House Consumer Privacy Bill of Rights [11]. Data manipulation is also recommended in order to minimize security risks. Indeed, the more data is collected, the more risk there is that the data may be lost, stolen, modified, exploited, or accessed by unauthorized parties.

43.3.1 Minimizing the Collection

Most companies tend to believe that they need to collect as much data as possible to have the ability to innovate around new uses of data. As IoT and DA applications are able to collect a significant amount of data in a continuous mode, the concept of minimizing the amount of data collected is especially relevant.

On the other hand, large volumes of data are vulnerable to hacking incidents (theft, modification, publication) because they constitute a more attractive target for data thieves.

Developers of IoT devices and users of DA services need to balance their goal of collecting as much information as is available against the effect that bulk collection might have on the privacy and security of their customers. The FTC has recommended that companies examine their business needs and data practices and develop policies that impose reasonable limits on the collection of data [12].

In practice, this means that businesses should pay attention to the nature and volume of data that they allow their devices or DA applications to collect. They should evaluate how they can limit their collection of data without affecting functionality, in order to limit the risk to the privacy and security of the individuals using these devices.

As part of the design of an IoT product or DA algorithm, a company should investigate whether it needs a particular data and whether this data can be deidentified. A company should examine its data practices and business needs and develop congruent policies and practices that impose reasonable limits on the collection and retention of consumer data.

There are many options. For example, in implementing its data policies and practices, the company may decide:

- Not to collect data
- To collect only the type of data necessary or the functionality of the product or service being offered
- To collect less sensitive data
- To deidentify the data collected
- To seek consumers' consent for collecting additional, unexpected categories of data

43.3.2 Minimizing the Retention

IoT and DA have the capacity to collect significant amounts of data in a continuous mode and over a long period. Collecting and retaining unnecessary data greatly increases the potential harm that could result from a data breach. Data that has not been collected or that has already been destroyed cannot fall into the wrong hands.

When evaluating the amazing data collection capability of IoT and DA in the framework of the FIPPs principles, it is clear that IoT business need to consider whether and how they should reduce the extent of their data retention in order to limit the privacy and security risks.

Best practices recommend limiting the collection of data to only the data needed for a specific purpose and safely disposing the data as soon as possible when the data is no longer needed for that specific purpose. For example, a recording device on a car could automatically delete old data after a certain amount of time or prevent individual data from being automatically synched with a central database.

43.3.3 Data Retention and Privacy

Retaining large amounts of data increases the potential harms associated with access to or misuse of data and increases the risk that the data will be used in a way that might surprise the customer.

For example, an IoT service provider could accumulate details of customers' use of the service over time, through the customer's use of one or several exercise and weight monitoring devices. The records of the changes in pulse, blood pressure, or weight, accumulated over a long period, can show patterns or anomalies, such as a sudden drop in weight. A sudden, significant drop in a person's weight could be interpreted as being symptomatic of a significant personal event or illness. The IoT service provider might be tempted to sell this information to interested third parties. The subscriber might be surprised to receive unexpectedly a steady flow of information concerning the treatment of depression. A similar situation arose in the case of the attempted sale of the customer list of a bankrupt magazine targeting gay minors. In that case, the FTC attempted to block the sale.[1] While this is not an IoT-specific case, it shows that the retention of information for long periods of time can have significant consequences if this information is shared with, or inadvertently disclosed to, third parties.

43.3.4 Data Retention and Security

IoT and DA, by nature, tend to collect a large amount of data over a long period of time. Larger databases are an attractive target. The more data is retained, the more attractive the database will be to data thieves. Conversely, data that has been deleted

[1] In connection with the bankruptcy of the XY magazine, a magazine for gay youth, the trustee attempts to sell in bankruptcy customer information accumulated over a 12-year period. At the time of the bankruptcy, the magazine had ceased to exist for more than 3 years. The Federal Trade Commission (FTC) attempted to block the sale arguing that the subscribers to the magazine likely did not contemplate that their personal details would be sold many years after they subscribed to the magazine. The FTC pointed out that the subscribers were likely to have moved on and the continued use of their information would have been contrary to their reasonable expectations. The FTC requested that the personal information accumulated by XY be deleted. The letter from David C. Vladeck, then director of the FTC Bureau of Consumer Protection to owners of the magazine XY, is available at http://www.ftc.gov/enforcement/cases-proceedings/closing-letters/letter-xy-magazine-xycom-regarding-use-sale-or

after serving its purpose (provided that the proper deletion methods were used) cannot be stolen or exploited. Thus, it is especially important in the IoT and DA context to ensure that the data collected is retained only for the minimum time necessary to accomplish the predefined purpose.

The FTC has repeatedly voiced its concern over the excessive and unnecessary retention of personal data. In several of its enforcement actions against companies that had suffered a major breach of security, the FTC blamed the devastating effect of the breach on the companies' excessive data retention periods. In its complaints and orders, the FTC noted that these companies could have mitigated the harm associated with a data breach by disposing of customer information they no longer had a business need to keep [13].

43.3.5 Application to IoT or DA Development

To avoid unnecessary or excessive data collection practices and reduce data retention duration, a business should make a conscious effort to develop and implement policies and practices that impose reasonable limits on the collection and retention of consumer data. These policies and practices should be developed in view of its own current and reasonably anticipated business needs.

Such an exercise is integral to a privacy-by-design approach. This is accomplished by evaluating:

- What types of data the device is collecting
- What types of data the device is transmitting to the business
- For what purpose(s) each category of data, and each data, is used
- For each purpose, whether the purpose could be accomplished with less data, or less specific data (e.g., if location data is necessary, would zip code be sufficient rather than a complete street address)
- How long data collected from the device should be stored
- Whether each data field should be kept for the same duration
- Whether deidentified data would be sufficient to achieve certain purposes

For example, assume that an IoT business designs a health band intended to collect information about how skin reacts to exposure to sunlight and sunrays. The device may need to collect geographic location, outside temperature, body temperature, and humidity level. However, the device may not need to collect precise geolocation information in order to work and may operate sufficiently with only zip code information rather than precise geolocation information such as street address.

If the device manufacturer believes that precise geolocation information might be useful for a future product feature that would enable users to find treatment options in their area, the company should evaluate whether it should wait to collect precise geolocation information until after it issues the product release containing the feature that requires such precise information.

If the company decides that it does require the precise geolocation information, it should provide a prominent disclosure about its collection and use of this information and obtain consumers' affirmative express consent. Finally, it should establish reasonable retention limits for the data collected.

43.4 DEIDENTIFICATION

In many cases, deidentified data might be as useful as data in identifiable form. The use of deidentified data may be a viable option in some circumstances. It might help minimize the potential risk to customers' privacy while preserving the value of the collected information for research purposes. Appropriately, deidentified data sets that are kept securely, and accompanied by strong accountability mechanisms, can reduce many privacy risks [14].

43.4.1 Benefits of Deidentification

When proper deidentification techniques are used in conjunction with reidentification risk management procedures, the collection, use, and disclosure of deidentified information have a number of advantages over the use of personally identifiable information. Most fundamentally, it reduces the risk of a privacy breach if the data is lost, stolen, or accessed by unauthorized persons because it is less likely that a person can be identified from information that has been properly deidentified [15].

43.4.2 Weakness of Deidentification

Deidentification is not always a silver bullet, and it should be implemented with care. Studies conducted by data scientists have shown that in some circumstances, data that had been deidentified could be reidentified. For example, a group of experts that had attempted to reidentify approximately 15,000 patient records that had been deidentified under the HIPAA standard [16] were able to reidentify some of the records. Using commercial data sources to reidentify the data, they were able to identify 0.013% of the individuals [17]. As technology improves, there is always a possibility that purportedly deidentified data could be reidentified [15].

43.4.3 Deidentification in IoT and DA

Because of the concerns and uncertainties described earlier, an IoT and DA business that intend to use deidentification should implement deidentification policies, procedures, and practices that are based on the assessment of the risks identified. These should be reviewed on a regular basis to ensure that they continue to be consistent with industry standards and best practices, technological advancements, legislative requirements, and emerging risks [15].

Businesses should also enter into written contracts with the individuals or organizations that are granted access to the deidentified information. These contracts should [15]:

- Prohibit the use of deidentified information, alone or with other information, to identify an individual.
- Place restrictions on any other use or subsequent disclosure of the deidentified information.
- Ensure that those who have access to the deidentified information are properly trained and understand their obligations in respect of such information.
- Require the recipient to notify the organization of any breach of the agreement.
- Set out the consequences of such a breach.

Further, the FTC has also advised that a company that maintains deidentified or anonymous data should:

- Take reasonable steps to deidentify the data, including by keeping up with technological developments.
- Publicly commit not to reidentify the data.
- Have enforceable contracts with any third parties with whom they share the data, requiring the third parties to commit not to reidentify the data.

43.4.4 Sensitivity of the Data

In considering reasonable collection and retention limits, it is also appropriate to consider the sensitivity of the data at issue: the more sensitive the data, the greater the potential harm if the data fell into the wrong hands or was used for purposes that the consumer would not expect [17].

In its Internet of Things Report, the FTC Staff recommends that companies should be careful to minimize their collection of sensitive data and limit such collection to data that meets their business goals and that is consistent with the customer's expectations [17].

43.4.5 Consumer's Consent

When discussing data minimization, in its Internet of Things Report, the FTC states that IoT companies have numerous options. They can decide not to collect data at all or collect only the fields of data necessary to the product or service being offered; other options include collecting less sensitive data or deidentifying the data that they collection.

However, if a company determines that none of these options work, it "can seek consumers' consent for collecting additional, unexpected data" [17]. This statement should be read with caution. While the FTC Report does not have the status of a Regulation, it clearly provides an indication of the current views of the FTC on this issue. It should be anticipated that if an IoT business were investigated for deficiencies in its personal data collection and handling practices, the FTC would be very interested in understanding the scope of the data collection, the disclosures made to the consumers, and the specific interaction between the IoT business and the consumer regarding the collection of data beyond the minimum necessary or expected. While it is not clear whether an actual consent would be required or an implied consent would suffice, it is clear however, that at a minimum, clear, and conspicuous disclosure would be required to ensure that the consent—whatever its form—is deemed valid.

43.5 DATA SECURITY

According to the FTC, a company's data security measures must be reasonable and appropriate in light of the sensitivity and volume of consumer information it holds, the size and complexity of its business, and the cost of available tools to improve security and reduce vulnerabilities [17]. It is clear that security must be built into IoT or DA applications. The nature, volume, and quality of the data collected and the continuous flow of this data within the IoT or DA service or ecosystem, such as other applications, devices, sensors, multiple clouds, and service providers, also expose each IoT device and DA service to numerous security vulnerabilities. A security flaw could enable unauthorized access and misuse of the data or facilitate attacks on other systems connected to the IoT or DA application. A hacker or a disgruntled employee could modify results or block access to certain key—perhaps vital—information.

43.5.1 Unique Vulnerabilities of IoT and DA

IoT and DA rely in great part on the transmittal of data to the cloud. Any security vulnerabilities could put the information stored on or transmitted at risk. If an IoT device or other devices connected to that device store sensitive data, such as account information or passwords, unauthorized access to the device might result in identity theft or fraud. As the use of IoT devices increases and the devices become more interconnected, the number and extent of vulnerabilities increases exponentially. Among other things, vulnerabilities could enable attackers to assemble large numbers of devices to use in a denial of service attack or to send malicious emails.

Unauthorized persons might exploit security vulnerabilities to create risks to physical safety in some cases. Consider, for example, the press reports concerning the decision taken by one public figure to deactivate the wireless capabilities of the device implanted to regulate his heartbeat to prevent any unauthorized access to the device [18].

These potential risks are exacerbated by the fact that securing connected IoT devices may be challenging for numerous reasons. Many companies may lack the knowledge or economic incentives to provide ongoing support or software security updates for their devices. Some IoT devices are inexpensive and essentially disposable, and companies may, therefore, decide that it does not make sense to support them. Some IoT or DA developers may not have experience in dealing with security issues. If vulnerability was discovered too late, it may be difficult or impossible to update the software or apply a patch to remedy it. In addition, if an update for an IoT device becomes available, many consumers may never hear about it. The combination of these problems may leave numerous devices unsupported or vulnerable.

Ensuring security is especially challenging, but it is essential because of the potential significant consequences of a breach of security. While security techniques evolve at the speed of light and each business will have to keep continually abreast of technical development in security products and security techniques, there are also best practices. Following these best practices will help provide a structure and goals to an otherwise daunting task.

43.5.2 How to Address Security Risks

IoT and DA developers, distributors, and service providers should ensure that their devices and services meet the optimum security for their offering under the circumstances. What constitutes reasonable security for a given device will depend on a number of factors, including the amount and sensitivity of data collected, the device functionalities, and the costs of remedying the security vulnerabilities. In many of its consent decrees and its publications, the FTC has identified its expectations regarding the development of an appropriate security program [19].

In its first case involving an Internet-connected device, against TRENDnet, the FTC detailed its expectation regarding IoT devices and services [20]. TRENDnet marketed a home monitoring camera connected to the Internet. The camera enabled homeowners to monitor remotely the activities in their home, for example, to monitor the whereabouts of a pet.

In its complaint, the FTC alleged, among other things, that the company transmitted user login credentials in clear text over the Internet, stored login credentials in clear text on users' mobile devices, and failed to test consumers' privacy settings to ensure that video feeds that were marked as "private" would in fact be private. Hackers were able to exploit these vulnerabilities to access live feeds from consumers' security cameras and conduct unauthorized surveillance of infants sleeping in their cribs and other activities normally conducted in the privacy of a home. The FTC consent order with TRENDnet required, among other things, that TRENDnet adopt the security program and security measures described in the section earlier.

43.5.3 Security

In its Internet of Things Report, the FTC summarized its recommendations with respect to security as follows [17]:

- Build security into the IoT devices and service. As part of the security-by-design process, consider conducting a privacy or security risk assessment, minimizing data collected and retained, and testing the security measures before launch.

- Train all employees about good security and ensure that security issues are addressed at the appropriate level of responsibility within the organization.
- Retain service providers that are capable of maintaining reasonable security and provide reasonable oversight for these service providers.
- When significant risks within the IoT systems are identified, implement a defense-in-depth approach, in which security measures are implemented at several levels.
- Consider implementing reasonable access control measures to limit the ability of an unauthorized person to access a consumer's device, data, or even the consumer's network.
- Continue to monitor IoT products and services throughout the life cycle and, to the extent feasible, patch known vulnerabilities.

43.5.4 Adoption of Proper Security Measures

The investigation of the numerous security breaches that have been disclosed over the past few years has revealed that numerous security flaws could have been avoided if security had been implemented from the beginning, at the time of the design and development, rather than as afterthought. The trend toward early focus on security issues has been named "security by design." "Security by design" means building security into a device or a service at the outset, as part of its design, development, and alpha testing, rather than as an afterthought.

Developers, distributors, and service providers should implement "security by design" by building security into their devices or applications at the outset. During the life of the IoT device or DA application, the business should continue with these same procedures and methods to continuously and regularly update its processes and procedures to meet the optimum security for its offering.

43.5.5 Defense-in-Depth Approach

If significant risks have been identified, the FTC recommends that a defense-in-depth approach, in which security measures are implemented at several levels, be considered [17]. For example, in many cases, an IoT device should not rely solely on the security of the customer's networks, such as passwords for their Wi-Fi routers, to protect the information on connected devices. IoT and DA businesses should take additional steps to secure the information, such as through encryption, when the data is in use, in transit, or in storage.

43.5.6 Reasonable Access Control

The Internet of Things Report also recommends that IoT developers and businesses consider implementing reasonable access control measures to limit the ability of an unauthorized person to access a consumer's device, data, and the consumer's network [17]. In some cases, the small size and limited processing power of many connected devices could inhibit encryption and other robust security measures.

Some connected devices are relatively inexpensive and may not have the capability to receive updates or patches to address vulnerability after they have been sold to a consumer. Strong authentication could be used to permit or restrict IoT devices from interacting with other devices or systems. The privileges associated with the validated identity determine the permissible interactions between the IoT devices and could prevent unauthorized access and interactions.

43.5.7 Appropriate Level of Responsibility within the Organization

As part of the security-by-design approach, the FTC recommends that security should be addressed at the appropriate level of responsibility within the organization, by appointing someone at the executive level to be responsible for security [17]. Security should also be a significant consideration in hiring decisions and in developing processes and mechanisms throughout the entire organization to improve security and instill security in each aspect of the design, development, support, and upgrade of a product.

43.5.8 Employee Training

Companies and their products and services are potentially subject to error or malignant acts of individuals who have access to the devices. In addition to external threats, there are internal vulnerabilities from the acts and omissions of employees and service providers.

Businesses should also train their employees about good security practices, recognizing that technological expertise does not necessarily equate to security expertise. They should train all employees about good security and ensure that security issues are addressed at the appropriate level of responsibility within the organization.

43.5.9 Service Providers

When retaining service providers, organizations should ensure that they choose individuals and entities that are capable of maintaining reasonable security. In addition, they should regularly monitor these service providers to ensure that they comply with the agreed-upon privacy and security requirements.

43.5.10 Product Monitoring: Vulnerability Patching

Companies should continue to monitor devices and applications throughout the life cycle and, to the extent feasible, patch known vulnerabilities.

Many IoT devices have a limited life cycle. At some point in the life of an IoT device, the customers are likely to use an out-of-date device that is vulnerable to significant, known, security, or privacy risks.

The IoT business should evaluate how to communicate the existence of these vulnerabilities and determine the extent to which the provision of security updates and software patches is practically and economically feasible. In this case, it is crucial to carefully balance the effect of the vulnerability against the feasibility and cost of providing the patch.

Businesses should also provide clear and conspicuous information about their ability to provide ongoing security updates and software patches. For example, they should disclose the length of time during which they plan to support and release software updates for a given product line. When they provide updates, they should also notify customers clearly and conspicuously—and repeatedly—about the existence of the known security risks and updates.

43.6 PROFILING ISSUES

43.6.1 Profiling

IoT and DA allow the collection of large volumes of granular data. These huge databases allow skilled data scientists to identify patterns and produce statistics and inferences. The resulting information may provide a wide range of personal details that could be combined to create the profile of an individual with, for example, information about the individual's family, friends, contacts, health, wealth, occupation, hobbies, religious or philosophical beliefs, and the like.

Most connected devices and DA applications allow to digitally monitor customers' private activities. The regular collection and variety of data collected may allow an entity to infer sensitive or intimate information through the observation of patterns. For example, the correlation of the calendar of religious holidays and the level of occupation of a home collected from a smart meter might allow inferences on a family's religion.

However, statistics and data are not perfect. There could be errors. For example, when utility sensors detect that two neighbors open their garage door at the same time every Sunday morning, it does not necessarily mean that they are doing the same thing. One could be going to his weekly tennis game and the other to his weekly service at the local church. Flawed analysis could infer that they are both very athletic, very religious, or that they have a personal relationship.

The large volume of data collected is likely to allow the creation of customer profiles that may include information beyond what the customer initially contemplated when purchasing the IoT device or signing up for the IoT service. With these profiles in hand, companies might be able to use certain data to make credit, insurance, and employment decisions.

43.6.2 Use of Information to Deny Credit, Insurance, or Employment

Businesses should be aware of the scope and meaning of the Fair Credit Reporting Act (FCRA) (15 U.S.C. §1681). The FCRA is a complex law that has been in effect since the early days of privacy in 1970.

Among other things, the FCRA imposes certain limits on the use of consumer data to make decisions about credit, insurance, employment, or for similar purposes.

Further, sections 15 U.S.C. §§1681e, 1681j of the FCRA impose numerous obligations on entities that qualify as consumer reporting agencies. Among other things, they are required to employ reasonable procedures to ensure the accuracy of their data and to give consumers access to their information.

Thus, businesses must be careful not to find themselves in a position where they are caught inadvertently violating the provisions of the FCRA.

43.7 RESEARCH AND ANALYTICS

The large volume of data collected and the databases created by combining all such data, provides the ability to conduct a variety of research, by taking advantage of the newest techniques used by data scientists, combining analytics, semantics, big data, and similar techniques and technologies. However, the use of data for DA or other research purposes is not necessarily contemplated by IoT users and owners of IoT devices or consumers; thus special precautions should be taken.

43.7.1 Uses Data for Research Purposes

Significant, valuable information may be obtained from the compilation and analysis of large volumes of data, such as that which is collected in the IoT context.

Today's businesses are eager to use the vast troves of data generated by connected devices to better understand their users. When conducting such complex, in-depth analysis, there is a significant risk that the ubiquitous collection and sharing of the data may lead to significant abuses. On the other end of the spectrum, some of these data analyses and compilation may lead to major discoveries in a wide variety of areas, such as in the medical field or in connection with urban management, such as the reduction of traffic congestion or better efficiency in handling emergencies.

Thus, there is a tension between allowing the analysis of this data and preventing it. From the standpoint of the FIPPs, the use of the data collected through a specific product or service for research and analytics purposes may be inconsistent with the knowledge, understanding, or expectations of the typical user of an IoT device.

In principle, data collected by a service must be used solely for the benefit of the consumer who has purchased the specific device or has subscribed to a particular service.

Thus, while it is tempting to use the large amount of data collected by IoT devices or other applications, this use, if uncontrolled, should be avoided, as it is likely to lead to uses of the collected data in ways that are inconsistent with consumers' expectations or relationship with a company.

To ensure that research and analysis is performed in a manner that is fair to both the service user and the provider of such services, businesses should ensure that they comply with the FFIPs, discussed earlier in this chapter, and among other things, provide users with clear and conspicuous notice and an opportunity to choose the manner in which their data may be used.

43.7.2 Data Deidentification

To the extent that IoT companies collect data that they will use for research purposes, they should deidentify the data where possible. Sound technical strategies for deidentifying the data should be coupled with administrative safeguards. For example, companies should publicly commit not to seek to reidentify data. They should also ensure that their service providers do the same, by requiring in written contract that these service providers follow specified guidelines or requirements.

43.8 IoT AND DA ABROAD

The legal issues and concerns surrounding the development of IoT and DA and the fast growing sales of IoT devices and DA services have also attracted the attention of regulators and standard setting organizations throughout the world.

43.8.1 European Union Article 29 Working Party

In September 2014, the Article 29 Working Party, the umbrella EU organization under which the data protection authorities of EU Member States meet and conduct join activities, published a document regarding its observations on the developments of IoT devices and services [21]. The opinion focused on the privacy and security issues raised by these devices and services.

Among other things, the Article 29 Working Party stressed the importance of giving users the ability to choose how or when their personal data could be used. The Article 29 Working Parting also indicated that, in its opinion, users must remain in complete control of their personal data throughout the product lifecycle. The Article 29 Working Party also pointed out that when an IoT business relies on consent as a basis for processing, the consent should be fully informed, freely given, and specific.

43.8.2 EU General Data Protection Regulation

Companies selling or providing services related to IoT devices should be aware of the provisions of the EU General Data Protection Regulation [3]. The GDPR replaces the 1995 EU Data Protection Directive, 95/46/EC, as of May 25, 2018. The GDPR will be the primary data protection law in the EU and the European Economic Area (EEA), that is, EU plus Norway, Lichtenstein, and Iceland.

The GDPR will apply to all entities that are established in the EU or EEA and process personal data, whether the processing occurs in the EU or not. In addition, it will apply to the processing of personal data of individuals who reside in the EU or the EEA when the processing is conducted by an entity that is established outside the EU or the EEA, if the processing relates to: (1) the offering of goods or services in the EU, whether payment is required or not; or (2) the monitoring of an individual's behavior that takes place within the EU.

The GDPR incorporates many of the principles found in the 1995 EU Data Protection Directive but contains numerous new concepts. For example, regarding consent, which is often a key to making legal the processing of personal data, the data controller should be able to demonstrate that such consent was given and that it was given freely, specific, informed, and unambiguous. In addition, companies will have significant documentation obligation, such as documenting their data collection activities, the measures taken to protect the security of the data, and much more.

Businesses that intend to sell or distribute their products throughout the EU and EEA should be aware of the details of the GDPR because the GDPR will significantly affect their operations to the extent that they include the processing of personal data of EU or EEA residents, whether or not the business is established in the EU or EEA.

43.8.3 Standard Setting Organizations

In addition to policy work conducted by government agencies and regulators, relevant standard organizations continue to be very active in attempting to create structures for the orderly development of IoT devices and DA business. For example, oneM2M, a global standards body, released a proposed security standard for IoT devices in 2014. The standard addresses such issues as authentication, identity management, and access control [22].

REFERENCES

[1] OECD Guidelines on the Protection of Privacy and Transborder Flows of Personal Data (1980) and (2013).

[2] Directive 95/46/EC of the European Parliament and of the Council of 24 October 1995 on the protection of individuals with regard to the processing of personal data and on the free movement of such data.

[3] Regulation (EU) 2016/679 of the European Parliament and of the Council of 27 April 2016, http://eur-lex.europa.eu/legal-content/EN/TXT/PDF/?uri=CELEX:32016R0679&from=EN (accessed August 31, 2016).

[4] Health Insurance Portability and Accountability Act of 1996, Pub. L. 104-191, 110 Stat. 1936.

[5] The White House, Consumer Data Privacy in a Networked World: A Framework for Protecting Privacy and Promoting Innovation in the Global Digital Economy (2012).

[6] FTC, Protecting Consumer Privacy in an Era of Rapid Change: Recommendations for Businesses and Policymakers (2012).

[7] FTC, Privacy Online: Fair Information Practices in the Electronic Marketplace: A Report to Congress (2000).

[8] Network Advertising Initiative (NAI) Code of Conduct 2013, Internet Advertising Bureau (IAB), and Interactive Advertising Privacy Principles (2008).

[9] FTC Staff Report, Mobile Privacy Disclosures: Building Trust Through Transparency 10–11 (2013) (“Mobile Disclosures Report”).

[10] Opening Remarks of FTC Chairwoman Edith Ramirez, Privacy and the IoT: Navigating Policy Issues, International Consumer Electronics Show, Las Vegas, Nevada, January 6, 2015.

[11] OECD: *OECD Guidelines Governing the Protection of Privacy and Transborder Flows of Personal Data* (2013).The White House, Consumer Data Privacy in a Networked World: A Framework for Protecting Privacy and Promoting Innovation in the Global Digital Economy (2012).

[12] FTC Privacy Report, Protecting Consumer Privacy in an Era of Rapid Change: Recommendations for Businesses and Policymakers (2012).

[13] CardSystems Solutions, Inc., No. C-4168, 2006 WL 2709787 (FTC Sept. 5, 2006) (consent order); DSW, Inc., No. C-4157, 2006 WL 752215 (FTC Mar. 7, 2006) (consent order); BJ's Wholesale Club, Inc., 140 FTC 465 (2005) (consent order).

[14] Arvind Narayanan and Edward Felten, No Silver Bullet: De-Identification Still Doesn't Work (July 9, 2014).

[15] Ann Cavoukian and Khaled El Emam, De-identification Protocols: Essential for Protecting Privacy (June 25, 2014).

[16] 45 C.F.R. § 165.514(b)(2).

[17] FTC Internet of Things Report.

[18] ABC News, October 2013 "Dick Cheney Feared Assassination via Medical Device Hacking" available at http://abcnews.go.com/US/vice-president-dick-cheney-feared-pacemaker-hacking/story?id=20621434 (accessed August 31, 2016).

[19] Credit Karma, Inc., File No. 132-3091 (Mar. 28, 2014) (consent); Fandango, LLC, File No. 132-3089 (Mar. 28, 2014) (consent); HTC America, Inc., No. C-4406 (July 2, 2013) (consent).

[20] TRENDnet, Inc., No. C-4426 (Feb. 7, 2014) (consent).

[21] Article 29 Data Protection Working Party: *Opinion 8/2014 on the Recent Developments on the Internet of Things* (September 16, 2014).

[22] oneM2M, Technical Specification, oneM2M Security Solutions, available at http://www.onem2m.org/images/files/deliverables/TS-0003-Security_Solutions-V-2014-08.pdf (accessed August 31, 2016).

44

IoT AND INNOVATION

WILLIAM KAO

Department of Engineering and Technology, University of California, Santa Cruz, CA, USA

44.1 INTRODUCTION

Innovation is the watchword or buzzword of our day. Everywhere you look people are talking about innovation, and everybody wants to be the next Apple, Google, Netflix, or Uber.

Today's challenging economic conditions allied with global competition provide a business landscape that is hostile to companies that stand still. The dramatic and seemingly incessant development of new technologies provides opportunities for new products, new services, and even new ways of doing business. Lastly, today's market is truly global, offering an unparalleled opportunity for businesses.

Innovation is about developing businesses in new ways that will maximize value from new ideas, thereby yielding benefits. It covers any improvement to your business from introducing new products and services to reengineering your business processes to repositioning your business in the market or developing a completely new business model.

44.2 WHAT IS INNOVATION?

The word innovation comes from the Latin root "innovatus" which means to renew or change [1, 2]. Innovation can mean many things: doing things differently, coming up with new ideas or products, being creative, thinking out of the box, taking risks to gain market share, etc. The ability of an organization to innovate depends not only in having the right tools and processes but also in having a good organization culture and stakeholder behavior.

Internet of Things and Data Analytics Handbook, First Edition. Edited by Hwaiyu Geng.

Companion website: www.wiley.com/go/Geng/iot_data_analytics_handbook/

Innovation can be defined as "exploiting new ideas leading to the creation of a new product, process or service." In innovation besides the invention of a new idea, it requires putting the idea into practice, adding value or improving quality, and finally bringing it to market. There is no innovation until an idea gets implemented and somebody takes risks to deliver something useful. It is the "profitable implementation of ideas."

44.2.1 Difference between Invention, Creativity, and Innovation

For many people it is unclear or confusing the difference between invention, creativity, and innovation [3].

Theodore Levitt, a well-known Harvard Business School marketing professor, says, "Creativity is thinking up new things. Innovation is doing new things." One can think of invention as the laying of an egg, while innovation is the laying, incubation, and hatching and even the sale of a chicken.

An *invention* is the creation of a new product, process, or service, but it may or may not have an economic motive, that is, it may not be commercialized.

Innovation entails the introduction of a new product, process, or service into the marketplace; it has an economic motive and results in commercialization.

Invention precedes innovation, while innovation follows invention:

$$\text{Innovation} = \text{invention} + \text{commercial exploitation (value)}$$

Besides the previous formula indicating the relation between innovation and invention, one can additionally describe the relationship between innovation and creativity as in Figure 44.1.

Note that innovation equals creativity times execution. That says that if you have creativity, but the execution is bad or faulty, innovation fails. As they say anything multiplied by zero results in zero.

44.2.2 Types of Innovation

For many years innovation was seen as the development of new products. Most innovation research was focused on improving product performance. There are however equal opportunities to create a competitive advantage in areas other than product and technology such as process, service, management, marketing, sales, and business model [4] (see Figure 44.2).

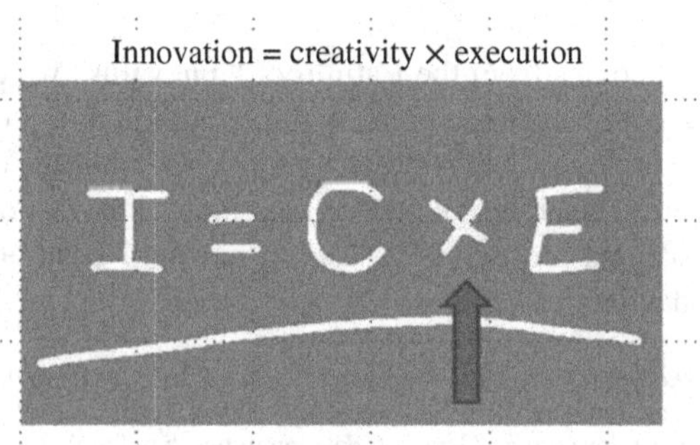

FIGURE 44.1 Relation between innovation and creativity.

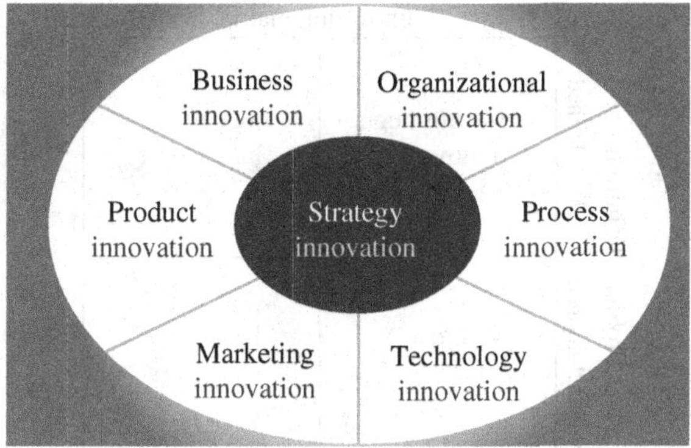

FIGURE 44.2 Areas of innovation.

44.2.3 The Innovation Management Matrix

By determining problem and domain definition (whether they are well defined or not), one can build a 2×2 Innovation Management Matrix encompassing four types of innovation: Basic, Breakthrough, Sustaining, and Disruptive[5] (see Figure 44.3):

1. **Basic research**: When you are trying to discover something truly new, neither the problem nor the domain is well defined. While most basic research happens in academic institutions, some businesses (e.g., IBM, Xerox, GE, Apple, Google) can excel at it as well.
2. **Breakthrough innovation**: Sometimes, even though the problem is well defined, organizations or even entire fields of endeavor can get stuck. Many firms are turning to open innovation platforms such as InnoCentive. Through a global network, seekers and solvers interact creating innovative solutions and a collective intelligence. It illustrates the concept of crowdsourcing allowing outsiders to solve problems that organizations are stuck on. Procter and Gamble has built its own Connect+Develop platform which allows them to benefit from expertise in a variety of domains across the world. Praveen Gupta, in his book *The Innovation Solution* [6], presents a Breakthrough Innovation Framework, consisting of a powerful Target, Explore, Develop, Optimize, and Commercialize(TEDOC) methodology and Return on Innovation.
3. **Sustaining innovation**: Whatever you do, you always want to get better at it. In this type of innovation, both the problem and domain are well defined. Large organizations tend to be very good at this type of innovation, because conventional R&D labs and outsourcing are well suited for it. Great sustaining innovators have good, clear product visions and are also great marketers. They see a need where no one else does.

   Apple's Steve Jobs was a superior innovator, and they improved the smartphone and the tablet computer to such an extent that it seemed like Apple created something completely new.

   Likewise in the auto industry Toyota makes cars just like other manufacturers but with better maintenance records.

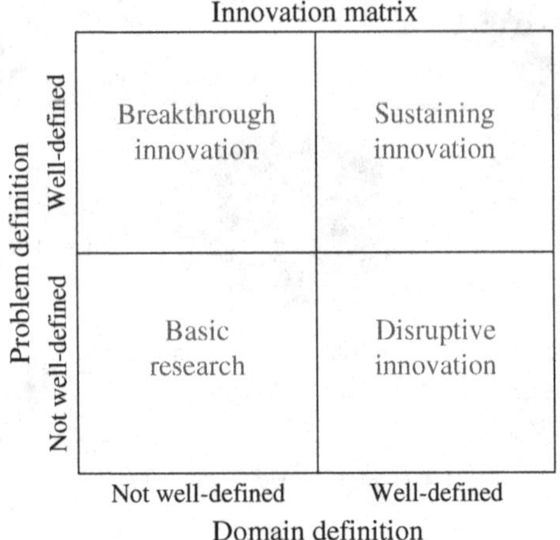

FIGURE 44.3 The innovation matrix.

4. **Disruptive innovation**: This is the most troublesome type. Disruptive innovations generally target light or nonconsumers of a category, so they require a new business model and have high failure rates. Once in awhile Google comes out with something that most people won't even understand and manage to fill needs consumers didn't even know they had.

   3M, the company that pioneered scotch tape and post-it notes, also came out with quite a few disruptive products.

Now let us take a look at another "innovative matrix" (see Figure 44.4) based on the relative levels of investment, risk, and return [7].

In the aforementioned matrix innovation is being structured along two orthogonal axes. On the horizontal axis, it is defined as either product innovation or process innovation. On the vertical axis, it is defined as either incremental innovation or disruptive innovation. The result, as shown in the diagram, is a square with four quadrants. Not every quadrant makes sense in every business climate, but none of the quadrants are inherently poor choices. The only low-return choice is no innovation at all.

In each quadrant, the relative level of investment, risk, and return for that type of innovation is given. For example, a disruptive process innovation (fourth quadrant) can return very large rewards, but to achieve it you have to make a large investment and accept a high level of risk. One should only make this kind of investment if the potential return is high. There may be considerable risk in achieving it, but the potential needs to be there. An incremental product innovation (first quadrant) is unlikely to generate large returns, but it is much less risky. It is usually a less expensive undertaking, so it often makes sense to do even if the returns are moderate.

First let us look at the horizontal axis. Here "product" in broad terms will include services as well as physical products.

Types of innovation

	Product	Process
Disruptive	High investment High risk High return	High investment High risk Highest return
Incremental	Low investment Low risk Moderate return	Moderate investment Moderate risk High return

FIGURE 44.4 Types of innovation.

Companies like Sony or Procter & Gamble sell physical products you can hold and touch, whether they be high-definition TVs or tubes of toothpaste. Companies like United Airlines or Bank of America sell services, not physical products.

Similarly, "process" in broad terms includes not only the methodology used to manufacture a physical product but also the channel used to deliver that product or service to the customer. Amazon's sales channel innovation, for example, was a new process that revolutionized retailing by making online purchasing safe, convenient, and less expensive than a traditional brick-and-mortar store. Similarly, Google developed a business model of giving away all of their services for free, funded entirely by advertising. This has certainly been one of the more successful models of the Internet era.

Now let us look at the chart's vertical axis with the terms incremental innovation or disruptive innovation. Incremental innovation is an improvement to product offerings in an existing market that incrementally improves a user's experience. Disruptive innovation opens new markets, throws established markets into turmoil, and has the ability to completely rearrange the structure of the marketplace. Companies that were formerly market leaders could be left behind once the dust settles. Much press has been devoted to the importance of disruptive innovation because large companies historically haven't been very good at capitalizing on it.

Xerox's PARC division made major innovations such as the Ethernet, the graphical user interface, and the mouse, but Xerox failed to capitalize on them, and Apple reaped most of the commercial benefits. We have seen many examples of market leaders falling by the wayside when they didn't react to fundamental changes in technology. Polaroid resolutely stuck with film technology long after the world moved digital. Disruptive innovation rightly gets a lot of attention because the cost of missing such sea changes in markets is catastrophic to those who ignore them. Managers use it as a rallying cry to stimulate innovation in their organizations. Incremental innovations are usually more important because disruptive innovations occur only rarely. And when they do, the first to market isn't always the one who reaps the rewards. A company that bases its future primarily on disruptive innovation (e.g., start-ups) is taking a long-shot bet. Large corporations let the venture capital world sort out the winners from the losers and often acquire rather than invent disruptive innovations. Far more often, incremental innovations are what drive a company's results. Take Apple, for example. Their most successful introductions of recent years, the iPod and iPhone, are both incremental innovations in existing markets. The iPod, the first digital audio player with a built-in hard drive, was a relative latecomer to the consumer digital audio market, which was then dominated by Sony. When the iPod was introduced, Apple had already launched iTunes, a software application reminiscent of Napster that allowed you to choose from millions of songs for the low price of 99 cents each.

When Apple introduced the iPhone in 2007, it was long after such other smartphones as the Treo from Palm and the BlackBerry from RIM. But once again, Apple successfully innovated incrementally. For the first time, mobile users could surf the Internet with an experience similar to a computer. And Apple extended their iTunes success with the equally successful iPhone applications store.

44.3 WHY IS INNOVATION IMPORTANT? DRIVERS AND BENEFITS

"Innovate or die." Besides the most obvious reason of making businesses more competitive, there are many other drivers of innovation [8]. Innovation creates growth and economic wealth, increases productivity, and avoids stagnation and product obsolescence. Innovation also reduces waste and environmental damage. It provides better goods and services at a cheaper price and creates more interesting work and a more challenging work environment for employees.

44.3.1 Strong Drivers for Innovation

- Increased competition
- Make use of new, available, useful technologies and exceed the competition in these technologies
- Shorter product life cycles
- Financial pressures to reduce costs
- Increase efficiency
- Industry and community needs for sustainable development
- Increased demand for accountability
- Market, economic, demographic, and social changes

- Rising customer expectations regarding service and quality
- Meet stricter and new regulations

44.3.2 Benefits of Innovation

- It increases value to your customers or clients through new or improved products or services.
- It improves the effectiveness of your operation through reengineering your business processes.
- It makes your organization stand out from "me-too" competitors.

44.4 HOW: THE INNOVATION PROCESS

Now that we have defined what is innovation and why it is so important to do innovation, we will discuss next how to do innovation in a methodical way. Two approaches/ways are presented: a "six-step systematic innovation approach" and a more detailed "nine-step innovation journey."

44.4.1 Six-Step Systematic Innovation Approach

In this systematic approach to innovation, six steps are identified [9]: investigation, preparation, incubation, illumination, verification, and finally application (see Figure 44.5). Brief explanations are given alongside.

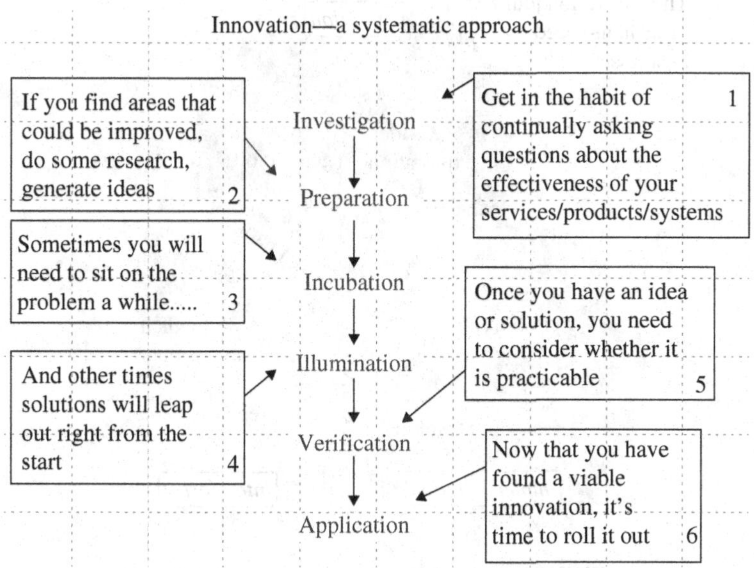

FIGURE 44.5 A systematic approach to do innovation.

44.4.2 Nine-Step Innovation Journey by Codexx

For businesses to survive [10, 11], let alone grow, they must be effective at innovation—at converting ideas to products and services in the market that deliver value to users and income to their creators. The journey from ideas to value is not a simple one. There are many steps along the way and many opportunities for good ideas to be lost, for poor ideas to reach the market, or for the journey itself to take longer than it really should.

The "Innovation Journey" model from the British company Codexx covers three phases (Initiation, Implementation, and Into Market) and a total of nine key steps from "ideas to value" (see Figure 44.6):

Phase 1: Initiation

1. Idea generation: Creating or capturing new ideas for innovative products
2. Idea exploration: Evaluating and testing the potential and feasibility of new ideas
3. Selection: Deciding which ideas to take further and which to shelf

Phase 2: Implementation

4. Prototype: Creating prototypes to further evaluate the concept
5. Develop: Development of the concept into a product
6. Go/no go: Decision on whether to proceed to market, delay, or cancel

Phase 3: Into market

7. Prepare for market: Engaging internal functions and partners to prepare for launch

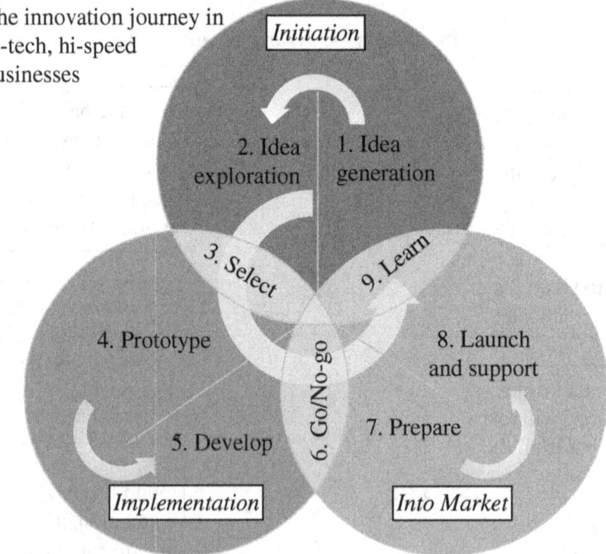

FIGURE 44.6 A nine-step innovation journey.

8. Launch and support: Launching the new product and supporting it in market
9. Learn: Learning from customers' use of the product to enhance current and future offerings

44.5 WHO DOES THE INNOVATION? GOOD INNOVATOR SKILLS

Innovation is everyone's responsibility, not just R&D. In order for an organization or company to succeed, everybody should be involved and engaged in innovation. Often creativity and innovation are thought to be belonging to only a select few individuals. Not true. We all have and can improve our ability to create and innovate. We already have these habits now and can focus on them and strengthen them.

44.5.1 Seven Traits of Highly Effective Innovators

There are seven essential traits and habits of the highly innovative people [12]: Curiosity, Resilience, Experimenting, Attentive, Thoughtful, Environmental, and Disciplined—the acronym spells out the word **CREATED**.

1. Innovators are **curious**. True curiosity is the wellspring of creative thought. Innovative people look at the ordinary and the everyday and ask why, how, when. They border on being obsessive in finding the true nature of the world, the people around them, and the better way of doing things. Their questioning disposition is often most prevalent in their work and domain of expertise. They enjoy wondering about things. The genesis of innovation is the ability to detect and crystallize problems. The innovative thinker opts for the slow process of incubating questions rather than favoring the quick and expedient answer.
2. Innovators are always **resilient**. Innovation is not always fun and "inspirational." It is often slow and difficult with failure just around every corner. Yet innovative people push through and they endure. Innovative people have a strong feeling for what is "right" and what is "better" which often prevents them from accepting easier solutions. Because of this continuous drive, they prevail. The ability to tolerate uncertainty, criticism, and failure yet relish the challenge of ultimately succeeding is a core attribute of highly successful innovators.
3. Innovators are constantly **experimenting**, with everything, with everyone, at any time. Innovative people love the unbeaten track: tinkering around with materials, ideas, actions, and possibilities. While they are serious about their goals and projects, they often show a playful and fun approach to probing for solutions and new possibilities. "What if, this could also be, let's try this, that looks like fun" are leitmotifs for innovation. Doing something without

immediately knowing why or to what end you are doing it is an essential creative exploration. Play, mess about, try out, and fool around. You never know what might be revealed.

4. Innovators are **attentive**. They pay attention diligently. Innovative people see the whole while noticing the details. They have a unique propensity for intense yet effortless concentration. They are able to immerse themselves in their experience for the pure enjoyment of the experience and become what has been dubbed "in the flow." While engrossed they are able to maintain perspective allow for, and even welcome, unexpected interferences and unanticipated patterns and truths to evolve. Where the serious problem solver is busily looking for things that she has already half decided are going to be relevant, and is therefore half-blind to everything else, the creator lowers the intensity of that motivational magnetic field, and so is able to look more carefully at what is there—and thus to spot the small clue that may give her the insight he needs. Too much careful thinking and not enough awareness is against the grain of the innovator.
5. Innovators are **thoughtful**. How people make use of their time, surroundings, and resources influences their ability to innovate. Most importantly, how they steer and their attention and thought processes are crucial. Innovators have learned how to "mind their mind." They know when to vest in deep concentration and when to leave some questions to the secret workings of the subconscious mind, when to ponder questions and possibilities, and when to think carefully and methodically. They use the time when their focus is sharpest wisely, and they allow and enjoy the semiautonomous play of images and metaphors that happens in states of reverie—as you are waking up or falling asleep, or doing something repetitive and habitual like showering or driving, for example. Innovators know when to keep on trying to figure something out and when to give up (for a while) and to relax. They are skilful orchestrators of their own states of mind and mental modes. Innovative people are able to segue smoothly and intuitively between focused and dreamy thinking, as appropriate, while less creative people tend to get stuck in one mental groove and can't get out.
6. Innovative people construct an optimal **environment**. Innovative people know that their physical and social environment can have a huge effect on their work and can make a big difference. They need different kinds of setting, support or challenge, at different times, and they regulate their social world so that it supports the kind of thinking that they need to do. They search for quiet spaces when needed, and they mix socially when the stimulus is called for, they withdraw into their private worlds at times, and at other times they partake fully in the pleasures and experience of the environment. Innovative people surround themselves with people who are going to support their creativity, whether emotionally, intellectually, or practically. They know how to use the rhythms of time to balance different kinds of thinking, and they know the worth of breaks and holidays.

7. Innovators are **disciplined**. Innovators possess a trait which is so important yet sometimes in seemingly tedious contradiction to innovation: discipline. Only through regular work at your interest and passion can it be innovative. Working through the highs and the troughs, the ups and downs, and the often natural inclination to "do something else" is the consistent fuel of the innovator. So keep going, keep at it, make the time, and stick to it. Only through regular work at your interest and passion can it be innovative.

44.5.2 Seven Skills of Effective Innovators

Jag Randhawa, a technology executive and author of *The Bright Idea Box*, discusses seven skills of effective innovators [13]: curiosity, listening, mining, borrowing, networking, problem solving, and writing:

1. **Curiosity**. Ask yourself why, what if, and how does it work? You may have worn shoes with Velcro straps. Velcro was invented by Swiss engineer George de Mestral in 1941. He removed burrs from his dog and decided to take a closer (curious) look at how they worked. The small hooks found at the end of the burr needles inspired him to create the now ubiquitous Velcro.
2. **Listening**. Listen to your customers, partners, stakeholders, and employees. Why are they calling and not calling, what are their most common request, and what are their feedbacks?
3. **Mining**. Frustrations represent opportunities. Can it be voided or done differently, and how would I solve it?
4. **Borrowing ideas**. From others, from nature (biomimicry). There is nothing new under the sun. How do competitors do it? Has another industry solved a similar problem? Who does it best? Learn from others.
5. **Networking**. Network with people with different skills (diversity). Take classes in a different field. Work with people in other cross-functional areas.
6. **Problem solving**. Keep a positive attitude. Persist, don't give up. Collaborate, share, and ask for help. Find solutions rather than complaining.
7. **Writing**. Write, doodle, and record every idea. Do not throw away ideas. Dedicate time to review old ideas. Some old ideas may be applicable now.

44.6 WHEN: IN A PRODUCT CYCLE WHEN DOES INNOVATION TAKES PART?

We mentioned in the section "WHO DOES THE INNOVATION?" that everybody in the organization and in their role can do innovation and have an impact in the bottom line of the company by continuously finding ways to improve their daily job such as increasing their work productivity, taking advantage of new technologies, minimizing waste, improving the channel, coming up with new business models, etc. So innovation can take part **anytime and anywhere** in the product cycle.

44.7 WHERE: INNOVATION AREAS IN IoT

In Parts II and III of this book, numerous articles have been presented on various Internet of Things (IoT) systems and their applications [14]. Some of the top "IoT innovations" will take place in the following 10 areas:

1. **Medical and healthcare systems**: IoT devices can be used to enable remote health monitoring and emergency notification systems. These health monitoring devices can range from blood pressure and heart rate monitors to advanced devices capable of monitoring specialized implants, such as pacemakers or advanced hearing aids. Specialized sensors can also be equipped within living spaces to monitor the health and general well-being of senior citizens while also ensuring that proper treatment is being administered and assisting people regain lost mobility via therapy as well. Other consumer devices to encourage healthy living, such as wearable heart monitors, are also a possibility with the IoT. More and more end-to-end health monitoring IoT platforms are coming up for antenatal and chronic patients, helping one manage health vitals and recurring medication requirements. Doctors can monitor the health of their patients on their smartphones after the patient gets discharged from the hospital.
2. **Consumer electronics and connected living**: It is now cost effective to add connectivity to a growing range of household appliances, providing benefits for both consumers and manufacturers. Connected washing machines and dishwashers, for example, could inform the householder when a cycle is complete remotely. Connected sprinkler systems and pet feeders can be remote controlled by householders while they are away from home. Using their smartphone, consumers could configure connected appliances to perform certain tasks at certain times.
3. **Environmental monitoring**: Environmental monitoring applications of the IoT typically use sensors to assist in environmental protection by monitoring air or water quality or atmospheric or soil conditions and can even include areas like monitoring the movements of wildlife and their habitats. Development of resource-constrained devices connected to the Internet also means that other applications like earthquake or tsunami early warning systems can also be used by emergency services to provide more effective aid. IoT devices in this application typically span a large geographic area and can also be mobile.
4. **Smart homes**: Home automation systems, like other building automation systems, are typically used to control lighting, heating, ventilation, air conditioning, appliances, communication systems, entertainment, and home security devices to improve convenience, comfort, energy efficiency, and security.
5. **Smart building and energy management**: IoT devices can be used to monitor and control the mechanical, electrical, and electronic systems used in various types of buildings (e.g., public and private, industrial, institutions, or

residential). Integration of sensing and actuation systems, connected to the Internet, is likely to optimize energy consumption as a whole. It is expected that IoT devices will be integrated into all forms of energy-consuming devices (switches, power outlets, bulbs, televisions, etc.) and be able to communicate with the utility supply company in order to effectively balance power generation and energy usage. Such devices would also offer the opportunity for users to remotely control their devices, or centrally manage them via a cloud-based interface, and enable advanced functions like scheduling (e.g., remotely powering on or off heating systems, controlling ovens, changing lighting conditions, etc.).

6. **Smart grid and industrial applications**: The IoT is especially relevant to the Smart Grid since it provides systems to gather and act on energy and power-related information in an automated fashion with the goal to improve the efficiency, reliability, economics, and sustainability of the production and distribution of electricity. Using Advanced Metering Infrastructure (AMI) devices connected to the Internet backbone, electric utilities can not only collect data from end-user connections but also manage other distribution automation devices like transformers and circuit breakers. Smart industrial management systems can also be integrated with the Smart Grid, thereby enabling real-time energy optimization. Measurements, automated controls, plant optimization, health and safety management, and other functions are provided by a large number of networked sensors.
7. **Urban infrastructure management**: Monitoring and controlling operations of urban and rural infrastructures like bridges, railway tracks, on- and offshore wind farms are key applications of the IoT. The IoT infrastructure can be used for monitoring any events or changes in structural conditions that can compromise safety and increase risk. It can also be used for scheduling repair and maintenance activities in an efficient manner by coordinating tasks between different service providers and users of these facilities. IoT devices can also be used to control critical infrastructure like bridges to provide access to ships. Usage of IoT devices for monitoring and operating infrastructure is likely to improve incident management and emergency response coordination, quality of service, and uptimes and reduce costs of operation in all infrastructure-related areas. Even areas such as waste management can benefit from automation and optimization that could be brought in by the IoT.
8. **Intelligent transportation**: The IoT can assist in integration of communications, control, and information processing across various transportation systems. Application of the IoT extends to all aspects of transportation systems, that is, the vehicle, the infrastructure, and the driver or user. Dynamic interaction between these components of a transport system enables inter- and intra-vehicular communication, smart traffic control, smart parking, electronic toll collection systems, logistic and fleet management, vehicle control, and safety and road assistance.

9. **Cyber security**: Concerns have been raised that the IoT is being developed rapidly without appropriate consideration of the profound security challenges involved and the regulatory changes that might be necessary. Given the concerns around information security with IoT, many companies are coming out with cyber-security products and services. Digital door locks, smartphone-controlled security systems, and automated home security products are on the rise.
10. **Manufacturing**: Network control and management of manufacturing equipment, asset and situation management, or manufacturing process control bring the IoT within the realm on industrial applications and smart manufacturing as well. The IoT intelligent systems enable rapid manufacturing of new products, dynamic response to product demands, and real-time optimization of manufacturing production and supply chain networks, by networking machinery, sensors, and control systems together.Digital control systems to automate process controls, operator tools, and service information systems to optimize plant safety and security are within the range and scope of the IoT. But it also extends itself to asset management via predictive maintenance, statistical evaluation, and measurements to maximize reliability. Smart industrial management systems can also be integrated with the Smart Grid, thereby enabling real-time energy optimization. Measurements, automated controls, plant optimization, health and safety management, and other functions are provided by a large number of networked sensors.

44.8 CONCLUSION

In this chapter we have covered the **what (what is innovation?), why (why do innovation? benefits), who (innovators, skills), where (areas to do innovation), when (when to do innovation?) and how (how to do innovation? process, lessons)** of Innovation.

We will conclude by sharing some lessons people have learned from doing innovation, as well as giving some notable quotes about innovation.

44.8.1 Some Innovation Lessons Worth Sharing

1. Innovation starts at the top [15]. Leaders need to create the vision and live the values. However, just a warning: "The best way to kill creativity in a team is letting the boss speak first."
2. Innovation can happen anywhere. Anyone can do it, but not everyone is good at it.
3. Innovation is a team sport. Operating in a silo is deadly. So collaborate.
4. Innovation is never easy, but it is always possible. Do it step by step, project by project. Small ideas need room and time to grow. Also need to see that progress is being made.

5. Innovation relies on trust. Listen to ideas. Reward bravery and risk. Just because it works for Google, it does not mean it will work for you. Need to create an innovation culture that fits.
6. The customer is not always right, but they do sit at the heart of all innovation.
7. Speed is mission critical.
8. Learn from failures. Fail early, fail fast, and fail inexpensively.

44.8.2 Some Notable Quotes (about Innovation)

1. No passion, no innovation.
2. Experiment fast. Learn, adapt, change.
3. Keep learning new ways, and take continuous feedback.
4. No escaping hard work.
5. No commitment, no innovation.
6. Collaborate. Embrace diversity.
7. See art everywhere and in everything.

REFERENCES

[1] What is innovation? Definition and meaning, http://www.businessdictionary.com/definition/innovation.html (accessed February 20, 2016).

[2] Innovation, http://en.wikipedia.org/wiki/Innovation (accessed August 13, 2016).

[3] What's the difference between creativity and innovation? http://www.innovationexcellence.com/blog/2012/08/04/whats-the-difference-between-creativity-and-innovation/#sthash.XHf1mZoK.dpuf (accessed February 20, 2016).

[4] The power of process at the fuzzy front end of innovation, https://innovationcrescendo.com/2013/09/24/the-power-of-process-at-the-fuzzy-front-end-of-innovation/ (areas of innovation) (accessed August 13, 2016).

[5] A better definition of innovation http://www.rdinsights.com/a-better-definition-of-innovation/# (accessed September 12, 2016).

[6] Praveen Gupta, *The Innovation Solution, Making Innovation More Pervasive, Predictable and Profitable*, Accelper Consulting/Createspace, Scotts Valley, CA, 2012.

[7] Stephen Hinch, The second tenet of innovation, http://stephenhinch.blogspot.com/2010/04/second-tenet-of-innovation.html (accessed August 13, 2016).

[8] Ram Charan and A. G. Lafley, Why innovation matters, http://www.fastcompany.com/874798/why-innovation-matters (accessed February 20, 2016).

[9] Innovation: a roadmap for homeless agencies—npfSynergy—2003, http://nfpsynergy.net/ (accessed February 20, 2016).

[10] Simon Bramwell and Alastair Ross, Innovation journey for technology—rich product businesses, February 2011. http://www.codexx.com/pdf/Innovation-Journey-Study-February-2011.pdf (accessed September 12, 2016).

[11] The innovation journey for technology businesses, http://www.codexx.com/tag/innovation-journey/ (accessed February 20, 2016).

[12] Gert Scholtz, The seven traits of highly effective innovators, https://www.linkedin.com/pulse/seven-traits-highly-effective-innovators-gert-j-scholtz?trk=pulse-det-nav_art (accessed August 13, 2016).

[13] Jag Randhawa, Innovator's Mindset: 7 Skills of Effective Innovators, http://www.slideshare.net/jagdeepr/innovators-mindset-7-skills-of-effective-innovators (accessed February 20, 2016).

[14] IoT Innovations at CES 2016: What to Expect, http://www.informationweek.com/iot/iot-innovations-at-ces-2016-what-to-expect-/d/d-id/1323734 (accessed February 20, 2016).

[15] 8 Lessons about innovation worth sharing, http://www.slideshare.net/ingosigge/creativity-innovation (accessed February 20, 2016).

FURTHER READING

Curtis R. Carlson and William W. Wilmot, *Innovation: The Five Disciplines for Creating What Customers Want*, Crown Business, New York, 2006.

Clayton Christensen, *Innovator's Dilemma: The Revolutionary Book that Will Change the Way You Do Business*, Harper Business, New York, 2011.

Marc de Jong, Nathan Marston, and Erik Roth, The eight essentials of innovation, *McKinsey Quarterly*, April 2015, pp. 36–47.

Peter F. Drucker, *Innovation and Entrepreneurship*, Harper & Row, New York, 1985.

Walter Isaacson, *The Innovators: How a Group of Hackers, Geniuses, and Geeks Created the Digital Revolution*, Simon & Schuster, New York, 2014.

Max Mckeown, *The Innovation Book*, Pearson, Harlow, UK/New York, 2014.

Andrew H. Van de Ven, *The Innovation Journey*, Oxford University Press, New York, 1999.

David Verduyn, 8 Steps of systematic innovation, https://www.youtube.com/watch?v=JF64tkWAUXI (accessed February 20, 2016).

45

INTERNET OF THINGS BUSINESS MODELS

HUBERT C.Y. CHAN
The Hong Kong Polytechnic University, Hong Kong, China

Almost all businesses are aware of the potential gains that the Internet of Things (IoT) has to offer but are unsure of how to approach it. This chapter proposes a business model that builds on the work by Turber et al. [1]. The model consists of three dimensions: “Who, Where, and Why.” “Who” describes collaborating partners, which builds the “Value Network.” “Where” describes sources of value co-creation rooted in the layer model of digitized objects, and “Why” describes how partners benefit from collaborating within the value network. With the intention of addressing “How” the proposed framework has integrated the IoT strategy category, tactics, and value chain elements. The framework was then validated through the case studies of some successful players of Hong Kong.

45.1 INTRODUCTION

IoT is an integrated part of Future Internet. According to the agreed protocol, any article can be connected and talk to each other. This can be achieved through a vast number of methods and technologies, including radio-frequency identification (RFID), near-field communication (NFC), infrared (IR) sensors, and many more. The IoT paradigm is a result of the convergence of three main visions: Internet oriented (middleware), things oriented (sensors) and semantic oriented (knowledge) [2]. However, the identification of the killer application and the definition of the underlying business

Internet of Things and Data Analytics Handbook, First Edition. Edited by Hwaiyu Geng.

Companion website: www.wiley.com/go/Geng/iot_data_analytics_handbook/

model in IoT are still missing [3]. Filling in this gap, the objectives of this article are to identify the killer applications or the value proposition and evaluate the business models through the case studies of some successful players. An IoT business model framework is proposed for this purpose.

45.2 IoT BUSINESS MODEL FRAMEWORK REVIEW

Business model is defined as the plan implemented by a company to generate revenue and make a profit from operations [4]. That being said, it is clear that a company that lacks a clear and organized plan to generate profit and revenue will likely be unsuccessful. Traditional business models are designed on a firm-centric basis; however due to the nature of the IoT ecosystem in which firms must collaborate with competitors and across industries, it is easy to see why traditional business models are not adequate. Moreover, fast-changing market environments in technology-related industries imply that companies must quickly adjust to market challenges in order to succeed. As a result, business model innovations are becoming "new routes to competitive advantage" [5]. Bucherer et al. [6] described some key issues when designing IoT business models, including "information between nodes and win–win information exchange for all stakeholders." Furthermore, Westerlund et al. [7] identified three contemporary challenges of the IoT, comprising the diversity of objects, the immaturity of innovation, and the unstructured ecosystems. The diversity of objects refers to a multitude of different types of connected objects and devices without commonly accepted or emerging standards. Immaturity of innovation implies that today's quintessential IoT innovations have not yet matured into products and services. Unstructured ecosystems mean the lack of defined underlying structures and governance, stakeholder roles, and value-creating logics. Despite of these challenges, several IoT business model frameworks exist, but there are still some major gaps in the IoT that need to be properly addressed.

Business models consist of several essential elements: "Who, What, How, and Why" [8]. "Who" refers to the target customer, "What" is the value proposition that is offered to the customer, "How" is the value chain to deliver the value proposition to the customer, and "Why" describes the underlying economic model to capture value. These four essential elements in Figure 45.1 must be addressed for a business model to be functional. This is the fundamental requirement in building our proposal.

As value creation in traditional product mindset shifts from solving existing needs in a reactive manner to addressing real-time and emergent needs in a predictive manner, filling out well-known frameworks and streaming established business models will not be enough [9]. Value creation in IoT can be classified into three layers [10]: manufacturing, supporting, and value creation. Manufacturing layer means that manufacturers or retailers can provide items such as sensors and terminal devices. Supporting layer collects data which can be utilized in the value creation processes, while the third layer uses IoT as a cocreative partner, because the network of things can think for itself. Chen Min [11] presents a more detailed four-layer architecture for IoT as shown in Figure 45.2:

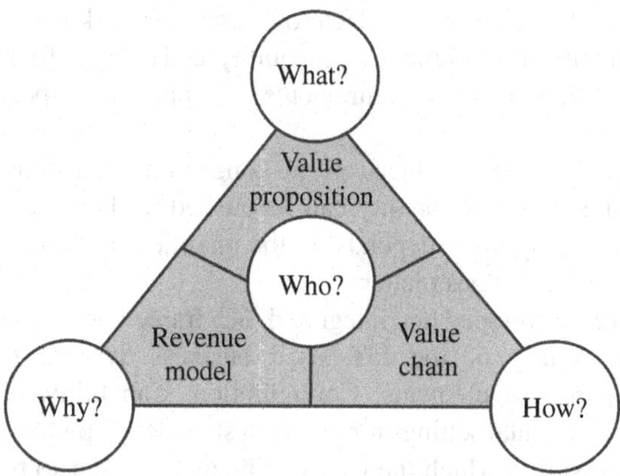

FIGURE 45.1 "The archetypal business model." Gassmann et al. [8]. Reproduced with permission of Springer.

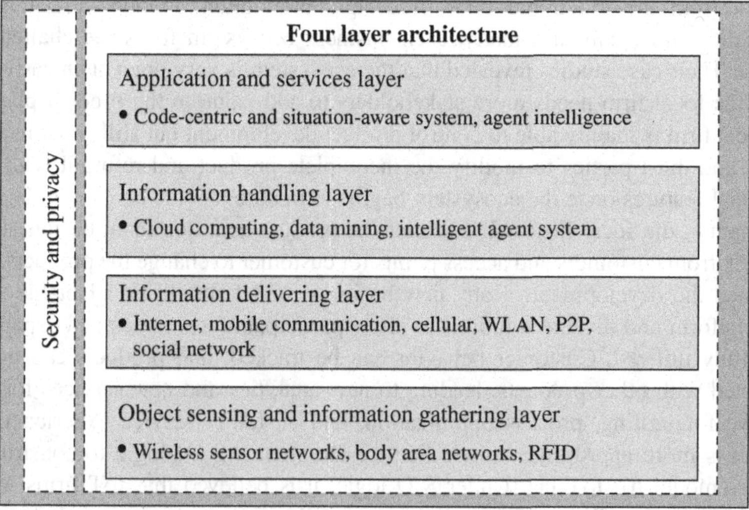

FIGURE 45.2 Architecture for Internet of Things. Chen [11]. Reproduced with permission of Springer.

- *Object sensing and information gathering:* The first step of enabling smart services is to collect contextual information about environment, "things," and objects of interest.
- *Information delivering:* Various wireless technologies such as wireless sensor networks (WSNs), body area networks (BANs), Wi-Fi, Bluetooth, Zigbee, GPRS, GSM, cellular and 3G, and so on can be used for delivering the information.
- *Information processing:* Pervasive and autonomic services are provided through ubiquitous machines in both "autonomic" and "smart" ways.

- *Application and smart services:* Heterogeneous network performance in terms of bandwidth utilization, computing capability, and energy efficiency is improved according to different users' requirements and application-specific design.

In this connected world, products are no longer one and done in a connected world. New features and functionality can be pushed to the customer on a regular basis. The success in any layer depends on the readiness and maturity of the other layers; thus the ecosystem does matter.

Rong et al. [12] developed an integrated 6C framework in order to improve systematic understanding of the IoT-based business system, namely, Context, Cooperation, Constructive elements, Configuration, Capability, and Change. The Context is the environmental settings for the ecosystem development. The Cooperation reflects the mechanism by which the partners interact in order to reach the strategic objectives. Constructive elements define the fundamental structure and supportive infrastructure of the ecosystem. Configuration seeks to identify the external relationships among partners. Capacity investigates the key success features of a supply network from the functional view of design, production, inbound logistics, and information management. Finally, each business ecosystem faces the challenge of Change. Their case studies revealed that the ecosystem is very open at an early stage where the focal firm needs more stakeholders to add value to the product platform. The focal firm is mainly able to control product development but still welcomes customers and third parties to modify the incomplete product and refine it with more functional features once the ecosystem begins to mature. As the ecosystem becomes very mature, the focal firm will consider the product as a dominant design and get feedback from customers and access points for customer to change the products. They suggested the development of any new business model should be adaptable via an open platform and diverse solutions to allow participants' resources and capabilities to be fully utilized. Customer behavior can be tracked, and products can now be connected with other products, leading to new analytics and new services for more effective forecasting, process optimization, and customer service experiences [9]. Thus, it is more appropriate to use Service-Dominant (S-D) logic to construct the business model for IoT [1]. Under S-D logic, it is believed that IoT firms will no longer simply be selling goods to the customer but will be acting as platforms for customers and competitors adding value upon. In S-D logic, the traditional firm-centric view is replaced by a network-centric view. Individual firms must work with market partners and customers to create "value creation networks," in which individual firms act as "organizers of value creation." Individual firms must be capable of making smart collaborations, as collaborations form the fundamental basis of the IoT. If a firm is unable or unwilling to be collaborative, it will not be competitive.

Turber et al. [1] posited that a business model framework in IoT consists of three dimensions: "Who, Where, and Why." "Who" describes collaborating partners, who build the "Value Network," and does not only include firms who create IoT products but also includes customers and stakeholders to reflect the network-centric sentiment that customers are cocreators and coproducers of value. "Where" describes sources of value cocreation rooted in the layer model of digitized objects. There are four places of opportunity for value to be added by the collaborators. They are the device, connectivity, service, and content layers. The device layer comprises a hardware and an operating system; the network layer involves transmission plus network standard and physical transport; the service layer provides direct interaction with users through application programs, while the content layer hosts the data, images, and information. "Why" describes the benefits for partners from collaborating within the value network, both monetary benefits and nonmonetary benefits as shown in Figure 45.3. Each dimension addresses the four-layered modular architecture of digitized products.

Westerlund et al. [7] provided three pillars for designing ecosystem business models required in the IoT context, namely, the drivers, value nodes, exchanges, and extracts of value. Drivers comprise both individual and shared motivations of diverse participants who fulfill a need to generate value, realize innovation, and make money in an ecosystem. Value nodes include various actors, activities, or (automated) processes that are linked with other nodes to create value. Value exchange refers to an exchange of value by different means, resources, knowledge, and information between and within different value nodes. Value extracts refers to part of ecosystem that extracts value or be monetized. These pillars are interconnected, and, in contrast to existing business model frameworks, they aim to explain the flows and action of a business model rather than components of the model. In fact, the "Who" of the framework of Turber et al. [1] are the drivers articulated by Westerlund et al. [7], the "What" reflects the value nodes, while the "Why" refers to the value extracts. However, the "How" that is the value exchange is not explicitly revealed.

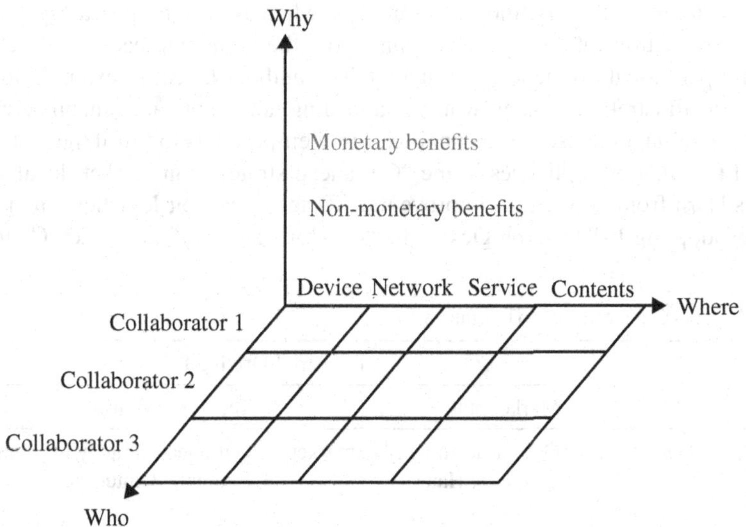

FIGURE 45.3 Framework design for a business model framework in the IoT context. Turber et al. [1, p. 24]. Reproduced with permission of Springer.

45.3 FRAMEWORK DEVELOPMENT

The Business Model Canvas [13] has been widely adopted for developing and documenting existing business model. It consists of nine elements: key partners, key activities, key resources, value propositions, customer relationships, channels, customer segments, cost structure, and revenue streams. However, it cannot fully address the network-centric sentiment and the connectivity nature of IoT characteristics. In the review of the framework posited by Turber et al. [1], it is the most comprehensive one, except the "competitive strategy," and "How" (value chain) elements need to be expanded as the service content increases particularly in IoT markets [14]. Strategy refers to the choice of business model through which the firm will compete in the marketplace. Business model refers to the logic of the firm, the way it operates, and how it creates value for its stakeholders, which in turn involves tactics which refer to the residual choices open to a firm by virtue of the business model it chooses to employ. We are going to explore how these elements are to be incorporated into our proposed framework.

45.3.1 Competitive Advantage

Competitive advantage can be gained with an effective strategy in grasping the emerging opportunities from IoT. Li et al. [15] proposed four categories of IoT strategies, as illustrated in Table 45.1: get-ahead strategy in market, catch-up strategy in market, get-ahead strategy in technology, and catch-up strategy in technology. Technology push is applying new technology to IoT application, while market pull is the demand in response to an identified and potential market need. "Get-ahead strategy" is a set of plans and action to enable the firm to stay head of other competitors so as to enjoy the first-mover advantages, whereas "catch-up strategy" is a set of plans and action following and learning from the industrial leaders' movements through operational efficiency and quality. The authors listed an example in each category as illustrations: Haier, which is a leading home appliance manufacturer in China, is facilitated to use IoT technologies in their products to fulfill the increasing demand for smarter appliances in the "Get-ahead strategy" in market; local supermarkets learn from Walmart to improve the efficiency of their logistics, orders, and sales by adopting IoT technologies in the "Catch-up strategy" in market; Cinterion,

TABLE 45.1 Categories of IoT Strategy

		Industrial Driving Force	
		Market pull	Technology push
Strategic intent	Get-ahead	Get-ahead strategy in market Example: Haier	Get-ahead strategy in technology Example: Cinterion
	Catch-up	Catch-up strategy in market Example: local supermarket	Catch-up strategy in technology Example: Junmp

Li et al. [15]. Reproduced with permission of Springer.

which is a supplier of wireless modules for the cellular Machine-to-Machine (M2M) communication, has successfully turned itself to be the world's leading player in this arena and enjoys a prestigious position among competitors in the "Get-ahead strategy" in technology; and Junmp, which is an RFID technology provider in China, produces RFID labels and readers through a cooperation agreement with the US Impinj (one of the leading innovators for RFID) in the "Catch-up strategy" in technology. Each strategy has its pros and cons, which depends on the firm's capability, resources availability, and business context.

Besides the strategy, we can study tactic of the focal firms to enrich our framework. There are six components in the digitally charged products derived by Fleisch et al. [16] for IoT general business model:

- Physical freemium: This component describes a physical asset that is sold together with a free digital service at no additional charge. Some percentage of customers will select premium charged services afterward.
- Digital add-on: A physical asset is sold very inexpensively at a thin margin. Over time, the customer can purchase or activate any number of digital services with a higher margin.
- Digital lock-in: This refers to a sensor-based, digital handshake that is deployed to limit compatibility, prevent counterfeits, and ensure warranties.
- Product as point of sales: Physical products become sites of digital sales and marketing services. The customer consumes services directly at the product or indirectly via a smartphone and identification technology.
- Object self-service: This component refers to the ability of things to independently place orders on the Internet.
- Remote usage and condition monitoring: "Smart" things can transmit data about their own status or their environment in real time. This makes it possible to detect errors preventatively and to monitor usage and the remaining inventory of consumables.

In fact, digital freemium is very common. An example is Apps, where digital asset is free at first, but customer may pay premium services afterward. All these components are the general tactics employed by most of the collaborators in the IoT ecosystem.

45.3.2 Value Chain

Value chains break down a firm's activities into a sequence of value-generating activities, starting from the conception and leading to end use. Value chains for IoT firms are more complicated than those of a traditional product; however the underlying concept remains the same. There are at least nine distinct product or service categories along the value chain in IoT in Table 45.2 [17]. However, the report does not group the categories into different layers of architecture.

TABLE 45.2 Various Layers of the IoT Value Chain [17]

Radios	Chips that provide connectivity based on various radio protocols
Sensors	Chips that can measure various environmental/electrical variables
Microcontrollers	Processors/Storage that allow low-cost intelligence on a chip
Modules	Combine radios, sensors, microcontrollers in a single package
Platform Software	Software that activates, monitors, analyzes device network
Application Software	Presents information in usable/analyzable format for end user
Device	Integrates modules with app software into a usable form factor
Airtime	Use of licensed or unlicensed spectrum for communications
Service	Deploying/Managing/Supporting IoT solution

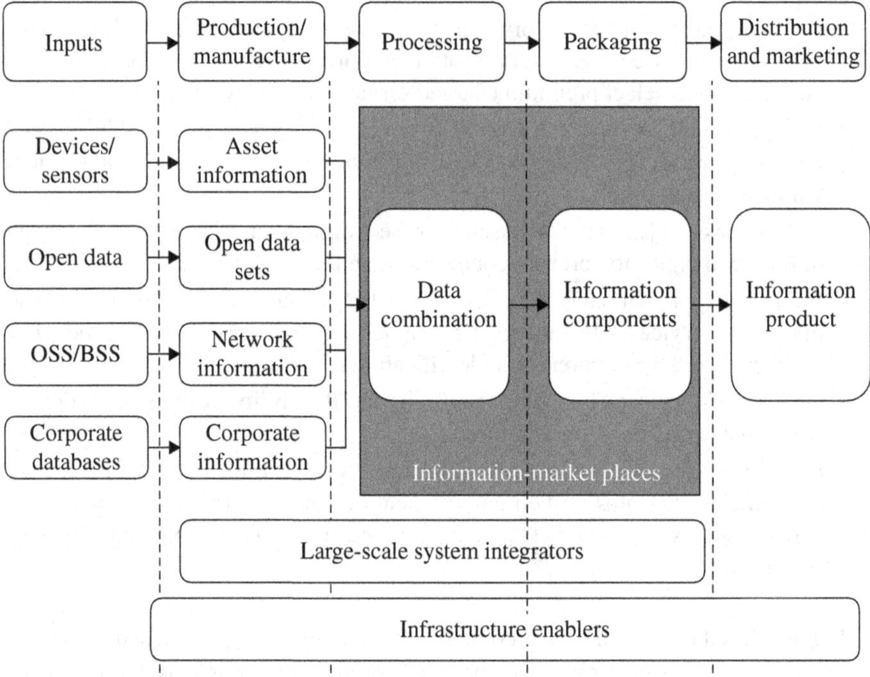

FIGURE 45.4 Information-driven value chain for IoT. Holler et al. [18]. Reproduced with permission of Elsevier.

Holler et al. [18] posited a more detail information-driven value chain for IoT. There are four inputs, as depicted in Figure 45.4. Each of these four inputs undergoes value addition through production/manufacture, processing, packaging, and distribution and marketing as a finished product. The raw data are collected via different types of sensors, actuators, open data, operating or business system, and corporate database. It will undergo processing and packaging transmitted through a wireless of fixed network prior to becoming useful information. Because of the variety, velocity, and volume of this Big Data, infrastructure enablers and large scale system integrator are

TABLE 45.3 Proposed IoT Business Model Framework

Company	Collaborator	Inputs	Network	Service/ Processing/ Packaging	Content/ Information Product	Benefits	Strategy	Tactic
	C1							
ABC	C2							
	C3							

required. However, different players along the value chain have to overcome the interoperability issue as mentioned before. In reality, many successful players have been forced to vertically integrate a full solution of hardware, software, and service designed for a vertical market because of lack of standardization of connected devices and very little commonality between applications in different vertical market.

45.3.3 Proposed Framework

Building on the framework of Turber et al. [1], I integrate the IoT strategy category and value chain into the following framework in Table 45.3. A two-dimensional model replaces a three-dimensional one where monetary and/or nonmonetary benefits are the value captured by the collaborators but not to the value nodes. We replace "Device" by "Input" since there are other types of input other than devices as shown in the IoT architecture. "Network" in IoT includes the "Production/Manufacturing" in the value chain. "Processing and Packaging" can be grouped under "Service." "Content" includes all the "Information Product." This grouping embraces all the layers proposed in Table 45.2 [17] and Figure 45.2 [11]. These two parameters are the value propositions which are the most important building block in business model for IoT [19]. A column specifying the category of IoT strategy (Table 45.1) [15—get-ahead strategy in market, catch-up strategy in market, get-ahead strategy in technology, and catch-up strategy in technology of each collaborator—has been added. Furthermore, a column reveals the tactic derived by Fleisch et al. [16] has been included.

45.4 CASE STUDIES

The qualitative analysis is based on multiple case studies which are adopted to explore the "why, what, and how" [20] essential elements of business model. A series of interviews with 2014 IoT Awardees [21], the secondary data of the Best Smart Hong Kong Award Winner 2015 (http://download.hkictawards.hk/v_files/HKICTA/doc_2015/categories/Category_Award_Booklet/08_SH_CategoryProgramme.pdf), and a real case study were used to illustrate the proposed framework. A summary table for each company based on the proposed framework is shown in the following.

45.4.1 Company: Hong Kong Communications Co. Ltd (http://www.hkc.com.hk)

Project Description: Work with a service provider (caregiver) to use Machine Learning in learning the resident's daily habits (duration spends in different rooms at different times). It had been found that the living habit of most of the elderly are pretty routine during weekdays, and the change will not be drastic if any. That means living pattern aligns with Gaussian (Normal) Distribution in different period of time during the day. The location of the resident inside the premise is recorded by motion sensors installed in different rooms. The floor layout of a real case study is depicted in Figure 45.5. There is a learning period of, say, 30 days initially; all the normal data will be sent to the system (Cloud) at the end of each day as training set. A software called Matlab k-means (http://www.mathworks.com/) was employed. Since most people will move around the different rooms for a short period of time, records of this numerous movements are not useful for the study; duration at any location of less than 3 minutes is regarded as noise and will be neglected. After the training period, the data at each location will appear as spherical clustering as shown in Figure 45.6. Any data falling within the clusters are normal, while those falling outside the clusters are anomaly (outliners are set at mean + two times standard deviation). The aggregated data of the last 30 days is used as moving data, and thus the program will adopt any update living habit as long as the deviation is not drastic. According to Figure 45.6, the residents will go to bed at around 00:30 and sleep for about 5 hours (300 minutes) normally. She will get up, go to the toilet, and then change clothes at the bedroom before going out. She will then come back to take a nap at around 15:00 for 20 minutes. The red dot was the last day of record showing a normal pattern falling within a cluster. If it falls outside the cluster, it will show up and send an alert to the caregiver.

Product/Service: A real-time monitoring of the living habit of a single living elderly with motion sensors for anomaly detection.

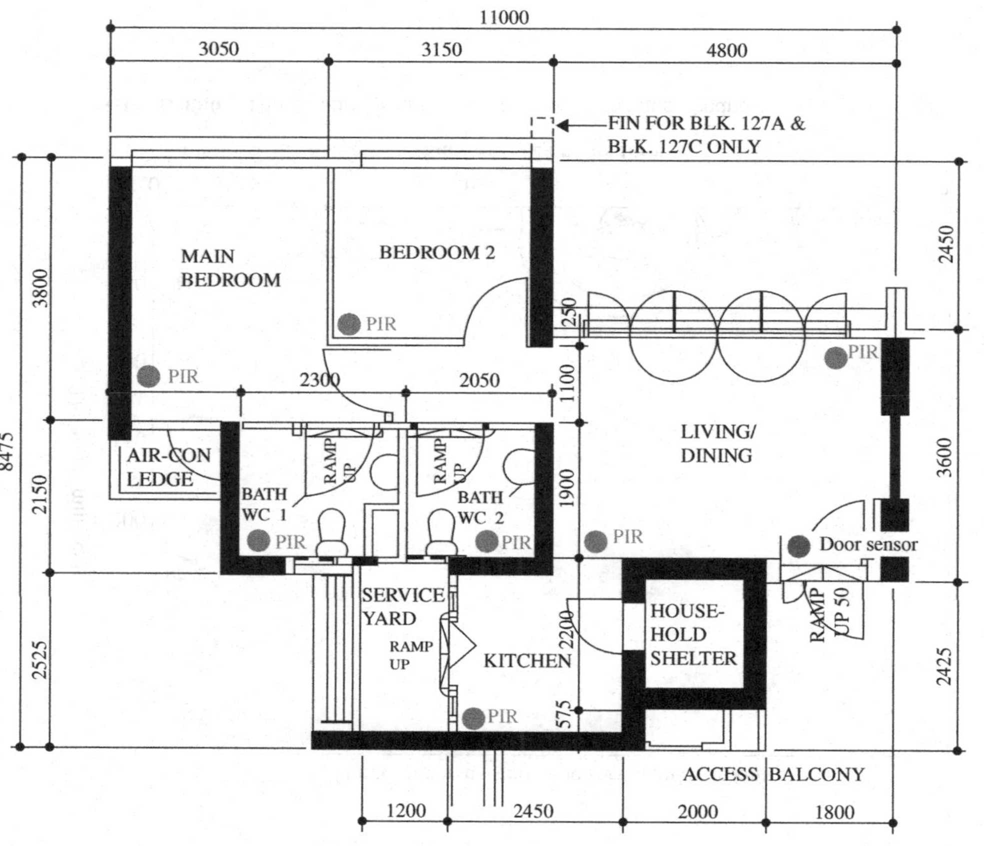

FIGURE 45.5 The position of the motion sensor (PIR) and a door sensor installed in the testing premise.

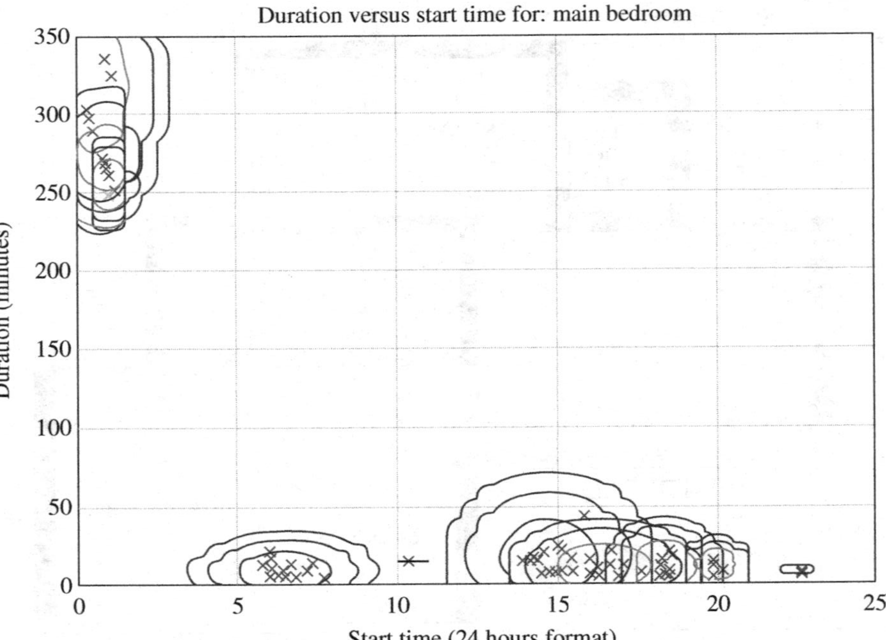

FIGURE 45.6 The duration against the start time at the main bedroom for over 30 days.

Collaborator	Inputs	Network	Service/Processing/ Packaging	Content/Information Product	Benefits	Strategy	Tactic
HK Comm	Motion sensors	ZigBee	Monitoring the resident living habit	Movement of the resident	Subscription service	Get-ahead strategy in technology	Low-cost setup
Resident	No need to wear any device at home	Subscribe a broadband		Disclose the movement inside the unit	Living safely and independently		
Caregiver		Internet	Call up the resident in case of anomaly detected	Movement information in Cloud	Automatic remote monitoring of the single living residents	Get-ahead strategy in market	Levering of IoT

45.4.2 Company: Digi Mobil Technology Ltd (www.gpsfinder.hk) and Rodsum Wireless Ltd (www.rodsum.com)

Project Description: The building cost of HK–Zhuhai–Macau Bridge project is as high as RMB76.2 billion and the progress is now in full swing. Since the construction is very close to the Hong Kong International Airport in Chek Lap Kok, the developers must efficiently control the heights of the construction machineries to prevent affecting the flight paths and causing danger. It is easier to control the heights on the land as compared with that at the sea because there are high and low tides at the sea, as well as the sailing vessels. The height restriction should meet the requirement from Civil Aviation Department; therefore, it is necessary to have a real-time monitoring system for precise height calculation. The system consists of five components, namely, a GPS antenna, a master controller, a sensor for measuring height and angle, a Draught Sensor, and an alarm system installed at different parts of the vessel. Together with the use of cloud server, if the height of the suspended platform of the engineering vessel exceeds the limit, the red light alarm will be turned on to remind the suspended platform operators to immediately lower the suspended platform back to the safe level. If the irregularities sustain, the system will automatically send messages and emails to alert engineer, manager, and Civil Aviation Department.

Product/Service: Equipment Height Real-time Monitoring System (EHRMS) for HK–Zhuhai–Macau Bridge Project.

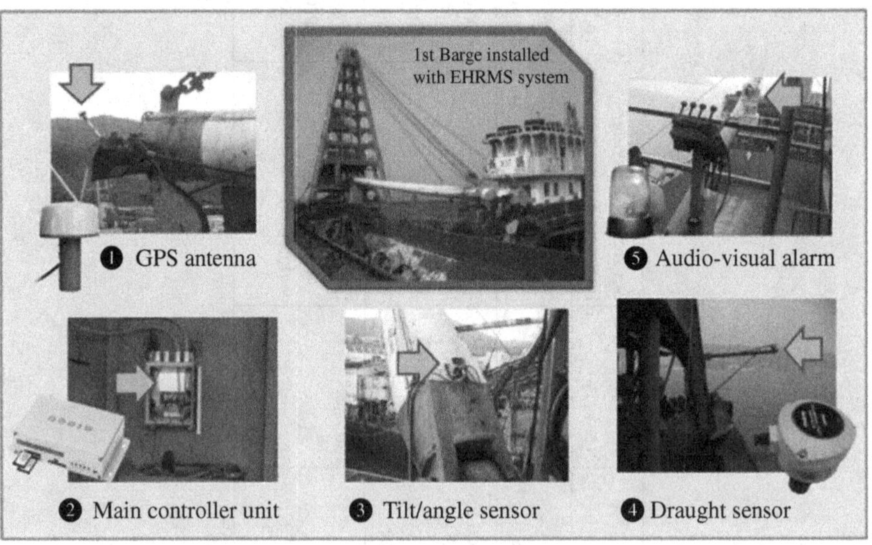

Once the sensor and the intranet device are installed on the suspended platform and working vessel, administrators and the government officers can monitor the height of the suspended platform anytime.

Collaborator	Inputs	Network	Service/Processing/ Packaging	Content/Information Product	Benefits	Strategy	Tactic
Construction boats	GPS, height and angle measurement device, alert system	Mobile network	Adjust the construction arm upon the receipt of alarm		Compliances		
Digi Mobil and Rodsum		Cloud service	Trigger the alert when the construction arm of the boat exceeds the limit; SMS and email to the boat management	GPS data, tide data from Observatory, height limit from Aviation Department, arm height and angle of each boat	Monthly charge for the height monitoring for 30 construction boats	Get-ahead strategy in technology	Digital Add-on and lock-in
Civil Aviation Department		Internet access			Safety monitoring	Get-ahead strategy in market	Remote Usage and Condition Monitoring

45.4.3 Company: Cenique Infotainment Group Ltd (www.cenique.com) and Hutchison Global Communications Ltd (www.hgc.com.hk)

Project Description: With the smart TV installed at the shopping mall, the camera on the TV can find out the effectiveness and ROI of digital signage and point-of-purchase displays through a Cloud Audience Analytics. It is a video "presence detection" solution providing shopper analytics about their profile such as gender, age, and so on in real time, automatically popping up the appropriate advertisement, and collecting and reporting insightful information for improving advertising effectiveness. Cenique provides the analytic software and gateway, while Hutchison provides the mobile network and Cloud facility.

Product/Service: IntelliSense Analytics.

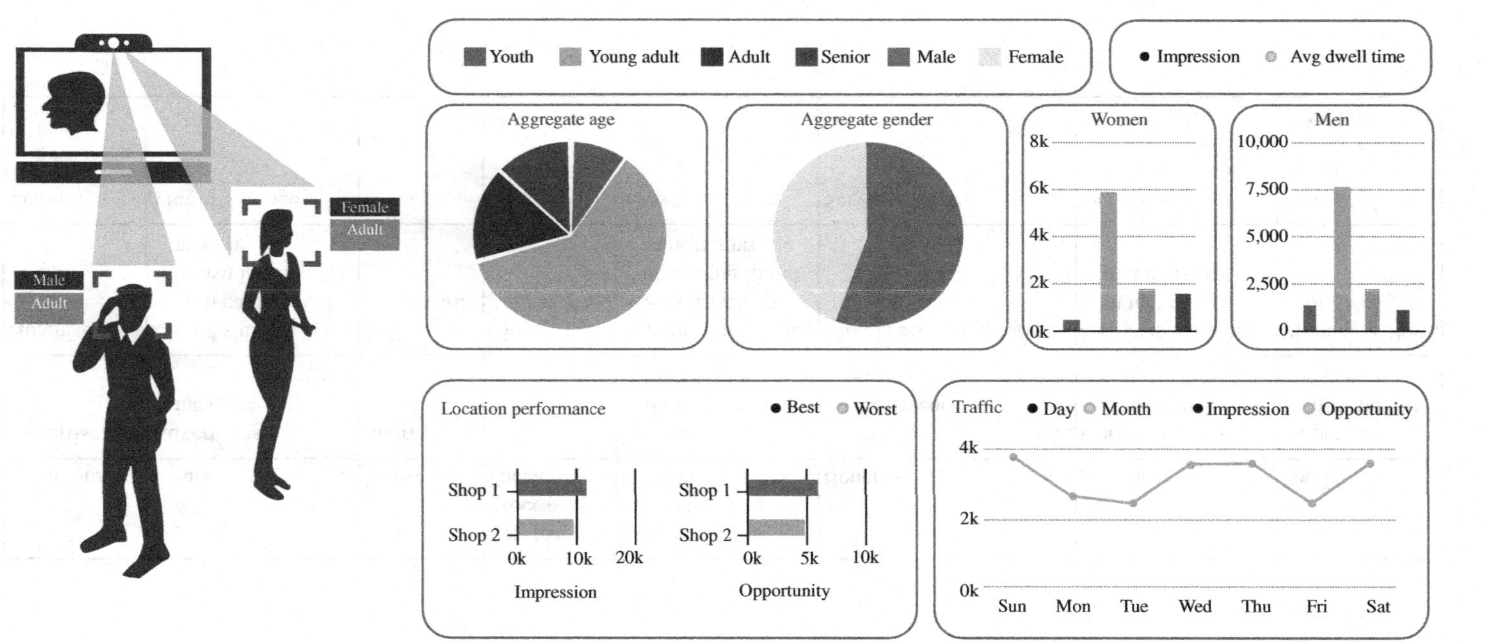

Male
Adult
Female
Adult
Youth
Young adult
Adult
Senior
Male
Female
Impression
Avg dwell time
Aggregate age
Aggregate gender
Women
8k
6k
4k
2k
0k
Men
10,000
7,500
5,000
2,500
0
Location performance
Best
Worst
Shop 1
Shop 2
0k
10k
20k
Impression
Shop 1
Shop 2
0k
5k
10k
Opportunity
Traffic
Day
Month
Impression
Opportunity
4k
2k
0k
Sun
Mon
Tue
Wed
Thu
Fri
Sat

Collaborator	Inputs	Network	Service/ Processing/ Packaging	Content/ Information Product	Benefits	Strategy	Tactic
Advertiser	Advertising message	Internet		Product and promotion information	Viewer profile, ROI, and effectiveness of the advertisement	Get-ahead strategy in market	Target customer information pushing
Hutchison	Digital signage with camera and transmission modem	Mobile network	Cloud service	Pop up appropriate Advertisement according to the viewer's profile	Cloud service platform	Get-ahead strategy in technology and market	Product as Point of Sales
Cenique	Viewer's image	Cloud	Data analytic	Viewer's profile	Service charges	Get-ahead strategy in technology	Remote Usage and Condition Monitoring

45.4.4 Company: Fukui Shell Nucleus Factory (http://www.fukuishell.com/)

Project Description: Unlike diamonds and gemstones which can now be sold online with confidence and have their own internationally recognized standards, the pearl industry is facing the challenges of identity, track and traceability, and authentication. Each pearl nucleus is embedded with an RFID tag, registered by technology called Metakaku™. By adding in data and information to the database along the cultivation period, each pearl shall have a record that tells its unique story about the pearl's true origin, cultivation period, and its current whereabouts. The RFID tag contains a unique Electronic Product Code (EPC) serial number via a unique serialization and the track-and-trace function provided by GS1, a nonprofit organization, industry-led global supply chain standards organization. A high-level authentication with a dynamic security key is done by a technology called AuthenTick™ developed by the Hong Kong R&D Centre for Logistics and Supply Chain Management (LSCM). The information is collected collaboratively throughout the value chain and is then stored in the database called the Global Pearl Database (GPD) developed by LSCM as well. The pearl enquirer can access the information by using the pearl search engine that is powered by the GS1's Search Directory and the GPD's information.

Product/Service: Metakaku, the technological enabler that allows unique identification for each pearl.

Collaborator	Inputs	Network	Service/Processing/ Packaging	Content/Information Product	Benefits	Strategy	Tactic
Fukui	RFID tag		Pearl nucleus embedded with RFID tag	Identity, track and traceability, and authentication	Technology enabler	Get-ahead strategy in technology	Increase product value via IoT benefit
GS1		Internet	Electronic Product Code Information Service (EPCIS) and search directory	A unique serialization of the numbers for Metakaku and searching function	Service platform	Get-ahead strategy in technology	Remote Usage and Condition Monitoring
LSCM		Internet	AuthenTick and Global Pearl Database (GPD)	Data collected throughout the value chain	Service platform	Get-ahead strategy in technology	Remote Usage and Condition Monitoring
Pearl enquirer	Pearl information via a RFID scanner	Internet		Authentication and product information	Product confidence		

45.5 DISCUSSION AND SUMMARY

In the business model framework proposed by Turber et al. [1], the context of the IoT is based on S-D logic. This "who, where, and why" framework clearly shows that in the IoT, the firms are not the only ones producing value for the customers to buy, but rather all collaborators have some role in the production of value. As shown in our framework earlier, all the collaborators have their own indispensible role in its IoT ecosystem.

Although the sample size is not big, the studies included user, vendor, system integration, and solution and service providers, while the applications are cross industries; thus the observations and the lessons from the interviews can validate the proposed framework with the following implications:

1. The technology breakthrough in miniaturization, energy efficiency of sensors enables a lot of portable devices and thus convenient and real-time IoT applications.
2. The development and deployment process is quite long or have gone through a Proof-of-Concept stage as there is no precedent in most of the applications. The advantage for the incumbents is high entry barrier if they can sustain the initial investment period.
3. Affordable wireless connectivity provides ubiquitous network connection for IoT devices.
4. As Cloud service is widely employed in most applications, it creates a common data storage platform for collaborators.
5. Sometimes external information is required in addition to the sensor data for holistic analysis. Some of these are open data such as weather, traffic, and geomapping.
6. Integration with client back-end system is required in most cases.
7. All the developers/Integrators adopted either a "get-ahead strategy in technology" or a "get-ahead strategy in market," while the clients mostly adopted a get-ahead strategy in market. The clients must buy in the idea, and the product/solution providers must accommodate more late-stage and postpurchase design changes for the fine-tuning after initial implementation.
8. The common tactics are digital or physical freemium, Product as Point of Sales, and Remote Usage and Condition Monitoring.
9. IoT applications may involve a lot of different appliances/devices with different control protocol.
10. After-sales services can be much more efficient compared with traditional system because of the Internet-connected nature of IoT. It can also share the workload of individual device through system optimization.
11. IoT adoption is not reaching a mass market yet; the network effect is confined to specific vertical market because a lot of customization, and thus the domain knowledge are required.

12. Retailers demonstrate the strengthening of customer relationships, marketing, and security through the adoption of IoT.
13. Although an open system which facilities connection to other devices through an open interface or API is feasible, closed system which aims to have customer purchase the whole system from a chosen solution provider is preferable.
14. The performance of the product shifts from the functionality of a discrete product to that of a broader system which consists of the connection network, data storage, system data analytics, and integration with the existing back-end management system.

With a rapidly growing industry and huge potential revenue over the horizon, companies can see the great potential in the IoT but struggle on how to approach it. Not much data exists on how business operates within an IoT context, and so companies are reluctant to take the big leap into the IoT. Just as soon as the business potential was realized, so was the need for network-centric IoT business model framework. This article addresses this issue by revealing the successful business models of the IoT Awardees. With that, we also testified that the value proposition, knowledge requirement, and thus the competitive advantages have been changing in the supply side, while a new era of IT-driven productivity growth is emerging in the demand side.

45.6 LIMITATIONS AND FUTURE RESEARCH

Pure sensor manufacturers or technology enablers are excluded in this study because I mainly focused on those businesses involving more than one collaborators or solution base to validate the proposed business model framework. The case studies are mainly from businesses in Hong Kong. Future study can be extended to other cities with different contexts. Although some of these case studies involved data analytic, the self-learning artificial intelligence application is not fully demonstrated. As the artificial intelligence or machine learning is indispensable in IoT applications, its impact on business model is a worthwhile topic for future research.

REFERENCES

[1] Turber, S., vom Brocke, J., Gassmann, O., and Flesich, E. (2014). Designing Business Models in the Era of Internet of Things. Proceedings of the 9th International Conference on Design Science Research in Information Systems and Technology, Miami, FL, May 22–24, 2014. (8462), pp. 17–31.

[2] Abdmeziem, R. and Tandjaoui, D. (2014). Internet of Things: Concept, Building Blocks, Applications and Challenges, January 28, 2014, pp. 1–17. http://arxiv.org/abs/1401.6877 (accessed September 12, 2016).

[3] Atzori, L., Iera, A., and Morabito, G. (2014). From "smart objects" to "social objects": the next evolutionary step of the Internet of Things. *IEEE Communications Magazine*, **52**(1): 97–105.

[4] Business Model. In Investopedia. Retrieved July 25, 2014 from http://www.investopedia.com/terms/b/businessmodel.asp (accessed August 22, 2016).

[5] Sun, Y., Yan, H., Lu, C., Bie, R., and Thomas, P. (2012). A holistic approach to visualizing business models for the Internet of Things. *Communications in Mobile Computing*, **1**(4): 1–7.

[6] Bucherer, E., Eisert, U., and Gassmann, O. (2012). Towards systematic business model innovation: lessons from product innovation management. *Creativity and Innovation Management*, **21**(2): 183–198.

[7] Westerlund, M., Leminen, S., and Rajahonka, M. (2014). Designing business models for the Internet of Things. *Technology Innovation Management Review*, **4**: 5–14.

[8] Gassmann, O., Frankenberger, K., and Csik, M. (2014). Revolutionizing the business model. In: Gassmann, O. and Schweitzer, F. (eds.) *Management of the Fuzzy Front End of Innovation*, Springer International Publishing, Cham, pp. 89–98.

[9] Hui, G. (2014). How the Internet of Things Changes Business Models, Harvard Business Review, July 29. https://hbr.org/2014/07/how-the-internet-of-things-changes-business-models (accessed September 12, 2016).

[10] Mejtoft, T. (2011). Internet of Things and Co-creation of Value. 2011 IEEE International Conference on Internet of Things and Cyber, Physical and Social Computing, Dalian, October 19–22, 2011, pp. 672–677.

[11] Chen, M. (2013). Towards smart city: M2M communications with software agent intelligence. *Multimedia Tools and Applications*, **67**: 167–178.

[12] Rong, K., Hu, G., Lin, Y., Shi, Y., and Guo, L. (2015). Understanding business ecosystem using a 6C framework in Internet-of-Things-based sectors. *International Journal of Production Economics*, **159**: 41–55.

[13] Strategyzer Business Model Generation. (2014). Available at http://www.businessmodelgeneration.com/canvas (accessed August 22, 2014).

[14] Kindstrom, D. (2010). Towards a service-based business model—key aspects for future competitive advantage. *European Management Journal*, **28**: 479–490.

[15] Li, Y., Hou, M., Liu, H., and Liu, Y. (2012). Towards a theoretical framework of strategic decision, supporting capability and information sharing under the context of Internet of Things. *Information Technology and Management*, **13**(4): 205–216.

[16] Fleisch, E., Weinberger, M., and Wortmann, F. (2015). Business Models and the Internet of Things. In: Ivana, P.Z., Krešimir, P., and Martin, S. (eds.) *Interoperability and Open-Source Solutions for the Internet of Things. Volume 9001*, Springer International Publishing, Cham, pp. 6–10.

[17] James, R. (2014). The Internet of Things: A study in Hype, Reality, Disruption, and Growth, Raymond James US Research, Technology & Communications, Industry Report, Raymond James & Associates, Saint Petersburg, FL.

[18] Holler, J., Tsiatsis, V., Mulligan, C., Avesand, S., Karnouskos, S., and Boyle, D. (2014). *From Machine-to-Machine to the Internet of Things: Introduction to a New Age of Intelligence*, Elsevier, Waltham, MA.

[19] Dijkman, R.M., Sprenkels, B., Peeters, T., and Janssen, A. (2015). Business models for the Internet of Things. *International Journal of Information Management*, **35**(6): 672–678.

[20] Yin, R.K., Bateman, P.G., and Moore, G.B. (1983). *Case Studies and Organizational Innovation: Strengthening the Connection*, COSMOS Corporation, Washington, DC.

[21] Communications Association of Hong Kong. (2015). 2015 Official Guide to ICT Industry in Hong Kong, Theme: The Internet of Things, Communications Association of Hong Kong, Hong Kong, pp. 60–91.

INDEX

Internet of Things and Data Analytics Handbook, First Edition. Edited by Hwaiyu Geng.

Companion website: www.wiley.com/go/Geng/iot_data_analytics_handbook/

Printed in the United States
By Bookmasters